Conversion Factors

Acceleration of gravity
$g = 9.80665$ m/s^2
$g = 980.665$ cm/s^2
$g = 32.174$ ft/s^2
1 ft/s$^2 = 0.304799$ m/s^2

Area
1 acre $= 4.046856 \times 10^3$ m^2
1 ft$^2 = 0.0929$ m^2
1 in$^2 = 6.4516 \times 10^{-4}$ m^2

Density
1 lb$_m$/ft$^3 = 16.0185$ kg/m^3
1 lb$_m$/gal $= 1.198264 \times 10^2$ kg/m^3
Density of dry air at 0°C, 760 mm Hg $= 1.2929$ g/L
1 kg mol ideal gas at 0°C, 760 mm Hg $= 22.414$ m^3

Diffusivity
1 ft^2/h $= 2.581 \times 10^{-5}$ m^2/s

Energy
1 Btu $= 1055$ J $= 1.055$ kJ
1 Btu $= 252.16$ cal
1 kcal $= 4.184$ kJ
1 J $= 1$ N m $= 1$ kg m^2/s^2
1 kW h $= 3.6 \times 10^3$ kJ

Enthalpy
1 Btu/lb$_m = 2.3258$ kJ/kg

Force
1 lb$_f = 4.4482$ N
1 N $= 1$ kg m/s^2
1 dyne $= 1$ g cm/s$^2 = 10^{-5}$ kg m/s^2

Heat flow
1 Btu/h $= 0.29307$ W
1 Btu/min $= 17.58$ W
1 kJ/h $= 2.778 \times 10^{-4}$ kW
1 J/s $= 1$ W

Heat flux
1 Btu/(h ft^2) $= 3.1546$ W/m^2

Heat transfer coefficient
1 Btu/(h ft^2 °F) $= 5.6783$ W/(m^2 K)
1 Btu/(h ft^2 °F) $= 1.3571 \times 10^{-4}$ cal/(s cm^2 °C)

Length
1 ft $= 0.3048$ m
1 micron $= 10^{-6}$ m $= 1$ μm
1 Å $= 10^{-10}$ m
1 in $= 2.54 \times 10^{-2}$ m
1 mile $= 1.609344 \times 10^3$ m

Mass
1 carat $= 2 \times 10^{-4}$ kg
1 lb$_m = 0.45359$ kg
1 lb$_m = 16$ oz $= 7000$ grains
1 ton (metric) $= 1000$ kg

Mass transfer coefficient
1 lb mol/(h ft^2 mol fraction) $= 1.3562 \times 10^{-3}$ kg mol/(s m^2 mol fraction)

Power
1 hp $= 0.7457$ kW
1 W $= 14.34$ cal/min
1 hp $= 550$ ft lb$_f$/s
1 Btu/h $= 0.29307$ W
1 hp $= 0.7068$ Btu/s
1 J/s $= 1$ W

Pressure
1 psia $= 6.895$ kPa
1 psia $= 6.895 \times 10^3$ N/m^2
1 bar $= 1 \times 10^5$ Pa $= 1 \times 10^5$ N/m^2
1 Pa $= 1$ N/m^2
1 mm Hg (0°C) $= 1.333224 \times 10^2$ N/m^2
1 atm $= 29.921$ in. Hg at 0°C
1 atm $= 33.90$ ft H$_2$O at 4°C
1 atm $= 14.696$ psia $= 1.01325 \times 10^5$ N/m^2
1 atm $= 1.01325$ bar
1 atm $= 760$ mm Hg at 0°C $= 1.01325 \times 10^5$ Pa
1 lb$_f$/ft$^2 = 4.788 \times 10^2$ dyne/cm$^2 = 47.88$ N/m^2

Specific heat
1 Btu/(lb$_m$ °F) $= 4.1865$ J/(g K)
1 Btu/(lb$_m$ °F) $= 1$ cal/(g °C)

Temperature
$T_{°F} = T_{°C} \times 1.8 + 32$
$T_{°C} = (T_{°F} - 32)/1.8$

Thermal conductivity
1 Btu/(h ft °F) $= 1.731$ W/(m K)
1 Btu in/(ft^2 h °F) $= 1.442279 \times 10^{-2}$ W/(m K)

Viscosity
1 lb$_m$/(ft h) $= 0.4134$ cp
1 lb$_m$/(ft s) $= 1488.16$ cp
1 cp $= 10^{-2}$ g/(cm s) $= 10^{-2}$ poise
1 cp $= 10^{-3}$ Pa s $= 10^{-3}$ kg/(m s) $= 10^{-3}$ N s/m^2
1 lb$_f$s/ft$^2 = 4.7879 \times 10^4$ cp
1 N s/m$^2 = 1$ Pa s
1 kg/(m s) $= 1$ Pa s

Volume
1 ft$^3 = 0.02832$ m^3
1 U.S. gal $= 3.785 \times 10^{-3}$ m^3
1 L $= 1000$ cm^3
1 m$^3 = 1000$ L
1 U.S. gal $= 4$ qt
1 ft$^3 = 7.481$ U.S. gal
1 British gal $= 1.20094$ U.S. gal

Work
1 hp h $= 0.7457$ kW h
1 hp h $= 2544.5$ Btu
1 ft lb$_f = 1.35582$ J

Food Science and Technology

IRFS&T

INTERNATIONAL REVIEW
OF
FOOD SCIENCE
&TECHNOLOGY

2009

Food Science and Nutrition

Global Supply and Sustainability

Legislation

Processing

Packaging

Contamination and Analytical Processes

Traceability

An Official Publication of the
International Union of
Food Science and Technology

IUFoST

2009 EDITION
available in September!
www.irfst-git.com

A Passion
For Communication
Since 1969

40 Years **GIT VERLAG**
A Wiley Company
www.gitverlag.com

Food Science and Technology

Edited by

Geoffrey Campbell-Platt

Professor Emeritus of Food Technology, University of Reading
President of IUFoST 2008–2010

WILEY-BLACKWELL

A John Wiley & Sons, Ltd., Publication

Library of Congress Cataloging-in-Publication Data

Food science and technology / edited by Geoffrey Campbell-Platt.
     p. ; cm.
   Includes bibliographical references and index.
   ISBN 978-0-632-06421-2 (hardback : alk. paper)
   1. Food industry and trade.   2. Biotechnology.   I. Campbell-Platt, Geoffrey.
II. International Union of Food Science and Technology.
   [DNLM: 1.   Food Technology.   2. Biotechnology.   3. Food Industry.
4. Nutritional Physiological Phenomena. TP 370 F6865 2009]
   TP370.F629 2009
   664–dc22   2009001743

A catalogue record for this book is available from the British Library.

Set in 9.5/12pt Palatino by Aptara® Inc., New Delhi, India

1   2009

Contents

Chapter 1

Professor Geoffrey Campbell-Platt
Professor Emeritus of Food
 Technology
University of Reading; President of IUFoST
 2008–2010
Whiteknights
Reading
RG6 6AP
United Kingdom

Chapter 2

Dr Richard A. Frazier
Senior Lecturer in Food
 Biochemistry
Department of Food Biosciences
University of Reading
Whiteknights
Reading
RG6 6AP
United Kingdom

Chapter 3

Professor Heinz-Dieter Isengard
University of Hohenheim
Institute of Food Science and
 Biotechnology
D-70593 Stuttgart
Germany

Professor Dietmar Breithaupt
University of Hohenheim
Institute of Food Chemistry
D-70593 Stuttgart
Germany

Chapter 4

Mr Brian C. Bryksa
Department of Food Science
University of Guelph
Guelph
Ontario N1G 2W1
Canada

Professor Rickey Y. Yada
Canada Research Chair in Food Protein Structure
Scientific Director, Advanced Foods and Materials
 Network (AFMNet)
Department of Food Science
University of Guelph
Guelph
Ontario N1G 2W1
Canada

Chapter 5

Professor Cherl-Ho Lee
Division of Food Bioscience and Technology
College of Life Sciences and Biotechnology
Korea University
1 Anamdong, Sungbukku,
Seoul
136-701 Korea

Chapter 6

Dr Tim Aldsworth
The University of Hertfordshire
College Lane Campus
Hatfield
AL10 9AB
United Kingdom

Professor Christine E.R. Dodd
  and **Professor Will Waites**
Division of Food Sciences
University of Nottingham
Sutton Bonington Campus
Loughborough
Leicestershire
LE12 5RD
United Kingdom

Chapter 7

Professor R. Paul Singh
Distinguished Professor of Food Engineering
Department of Biological and Agricultural
  Engineering
Department of Food Science and Technology
University of California
One Shields Avenue Davis
CA 95616
USA

Chapter 8

Professor Keshavan Niranjan
Professor of Food Bioprocessing
Editor, Journal of Food Engineering
University of Reading
Whiteknights
PO Box 226
Reading
RG6 6AP
United Kingdom

Professor Gustavo Fidel Gutiérrez-López
Professor of Food Engineering
Head, PhD Program in Food Science and Technology
Escuela Nacional de Ciencias Biológicas
Instituto Politécnico Nacional
Carpio y Plan de Ayala S/N
Santo Tomás, 11340
Mexico, DF
México

Chapter 9

Dr Jianshe Chen
Department of Food Science and Nutrition
University of Leeds
Leeds
LS2 9JT
United Kingdom

Dr Andrew Rosenthal
Nutrition and Food Science Group
School of Life Sciences
Oxford Brookes University
Gipsy Lane Campus
Oxford
OX3 0BP
United Kingdom

Chapter 10

Professor R. Paul Singh
Distinguished Professor of Food
  Engineering
Department of Biological and Agricultural
  Engineering
Department of Food Science and Technology
University of California
One Shields Avenue Davis
CA 95616
USA

Chapter 11

Professor Gordon L. Robertson
University of Queensland and
Food • Packaging • Environment
6066 Lugano Drive
Hope Island
QLD 4212
Australia

Chapter 12

Professor C. Jeya Henry
Professor of Food Science and Human Nutrition
School of Life Sciences
Oxford Brookes University
Gipsy Lane
Oxford
OX3 OBP
United Kingdom

Ms Lis Ahlström
Researcher
School of Life Sciences
Oxford Brookes University
Gipsy Lane
Oxford
OX3 0BP
United Kingdom

Chapters 13 and 14

Dr Herbert Stone and **Dr Rebecca N. Bleibaum**
Tragon Corporation
350 Bridge Parkway
Redwood Shores
CA 94065-1061
USA

Chapter 15

Dr David Jukes
Senior Lecturer in Food Regulation
Department of Food Biosciences
University of Reading
Whiteknights
Reading
RG6 6AP
United Kingdom

Chapter 16

Dr Gerald G. Moy
GEMS/Food Manager
Department of Food Safety, Zoonoses and
    Foodborne Disease
World Health Organization
Geneva
Switzerland

Chapter 17

Dr Michael Bourlakis
Senior Lecturer
Brunel University
Business School
Elliot Jaques Building
Uxbridge
Middlesex
UB8 3PH
United Kingdom

Professor David B. Grant
Logistics Institute
Business School
University of Hull
Kingston upon Hull
HU6 7RX
United Kingdom

Dr Paul Weightman
School of Agriculture, Food and Rural
    Development
Newcastle University
Agriculture Building
Newcastle upon Tyne
NE1 7RU
United Kingdom

Chapter 18

Professor Takahide Yamaguchi
Professor of Management
Graduate School of
    Accountancy
University of Hyogo
Kobe-Gakuentoshi Campus
Kobe, 651-2197
Japan

Professor Ray Winger
Professor of Food Technology
Institute of Food, Nutrition and
    Human Health
Massey University
Private Bag 102 904
North Shore Mail Centre
Albany
Auckland
New Zealand

Dr Sue H.A. Hill and **Professor Jeremy D. Selman**
Managing Editor and Head of Editorial and
    Production
and Managing Director
International Food Information Service
    (IFIS Publishing)
Lane End House
Shinfield Road
Shinfield
Reading
RG2 9BB
United Kingdom

Geoffrey Campbell-Platt

Food science and technology is the understanding and application of science to satisfy the needs of society for sustainable food quality, safety and security.

At several universities worldwide, degree programmes in food science and technology have been developed in the past half-century. This followed the lead of the University of Strathclyde (then the Royal College of Science and Technology) in Glasgow, Scotland, under the leadership of the first Professor of Food Science, who also became President of the International Union of Food Science and Technology (IUFoST), the late John Hawthorn.

The aim of these courses has been to provide food science and technology graduates with the ability, through multidisciplinary studies, to understand and integrate the scientific disciplines relevant to food. They would then be able to extend their knowledge and understanding of food through a scientific approach, and to be able to apply and communicate that knowledge to meet the needs of society, industry and the consumer for sustainable food quality, safety and security of supply.

1.1 Food science and technology course elements

Students studying food science and technology in higher education need to have undertaken courses in the basic scientific disciplines of chemistry, biology, mathematics, statistics and physics. These are developed in food science and technology degree programmes through course elements in Food Chemistry, Food Analysis, Food Biochemistry, Food Biotechnology, Food Microbiology, Numerical Procedures and Food Physics. These are all covered by chapters in this book, followed by chapters covering Food Processing, Food Engineering and Packaging. Further courses are required in Nutrition, Sensory Evaluation, Statistical Techniques, and Quality Assurance and Legislation. Regulatory Toxicology and Food Safety is addressed, as is Food Business Management. Other course elements in Food Marketing and Product Development are included, together with chapters on Information Technology, and Communication and Transferable Skills.

Food science and technology are science-based courses, requiring a good grounding in science and the use of laboratory and pilot-plant facilities, to reinforce the theoretical knowledge acquired. As well as acquiring practical laboratory and observation skills, laboratory experiments need to be written up, developing important reporting and interpretation skills. Universities therefore require up-to-date facilities for chemical, microbiological laboratory exercises, and processing pilot-plant facilities for teaching the principles of unit processing and engineering operations, as well as sufficient well-qualified staff to teach the range of disciplines covered in this book.

1.2 Evolution of the book

The book has evolved from a working group of the Committee of University Professors of Food Science and Technology (CUPFST), United Kingdom, who sought to agree a framework of common course elements for the various food science and technology courses established in the UK. Newer universities advised that each course element should be based on outcomes, which should be achieved on successful completion, and it is these outcome headings that have largely been used as subject headings in each chapter of this book. This approach is popular internationally and is used by professional institutes such as the Institute of Food Science and Technology (IFST) in the UK, and the book has evolved in consultation with the recommended Education Standards for Food Science of the Institute of Food Technologists (IFT) in the USA.

The IFT recognises food science as the discipline in which engineering, biological and physical sciences are used to study the nature of foods, the causes of deterioration, the principles underlying food processing, and the improvement of foods for the consuming public. Food technology is recognised as the application of food science to the selection, preservation, processing, packaging, distribution and use of safe, nutritious and wholesome food. In short, it could be said that the food scientist analyses and takes apart food materials, whilst the food technologist puts all that knowledge into use in producing safe, desired food products. In practice, as recognised throughout the world, the terms are often used interchangeably, and practising food scientists and technologists have to both understand the nature of food materials and produce safe, nutritious food products.

It is understood, and desirable, that the various food science and technology courses offered will vary, reflecting particular research interests and expertise, in different institutions, and students will want to develop their own interests through specific module choices or individual research projects. However, the purpose of establishing the core competencies, reflected in the chapters of this book, is to recognise what a food science or food technology graduate can be expected to achieve as a minimum, so that employers and regulators know what to expect of a qualified graduate, who could then expect, after suitable relevant experience, to become a member of a professional body, such as IFT or IFST, or a Chartered Scientist.

1.3 Food safety assurance

In our increasingly interdependent globalised world, food safety is an implied term in the 'food purchasing or food service' consumer contract, which often appears to be addressed publicly only when something goes wrong. In fact, food control agencies and food retailers require processors and manufacturers to apply Hazard Analysis Critical Control Points (HACCP) to all their processes. This, combined with good practices, such as Good Manufacturing Practice (GMP), and traceability, build quality and safety assurance into the food chain, which is inherently better with the very large number of food items produced and eaten frequently, and when individual item or destructive testing can only give a limited picture of the total production. Both HACCP and GMP require good teamwork by all involved in food processing, and it is the multidisciplinary-trained food scientist or technologist who usually is called upon to lead and guide these operations.

In our modern world where food ethics are to the fore, in terms of sustainable production practices, care of our environment, fair-trade, packaging recycling and climate-change concerns, food scientists and technologists will have an increasing role to play, in keeping abreast of these issues and the science that can be applied to help address them. Food scientists, to be successful, already need good interpersonal, communication and presentation skills, which may be learned through example, mentoring and practice in as many different situations as possible; in the future, these skills promise to be in even greater demand, as scientists engage with increasingly demanding members of the public.

1.4 The International Union of Food Science and Technology (IUFoST)

IUFoST is the international body representing some 65 member countries and some 200,000 food scientists and technologists worldwide. IUFoST organises World Congresses of Food Science and Technology in different locations around the world, normally every 2 years, at which the latest research and ideas are shared, and the opportunity is provided for young food scientists to present papers and posters and to interact with established world experts. Higher education in food science and technology has been of great interest for several years, with many developing countries looking for guidance in establishing courses in the subject, or to align them more closely with others, to help graduates move more successfully between countries and regions. IUFoST is also helping the development of Distance Education, where people are in employment and not able to attend normal university courses. IUFoST therefore sees the publication of this book as an important part of its contribution to helping internationally in sharing knowledge and good practice.

IUFoST has also established the International Academy of Food Science and Technology (IAFoST), to which eminent food scientists can be elected by peer review, and are designated as Fellows of IAFoST. The Fellows have acted as lead authors and advisers in the increasing range of authoritative Scientific Information Bulletins published by IUFoST, through its Scientific Council, which help summarise key food issues to a wider audience.

1.5 The book

In writing this book, we have been honoured to have the 20 chapters written by 30 eminent authors, from 10 different countries. All authors are experts in their respective fields, and together represent 15 of the world's leading universities in food science and technology, as well as four leading international organisations. We are particularly honoured that several of the authors are distinguished Fellows of IAFoST, so helping directly to inspire younger potential food scientists and technologists through this textbook for students.

It is therefore hoped that this book is adopted widely, providing tutors and students with the basic content of the core components of food science and technology degrees, while providing guidance through references to further knowledge and for more advanced study. If this work provides the opportunity to help students worldwide in sharing a common ideal while developing their own interests and expertise, the original aim of Professor John Hawthorn in developing this vital subject, so essential for all of us, from Scotland to a worldwide discipline, will have been achieved.

Supplementary material is available at www.wiley.com/go/campbellplatt

Richard A. Frazier

Key points

- Carbohydrate chemistry: structures, properties and reactions of major monosaccharides, oligosaccharides and polysaccharides in foods.
- Proteins: chemistry of the amino acids and their role in protein structure, a description of the major forces that stabilize protein structure and how they are disrupted during protein denaturation.
- Lipids: structure and nomenclature, polymorphism of triglycerides, oil and fat processing (hydrogenation and interesterification), and lipid oxidation.
- Chemistry of minor components in foods: permitted additives, vitamins and minerals.
- Role of water in foods: water activity, its determination and the importance for microbial growth, chemical reactivity and food texture.
- Physical chemistry of dispersed systems: solutions, lyophilic and lyophobic dispersions, colloidal interactions and the DLVO theory, foams and emulsions.
- Chemical aspects of organoleptic properties of foods.

2.1 Introduction

Food chemistry is a fascinating branch of applied science that combines most of the sub-disciplines of traditional chemistry (organic, inorganic and physical chemistry) together with elements of biochemistry and human physiology. Food chemists attempt to define the composition and properties of food, and understand the chemical changes undergone during production, storage and consumption, and how these might be controlled. Foods are fundamentally biological substances and are highly variable and complex; therefore, food chemistry is a constantly evolving and expanding field of knowledge that underpins other areas of food science and technology. This chapter cannot hope to encompass all of the intricacies and details of food chemistry, but instead attempts to provide an overview of the fundamental areas that constitute this important area of science. To delve deeper, the reader is encouraged to refer to one or more of the excellent texts relating to food chemistry that are listed as further reading at the end of this chapter.

2.2 Carbohydrates

Carbohydrate is the collective name for polyhydroxyaldehydes and polyhydroxyketones, and these compounds form a major class of biomolecules that perform several functions in vivo, including the storage and transport of energy. Indeed, carbohydrates are the major source of energy in our diet. The name carbohydrate derives from their general empirical formula, which is $(CH_2O)_n$; however, the carbohydrate group contains several derivatives and closely related compounds that do not fit this general empirical formula but are still considered to be carbohydrates. There are three distinct classes of carbohydrates: *monosaccharides* (1 structural unit), *oligosaccharides* (2–10 structural units) and *polysaccharides* (more than 10 structural units).

2.2.1 Monosaccharides

The monosaccharides are also termed *simple sugars*, are given the suffix *-ose* and classified as *aldoses* or *ketoses* depending on whether they contain an aldehyde or ketone group. The most common monosaccharides are either *pentoses* (containing a chain of five carbon atoms) or *hexoses* (containing a chain of six carbon atoms). Each carbon atom carries a hydroxyl group, with the exception of the atom that forms the carbonyl group, which is also known as the *reducing group*.

Simple sugars are *optically active* compounds and can contain several asymmetrical carbon atoms. This leads to the possibility for the formation of multiple *stereoisomers* or *enantiomers* of the same basic structure. To simplify matters, monosaccharides are assigned optical configurations with respect to comparison of their highest numbered asymmetric carbon atom to the configuration of D-glyceraldehyde or L-glyceraldehyde (see Fig. 2.1). By convention, the carbon atoms in the monosaccharide molecule are numbered such that the reducing group carries the lowest possible number; therefore, in aldoses the reducing group carbon is always numbered 1 and in ketoses the numbering is started from the end of the carbon chain closest to the reducing group. Most naturally occurring monosaccharides belong to the

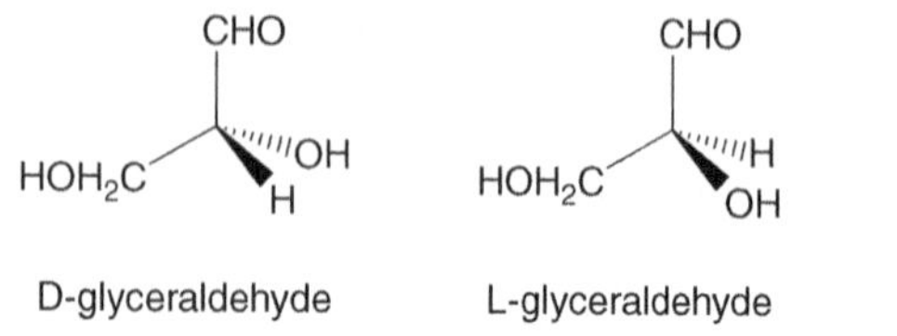

Figure 2.1 The D and L stereoisomers of glyceraldehyde.

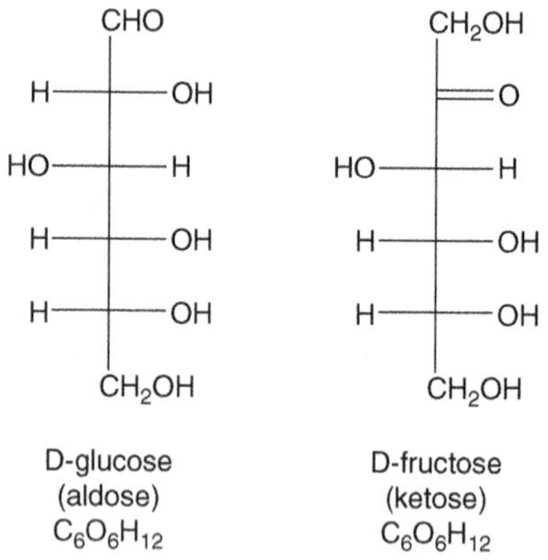

Figure 2.2 Fischer projections of the structures of D-glucose and D-fructose.

D-series, i.e. their highest numbered carbon has a similar optical configuration to D-glyceraldehyde.

The stereochemistry of the monosaccharides is depicted using the *Fischer projection* as shown for D-glucose and D-fructose in Fig. 2.2. All bonds are depicted as horizontal or vertical lines; all horizontal bonds project toward the viewer, while vertical bonds project away from the viewer. The carbon chain is depicted vertically with the C1 carbon at the top.

Aldoses and ketoses commonly exist in equilibrium between their *open-chain* form and cyclic structures in aqueous solution. Cyclic structures form through either a *hemiacetal* or a *hemiketal* linkage between the reducing group and an alcohol group of the same sugar. In this way sugars form either a five-membered *furanose* ring or a six-membered *pyranose* ring as shown in Fig. 2.3 for D-glucose. The formation of a furanose or pyranose introduces an additional asymmetric carbon; hence two *anomers* are formed (α-anomer and β-anomer) from each distinct open-chain monosaccharide. The interconversion between these two anomers is called *mutarotation*.

The cyclic structures of carbohydrates are commonly shown as *Haworth projections* to depict their three-dimensional structure. However, this projection does not account for the tetrahedral geometry of carbon. This is most significant for the six-membered pyranose ring, which may adopt either a *chair* or a *boat* conformation as depicted in Fig. 2.4. Of these, the chair conformation is favoured due to its greater thermodynamic stability. Within this conformation the bulky CH_2OH group is usually found in an equatorial position to reduce steric interactions.

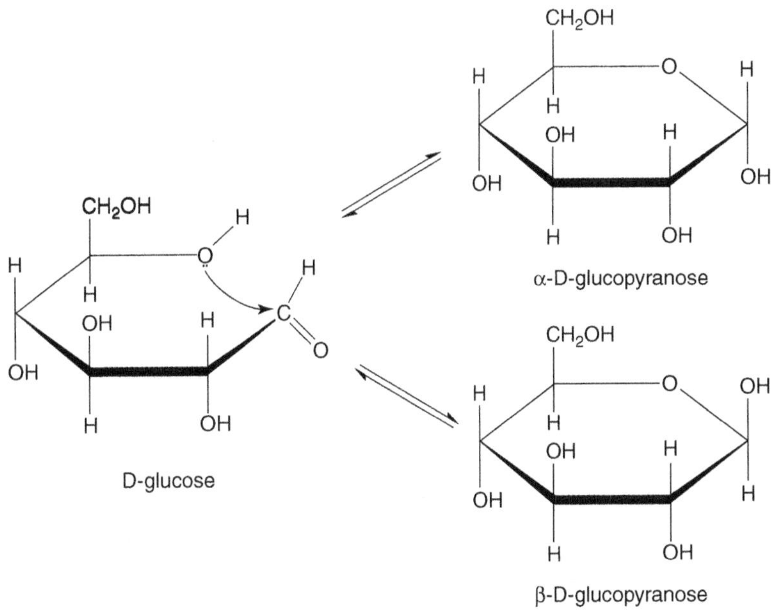

CH₂OH

α-D-glucopyranose

D-glucose

CH₂OH

β-D-glucopyranose

Figure 2.3 The formation of a hemiacetal linkage between the C1 carbon of D-glucose and the hydroxyl group of its C5 carbon leading to two anomers of D-glucopyranose. The rings are depicted as Haworth projections.

2.2.2 Oligosaccharides

Oligosaccharides contain 2–10 sugar units and are water soluble. The most significant types of oligosaccharide occurring in foods are disaccharides, which are formed by the condensation (i.e. water is eliminated) of two monosaccharide units to form a *glycosidic bond*. A glycosidic bond is that between the hemiacetal group of a saccharide and the hydroxyl group of another compound, which may or may not be itself a saccharide. Disaccharides can be homogeneous or heterogeneous and fall into two types:

1 *Non-reducing sugars* in which the monosaccharide units are joined by a glycosidic bond formed between their reducing groups (e.g. sucrose and trehalose). This inhibits further bonding to other saccharide units.

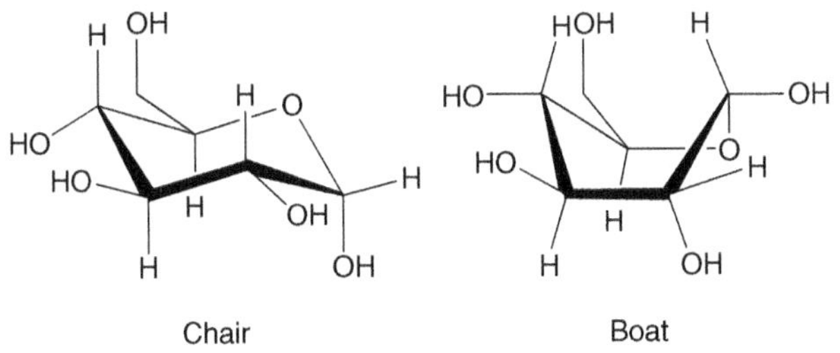

Chair   Boat

Figure 2.4 Chair and boat conformations of α-D-glucopyranose.

2 *Reducing sugars* in which the glycosidic bond links the reducing group of one monosaccharide unit to the non-reducing alcoholic hydroxyl of the second monosaccharide unit (e.g. lactose and maltose). A reducing sugar is any sugar that, in basic solution, forms an aldehyde or ketone allowing it to act as a reducing agent, and therefore includes all monosaccharides.

Of the disaccharides, sucrose, trehalose and lactose are found free in nature, whereas others are found as *glycosides* (in which a sugar group is bonded through its anomeric carbon to another group, e.g. a phenolic group, via an O-glycosidic bond) or as building blocks for polysaccharides (such as maltose in starch), which can be released by hydrolysis. Probably the three most significant disaccharides in food are sucrose, lactose and maltose, whose structures are depicted in Fig. 2.5.

Sucrose is the substance known commonly in households as sugar and is found in many plant fruits and saps. It is isolated commercially from sugar cane or the roots of sugar beet. Sucrose is composed of an α-D-glucose residue linked to a β-D-fructose residue and is a non-reducing sugar. Its systematic name is α-D-glucopyranosyl-(1↔2)-β-D-fructofuranoside (having the suffix -*oside*, because it is a non-reducing sugar). It is the sweetest tasting of the disaccharides and is an important source of energy.

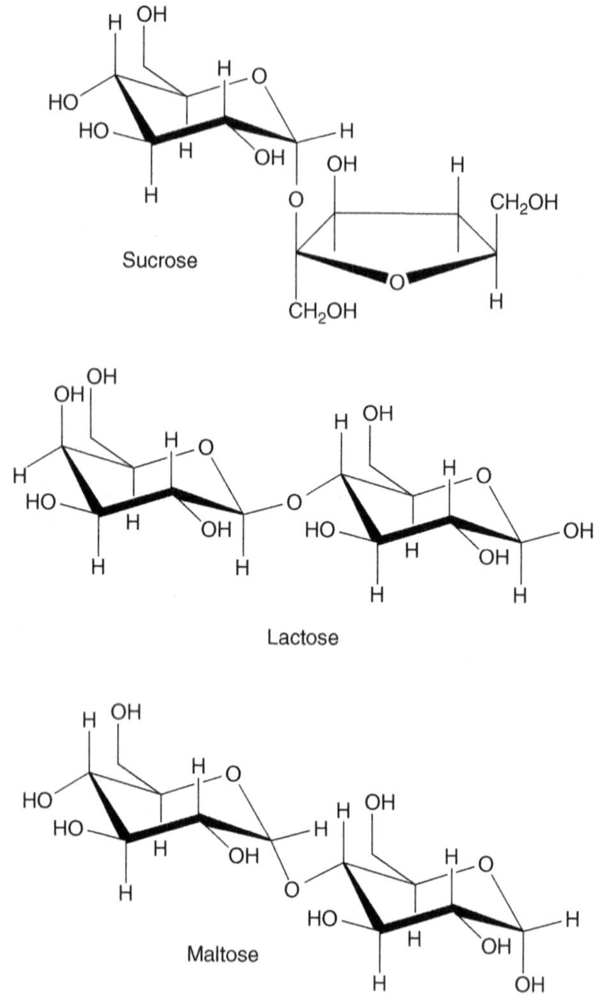

Figure 2.5 The structures of some common disaccharides: sucrose, lactose and maltose.

Lactose is found in mammalian milk and its systematic name is β-D-galactopyranosyl-(1↔4)-β-D-glucopyranose. To aid the digestion of lactose, the intestinal villi of infant mammals secrete an enzyme called lactase (β-D-galactosidase), which cleaves the molecule into its two subunits β-D-glucose and β-D-galactose. In most mammals the production of lactase gradually reduces with maturity into adulthood, leading to the inability to digest lactose and so-called *lactose intolerance*. However, in cultures where cattle, goats and sheep are milked for food there has evolved a gene for lifelong lactase production.

Maltose is formed by the enzymatic hydrolysis of starch and is an important component of the barley malt used to brew beer. It is a homogeneous disaccharide consisting of two units of glucose joined with an α(1→4) linkage, and is systematically named 4-O-α-D-glucopyranosyl-D-glucose. Maltose is a reducing sugar and the addition of further glucose unit yields a series of oligosaccharides known as maltodextrins or simply dextrins.

2.2.3 Polysaccharides

Polysaccharides are built of repeat units of monosaccharides and are systematically named with the suffix *-an*. The generic name for polysaccharides is *glycan* and these can be *homoglycans* consisting of the one type of monosaccharide or *heteroglycans* consisting of two or more types of monosaccharide.

Polysaccharides have three main functions in both animals and plants: as sources of energy, as structural components of cells, and as water-binders. Plant and animal cells store energy in the form of *glucans*, which are polymers of glucose such as starch (in plants) and glycogen (in animals). The most abundant structural polysaccharide is cellulose, which is also a glucan and is found in plants. Water-binding substances in plants include agar, pectin and alginate.

Polysaccharides occur as several structural types: *linear* (e.g. amylose, cellulose), *branched* (e.g. amylopectin, glycogen), *interrupted* (e.g. pectin), *block* (e.g. alginate) or *alternate repeat* (e.g. agar, carrageenan). According to the geometry of the glycosidic linkages, polysaccharide chains can form various conformations, such as *disordered random coil, extended ribbons, buckled ribbons* or *helices*. One of the most important properties of a great number of polysaccharides in foods is that they are able to form aqueous gels and thereby contribute to food structure and textural properties (e.g. mouth-feel).

2.2.3.1 Starch

Starch occurs in the form of semi-crystalline granules ranging in size from 2 to 100 μm, and consists of two types of glucan: *amylose* and *amylopectin*. Amylose is a linear polymer of α(1→4) linked α-D-glucopyranose and constitutes 20–25% of most starches. Amylopectin is a randomly branched polymer of α-D-glucopyranose consisting of linear chains with α(1→4) linkages with 4–5% of glucose units also being involved in α(1→6) linked branches. On average the length of linear chains in amylopectin is about 20–25 units. The chemical structures of amylose and amylopectin are shown in Fig. 2.6.

Amylose molecules contain in the region of 10^3 glucose units and form helix structures which entrap

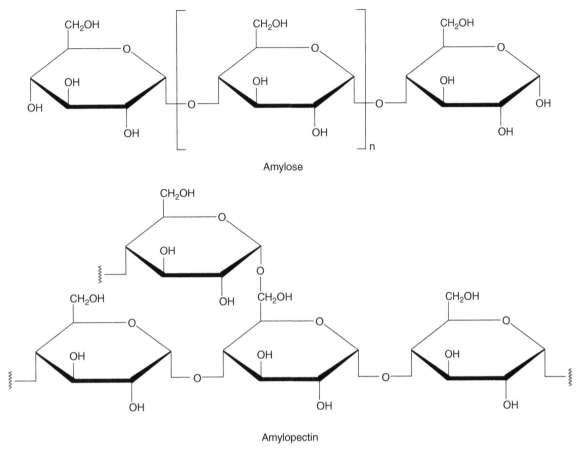

Figure 2.6 Chemical structures of amylose and amylopectin.

other molecules such as organic alcohols or fatty acids to form clathrates or helical inclusion compounds. Indeed, the blue colour that results when iodine solution is used to test for starch is thought to be due to the formation of an inclusion compound.

Amylopectin is a much larger molecule than amylose, containing approximately 10^6 glucose units per molecule, and forms a complex structure. This structure is described by the *cluster model* and has three types of chain (see Fig. 2.7): *A chains* that are unbranched and contain only $\alpha(1{\to}4)$ linkages, *B chains* that contain $\alpha(1{\to}4)$ and $\alpha(1{\to}6)$ linkages, and *C chains* that contain $\alpha(1{\to}4)$ and $\alpha(1{\to}6)$ linkages plus a reducing group. The linear A chains in this structure form clusters that are crystalline in nature, whereas the branched B chains give amorphous regions.

Starch granules undergo a process called *gelatinization* if heated above their gelatinization temperature (55–70°C depending on the starch source) in the presence of water. During gelatinization granules first begin to imbibe water and swell, and as a conse-

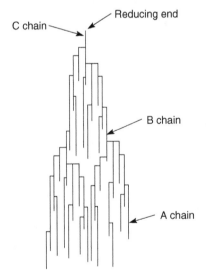

Figure 2.7 Amylopectin cluster model.

quence they progressively lose their organized structure (detected as a loss of *birefringence*). As time progresses, granules become increasingly permeable to water and solutes, thus swelling further and causing the viscosity of the aqueous suspension to increase sharply. Swollen starch granules leach amylose, which further increases viscosity to the extent that a paste is formed. As this paste is allowed to cool, hydrogen-bonding interactions between amylopectin and amylose lead to the formation of a gel-like structure.

Prolonged storage of a starch gel leads to the onset of a process termed *retrogradation*, during which amylose molecules associate together to form crystalline aggregates and starch gels undergo shrinkage and syneresis. Retrogradation can be viewed as a return from a solvated, dispersed, amorphous state to an insoluble, aggregated or crystalline condition. To avoid retrogradation in food products, *waxy starches* can be used that contain only amylopectin. Chemically modified starches are also available that have been depolymerized (i.e. partially hydrolysed), esterified or crosslinked to tailor their properties for particular end uses.

2.2.3.2 Glycogen

The polysaccharide that animals use for the short-term storage of food energy in the liver and muscles is known as *glycogen*. Glycogen is similar in structure to amylopectin, but has much higher molecular weight and a higher degree of branching. Branching aids the rapid release of glucose since the enzymes that release glucose attack on the non-reducing ends, cleaving one glucose molecule at a time. More branching equates to more non-reducing ends meaning more rapid release of energy. The metabolism of glycogen continues post-mortem, which means that by the time meat reaches the consumer it has lost all of its glycogen.

2.2.3.3 Cellulose

The most abundant structural polysaccharide is *cellulose*. Indeed, there is so much cellulose in the cell walls of plants that it is the most abundant of all biological molecules. Cellulose is a linear polymer of $\beta(1\rightarrow4)$ linked glucopyranose residues. The β-linkage in cellulose is not susceptible to attack by salivary amylases that break down starch α-linkages, and therefore cellulose forms a major part of *dietary fibre*. Dietary fibre is not digested by enzymes in the small intestine and is hence utilized by colonic microflora

via fermentation processes. So-called *hemicelluloses*, including *xylans*, which are major constituents of cereal bran, are another major component of dietary fibre.

2.2.3.4 Pectins

Pectins are mainly used in food as gelling agents. Pectins are heteroglycans and have complex structures that are based on a polygalacturonan backbone of $\alpha(1\rightarrow4)$-linked D-galacturonic acid residues, some of which are methylated. Into this backbone, there are regions where D-galacturonic acid is replaced by L-rhamnose, bonded via $(1\rightarrow2)$ linkages to give an overall rhamnogalacturonan chain. Pectins are characterized by *smooth regions* that are free of L-rhamnose residues and *hairy regions* consisting of both D-galacturonic acid and L-rhamnose residues. The hairy regions are so-called because they carry side chains of neutral sugars including mainly D-galactose, L-arabinose and D-xylose, with the types and proportions of neutral sugars varying with the origin of pectin.

As stated above, pectins are mainly applied in foods for their gelling properties, especially in jams and preserves. *Gels* consist of a three-dimensional polymeric network of chains that entrap water. Pectin gels are stabilized by *junction zones*, which are crystalline regions where smooth regions align themselves and interact. The hairy regions of pectin disrupt these junction zones, preventing extensive aggregation that could lead to precipitation as occurs during amylose retrogradation.

2.2.3.5 Gums

Distinct from those polysaccharides that form gels are a group of polysaccharides that are called *gums*. Gums have a high affinity for water and give high-viscosity aqueous solutions, but are not able to form gels. The reason for this is that all gums possess structures that incorporate a very high degree of branching or highly interrupted chains. This prevents the formation of junction zones (such as in pectins) that are a feature of polysaccharide gels. A notable gum that is commonly employed in foods is xanthan gum, which is secreted by *Xanthomonas campestris* and has a backbone of $\beta(1\rightarrow4)$-linked glucopyranose with trisaccharide branch points every five residues.

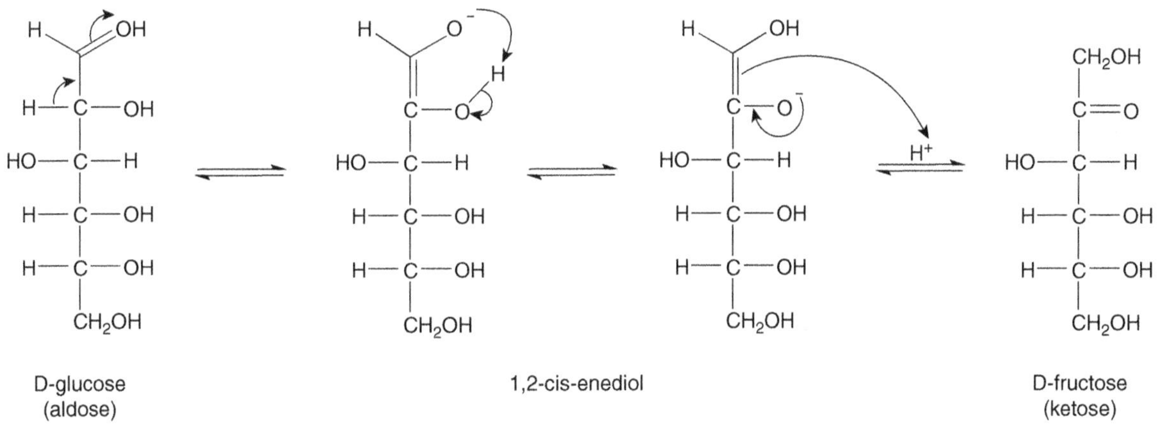

Figure 2.8 The isomerization of sugars to form an enediol intermediate prior to caramelization.

2.2.4 Reactions of carbohydrates

2.2.4.1 Caramelization

When a concentrated solution of sugars is heated to temperatures above 100°C, various thermal decomposition reactions can occur leading to formation of flavour compounds and brown-coloured products. This process, which particularly occurs during the melting of sugars, is called *caramelization*. Caramelization is a *non-enzymic browning reaction* like the Maillard reaction discussed below.

During caramelization, the first reaction step is the reversible isomerization of aldoses or ketoses in their open chain forms to form an *enediol* intermediate (Fig. 2.8). This intermediate can then dehydrate to form a series of degradation products – in the case of hexoses the main product is *5-hydroxymethyl-*

2-furaldehyde (HMF), whereas pentoses yield mainly *2-furaldehyde* (furfural). HMF and furfural are considered useful indicators of accurate storage temperature of food samples.

2.2.4.2 Maillard browning

The *Maillard reaction* is the chemical reaction between an amino acid and a reducing sugar that, via the formation of a pool of reactive intermediates, leads to the formation of flavour compounds and melanoidin pigments (non-enzymic browning). The initial step in this complex series and network of reactions is the condensation of reducing sugar and amino acid. The reactive carbonyl group of the sugar reacts with the nucleophilic amino group of the amino acid to form an *Amadori compound* as shown in Fig. 2.9. This

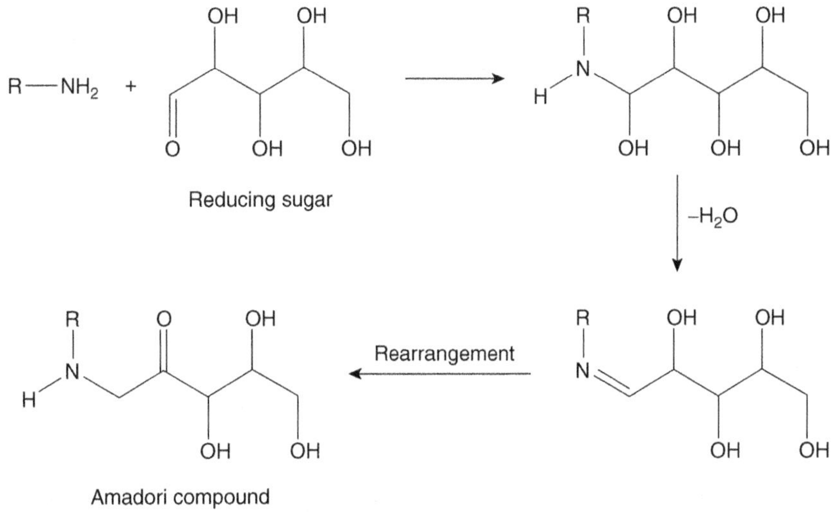

Figure 2.9 The formation of an Amadori compound during the initial stage of the Maillard reaction.

reaction normally requires heat (usually >100°C), is promoted by low moisture content, and is accelerated in an alkaline environment as the amino groups are deprotonated and hence have an increased nucleophilicity. Various reducing sugars have differing rates of reaction in the Maillard reaction; pentoses such as ribose, xylose and arabinose are more reactive than hexoses such as glucose, fructose and galactose. Different sugars give different breakdown products and hence unique flavour and colour.

2.2.4.3 Toxic sugar derivatives

The Maillard reaction, while desirable in many respects, does have certain implications for the loss of essential amino acids (cysteine and methionine), the formation of mutagenic compounds and the formation of compounds that can cause protein cross-linking, which is implicated in diabetes. The most concerning aspect is the potential for toxic sugar derivatives with mutagenic properties, primarily the group of compounds called *heterocyclic amines*. These are particularly associated with cooked meat, especially that which has been grilled at high temperature for long cooking times. In recent times the formation of *acrylamide* has been an issue of concern in potato-based snack foods.

2.3 Proteins

Proteins are polymers of amino acids linked together by peptide bonds. They can also be referred to as *polypeptides*. Proteins are key constituents of food, contributing towards organoleptic properties (particularly texture) and nutritive value. Proteins participate in tissue building and are therefore abundant in muscle and plant tissues.

2.3.1 Amino acids – the building blocks of proteins

2.3.1.1 Amino acid structure

The general structure of an *amino acid* is depicted in Fig. 2.10, and consists of an amino group (NH₂), a carboxyl group (COOH), a hydrogen atom and a distinctive R group all bonded to a single carbon atom, called the α-*carbon*. The R group is called the *side chain* and determines the identity of the amino acid.

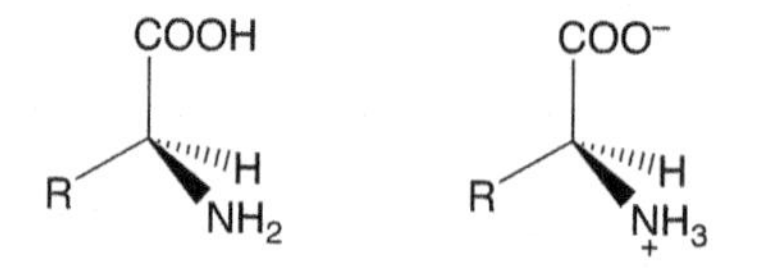

Figure 2.10 The general structure of an amino acid.

Amino acids in solution at neutral pH are predominantly *zwitterions*. The ionization state varies with pH: at acidic pH, the carboxyl group is un-ionized and the amino group is ionized; at alkaline pH, the carboxyl group is ionized and the amino group is un-ionized.

There are 20 different amino acids that are commonly found in proteins. The R group is different in each case and can be classified according to several criteria into four main types: *basic*, *non-polar* (hydrophobic), *polar* (uncharged) and *acidic*. Tables 2.1, 2.2 and 2.3 categorize the amino acids according to these types. The four different functional groups of amino acids are arranged in a tetrahedral array around the α-carbon atom; therefore, all amino acids are optically active apart from glycine. Of the possible L- or D-isomers, proteins contain only L-isomers of amino acids.

Some proteins contain *non-standard amino acids* in addition to the 20 standard amino acids (Fig. 2.11). These are formed by modification of a standard amino acid following its incorporation into the polypeptide chain (*post-translational modification*). Two examples that are encountered often in food proteins are *hydroxyproline* and *O-phosphoserine*. Hydroxyproline occurs in collagen and O-phosphoserine occurs in caseins.

2.3.1.2 Peptide bonds

The *peptide bond* is the covalent bond between amino acids that links them to form peptides and polypeptides (Fig. 2.12). A peptide bond is formed between the α-carboxyl group and the α-amino group of two amino acids by a condensation (or dehydration synthesis) reaction with the loss of water. Peptides are compounds formed by linking small numbers of amino acids (up to 50). A polypeptide is a chain of 50–100 amino acid residues. A protein is a polypeptide chain of 100+ amino acid residues and has a positively charged nitrogen-containing amino group at one end (*N-terminus*) and a negatively charged carboxyl group at its other end (*C-terminus*).

Table 2.1 Basic and acidic amino acids.

Amino acid	Single letter code	Structural formula	Amino acid	Single letter code	Structural formula
Basic: Arginine (Arg)	R		**Acidic:** Aspartic acid (Asp)	D	
Histidine (His)	H		Glutamic acid (Glu)	E	
Lysine (Lys)	K				

Table 2.2 Non-polar amino acids.

Amino acid	Single letter code	Structural formula	Amino acid	Single letter code	Structural formula
Alanine (Ala)	A		Phenylalanine (Phe)	F	
Isoleucine (Ile)	I		Proline (Pro)	P	
Leucine (Leu)	L		Tryptophan (Trp)	W	
Methionine (Met)	M		Valine (Val)	V	

Table 2.3 Polar amino acids.

Amino acid	Single letter code	Structural formula	Amino acid	Single letter code	Structural formula
Asparagine (Asn)	N		Serine (Ser)	S	
Cysteine (Cys)	C		Threonine (Thr)	T	
Glutamine (Gln)	Q		Tyrosine (Tyr)	Y	
Glycine (Gly)	G				

A special feature of the peptide bond is its *partial double bond character*. This arises because the peptide bond is stabilized by *resonance hybridization* between two structures, one single bonded between the carbon and nitrogen atoms, the other double bonded. As a consequence, the peptide bond is planar and stable. This has implications for the possible conformations adopted by a polypeptide chain, since no rotation is possible around the peptide bond. However, rotation is possible around bonds between the α-carbons and the amino nitrogen and carbonyl carbon of their residue.

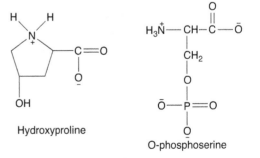

Hydroxyproline

O-phosphoserine

Figure 2.11 Chemical structures of some non-standard amino acids common in food proteins.

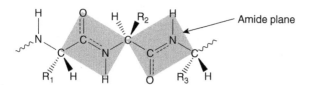

Figure 2.12 Peptide bonds in a polypeptide. The partial double bond character is represented by the dashed double bonds. The shaded boxes highlight atoms that exist within the same plane.

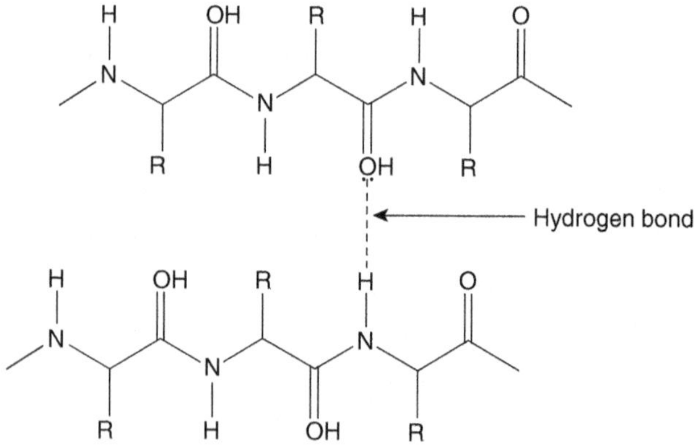

Figure 2.13 Hydrogen bonding between two polypeptides.

2.3.2 Molecular structure of proteins

2.3.2.1 Primary structure

The *primary structure* of a protein is simply the sequence of amino acids listed from the N-terminal amino acid. There are more than a billion possible sequences of the 20 amino acids and every protein will have a unique primary structure which determines how the protein folds into a three-dimensional conformation. If we compare the primary sequence Leu-Val-Phe-Gly-Arg-Cys-Glu-Leu-Ala-Ala with Gly-Leu-Arg-Phe-Cys-Val-Ala-Glu-Ala-Leu, these two peptides have the same number of amino acids, the same kinds of amino acids, but have different primary structures.

2.3.2.2 Secondary structure

The secondary structure of a protein describes the arrangement of the protein backbone (polypeptide chain) due to *hydrogen bonding* between its amino acid residues. Hydrogen bonding can occur between an amide hydrogen atom and a lone pair of electrons on a carbonyl oxygen atom, as shown in Fig. 2.13.

The peptide bond is planar, offering no rotation around its axis. This leaves only two bonds within each amino acid residue that have free rotation, namely the α-carbon to amino nitrogen and α-carbon to carboxyl carbon bonds. The rotations around these bonds are represented by the *dihedral angles* ϕ (phi) and ψ (psi), as shown for a tripeptide of alanine in Fig. 2.14. Ramachandran plotted ϕ and ψ combinations from known protein structures and found that there are certain sterically favourable combinations that form the basis for the preferred secondary structures. He also found that unfavourable orbital overlap precludes some combinations: $\phi = 0°$ and $\psi = 180°$; $\phi = 180°$ and $\psi = 0°$; $\phi = 0°$ and $\psi = 0°$.

Two kinds of hydrogen bonded secondary structures occur frequently with features that repeat at regular intervals. These *periodic structures* are the α-*helix* and the β-*pleated sheet*. The β-*pleated sheet* can give a two-dimensional array and can involve more than one polypeptide chain.

α-Helix

The α-helix is a coiled rod-like structure and involves a single polypeptide chain. The 'α' denotes that if you were to view the helix down its axis then you would note that it spirals clockwise away from you. The α-helix is stabilized by hydrogen bonding parallel to the helix axis and the carbonyl group of each residue is hydrogen bonded to the amide group of the residue that is four residues away if counting from the N-terminus. There are 3.6 residues for each turn of the helix and the dihedral angles are $\phi = -57°$ and $\psi = -48°$. The R group of each residue protrudes from the helix and plays no role in the formation of

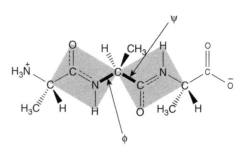

Figure 2.14 Bonds adjacent to peptide bonds with free rotation are depicted in bold with their respective dihedral angles ϕ and ψ.

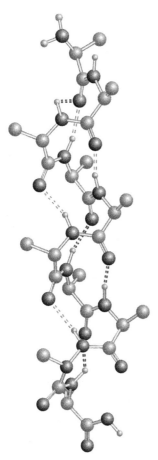

β-Pleated sheet

In β-pleated sheets the peptide backbone is almost completely extended (termed a β-*strand*) and hydrogen bonding is perpendicular to the direction of the polypeptide chain. Hydrogen bonds form between different parts of a single chain that is doubled back on itself (*intrachain bonds*) or between different chains (*interchain bond*) giving rise to a repeated zigzag structure (see Fig. 2.16). β-Sheets can be either *parallel* (where the β-strands run in the same direction) or *antiparallel* (where the β-strands run in opposite directions). The dihedral angles are $\phi = -119°$ and $\psi = +113°$ for pure parallel, and $\phi = -139°$ and $\psi = +135°$ for pure antiparallel β-sheets.

β-Turns

β-*Turns* are essentially hairpin turns in the polypeptide chain that allow it to reverse direction. In a β-turn the carbonyl oxygen of one residue is hydrogen bonded to the amide proton of a residue three

Figure 2.15 Ball and stick representation of an α-helix structure showing the position of hydrogen bonds between amide hydrogens (small white spheres) and carbonyl oxygens (larger dark spheres).

hydrogen bonding as part of the α-helix structure. In the illustration of an α-helix structure shown in Fig. 2.15, the hydrogen bonds are shown as dotted lines between backbone amide hydrogens and backbone carbonyl oxygens.

Proteins contain varying amounts of α-helix structure. Properties of α-helices include strength and low solubility in water. These properties arise because all amide hydrogen and carbonyl oxygen is involved in hydrogen bonds. Multiple strands of α-helix may entwine to make a protofibril such as in the muscle protein myosin.

All amino acids can be found in α-helix structure apart from *proline*, which disrupts α-helix. This is because its cyclic structure causes a bend in the backbone as a result of the restricted C–N bond rotation. This prevents the α-amino group from participating in intrachain hydrogen bonding.

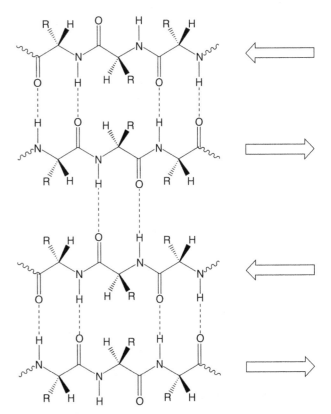

Figure 2.16 Hydrogen bonding within an antiparallel β-sheet structure. The arrow shows the direction of the polypeptide chain.

residues away. Proline and glycine are prevalent in β-turns. β-Turns are estimated to comprise between a quarter and a third of all residues in proteins and they are commonly found to link two strands of anti-parallel β-sheet.

Collagen triple helix

Collagen is a component of bone and connective tissue and is organized as strong water-insoluble fibres. It has a unique periodic structure comprising of three polypeptide chains wrapped around each other in a repeat sequence of X-Pro-Gly or X-Hyp-Gly (where X can be any amino acid). Proline and hydroxyproline make up to 30% of the residues in collagen, and hydroxylysine is also present. Every third position in the *collagen triple helix* is Gly because every third residue must sit inside the helix and only Gly is small enough.

The individual collagen chains are also helices and the three strands are held together by hydrogen bonds involving hydroxyproline and hydroxylysine residues. The molecular weight of the triple-stranded array is approximately 300,000 Daltons, involving approximately 800 amino acid residues. Intra- and intermolecular cross-linking stabilizes the collagen triple helix structure, especially covalent bonds between lysine and histidine. The amount of cross-linking increases with age. A major role for vitamin C (L-ascorbic acid) in vivo is in making collagen: proline and lysine in collagen are converted to 4-hydroxyproline and 5-hydroxylysine using this vitamin. Scurvy is a disease arising from a deficiency of vitamin C, and results in skin lesions, bleeding gums and fragile blood vessels.

2.3.2.3 Tertiary structure

The *tertiary structure* of a protein is the three-dimensional arrangement of all atoms within the molecule, and takes into account the conformations of side chains and the arrangement of helical and pleated sheet sections with respect to each other. Proteins *fold* to make the most stable structure and this structure will generally minimize solvent contact with residues of opposing polarity and hence minimize overall free energy. Therefore, in aqueous solution, proteins generally exist with their hydrophobic residues to the inside and their hydrophilic residues to the outside of their three-dimensional conformation.

There are two categories of tertiary structures:

- *Fibrous proteins* – overall shape is a long rod; mechanically strong; usually play a structural role in nature; relatively insoluble in water and unaffected by moderate changes in temperature and pH.
- *Globular proteins* – helical and pleated sheet sections fold back on each other; interactions between side chains important for protein folding; polar residues face surface and interact with solvent; non-polar residues face interior and interact with each other; structure is not static; generally more sensitive to temperature and pH change than their fibrous counterparts.

The tertiary structure of a protein is held together by interactions between the side chains. These can be through *non-covalent* interactions or *covalent* bonds. The most common non-covalent interactions are *electrostatic* (ionic bonds, salt bridges, ion pairing), *hydrogen bonds*, *hydrophobic interactions* and *van der Waals dispersion forces*. Covalent bonds in protein structure are primarily *disulphide bonds* (sulphur bridges) between cysteine residues, although other types of covalent bond can form between residues.

Electrostatic interactions

Some amino acids contain an extra carboxyl group (aspartic acid and glutamic acid) or an extra amino group (lysine, arginine, histidine). These groups can be ionized and therefore an ionic bond could be formed between the negative and the positive group if the chains folded in such a way that they were close to each other.

Hydrogen bonds

Hydrogen bonds can form between side chains since many amino acids contain groups in their side chains which have a hydrogen atom attached to either an oxygen or a nitrogen atom. This is a classic situation where hydrogen bonding can occur. For example, the amino acid serine contains a hydroxyl group in its side chain; therefore, hydrogen bonding could occur between two serine residues in different parts of a folded chain.

Hydrophobic interactions

Non-polar molecules or groups tend to cluster together in water; these associations are called hydrophobic interactions. The driving force for hydrophobic interactions is not the attraction of the non-polar molecules for one another, but is due to entropic factors relating to the strength of hydrogen bonding between water molecules.

Van der Waals dispersion forces

Several amino acids have quite large hydrocarbon groups in their side chains (e.g. leucine, isoleucine and phenylalanine). Temporary fluctuating dipoles in one of these groups could induce opposite dipoles in another group on a nearby folded chain. The dispersion forces set up would be enough to hold the folded structure together, although van der Waals forces are weaker and less specific than electrostatic and hydrogen bonds.

Disulphide bonding

If two cysteine side chains are oriented next to each other because of folding in the peptide chain, they can react to form a covalent bond called a *disulphide bond* or a *sulphur bridge*.

2.3.2.4 Quaternary structure

Not all proteins possess *quaternary structure*; it is only a property of proteins that consist of more than one polypeptide chain. Each chain is a subunit of the *oligomer* (protein), which is commonly a dimer, trimer or tetramer. Haemoglobin has quaternary structure. It is a tetramer consisting of two α- and two β-chains. The chains are similar to myoglobin and haemoglobin is able to bind four oxygen atoms through positive cooperativity.

2.3.3 Denaturation of proteins

The forces that stabilize the secondary, tertiary and quaternary structures of proteins can be disrupted through various chemical or physical treatments. This disruption of the native protein structure is defined as protein *denaturation*, which is an important process that may occur during the processing of foods. Denaturation is a change in a protein which causes an alteration in its physical and/or biological properties without rupture of its peptide bonds. It is generally observed as unfolding of the protein molecule from its uniquely ordered structure to a randomly ordered peptide chain. In the case of globular proteins, the denaturing process is often followed by *aggregation*, since previously buried hydrophobic residues are exposed to solution.

Denaturation is accompanied by a loss of native biological activity, but also affects physical properties. Some important consequences of protein denaturation are:

- Loss of biological activity (e.g. enzyme activity).
- Loss of solubility and changes to water-binding capacity.
- Increased intrinsic viscosity.
- Increased susceptibility to proteolysis.

Denaturation can be reversible, but if disulphide bonds are broken the denaturation process is often considered irreversible. Different proteins have different susceptibilities to denaturation since their individual structures are different. There are various denaturing agents that can destabilize protein structures that are categorized as physical agents or chemical agents.

Physical agents include heat, mechanical treatment, hydrostatic pressure, irradiation, and adsorption at interfaces. *Heat* is the most commonly encountered physical agent and is able to destabilize many bonds within proteins, including electrostatic bonds, hydrogen bonds and van der Waals interactions. Heat denaturation is useful in food processing since it tends to lead to improvement of sensory properties and protein digestibility, and can be used to manipulate foaming and emulsifying properties. Heating also promotes the participation of proteins in the Maillard reaction, which leads to the loss of nutritionally available lysine residues.

Chemical agents to denature proteins include acids, alkalis, metals, organic solvents and various organic solutes. Exposure to *acids* or *alkalis* (i.e. pH changes) affects the overall net charge on a protein, which will change the extent of electrostatic interactions, both attractive and repulsive. Most proteins are stable within a pH range around their isoelectric point (zero net charge) and the effects of acids or alkalis are normally reversible.

The presence of *organic solvents* weakens hydrophobic interactions since non-polar side chains become more soluble. *Organic solutes* can have a variety of effects. Urea alters the structure of water in such a way as to weaken hydrophobic interactions, leading to protein unfolding. Sodium dodecyl sulphate (SDS) is an anionic detergent that binds irreversibly to charged groups within a protein, inducing a large net negative charge that increases electrostatic repulsion, leading to unfolding. Reducing agents, such as mercaptoethanol and dithiothreitol, break disulphide bonds in proteins.

2.3.4 Post-translational modification

Post-translational modification of a protein is a chemical change that has occurred after the protein was synthesized by the body. Post-translational modification can create new functional families of proteins by attachment of biochemical functional groups to reactive groups on amino acid side chains, such as phosphate (*phosphoproteins*), various lipids (*lipoproteins*) and carbohydrates (*glycoproteins*). More simple modifications of amino acid side chains are also possible, such as the hydroxylation of lysine and proline encountered in collagen due to the action of vitamin C.

Enzymes also cause post-translational modification, such as peptide bond cleavage by specific proteases. One example of particular relevance in food production is the action of chymosin on casein. Chymosin cleaves the peptide bond between phenylalanine and methionine in κ-casein, which is used to bring about extensive precipitation and curd formation during cheese making.

2.3.5 Nutritional properties of proteins

Food proteins have an important nutritional role and are primarily used by the body to supply nitrogen and amino acids from which the body synthesizes its own proteins. Within the gastrointestinal tract, hydrolytic enzymes break down food proteins into their component amino acids, which are then used by the body to synthesize other substances. The liver balances the pattern of amino acid supply against the needs of synthesis.

In terms of nutritive value, proteins are classified according to their content of non-essential and essential amino acids. *Non-essential amino acids* are synthesized by the body and require only an adequate supply of amino nitrogen and carbohydrate. However, humans cannot synthesize some *essential amino acids* and these must be supplied by our diet. The essential amino acids are: histidine, isoleucine, leucine, lysine, methionine, phenylalanine, threonine, tryptophan and valine.

Protein efficiency ratio (PER) is used as a measure of how well food protein sources supply essential amino acids. Human breast milk is treated as a standard and is given a PER score of 100%. In general,

animal protein foods (eggs, milk and meat) are very efficient sources, whereas plant protein foods are less efficient since they are usually deficient in either lysine or methionine. For this reason, vegetarians need to consume a balanced diet of plant products to ensure sufficient supply of these two amino acids.

2.4 Lipids

Lipids are a group of molecules that contribute to the structure of living cells and are also used in the body for the purpose of energy storage. Dietary lipids have important roles for provision of energy and as carriers of fat-soluble vitamins. Generally speaking, all lipids are soluble in non-polar organic solvents and have low solubility in water. Dietary lipids are commonly referred to as *oils* and *fats*. Edible oils are liquid at room temperature, whereas fats are solid or semi-solid at room temperature. The lipids found in oils and fats are chemically very diverse, but are predominantly long-chain *fatty acid esters*. Other lipid types encountered in foods are also either fatty acids or derivatives of fatty acids, and include *triglycerides*, *phospholipids*, *sterols* and *tocopherols*.

Lipids can be broadly classified into three main groups:

- *Simple lipids* yield two classes of product when hydrolyzed, e.g. glycerides (acylglycerols) which are hydrolyzed to give glycerol and a fatty acid.
- *Complex lipids* yield three or more classes of product when hydrolyzed, e.g. phospholipids, which are hydrolyzed to give alcohols, fatty acids and phosphoric acid.
- *Derived lipids* are non-hydrolyzable and do not fit into either of the above classes, e.g. sterol, tocopherol and vitamin A.

2.4.1 Lipid structure and nomenclature

2.4.1.1 Fatty acids

A *fatty acid* is a carboxylic acid having a long unbranched aliphatic tail or chain, and can be described chemically as an *aliphatic monocarboxylic acid*. The aliphatic chain can be either *saturated* (no double bonds between carbons) or *unsaturated* (one or more double bonds between carbons). Saturated

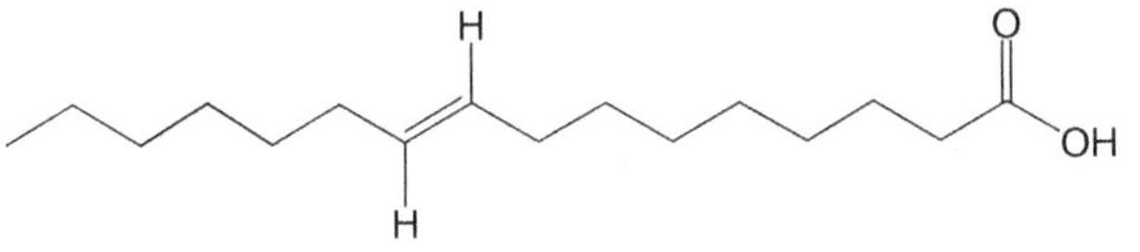

trans-9-hexadecenoic acid

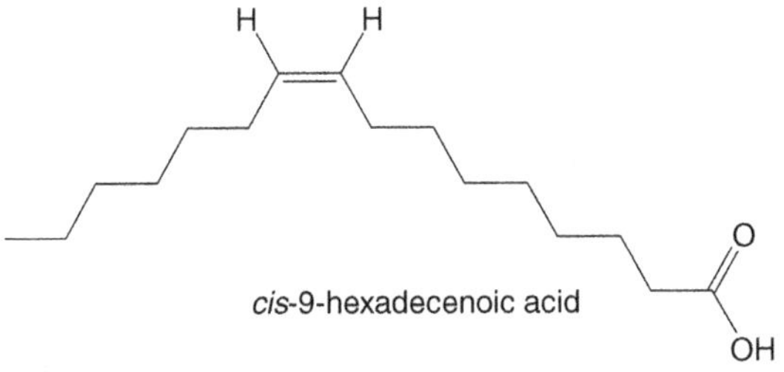

cis-9-hexadecenoic acid

Figure 2.17 *Cis* and *trans* double bond configurations in a fatty acid chain.

fatty acids have the general chemical structure $CH_3(CH_2)_{n-2}CO_2H$, and commonly contain an even number of carbon atoms from $n = 4$ to $n = 20$. In unsaturated fatty acids the double bonds can adopt either a *cis*- or a *trans*-configuration as illustrated in Fig. 2.17. Unsaturated acids can contain one (*monounsaturated*) or several (*polyunsaturated*) double bonds.

There are several conventions for the naming of fatty acids; they can be referred to by their *systematic name* or by a *trivial name*, and it is important to be familiar with both names. A shorthand notation for fatty acids is often employed, which takes the form of a *lipid number*, C:D, where C is the number of

carbon atoms and D is the number of double bonds in the fatty acid. This notation can be too general for unsaturated fatty acids since the position of the double bond is not specified; therefore, the lipid number is usually paired with a *delta-n* notation such that each double bond is indicated by Δ^n, where the double bond is located on the nth carbon–carbon bond, counting from the carboxylic acid end. Each double bond is preceded by a *cis*- or *trans*- prefix, indicating the configuration of the bond. Table 2.4 summarizes names and notations for some of the most common fatty acids in foods.

2.4.1.2 Triglycerides

While fatty acids are the most common structural component of lipids, oils and fats are largely composed of mixtures of *triglycerides*. Triglycerides are also known as *triacylglycerols* and are esters of three fatty acids with glycerol. A typical triglyceride structure is depicted in Fig. 2.18. The three fatty acid residues may or may not be the same, i.e. they can be simple or mixed triglycerides. Natural oils and fats will therefore contain a characteristic profile of different fatty acids dependent on their source. For example, fish oils are rich in long-chain polyunsaturated fatty acids (PUFAs) with up to six double bonds, while many vegetable oils are rich in oleic and linoleic acids. Vegetable oils are important in the diet since the body is unable to synthesize linoleic acid, which is an important precursor of *prostaglandins*, a class of hormones that are involved in inflammation and smooth muscle contraction.

Table 2.4 Fatty acid nomenclature.

Systematic name	Trivial name	Lipid number
Butanoic acid	Butyric acid	4:0
Hexanoic acid	Caproic acid	6:0
Octanoic acid	Caprylic acid	8:0
Decanoic acid	Capric acid	10:0
Dodecanoic acid	Lauric acid	12:0
Tetradecanoic acid	Myristic acid	14:0
Hexadecanoic acid	Palmitic acid	16:0
cis-9-Hexadecenoic acid	Palmitoleic acid	16:1, *cis*-Δ^9
Octadecanoic acid	Stearic acid	18:0
cis-9-Octadecenoic acid	Oleic acid	18:1, *cis*-Δ^9
cis, cis-9,12-Octadecadienoic acid	Linoleic acid	18:2, *cis,cis*-$\Delta^{9,12}$
all-*cis*-9,12,15-Octadecatrienoic acid	Linolenic acid	18:3, *cis,cis,cis*-$\Delta^{9,12,15}$
Eicosanoic acid	Arachidic acid	20:0
all-*cis*-5,8,11,14-Eicosatetraenoic acid	Arachidonic acid	20:4, *cis,cis,cis,cis*-$\Delta^{5,8,11,14}$
cis-13-Docosenoic acid	Erucic acid	22:1, *cis*-Δ^{13}

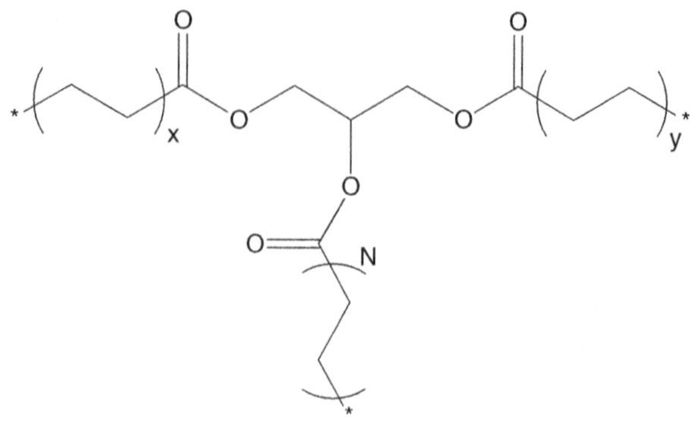

Figure 2.18 The general structure of a triglyceride.

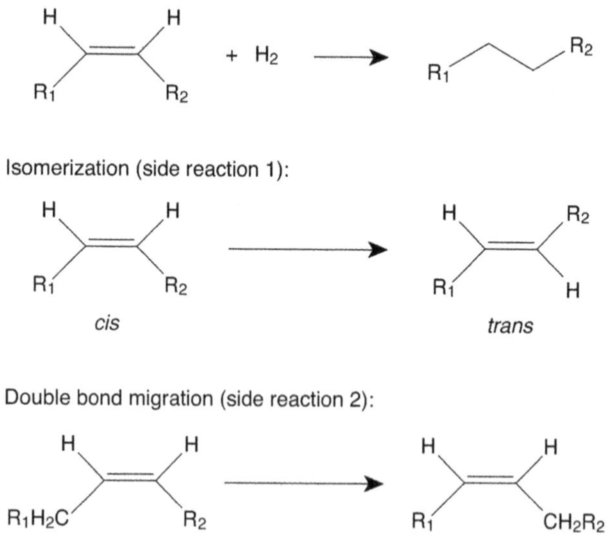

2.4.2 Polymorphism

Polymorphism is an important property of triglyceride crystallization that influences the melting properties of triglycerides. Polymorphic forms have the same chemical composition, but differ in their crystalline structure. Each has a characteristic melting point and there are three basic polymorphic forms: α, β, and β'. The most stable form is β and the least stable is α, and these can be present together within the same sample of a fat. Crystals of β' fats tend to be small and needle-like, and therefore form better emulsions than the other polymorphic forms.

It is possible to transform a fat from one polymorphic form to another through melting and recrystallization. If a triglyceride is melted and then cooled rapidly it will adopt the α-form. Slow heating will then give a liquid fat that recrystallizes into the β'-form. Repeating this second step yields the stable β-form.

2.4.3 Oil and fat processing

2.4.3.1 Hydrogenation of lipids

Hydrogenation is an important industrial process to convert liquid oils into semi-solid fats for the production of margarine or shortenings. Hydrogenation also increases oxidative stability since unsaturated fatty acids are converted to saturated fatty acids. Briefly, during hydrogenation oils are exposed to hydrogen gas under conditions of high temperature (150–180°C) and pressure (2–10 atm) in the presence of a nickel catalyst. The main reaction is given in Fig. 2.19 alongside the two possible side reactions that may occur: *isomerization* and *double bond migration*.

Figure 2.19 Reactions that occur during the hydrogenation of oils and fats.

The isomerization from *cis*- to *trans*-fat is undesirable since the nutritional quality of the fat is reduced. Indeed, the consumption of *trans*-fat arising from partial hydrogenation is linked to an increased risk of coronary heart disease. There has consequently been a worldwide drive to eliminate the consumption of *trans*-fat in the diet. Double bond migration is also linked to a reduction of nutritional quality. Given the occurrence of side reactions, it is important to optimize the *selectivity* of hydrogenation by use of optimal catalysts and conditions.

2.4.3.2 Interesterification

Naturally occurring fats do not contain a random distribution of fatty acids amongst their triglycerides. This is important since the physical characteristics of fats are affected by the fatty acid distribution in triglycerides as well as the overall nature of the fatty acids present. To improve the physical consistency of fats it is possible to perform *interesterification*, which rearranges the fatty acids so that they become distributed randomly among the triglyceride molecules.

Interesterification can be achieved through heating fats at high temperature ($<200°$C) for long periods, but it is more efficient to employ catalysts that speed up the process (30 min) and lower the temperature required (50°C). Sodium methoxide is the most

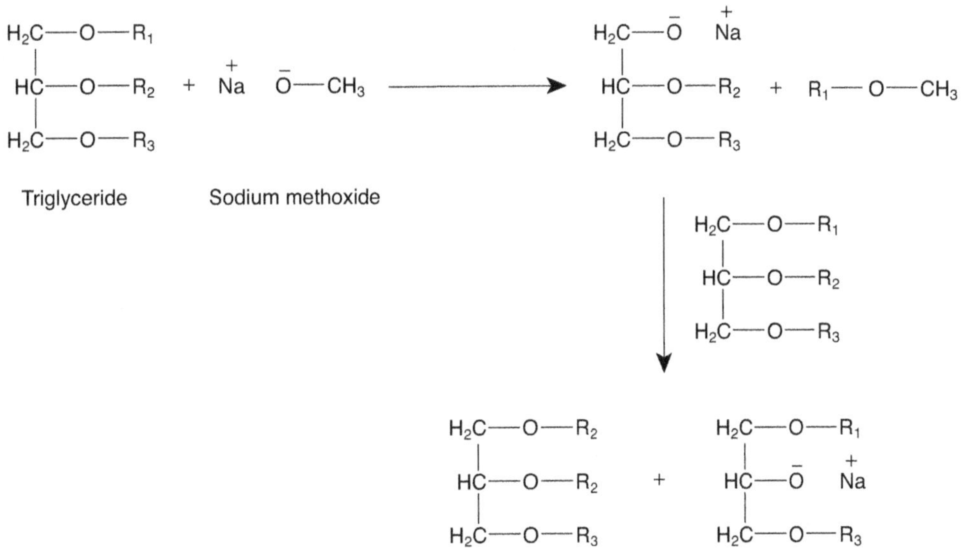

Figure 2.20 shows the reaction scheme for sodium methoxide catalyzed interesterification.

Figure 2.20 Sodium methoxide catalyzed interesterification of triglycerides.

popular catalyst for this process which is shown in Fig. 2.20.

In addition to chemically catalyzed interesterification, it is possible to use enzymes to catalyze a similar process known as *transesterification*. In this case fungal lipases are used to modify palm oils that are rich in 1,3-dipalmityl, 2-oleyl glycerol (POP triglycerides) to yield a fat with an identical profile of POP, POS (1-palmityl, 2-oleyl, 3-stearyl glycerol) and SOS (1,3-distearyl, 2-oleyl glycerol) triglycerides to cocoa butter, which is more expensive (hence value is added).

2.4.4 Lipid oxidation

2.4.4.1 Mechanism

Lipid oxidation is a major cause of food spoilage and causes the generation of off-flavours and off-odours that are termed *rancid*. The fundamental mechanism of lipid oxidation is that of *autoxidation*, which comprises three steps of *initiation*, *propagation* and *termination* as shown in Fig. 2.21. The initiation step involves the generation of highly reactive *free radicals* (molecules having unpaired electrons). These then react with atmospheric oxygen to generate *peroxy radicals* (ROO·), and a chain reaction is set in motion until terminated by the formation of non-radical products.

Autoxidation is accelerated at higher temperature and is more rapid for fats containing polyunsaturated fatty acids. Autoxidation is also sensitive to small concentrations of *antioxidants* or *pro-oxidants*. Pro-oxidants are predominantly metal ions (e.g. iron) that increase the rate of lipid oxidation, either through free radical initiation, acceleration of hydroperoxide (ROOH) decomposition, or activation of molecular oxygen (O_2) to give reactive singlet oxygen ($^1O_2$) and peroxy radicals.

2.4.5 Antioxidants

Antioxidants are substances that can retard the autoxidation of lipids. Both natural and synthetic antioxidants are available as permitted food additives,

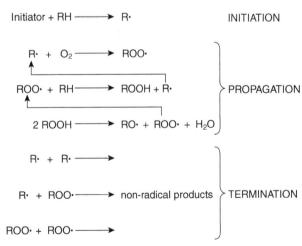

Figure 2.21 The reactions occurring during the autoxidation of lipids.

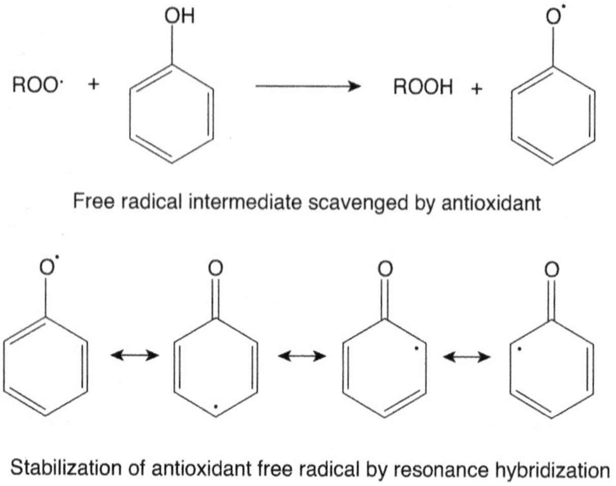

Free radical intermediate scavenged by antioxidant

Stabilization of antioxidant free radical by resonance hybridization

Figure 2.22 Free radical scavenging by a phenol ring.

and these are mainly phenolic compounds bearing various ring substitutions. The mechanisms of antioxidant activity are a topic of active research, particularly for naturally occurring polyphenols that are abundant in many plant-derived foods. In the case of most synthetic antioxidants, the antioxidant acts as a *radical scavenger* to block the propagation step of autoxidation (see Fig. 2.22).

2.5 Minor components of foods

2.5.1 Permitted additives

The naturally occurring components of foods have a wide range of functional properties that contribute to overall product quality. However, in some cases we may wish to use an additive to enhance food quality, be it for better appearance, texture, flavour, nutritive value or shelf-life. The use of such food additives, especially those that are synthetic in origin, is strictly regulated by government legislation. Briefly summarized here are some of the colours and preservatives that are used as food additives.

2.5.1.1 Colours

Colour is a key sensory component of food and therefore many processed foods contain added colorants. These colorants can be naturally occurring pigments or synthetic dyes, but in recent years there have been moves away from the use of synthetic dyes due to consumer demands.

There are several naturally occurring pigments that are used in foods which are derived from plants, insects and bacteria, including chlorophylls, carotenoids and anthocyanins. *Chlorophylls* are green pigments present in leafy vegetables, fruit, algae and photosynthetic bacteria. Chlorophylls are unstable to heat and insoluble in water; therefore, chlorophyll derivatives (e.g. copper chlorophyllin) are used as added colorants.

Carotenoid pigments give yellow and orange colours in fruits and vegetables, and can be subdivided into carotenes (hydrocarbons) and xanthophylls (contain oxygen). β-Carotene gives the orange colour to carrots, while lycopene occurs in tomatoes and astaxanthin is the molecule responsible for the pink colour of salmon. Carotenoids are of interest for other than their colour since they are generally found to possess antioxidant properties.

Anthocyanins are polyphenolic compounds found in flowers, fruit and vegetables, and give rise to red, violet and blue colours. Fruits such as blackcurrants, blackberries, blueberries, raspberries, strawberries and grapes are particularly rich in anthocyanins. Anthocyanins are the glycoside form of anthocyanidins and, like carotenoids, are noted for their antioxidant properties. Anthocyanins find widespread use as food colours in confectionery and soft drinks, but are of limited use in some foods as their colour is not stable outside of the acid pH range.

Synthetic dyes or artificial food colorants are commonly *azo dyes* (e.g. carmoisine, amaranth), whose colour originates from the azo group (R_1-N=N-R_2). The R-groups in azo dyes are normally aromatic systems, giving a conjugated double bond system that allows for a range of colours (yellow, orange, red, brown). Other synthetic dyes are triarylmethanes (green S, brilliant blue FCF), xanthenes (erythrosine) and quinolines (quinoline yellow).

2.5.1.2 Preservatives

Preservatives are added to food either to prevent rancidity or to prevent microbial growth. Antioxidants have been discussed earlier and can be naturally occurring or synthetic in origin; synthetic antioxidants include butylated hydroxyanisole (BHA), butylated hydroxytoluene (BHT) and propyl gallate. Antimicrobial agents are used to prevent spoilage by bacteria, yeast and moulds, and include sulphur dioxide (in the form of SO_2 generating sulphites), benzoic acid (or benzoates), sorbic acid

Table 2.5 Classification of vitamins according to solubility.

Water-soluble	Fat-soluble
Thiamin (vitamin B_1)	Retinol (vitamin A)
Riboflavin (vitamin B_2)	Cholecalciferol (vitamin D)
Niacin (nicotinic acid, nicotinamide)	α-Tocopherol (vitamin E)
Pantothenic acid (vitamin B_5)	Phylloquinone (menaquinones, vitamin K)
Pyridoxine (vitamin B_6)	
Biotin (vitamin B_7)	
Folic acid (vitamin B_9)	
Cobalamin (vitamin B_{12})	
L-Ascorbic acid (vitamin C)	

(or sorbates), nisin (a polypeptide antibiotic) and nitrites.

2.5.2 Vitamins

The *vitamins* are a group of organic nutrients that cannot be synthesized in sufficient quantities by our bodies and therefore must be obtained from the diet. Vitamins are impossible to classify in structural terms since they are widely disparate in structure and are defined only by their biological and chemical activity. Each vitamin constitutes a group of *vitamer* compounds, which show the biological activity of a particular vitamin. The only broad chemical classification of vitamins is by solubility into two groups: *water-soluble vitamins* and *fat-soluble vitamins* (see Table 2.5). Water-soluble vitamins are easily absorbed by the body and excreted. Fat-soluble vitamins are absorbed through the intestinal tract with the help of lipids.

Vitamins have many biochemical functions: hormones, antioxidants, cell signalling, tissue growth, etc. Most vitamins function as precursors for enzyme cofactor biomolecules (coenzymes) that help act as catalysts and substrates in metabolism. When acting as part of a catalyst, vitamins are bound to enzymes and are called *prosthetic groups*.

2.5.3 Minerals

Minerals in foods include a range of inorganic elements that are required by living organisms to support biochemical processes, including building bones and teeth, transmitting nerve signals, energy conversion from food and vitamin biosynthesis. Minerals encompass all elements that are required as essential nutrients other than carbon, hydrogen, nitrogen and oxygen, which are all present in common organic molecules. Minerals can be found in varying amounts in a variety of foods such as meat, fish, cereals, milk and dairy products, vegetables, fruit and nuts. There are two types of mineral: *bulk minerals* (macro-minerals or essential minerals) and *trace minerals*. The distinction between these two types is that the body needs larger amounts (>200 mg/day) of bulk minerals than it does trace minerals. Trace minerals can be hazardous if consumed in excess.

The bulk minerals include calcium, chloride, magnesium, phosphorus, potassium and sodium. Trace minerals include cobalt, copper, fluorine, iodine, iron, manganese, molybdenum, nickel, selenium, sulphur and zinc. Some important food sources of certain minerals are: dairy products and green leafy vegetables for calcium; nuts, soy beans and cocoa for magnesium; table salt, olives, milk and spinach for sodium; legumes, potato skin, tomatoes and bananas for potassium; table salt for chloride; meat, eggs and legumes for sulphur; red meat, green leafy vegetables, fish, eggs, dried fruits, beans and whole grains for iron.

2.6 Water in foods

2.6.1 Water activity

Water is the most abundant component of food and is important to the stability of most food products. Water is a controlling factor in microbial spoilage, and also contributes to chemical and physical stability. For this reason the control of water content is commonly used as a preservation technique, such as by spray drying or freeze drying.

One of the most important proximate analyses of food raw materials and products is to determine moisture or *water content*. This is a quantitative analysis that determines the total amount of water present by drying methods, Karl Fischer titration, near infrared spectroscopy, NMR spectroscopy, etc. However, microbial responses and chemical reactions in foods cannot be reliably predicted by a direct relationship to water content alone, since different foods having the same water content can have very different properties. This is because there are differences in the strength with which water associates with other components in different foods.

Chemically bound water cannot be used to support microbial growth, so it is difficult to assess using only water content values how much water is available to microbes or to influence other aspects of product quality (e.g. rheological properties). Therefore, the influence of water on certain properties is better expressed also in terms of *water activity* (a_w), which is an expression of the reactivity of water in a food, and indicates how tightly water is structurally or chemically bound. Water activity is a thermodynamic concept related to the chemical potential (or Gibbs free energy) of water in solution (μ_w), the chemical potential of pure water (μ_w^o), the gas constant ($R = 8.314$ J mol^{-1} K^{-1}) and temperature (T) by the following expression:

$$\mu_w = \mu_w^o + RT \ln a_w$$

The chemical potential of unbound or free water in solution is equal to the chemical potential of pure water; thus, by the above expression for free water $a_w = 1$ (since ln 1 = 0). The full range of water activity is $0 \leq a_w \leq 1$ and it has no units. Since it is a thermodynamic concept, to measure a_w a system being measured must be in equilibrium, the temperature must be defined, and a standard state must be specified (pure water).

There is no method available to directly measure a_w. However, a_w can be defined more simply in terms of the vapour pressure of water above a sample of a food (p) and the vapour pressure of pure water at the same temperature (p_0):

$$a_w = \frac{p}{p_0}$$

The value of a_w for a sample can therefore be determined from the *equilibrium relative humidity* (ERH, %) of air surrounding the sample in a sealed measurement chamber:

$$a_w = \frac{ERH}{100}$$

Relative humidity sensors measure ERH through changes in electrical resistance or capacitance of the sensor material.

2.6.2 Microbial growth, chemical reactivity and food texture

Control of a_w is very important in the food industry, since it strongly influences microbial growth, the rate of chemical reactions and food texture. The growth of most bacteria, including pathogenic *Salmonella*, *Escherichia* and *Clostridia*, is inhibited below $a_w = 0.91$. Most yeasts are inhibited below $a_w = 0.87$ and most moulds are inhibited below $a_w = 0.80$. Below $a_w = 0.60$ there is almost no possibility for microbial proliferation. Highly perishable foods (fruits, vegetables, meat, fish and milk) are found to have $a_w > 0.95$.

In relation to *chemical reactivity*, water can act as a solvent or reactant, or alter the mobility of reactants by affecting viscosity. Water activity thus influences non-enzymic browning, lipid oxidization, degradation of vitamins, enzymatic hydrolysis, protein denaturation, and starch gelatinization and retrogradation. It is found that the relative rates of certain reactions are optimal over particular windows of a_w, as is illustrated in Fig. 2.23 for enzymatic hydrolysis, lipid oxidation and browning reactions.

Textural properties of foods show large variations with changes in a_w. Foods with high a_w normally are described as moist, juicy, tender or chewy. When the a_w of these products is lowered, undesirable textural attributes are observed, such as hardness, dryness,

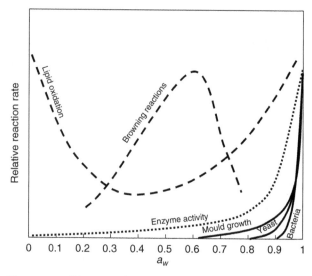

Figure 2.23 Water activity vs stability.

staleness and toughness. Foods with low a_w normally are described in terms of crispness and crunchiness, while these products at higher a_w lose these attributes. Water activity is therefore a critical factor for the sensory acceptability of many foods.

2.6.3 Sorption isotherms

The process of *sorption* encompasses the action of both adsorption and absorption. *Adsorption* is a surface phenomenon relating to the physical adherence or chemical bonding of molecules onto the surface of another molecule or material. *Absorption* involves the incorporation of a substance in one state into another of a different state (e.g. a liquid can be absorbed by a solid; a gas can be absorbed by a liquid).

Sorption isotherms are plots of the relationship at a constant temperature between water content (g H_2O/g dry matter) and a_w. It is important to note that sorption isotherms are temperature dependent. Figure 2.24 shows a generalized plot of a water sorption isotherm. This plot shows the typical sigmoidal shape observed for most foods. Also depicted is the *hysteresis* between the individual *desorption* (removal of water) and *resorption* (addition of water) isotherms, which is a phenomenon that is typically encountered due primarily to physical changes undergone by a food during the addition or removal of water. The practical consequence is that for any given a_w a food will contain more water during desorption than during resorption.

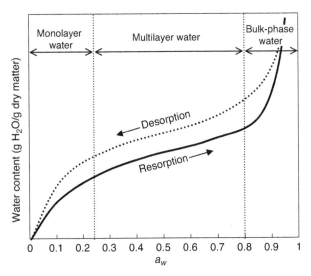

Figure 2.24 A generalized plot of a water sorption isotherm.

There are three zones shown in Fig. 2.24, which denote the type of water sorption: *monolayer water* is that which is mostly strongly bound to the surface of the food, and represents water bound to accessible polar groups on the food surface; *multilayer water* represents additional layers of adsorption around polar groups through hydrogen bonding; *bulk-phase water* is the least strongly bound and therefore most mobile water in the food. Bulk-phase water is the most easily removed water from a food and is typically available as a solvent.

2.7 Physical chemistry of dispersed systems

The application of physical chemistry to foods is wide and far reaching. It encompasses concepts of thermodynamic equilibria, chemical bonding, interaction forces and reaction kinetics, the fundamentals of which are dealt with in texts dedicated to physical chemistry and the physical chemistry of foods. Aspects such as structure/function relationships in food macromolecules (i.e. polysaccharides and proteins) are touched on in other sections of this chapter. In this section, the physical chemistry of solutions and dispersed systems is discussed.

2.7.1 Solutions

A *solution* is a homogeneous mixture composed of two or more substances. In a simple solution, a *solute* is dissolved in a *solvent*. The most familiar types of solution are those where a solid is dissolved in a liquid (e.g. salt in water). However, gases may dissolve in liquids (e.g. carbon dioxide in water) and liquids may dissolve in other liquids (e.g. ethanol in water). Solutions are present in almost all foods; otherwise the food itself is in the form of a solution. The property of a solute that dissolves in a solvent is called *solubility*.

2.7.1.1 Solvents

Liquid solvents are classified as *polar* or *non-polar*. Polar solvents are molecules that contain *dipoles* (usually due to an electronegative atom such as oxygen or nitrogen that draws electrons towards it). Water is the most common polar solvent, whereas hydrocarbons like *n*-hexane are examples of non-polar solvents. In general, polar solutes will be more soluble in polar solvents and less soluble (or insoluble) in non-polar solvents. Similarly, non-polar solutes have greatest solubility in non-polar solvents.

2.7.1.2 Solvation

Solvation involves different types of intermolecular interactions: hydrogen bonding, ion–dipole and dipole–dipole attractions or van der Waals forces. Hydrogen bonding, ion–dipole and dipole–dipole interactions occur only in polar solvents; therefore, ionic solutes (e.g. salts) will only dissolve in polar solvents. The solvation process will only be thermodynamically favoured if the free energy of the system decreases. In other words, the solvent must stabilize the solute for solvation to occur, i.e. the solute molecules are preferentially surrounded by solvent molecules rather than by other solute molecules.

If the solvent is water, the process of solvation is called *hydration*. Generally, ions or ionic groups are most soluble in water and are highly hydrated through ion–dipole interactions. Polar groups, especially those that can form hydrogen bonds, are said to be hydrophilic and are also highly hydrated, although not as strongly as ions. Non-polar groups are hydrophobic and cannot participate in hydrogen bonding; their presence in water causes breaking of hydrogen bonds, which is energetically unfavourable, leading to solute aggregation to minimize the disruption to hydrogen bonding. This is the reason why denatured proteins that have exposed non-polar regions will tend to aggregate and precipitate from water.

2.7.1.3 Factors affecting solubility

A *saturated solution* is formed when no more solute can be dissolved in a solvent. However, solubility is a thermodynamic property that relies on an equilibrium state. Therefore, the concentration at which a solution becomes saturated can be influenced significantly by different environmental factors, such as temperature and pressure, although the pressure dependence of solubility is negligible for solids and liquids (condensed phases). Solubility may also strongly depend on the presence of other solutes in solution.

The solubility of solid solutes is normally increased by increasing the temperature of the solvent. It is therefore possible for some solute–solvent combinations to form a *supersaturated* solution by increasing the temperature to dissolve more solute. Most gases are less soluble at higher temperatures, while the solubility of liquids in liquids is generally less temperature-sensitive than that of solids or gases.

2.7.1.4 Partitioning

Often in foods there are present immiscible solvents or phases (e.g. water and oil) in which a solute may have limited or differing solubility. This is very common for flavour molecule solutes and various food additives. In these cases the solute partitions itself between the two solvents according to its *partition ratio* (K_D) for the two phases:

$$K_D = \frac{c_1}{c_2}$$

where c_1 and c_2 are the solute concentrations in phase 1 and phase 2, respectively. Essentially, the higher the solubility in phase 1 relative to phase 2, the higher will be the partition ratio. The concept of partitioning is used extensively in analytical chemistry, particularly for the separation of complex chemical mixtures using chromatography.

2.7.2 Dispersed systems

The majority of foods are *dispersed systems*; they are non-homogeneous (*heterogeneous*) mixtures. Heterogeneous systems have structural elements that can vary considerably for the same overall chemical composition depending upon how they were created. Dispersed systems are generally composed of one or more dispersed phases and a *continuous phase*. The *dispersed phase* is usually a particle, crystal, fibre or particle aggregate. Dispersions are called colloids if the particles are larger than molecules and too small to be visible (10^{-8} to 10^{-5} m).

There are two main types of dispersed system: those that are *lyophilic* (solvent liking) and those that are *lyophobic* (solvent hating). Lyophilic dispersions are in thermodynamic equilibrium and form spontaneously on mixing. These include *macromolecular solutions*, a macromolecule being a polymer such as a polysaccharide or protein where the molecules are large enough to be considered particles, and *association colloids*. Association colloids such as surfactant micelles occur when small molecules associate to form a larger structure in order to optimize their interaction with solvent.

Lyophobic dispersions do not form spontaneously, i.e. energy input is required. In this type of dispersion, the dispersed and continuous phases are *immiscible*. Notable lyophobic dispersions encountered in foods include *foams* (gas dispersed in liquid or solid), *emulsions* (liquid dispersed in liquid) and *gels* (solid dispersed in liquid). The stability of lyophobic dispersions is dependent upon colloidal interactions and the stabilization of interfaces, as discussed in the following sections.

2.7.2.1 Colloidal interactions

Colloidal interaction forces primarily act perpendicular to the surface of colloid particles. Colloid particles are essentially hard particles that cannot overlap; hence the most fundamental colloidal interaction force is *excluded volume repulsion*. However, two other forces are essential to our understanding of colloidal interaction and colloid stability; these are *van der Waals attraction* and *electrostatic repulsion*. These interaction forces are best discussed in terms of the *DLVO theory*, which is named for its originators Deryagin, Landau, Verwey and Overbeek.

Van der Waals attractive forces act between all molecules. The forces are due to interaction between two dipoles that are either permanent or induced. In the case of induced dipole–induced dipole interactions, fluctuations in electron density give rise to a temporary dipole in a particle, which itself induces a dipole in neighbouring particles. The temporary dipole and the induced dipoles are then attracted to each other by these *dispersion forces* or *London forces*. Van der Waals forces are weak and only act over short distances.

All aqueous surfaces carry an electric charge or *electrostatic surface potential* that causes ions of opposite charge (*counter ions*) to accumulate at the surface to form an *electrical double layer*. The electrical potential of the electrical double layer is often described as its *zeta potential*. As the surfaces of two particles in an aqueous medium approach one another, their electrical double layers will begin to overlap, and it is this overlap that gives rise to electrostatic repulsive forces.

According to DLVO theory, the stability of a colloidal system is determined by the sum of the van der Waals attractive and electrostatic repulsive forces that exist between particles as they approach each other (see Fig. 2.25). An energy barrier resulting from electrostatic repulsion prevents two particles approach-

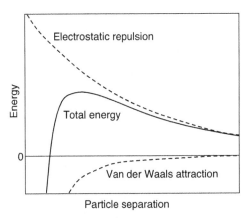

Figure 2.25 The variation of free energy with particle separation as described by the DLVO theory.

ing one another and adhering together; however, if the particles collide with sufficient energy to overcome that barrier, the van der Waals attractive force will pull them into contact. Therefore, if colloid particles have a sufficiently high electrostatic repulsion, a colloidal dispersion will resist flocculation and the colloid will be stable, but if electrostatic repulsion is weak or non-existent, then flocculation will eventually take place. Electrostatic repulsion can be weakened by the addition of a salt to the system or by changing the pH to neutralize surface potentials.

2.7.2.2 Foams and emulsions

Foams and emulsions are perhaps the most commonly encountered lyophobic dispersions in foods. As has been mentioned earlier, foams comprise a gas dispersed in a liquid. In foods, foams are commonly found as air-in-water dispersions. Emulsions are dispersions of two immiscible liquids. There are two types of emulsion common in foods: oil-in-water (e.g. milk, cream, mayonnaise) and water-in-oil (e.g. butter, margarine).

To create a foam or emulsion requires the input of energy since the interfacial area between the two immiscible phases is increased, which increases *interfacial free energy* or *surface tension*. The increase in interfacial free energy makes foams and emulsions inherently unstable systems; therefore, some means of stabilization is required via a reduction in surface tension by the use of *surfactants*.

Surface tension is the force that acts to minimize the interfacial area between two immiscible phases. The existence of surface tension is the reason that water droplets have a spherical shape, since this is the geometry that gives the smallest surface area for any given volume. Surface tension also causes the coalescence of small droplets into larger droplets when they collide. Surface tension is defined as a force per unit length and is given the symbol γ and the unit $N\,m^{-1}$. Surface tension acts in the same direction as the interface. Surface tension is equivalent to the interfacial free energy, which is the work required to increase the area of an interface, and is given the unit $J\,m^{-2}$. Since $1\,J = 1\,N\,m$, surface tension and interfacial free energy are identical parameters.

Surfactants are essential to the formation of foams and emulsions since they act to reduce interfacial free energy or surface tension. Surfactants are *amphiphilic*, which means that they contain both hydrophilic and lipophilic portions within the same molecule. This property means that surfactants tend to adsorb at interfaces, i.e. they are *surface active*. The term surfactant is normally associated with small-molecule surfactants, also called *detergents* or *emulsifiers*; however, proteins can also act as surfactants and are generally better at stabilizing foams and emulsions since they form intermolecular associations that stabilize their surface films.

Surfactants are characterized by their *hydrophilic–lipophilic balance* (HLB) values. An HLB value of 0 corresponds to a completely lipophilic molecule and a value of 20 corresponds to a completely hydrophilic molecule. The HLB value can be used to predict the surfactant properties of a molecule; surfactants with HLB values from 4 to 6 are good water-in-oil emulsifiers (e.g. sorbitan mono stearate); surfactants with HLB values from 8 to 18 are good oil-in-water emulsifiers (e.g. lactoyl monopalmitate, sorbitan mono laurate, lecithin, poly(oxyethylene) sorbitan mono oleate).

2.8 Chemical aspects of organoleptic properties

2.8.1 Taste and odour reception

Taste is detected in the mouth and is a form of *chemoreception*, which is the transduction of a chemical signal into an action potential (or nerve impulse). Individual *taste buds* (which each contain approximately 100 taste receptor cells) are concentrated on the upper surface of the tongue and respond to each of the primary tastes (sweetness, bitterness, sourness and saltiness).

Taste is only one component of the sensation of *flavour* in the mouth. The *odour* or *aroma* of food is detected by the *olfactory receptors* on the surface of the *olfactory epithelium* of the nose. Olfactory receptors detect volatile aroma compounds, whereas taste compounds are normally polar, water-soluble and nonvolatile.

2.8.2 Primary tastes

2.8.2.1 Sweetness

Sweet-tasting compounds are characterized by a *glycophore*, which is able to bind to G-protein coupled transmembrane receptors on the tongue. According to the AH-B theory of sweetness this glycophore must contain a hydrogen bond donor (AH) and a Lewis base (B) separated by about 0.3 nanometres. The AH-B unit binds with a corresponding AH-B unit on the sweetness receptor to produce the sensation of sweetness. The AH-B theory provides a satisfactory explanation for the sweetness of most sugars; however, this theory was later refined to the AH-B-X theory to explain why some groups of compounds, particularly amino acids and non-sugar sweeteners, show a correlation between hydrophobicity and sweetness. Therefore, the AH-B-X theory proposes that a compound must have a third binding site (X) that could interact with a hydrophobic site on the sweetness receptor via London dispersion forces.

2.8.2.2 Bitterness

The bitter taste is perceived by many to be unpleasant. Some examples of bitter foods and beverages include coffee, dark chocolate, beer, citrus peel and many plants in the Brassicaceae family (cabbage). Quinine is a well-known example of a bitter-tasting compound and is used in tonic water. Quinine is an alkaloid, as are nicotine, caffeine and theobromine, which also all give bitter tastes. Many alkaloids have toxic effects on animals, and it is likely that the ability to perceive bitterness evolved as a defence mechanism. Research has shown that TAS2Rs (taste receptors, type 2) coupled to the G-protein gustducin are responsible for the human ability to taste bitter substances.

2.8.2.3 Sourness

Sourness is related to acidity and is sensed by hydrogen ion channels in the tongue, which detect the concentration of hydronium ions (H_3O^+ ions) that are formed from acids and water in the mouth.

2.8.2.4 Saltiness

Saltiness is a taste mainly produced by the presence of sodium ions, although other alkali metal ions also taste salty. However, the larger the ions are, the less salty is the taste. Potassium ions most closely resemble the size of sodium and are therefore quite alike in saltiness; hence potassium chloride is the principal ingredient in salt substitutes. Saltiness is detected by the passage of ions through ion channels in the tongue.

2.8.3 Secondary tastes

2.8.3.1 Meatiness (umami)

Meatiness or *umami* is the name for the taste produced by compounds such as *glutamate*, which are found in meat, cheeses and soy sauce. The umami sensation is due to binding to glutamate receptors on the tongue. The food additive *monosodium glutamate* (MSG) produces a strong umami taste. Umami taste is also caused by the nucleotides 5′-*inosine monophosphate* (IMP) and 5′-*guanosine monophosphate* (GMP). These compounds are naturally present in many protein-rich foods.

2.8.3.2 Astringency

Astringency is a dry mouth-feel in the oral cavity that is most associated with polyphenolic compounds. It is believed that astringency occurs due to binding of polyphenols to salivary proline-rich proteins leading to precipitation. Red wine and tea are most associated with astringency.

2.8.3.3 Pungency

Substances such as ethanol and capsaicin cause a burning sensation by inducing a nerve reaction together with normal taste reception. Pungency is essentially a response to chemical irritation.

Further reading and references

Belitz, H.-D. and Grosch, W. (1999) *Food Chemistry*, 2nd edn. Springer, Berlin.

Coultate, T.P. (2002) *Food: The Chemistry of its Components*, 4th edn. The Royal Society of Chemistry, Cambridge.

Damodaran, S., Parkin, K.L. and Fennema, O.R. (2007) *Fennema's Food Chemistry*, 4th edn. CRC Press, Boca Raton.

Walstra, P. (2003) *Physical Chemistry of Foods*. Marcel Dekker, New York.

Supplementary material is available at www.wiley.com/go/campbellplatt

Heinz Dieter Isengard and Dietmar Breithaupt

Key points

- This chapter determines the basic properties of foodstuffs and outlines the most important instrumental methods in the area of food analysis.
- Nitrogen is determined particularly for the estimation of protein content. Analyses can be carried out by the Kjeldahl and the Dumas methods. Fat content is determined after extraction of the fat from the matrix. A classical method is the extraction by a Soxhlet apparatus. Dry matter and water content have particularly high importance for food. Methods to determine this property are discussed broadly.
- Further emphasis is put on various sorts of titration, namely acid-basic, redox, complexation and precipitation techniques. In this context, pH measurement is addressed. Enzymatic analysis is also mentioned.
- Chromotographic techniques aim at the separation of mixtures into components using the different affinity of the components to a stationary and a mobile phase. The major chromatographic techniques described are: thin layer and paper chromatography (TLC and PC), high performance liquid chromatography (HPLC), ion chromatography (IC), gas chromatography (GC) and capillary electrophoresis (CE).
- Spectrometry uses the interaction of light with the material. Techniques using this interaction are described in the sections on spectroscopy in the ultraviolet and visible wavelength domain (UV/vis), fluorescence spectroscopy, infrared (IR) and near infrared (NIR) spectroscopy, microwave resonance spectroscopy, atomic absorption spectroscopy (AAS) and nuclear magnetic resonance (NMR) spectroscopy. The differentiation between different isotopes of an element is addressed in the section on isotope ratio-mass spectroscopy.
- One section is also dedicated to the measurement of the caloric value of food (calorimetry).

3.1 Macro analysis

3.1.1 Sampling techniques

When a material is to be examined, in most cases the whole bulk will not be analysed, particularly if it is changed by the measurement(s) and will, consequently, no longer be available for other purposes after analysis. Usually samples will be taken from the bulk for examination. The result of the analyses should be valid for the whole bulk. It is therefore important that the sample be representative for the material. This is no problem for homogeneous material, provided that the sample collection itself does not change the composition of the material. Many food materials are, however, heterogeneous and "sampling" is therefore an important first step of analysis. Even if the precision and accuracy of the analysis is very good, the result may be erroneous and inaccurate when the test sample does not represent the material.

As foods occur in very different forms, it is not possible to give advice how to proceed in every situation. Only general hints can be given as follows.

The test material should be made as homogeneous as possible. Liquids should be homogenised before taking samples. The particle size of solids should be made as small as possible, for instance by grinding, and the material should then be thoroughly mixed. These operations must, however, not change the composition of the material. Thus, grinding or milling may cause a loss of volatile compounds. Also, the water content may be changed, and even if water content itself is not part of the analysis, the results for other analytes will then be incorrect because concentrations or mass concentrations are often referred to the original mass (which has been changed by the sample preparation). Some components may react with the material of the sample container or be adsorbed onto its surface.

Samples should be taken from different positions within the test material. The sample size must be sufficient for all intended determinations, including replicates and possibly repetitions of the tests. If a sample is not analysed immediately it must be stored in a way that its composition will not change. If it is kept as a proof it must be sealed.

A certain indication that the samples are representative for the whole bulk can be concluded from the standard deviations. If the standard deviation of the replicate results of the single samples is in the same order as the standard deviation of the mean of the different sample values, the sample results may be regarded as representative. When the heterogeneity of the material is very high, the mean value of more samples must be taken into account for this reason.

3.1.2 The Kjeldahl method for determining nitrogen

The Kjeldahl method allows the determination of "crude protein". "Crude" means that the result is

only approximate and does not reflect the exact protein content of the sample. The method is based on digestion of the sample. The nitrogen contained in proteins is converted into ammonium ions which are then determined. From this result the protein content of the sample is recalculated.

The digestion is carried out in a special apparatus (see Colour Plate 1) with concentrated sulphuric acid at about 400°C in the presence of potassium sulphate to elevate the boiling point. For more rapid reaction, a catalyst such as copper sulphate or titanium dioxide is added. These catalysts are usually added to the sample in a mixture with potassium sulphate, for example 3 g TiO_2 + 3 g $CuSO_4$ + 100 g K_2SO_4. Carbon and hydrogen are oxidised, whereas most of the nitrogen in the sample is reduced to ammonium. This is particularly the case for nitrogen in peptide bonds, amino and amido groups.

After the end of digestion – when the solution should be clear and have a bluish or greenish colour – the solution is cooled to room temperature and carefully diluted with water. The whole solution or an aliquot is transferred into a distillation unit. Sodium hydroxide solution (approximately 50%) is added. The ammonia formed is transferred by steam distillation into a flask that contains an excess of acid for neutralisation. This can be a known amount of a standard solution of a strong acid like HCl. The part of it that has not been neutralised by NH_3 is then titrated with a standard solution of a base. The amount of ammonia is calculated from the difference. Another possibility is the use of weak boric acid to absorb the ammonia:

$$NH_3 + H_3BO_3 \rightarrow NH_4^+ + H_2BO_3^- \quad (3.1)$$

Quantification is then carried out by titration with a standard solution of a strong acid like HCl:

$$H_2BO_3^- + H_3O^+ + Cl^- \rightarrow H_3BO_3 + H_2O + Cl^- \quad (3.2)$$

The sum of Equations 3.1 and 3.2 gives:

$$NH_3 + H_3O^+ \rightarrow NH_4^+ + H_2O \quad (3.3)$$

The amount of ammonia can thus be calculated from the consumption of the standard acid solution used.

As the purpose of the analysis is to determine the protein content of the sample, a relation between the measured amount of ammonia (produced from the sample) and the protein content of the sample must be established.

The mass percentage of nitrogen is very similar in different proteins and is about 16%. To calculate the protein content of a sample from the nitrogen content, the nitrogen value has therefore to be multiplied by a conversion factor of 6.25. This is a good average for many proteins like those occurring in meat, fish and eggs. However, if the protein contains amino acids with – compared to the average – a significantly higher percentage of nitrogen, like lysine, arginine or asparagine, or a lower percentage of nitrogen, like leucine, tyrosine or glutamic acid, the conversion factor has to be lower or higher, respectively. Thus, for example, the conversion factor for gelatine is 5.55 and 6.38 for milk and dairy products.

The choice of the "correct" conversion factor is not the only uncertainty of the Kjeldahl method. Beside the nitrogen of proteins, that of other compounds may also be converted into ammonium ions by the digestion. This concerns ammonium salts, free amino acids, nucleic acids, nucleotides, vitamins and others. Instead of "crude protein" the result of a Kjeldahl analysis is therefore often given as "total nitrogen, calculated as protein" and the conversion factor applied is indicated. This, however, may also not be correct, since some nitrogen compounds in the sample will not be converted into ammonia. This concerns, for instance, those with nitro or azo groups. As such compounds are usually rare, the error will not be very important.

A further technique for determining protein content is the Dumas method (see section 3.2.11).

3.1.3 The Soxhlet and Gerber methods for determining lipids

3.1.3.1 The Soxhlet method

The Soxhlet method is a method to determine the crude lipid content by extracting the lipids from the sample with consecutive gravimetric measurement.

A weighed amount of the sample is placed in a porous thimble ("Soxhlet thimble") which is itself placed in a container department of an apparatus for continuous extraction, the Soxhlet extractor (see Colour Plate 2). The dry distillation flask is weighed before the extracting solvent is poured in. This is a non-polar liquid, usually petroleum ether or diethyl ether. The flask with the solvent is heated and the vapour is condensed in a reflux condenser. The liquid then drops into the thimble where it extracts the lipids from the sample. The thimble container is connected by a siphon to the distillation flask. When the liquid in the thimble container has reached a certain level the solvent, which now contains lipids from the sample, flows back into the distillation flask. There the lipids are kept back while the solvent is continuously distilled into the thimble. The extraction process is thus repeated several times. When extraction is assumed to be finished, the distillation flask that now contains the lipids dissolved in the extraction liquid is removed. The solvent is removed, usually in a rotary evaporator. The flask with the lipids as residue is dried to eliminate the last solvent traces, cooled and weighed. The difference between the dry weight and the original weight is the amount of lipids extracted from the sample. The quotient of this mass and the sample mass is the relative crude lipid content; multiplied by 100% it gives the crude lipid content in percentage by mass.

The method described above covers lipids that are free and directly accessible for extraction. If, however, lipids bound to proteins or carbohydrates should also be detected, these must be digested prior to extraction. In the Weibull–Stoldt method this is carried out by boiling the sample with hydrochloric acid (12–14%). The still hot assay is filtered through a moist pleated filter. The filter is then extracted according to the Soxhlet method. It should be mentioned that the lipids will most probably undergo alterations such as hydrolysis during the digestion process. The lipid fraction obtained by Soxhlet extraction is therefore not identical to the original composition in the sample.

3.1.3.2 The Gerber method

A particular method of lipid determination exists for dairy products, the Gerber method. Concentrated sulphuric acid is placed in a specially designed and calibrated tube, a Gerber tube or butyrometer, which has a graduation. The sample is added and the mixture is cautiously shaken. This makes the temperature rise and thus causes the lipids to liquefy. If necessary, additional heating is possible. For better phase separation, 1-pentanol or isoamyl alcohol can be added. The mixture is centrifuged and the tube brought to a standardised temperature (65°C ± 2°C). The lipids form the upper layer. The volume of the lipids is read on the graduation scale. Because of the calibration of the tube, the scale indicates directly the relative lipid content.

3.1.4 Dry weight and water content

Water is present in every foodstuff, possible water contents ranging from extremely low values in dried products to extremely high values in beverages. Water content is of utmost significance in many respects. Physical properties like conductivity for heat and electrical current, density and particularly rheological behaviour depend on the water content of a product. This has an influence on the design of technological processes. As the moisture of a substance can change over the course of time, information on contents is often related to dry matter. It is obvious that correct information on the composition related to dry matter depends strongly on the precision with which moisture is analysed. Free water is necessary for microbiological life and for most of the enzymatic activities and, thus, water content influences – via water activity – the stability and the shelf life of foodstuffs. Storage volume and mass depend on the amount of water in the product and so do transport costs. As water is relatively cheap, its presence particularly in expensive products is interesting from a commercial point of view and for this and other reasons legal regulations and limits exist. Certified reference materials contain certain components in a guaranteed amount or concentration. These values depend on the water content of the material, which must therefore be determined with high accuracy and precision. The determination of water content is, consequently, certainly the most frequent analysis performed on foodstuffs.

Several methods exist to determine the water content of foodstuffs. They may be classified into different groups. Direct methods are those that aim at a quantitative determination of the water itself. The physical techniques among these measure the amount of water obtained or the mass loss observed after separation of water from the other components of a product. Chemical methods are based on a selective reaction of the water in the sample. Indirect methods may either determine a macroscopic property of the sample which depends on its water content or measure the response of the water molecules in the sample to a physical influence.

3.1.4.1 Direct methods based on physical separation of water

One possibility to separate water from the other components in the product is to place the sample in a desiccator close to a very hygroscopic substance like diphosphorus pentoxide or to use a molecular sieve. The difference in mass before and after the transfer of the water is measured. This process, of course, leads principally only to a distribution of the water in an equilibrium which depends on the difference in hygroscopicity of the two products competing for the water and the volume of air in the apparatus. A part of the water will therefore remain in the sample.

Water can also be separated by distillation. Compounds forming an azeotropic mixture with water, like toluene or xylene, which separates again after condensation, are often used. The water obtained is usually measured by volume.

The most frequently applied method is based on the mass loss that the product undergoes by a heating process. These drying techniques with a convective heating principle comprise ordinary oven drying and vacuum oven drying. It is important to be aware that drying techniques do not measure the water content as such. The result is a mass loss under the conditions applied. These conditions can principally be freely chosen and the results are, consequently, variable. Even the results of official methods with a certain parameter set are only a convention by definition and do not necessarily reflect the true water content. Drying to a constant mass is often required, but a real constancy is only achieved in rare cases. Tightly bound water escapes detection, but a distinction between "free" and "bound" water is nevertheless usually not possible. The mass loss is caused by not only water but also all the substances volatile under the drying conditions, either already contained in the original sample or produced by the heating process. The application of low pressure in vacuum ovens reduces the danger of producing volatile decomposition compounds but does not allow a distinction to be made between water and other volatile substances already present in the product. The results of drying methods should therefore not be termed water content. The most adequate term is mass loss on drying (mentioning the drying conditions), but the expression moisture or moisture content is common as a compromise.

To shorten the long determination times in drying ovens with convective heating principle, more efficient heating sources have been introduced. In such dryers the sample is exposed, on the pan of a built-in balance, to infrared (or "halogen") or microwave radiation and the mass loss is registered. The more intensive way of heating in these dryers as compared with usual drying ovens makes samples even more susceptible to decomposition reactions, resulting in the production of volatile matter and, thus, pretending a higher water content of the sample. The results can vary very broadly depending on the drying parameters applied.

Mass-loss results can, however, be matched with the results of another method, particularly a reference method, by adjusting the parameters in an appropriate way. In these cases the two errors by leaving a part of the water undetected and by determining other volatile substances as water compensate each other. Such calibrations are particularly relevant for the rapid techniques mentioned and must be established for every type of product in a specific way. All the parameters like drying mode or programme including temperature and time, stop criterion, sample size and sample distribution on the pan of the balance and, in some situations, even the time interval between consecutive measurements must be considered.

3.1.4.2 Karl Fischer titration as a direct method based on a chemical reaction

By far the most important chemical determination method is Karl Fischer titration. It is based on a two-step reaction. In the first step an alcohol ROH (normally methanol) is esterified with sulphur dioxide; to obtain a quantitative reaction, the ester is neutralised by a base Z to yield alkyl sulphite (Eq. 3.4). The "classical" pyridine has been replaced by other bases like imidazole in modern reagents. In the second step alkyl sulphite is oxidised by iodine to give alkyl sulphate in a reaction that requires water; the base, again, provides for a quantitative reaction (Eq. 3.5):

$$ROH + SO_2 + Z \rightarrow ZH^+ + ROSO_2^- \quad (3.4)$$

$$ZH^+ + ROSO_2^- + I_2 + H_2O + 2\,Z$$
$$\rightarrow 3\,ZH^+ + ROSO_3^- + 2\,I^- \quad (3.5)$$

The overall reaction is:

$$3\,Z + ROH + I_2 + H_2O \rightarrow 3\,ZH^+ + ROSO_3^- + 2\,I^-$$
$$(3.6)$$

The consumption of iodine is measured. In the coulometric variation of the Karl Fischer titration, iodine is formed from iodide in the titration cell by anodic oxidation. In the volumetric variation, which

is more relevant in food analysis, iodine is added in a solution. The sample is placed in the titration cell which contains the working medium which has been titrated to dryness before the addition of the sample. In the so-called one-component technique, this working medium consists of methanol and the titration solution then contains all the other chemical components, iodine, sulphur dioxide and the base, dissolved in an appropriate solvent. In the two-component technique, the working medium contains sulphur dioxide and the base dissolved in methanol; the titration agent is a methanolic solution of iodine. The water equivalent of the respective titrating solution is determined by titrating standards with known water content. The end-point indication, both in the coulometric and the volumetric varieties, is based on an electrochemical effect. Two platinum electrodes, submerged in the working medium in the titration cell (see Colour Plates 3 and 4), are polarised either by a constant current (bipotentiometric or voltametric technique) or by a constant voltage (biamperimetric technique) and the voltage or the current respectively to maintain this situation is monitored. After the water of the sample is consumed, iodine can no longer react and the redox couple iodine/iodide is then present and renders the respective oxidation and reduction possible. This makes the voltage necessary to maintain the constant current drop abruptly (voltametric technique) or makes the current resulting from the constant voltage rise abruptly (biamperimetric technique). This sudden change is used to indicate the end point. When the voltage remains below (respectively the current above) a certain chosen value for a certain time, the determination is completed (see Fig. 3.1). This so-called stop delay time is important in order to allow for the detection of water that may not be immediately available, particularly when samples are analysed which are not or not completely soluble in the working medium. In these cases water reaches the working medium by diffusion and extraction processes only with a certain delay.

A major aspect of the Karl Fischer titration is that water needs to be in direct contact with the reagents, a fact that may cause problems with insoluble samples. Several measures exist, however, to provide for a practically complete detection of the water. They include:

- a long stop delay time;
- external extraction of the water and titration of an aliquot of this solution;

KFT Ipol

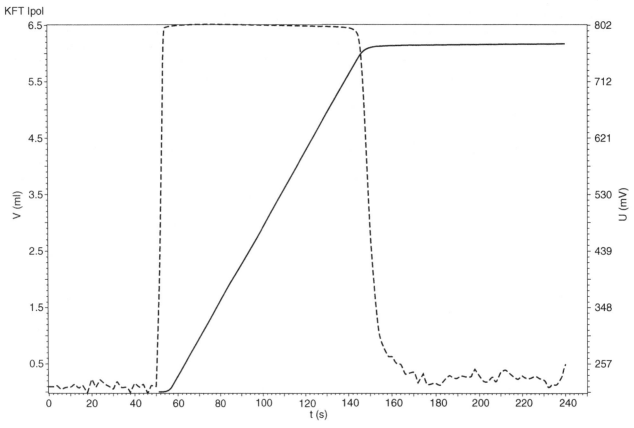

Figure 3.1 Karl Fischer titration curves (using the voltametric technique) showing voltage versus time (U/t) (dashed line) and volume versus time (V/t) (solid line).

■ internal extraction in the titration vessel prior to the start of the titration;

■ reducing of the particle size by external sample preparation or by using a homogeniser in the titration cell (see Colour Plate 3);

■ working at elevated temperature, even up to the boiling point of the working medium;

■ the addition of solvents to the working medium;

■ replacement of methanol by other alcohols in order to change the polarity.

When parameters can be found to prevent the sample from being dissolved in the working medium, the surface water can be determined selectively. When, in a second assay, conditions are applied to dissolve the sample completely or to set the water free by appropriate techniques, the total water content can be determined. The difference between the results is the "interior" or "bound" water of the sample.

Automatic titrations in series are possible by the use of automatic sample changers.

3.1.4.3 Indirect methods based on the measurement of a macroscopic sample property that depends on its water content

The water content affects various properties of the sample, for instance density. If only approximate results are demanded and if the composition of the product is simple and differs only in water content, a calibration of the water content against density can be established. In a similar way, polarimetry and refractometry may also serve as methods to determine water content for solutions that differ only in water content. A wide field is the application of electric properties of samples to measure their water content. These techniques comprise the determination of conductivity, resistance, capacitance or permittivity. As these properties do not, however, depend exclusively on the water content of the product, a calibration is inevitable.

Water content and water activity are correlated via sorption isotherms which are very product-specific. If the isotherm is known for the product in question and if the water activity is measured, water content can be read from the isotherm.

3.1.4.4 Indirect methods based on the measurement of the response of the water molecules to a physical influence

These extremely rapid methods that can even be adapted to serve as in-line or at-line techniques comprise low-resolution (or time-domain) nuclear magnetic resonance (LR-NMR or TD-NMR) (see section 3.2.8), near infrared (NIR) (see section 3.2.1.4) and microwave (MW) (see section 3.2.1.5) spectroscopy. The response of each water molecule may, however, be different depending on the bonding state within a given product. As the distribution of these bonding states is again different from one product to another, a very product-specific calibration against a reference method is necessary. In NIR spectroscopy the spectra are very complex. For their evaluation chemometric techniques are therefore necessary.

3.1.5 Ash determination

Ash is determined in order to obtain information on the mineral content of the food. Two principles are used: dry ashing and wet ashing.

3.1.5.1 Dry ashing

The sample is weighed into quartz, porcelain or platinum crucibles. Quartz is fragile and not resistant to alkali, hydrofluoric acid and phosphoric acid and has a low conductibility for heat. Porcelain has similar properties but a higher sensibility to rapid temperature changes and can set silicon compounds free. Platinum is mechanically stable and chemically inert against most influences. Some elemental metals, however, can cause corrosion. Platinum crucibles should therefore, particularly in the heat, not be brought into contact with metal surfaces and should be touched only with platinum-tipped tongs.

Liquid samples may be heated at about 100°C to concentrate them into a smaller volume. The crucibles with the samples are heated at about 550°C in a muffle furnace for several hours. Organic matter is burnt and inorganic elements, particularly metals, are transferred into oxides and salts like carbonates or phosphates. After cooling the crucibles are reweighed. The mass loss is the ash or "mineral content". Of course, the compounds formed during ash-

ing were not contained in this form in the sample. The mass loss is sometimes related to dry matter. In this case the water content (see section 3.1.4) must also be determined and be taken into account. Ashing can be accelerated and improved by adding ethanol or hydrogen peroxide solution to the sample. For meat samples an addition of magnesium acetate has proved to be advantageous. As this leads to the formation of magnesium oxide, a blank with the same amount of magnesium acetate must be analysed in parallel and the result of this assay should be accounted for.

Special applications exist. Thus the degree of fineness of flour is determined by ashing at 900°C. The so-called flour type is defined as milligrammes of ash per 100 g dry matter.

In some situations it is advantageous to produce a so-called sulphate ash, for instance if lead must not be lost in the form of lead chloride (which may be formed in the standard procedure). In this case 5 ml of 10% sulphuric acid per 5 g is added to the sample. After cooling the ash, 2–3 ml of 10% sulphuric acid is added and heating at 550°C is repeated until a constant mass is reached.

Several categories of ashes can be distinguished. (Water-) soluble ash is the part that is soluble in hot water. (Water-) insoluble ash is the remaining part. Acid-insoluble ash is the residue after heating the ash with dilute hydrochloric acid. Since it contains much silicon dioxide, it is often called "sand".

The ash can be characterised by its "alkalinity". This may be relevant for fruit juices or wine. The value allows an estimation of the fruit content. A certain volume of a standard acid solution, e.g. 0.1 M HCl, is added to the ash, the mixture is heated to boiling, cooled down and titrated back with a standard hydroxide solution. The ash alkalinity is defined as the amount of sodium hydroxide in millimoles that corresponds to the basic components in the ash dissolved or dispersed in 1 l of 0.1 N acid.

Ashes are often the starting material for elemental analyses.

3.1.5.2 Wet ashing or wet digestion

This technique is usually used for digesting organic material and not so much for determining ash or mineral content.

The sample is heated with a volatile oxidising agent under reflux or in a closed vessel under pressure. The reagents added must be accounted for when calculating the result. This can be done by doing a blank analysis without sample. Because of the chemical aggressiveness of the reagents tetrafluoroethylene vessels are often used and very cautious working is necessary to avoid explosions. Common oxidising agents are: mixtures of concentrated sulphuric acid and concentrated nitric acid, sometimes with the addition of potassium permanganate or hydrogen peroxide solution (50%), mixtures of concentrated hydrochloric acid and concentrated nitric acid, mixtures of 60% perchloric acid and concentrated nitric acid, and mixtures of chloric acid and perchloric acid, chloric acid and concentrated nitric acid. In some cases ultraviolet radiation in the presence of hydrogen peroxide is sufficient. The application of microwaves is also very effective. For ash determination the liquid obtained has to be concentrated to dryness. The values obtained for ash differ from those received by dry ashing, because the chemical reactions during the process are different.

3.1.6 pH and pH electrodes

3.1.6.1 The pH value

The pH value or pH is defined as

$$pH = -\log[H_3O^+]. \tag{3.7}$$

It is a measure of the acidity or alkalinity of an aqueous system. In every aqueous system the equilibrium

$$2\,H_2O \rightleftarrows H_3O^+ + OH^- \tag{3.8}$$

exists. It lies very much on the right side. Applying the law of mass-action and considering the concentration of water, $[H_2O]$, in an aqueous system as constant, we find the ionic product to be

$$[H_3O^+]\cdot[OH^-] = 10^{-14}\ mol^2/l^2. \tag{3.9}$$

In neutral water the concentration of hydroxonium and hydroxide ions is equal:

$$[H_3O^+] = [OH^-] = 10^{-7}\ mol/l. \tag{3.10}$$

The pH value of a neutral aqueous system is therefore 7. Acid solutions are characterised by $[H_3O^+] > [OH^-]$ and, thus, pH < 7; basic solutions are alkaline and characterised by $[H_3O^+] < [OH^-]$ and, thus, pH > 7.

3.1.6.2 pH measurement

Measurement of pH is carried out with so-called glass electrodes. The potential difference between two electrodes is measured. One of these has a constant and known potential. The other one is the glass electrode in the true sense (see below). Its potential is dependent on the concentration of hydroxonium ions and, thus, the pH in the surrounding medium. Usually both electrodes are combined in one instrument, which is (not quite correctly) commonly called "glass electrode" (see Colour Plate 5). This device contains two electrodes and is submerged in the medium which is to be analysed for pH. The electrode with the constant potential, the reference electrode, may be a silver wire in a cell containing a highly concentrated or saturated potassium chloride solution and a silver chloride residue. The silver wire is connected to a potentiometer. The glass electrode contains the same metal (silver) and the same electrolyte solution as the reference electrode. The metal wire, the reference junction, is also connected to the potentiometer. The surface of the glass electrode is a very thin spherical glass membrane which swells in aqueous surroundings to form a gel. Several potentials exist between the various layers (internal solution, inner hydrated gel layer, glass, outer hydrated gel layer, external solution). The potential E of the reference junction depends on the pH to be measured:

$$\begin{aligned} E &= E_0 + \frac{R\cdot T}{n\cdot F}\cdot(pH_i - pH) \\ &= E_0 + 0.059\cdot(pH_i - pH)\ (\text{at room temperature}) \end{aligned} \tag{3.11}$$

where

E_0 is the sum of potentials of the various layers,
R is the general gas constant (8.314 J/K ·mol),
T is temperature (with $T = 298$ K the factor 0.059 is obtained),
F is the Faraday constant (96,485 A·s/mol),
n is the number of electrons involved in the molecular reaction (in this case oxidation or reduction of hydrogen),
pH_i is the pH of the solution in the glass electrode.

E_0 is constant for a given instrument (and usually close to 0 V). pH_i is also constant and the potential E is, thus, only dependent on pH. The correlation between E_0 and pH can be established by measuring solutions with a known pH (standard buffer solutions). This is the calibration of glass electrodes that is necessary before taking measurements.

3.1.7 Titrations

Titrations are techniques to determine the amount or the concentration of an analyte, which in this context is also called the titrand, by adding standard solutions of a reagent, which is here also called titrant, to the sample (see Colour Plate 6). The reagent volume is measured. The technique is therefore also called volumetric analysis. This method can be applied if the following conditions are fulfilled: the reaction of the reagent with the analyte is stoichiometrically known; the reaction is practically quantitative; the reaction rate is high; and the end point can be indicated clearly. According to the type of reaction between reagent and analyte, different sorts of titrations are distinguished.

Using sample changers, several samples can be titrated automatically in series (see Colour Plate 7).

3.1.7.1 Acid-base titrations or neutralisation titrations

The amount or concentration of an acid or a base is analysed. During titration the pH value (see section 3.1.6) changes with the volume of reagent added. The so-called neutralisation curve (pH versus volume) has a point of inflection at the equivalent point. The pH value changes most rapidly upon reagent addition at the equivalent point and the curve has the steepest slope. The pH can be monitored with a pH electrode (see section 3.1.6).

The end-point indication is also possible with an acid-base indicator. Such indicators change their structure depending on the pH. For instance, an indicator molecule HX may be able to dissociate a proton to give X^- in which a rearrangement of mesomeric electron systems gives X^- another colour than HX, which makes the two forms of the indicator visually distinguishable. HX and X^- are in a pH-dependent equilibrium with each other thus:

$$HX \text{ (colour I)} + H_2O \rightleftarrows H_3O^+ + X^- \text{ (colour II)}$$
$$(3.12)$$

In a solution with high hydroxonium concentration (low pH) colour I prevails, whereas colour II is vis-

ible in solutions with low hydroxonium concentration. The pH, where the colour changes, depends on the dissociation constant K_{Ind} of the indicator, also called the indicator constant:

$$K_{Ind} = \frac{[H_3O^+] \cdot [X^-]}{[HX]} \qquad (3.13)$$

For high values of $[H_3O^+]$, $[X^-]$ must be low and $[HX]$ must be high, and colour I is then visible. For low values of $[H_3O^+]$ colour II is visible. The point of colour change or transition point, t.p., where $[X^-] = [HX]$, is at the hydroxonium concentration $[H_3O^+]_{t.p.}$ that corresponds numerically to the indicator constant. The pH value at the transition point $pH_{t.p.}$ corresponds to the pK value of the indicator pK_{Ind} ($pK_{Ind} = -\log K_{Ind}$). To make a distinction between the two colours really visible to the human eye, one of the concentrations should be approximately ten times bigger than the other. This means that the colour changes within a transition interval of approximately two pH units. This is, however, usually sufficient, because the slope of the pH curve (pH versus the added volume of reagent) is very steep around the equivalent point (as mentioned above). The indicator must, however, be chosen in a way that its indicator constant lies close to the hydroxonium concentration of the equivalence point.

When strong acids are titrated with hydroxide solutions and vice versa, the neutralisation curve is extremely steep near the equivalence (inflection) point, which then corresponds to a pH value of 7, because the chemical reaction is in all these cases the same:

$$H_3O^+ + OH^- \rightarrow 2\,H_2O \qquad (3.14)$$

Any acid-base indicator with a pK value between 4 and 10 can be used. The choice of the indicator may become more important when weak acids are titrated with hydroxide solutions or weak bases are titrated with strong acids. The equivalence pH value is then not 7 and the neutralisation curve is less steep there. This is due to the buffering capacity of weak acids (bases) and their corresponding anions (cations) like CH_3COOH/CH_3COO^- or NH_3/NH_4^+. The indicator must then be better adapted to the specific situation. When a weak acid is titrated with a hydroxide solution, the equivalence point is shifted to a basic pH.

Some examples of the application of acid-base titrations in the food domain are:

- crude protein determination according to the Kjeldahl method (see section 3.1.2);
- test for acidity of flour, milk, mash and must;
- determination of the saponification number of fats and oils;
- determination of the acid value of fats and oils.

3.1.7.2 Redox titrations

Redox titrations allow the determination of oxidisable or reducible substances. Indication is possible:

- if the reagent or the analyte changes colour after the redox reaction;
- by potential measurement against a reference electrode; or
- by means of a redox indicator with a redox potential between those of the systems involved and which has different colours in the reduced and the oxidised form.

For redox reactions, the Nernst equation is of fundamental importance:

$$E = E_0' + \frac{R \cdot T}{n \cdot F} \cdot \ln \frac{\prod_i c_i^{v_i}(\text{ox})}{\prod_j c_j^{v_j}(\text{red})} \qquad (3.15)$$

or in common (decimal) logarithmic form:

$$E = E_0 + \frac{R \cdot T}{n \cdot F} \cdot \log \frac{\prod_i c_i^{v_i}(\text{ox})}{\prod_j c_j^{v_j}(\text{red})} \qquad (3.16)$$

where

E is the potential of a conductor submerged in a solution in which a redox reaction is run (the conductor may itself be involved in the reaction),

E_0' is the so-called (natural) normal potential of the system (the potential for the case that the argument of the logarithmic function is 1, which is achieved when all the c_i and c_j are 1 mol/l),

E_0 is the corresponding (decimal) normal potential when the common (decimal) logarithm (containing the recalculation factor from ln to log) is used,

R is the general gas constant (8.314 J/K·mol),

T is temperature (in K),

F is the Faraday constant (96,485 A·s/mol),

n is the number of electrons exchanged between the reactants on the molecular level,

c_i is the concentration (or more exactly the activity) of the reactant i on the "oxidised side" of the reaction equation,

c_j is the concentration (or more exactly the activity) of the reactant j on the "reduced side" of the reaction equation,

v_i is the stoichiometric number of the reactant i on the "oxidised side" of the reaction equation,

v_j is the stoichiometric number of the reactant j on the "reduced side" of the reaction equation.

The following example shows an application.

Permanganate ions can be reduced to manganese(II) ions in acidic solution:

$$\text{MnO}_4^- + 8\,\text{H}_3\text{O}^+ + 5\,\text{e}^- \rightleftharpoons \text{Mn}^{2+} + 12\,\text{H}_2\text{O} \qquad (3.17)$$

The Nernst equation for this system is:

$$E_{\text{MnO}_4^-} = E_{0,\,\text{MnO}_4^-} + \frac{R \cdot T}{5 \cdot F} \cdot \log \frac{[\text{MnO}_4^-] \cdot [\text{H}_3\text{O}^+]^8}{[\text{Mn}^{2+}]} \qquad (3.18)$$

Normally, the denominator of the logarithmic argument should contain the factor $[\text{H}_2\text{O}]^{12}$. However, the concentration of water in a diluted aqueous system is regarded as constant, and this value is included in the normal potential, $E_{0,\,\text{MnO}_4^-}$, which has a value of 1.51 V. From the equation it is obvious that the redox potential of permanganate depends very much on the pH value of the solution.

Iodide can be oxidised to iodine:

$$2\,\text{I}^- \rightleftharpoons \text{I}_2 + 2\,\text{e}^- \qquad (3.19)$$

The Nernst equation for this system is:

$$E_{\text{I}_2} = E_{0,\,\text{I}_2} + \frac{R \cdot T}{2 \cdot F} \cdot \log \frac{[\text{I}_2]}{[\text{I}^-]^2} \qquad (3.20)$$

$E_{0,\,\text{I}_2}$ has a value of 0.58 V. The redox potential of the permanganate system is higher than that of the iodine system, provided that the concentrations are in a "normal" range. This means that the position of the equilibrium

$$2\,MnO_4^- + 16\,H_3O^+ 10\,I^-$$
$$\rightleftharpoons 2\,Mn^{2+} + 5\,I_2 + 24\,H_2O \quad (3.21)$$

is clearly on the right side and that iodide can be oxidised by permanganate in acid solution. The inverse reaction is practically not possible.

Iodine can, however, oxidise thiosulphate to give tetrathionate according to the equation

$$I_2 + 2\,S_2O_3^{2-} \rightleftharpoons 2\,I^- + S_4O_6^{2-}, \quad (3.22)$$

because $E_{0,\,S_4O_6^{2-}}$ is only 0.08 V.

Some examples of the application of redox titrations in the food domain are:

- determination of the iodine number of fats and oils;
- determination of the peroxide value of fats and oils;
- determination of reducing sugars.

3.1.7.3 Complexation titrations

With this sort of titration, usually metal ions in aqueous solutions are determined. The technique is based on the different stability of metal complexes.

Before starting the titration, a complexing agent, for instance X, is added to the solution to be titrated. It will form a complex with the metal ion, for instance Me^{2+}. The agent and the complex must have different colours:

$$Me^{2+} + X\,(colour\ I) \rightleftharpoons [MeX]^{2+}\,(colour\ II) \quad (3.23)$$

X may have an electrical charge, and also more than one complexing agent may be coordinated to the metal ion. The complex will have then another charge than indicated in Eq. 3.23. The equilibrium is on the right side, and the solution has therefore the colour II before the start of the titration.

This solution is titrated with a standard solution of another complexing agent, which is colourless and forms a colourless complex with the metal ion that is more stable than the metal-X complex. Usually, a salt of ethylenediamine tetra-acetic acid is used, mostly the disodium salt Na_2H_2EDTA:

$$Me^{2+} + H_2EDTA^{2-}\,(colourless) + 2\,H_2O$$
$$\rightleftharpoons [MeEDTA]^{2-}\,(colourless) + 2\,H_3O^+$$
$$(3.24)$$

The equilibrium lies practically completely on the right side. As the metal ions are, thus, bound in the complex, metal ions complexed by X according to Eq. 3.23 are now set free (to maintain this equilibrium). These are then also complexed according to Eq. 3.24. By this continued process MeX^{2+} eventually disappears and X is liberated. This leads to a change in colour of the solution from II to I. This indicates that all the metal ions are now complexed by EDTA.

This technique can be used to measure the hardness of water which is defined by the concentration of calcium and magnesium ions.

3.1.7.4 Precipitation titrations

In this technique ions are determined by adding standard solutions of an ion that forms a precipitate with the analyte. An indicator is added that also reacts with the reagent after the analyte has practically completely precipitated.

Thus, chloride can be titrated with a standard solution of silver nitrate. As indicator some drops of sodium chromate solution are added. This is the method according to Mohr. First, chloride forms the nearly insoluble and colourless silver chloride:

$$Cl^- + Ag^+ \rightarrow AgCl \downarrow \quad (3.25)$$

When practically all the chloride ions are precipitated, additionally added silver ions will react with chromate to form the reddish-brown silver chromate:

$$Ag^+ + CrO_4^{2-} \rightarrow Ag_2CrO_4 \downarrow \quad (3.26)$$

Another possibility is the method according to Volhard. Here, an excess of a standard solution of silver nitrate is added. The excess is determined by titrating with a standard solution of an alkali or ammonium thiocyanate, which yields a precipitate of silver thiocyanate:

$$Ag^+ + SCN^- \rightarrow AgSCN \downarrow \quad (3.27)$$

As indicator, some drops of iron(III) chloride solution are added. When silver has been practically completely precipitated, the first excess of thiocyanate leads to a red colour of an iron(III) complex:

$$[Fe(H_2O)_6]^{3+} + 3\,SCN^- \rightarrow [Fe(H_2O)_3(SCN)_3] + 3\,H_2O$$
$$(3.28)$$

Colorimetry is based on the comparison of solutions. A solution with known concentration in a cuvette is taken as standard. To this, a solution with unknown concentration is compared. The thickness or optical path-length of this latter solution is varied until the colour corresponds visually to the standard solution. This variation can be realised by using wedge-shaped cuvettes or by introducing glass objects into the cuvette which shorten the path-length. When the colour of standard and sample solutions correspond to each other, the concentration can be calculated from the known concentration and the two path-lengths using the law of Lambert and Beer (see UV/vis spectroscopy, section 3.2.1.1).

The visual technique of colorimetry is practically completely replaced by photometry which uses photometers to measure light absorption quantitatively. This technique is described in 3.2.1.1.

Enzymes are specific catalysts for certain reactions. This can be used for specific analyses of certain substances. Enzymes do not change the equilibrium constant of a reaction. They only accelerate the reaching of the equilibrium, no matter from which side. This fact has consequences for the nomenclature. Only one name (and for only one enzyme!) is given for both reaction directions, even when the equilibrium lies extremely on one side. This is done after a certain principle. All the reactions (and equilibria) and thus the enzymes are classified in six groups and only one reaction direction is considered for nomenclature, no matter where the equilibrium lies in a specific case. This leads to situations in which an enzyme is named after a reaction which does not really happen or only minimally, but rather the reverse reaction. Examples of this situation are given below.

A very frequently used principle in enzymatic analysis is the determination of reducible or oxidisable substances that react with nicotinamide adenine dinucleotide or the corresponding phosphate in the oxidised (NAD^+, $NADP^+$) or the reduced ($NADH + H^+$, $NADPH + H^+$) form. The oxidised and reduced forms can easily be distinguished and determined quantitatively by photometry (see section 3.2.1.1). Both have an absorption maximum at 260 nm, and the reduced forms NADH and NADPH have an additional maximum at 340 nm due to the quinoid structural element that is formed from nicotinamide by reduction.

An example is the determination of ethanol using the enzyme alcohol dehydrogenase (ADH):

$$CH_3CH_2OH + NAD^+ \rightarrow CH_3CHO + NADH + H^+ \text{ (catalysed by ADH)} \quad (3.29)$$

As the molecular ratio of ethanol and NADH is known (1:1), the original concentration of ethanol in the sample can easily be calculated from the concentration of NADH formed. For this purpose, a calibration curve (absorption at 340 nm versus NADH concentration) must have been established.

This principle can also be used for indirect analyses. An example is the determination of citrate. Citrate is first split into oxalacetate and pyruvate using citrate lyase (CL) as catalyst (Eq. 3.30). Oxalacetate is then reduced to form malate in the presence of malate dehydrogenase (MDH) (Eq. 3.31). A part of oxalacetate is decarboxylated and yields pyruvate (Eq. 3.32). This is also reduced to form lactate with lactate dehydrogenase (LDH) as catalyst (Eq. 3.33).

$$^-OOC–CH_2–C(OH)(COO^-)–CH2–COO^- \rightarrow\ ^-OOC–CH_2–CO–COO^- + CH_3–COO^- \text{ (catalysed by CL)} \quad (3.30)$$

$$^-OOC–CH_2–CO–COO^- + NADH + H^+ \rightarrow\ ^-OOC–CH_2–CHOH–COO^- + NAD^+ \text{ (catalysed by MDH)} \quad (3.31)$$

$$^-OOC–CH_2–C(OH)(COO^-)–CH2–COO^- + H^+ \rightarrow CH_3–CO–COO^- \quad (3.32)$$

$$CH_3–CO–COO^- + NADH + H^+ \rightarrow CH_3–CHOH–COO^- + NAD^+ \text{ (catalysed by LDH)} \quad (3.33)$$

From the decrease of absorption at 340 nm due to the loss of NADH, which had been added before the analysis, the original citrate concentration can be calculated.

MDH and LDH are called after the reverse reaction. Generally, NAD^+ is regarded as acceptor and the (other) substrate as donor of protons and electrons in reactions catalysed by oxido-reductases (see also ADH above), even if the reaction direction is essentially the opposite (see above). The oxido-reductases are one of the six enzyme groups mentioned above.

3.2 Instrumental methods

3.2.1 Spectroscopy

Spectroscopic methods utilise physicochemical interactions between electromagnetic radiation and molecules or atoms. For spectroscopic methods, only particular wavelengths are of importance. Usually, wavelengths (λ) are given in nanometres, which is 10^{-9} m. From 200–780 nm, energy absorption causes electron transition in a molecule (UV/vis spectroscopy). Absorption of radiation with wavelengths of 800–10^6 nm, the infrared (IR) region, causes the molecule to vibrate. The resulting characteristic molecular vibration spectra are used in IR spectroscopy to identify certain functional groups. Wavelengths between 10^6 and 10^9 nm (microwaves) cause molecular rotations. Thus, the overall energy of an organic molecule is composed of the electronic, the vibrational and the rotational energies.

3.2.1.1 UV/vis spectroscopy (photometry)

UV/vis spectroscopy is based on the transition of outer electrons of molecules induced by absorption of UV light (200–400 nm) or visible light (400–780 nm). UV light possesses enough energy to raise σ- or non-binding n-electrons to excited stages (e.g. σ–σ* (130 nm) or n–σ* transition (200 nm)) whereas visible light is responsible for transition of n- or Π-electrons belonging to conjugated systems (e.g. Π–Π* or n–Π* transitions). In consequence, molecules comprising no non-binding electrons (e.g. with an absence of oxygen, nitrogen or halogen atoms) or no double bonds (e.g. alkanes) are mostly not suitable for identification or detection by UV/vis spectroscopy.

Quantification is based on Lambert–Beer's law (Eq. 3.34) which connects the read-out of a spectrometer (absorbance, A) with physicochemical characteristics of the compound (ε_λ = molar extinction coefficient, determined at a specific wavelength λ) with the concentration (c):

$$A = \log I_0/I = \varepsilon_\lambda (\text{L} \cdot \text{mol}^{-1} \cdot \text{cm}^{-1}) \cdot c[\text{mol} \cdot \text{L}^{-1}] \cdot d[\text{cm}]$$
(3.34)

where

d is the length of the photometer cell (cuvette),
I_0 is the intensity of the incident light,
I is the intensity of the transmitted light.

In each case absorption of a photon from light with appropriate energy causes elevation of a molecule from the ground state to an elevated level. These electronic transition phenomena of molecules provide typical absorption spectra comprising one or more absorption bands. They are obtained by plotting the respective absorbance against λ (see Fig. 3.2). Their characteristic maxima as well as their shape are useful for compound identification by comparison with spectra libraries. They also serve for finding a suitable wavelength for a photometric detector used in chromatography.

UV/vis spectrophotometry is one of the most widely used techniques in food analysis. For example, amino acids can be detected as blue derivatives after reaction with ninhydrin. Glucose, fructose and sucrose are detectable after separation by thin layer chromatography and derivatisation with diphenyl amine/aniline as brownish or bluish spots. Highly sensitive reactions are described for minerals: iron, a typical component in most natural waters, may be detected photometrically as 1,10-phenanthroline-Fe(II)-complex at 510 nm. UV spectroscopy is used for detection of NADH (340 nm), a typical product of enzymatic measurements of food constituents, such as organic acids, ethanol, sugars and others (see section 3.1.9).

3.2.1.2 Fluorescence

Fluorescence is a phenomenon that occurs when light is emitted from a molecule which is in an elevated energy state. Electrons in the elevated level must be in the singlet state (S_1), i.e. one electron of a pair of non-binding electrons is spinning opposed to the other. Relaxation of the electron from the excited to the ground level (S_0) comes along with emission of fluorescent light. In general, the wavelength of the emitted light is longer (corresponding to a lower energy) than that which caused excitation of electrons.

In general, molecules that fluoresce are inelastic and possess no possibility to reach the ground level by vibrational relaxation ("loose-bolt effect"). They mostly contain either an electron-donating group or multiple conjugated double bonds (e.g. PAKs). The presence of groups that tend to withdraw electrons (nitro groups) usually destroys fluorescence.

Fluorescence spectroscopy provides an outstanding sensitivity since only the fluorescent component is regarded. Comparison of the sensitivity achievable by UV/vis or by fluorimetric detection, respectively,

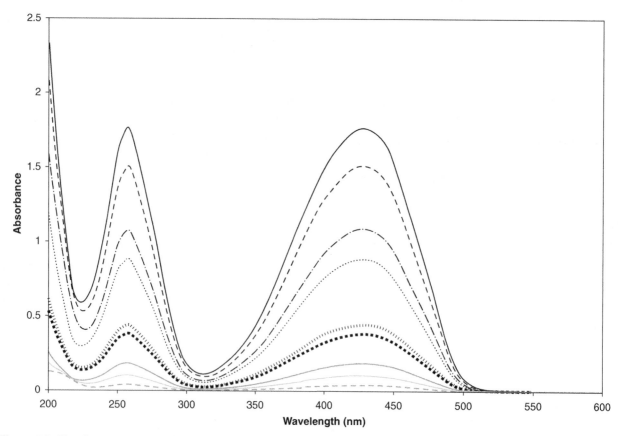

Figure 3.2 Absorbance spectrum of solutions of tartrazine in different concentrations.

reveals an increase by a factor of 1000. Thus, conversion of non-fluorescent compounds into fluorescent derivatives is a common procedure in the determination of food components (e.g. conversion of vitamin B_1 into fluorescent thiochrome by oxidation). Typical compounds showing intense fluorescence without derivatisation are quinine, riboflavin (vitamin B_2) and several aflatoxins produced by *Aspergillus* ssp. and other moulds found on nuts and cereals.

3.2.1.3 IR spectroscopy

Infrared (IR) radiation provides insufficient energy to cause electron transition, but results in transitions in the vibrational and rotational states in the ground electronic state of a molecule. The absorption of IR radiation results mainly in the stretching and contracting of bonds and in the changing of bond angles of molecules. The IR region of the electromagnetic spectrum can be divided into three parts: near infrared (NIR, 800–2500 nm), mid infrared (MIR, $2.5 \cdot 10^3$–$50 \cdot 10^3$ nm) and far infrared (FIR,

$50 \cdot 10^3$–10^6 nm). With the exception of homonuclear diatomic molecules (e.g. O_2) all molecules absorb IR radiation. Thus, IR spectroscopy is one of the most applied methods in analytical chemistry.

MIR is a tool for structural analysis of molecules. It can, however, also be used to check identity, and thus also the authenticity, of substances by comparing the spectrum of a sample with the spectrum of a reference substance (see Colour Plate 8).

3.2.1.4 Near infrared spectroscopy

In NIR spectroscopy, the light absorbance is measured either in a transmission mode for solutions or in a reflectance mode for solids, or even in a combination of both (transflectance) for dispersions. Measurements are possible in cuvettes, through walls of containers (like glass bottles, bottoms of dishes (see Colour Plate 9) or plastic bags) or even at a distance from the spectrometer using probes connected by a bundle of fibre-optic light guides with the spectrometer (see Colour Plate 10). This allows measurements in dangerous environments or online. While "classical" IR spectroscopy is based on the first harmonic oscillations of chemical bonds in molecules, NIR spectroscopy registers overtones of these and combination oscillations. These are much weaker and therefore do not necessitate very thin layers or high dilution rates as required in IR spectroscopy. The disadvantage is the great number of bands in this wavelength area, which cannot be separated into single peaks. Practically every substance in a sample contributes to the NIR spectrum, which becomes more or less a continuum consisting of an overlay of an enormous number of peaks that cannot be attributed precisely to a certain component, although water gives signals at 1450 and 1940 nm. The measurements are also influenced by temperature and by the colour and the particle size of the sample. Therefore analyses need a very product-specific calibration against a reference method based on a great number of single measurements in the whole range of the expected water contents.

Originally, NIR spectrometers worked with several wavelength filters, still a widely used technique for many applications, particularly for the determination of water in food. The values obtained for each wavelength are given a mathematical factor with a statistical weight, which is calculated by empirical mathematical approximation to match the result (the sum of these values, plus a constant factor) with the result obtained by a reference method. The quality of the evaluation increases with the number of filters used. An important step forward was the exploitation of the entire NIR wavelength range by the application of Fourier-transform NIR (FT-NIR) and the use of chemometric evaluation methods. NIR spectroscopy has the advantage that different components and properties of the sample can be measured (after respective calibration) simultaneously.

The NIR spectroscopy technique can be used for qualitative analyses, for instance quality or authenticity control of raw materials (see Colour Plate 11) and for quantitative analyses to measure the content of certain components or properties of a sample.

3.2.1.5 Microwave spectroscopy

The velocity of microwaves depends on the dielectric properties of the material they cross. As the dielectric constant increases, the propagation rate becomes smaller and with it the wavelength. The small dipoles of water molecules can easily be oriented in the rapidly oscillating electromagnetic field. This results in an extraordinarily high dielectric constant and explains the great effect of water on the microwave wavelength. The water molecules cannot, however, follow the oscillations of the field exactly; a small time lag occurs, resulting in partial conversion of the field energy into translational energy, and thus in a gradual damping of the amplitude of the microwaves.

The sample is placed between the emitter and the receiver of the microwaves. For physical reasons no metallic or other highly conductive substances must be present between these microwave antennae. Measurement of the shift of the wavelength and the attenuation of the amplitude of the waves (and thus the microwave energy) after passing through the sample can be used to determine the amount of water present. Several wavelengths may be used in quick succession for the same sample. The mean value of these measurements is more reliable than the result obtained with a single wavelength. The effect depends on the interaction of the waves with the water molecules and thus their number. This means that the measurement is dependent on the concentration of water as well as on the thickness and density of the product layer. These other properties must therefore either be kept constant (as parameters of the calibration) or be measured and accounted for.

Only freely movable water can be measured accurately, as crystallised or tightly bound water molecules cannot be oriented in the field in the same way. Different intermediate states exist, an aspect that must be covered by the calibration, which must therefore be product-specific.

The principle of the microwave resonator method is mainly the same as described for microwave spectroscopy. Here, too, the wavelength shift and the attenuation of the microwave energy are monitored. The difference is that the measurement is carried out in a resonator chamber, in which standing microwaves are produced. Their frequency equals the resonance frequency of the chamber. The plot of the measured diode signal against the frequency has the form of a slim peak with the maximum at the resonance frequency. This frequency changes as soon as a sample is brought into the resonator chamber, the shift being dependent on the amount of water in the product. At the same time the peak becomes lower and broader. These quantities are evaluated mathematically, allowing calculation of the water content and the packing density independently of each other.

A product-specific calibration is also necessary. See Colour Plate 12.

3.2.2 Chromatography

3.2.2.1 General observations

Chromatography aims at the separation of mixtures into their components. It utilises the different affinity of these components to two separate phases. One of these phases is stationary (and is therefore solid or liquid), the other mobile (liquid or gaseous). The mobile phase moves through or over the stationary phase and also transports the sample to be separated into its components and thus analysed. Various techniques exist. The kind of the affinity of the components to the two phases can be different. Components with a high affinity to the stationary phase will be retained more and therefore move more slowly. They will, during analysis, cover a shorter way or take a longer time to pass through the stationary phase, depending on whether the analysis is ended after a given time (like in thin layer chromatography, see section 3.2.2.2) or after all the components have passed through the stationary phase (like in high performance liquid chromatography, see section 3.2.2.3, and gas chromatography, see section 3.2.4). Respectively, components with a higher "preference" for the mobile phase will have a higher flow speed through the analytical device.

Various criteria can serve to classify the different chromatographic techniques. For example, according to the nature of the stationary phase we have paper chromatography (PC) or thin layer chromatography (TLC). The term gas chromatography (GC) indicates that the mobile phase is a gas. Column chromatography (CC) means that the separation process takes place in a column. The state of aggregation of the two phases used can also serve to distinguish the techniques. Mentioning the mobile phase first, we have liquid–solid (LSC), liquid–liquid (LLC), gaseous–solid (GSC) and gaseous–liquid (GLC) chromatography. The reason for the difference in affinity of the components to the two phases can also be referred to. Thus, we have adsorption chromatography (the separation criterion is the different adsorption to the stationary phase), partition chromatography (the analytes are differently partitioned between the mobile phase and a liquid stationary phase), ion chromatography (the distribution between the phases includes ionic interactions), chiral chromatog-

raphy (optical isomers are retained differently by the stationary phase) and affinity chromatography (specific interactions are utilised like those of enzymes with substrates or antigens with antibodies). All these criteria overlap. A certain gas chromatographic technique may involve a column chromatography, a GLC and a partition chromatography.

At the end of the separation process the components have to be identified. The qualitative identification is possible by comparing the time the substance needs to pass through the device, the so-called retention time, with the time a known substance needs. This applies to techniques where all the components must pass through the chromatographic apparatus. If, however, the analysis is stopped after a certain time, the distance the component in question has covered is compared with the one covered by a known substance. When the retention time or, respectively, the covered distance of the substance to identify is identical with that of the known substance, there is a high probability that the substances are the same. The quantitative analysis is carried out using a detector which is placed at the end of the chromatographic system. It gives a signal that is proportional to the amount or concentration of the component which passes through the detector. The relationship between signal and concentration must be established with known concentrations of the analyte. Thus, a calibration curve (usually a straight line) is obtained.

3.2.2.2 Thin layer chromatography (TLC) and paper chromatography (PC)

The stationary phase in PC is a special paper; in TLC it is a thin layer on a support of a glass plate or a metal foil (usually aluminium). The size is usually in the range of 20 by 20 cm. The material of the thin layer (thickness approximately 0.25 mm) is often a powder (particle size about 12 μm, pore size approximately 6 nm) of cellulose, aluminium oxide or silica gel. The surfaces of these materials are relatively polar, and the mobile phase will therefore be a less polar or non-polar liquid. This combination is the so-called normal-phase technique. The surface of the stationary phase can, however, be modified by reacting the OH groups at the surface with halogen alkyl silanes. By condensation (release of hydrogen halides) the surface is covered with alkyl chains and becomes, thus, non-polar. The mobile phase will then be a more polar liquid. This is the reversed-phase

(RP) technique. Depending on the length of the alkyl chains, terms such as RP8 or RP18 are used.

The solutions to be analysed (about 1 µl) are applied to the stationary phase as little spots by capillaries along a starting line about 1–2 cm from the bottom. Some of these solutions may be solutions of known substances. The paper, plate or foil is then brought into a chamber which contains the mobile phase. This can be a mixture of several components. This solvent system moves into and through the thin layer mobile phase by capillary forces. It reaches the starting line with the points of application and then transports the components with a speed depending on the relative affinity of the single compounds to the stationary and the mobile phase. When the solvent front has nearly reached the upper end of the plate or foil, the analysis is stopped by taking the stationary phase out of the chamber and marking the solvent front. The original spots will now have moved a certain distance. If the solution contained several substances, several spots in a vertical line will appear.

To characterise a specific substance X, the distance of the spot from the starting line, d_X, relative to the distance covered by the solvent system, d_S, can be used. This is the retention factor, R_f (see Eq. 3.35):

$$R_f(X) = \frac{d_X}{d_S} \qquad (3.35)$$

where

$R_f(X)$ is the retention factor of X,
d_X is the distance covered by X,
d_S is the distance covered by the solvent system.

The R_f values depend strongly, of course, on the stationary phase and the mobile phase used. Only for these conditions are they characteristic for a certain compound. R_f values lie in the range between 0 (substance is absolutely insoluble in the solvent system) and 1 (substance is ideally soluble and has no affinity to the stationary phase). In order to provide for a good separation, the phases should be chosen in a way that the R_f values are clearly different from 0 and 1. R_f values serve for qualitative analyses.

In most situations the compounds to be separated will not be visible. In some cases they may be fluorescent and can be observed under an ultraviolet lamp. In other cases a derivatisation of the substances is necessary in order to obtain coloured derivatives. This can be done by spraying the dried paper, plate or foil with a reagent solution. An example is the reaction of amino acids with ninhydrin to obtain blue spots (but for proline and hydroxyproline, yielding yellow derivatives). Another possibility is the prederivatisation of the analytes. Amino acids give fluorescent derivatives with dansyl chloride which can be observed after the chromatographic run under a UV lamp. Of course, the derivatives have other R_f values than the original analytes.

For quantitative analyses the deepness in colour of the spots can be measured by densitometers. This method must be calibrated with spots obtained with known amounts of substance. Another possibility is to scratch the spots off the plate and analyse them quantitatively by a classical method.

3.2.2.3 High performance liquid chromatography (HPLC)

This technique serves for the separation of mixtures which are liquid or can be dissolved completely.

The stationary phase is similar to that used in thin layer chromatography. The particle size is usually in the range of 3–10 µm. It is packed in a column (glass or metal) with an inner diameter of usually about 5 mm and a length of 10–75 cm (usually about 25 cm) (see Colour Plates 13 and 14). The mobile phase, which in this technique is often called eluent or eluant, is pumped through the column. Because of the small particle size of the stationary phase, high pressures are necessary (3–40 MPa or 30–400 bar). The sample solution is injected through a valve into the eluent flow. The sample injected must be a clear solution to avoid clogging of the column material by dispersed particles. It should therefore be membrane filtrated before injection.

Liquid chromatography plays an important role in modern food analysis, since many target compounds are poorly volatile (lipids, vitamins, antioxidants), and are therefore not accessible to GC techniques. In modern HPLC high pressure is used to force the flow of a solvent through a column filled with adsorption material. In principle, two different systems can be characterised by different stationary (column) and mobile (solvent) phases: normal phase chromatography uses polar stationary phases (SiO_2 or Al_2O_3) and non-polar mobile phases (hexane, ethyl acetate), whereas reversed phase (RP) chromatography uses non-polar stationary phases (RP-8, RP-18, RP-30) and relatively polar mobile phases (methanol,

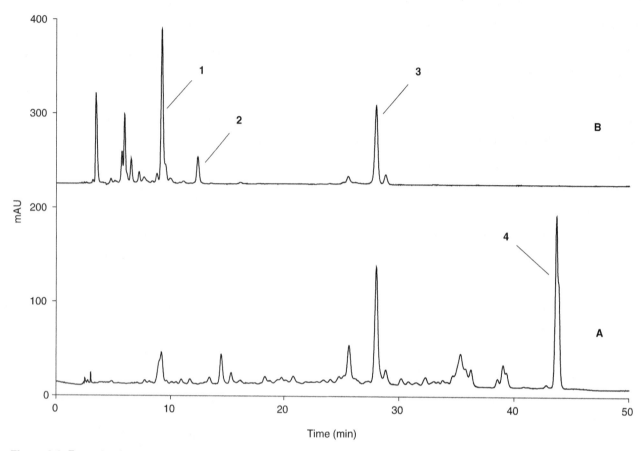

Figure 3.3 Example of a separation of phytonutrients by HPLC. Representative HPLC chromatogram (diode array detector, DAD, 450 nm) of tomato (**A**) and spinach (**B**) extract, using an RP-C30-column for separation and a gradient consisting of methanol, *tert*.-butyl methyl ether, and water for separation. The peak assignment of the main carotenoids is as follows: lutein (**1**), zeaxanthin (**2**), β-carotene (**3**), lycopene (**4**).

acetonitrile, water). Actually, most applications for food components are performed with RP systems.

The type of detector used depends on the analyte under investigation (see Fig. 3.3). If the analyte absorbs light in the UV/vis region, a UV detector or a diode array detector (DAD) may be used. If the analyte shows fluorescence – with or without derivatisation – sensitive detection is possible using a fluorescence detector. Furthermore, the refractive index (RI) detector is applicable to all compounds changing the refractive index of the solvent; however, the sensitivity of this detector is rather poor. HPLC is actually used for determination of sugars (RI detector), preservatives (UV detector), food dyes (UV/Vis detector), sweeteners (UV detector) and other non-volatile food additives.

Chromatographic methods useful for the isolation and clean-up of proteins are summed under the designation "fast protein liquid chromatography"

(FPLC). The following special FPLC techniques are in use:

- Size-exclusion chromatography (SEC) is applied for the separation of proteins of different molecular weight and shape. Small molecules penetrate pores of a resin and are retained, while large molecules are not retained and elute earlier from the column.
- Hydrophobic interaction chromatography (HIC) is based on hydrophobic interaction between the stationary phase and proteins. Usually, the protein is dissolved in a buffer containing high concentrations of ammonium sulphate. Proteins are eluted as the salt concentration in the buffer is decreased.
- Ion-exchange chromatography (IEX) separates proteins based on differences between their pH-dependent total charges. The respective protein

must have a charge opposite to that of the functional group attached to the stationary phase. Elution is achieved by increasing the ionic strength of the eluent to break the ionic interaction.

3.2.2.4 Ion chromatography

Ion chromatography (IC) is a special type of liquid chromatography using a column filled with a cation or an anion exchange material: ionic groups are bound to an inert material, e.g. polystyrene, cellulose or silica gel (see Colour Plate 15). Typical groups are carboxyl or sulfonyl groups for the exchange of cations and quaternary amines for anions. Ionic analytes passing through the column interact with the groups bound to the stationary phase and are retained according to the strength of the electrostatic interaction. Thus, separation of ions is achieved. For detection, a conductivity detector is used. The initial conductivity of the eluent is lowered by neutralisation applying a suppressor column. IC is used in food chemistry mainly for quantitative determination of ions such as Na^+, K^+, Mg^{2+}, Ca^{2+}, Cl^-, NO_2^-, NO_3^- or SO_4^{2-} in drinking water or as an effective clean-up step in sample preparation.

3.2.3 Capillary electrophoresis

Electrophoretic techniques are based on the fact that charged molecules are attracted by an electrode of opposite charge in an electric field. Anions (−) move towards the anode, while cations (+) move towards the cathode. Charged molecules such as proteins, peptides, nucleic acids and other biopolymers migrate to the respective electrode and form a band which can be visualised by staining with dyes (e.g. Coomassie Blue) or silver ions. Transferring the principles of electrophoretic separation to the micro scale results in development of capillary electrophoresis (CE), which uses – untreated or coated – fused silica capillaries filled with a buffer solution or an appropriate gel and a suitable detector based on UV spectroscopy, fluorescence or electroconductivity. For example, DNA fragments may be separated in a borate buffer and are detected at 260 nm. In recent years, neutral molecules could be separated by CE, too. The addition of a surfactant such as sodium dodecyl sulphate results in the formation of micelles burying the hydrophobic neutral analyte in their core, and exhibiting the hydrophilic negatively charged surfactant molecules to the outer surface. These micelles are separated according to electrophoretic principles. This technique has been called micellar electrokinetic capillary chromatography (MECC).

3.2.4 Gas chromatography and mass spectroscopy (detection and structural analysis)

Gas chromatographic (GC) techniques use an inert gas (hydrogen, helium, nitrogen, argon) for the transport of volatile analytes through a column, which is placed in a heatable furnace. Typically, temperature programmes range from 50 to 300°C. Two types of columns are used in food analysis: packed columns and capillary columns. Due to the enhanced separation efficiency, most applications actually use the latter system. The stationary phase of a typical capillary column is a non-polar or a polar thin film (0.3–2 μm) of various materials, spread over the inside of a narrow silica column.

For detection, various systems have been developed. The most versatile detector is the flame-ionisation detector (FID), which is able to detect all compounds comprising C–H or C–C bonds. The carrier gas stream eluting from the column is burnt in an H_2/air-flame. This process generates electrons, which are detected by a counting electrode, placed in an electric field. Another widely spread detector is the electron capture detector (ECD), which is applied for the sensitive detection of compounds comprising halogens, sulphur or nitro groups (especially pesticides). The detection principle is based on the capture of electrons emitted from a radioactive source as Ni^{63} (beta radiator). In recent years, coupling GC with mass spectrometry (GC-MS) has become a valuable tool for both structural analysis and selective compound detection. Special interfaces have been developed to remove the excess of carrier gas eluting from the GC column. Several types of mass spectrometers are available (e.g. quadruple-MS or time-of-flight-MS) (see Colour Plate 16). Bombardment of analytes eluting from the GC column with electrons (70 eV) in the ion source results in formation of radical cations ($M^{+\bullet}$) which further decay and reveal daughter ions. This fragmentation pattern allows for structural elucidation (library search). Selecting a definite mass allows for quantitative determinations too.

One of the main applications of GC in food analysis is determination of fatty acids, converted to fatty acid methyl esters (FAMEs) by hydrolysis of triacylglycerides and subsequent derivatisation with BF_3/MeOH. The fatty acid pattern allows for identification of fats and oils and of adulterations. Pesticide and aroma compound analyses are further ambitious applications for GC.

Isotope ratio-MS (IR-MS) is a technique that measures the relative stable isotopic abundance of a small number of elements. Since samples are combusted in a pyrolysis interface, IR-MS analyses actually small variations in the isotopic composition of a number of gaseous molecules. Usually, the ratio of $^{13}C/^{12}C$ is measured as CO_2, $^{15}N/^{14}N$ as N_2, D/H as D_2/H_2, and $^{18}O/^{16}O$ as CO_2 or CO. Results are expressed as δ values relative to a commercially available reference material (e.g. PDB, PeeDeeBelemnite, a fossil $CaCO_3$). For determination of the $^{13}C/^{12}C$ ratio the following equation is used:

$$\delta^{13}C_{PDB} = ((R_S/R_{SD}) - 1) \times 1000\,(\text{‰}) \qquad (3.36)$$

The symbol "R" defines the respective isotopic ratio of an unknown sample (S) and a standard (SD). IR-MS can be used to determine isotopes at their natural abundance level or to detect stable isotope ratios of molecules artificially enriched in one isotope. The latter technique is mostly applied in biochemical research.

In food chemistry, IR-MS is applied to determine the geographical origin of food (wine, milk, butter, cheese), to differentiate between artificial and natural compounds (flavours) or to state the possible source of environmental contaminations in soil and waste water. For instance, the isotopic ratios determined in apple juice can indicate an adulteration with corn syrup. Ethanol originating from sugarcane can be distinguished from synthetic ethanol and from ethanol made from European sugar beets, thus allowing the verification of rum provenance. The ratios of $^{13}C/^{12}C$, $^{15}N/^{14}N$ and others were used to indicate the geographic origin of biological material such as ivory. Stable isotope determinations have recently been incorporated into international regulations such as European Union requirements for food products.

Atomic absorption spectroscopy (AAS) provides the possibility to determine trace amounts of metals in mostly liquid sample matrices (see Colour Plate 17).

AAS is based on the fact that atoms absorb only radiation with characteristic energy. If an element present in a sample is vaporised, specific wavelengths – emitted by a lamp – cause excitation of electrons to an elevated state, resulting in reduction of the initial light intensity. This reduction is proportional to the element concentration in the sample and is used for quantitative determinations. As radiation sources, two types of lamps are used: hollow cathode lamps (HCL) and electrodeless discharge lamps (EDL). Usually, for determination of each element another lamp is required. Hollow cathode lamps are available from several manufacturers either as single or as multiple element lamps. Both types of lamps emit characteristic lines of the respective atomic species. For determination of alkali metals, no lamp is needed to excite electrons. Atomisation in the flame provides enough energy to initiate visible light emission during the return of electrons from an excited to the ground level. This is the basis of simple flame photometry (atom emission spectroscopy, AES), to date applied for the determination of alkali metals (sodium: 589.6 nm and potassium: 766.5 nm) in drinking water.

For the determination of elements easily forming hydrides (arsenic, antimony, selenium, tellurium, tin) the hydride technique was developed: after reduction with $NaBH_4$, volatile hydrides are purged with an inert gas in a heated cuvette and analysed based on AAS principles. Due to its high vapour pressure, mercury is the only element that is detectable in its metallic form.

The combination of two or even more separate analytical techniques via appropriate interfaces creates hyphenated methods, such as GC-MS (gas chromatography-mass spectrometry), LC-MS (liquid chromatography-mass spectrometry), LC-IR (liquid chromatography-infrared-spectroscopy) and LC-NMR (liquid chromatography-nuclear magnetic resonance). The purpose of coupling is the need for unambiguous detection of the compounds separated by each of the chromatographic techniques. Furthermore, it is possible to determine additional information about an analyte. Nowadays GC-MS is the most popular of all hyphenated techniques since it combines the separation efficiency of GC with the high degree of structural information provided by MS. Using MS as a GC detector offers the advantage of using isotope-labelled compounds as internal standards, a method facilitating modern isotope dilution analysis. GC-MS and LC-MS are indispensable techniques for the determination of pesticides or environmental contaminations (PCBs, PAKs) in complex matrices.

3.2.8 Nuclear magnetic resonance (NMR) (structural analysis, state of water and fat)

The hydrogen nuclei of the compounds contained in a sample have a nuclear spin. Placed in a magnetic field they precess around the field axis with the so-called Larmor frequency, which depends on the nature of the nucleus and which is proportional to the magnetic field strength. These spins can be excited and synchronised by using a short, strong radio frequency pulse, resulting in an oscillating magnetic field that induces an alternating voltage, the NMR (nuclear magnetic resonance) signal. The magnitude of these oscillations, which is proportional to the number of hydrogen atoms in the sample, can be measured. After switching off the aligning radio frequency pulse, a relaxation is observed as the nuclear spins fall back into their original state, resulting in a decay of the NMR signal. The rate of this decay depends very much on the surroundings of the hydrogen atoms. In a solid environment the oscillations are heavily damped, and the decay is very rapid. After about 70 μs the NMR signal has disappeared. In a liquid environment, however, the amplitude of the NMR signal may still have 99% of its original value at that time. This allows a distinction to be made between solids, oils and water based on the different decay times, as well as to measure the amount of these different components in the sample (see Colour Plate 18).

Water in food samples is not determined as such, but rather hydrogen atoms in a certain environment. Free water is easily detected, while strongly bound water does not fall into this category. All the transitory states may occur. This makes a product-specific calibration necessary to correlate the NMR results with those of a reference method. For practical reasons the water content should be below about 15% because of the long relaxation time of hydrogen nuclei in free water.

Single-sided NMR spectrometers have also been developed. They do not use homogeneous magnetic fields. They are placed directly on the (big) sample.

3.2.9 Immunoassays and ELISA techniques

Immunochemical methods are based on the non-covalent binding of an antigen by an antibody. Antibodies (immunoglobulins) are commercially produced in living organisms (rat, rabbit, horse, goat) against a foreign molecule (the analyte, chemically bound to a protein), which is injected in the muscle. The antibody in the crude plasma or the isolated antibody itself will recognise the analyte and may be employed as a highly specific detector for the respective analyte.

Due to different detection principles, several immunochemical systems are distinguishable: fluorescence (FIA), radio (RIA) and enzyme (EIA) immunoassays. In previous systems, radiolabelled molecules were used to detect the antigen–antibody complex (radioimmunoassay). However, the exposure to radioactive tracers complicates this system. In the so-called competitive ELISA (enzyme-linked immunoabsorbent assay), antibodies are immobilised on a solid carrier. In a second step, enzyme-linked antigens and free antigens of a sample are added. The enzyme-linked and the free antigens compete for free binding sites. The higher the antigen concentration in the sample is, the less enzyme-linked antigens bind to the antibodies fixed on the carrier. Quantification is based on a subsequent enzymatic reaction (e.g. horseradish peroxidase, alkaline phosphatase), where only the bound antigen–antibody enzyme complex converts an added substrate to a dye, which is measured photometrically.

Numerous variants of ELISA have been developed and are in use in food analyses. Immunochemical techniques are actually used to detect traces of nut proteins not labelled on food packaging. Due to the presence of foreign proteins, adulterations of food may be detected in a qualitative and quantitative manner (e.g. bovine milk proteins in cheese made from goat's milk). Furthermore, sensitive ready-to-use kits for the detection of aflatoxins, pesticides (atrazin) and vitamins (folic acid) are available on the market.

3.2.10 Thermal analysis (bomb calorimetry, differential scanning calorimetry)

The total energy of food can be determined in two different ways. The first is by calculating the energy mathematically on the basis of quantitative analysis of the main food constituents (fat, protein, carbohydrates, organic acids, ethanol). The second provides a direct method to determine the total energy experimentally: an aliquot of the food is oxidised (burnt) in a sealable metal tube or a so-called calorimetric bomb (see Colour Plate 19) under a high pressure of oxygen, resulting in the formation of carbon dioxide, water and other gases of low molecular weight. The increase in the temperature of the water surrounding the tube is used for calculation of the total energy (bomb calorimetry). However, this method may oxidise food constituents that are not bio-available during human digestion (e.g. dietary fibre). A distinction must therefore be made between the physical or gross calorific value and the physiological calorific value of food. Thus, this method is actually not accepted in all countries.

Different techniques of bomb calorimetry are common: isothermal and adiabatic methods. A particular technique has been developed that works without surrounding water. This is the isoperibolic (or "double-dry") method (see Colour Plate 19).

Differential scanning calorimetry (DSC) measures the difference in heating power needed to heat a food sample and a reference compound in a furnace, while temperatures of both are kept identical. Temperature is continuously changed and the heat effect is measured. To achieve equal temperature, two heaters are needed. One of the most frequent uses of DSC is the purity analysis of pharmaceuticals. Due to different melting or transition temperatures, DSC is further used to characterise mixtures of plastic waste. This method serves also to measure transition energies and temperatures between different states, e.g. states of aggregation, crystal forms, different hydrates, glass transition, decarboxylation or other degradation reactions. The technique can be combined with mass measurements (thermogravimetric analysis, TGA) and also mass spectroscopy (TGA/MS).

3.2.11 The combustion method for nitrogen (Dumas)

Determination of nitrogen concentration allows for the calculation of the protein content of food. The Dumas method is based on the determination of nitrogen generated by high-temperature combustion of nitrogen-containing food compounds. The entire sample is combusted in a steam of oxygen at 900°C, producing oxides of nitrogen (NO_x), water, sulphur dioxide and carbon dioxide. Helium is used as a carrier gas to transport all combustion products. The oxides of nitrogen are detected by a thermal conductivity detector after reduction to nitrogen on a hot copper catalyst. By-products of the combustion are removed by adsorption. Currently, several automated systems are available. The method is said to replace the traditional Kjeldahl method for protein determination. For determination of nitrogen in animal feed the method is already in use.

Further reading and references

Baltes, W. (ed.) (1990) *Rapid Methods for Food and Food Raw Material*. Behr's, Hamburg.

Barker, P.J. (1990) Low-resolution NMR. In: *Rapid Methods for Food and Food Raw Material* (ed. W. Baltes). Behr's, Hamburg.

Burns, D.A. and Ciurczak, E.W. (eds) (1992) *Handbook of Near-Infrared Analysis*. Marcel Dekker, New York.

Engelhardt, H. (ed.) (1986) *Practice of High Performance Liquid Chromatography, Applications, Equipment and Quantitative Analysis*. Springer, Berlin.

Field, L.D. (1989) *Analytical NMR*. John Wiley, New York.

Fried, B. and Sherma, J. (1999) *Thin-Layer Chromatography*, 4th edn. Marcel Dekker, New York.

Fritz, J.S. and Gjerde, D.T. (2000) *Ion Chromatography*. Wiley-VCH, Weinheim.

Fung, D.Y.C. and Matthews, R.F. (1991) *Instrumental Methods for Quality Assurance in Foods*. Marcel Dekker, New York.

Goldsby, R.A., Kindt, T.J., Osborne, B.A. and Kuby, J. (2004) *Immunology*. W.H. Freeman, New York.

Gordon, M.H. (ed.) (1990) *Principles and Applications of Gas Chromatography in Food Analysis*. Ellis Horwood, New York.

Gruenwedel, D.W. and Whitaker, J.R. (1984–1987) *Food Analysis, Principles and Techniques* (8 volumes). Marcel Dekker, New York.

Grünke, S. and Wünsch, G. (2000) Kinetics and stoichiometry in the Karl Fischer solution. *Fresenius' Journal of Analytical Chemistry*, **368**, 139–47.

Guthausen, G., Todt, H., Burk, W., Schmalbein, D., Guthausen, A. and Kamlowski, A. (2006a) Single-sided NMR in foods. In: *Modern Magnetic Resonance* (ed. G.A. Webb), pp. 1873–5. Springer, Berlin.

Guthausen, G., Todt, H., Burk, W., Schmalbein, D. and Kamlowski, A. (2006b) Time-domain NMR in quality control: more advanced methods. In: *Modern Magnetic Resonance* (ed. G.A. Webb), pp. 1713–16. Springer, Berlin.

Handley, A.J. and Adland, E.R. (2001) *Gas Chromatographic Techniques and Applications*. Sheffield Academic Press, Sheffield.

Hauschild, T. (2005) Density and moisture measurements using microwave resonators. In: *Electromagnetic Aquametry, Electromagnetic Wave Interaction with Water and Moist Substances* (ed. K. Kupfer). Springer, Heidelberg.

Heinze, P. and Isengard, H.-D. (2001) Determination of the water content in different sugar syrups by halogen drying. *Food Control*, **12**, 483–6.

Hirschfeld, T.B. and Stark, E.W. (1984) *Near Infrared Analysis of Foodstuffs – Analysis of Food and Beverages*. Academic Press, New York.

International Union of Biochemistry and Molecular Biology (IUBMB) (1992) *Enzyme Nomenclature 1992*. Academic Press, New York.

Isengard, H.-D. (1995) Rapid water determination in foodstuffs. *Trends in Food Science and Technology*, **6**, 155–62.

Isengard, H.-D. (2001) Water content, one of the most important properties of food. *Food Control*, **12**, 395–400.

Isengard, H.-D. (2008) Water determination – scientific and economic dimensions. *Food Chemistry*, **106**, 1379–84.

Isengard, H.-D. (2008) The influence of reference methods on the calibration of indirect methods. In: *Nondestructive Testing of Food Quality* (eds J. Irudayaraj and C. Reh), pp. 33–43. Blackwell Publishing, Oxford, and the Institute of Food Technologists (IFT Press), Ames, Iowa.

Isengard, H.-D. and Färber, J.-M. (1999) Hidden parameters of infrared drying for determining low water contents in instant powders. *Talanta*, **50**, 239–46.

Isengard, H.-D. and Heinze, P. (2003) Determination of total water and surface water in sugars. *Food Chemistry*, **82**, 169–72.

Isengard, H.-D. and Nowotny, M. (1991) Dispergierung als Vorbereitung für die Karl-Fischer-Titration. *Deutsche Lebensmittel-Rundschau*, **87**, 176–80.

Isengard, H.-D. and Präger, H. (2003) Water determination in products with high sugar content by infrared drying. *Food Chemistry*, **82**, 161–7.

Isengard, H.-D. and Schmitt, K. (1995) Karl Fischer titration at elevated temperatures. *Mikrochimica Acta*, **120**, 329–37.

Isengard, H.-D. and Striffler, U. (1992) Karl Fischer titration in boiling methanol. *Fresenius' Journal of Analytical Chemistry*, **342**, 287–91.

Isengard, H.-D. and Walter, M. (1998) Can the true water content in dairy products be determined accurately by microwave drying? *Zeitschrift für Lebensmittel-Untersuchung und -Forschung*, **207**, 377–80.

Isengard, H.-D., Kling, R. and Reh, C.T. (2006) Proposal of a new reference method to determine the water content of dried dairy products. *Food Chemistry*, **96**, 418–22.

James, C.S. (1995) *Analytical Chemistry of Foods*. Chapman & Hall, Glasgow.

Kellner, R., Mermet, J.-M., Otto, M. and Widmer, H.M. (1998) *Analytical Chemistry*. Wiley-VCH, Weinheim.

Köstler, M. and Isengard, H.-D. (2001) Quality control of raw materials using NIR spectroscopy in the food industry. *G.I.T. Laboratory Journal*, **5**, 162–4.

Kraszewski, A. (1980) Microwave aquametry. *Journal of Microwave Power*, **15**, 207–310.

Kress-Rogers, E. and Kent, M. (1987) Microwave measurement of powder moisture and density. *Journal of Food Engineering*, **6**, 345–76.

Lough, W.J. and Wainer, I.W. (1995) *High Performance Liquid Chromatography, Fundamental Principles and Practice*. Blackie Academic & Professional, Glasgow.

Matissek, R. and Wittkowski, R. (eds) (1992) *High Performance Liquid Chromatography in Food Control and Research*. Behr's, Hamburg.

Matissek, R., Schnepel, F.-M. and Steiner, G. (2006) *Lebensmittelanalytik*, 3rd edn. Springer, Berlin.

Meyer, W. and Schilz, W. (1980) A microwave method for density-independent determination of moisture content of solids. *Journal of Physics D: Applied Physics*, **13**, 1823–30.

Nielsen, S.S. (1998) *Food Analysis*, 2nd edn. Aspen, Gaithersburg, Maryland.

Osborne, B.G. and Fearn, T. (1988) *Near Infrared Spectroscopy in Food Analysis*. Longman Scientific & Technical, New York.

Paré, J.R.C. and Bélanger, J.M.R. (eds) (1997) *Instrumental Methods in Food Analysis*. Elsevier, Amsterdam.

Pomeranz, Y. and Meloan, C.E. (1994) *Food Analysis, Theory and Practice*, 3rd edn. Chapman & Hall, New York.

Rückold, S., Grobecker, K.H. and Isengard, H.-D. (2000) Determination of the contents of water and moisture in milk powder. *Fresenius' Journal of Analytical Chemistry*, **368**, 522–7.

Rückold, S., Grobecker, K.H. and Isengard, H.-D. (2001a) Water as a source of errors in reference materials. *Fresenius' Journal of Analytical Chemistry*, **370**, 189–93.

Rückold, S., Grobecker, K.H. and Isengard, H.-D. (2001b) The effects of drying on biological matrices and the consequences for reference materials. *Food Control*, **12**, 401–407.

Rückold, S., Isengard, H.-D., Hanss, J. and Grobecker, K.H. (2003) The energy of interaction between water and surfaces of biological reference materials. *Food Chemistry*, **82**, 51–9.

Rudi, T., Guthausen, G., Burk, W., Reh, C.T. and Isengard, H.-D. (2008) Simultaneous determination of fat and water content in caramel using time-domain NMR. *Food Chemistry*, **106**, 1379–84.

Sandra, P. and Bicchi, C. (eds) (1987) *Capillary Gas Chromatography in Essential Oil Analysis*. Dr. Alfred Huethig, Heidelberg.

Schmitt, K. and Isengard, H.-D. (1998) Karl Fischer titration – a method for determining the true water content of cereals. *Fresenius Journal of Analytical Chemistry*, **360**, 465–9.

Scholz, E. (1984) *Karl Fischer Titration*. Springer, Berlin.

Schwedt, G. (2007) *Taschenatlas der Analytik*, 3rd edn. Wiley-VCH, Weinheim.

Sherma, J. and Fried, B. (eds) (2003) *Handbook of Thin-Layer Chromatography*, 3rd edn. Marcel Dekker, New York.

Skoog, D.A., West, D.M. , Holler, F.J. and Crouch, S.R. (2000) *Analytical Chemistry, an Introduction*, 7th edn. Saunders College Publishing, Philadelphia.

Tanai, T. (1999) *HPLC, a Practical Guide*. The Royal Society of Chemistry, Cambridge.

Todt, H., Guthausen, G., Burk, W., Schmalbein, D., and Kamlowski, A. (2006) *Time-Domain NMR in Quality Control: Standard Applications in Food*. Springer Netherlands, Dordrecht.

Todt, H., Burk, W., Guthausen, G., Guthausen, A., Kamlowski, A. and Schmalbein, D. (2001) Quality control with time-domain NMR. *European Journal of Lipid Science and Technology*, **103**, 835–40.

Weaver, C.M. and Daniel, J.R. (2003) *The Food Chemistry Laboratory, a Manual for Experimental Foods, Dietetics, and Food Scientists*. CRC Press, Boca Raton.

Weston, A. and Brown, P.R. (1997) *HPLC and CE, Principles and Practice*. Academic Press, San Diego.

Wrolstad, R.E., Acree, T.E., Decker, E.A., *et al.* (2005) *Handbook of Food Analytical Chemistry – Water, Proteins, Enzymes, Lipids, and Carbohydrates*. John Wiley, Hoboken, New Jersey

Yazgan, S., Bernreuther, A., Ulberth, F. and Isengard, H.-D. (2006) Water – an important parameter for the preparation and proper use of certified reference materials. *Food Chemistry*, **96**, 411–17.

Supplementary material is available at www.wiley.com/go/campbellplatt

Food biochemistry

Brian C. Bryksa and Rickey Y. Yada

Key points

- This chapter focuses on basic biochemical concepts relevant to foods in six main categories: carbohydrates, proteins, lipids, nucleic acids, enzymes, and food processing and storage.
- In addition to sweetening, carbohydrates are important to food processes such as gelation, emulsification, flavour encapsulation, flavour binding, coloration and flavour production via browning reactions, and control of humidity and water activity.
- The genetic information encoded by DNA serves the singular purpose of building proteins, of which there are two types: globular and fibrous proteins.
- Globular proteins include enzymes, transport proteins and receptor proteins while structural and motility-related (muscle) proteins are fibrous.
- Triglycerides comprise 98% of food lipids, serving critical roles as energy sources as well as contributing to mouthfeel and satiety.
- The advent of genetic engineering and identification has allowed for the tailoring of specific food characteristics as well as verification of authenticity.
- Food processing and storage serve to provide or preserve food properties such as colour, texture, flavour and nutritional value, as well as to provide chemical and microbial safety barriers, and extended stabilities.

4.1 Introduction

By its nature, food science is an interdisciplinary field requiring an understanding of microbiology, chemistry, biology, biochemistry, and engineering as they relate to food systems. While it is to be expected that the exciting and seductive fields of product development and sensory analysis are most coveted by students of food science, an understanding of biochemical processes underlies the ability to control many aspects of food characteristics and stabilities. This chapter focuses on the fundamental components which comprise most foods and the basic principles of biochemical processes pertinent to foods. As entire text books have been written on the subject of food biochemistry, this chapter is intended to be a condensed compilation of basic biochemical concepts relevant to foods in six main categories: carbohydrates, proteins, lipids, nucleic acids, enzymes, and food processing and storage.

4.2 Carbohydrates

Carbohydrates are one of the four major classes of biomolecules and comprise the majority of organic mass on earth. The term 'carbohydrate' literally means 'carbon hydrate' which is reflected in the basic building block unit of simple carbohydrates, i.e. $(CH_2O)_n$. Their biologically important roles are energy storage (e.g. plant starch, animal glycogen), energy transmission (e.g. ATP, many metabolic intermediates), structural components (e.g. plant cellulose, arthropod chitin), and intra- and extracellular communication (e.g. egg–sperm binding, immune system recognition). Critical for the food industry, carbohydrates make up three quarters of the dry weight of all land plants and seaweeds, thus serving as the primary nutritive energy sources from foods like grains, fruits, and vegetables as well as being important ingredients for many formulated or processed foods. In addition to sweetening agents they are involved in food functions such as gelation, emulsification, flavour encapsulation, flavour binding, coloration and flavour production via browning reactions, and control of humidity and water activity.

4.2.1 Nomenclature and structures

The basic unit of a carbohydrate is termed a *monosaccharide*. Two monosaccharides joined together are called *disaccharides*; three are called *trisaccharides*, and so on. Two to ten monosaccharides in a chain are collectively known as *oligosaccharides* while ten or more are termed *polysaccharides*.

The simplest carbohydrates, monosaccharides, are aldehydes or ketones having two or more hydroxyl (-OH) groups and at least three carbons (*trioses*). The two carbohydrate trioses are glyceraldehyde (an *aldose*) and dihydroxyacetone (a *ketose*). The term aldose indicates the presence of an aldehyde group while ketose indicates that a ketone group is contained (see Fig. 4.1).

In nomenclature, carbons are numbered from the carbonyl carbon. Glyceraldehyde contains an asymmetric carbon; thus both D- and L-*stereoisomers* exist for this carbohydrate building block, with D- and L- referring to the configuration of the asymmetric carbon farthest from the carbonyl group (ketone or aldehyde). The D- and L- forms of glyceraldehyde are said to be *enantiomers* – mirror images of each other. Sugars differing in configuration at a single asymmet-

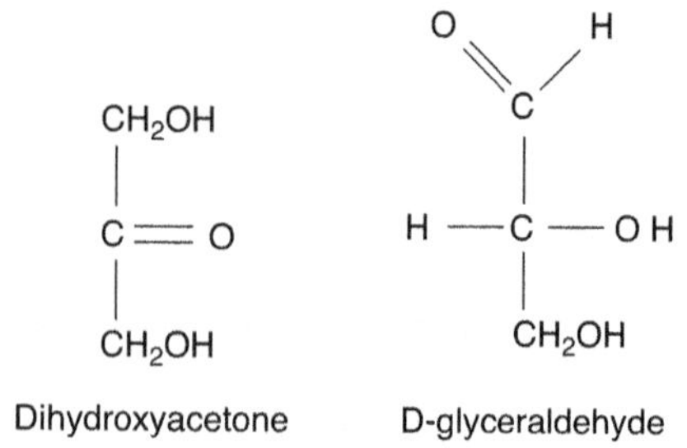

Figure 4.1 The structures of dihydroxyacetone and D-glyceraldehyde.

ric carbon, like D-mannose and D-glucose, are termed *epimers* (Fig. 4.2).

The common food-related monosaccharides glucose, mannose, and galactose are aldoses since they contain an aldehyde group, while fructose is a ketose having a ketone group. Figure 4.2 shows open chain structures; however, sugars in solution, like glucose and fructose, exist as closed ring structures. The aldose sugars, like glucose, cyclize to form a six-membered ring called a pyranose, a term that refers to its structural similarity to pyran. Ketose sugars, e.g. fructose, cyclize to form *furanoses*, five-membered rings structurally related to *furan* (Fig. 4.3).

The aldehyde of an aldose can react with an alcohol to form a *hemiketal*. Cyclization results from such reactions occurring in an intramolecular manner. Similarly, a ketone of a ketose can also react with an alcohol to form a hemiketal.

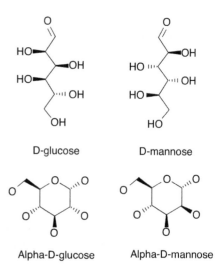

Figure 4.2 The structures of D-glucose and D-mannose, epimers of each other. The linear sugar structures (top) are shown in a manner which highlights that glucose and mannose are near-mirror images; however, note that the two sugars have opposite configurations at asymmetric centre C2.

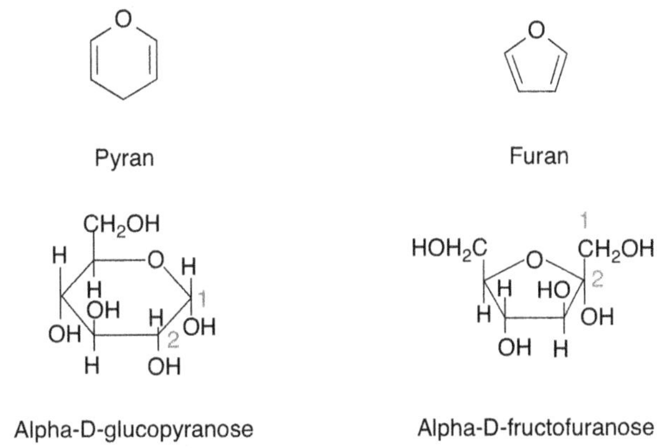

Pyran

Furan

CH₂OH

Alpha-D-glucopyranose

Alpha-D-fructofuranose

Figure 4.3 The five- and six-member ring structures of furanoses and pyranoses, respectively, represented here by fructose and glucose.

Upon formation of a cyclic structure, pentoses and hexoses contain an additional asymmetric carbon. In glucose, the open chain carbonyl carbon C-1 becomes an asymmetric carbon in the ring configuration, thus allowing for two distinct structural forms termed anomers; α- and β-glucopyranose. The α and β conventions refer to the configuration of the hydroxyl group attached to the carbonyl carbon, where α indicates the hydroxyl is below the plane of the ring structure and β indicates it is above the ring plane, respectively drawn as 'down' and 'up'.

The carbonyl carbon is designated the anomeric carbon atom corresponding to C-1 in glucose and C-2 in fructose. Interconversion between the α and β forms occurs by a process known as *mutarotation*. The switch between anomers takes place via the open chain form. As an example, 100% β-D-glucose, or 100% α-D-glucose, dissolved in water will equi-

librate after a few hours at 36% α-D-glucose, 64% β-D-glucose, and far less than 1% open chain. Such changes can be detected by measuring changes in absorption of optically polarized light.

The structures of carbohydrates contain several hydroxyl groups (-OH) in every molecule, a structural feature that imparts a high capacity for hydrogen bonding. This in turn makes carbohydrates very hydrophilic, a property that allows them to serve as a means of moisture control in foods. The ability of a substance to bind water is termed *humectancy*, one of the most important properties of carbohydrates in foods. Sometimes it is desirable to control water's ability to enter and exit foods. For example, to avoid stickiness in baked icings one needs to limit the entry of water. Maltose and lactose (discussed below) have low humectancies; therefore, they allow for sweetness as well as the desired texture. Hygroscopic sugars like corn syrup and invert sugar (discussed below) are more suited for use in foods where water loss must be prevented, e.g. baked goods. In addition to the presence of hydroxyl groups, humectancy is also dependent on the overall structures of carbohydrates. Glucose and fructose contain the same number of hydroxyl units; however, fructose binds more water than glucose.

4.2.2 Sugar derivatives – glycosides

Most chemical reactions of carbohydrates occur via their hydroxyl and carbonyl groups. Under acidic conditions, the carbonyl carbon of a sugar can react with the hydroxyl of an alcohol like methanol (wood alcohol) to form α- and β-glucopyranoside (see below). In the case of methanol (CH₃-OH) the

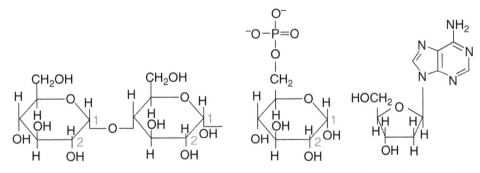

Figure 4.4 Left shows the structure of maltose, two glucose units joined by an α-1,4-glycosidic linkage, a type of O-glycosidic linkage. Middle is another example of an O-glycosidic linkage, that between C6 of glucose and inorganic phosphate in glucose 6-phosphate. Right is deoxyadenosine, a component of DNA which contains an N-glycosidic linkage between deoxyribofuranose and the base adenine.

R-group (methyl group, CH_3-) is referred to as an *aglycon*. Such carbonyl carbon–oxygen bonds are called *O*-glycosidic bonds. The individual sugar residues of polysaccharides such as starch and glycogen are joined by *O-glycosidic bonds* between the carbonyl carbon of one sugar and the alcohol of another.

Two other important examples of glycosidic bonds are those between sugar carbonyl groups and amines (e.g. DNA and RNA building blocks), as well as sugar carbonyl bonds to phosphate (e.g. phosphorylated metabolic intermediates). A glycosidic bond between a carbonyl carbon and the nitrogen of an amine group (R-NH) is termed an *N*-glycosidic bond and the compound is an amino-glycoside (Fig. 4.4). Similarly, reactions of carbonyl carbons with thiols (R-SH) produce thio-glycosides.

4.2.3 Food disaccharides

The food carbohydrates sucrose, lactose, and maltose are all disaccharides; thus they consist of two monosaccharides joined by *O*-glycosidic bonds. These common sugars are three principal disaccharides of the food industry. Although sucrose is enormously important economically for direct use as a sweetening agent and indirectly as a carbon source in fermentations, it exists only in low concentrations throughout most of the plant kingdom with the exception of cane and beet, the main sources of sucrose for the food industry. Sucrose is composed of one glucose and one fructose joined between glucose C1 and fructose C2. The glycosyl bond is α for glucose and β for fructose; thus it is written as α-glucopyranosyl-(1-2)-β-fructofuranoside. Unlike most carbohydrate molecules, the sucrose α to β bond configuration has both carbonyl carbons involved in the glycosyl bond. Normally, aldehydes and ketones can act as reducing agents; thus sucrose is a *non-reducing sugar* as no free aldehyde is present in sucrose. The enzyme responsible for catalyzing the hydrolysis of sucrose to glucose and fructose is *sucrase*. Sucrose treated with sucrase is sometimes called invert sugar because the products have an opposite, or inverted, optical activity, and thus sucrase is also called *invertase*.

The disaccharide lactose is made up of galactose and glucose bound by a β-1,4 glycosidic bond. Often referred to as milk sugar, its complete name is β-galactopyranosyl-(1-4)-α-glucopyranose, and the presence of a free hemiacetal group, i.e. not part of the glycosyl bond, at C1 of the glucose moiety makes it a reducing sugar. Lactose is the primary source of carbohydrates for infant mammals and it is of particular importance among disaccharides in terms of its hydrolysis. Lactose is hydrolyzed by lactase in mammals and by β-galactosidase in bacteria. All normal, young mammals produce lactase for digestion of milk sugar; however, most humans lose lactase production by adulthood. All dairy products containing milk or milk ingredients contain lactose unless it has been consumed by lactase added for this purpose to unfermented dairy products, or by lactic acid bacteria during fermentation. Value-added milk that has had most of its lactose hydrolyzed by treatment with lactase is widely available for purchase. Also, fermented dairy foods like yoghurt and cheese contain less lactose post-fermentation compared to the starting materials because the lactose is converted into lactic acid by bacteria. For example, old cheddar cheese contains virtually no remaining lactose. Lactase deficiency in humans results in lactose intolerance, a clinical condition that is important in societies that consume large amounts of dairy products. Normally, lactose would be digested to glucose and galactose by lactase in the small intestine where sugars are normally broken down to monosaccharides, the only sugars that are absorbed. A build-up of sugar in the small intestine results in fluid influx due to increased osmolality (Lomer *et al.*, 2008). When sugar, including lactose, reaches the lower gut it is anaerobically fermented by bacteria producing gases and short chain acids which further irritate the intestinal lining. Bloating, cramping, and diarrhoea result, proportional to the level of lactase deficiency and the amount of lactose consumed. With regard to health, such conditions are detrimental in terms of nutrient absorption efficiency and hydration. It is thought that humans have evolved the ability to digest lactose through adulthood over thousands of years since dairy farming practices were begun.

Maltose is composed simply of two glucose units joined by an α-1,4 glycosidic linkage. Maltose is derived from starch by treatment with β-amylase which releases maltose units from the non-reducing (C4-presenting) end of the poly-glucose chain. As β-amylase releases maltose the sweetness of the reaction mixture increases, although maltose has limited applications as a sweetener in foods. The term malt in the context of beer making is derived from barley or other grains steeped in water. Upon germination, β-amylase is produced resulting in hydrolysis of starch to maltose. Maltose itself is a reducing sugar due to the free hemiacetal group and upon release by β-amylase it can be further hydrolyzed by α-amylase, yielding free glucose.

4.2.4 Browning reactions of carbohydrates

Foods are affected by browning reactions frequently, sometimes detrimentally and other times purposefully to impart a desired flavour and/or colour. In general, these reactions can be classified into three categories: oxidative/enzymatic browning, caramelization, and non-oxidative/non-enzymatic/Maillard browning. *Oxidative browning*, or enzymatic browning, is discussed in section 4.3.8. The latter two types of browning involve carbohydrate reactions.

Caramelization involves a complex group of reactions that are the result of direct heating of carbohydrates, particularly sugars and sugar syrups. Essentially, dehydration reactions result in the formation of double bonds which absorb different light wavelengths. Also, anomeric shifts, ring size alterations, and breakage of glycosidic bonds result from *thermolysis* which causes dehydration to form anhydro rings or the introduction of double bonds into sugar rings which lead to furans (Eskin, 1990). The formation of conjugated double bonds results in the absorption of light and the production of colour. Condensation will occur with unsaturated rings to polymerize ring systems, which in turn results in useful colours and flavours. Two important roles of caramelization in the food industry are caramel flavour and colour production, processes in which sucrose is heated in solution with acid or acid ammonium salts to produce a variety of products in food, candies, and beverages. There are three commercial types of caramel colours: (1) acid fast caramel, used in cola drinks, is made using ammonium bisulphite catalyst; (2) brewers' colour, found in beer, is made from sucrose in the presence of ammonium ion; (3) bakers' colour, in baked goods, results from direct pyrolysis of sucrose to give burnt sugar colour. Certain pyrolytic reactions result in the production of unsaturated ring systems that have unique tastes, and fragrances result from certain pyrolytic reactions, e.g. maltol contributes 'cotton candy' odour and is a flavour compound in baked bread, coffee, cocoa, etc., while furaneol imparts a 'strawberry' flavour and contributes to the odours of roasted products, including beef and coffee among others (Ko *et al.*, 2006) (Fig. 4.5).

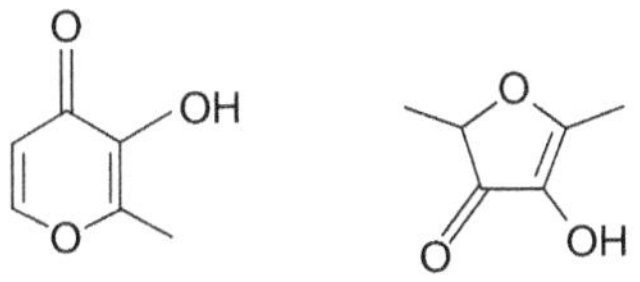

Figure 4.5 The structures of maltol (left) and furaneol (right).

The *Maillard reaction* is one of the most important reactions encountered in food systems and it is also called non-enzymatic or non-oxidative browning. Overall, this type of carbohydrate browning involves reducing sugars and amino acids, or other nitrogen-containing compounds, reacting to produce N-glycosides displaying red-brown to very dark brown colours, caramel-like aromas, and colloidal and insoluble melanoidins. There are a complex array of possible reactions that can take place via Maillard chemistry and the aromas, flavours, and colours can be desirable or undesirable (BeMiller and Whistler, 1996).

Initially, the carbonyl carbon of an open chain reducing sugar reacts with a non-ionized amino compound (often a free amino acid or a lysine side chain). This is then followed by the loss of water and ring closure to form a glycosylamine. The glycosylamine then undergoes the Amadori rearrangement to produce 1-amino-2-keto sugar (see Fig. 4.6). If the initial sugar is ketose, the glycosylamine undergoes a reverse-Amadori (Heyns) rearrangement to form a 2-amino aldose. Amadori compounds may be degraded in two ways, via either 3-deoxyhexosone or methyl α-dicarbonyl, both producing melanoidin pigments.

Glycoside reactions of this type with amino acids are irreversible; thus the reactants are 'lost', an important fact from a nutritional standpoint since lysine is

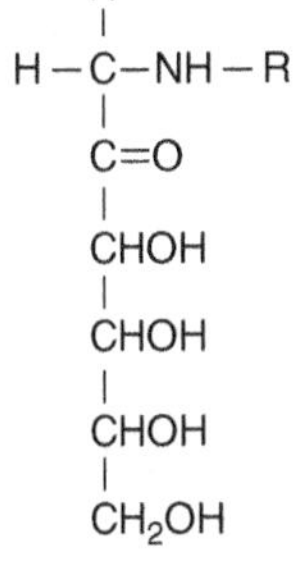

Figure 4.6 The product of the Amadori reaction, 1-amino-1-deoxyketose, which can further rearrange, dehydrate, and deaminate resulting in various possible aldehyde compounds.

a nutritionally essential amino acid and its side chain can be free to react in the Maillard reaction even when part of a polypeptide/protein. Other amino acids that may be lost due to the Maillard reaction include the basic amino acids L-arginine and L-histidine.

4.2.5 Glycans

Polysaccharides, or glycans, are made up of glycosyl units in a linear or branched structure. The three major food-related glycans are amylose, amylopectin, and cellulose. These are all chains of D-glucose; however, they are structurally distinct based on the types of glycosidic linkages that join the glucose units and the amount of branching in their respective structures. Both amylose and amylopectin are forms of starch, the energy storage molecules of plants, and cellulose is the structural carbohydrate that provides structural rigidity to plants.

Most energy derived from carbohydrate food sources is from *starch* which is critical, both structurally and nutritionally, for many foods, especially flour-based foods, tubers, cereal grains, corn, and rice. Starch can be both linear (amylose) and branched (amylopectin). Amylose glucose units are joined only by α-1,4 linkages and it usually contains 200–3000 units. The chains form helices such that the interior is lipophilic and the exterior is hydrophilic – the hydrophilic hydroxyl groups (-OH) of the glucose units orient to the exterior. Amylopectin also contains α-1,4 linkages with branch points at α-1,6 linkages. Branch points occur approximately every 20–30 α-linkages. The branched molecules of amylopectin produce bulkier structures than amylose. Normal starches contain approximately 25% amylose, although amylose contents as high as 85% are possible. By contrast, starches containing only amylopectin occur and are termed waxy starches.

The synthesis of starch (and glycogen in animals) is not energetically favourable (Stryer, 1996); therefore, glucose molecules must first be converted to an activated precursor prior to being added to the ends of starch molecules. *ADP-glucose pyrophosphorylase* (AGPase) catalyzes the reaction of glucose-1-phosphate with ATP to form ADP-glucose (activated glucose). ADP-glucose is then used as substrate by *starch synthase enzymes* that add glucose units to the end of a growing polymer chain, releasing ADP in the process. Branches in amylopectin are introduced by starch-branching enzymes that hydrolyze 1,4-glycosidic bonds, replacing them with 1,6-glycosidic bonds using other glucose units. Sucrose, the other plant energy storage molecule, is also synthesized using an activated glucose, except that it is joined to fructose-6-phosphate instead of a starch end yielding the disaccharide. This synthesis is catalyzed by sucrose 6-phosphate synthase.

In the digestion of starch, three enzymes that hydrolyze starch are α-amylase, β-amylase, and glucoamylase. α-Amylase cleaves starch molecules internally and thus it is an *endo*-enzyme. As discussed earlier, it hydrolyzes α-1,4 glucan linkages randomly, but not α-1,6 bonds, and it causes a rapid decrease in viscosity; therefore, α-amylase is also known as the liquefying enzyme. *Reducing power* of starch solutions increases upon α-amylase treatment since additional reducing ends are exposed upon cleavage events. β-Amylase is an *exo-enzyme*, i.e. hydrolysis occurs sequentially from end units, removing maltose units from the non-reducing end. Since maltose increases sweetness, β-amylase is known as the saccharifying enzyme. β-Amylase does not hydrolyze α-1,6 bonds of amylopectin and cleavage stops two or three glucose units from the branch point, resulting in residues referred to as *limit dextrin*. *Glucoamylase* (amyloglucosidase) can hydrolyze both α-1,4 and α-1,6 bonds to yield glucose. The most important food functions of the starch-hydrolyzing enzymes are to provide sugars for fermentation, reducing sugars in non-enzymic browning, and to alter texture, mouthfeel, moistness, and sweetness of affected foods.

Starch exists in packets called granules that are deposited in organelles known as amyloplasts. The size and shape of starch granules vary with the plant source, properties that can be used to identify starch sources. All granules contain a cleft called the hilium which serves as a nucleation point around which the granule develops as part of plant energy storage. Granules vary in size from 2 to 130 μm and they have a crystalline structure such that the starch molecules align radially within the crystals. Spherocrystalline structures like starch granules oriented in different configurations produce differential patterns when exposed to polarized light, an optical property referred to as *birefringence*.

Intact starch granules are insoluble in cold water. Granules can, however, reversibly absorb water, thereby swelling slightly and returning to original size if re-dried. If the temperature of the system/suspension is increased, starch molecule vibrations increase, causing breakage of starch inter-molecular bonds and resulting in increased hydrogen bonding to water. This penetration of water increases the separation of segments and decreases crystalline regions. Continued heating results in a complete loss of crystallinity and a loss of birefringence such that the uptake of water is irreversible; this is called the *gelatinization point* or *gelatinization temperature*. Amy-lose molecules, due to their linearity, can diffuse from the granule during the early stages of gelatinization and emerge in the extragranular solution, the aque-ous solution between starch granules.

Gelatinization usually occurs over a narrow tem-perature range and is dependent on the source, and composition, of the starch (Zobel and Steven, 1995). During gelatinization, granules swell extensively to form a thick paste in which almost all the water has entered the granules. Extensive H-bonding with wa-ter causes swelling and the granules push tightly against each other. Highly swollen granules can be easily ruptured and disintegrated by mild stirring, causing a large decrease in paste viscosity. Upon re-cooling of a heated starch solution, starch polymers reassociate under the lower molecular kinetic energy conditions. The reassociation of starch chains yield-ing a more ordered structure, referred to as *retrogra-dation*, results in the formation of crystalline aggre-gates and a rubbery, gel-like texture. Amylose tends to form a better gel due to its linear structure which allows for hydrogen bond formation more so than amylopectin. Upon retrogradation, some water may be excluded from the gel structure, a process known as *syneresis* or *weeping*.

Starch gelatinization, starch solution viscosity, and gel characteristics depend not only on tempera-ture but also on the types, and amounts, of other constituents. Although water controls reactions and physical behaviour in foods, it is not the total amount of water that is important, but rather the availability of the water to participate in these reactions/interactions, termed *water activity*. Water activity is influenced by salts, sugars, and other strong water-binding agents which, if present in high amounts, will lower water activity, thereby limiting or preventing gelatinization due to increased com-petition for water. A high sugar concentration will decrease the rate of starch gelatinization, peak vis-cosity, and gel strength. Disaccharides have a greater effect than monosaccharides in delaying gelatiniza-tion and decreasing peak viscosity (Gunaratne *et al.*, 2007). Sugar decreases gel strength by interfering

with the formation of junctions. Lipids (mono, di, and triacylglycerols) also affect starch gelatinization since fats complex with amylose which retards the swelling of granules. Fatty acids or fatty acid compo-nents of monoacylglycerols can form inclusion com-plexes with helical amylose structures and outside portions of amylopectin. These complexes are less easily leached from the granules, interfere with junc-tion zones, and also prevent staling (basically ret-rogradation). Acidic conditions affect starch solutions via starch acid hydrolysis. Hydrolyzed starch cannot bind water as well; therefore the addition of acid re-sults in decreased viscosity, e.g. salad dressings, fruit pie fillings. To avoid starch hydrolysis, cross-linked starches are commonly used.

Modified starches are starches that have been chem-ically or physically altered in order to obtain a de-sired property. The following are selected common examples of such modifications. In order to ob-tain a starch that is dispersible in cold water, pre-gelatinized starches are made by heating to temper-atures just below the gelatinization temperature so that the granules swell but do not break. The slurry is then drum dried and the resulting product de-velops viscosity with little heat or time upon dis-persion in cold water. Instant gravies and puddings are examples of foods that depend on pre-gelatinized starches. Chemical modification can also affect the functional properties of starches. For example, ret-rogradation of starches can be prevented by adding charged groups to starch molecules, thereby reducing polymer–polymer hydrogen bonding. Acid-modified starches, produced by holding starch granules below gelatinization temperature in an acidic medium (the acid hydrolyzes glycosidic linkages in starch without breaking up granules) followed by drying, are used in candy manufacture because easy-to-handle, low-viscosity fluids are produced which can be poured into moulds but set into a firm gel upon cooling or aging, e.g. starch gum candies. Lastly, cross-linked starches are produced by forming a chemical bridge between adjacent starch chains via covalent link-ages. The cross-links prevent starch granules from swelling normally and provide greater stability to heat, agitation, and damage from hydrolysis, thereby reducing granule rupturing. PO_4 is used to link hy-droxyl groups (-OH) of different starch chains. These starches are used in baby goods, salad dressing, fruit pie filling, and cream-style corn where they func-tion as thickeners and stabilizers. They also provide resistance to gelling and retrogradation, show good

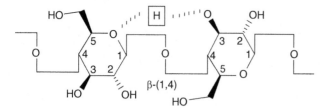

Figure 4.7 Cellulose is bound by beta-1,4 linkages with every other glucose residue flipped 180 degrees. Structural rigidity is provided by hydrogen bonding between the ring O and C3 OH.

freeze–thaw stability, and do not undergo syneresis or weep on standing.

4.2.6 Cellulose

Cellulose is the most abundant polysaccharide on the planet, accounting for almost half of all the carbon in the biosphere. Although starch is the most important food carbohydrate in terms of nutrition and functionality, cellulose is also important since it is the primary structural component of plant cell walls, and therefore it is ubiquitous in terms of unrefined, plant-based foods and ingredients. Cellulose is a *homoglucan* composed of β-D-glucose linear chains joined by β-(1,4) glycosidic linkages. A high level of hydrogen bonding between adjacent glucose units, which are rotated 180 degrees relative to one another every other residue, within the linear glucosyl chain provides rigidity and structural strength (Fig. 4.7).

Crystalline regions of ordered, tightly associated chains are formed due to the extensive hydrogen bonding. Less ordered (amorphous) regions are connected to the crystalline regions as well. Cellulose fibres are cross-linked by other polysaccharides, thus reinforcing the strength of the fibres, e.g. lignin in wood. Various modifications to cellulose yield cellulose-based food ingredients with desirable functions. The amorphous region of cellulose is easily penetrated and acted upon by solvents and chemical reagents first. Microcrystalline cellulose is produced by acid hydrolysis of the amorphous regions, leaving tiny, acid-resistant crystalline regions. This type of cellulose is non-metabolizable, and acts as a bulking and rheological control agent in low-calorie foods.

Derivatives of cellulose can be made through chemical modification under strongly basic conditions where substituents, e.g. methyl or propylene, react and bind at sugar hydroxyl groups. The resultant derivatives are ethers (oxygen bridges) joining sugar residues and the substituent (Coffee *et al.*, 1995):

$$\text{cellulose-OH} + CH_3Cl \rightarrow \text{cellulose-O-CH}_3 + NaCl$$
$$+ H_2O \text{ (in the presence of NaOH)}$$

A major reason to use cellulose derivatives is to add bulk to food products. Two important food-related derivatives of cellulose are carboxymethylcellulose (CMC) and methylcellulose (MC). CMC acts as a gum to help solubilize common food proteins such as gelatin, casein, and soy proteins. CMC–protein complexes contribute to an increase in viscosity. CMC acts as a binder and thickener in pie filling, puddings, custards, and cheese spreads. The water-binding capacity of CMC is useful in ice cream and other frozen desserts by preventing ice crystal growth. CMC also retards sugar crystal growth in confectionaries, glazes, and syrups. It helps stabilize emulsions in salad dressings and is used in dietetic foods to provide bulk, body, and mouthfeel normally contributed by sucrose. In low-calorie carbonated beverages CMC retains CO_2. CMC is used mainly to increase viscosity; however, viscosity decreases with increasing temperature. It is stable at pH 5–10, with a maximum stability at pH 7–9.

Methylcellulose, another cellulose ether derivative, exhibits thermo gelation: when heated it forms a gel that reverts, or melts, upon cooling (Coffee *et al.*, 1995). When heated, the viscosity initially decreases, but then rapidly increases and the gel develops due to interpolymer hydrophobic bonding. Methylcellulose is not digestible and therefore has no caloric value. In baked goods, methyl cellulose increases water absorption and water retention; in deep-fat-fried foods it decreases oil absorption; in dietetic foods it acts as a syneresis inhibitor and bulking agent; in frozen foods it prevents syneresis; while in salad dressings it acts as a thickener and stabilizer of emulsions.

Hemicellulose includes a class of polymers that yield pentoses, glucuronic acids, and some deoxy sugars upon hydrolysis of cellulose. Hemicellulose is used in baked goods to improve the water binding of flour, and it greatly retards staling. It is also a source of dietary fibre, although the absorption of some vitamins and minerals may decrease as a result of binding.

Pectin substances include polymers composed mainly of α-(1,4)-D-galacturonopyranosyl units, and constitute the middle lamella of plant cells. The nature of pectin compounds is responsible for the texture in fruits and vegetables. Texture changes during ripening are largely due to enzymatic breakdown of middle lamellar substances, including pectins which act as intercellular cement (Eskin, 1990). Commercial importance of pectins is largely due to their ability to form sugar and acid-containing, or sugar and calcium-containing, spreadable gels. Various pectic substances exist, and their structural differences are due to methyl ester content or degree of esterification (DE), defined as the ratio of esterified residues to total residues of D-galacturonic acid. Pectinic acids are less highly methylated pectic substances derived from enzymatic action by protopectinase and pectin methylesterase on protopectin, the pectic substance found in the flesh of immature fruits and vegetables. Carboxylic acid groups of non-esterified galacturonic acid residues experience charge–charge repulsion due to their negative charge states at physiological pH. Upon acidification (as in fruit jam making) the acid groups become protonated (neutral charge), thereby lowering inter-pectin chain repulsive forces. If hydration of the pectin chains is lowered by the addition of sucrose (hygroscopic sugar) then junction zones form between unbranched pectin chains (BeMiller and Whistler, 1996) in the formation of a gel. Low DE (low methoxy) pectins can gel in the absence of sugars, but require the divalent ion calcium to bridge pectinic acid chains. Such pectins are used in low-sugar dietetic jams and jellies.

A gum is defined as a water-soluble polysaccharide extractable from land/marine plants or microorganisms that can contribute viscosity or gelling ability to its dispersion. For example, *guar gum* is a galactomannan with β-(1,4)-D mannosyl backbone units, half of which have branch points of single α-1,6-D-galactosyl units. Guar gum hydrates in cold water to give a viscous, thixotropic mixture. It has the highest viscosity of all natural gums that is degraded at high temperatures and is usually used at levels of 1% or less. When used in cheeses, guar gum eliminates syneresis (i.e. exudation of water), in ice cream it gives body, in baked goods it promotes longer shelf-life, in meat

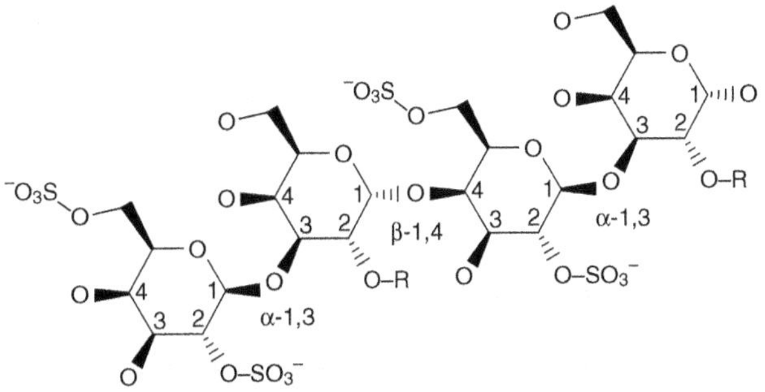

Figure 4.8 Carrageenans have alternating β- and α-galactose residues joined by alternating α-1,3 and β-1,4 glycosidic links. Lambda carrageenan is shown above where R is either H or SO3⁻.

products it improves casing stuffing, and in dressings and sauces it increases viscosity, and gives a pleasant mouthfeel.

Carrageenan, another gum, is a linear chain of D-galactopyranosyl units joined by alternating α-1,3- and β-1,4-glycosidic linkages. Most galactosyl units contain sulphate esters at C2 or C6 (BeMiller and Whistler, 1996). Carrageenans of commercial importance consist of three polymer types including iota, kappa, and lambda, with the two former carrageenans being the most important in the food industry. They are alkali-extracted from seaweed which yields a blend of carrageenans whose gelling properties are dependent on the associating cation, e.g. potassium results in a firm gel while sodium results in water solubility. Carrageenan is used in water-based and milk-based systems to stabilize suspensions and it acts synergistically with other gums, e.g. locust bean gum, to increase viscosity, gel strength, and gel elasticity. The structure of lambda carrageenan is shown in Fig. 4.8 to illustrate the relatively complex bond combinations characteristic of the many food gums.

As outlined above, the characteristics of carbohydrates in both their natural states and as processed food ingredients determine the properties of many foods. Understanding these relationships leads to better food products, and a basic understanding of how the body integrates this most important energy source is also important from a nutrition standpoint. *Glycolysis* is a fundamental pathway of metabolism consisting of a series of reactions where glucose is converted to pyruvate via nine enzymatically

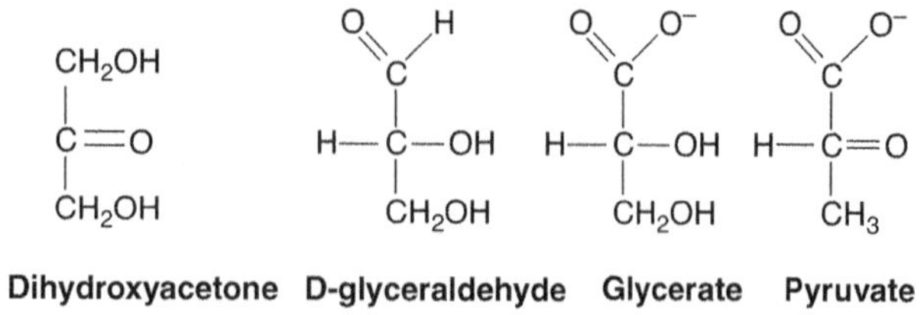

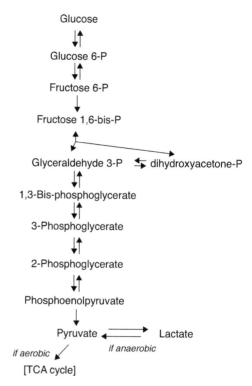

Figure 4.9 The four carbon intermediates of glycolysis are derivatives of four 3-C molecules: dihydroxyacetone, glyceraldehyde, glycerate, and pyruvate. Glucose and fructose are the base molecules for 6-C intemediates.

catalyzed reactions. Glycolytic processing of each glucose molecule results in a modest gain of only two *ATP* (the universal energy currency). As seen below, this gain is small, but creation of pyruvate feeds another metabolic pathway called the *tricarboxylic acid cycle* (TCA cycle) which yields another two ATP. Glycolysis and the TCA cycle also generate fully reduced forms of nicotinamide adenine dinucleotide (NAD$^+$; NADH) and flavin adenine dinucleotide (FAD; FADH$_2$) which drive oxidative phosphorylation of ADP such that a net 30 ATP are generated for each glucose initially entering glycolysis (Stryer, 1996).

4.2.10 Glycolysis

Intermediates of glycolysis contain either six or three carbons. The C6 molecules of glycolysis are glucose initially, followed by phosphorylated glucose, phosphorylated fructose, and double-phosphorylated fructose. The C3 molecules are all derivatives of dihydroxyacetone, glyceraldehyde, glycerate, or pyruvate (Fig. 4.9).

Glycolysis takes place in the cytosol and the overall glycolytic pathway is presented in Fig. 4.10.

The initial step of glucose-based energy derivation is the phosphorylation of glucose by the enzyme *hexokinase* at the expense of using one ATP yielding glucose-6-phosphate (G6P). Next, *phosphoglucose isomerase* converts G6P to fructose-6-phosphate (F6P). F6P undergoes the second phosphorylation event by *phosphofructokinase* yielding fructose-1.6-bisphosphate (FBP) at the expense of a second ATP. The critical control point of glycolysis is at this second phosphorylation step since *phosphofructokinase* is inhibited by high levels of ATP, acidity (H$^+$), and citrate (from the TCA cycle). Thus far, glycolysis is net −2 ATP. However, the activated aldose sugar FBP is then split by *aldolase* giving two phosphorylated trioses: dihydroxyacetone phosphate (DHAP) and glyceraldehyde-3-phosphate (GAP). DHAP is con-

Figure 4.10 The intermediates of glycolysis. All glycolytic intermediates are phosphorylated. The 3-C reactions beginning with glyceraldehyde 3-P occur twice per glucose entering glycolysis.

verted to a second molecule of GAP by *triose phosphate isomerase*, and thus further glycolytic reactions occur twice for every glucose entering glycolysis. The subsequent steps of glycolysis result in some net energy gain. *Glyceraldehyde-3-phosphate dehydrogenase* catalyzes coupled reactions where GAP is further phosphorylated with inorganic phosphate (P$_i$) and NAD$^+$ is reduced giving as products NADH, H$^+$, and 1,3-bisphosphoglycerate (1,3-BPG). Catalyzed by *phosphoglycerate kinase*, 1,3-BPG is dephosphorylated by ADP yielding ATP and 3-phosphoglycerate (3PG), which is then converted to 2PG by phosphoglycerate mutase. *Enolase* then catalyzes the dehydration of 2PG, giving phosphoenol pyruvate (PEP) which transfers its phosphate to ADP, yielding ATP and pyruvate, a reaction catalyzed by *pyruvate kinase*. As noted above, pyruvate subsequently feeds into the TCA cycle for further energy production.

4.2.11 The TCA cycle

Pyruvate resulting from glycolysis in the cytosol under aerobic conditions undergoes oxidative decarboxylation by pyruvate dehydrogenase inside the

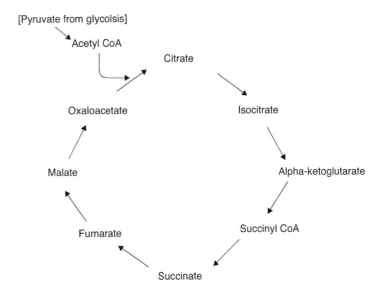

Figure 4.11 The steps of the tricarboxylic acid cycle.

mitochondria. Oxidation of pyruvate results in electron transfer to NAD^+, acetylation (2C) of coenzyme A (CoA), and oxidation of pyruvate C3, thereby producing acetyl-CoA, CO_2, and NADH. Many other energy source biomolecules are converted to acetyl-CoA as well and it is this intermediate which enters the tricarboxylic acid cycle (TCA) for completion of fuel carbon's oxidation. An overview of the TCA cycle is as follows, with the cycle skeleton depicted in Fig. 4.11.

The initial step of this oxidative cycle involves the recycling of oxaloacetate (OAA) (C4) by *citrate synthase*-catalyzed reaction with acetyl-CoA (C2) yielding citrate (C6) which contains three carboxylic acid groups. Isomerization to isocitrate by aconitase occurs next, thus allowing for the first of two decarboxylations. Isocitrate reduces NAD^+ and is then decarboxylated to α-ketoglutarate by *isocitrate dehydrogenase*. Next, *α-ketoglutarate dehydrogenase complex* catalyzes a similar reaction to the pyruvate → acetyl CoA reaction leading into the TCA cycle; NAD^+ is reduced and α-ketoglutarate is decarboxylated, giving succinyl CoA, CO_2, and NADH. Succinyl CoA is then cleaved by succinyl CoA synthetase in a reaction coupled to phosphorylation of GDP yielding GTP which, among other functions, can transfer its phosphoryl group to ADP. The final three steps of the cycle involve two redox reactions along the way to returning to oxaloacetate. *Succinate dehydrogenase* catalyzes the reduction of FAD to $FADH_2$ in the oxidation of succinate to give fumarate. Fumarate is then hydrated yielding malate by *fumarase*. Finally, *malate dehydrogenase* catalyzes the oxidation of malate to oxaloacetate and the concomitant reduction of NAD^+ to NADH,

thus completing the cycle. The NADH and $FADH_2$ generated by glycolysis and the TCA cycle are subsequently part of oxidative phosphorylation where they transfer their electrons to O_2 in a series of electron transfer reactions whose high energy potential is used to drive phosphorylation yielding ATP.

4.3 Proteins

The genetic information encoded by DNA within chromosomes contains all the information needed for an organism's bioprocesses, with the exclusive function of building only one type of biomolecule: proteins. Proteins are polymers of *amino acids*, of which there are 20, and the amino acids are joined by *peptide bonds* in a non-super imposable fashion – that is, X-Y-Z is not super imposable on Z-Y-X by any rotation or translation. Thus an enormous number of potential different proteins exist, considering the possibilities in protein length, amino acid order, and amino acid composition/content.

4.3.1 Amino acids

Amino acids, the building blocks of proteins, share an N-C_α-C backbone consisting of a carbon atom (C_α) that is covalently bonded to an amino group, a carboxylic acid group, a hydrogen atom, and one of 20 substituents termed *R groups* or *side chains*, hence the general formula $^+NH_3$-CHR-COO^- which describes 19 of the 20 amino acids, with the exception of proline. Due to R group differences, the 20 amino acids can be divided into three categories: non-polar,

Table 4.1 Categorizing the 20 amino acids and their side chain pK values.

Amino acid	Three-letter abbreviation	Side chain pK	Ionizable group
Non-polar amino acid			
Glycine	Gly	—	—
Alanine	Ala	—	—
Valine	Val	—	—
Leucine	Leu	—	—
Isoleucine	Ile	—	—
Phenylalanine	Phe	—	—
Tryptophan	Trp	—	—
Methionine	Met	—	—
Proline	Pro	—	—
Polar amino acid			
Serine	Ser	—	—
Threonine	Thr	—	—
Asparagine	Asn	—	—
Glutamine	Gln	—	—
Ionizable amino acid			
Aspartic acid	Asp	3.9	–COOH
Glutamic acid	Glu	4.4	–COOH
Histidine	His	6	Imidazole -C=N-
Lysine	Lys	10.5	$-NH_2$
Arginine	Arg	12.5	C=NH
Tyrosine	Tyr*	10.1	Phenyl-OH
Cysteine	Cys*	8.3	-SH

*Tyr and Cys are weak acids and are not charged under normal conditions.

polar, and ionizable, listed in Table 4.1. Note that proline is an exceptional amino acid in that its side chain is covalently bound to both the α-carbon and the backbone nitrogen. At neutral pH, most free amino acids are *zwitterionic*, i.e. they are dipolar ions.

Amino acids (aa) can be linked together with covalent bonds between the backbone carboxyl carbon of aa_1 and the amino nitrogen of aa_2 (...-N-C-C-N-C-C-...) and this bond is called the *peptide bond*. Two or more amino acids joined together are called a *polypeptide*, while molecules that consist of one or more polypeptides are termed *proteins*. Although the amino groups and carboxyl groups of free amino acids are ionizable, residues within a protein are covalently linked together, and therefore are not available for proton exchange. Thus, the charge of a protein is determined by the charge states of the ionizable amino acid R-groups that make up the polypeptide, namely Asp, Glu, His, Lys, Arg, Cys, and Tyr. The charge state of an ionizable molecule is dictated by its *pK* value, the pH at which an equal number of protonated and deprotonated species exist resulting in a net charge of zero. If pH is above pK of a particular molecular species then the molecules

will be deprotonated and if pH is lower than pK then the molecules will be protonated. Table 4.1 shows the side chain pK values of each amino acid containing ionizable side chains.

If a given polypeptide is at pH above the side chain pK values for Tyr, Cys, Asp, and Glu then the side chains will be deprotonated, having a net negative charge because these are acidic amino acids (note: Tyr and Cys require pH above physiologic pH to act as acids). These acidic residues have neutral charges below their respective pK values. By contrast Lys, Arg, and His are basic amino acids, and therefore they are neutral above their respective pK values while they are positively charged below their pKs. The pH at which a protein has zero net charge is called *pI* or the *isoelectric point*.

The interiors of protein structures are often highly hydrophobic. In such non-aqueous microenvironments, proton exchange is inhibited, resulting in side chain pK values that may differ greatly from those for free amino acids in solution. Importantly, a positively charged side chain can form a salt bridge with a negatively charged side chain. For example, lysine and aspartate typically have opposite charges under the same conditions and if the side chains are proximate then the negatively charged carboxylate of Asp can salt-bridge to the positively charged ammonium of Lys. Another important inter-residue interaction is covalent bonding between cysteine side chains. Under oxidizing conditions, the sulphhydryl groups of cysteine side chains (-S-H) can form a thiol covalent bond (-S-S-) also known as a *disulphide bond*.

4.3.2 Nutrition

Protein digestion, especially from animal sources, begins with cooking as heat causes proteins to unfold, allowing for higher digestibility. Chewing provides mechanical breakdown of proteins. In the stomach and upper intestine, two types of *proteases* (enzymes that hydrolyze peptide bonds) act on dietary proteins. *Endopeptidases* are proteases that cleave interior peptide bonds of polypeptide chains while *exopeptidases* are proteases that cleave at the ends of proteins exclusively. Protein digestion would be very inefficient in the absence of one protease type since interior bonds would be buried within proteins and thus unavailable to exopeptidases in the absence of endopeptidase, and exopeptidases are required for release of individual amino acids. In the stomach, pepsin, an acid protease, functions at extremely low pH to release peptides from muscle and collagen proteins, while in the upper intestine, serine proteases trypsin and chymotrypsin further digest peptides, yielding free amino acids for absorption into the blood (Champe *et al.*, 2005).

In terms of diet, protein intake provides the body with amino acids used for building muscle as well as for energy derivation via the TCA cycle directly, or via gluconeogenesis for glucogenic amino acids (amino acids capable of conversion to glucose). Eight amino acids are important in the maintenance and growing of muscle fibre. These eight amino acids, lysine, methionine, phenylalanine, threonine, tryptophan, valine, leucine, and isoleucine, are termed the *essential amino acids* because humans do not synthesize them in sufficient quantities, if at all, for maintenance of optimal health. Infants also require dietary histidine as well as those listed above. In addition to the presence of essential amino acids, their relative quantities are also important since our bodies require varying amounts of each. The essential amino acid present in the smallest amount is termed *limiting amino acid*. Animal protein sources provide all essential amino acids for the human diet, with egg protein

rating as the ideal, or perfect, protein since its amino acid profile (the relative quantities of amino acids) is optimal for human dietary needs. By contrast, most cereals are deficient in lysine, while oilseeds and nuts are deficient in lysine as well as methionine.

4.3.3 Protein synthesis

Although the genetic code uses only four bases (A, T, G, C), different three-base combinations of these bases make 64 different codons that encode for the 20 amino acids as well as the signal 'stop'. Included are degenenerate codons: different codons that encode for the same amino acid or for the message 'stop'. When the body signals that a protein's synthesis is needed it 'turns on' the corresponding gene (DNA) by transcribing it into messenger RNA (*mRNA*) which is essentially a copy of the gene. mRNA is then used as a template for protein synthesis at the ribosome by the process known as *translation* and the number of mRNA copies available will determine the amount of protein that will be made. Available to the ribosome are a pool of transfer RNAs (*tRNA*), each carrying one specific amino acid at its free hydroxyl (3′) end. Each tRNA also contains an anticodon loop that is specific for binding to mRNA codons. As tRNAs bind at the *ribosome* sequentially, the polypeptide chain is formed. Critical is the state of the amino acid backbone carboxyl group: attachment of the amino acid to the tRNA by various synthetases involves the activation of the amino acid carboxyl by ATP; thus it is able to attach to the amino end of preceding, neighbouring amino acids in the chain.

Since primary structure is a major determinant of a protein's physico-chemical properties, changes made to amino acid residues can alter the functionality of the greater molecule. Thus, one needs to be aware of a plethora of protein modifications collectively known as *post-translation modifications*. Proteins can be altered in many ways, both chemically and structurally. Examples of common chemical changes post-translation include deamination of Arg yielding citrulline, deamidation of Gln → Glu and Asn → Asp, amidation of the C-terminus, phosphorylation of Ser, His, Tyr, or Thr, and glycosylation (adding sugar residues) to Asn, Ser, or Thr. In terms of nutrition, modifications of amino acids are important since they result in lower bioavailability of the affected amino acids. Structural changes include cleavage at a specific site on a protein and disulphide bond formation between Cys residues.

Polypeptides fold such that *free energy* is minimized. Hydrophobic residues will be in a lower energy state when sequestered from aqueous environments, in proximity to other hydrophobic side chains in the interior of proteins. Aliphatic and aromatic groups attract via hydrophobic interactions, forces of relatively low energy individually.

Protein structures are conceptualized in terms of four categories of structure: primary, secondary, tertiary, and quaternary structures. Primary structure refers to the amino acid sequence, while secondary structures are the structures formed by amino acid chains, e.g. α-helix, β-sheet, random coil. The tertiary structure of a protein refers to the way in which secondary structures, e.g. helices, sheets, and random coils, pack together in three dimensions. Quaternary structure refers to the association of tertiary structures, e.g. two associated subunits, each a separate polypeptide chain.

The way that a polypeptide chain folds is determined by the constituent amino acids. For example, α-helix formation requires that the polypeptide chain 'twists', and thus peptide bonds between residues must allow for accommodating angles between residues. In the cases of β-sheets, individual strands hydrogen-bond together and thus residues' side chains capable of H-bonds must line up to form the sheet of β-strands. Amino acid composition and order also influences tertiary and quaternary structures via salt bridges and disulphide bonds. Such interactions may take place between residues of different regions of the primary structure, thus influencing the overall folding of tertiary or quaternary structures.

The overall shape of a protein falls into one of two all-encompassing categories: *globular* and *fibrous*. Globular proteins include enzymes, transport proteins, and receptor proteins (Voet and Voet, 1995). They have an overall spherical shape as their compact structures have relatively low length-to-width ratios. Globular proteins contain a mixture of secondary structure types that fold together into compact protein structures which are generally highly soluble. Fibrous proteins, by contrast, are elongated structures that are simple compared to globular proteins. They include structural (e.g. keratin of hair) and motile (e.g. myosin of muscle) protein functions.

Not surprisingly, proteins that exist in aqueous environments have hydrophilic exteriors. However, there are proteins that must exhibit both hydrophilic and hydrophobic character, and these are membrane proteins. Membranes of living systems are made up of lipid bilayer, and hence proteins that are located, or anchored, in membranes must be stable and active under non-aqueous conditions. Aside from the compartmentalizing function of the membrane itself, membrane processes are carried out by membrane proteins. Membranes are comprised of 18–75% protein (Stryer, 1996). Such proteins may either span the width of the membrane, thus having a large hydrophobic region, or anchor in the membrane. The extraction of membrane proteins often requires the use of mild detergents to help maintain solubility by keeping the hydrophobic domains from aggregating.

The 3D form taken by a protein under its natural conditions is referred to as its *native structure*. When a protein's native structure is altered (without peptide bond cleavage) its new structure, or conformation, is called *denatured*, and if it is impossible to regain the native structure from the denatured form then it is considered to be *irreversibly denatured*. Generally, protein unfolding results in loss of solubility to varying degrees for different proteins due to exposure of the hydrophobic core. Exposed hydrophobic regions tend to associate via hydrophobic interactions, causing aggregation and subsequent precipitation. Additionally, there is a critical amount of water associated with the exterior of a given protein that maintains its solubility. When the protein water requirement is fulfilled, i.e. the *hydration shells* (or 'solvation layers') contain bound water, then solubility is maintained because the attractive forces between proteins remain dampened. When a soluble protein is introduced into an organic solvent, then water is stripped from its solvation layers. Upon reintroduction to an aqueous environment solubility may not be regained if the 'permanent water' layer was stripped and the protein would remain as precipitate.

Denaturation often arises from heating protein beyond its melting temperature (T_m) or changing pH to a value where a particular protein is unstable. Heating provides heat energy, resulting in 'breakage' of low-energy bonds (hydrogen and van der Waals interactions) required for the native conformation, thereby allowing the protein to unfold. Heating can also cause breakdown of certain amino acid

R-groups. Disulphide bonds can be broken via release of hydrogen sulphide, serine can dehydrate, and glutamine and asparagine can deamidate.

Exposure of a protein to pH values resulting in ionization of proximate side chains of like charge will cause charge–charge repulsion (a high energy state), thus triggering a conformational change to a lower energy state (separation of like charges). This process is called *pH denaturation*. For example, the digestive enzyme pepsin is only stable under acidic conditions. Its structure contains several aspartate residues that become ionized (deprotonated) at neutral pH, resulting in an irreversible conformational change. Reversion to the native form is not possible essentially because reprotonation of the ionizable side chains is not in itself a driving force to regaining the native form. Another denaturation scenario arises when disulphide bonds are reduced (cleaved) either by lowering pH (-S-S- → -SH) or by exposure to reducing agents. Often proteins whose native conformations involve disulphide bonds are irreversibly denatured upon reduction of these linkages.

Another important method of protein denaturation is through mechanical treatments, particularly important in the food industry. Proteins having hydrophobic regions sequestered towards the centre of the structure can be denatured when exposed to *shear forces*. Kneading and rolling dough for baked goods as well as whipping for foam production result in exposing proteins to shear forces, thereby causing denaturation principally due to α-helix disruption.

Proteins also denature at interfaces where high energy boundaries between two phases drive the unfolding of a protein molecule which lowers the overall energy of the system. For example, foams consist of water and air phases and emulsions have water and oil phases. The interfaces where the respective phases meet are not stable since air/oil are hydrophobic. Figure 4.12 shows the adsorption of emulsifying proteins to a dairy fat globule. Proteins that are present tend to migrate to the interfaces such that hydrophilic portions orient to the water phase and hydrophobic portions contact the air/oil, which results in lowering the energy of the system and stabilizing the foam or emulsion.

Other sources of denaturation include irradiation, organic solvents, and introduction or exclusion of metals. Also, some proteins, especially enzymes, have a requirement for ions for optimal functioning.

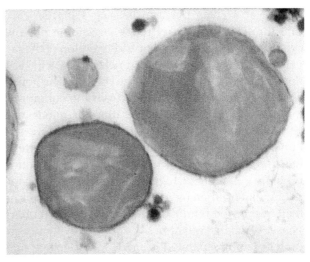

Figure 4.12 Cross section of a fat globule in an emulsion, visualized by transmission electron microscopy, showing the darkly stained proteinaceous membrane that forms around the periphery as the hydrophobic portions of proteins adsorb to fat interfaces during emulsification. Fat globules are approximately 1 μm in diameter, membrane thickness is approximately 10 nm. (Image courtesy of Prof. H. D. Goff, University of Guelph.)

4.3.6 Protein purification

Critical to the study of food protein structure and function is the ability to purify protein. Separating a particular protein from a complex mixture of many proteins, often hundreds or thousands, is done by taking advantage of differing biochemical characteristics such as charge, pI, mass, molecular shape and size, hydrophobicity, ligand affinity, and enzymatic activity, among others.

Two critical categories of purification techniques typically employed in protein purification schemes are *ion exchange* and *size exclusion*. In practical terms of purification, *chromatography* is the most important technique used. *Preparative chromatography* involves the separation of proteins (or other substances) dissolved in a mobile phase based on their differential interactions with a stationary phase. The mobile phase can be hydrophobic (organic solvent) or hydrophilic (aqueous buffer). The stationary phase can be hydrophobic or hydrophilic and/or charged or neutral and/or an immobilized ligand or molecule with specific affinity for proteins.

The simplest example of chromatography is based on ion exchange where the stationary phase consists of charged molecules immobilized on a solid polymer packed into a column. If a target protein is anionic at a given pH then passing a protein mixture through a stationary phase containing positive charges will result in binding and retention of the target protein, while cationic, and neutral, proteins will flow through the anion exchange column. The bound anionic proteins can be eluted from the column by introducing increasing concentrations of anions (eluent) that compete for the positive charges of the stationary phase. Weakly bound proteins will elute at lower concentrations of eluent and strongly bound proteins will elute at higher concentrations of competing eluent, thereby separating proteins based on their respective binding affinities for the stationary phase. Very large quantities to very small amounts of protein can be separated in single ion exchange runs depending on the equipment scale chosen.

In the case of *size exclusion chromatography*, column packing consists of inert, spherical packing material (beads) containing specific pore sizes. The highest molecular mass of protein capable of passing through the pores is termed the *exclusion limit*. Proteins of mass in excess of this limit may only pass around the packing material, thus taking the shortest path through the column, while smaller proteins that enter and exit beads take a more convoluted, longer path. Hence, larger proteins elute before smaller proteins in size exclusion chromatography, thereby allowing for the separation of many proteins within a mixture based on size. The resolution for separation of masses varies with the quality of setup purchased and length of the column. Generally, only milligram quantities are separated by this chromatographic technique. However, filtration devices with specific pore sizes, e.g. ultrafiltration, can be used for bulk separation of proteins where only two fractions are obtained, flow-through and retentate, yet with the advantage of processing relatively large quantities of protein per run.

4.3.7 Protein analysis techniques

The most basic type of protein analysis is that of determining the *total protein* content of substances. This measurement is based on the average nitrogen content of proteins, 16%. Organic nitrogen (including that from protein) reacts with sulphuric acid when heated to give ammonium sulphate. This product is then converted to ammonium hydroxide which can be quantified by titration, indicating the quantity of nitrogen present and hence the approximate amount of protein in a food.

To determine the size of proteins, denaturing *electrophoresis* is employed. Briefly, when proteins are passed through a polymerized acrylamide gel, smaller proteins will travel more quickly than larger proteins, allowing for a means for their separation. The motive force that makes proteins pass through the gel is provided by applying a voltage across the gel such that negative molecules migrate down the gel toward the anode. This technique is called polyacrylamide gel electrophoresis (PAGE). If charge amounts and size of proteins were both variable then it would be impossible to determine a protein's molecular weight by PAGE. To isolate mass as the only variable among proteins in a sample, the strong detergent sodium dodecyl sulphate (SDS) is added to completely unfold proteins and to normalize charge to mass ratio. The reducing agent β-mercaptoethanol is also added to reduce disulphide bonds to help completely unfold proteins. This ensures that both compact structure proteins and bulky structure proteins migrate through the gel at a rate only proportional to their masses and unaffected by native molecular shape or native charge. The SDS-PAGE gel in Fig. 4.13 is original and unpublished. Lanes 1, 2, 4–8 represent HPLC fractions from an aspartic protease purification.

A standard curve for known molecular weight proteins consisting of relative mobility versus log (molecular weight) is used to calculate the mass of any distinct protein in a sample. Subsequently, individual bands can be excised and identified by amino acid sequencing (Edman, 1970) or mass spectrometry (Nesvizhskii, 2007).

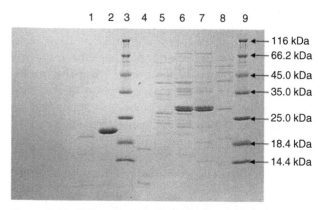

Figure 4.13 12% acrylamide SDS-PAGE stained with Coomassie brilliant blue R-250, a protein-binding dye. Lanes 3 and 9 contain molecular weight standards for the determination of unknown sample bands' masses.

4.3.8 Oxidative browning

Oxidative browning, or enzymatic browning, is the browning associated with cut or bruised apples, bananas, pears, and lettuce. The oxidation involves phenolase, also called polyphenol oxidase (PPO), an enzyme normally compartmentalized such that oxygen is sequestered from PPO. Upon injury or cutting of such produce, PPO principally acts on tyrosine residues in the presence of oxygen via an initial hydroxylation reaction of the phenolic ring followed by the oxidation itself. Oxidative browning is of commercial significance particularly in fruits and vegetables, due to the undesirability of the brown colour associated with bruising and decomposition resulting from such reactions (Kays, 1991). Other phenolic compounds can also be oxidized by phenolase to produce desirable brown pigmentation in products such as raisins, prunes, dates, cider, and tea.

4.4 Lipids

Lipids are one of the three major classes of biomolecules, representing a diverse group of compounds characterized by total insolubility, or low solubility, in aqueous environments that include the building blocks of biological membranes of all living organisms, substituents of lipoproteins, and the energy storage form of all animals. The insolubility results in lipids' compartmentalization in the body in membranes or in adipocytes (fat cells), the need for conjugation to soluble protein carriers for transport through aqueous environments, or the need for attachment of polar compounds to increase the overall solubility.

4.4.1 Fatty acids

The structure of a simple *fatty acid* (FA) is a hydrocarbon chain with a carboxylic acid group at one end: $CH_3\text{-}[CH_2]_n\text{-}COOH$. Fatty acids are named according to the length of the hydrocarbon chain. For example, a four carbon FA is butanoic acid; a five carbon FA is pentanoic acid; a six carbon FA is hexanoic acid, and so on. These would have the designations 4:0, 5:0, and 6:0, respectively, designating the number of carbons left of the colon and the number of double bonds to the right of the colon. The presence of a double bond (unsaturated) in a fatty acid is written as -enoic in place of -anoic. For example, hexadecanoic acid (16:0) becomes hexade*cenoic* acid (16:1), 16:2 is termed hexadeca*dienoic* acid having two double bonds, and 16:3 is hexadeca*trienoic* acid. If 16:1 has its double bond between C7 and C8 then it is written as 7-hexadecenoic acid, where carbon number is counted from the carboxylic acid carbon (C1).

The methyl end of the fatty acid is termed the *omega carbon* (ω); thus when designating unsaturated fatty acids using the omega convention, 9,12-octadecadienoic acid (18:2) becomes 18:2 ω-6 because the first double bond is six carbons from the ω carbon. Table 4.2 shows some common fatty acids, their lengths, and double bond characteristics.

Double bonds separated by one or more methylene groups are called unconjugated double bonds whereas ones not separated by a methylene group, i.e. (...-CH=CH-CH=CH-...), are termed *conjugated*

Table 4.2 Selected food fatty acid names, lengths, and double bonds. (Adapted from Nawar, 1996.)

Fatty acid name	Systematic name	Number of carbons	Abbreviation
Butyric	Butanoic	4	4:0
Lauric	Dodecanoic	12	12:0
Myristic	Tetradecanoic	14	14:0
Palmitic	Hexadecanoic	16	16:0
Stearic	Octadecanoic	18	18:0
Oleic	9-Octadecenoic	18	18:1 (n-9)
Linoleic	9,12-Octadecadienoic	18	18:2 (n-6)
Linolenic	9,12,15-Octadecatrienoic	18	18:3 (n-3)
Arachidic	Eicosanoic	20	20:0
Arachidonic	5,8,11,15-Eicosatetraenoic	20	20:4 (n-6)
EPA	5,8,11,14,17-Eicosapentaenoic	20	20:5 (n-3)
DHA	4,7,10,13,16,19-Docosahexaenoic	22	22:6 (n-6)

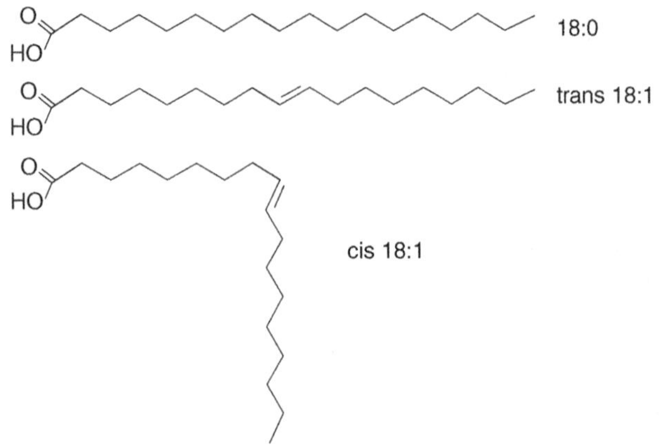

Figure 4.14 Representation of stearic acid, cis-oleic acid, and trans-oleic acid. Trans fatty acids, produced inadvertently during hydrogenation of unsaturated oils, have lower steric interference with respect to packing compared to cis double bonds.

double bonds. Lastly, the configuration of double bonds can be either *cis* or *trans* (Fig. 4.14). Geometrically, the cis configuration is the naturally occurring form and is much more bulky than the trans form as well as being susceptible to oxidation. The trans configuration has a configuration that is more linear in three dimensions, displaying properties like a saturated chain, and it is not found in nature.

Humans cannot synthesize two fatty acids due to a lack of the enzymes responsible for creating double bonds beyond C9 of the FA chain. They are the ω-3 FA linoleic acid (18:2) and the ω-6 FA linolenic acid (18:3).

Chain length and degree of unsaturation (number of double bonds) dictates the melting point and stability of FAs, two critical food lipid properties. Melting point, the temperature at which a lipid converts from a solid to a liquid, is generally lower for FAs of shorter length and/or more double bonds, e.g. stearic acid, 18:0, melts at higher temperature (70°C) than oleic acid, 18:1 (13°C). Additionally, the presence of trans FAs raises melting temperature since they experience less steric hindrance compared to cis FAs. In terms of stability, double bonds are susceptible to oxidation reactions, and thus stability decreases with increasing degree of unsaturation.

4.4.2 Triglycerides and phospholipids

Triglycerides (TG) are the lipid form that is commonly referred to as fat (solid at room temperature) and oil (liquid at room temperature). TGs comprise 98% of food lipids as they are the major energy storage form in animals and seeds as well as selected fruits like avocado and olive. TGs contain about six times more energy than glycogen or starch by weight and are stored in droplet form in the cytoplasm of fat cells. In terms of foods, TGs are important energy sources, and they contribute to mouthfeel and satiety. The fat-soluble vitamins A, D, E, and K require fat consumption for adequate levels to be absorbed. TGs also play critical roles in the structures of high-fat food emulsions like ice cream and chocolate. For example, in the making of ice cream liquid fat globules partially aggregate until a continuous network forms (partial coalescence), yielding a 'solid' product. The individual TG globules do not fully aggregate and are cemented to one another at globule interfaces due to interacting fat crystal networks (Goff, 1997).

TGs are molecules made up of a glycerol backbone to which three fatty acids are bound via ester bonds; thus they are also called *triacylglycerols*. TGs can have two or three identical FAs or three different FAs, and thus a wide range of properties exist among TGs. The order in which different FAs are attached to the glycerol backbone also varies. For example, oil seeds generally have a preference for unsaturated FAs at the sn-2 position and saturated FAs occur almost exclusively at the outer positions. Animal fats more frequently contain saturated FAs at sn-2, and they generally contain 16:0 at sn-1 and 14:0 at sn-2 (Nawar, 1996). As shown in Fig. 4.15, FAs within TGs are numbered according to their stereospecific positions relative to the glycerol backbone and each is assigned a stereospecific number, sn-1, sn-2, and sn-3. The sn-2 position is conventionally shown on the left side.

Another class of lipids is that of the *phospholipids* (PL) of which there are two types: *glycerophospholipids* (GPL) and *sphingophospholipids* (SPL). Most phospholipids are of the glycero- type, i.e. they contain a

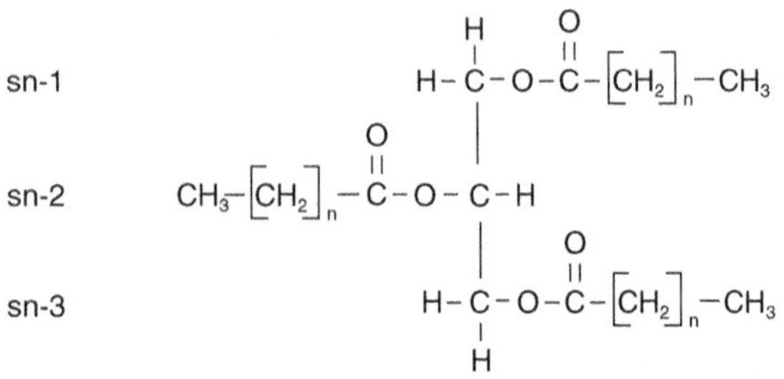

Figure 4.15 The general structure of a triglyceride, where n indicates the number of methylene groups between the carbonyl carbon and the omega carbon of an FA chain.

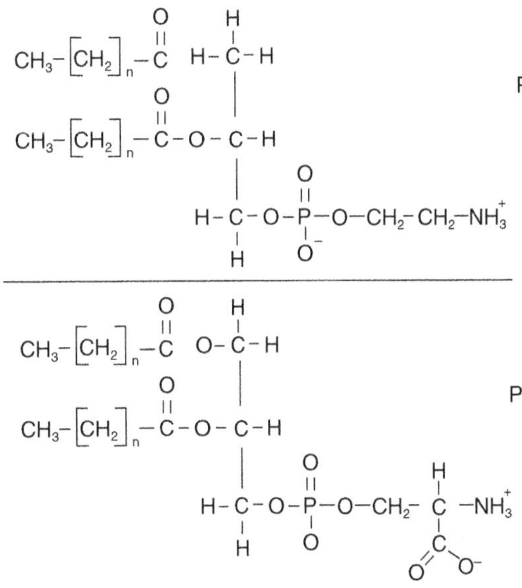

Figure 4.16 The structures of two common phospholipid types, where n is variable.

glycerol backbone like TGs. SPLs contain a sphinganine backbone, an amino alcohol, rather than glycerol. Phospholipids in general are structurally and functionally different from TGs in that they contain only two FA chains (diacylglyceride), with the third backbone position containing a phosphodiester bridge (attached via PO_4^-) to various substituents such as serine, choline, and ethanolamine. Figure 4.16 shows the basic structures of two representative PL types.

Substituents, including the phosphate bridge, are highly polar, conferring an amphipathic character to PLs, the critical property that allows, and causes, membrane structures to form as a *phospholipid bilayer*. Note that PL 'tails', the FA component, are highly hydrophobic and thus orient towards other FAs in aqueous environments, while the phosphodiester ends interact favourably with water, on each side of the bilayer, thereby minimizing the system energy. Hydrophobic forces and van der Waals interactions between hydrocarbon chains within the bilayer interior stabilize the structure. The phenomenon of bilayer formation is the basic process requisite for all known life forms because this is the basis for all organisms' membrane structures. The exclusion of water from contacting the hydrophobic PL moieties, while energetically favourable, causes bilayers to form lipid vesicles called *liposomes*. Lipid vesicles form spontaneously and in vitro formation is aided by high-frequency agitation of PLs in water. Liposomes may be used for the study of membrane properties and substance permeabilities, drug delivery,

and, in terms of foods, for microencapsulation of various food ingredients. For example, encapsulation of vitamin C significantly improves shelf-life by about 2 months when degradative components like copper, lysine, and ascorbate oxidase are present by providing a physical barrier and thereby preventing reactions. Also, the release of ingredients encapsulated by liposomes can be controlled to a particular temperature – the temperature at which the liposome bilayer breaks down (the melting point of the PLs), thus releasing its contents (Gouin, 2004).

4.4.3 Food lipid degradation

Lysis and oxidation are the principal ways in which TGs are degraded in foods. *Lipolysis* refers to the hydrolysis of the ester linkage between the glycerol backbone and FAs, thereby releasing free FAs. Lipases cleave FAs at the outer (sn-1 and sn-3) position on the glycerol backbone of TGs. Free FAs are more susceptible to oxidation, and they are more volatile, after release from the glycerol backbone than as part of TGs. Although free FAs are essentially absent in living animal muscle, they are released by *lipases* (lipolytic enzyme) activity after slaughter. To minimize the activity of lipases, temperature is controlled and processing must be done promptly. By contrast, oil seeds undergo lipolysis pre-harvest as a normal process, and therefore contain a significant amount of free FAs. The resulting acidity is neutralized with sodium hydroxide after extraction of the oil component of seeds. In terms of dairy products such as cheeses and yoghurts, and bread, controlled lipolysis is used as a means of producing desired odours and flavours via microbial and endogenous lipases. However, lipolysis is also responsible for the development of rancid flavour in milk resulting from the release of short chain FAs. Deep frying also produces undesirable lipolysis due to the high heat and introduction of water from foods cooked in the oil medium.

Phospholipids are referred to as lecithin, a natural emulsifier and surfactant broadly used across myriad processed food products. The source of particular lecithin influences the proportion of PLs and thus its properties. Phospholipids are subject to the lipolytic activities of *phospholipases*. There are four phospholipase cleavage sites: sn-1, sn-2, sn-3, and sn-4, each side of the phosphodiester. Different phospholipases have specificities for one of the four cleavage sites and thus can release an FA or the PLs substituent. Unlike lipases, phospholipases can cleave the middle position of the glycerol backbone, thereby releasing the sn-2 FA. For example, phospholipase A_2 activity results in the production of lysolethicin (hydroxyl at sn-2) via release of the sn-2 FA which may be subsequently oxidized.

Lipid oxidation, a major cause of food spoilage, occurs via two principal routes: autoxidation and lipoxygenase activity. The oxidized by-products of lipid oxidation result in off-flavours and off-odours associated especially with rancidity. In addition, oxidation of essential fatty acids lowers the nutritive value of the food. By contrast, in the case of cheeses and dried foods, certain oxidized lipid products are desirable.

Autoxidation occurs in three steps beginning with the removal of a hydrogen atom from a FA chain yielding a free radical, a thermodynamically unfavourable process that requires initiation by an existing free radical, catalysis by a metal, or light exposure (singlet oxygen). Next, the FA free radical reacts with molecular oxygen, giving a peroxy radical (ROO·). This second step is self-propagating in that peroxy radicals react with other FAs forming more free radicals as part of a chain reaction. Termination of individual reactions occurs via reaction with other free radicals which results in stable compounds. Unsaturated FAs are most susceptible to autoxidation because free radical formation is aided by the presence of double bonds, e.g. linoleic acid (18:2) is 20 times more susceptible to oxidation than oleic acid (18:1).

To prevent or slow oxidation of lipids, antioxidants are added to food products. Essentially, antioxidants work by reacting with free radicals, becoming free radicals themselves. Antioxidants are quenched by reaction with other free radicals instead of reaction with molecular oxygen, thereby preventing propagation of the free radical chain reaction. An extensive review on food antioxidants in terms of their effectiveness and modes of action was recently published by Laguerre and colleagues (Laguerre *et al.*, 2007).

Lipid oxidation by enzymatic action is catalyzed by a group of enzymes termed *lipoxygenases* (LOX), whose activities are most important in legumes and cereals. Similar to non-enzymatic oxidation, the first step in lipid oxidation is the formation of an FA free radical. The second major step in the overall reaction mechanism is reaction of the free radical with O_2 yielding a hydroperoxide product (Klinman, 2007). For soybean LOX the LOX-catalyzed reaction is specific for position 11 of linoleic acid (double bond), resulting in both 9- and 13- hydroperoxide products. In addition to rancid off-flavours, LOX can also cause deleterious effects on vitamins and colour compounds.

4.4.4 Triglyceride metabolism

The derivation of energy from triglycerides stored in adipocytes occurs by a process called β-oxidation. However, first fatty acids must be released from the glycerol backbone by adipose cell lipase in fat cells which releases all three fatty acids, leaving glycerol to be converted to one of multiple glycolytic intermediates. TGs in the digestive tract are acted upon by pancreatic lipases which release FAs at sn-1 and sn-3; free FAs and the remaining monoacylglycerol are absorbed by intestinal mucosal cells where triglycerides are resynthesized and then packaged as chylomicrons and released into the blood stream. Once at tissues where energy will be derived from the chylomicrons, lipoprotein lipase releases all three FAs similar to the adipocyte scenario above. Free FAs are activated by linking to CoA, catalyzed by acyl CoA synthase using ATP. At the mitochondrial membrane, the FA is then transferred from FA-CoA to carnitine, a lysine derivative, and the new complex (acyl carnitine) is then transported across the membrane into the mitochondria where the FA is rejoined to CoA (giving acyl CoA). A series of successive oxidative steps that result in the transfer of electrons (reduction) to electron carriers, e.g. NADH, $FADH_2$, that subsequently enter the electron transport chain yielding many ATPs, e.g. palmitic acid, yields a net total of 109 ATP.

4.4.5 Cholesterol

Cholesterol is one of the most important lipid structures as it controls the fluidity of membranes and it is the precursor to all steroid hormones. It is synthesized in all eukaryotes; however, it is only present in appreciable amounts in animals. Plants instead contain phytosterols, highly similar molecules structurally which have many nutraceutical applications (Kritchevsky and Chen, 2005). Cholesterol's structure is made up of adjacent carbon rings; a 27 carbon structure synthesized by a liver pathway consisting of successive polymerization and cyclization reactions beginning with the single, two carbon start material acetyl CoA. Since cholesterol is insoluble in aqueous environments it must be transported through the blood as a lipoprotein within heterogeneous lipid structures called chylomicrons which have polar lipid shells to maintain solubility. The densities of lipoproteins vary greatly; imbalance in low-density lipoprotein metabolism is a critical risk factor for heart disease development.

4.4.6 Steroid hormones and prostaglandins

An important lipid type not necessarily associated with food science readily is that of hormones. Hormones are messenger molecules in the form of peptides or lipids found in all plants and animals which control a vast number of cellular and organ processes. Specifically, *steroid hormones* and *prostaglandins* are hormone lipids that are fat soluble and some are used artificially for yield increases in the animal food chain. Also, these molecules are released into the environment via non-food uses where they can be taken up by food organisms. Fat-soluble molecules, if stable, can accumulate in adipose tissue, thereby presenting the possibility of biomagnification. This phenomenon can involve a direct route such as the accumulation of steroid estrogen-like compounds in food animals (Moutsatsou, 2007) or an indirect effect on hormone levels of organisms via accumulation of organic compounds that alter hormone production (Verreault et al., 2004).

Prostaglandins (PGs) are 20 carbon lipid molecules that contain a 5-member ring which have relatively short half lives and their functions include inflammation control and blood pressure control. Dietary essential fatty acids are critical in the biosynthesis of PGs. For example, arachidonate (20:4) is a direct precursor to PGs containing two non-ring double bonds, i.e. PG class PGE_2 which is acted upon by cyclo-oxygenase (COX). The role of PGs in maintaining good health highlights the importance of essential fatty acid intake through diet as well as the importance of protecting essential fatty acids (Moreno et al., 2001) and antioxidants (Boehme and Branen, 1977) from degradation in food products.

4.4.7 Terpenoids

A lipid type important to flavours and aromas of seasonings, herbs, and fruits is that of the *terpenoids*, a diverse and complex chemical group. Like steroids, most terpenoids are multicyclic compounds made via successive polymerization and cyclization reactions. All terpenoids are derived from isoprene, a five carbon hydrocarbon containing two double bonds. Terpenoids, especially C10 (monoterpenes) and C15 (sesquiterpenes), are components of the flavor profiles of most soft fruit in varying concentrations (Maarse, 1991). In some fruit species, they are of great importance for the characteristic flavour and aroma. For example, citrus fruits are high in terpenoids, as is mango (Aharoni et al., 2004).

4.5 Nucleic acids

4.5.1 DNA structure

DNA (deoxyribonucleic acid) and RNA (ribonucleic acid) are composed of a nitrogenous base, a sugar, and a phosphate. The bases, of which there are only four, make up the basis for all of life's information code, while the sugar (ribose or deoxyribose) and phosphate play structural roles. The four bases are derivatives of two parent nitrogenous bases: adenine (A) and guanine (G) are purines while thymine (T) and cytosine (C) are pyrimidines. A *nucleoside* is a molecule made up of a purine or pyrimidine bonded to a sugar; thus DNA is made up of polymers of deoxyadenosine, deoxyguanosine, deoxythymidine, and deoxycytidine (Fig. 4.17). Adding one or more phosphate groups to a nucleoside gives a *nucleotide*, e.g. deoxyadenosine 5'-triphosphate (dATP). The four building blocks of DNA are thus dATP, dGTP, dTTP, and dCTP which are bonded together by

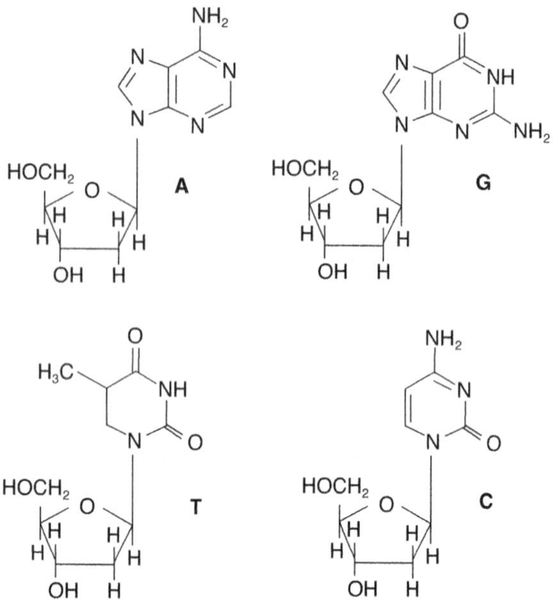

Figure 4.17 Deoxyadenosine (A), deoxyguanosine (G), deoxythymidine (T), and deoxycytidine (C) are the four sugar bases of DNA; the four structures that encode for every gene in nature. 'Deoxy' refers to the lack of a hydroxyl at C2 of the sugar (deoxyribose) ring.

DNA polymerase, resulting in a phosphodiester bridge (and release of PP_i) between the 3' position of one nucleotide and the 5' position of the next nucleotide (the prime symbol denotes a carbon within the sugar portion of the nucleotide).

DNA molecules exist as two-stranded complexes held together by hydrogen bonds between opposing strands' bases with purine-pyrimidine H-bond pairings – *base pairs* (bp). Specifically, A-T and G-C H-bond pairs form exclusively and are called *complementary*. Two-stranded DNA spontaneously forms a helix, hence the term '*double helix*'. DNA molecules can be extremely long, containing many thousands of base pairs and having molecular weights in the millions, or billions, of Daltons, compared to the largest known protein at 3 million Daltons.

Within the long strands of DNA are regions that encode for proteins called *genes*, pieces of DNA that encode for specific proteins. Within genes are three-nucleotide units called *codons* which each encode for a specific amino acid; thus genes are a string of codons. When an organism requires more of a particular protein, *transcription* is initiated where the DNA is copied in the form of messenger RNA (mRNA), which is then used as a template for *translation* of the transcript at the ribosome into a polypeptide based on the original DNA sequence of codons.

4.5.2 Food DNA manipulation

In the last half of the twentieth century, enormous leaps were realized in understanding how proteins were encoded for in the form of DNA, how proteins were made and how they regulated natural processes, and how to manipulate the encoded information. This culminated in the tailoring of specific characteristics, or genetic engineering, to better serve human food needs. Upon the solving of DNA's structure and the ability to sequence DNA, the abilities to alter specific codons and to amplify DNA were critical advances towards genetic engineering. Furthermore, the isolation and commercialization of enzymes that can be used to either cut DNA at specific sites (nucleases) or attach pieces of DNA together (ligases) have spawned two modern food science areas highlighted here: protein engineering and DNA-based food authentication.

Protein engineering has allowed for the ability to study relationships between protein structure and protein function in the isolation of controlled experiments. Critical was the invention of the polymerase chain reaction (PCR) which allows the amplification of any specific region of DNA simply by cycling temperature in the presence of polymerase enzyme and two small pieces of DNA called *primers* which hybridize (match in sequence) to two ends of the DNA strand of interest. Essentially, copies of the DNA strand are made by polymerase from one primer to the other in successive rounds of copying. The amplification is attained due to a greater and greater number of copies available for further subsequent copying in each cycle of PCR, thereby resulting in an exponential growth of copies. Today, primers are synthesized at low cost commercially.

If a mutation (i.e. an alternative nucleotide) is introduced into a PCR primer such that a codon encodes for an alternative amino acid then the resultant copy DNA sequence will encode for a protein having a different amino acid at the specified position. Copy DNA (cDNA) can be manipulated by engineering nuclease cut sites at each cDNA end that match identical cut sites in commercially available *vectors*, i.e. DNA used to transfer a recombinant gene from one organism to another. The (mutant) copy DNA is ligated into an expression vector and the resultant DNA construct is said to be *recombinant*. Expression vectors are often *plasmids* which are circular DNA vectors capable of replication in bacteria and simple eukaryotes like yeast. Upon recombinant expression (translation) in the chosen organism in small or large quantities, the target protein can be isolated for further study of its function and structure.

Beyond studying structure–function relationships of proteins, genetically engineered food plants have been created and a classic example is that of the Flavr savr™ tomato, originally available for consumption in 1994 (Martineau, 2001). Essentially, the engineered tomato was made to contain a so-called non-sense gene that encoded for mRNA which would base-pair match to the natural mRNA of the gene for polygalacturonase, an enzyme responsible for the breakdown of a cell wall component during ripening. The natural polygalacturonase mRNA and the non-sense, recombinant mRNA are expressed and bind to each other which would physically prevent translation of the polygalacturonase mRNA at the ribosome. The result was slowed spoilage of the engineered tomato, allowing the producer to vine-ripen the Flavr savr™ without spoilage during subsequent transport to market and resulting in superior flavour and appearance relative to natural tomatoes picked green to avoid transport spoilage. The genetic modifications to the Flavr savr™ resulted in no significant changes in micro- and macronutrients, pH, acidity, or sugar content relative to non-transgenic tomatoes.

Another food science area enhanced by DNA technology is that of *food authentication*. Fragments of DNA can be identified by use of *DNA probes*, short pieces of DNA complementary to the region of interest that contain a detectable feature. Detection of the probe can be achieved by attaching a fluorescent group or other specifically reactive group to the probe, or by synthesizing the probe from radioactive (^{32}P) DNA. When test DNA from a crop or food ingredient is immobilized on a piece of nitrocellulose (blotting), it can be tested for the presence of a DNA fragment of interest by treatment of the blot with a complementary probe. The fragment of interest could be a fragment characteristic from known samples, i.e. a DNA fragment that is only present in confirmed crop standards.

An early example of authentication using DNA technology involved an attempt to distinguish between chicken, pig, goat, sheep, and beef cooked meats by probing with DNA from each respective animal (Chicuni *et al.*, 1990). The results were that pig and chicken were specific to their own species whereas the ruminants' DNA was cross-reactive. This presented confirmation that DNA technology could be used, at least within limits, to distinguish heat-treated meat products and exclude the possibility of the presence of particular ingredients. One limitation of the above methodology was that the matching DNA sequences were present across ruminant species, thereby lowering the resolution of the assay.

A more modern way to improve resolution in terms of species identification has been to PCR amplify, and DNA sequence, a highly conserved gene such as metabolic protein cytochrome b, then compare the variability of test sample sequences with known rates of variation among species. Such 'genetic distance' testing is common and a current example is that of tuna species identification (Michelini *et al.*, 2007). Such tests can be done for high heat-processed samples, at a low cost, and can be highly automated for steps post sample collection. The information from these types of authentication tests can aid in quality control in processing operations.

4.6 Enzymology

4.6.1 Introduction to enzyme reactions

A basic definition of an enzyme could be 'A protein folded in such a way that interaction between itself and reactants results in a specific reaction at a fast relative rate'. The majority of chemical reactions in living systems are catalyzed by proteins called *enzymes*. Enzymes cause enormous increases in reaction rates by lowering reactions' energy barriers through optimal orientation of reactants often using temporary bond formation between substrate and enzyme. Enzymes work in both polar (cytoplasm, extracellular space) and non-polar environments (adipocytes, membranes) and their structural designs vary accordingly. The potential for inactivation by pH change, temperature increase, UV-bleaching, etc. follows the same denaturation principles as outlined in section 4.3.

4.6.2 Basics of enzyme reaction energetics

At constant pressure, free energy is a function of both the change in internal energy of the system and changes in entropy. Reactions can be spontaneous even when energy of products is higher than energy of reactants since heat can be absorbed from surroundings; thus system energy is not a predictor of reaction behaviour. In addition, entropy is not easily, directly measured. Biochemical reactions are instead described energetically in terms of free energy change. Reactions that do not occur spontaneously are inhibited by a free energy barrier too large to easily overcome and the equilibrium, expressed as $K_{eq} = $ [products]/[reactants], will favour reactants or be zero. The *transition state* (TS) is the highest free energy state occurring going from reactants to products. It is at this step that enzymes are most able to accelerate a given reaction since this is the most important barrier restricting the reaction rate. Reactions with a net decrease in free energy are spontaneous, and change in free energy depends only on the difference between product free energy and reactant free energy; free energy is independent of intermediate reaction steps or changes thereof.

For most reactions, the initial step of an enzyme catalyzed reaction is the binding of substrate to form enzyme-substrate complex (ES), where E and S are in equilibrium with ES and the reaction is reversible. Substrate binds in the *active site*, the part of the enzyme where the amino acid residues directly involved in positioning and bond breakage/formation are located; these particular active site amino acids are called the *catalytic residues*. The second general step of enzyme catalysis is that of product (P) formation and release of the enzyme, an irreversible step where the supply of ES complex is not limiting to the second step. Overall, we have:

$$E + S \leftrightarrow ES \rightarrow E + P$$

4.6.3 Reaction rates

Rate is determined by measuring the rate of reactant disappearance or the rate of product appearance, usually spectroscopically, at various substrate concentrations. The rate of a reaction is given by the Michaelis–Menten equation:

$$V = [V_{max}][s]/[s] + [K_m]$$

where

V_{max} is defined as the theoretical maximum speed attained as substrate concentration approaches infinity,

K_m is the substrate concentration at which half of V_{max} is attained.

K_m is often referred to as the binding constant because it is an indicator of the enzyme's relative affinity for substrate, i.e. low K_m indicates high affinity. If one plots V vs. s then the rate can be calculated by non-linear regression of the direct plot or alternatively various linear transformations are possible. Enzymes only follow Michaelis–Menten behaviour if the enzyme is saturated with substrate; thus velocity, v, must be measured near the beginning of reactions for each concentration of substrate, and $[E] \ll [S]$, thus ensuring that the ES $\rightarrow$ E + S irreversible step is the rate limiting step being measured.

4.6.4 Inhibition

Factors affecting enzyme catalyzed reaction rates include the presence of *competitive inhibitors*, molecules that bind in the active site of an enzyme but do not undergo conversion to products, thereby competing directly with substrate for the active site and slowing or stopping catalysis. An example of an irreversible inhibitor is that of the aspartic proteinases (AP), a class of enzymes that hydrolyze peptides and proteins in the stomachs of animals, also used in cheese making, among other functions. The inhibitor pepstatin binds in pepsin-like AP active sites with a binding constant very heavily favoured towards the ES complex due to the formation of a highly stable, non-hydrolyzable complex with catalytic residues (Tanaka and Yada, 2004). Unlike competitive inhibitors, non-competitive inhibitors reduce reaction rates via suppression of V_{max} due to distortion of optimal enzyme structure. A third mode of inhibition is that of allosteric inhibition where either substrate binds at an active site in an enzyme with multiple active sites, or a separate regulatory molecule binds to a site other than an active site, causing a change in the enzyme's affinity for substrate. Allosteric enzymes do not follow Michaelis–Menten behaviour (Stryer, 1996).

4.6.5 Catalysis in solvents

Many enzymes that can be isolated in an active form for in vitro use or study tend to be aqueous proteins simply because they remain soluble during purification without the use of detergents. Some food enzymes such as lipases have been found to remain active at high organic solvent concentrations. Interest in testing enzymatic activity in non-aqueous environments is in part for the discovery of novel specificities or activities not normally associated with a particular enzyme under native conditions. As a further step to this end, chemical modifications to the exterior of lipase to alter its hydrophobicity was shown to improve lipase activity and stability, and alter specificity in organic solvents (Salleh *et al.*, 2002). An example of a practical food application is that of manufacturing emulsifiers from triglycerides and propylene glycol using lipase as a catalyst in organic solvents as well as catalytic rate improvement by immobilizing the enzyme (Liu *et al.*, 1998). Such technologies can contribute to the manufacture of food ingredients like emulsifiers to yield highly consistent ingredient properties.

4.6.6 Biosensors

Biosensors are a technology that combines the use of a biological/biochemical component with a detection device. The bio component can be a tissue, microorganism, organelle, cell receptor, enzyme, antibody, nucleic acid, or natural product. The nature of the sensor can be optical, electrochemical, thermometric, piezoelectric/voltage inducement, magnetic, or micromechanical. The three main classes of biological elements used in biosensors are enzymes, antibodies, and nucleic acids (Lazcka *et al.*, 2007). Applicable to the area of food is the detection of pesticide residues in complex mixtures. Immobilized organophosphorus hydrolase (OPL) is an enzyme that catalyzes the hydrolysis of organophosphate (found in insecticides and pesticides). OPL has been used for organophosphate detection by cross-linking it to the surface of a pH electrode. The result was an enzyme-based biosensor that detected organophosphate by inducing a change in voltage in a reproducible manner (Mulchandani *et al.*, 1998).

A consideration in the use of enzyme biosensors is the use of low availability enzymes that are difficult to prepare in sufficient quantities such that time and cost could be prohibitive. As a solution to such a situation, an enzyme biosensor could be recombinantly expressed, purified, and subsequently immobilized from a suitable microorganism, thereby providing the necessary quantity of enzyme biosensor from a reliable source (Lei *et al.*, 2006).

4.7 Food processing and storage

The biochemical effects of heat, pH, oxygen, and light on food components have been discussed in the respective sections above so the focus of this section will be on selected topics related to processing and storage effects. Food processing and storage serves the important purposes of ensuring that food products or ingredients have the desired properties, e.g. bleaching flour, ensuring the safety of the processed product, and attaining the expected shelf-life of the product. Processing, packaging, and storage must also aim to protect the desirability of the food by preserving, when possible, properties like colour, texture, and flavour as well as nutritional value.

4.7.1 Irradiation

Irradiation of foods results in a large reduction in microbial contamination and insect pests while causing very small changes in the nutritionally important food components. Some controversy and skepticism exists regarding this processing method, so to ensure consumer acceptance and confidence, reliable and sensitive detection methods have been researched. Irradiated foods are distinguishable from those that are not irradiated for validation purposes, allowing for confirmation of labeling accuracy. Such validation tests can involve electron spin resonance spectroscopy, luminescence, fluorescence, gas chromatography, and DNA-based methods, depending on the food product in question. For example, chicken, pork, and beef can be tested for irradiation treatment by analyzing fatty acid by-product fragments formed during the irradiation process (Delincee, 2002). 1-Tetradecene and pentadecane are produced by breakage of the hydrocarbon chain of palmitic acid, while 1-hexadecadiene and heptadecene originate from oleic acid (Rahman *et al.*, 1995). These compounds can be identified by gas chromatography (GC). GC is a powerful analytical technique for the separation of compounds from simple or very complex mixtures. GC involves the injection of gas or liquid sample which is transported through a column (stationary phase) by an inert carrier gas (mobile phase). There are various types of columns that can be used in GC, but the important point is that the stationary phase interacts differentially with sample components, thereby causing their separation as they pass through. Of the detection options for GC, flame ionization detection (FID) is well suited for hydrocarbon detection as it is sensitive to nanogram quantities and hydrocarbons ionize well by this detection method.

4.7.2 Active packaging

Active packaging (AP) is a term used to describe packaging that performs a function beyond the passive barrier role of traditional packages. Roles of APs include oxygen scavenging, moisture control, carbon dioxide and ethanol generation, and antimicrobial roles (Suppakul *et al.*, 2003). The prevention of food-borne pathogen growth in foods is critical in the avoidance of food recalls, most of which result from post-processing practices. Packaging that actively aids in reducing microbial/pathogen load has been a prime focus. A novel, packaging film was made to be both edible and antipathogenic for use in meat products. The edible film in question was a low pH whey protein isolate-based film containing between 0.5% and 1.5% para-aminobenzoic acid (PABA) or sorbic acid (SA). pH was lowered to 5.2 by the addition of organic acids acetic acid and lactic acid. Whey protein films form due to disulphide bond, H-bond, and hydrophobic interactions formed between whey protein molecules upon heat denaturation which exposes the internal SH and hydrophobic groups (Cagri *et al.*, 2001). All of the concentrations of PABA and SA showed antimicrobial activity, with stronger action noted for the higher concentrations.

4.8 Summary

Understanding the food components that comprise foods and their biochemistry is critical for the study of food processes. The encoding of life's processes takes place through protein-mediated reactions. Both the preservation of fresh produce and the incubation of fermented products depend on the control of respiration. Certain amino acids, and fatty acids, are designated 'essential' because the body cannot synthesize them in adequate amounts, and they therefore must be ingested via dietary sources. Unique flavours and colours result from chemical or enzymatic reactions to otherwise common carbohydrates. The most important food source for energy derivation from carbohydrates takes place via glycolysis, whereas in addition lipid- and amino acid-derived energy sources of energy, the citric acid cycle, and oxidative phosphorylation are responsible for most energy production. Structure- and function-based studies have been important for identifying critical amino acids, and entire genes, in functionalities of food proteins. Functional proteins can be utilized as biosensors and packaging aids towards the goal of extending food product shelf-life and quality. Thus, biochemical processes are central to food applications, and their causes and effects are ubiquitous among food systems.

Further reading and references

Aharoni, A., Giri, A.P., Verstappen, F.W.A., *et al.* (2004) Gain and loss of fruit flavor compounds produced by wild and cultivated strawberry species. *The Plant Cell*, **16**, 3110–31.

BeMiller, J.N. and Whistler, R.L. (1996) Carbohydrates. In: *Food Chemistry*, 3rd edn (ed. O.R. Fennema), pp. 216–17. Marcel Dekker, New York.

Boehme, M.A. and Branen, A.L. (1977) Effects of food antioxidants on prostaglandin biosynthesis. *Journal of Food Science*, **42**, 1243–6.

Cagri, A., Ustunol, Z. and Ryser, E.T. (2001) Antimicrobial, mechanical, and moisture barrier properties of low pH whey protein-based edible films containing p-aminobenzoic or sorbic acids. *Journal of Food Science*, **66**, 865–70.

Champe, P.C., Harvey, R.A. and Ferrier, D.R. (2005) *Biochemistry*, 3rd edn, pp. 243–51. Lippincott Williams and Wilkins, Baltimore.

Chicuni, K., Ozutzumi, K., Koishikawa, T. and Kato, S. (1990) Species identification of cooked meats by DNA hybridization assay. *Meat Science*, **27**, 119–28.

Coffee, D.G., Bell, D.A. and Henderson, A. (1995) Cellulose and cellulose derivatives. In: *Food Polysaccharides and their Applications* (ed. A.M. Steven), pp. 127–39. Marcel Dekker, New York.

Delincee, H. (2002) Analytical methods to identify irradiated food – a review. *Radiation Physics and Chemistry*, **63**, 455–8.

Edman, P. (1970) Sequence determination. *Molecular Biology, Biochemistry, and Biophysics*, **8**, 211–55.

Eskin, N. (1990) *Biochemistry of Foods*, 2nd edn, pp. 268–72. Academic Press, San Diego.

Goff, H.D. (1997) Partial coalescence and structure formation in dairy emulsions. In: *Food Proteins and Lipids* (ed. S. Damodran), pp. 137–47. Plenum Press, New York.

Gouin, S. (2004) Microencapsulation: industrial appraisal of existing technologies and trends. *Trends in Food Science and Technology*, **15**, 330–47.

Gunaratne, A., Ranaweera, S. and Corke, H. (2007) Thermal, pasting, and gelling properties of wheat and potato starches in the presence of sucrose, glucose, glycerol, and hydroxypropyl beta-cyclodextrin. *Carbohydrate Polymers*, **70**, 112–22.

Kays, S.J. (1991) *Postharvest Physiology of Perishable Plant Products*. Van Nostrand Reinhold, New York.

Klinman, J. (2007) How do enzymes activate oxygen without inactivating themselves? *Accounts of Chemical Research*, **40**, 325–33.

Ko, H.S., Kim, T.H., Cho, I.H., Yang, J., Kim, Y. and Lee, H.J. (2006) Aroma active compounds of bulgogi. *Journal of Food Science*, **70**, 517–22.

Kritchevsky, D. and Chen, S.C. (2005) Phytosterols – health benefits and potential concerns: a review. *Nutrition Research*, **25**, 413–28.

Laguerre, M., Lecomte, J. and Villeneuve, P. (2007) Evaluation of the ability of antioxidants to counteract lipid oxidation: existing methods, new trends and challenges. *Progress in Lipid Research*, **46**, 244–82.

Lazcka, O., Del Campo, F.J. and Munoz, F.X. (2007) Pathogen detection: a perspective of traditional methods and biosensors. *Biosensors and Bioelectronics*, **22**, 1205–17.

Lei, Y., Chen W. and Mulchandani, A. (2006) Microbial biosensors. *Analytica Chimica Acta*, **568**, 200–210.

Liu, K., Chen, S. and Shaw, J. (1998) Lipase-catalyzed transesterification of propylene glycol with triglyceride in organic solvents. *Journal of Agriculture and Food Chemistry*, **46**, 3835–8.

Lomer, M., Parkes, G. and Sanderson, J. (2008) Review article: lactose intolerance in clinical practice – myths and realities. *Alimentary Pharmacology and Therapeutics*, **27**(2), 93–103.

Maarse, H. (1991) *Volatile Compounds in Foods and Beverages*. Marcel Dekker, New York.

Martineau, B. (2001) *First Fruit: The Creation of the Flavr savr™ Tomato and the Birth of Genetically Engineered Food*. McGraw-Hill, New York.

Michelini, E., Cevenini, L., Mezzanotte, L., *et al.* (2007) One-step triplex-polymerase chain reaction assay for the authentication of yellowfin (*Thunnus albacares*), bigeye (*Thunnus obesus*), and skipjack (*Katsuwonus pelamis*) tuna DNA from fresh, frozen, and canned tuna samples. *Journal of Agriculture and Food Chemistry*, **55**, 7638–47.

Moreno, J.J., Carbonell, T., Sánchez, T., Miret, S. and Mitjavila, M.T. (2001) Olive oil decreases both oxidative stress and the production of arachidonic acid metabolites by the prostaglandin G/H synthase pathway in rat macrophages. *Journal of Nutrition*, **131**, 2145–9.

Moutsatsou, P. (2007) The spectrum of phytoestrogens in nature: our knowledge is expanding. *Hormones (Athens)*, **6**, 173–93.

Mulchandani, A., Mulchandani, P. and Chen, W. (1998) Enzyme biosensor for determination of organophosphates. *Field Analytical Chemistry and Technology*, **2**, 363–9.

Nawar, W.W. (1996) Lipids. In: *Food Chemistry*, 3rd edn (ed. O.R. Fennema), pp. 237–43. Marcel Dekker, New York.

Nesvizhskii, A.I. (2007) Protein identification by tandem mass spectrometry and sequence database searching. *Methods in Molecular Biology*, **367**, 87–119.

Rahman, R., Haque, A.K.M.M. and Sumar, S. (1995) Chemical and biological methods for the identification of irradiated foodstuffs. *Nutrition and Food Science*, **95**, 4–11.

Salleh, A.B., Basri, M., Taib, M., *et al.* (2002) Modified enzymes for reactions in organic solvents. *Applied Biochemistry and Biotechnology*, **102**, 349–57.

Steven, A.M. (1995) *Food Polysaccharides and their Applications*. Marcel Dekker, New York.

Stryer, L. (1996) *Biochemistry*, 4th edn. W.H. Freeman, New York.

Suppakul, P., Miltz, J., Sonneveld, K. and Bigger, S.W. (2003) Active packaging technologies with an emphasis on antimicrobial packaging and its applications. *Journal of Food Science*, **68**, 408–20.

Tanaka, T. and Yada, R.Y. (2004) Redesign of catalytic center of an enzyme: aspartic to serine proteinase. *Biochemical Biophysical Research Communications*, **323**, 947–53.

Verreault, J., Skaare, J.U., Jenssen, B.M. and Gabrielsen, G.W. (2004) Effects of organochlorine contaminants on thyroid hormone levels in Arctic breeding glaucous gulls, *Larus hyperboreus*. *Environmental Health Perspectives*, **112**, 532–7.

Voet, D. and Voet, J.G. (1995) *Biochemistry*, 3rd edn. John Wiley, Hoboken.

Zobel, H.F. and Steven, A.M. (1995) Starch: structure, analysis, and application. In: *Food Polysaccharides and their Applications* (ed. A.M. Steven), pp. 27–31. Marcel Dekker, New York.

Cherl-Ho Lee

Key points

- The history of food biotechnology including alcohol fermentation, acid fermentation, bread fermentation and amino acid/peptide fermentation throughout the world.
- Recent developments in enzyme technology and the industrial production of amino acids, nucleic acids and organic acids.
- The basis of genetic engineering and tissue culture used in modern food biotechnology.

5.1 History of food biotechnology

Biotechnology has been broadly defined as the utilization of biologically derived molecules, structures, cells or organisms to carry out a specific process (Wasserman *et al.*, 1988). Many conventional food processing techniques utilize living organisms and bioactive molecules especially in the brewing and fermentation industries. Alcoholic food and drinks, cheese, yogurt, lactic acid fermented vegetables, soybean sauce and fish sauce have been made for thousand years by using naturally occurring microorganisms which grow in a specific environmental condition. Plant enzymes such as malt have been used in the brewing industry long before man acquired the knowledge of enzyme chemistry.

Traditional food fermentation technologies are based on the natural process whereby wet foodstuff undergoes microbial degradation, and when it is edible we call it fermented food and when it is not we call it spoiled or putrid food. Man has acquired

fermentation skills over a long time and has developed unique technologies suitable for the specific environment and raw materials available in different regions of the world. The first product of fermentation man discovered was alcoholic fermented fruits, which contain sugar that is fermented by natural yeast to make alcohol. More sophisticated fermentation skills using cereals to make alcohol were developed later; beer in Egypt and rice wine in Northeast Asia, both in around 4000 BC (Owades, 1992; Lee, 2001). The oldest known written recipe was found on a 4000-year-old Mesopotamian clay tablet, for beer (Owades, 1992). The Babylonians made 16 kinds of beer, using barley, wheat and honey. The Chinese book *Shijing* (1100—600 BC) has a poem describing a 'thousand wines of Yao', a legendary nation of China in around 2300 BC. It appears that the fermentation technology in Northeast Asia must have been invented by the littoral foragers of the primitive Pottery Age (8000–3000 BC) before agriculture began (Lee, 2001).

Drying and fermentation were the most important food preservation technologies until the industrial revolution in the seventeenth century in most regions of the globe from temperate to tropical regions. The world of microorganisms was opened to human beings with the invention of the microscope by Antonie van Leeuwenhoek (1632–1723), and the scientific control of fermentation began with the studies of Louis Pasteur (1822–1895). Pasteur showed that good wine batches had certain types of ferment (microorganism), and bad batches had other types of ferment. By heating juices at 63°C for 30 min he could kill the bad ferments and after cooling the juice he could consistently produce satisfactory wine by inoculating ferments from good wine batches into the juice. This idea was also applied in the pasteurization process of milk, which contributed greatly to the improvement of food hygiene.

Enzymes, the biocatalysts, have been known since the early seventeenth century from observation of their role in digestion and fermentation processes. However, isolation of the crystalline form of enzyme was first achieved with urease in 1926. Thereafter amylase, carboxypeptidase, papain and pepsin were isolated from plants, animals and microorganisms. With the development of enzyme technology, the conventional industrial enzymes originating from plants and animals were substituted by microbial enzymes. Chymotrypsin from microorganisms containing the milk clotting enzyme chymosin has partially replaced rennet from the stomach lining of young calves. When genetic engineering techniques developed in the 1970s, the first application of GMO in food was the production of food enzymes.

Using GMO microorganisms, numerous food enzymes have been developed which have higher activity and are more tolerant of extreme working conditions such as high temperature. The production of biotech crops, especially corn and soybean, has increased rapidly since their first marketing in 1995. The cultivation area of GM crops reached over 120 million hectares in a total of 23 countries in 2007 (James, 2007). Table 5.1 summarizes the milestone events of food biotechnology through history.

5.2 Traditional fermentation technology

The traditional fermented foods of the world can be classified by the materials obtained from bioconversion, such as alcohol fermentation, acid fermentation, carbon dioxide (bread) fermentation and amino acid/peptide fermentation from protein (Steinkraus, 1993; Lee, 2001). Depending on the raw materials used numerous varieties of fermented food are made in each class of fermentation (Steinkraus, 1983). For example, in alcohol fermentation, wine from grape, cider from apple, toddys from palm sap, beer from barley or corn, *chongju* from rice and even *mayuchu* from horse milk are made. Furthermore, by distillation, brandy, rum, vodka, whisky and soju are produced. Figure 5.1 shows a traditional fermented food map of the world.

The traditional societies can be divided by the indigenous fermented foods they produce; for example, the cheese/yogurt culture of the Middle East, Northern Africa and Europe, fish sauce culture of Southeast Asia and soybean sauce culture of Northeast Asia. These products are made by the breakdown of proteins to produce the umami (meaty) taste of amino acids and peptides, and form the basic flavor of the meals and condiments characterizing the dietary culture of these different regions.

5.2.1 Alcohol fermentation

Alcoholic beverages have played an important role in human spiritual and cultural life both in Western and Eastern societies. Unlike in Europe and the Middle East, where most of the indigenous alcoholic beverages are produced from fruits, in the Asia-Pacific region alcoholic beverages are produced from cereals and serve as an important source of nutrients. European beer uses barley malt as the major raw material, while Asian alcoholic beverages utilize mold-grown *nuruk* made from rice or wheat as the fermentation starter.

Alcohol is produced from glucose by the action of the alcohol producing yeast, *Saccharomyces cerevisiae*. In 1810 Gay-Lussac established the fermentation formula:

$$C_6H_{12}O_6 \rightarrow 2\,C_2H_5OH + 2\,CO_2$$

The general processes of alcohol fermentation of wine, beer and rice wine are presented as follows:

$$\text{Fruit juices} \xrightarrow{\text{yeasts}} \text{Wine}$$

$$\text{Barley} \xrightarrow{\text{sprouting}} \text{Malt} \xrightarrow{\text{yeasts}} \text{Beer}$$

$$\text{Rice} \xrightarrow{\text{molds}} \textit{Nuruk} \xrightarrow{\text{yeasts}} \textit{Chongju} \text{ (rice wine)}$$

Table 5.1 Milestones in food biotechnology.

Date	Milestone in food biotechnology
6000 BC	Use of earthenware for cooking and storage in Northeast Asia Yeasts employed to make wine and beer in the Middle East
4000 BC	Leavened bread produced with the aid of yeasts in Egypt Fungal fermentation of cereals in earthen jars in Northeast Asia Salt fermentation of marine products and plants in earthen jars Curdling of milk in skin bags for cheese making in the Middle East
2000 BC	Fermentation starter 'nuruk' employed for rice wine 'Thousand rice-wines in Yao' of China Use of soybean as food in South Manchuria and the Korean Peninsula
200 BC	*Bacillus subtilis* employed to ferment soybean 'shi'
1680	Antoni van Leeuwenhoek invented the microscope and discovered microbes
1857	Louis Pasteur discovered anaerobic fermentation Pasteurization process began
1876	Pasteur proved microbial action on beer fermentation
1897	Buchner discovered that enzymes in yeast juice convert sugar into alcohol
1904	Pure cultured fermentation starter 'koji' developed in Japan
1912	Industrial chemicals (acetone, butanol, glycerol) obtained from bacteria
1928	Alexander Fleming discovered penicillin
1953	Industrial production of glutamate by soil bacteria Double helix structure of DNA revealed by Watson and Crick
1960	Industrial enzyme production from microorganisms
1965	Borlaug's Green revolution
1973	DNA recombination by Cohen and Boyer
1975	Hybridomas which make monoclonal antibodies first created
1976	US NIH guidelines on genetic engineering
1982	Genetic engineered insulin approved for use in diabetics in the US and UK First approval for release of GM microbes into the environment
1994	Introduction of GM Flavr Savr Tomato on market by Calgene, Inc. Herbicide-tolerant Round-Up Ready GM soybean by Monsanto Co.
1996	Herbicide-tolerant and insect-resistant YieldGard GM corn on market
2000	Development of Golden Rice™

Malt contains amylases and is able to break down starch into fermentable sugars. The fermentation starter, *nuruk* in Korea, is made by growing molds on cereals, either raw or cooked, to digest starch into sugars, which are then consumed by the yeast to produce alcohol. Rice-wine fermentation is therefore called a two-step fermentation process. Table 5.2 summarizes the names of cereal fermentation starters used in Asia-Pacific regions, their ingredients and the microorganisms involved.

5.2.1.1 Wine

Wine is the fermented product of grape, mainly of cultivars of *Vitis vinifera*. Fermented juices of many other fruits, for example apples, berries, peaches and even herbs, are also called wine. The distinctive character of various wines depends on the composition of the raw material, the nature of the fermentation process, and processing and aging treatments. Table wines with excess carbon dioxide include white,

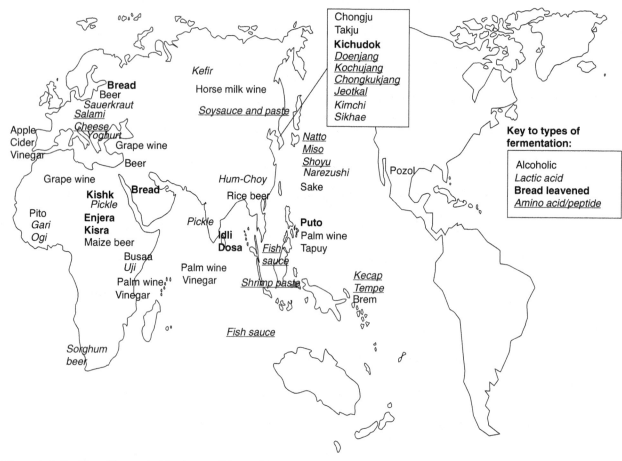

Figure 5.1 Traditional fermented food map of the world.

Table 5.2 Names of fermentation starters in different countries and the major ingredients used (Lee, 1998).

Country	Name	Ingredients commonly used	Shape	Microorganisms
China	Chu	Wheat, barley, millet, rice (whole grain, grits, flour or cake)	Granular	Rhizopus Amylomyces
Korea	Nuruk	Wheat, rice, barley (whole grain, grits or flour)	Large cake	Aspergillus Rhizopus Yeasts
Japan	Koji	Wheat, rice (whole grain, grits or flour)	Granular	Aspergillus
Indonesia	Ragi	Rice (flour)	Small cake	Amylomyces Endomycopsis
Malaysia	Ragi	Rice (flour)	Small cake	No data available
Philippines	Bubod	Rice, glutinous rice (flour)	Small cake	Mucor Rhizopus Saccharomyces
Thailand	Loogpang	Bran	Powder	Amylomyces Aspergillus
India	Marchaa	Rice	Flat cake	Hansenula Mucor Rhizopus

pink and red sparking wines, with or without Muscat flavor (champagne, spumante, Sekt, etc.). Non-sparkling white table wines are dry or sweet with regional, varietal and proprietary names (Riesling, Chardonnay, Chablis, Sauternes, etc.) Pink and red table wines are among the most important in terms of volume or consumer demand and have varietal, regional and proprietary names (Cabernet Sauvignon, Pinot noir, Burgundy, Bordeaux, etc.). There are also numerous dessert wines named after the grape variety from which they are produced, the processes and the added flavors and herbs (Amerine *et al.*, 1980).

Wines are produced in many different ways. In general, grapes are washed, removed from the stem and macerated. The must is stored in a fermentation vat. The normal flora of ripe grapes and the winery contain sufficient yeast to initiate fermentation. However, the addition of pure actively growing yeast cultures or pressed yeast is common and desirable, and numerous strains of yeast for wine making are available in the market. To prevent growth of undesirable microorganisms 25–100 mg/L sulfur dioxide (in the form of potassium metabisulfite) is added to the must about 2 hours before the yeast. Sulfur dioxide acts as a selective antiseptic for bacteria and wild yeasts and permits the more rapid growth of the added yeast.

The optimum temperature for fermentation differs in red and white wines; red musts are fermented at up to 26.7°C and whites to 15°C. Red musts are fermented on the skins until the maximum color is extracted, often 5–10 days. The solid material, now called the marc or pomace, is transferred to the press. The fermentation of white musts, at the lower temperature, may last several weeks. A basket press or Willmes press in the larger wineries is often used for the filtration. For fining of the filtrate gelatine solution or egg white is added to combine with tannins and related compounds and precipitate. In order to remove excess potassium acid, tartrate wines are stored at a low temperature of approximately –2°C. The clarified wines are bottled and stored in a cool place for aging.

The maximum sugar accumulated by grapes during ripening depends on the variety of grape and the climate of the region where the grapes are grown. For most varieties ca. 15–25% sugar is reached, but late-harvest fruits may have 30–40% sugar. The actual percentage of alcohol in the finished wine depends not only on the sugar content of the must (crushed grapes) but also on the completeness of the fermentation, and in the case of dessert wines, the amount of alcohol added during or after the fermentation. At least 9 vol % of alcohol is needed to prevent rapid acetification of the finished product. When the sugar content of the grapes is too low to attain this percentage alcohol (1% sugar yields 0.55% alcohol according to the fermentation formula), sugar or grape concentrate may be added. The process, called chaptalization, is normal in cool-climate regions (eastern United States, Germany, France (Burgundy), etc.) but is prohibited in other areas (California, Spain, Italy, etc.).

5.2.1.2 Beer

Beer is an alcoholic beverage derived from barley malt, with or without other cereal grains (rice, corn, sorghum and wheat), and flavored with hops. Malt is made by three steps: the steeping of barley, germination and drying. By keeping steeped barley (45% moisture) in a humid and dark place for 4–6 days the barley starts to germinate and produce starches, splitting enzymes, α-amylase and β-amylase, and also protease and cellulase. When the modification is completed, the malt is dried. The drying stabilizes the malt and allows beer to be produced year round anywhere, even where barley is not grown. This differs from wine, which is only produced seasonally and only near grape-growing areas.

Beer production involves three distinct stages: brewing, fermentation and finishing. An extract of the crushed malt and the grains selected is prepared to make wort. This step takes about 4–10 hours at around 50°C and is often referred to as brewing. The mashing temperature and duration vary with the mashing systems. Mashing converts insoluble starch into fermentable sugars, maltose and glucose, and proteins into peptides and amino acids. The converted mash is separated into liquid (wort) and insoluble husk by filtration. The clear wort is boiled in a kettle. During the heating period hops are added. The insoluble humulones in the hops undergo a chemical rearrangement during this process to form isohumulones, which are soluble in water and impart to beer a palate-cleansing bitterness that provides beer with its unusual property of drinkability (Owades, 1992).

The next stage is fermentation, the conversion of wort by yeast into beer. Yeast is added to a cool wort containing oxygen, fermentable sugars and various nutrients including amino acids. The two main types of beer, lager and ale, are fermented with different strains of yeast. Lager is produced by bottom-fermenting *Saccharomyces uvarum* (*carlsbergensis*) at fermentation temperatures between 7 and 15°C, and at the end of fermentation, these yeasts flocculate and collect at the bottom of the fermenter. For the production of ale, top-fermenting yeast, *Saccharomyces cerevisiae*, is used at fermentation temperatures between 18 and 22°C. *Saccharomyces cerevisiae* is less flocculent and collected for reuse from the surface of the fermenting wort (Russell and Stewart, 1995). The differentiation of lagers and ales on the basis of bottom and top cropping has become less distinct with the advent of vertical conical bottom fermenters and centrifuges.

The fermented beer may be finished in several ways. The simplest and most widely used method is merely to transfer the beer to another tank, chilling it en route, and keeping it for 7–14 days, which is called 'rest'. During this period much of the still suspended yeast settles, and some harsh, sulfury notes and undesirable flavor compounds, notably diacetyl, are removed. After finishing, the beer is filtered, always in the cold, through diatomaceous earth as the filter medium. For sterile filtration to remove all yeast and lactobacilli, another diatomaceous earth filter, cotton fibers, a porous plastic sheet or a ceramic filter are used as the retaining barrier. If packaged beer has not been sterile-filtered, it must be pasteurized, as beer is a fertile medium for many microbes. The pasteurization may be done just before filling (bulk pasteurization) or after filling in long tunnels with hot-water sprays (tunnel pasteurization). Bulk pasteurization takes about a minute, tunnel pasteurization about an hour.

5.2.1.3 Rice wine

Rice wine is a generic name for alcoholic beverages made from cereals, mainly rice, in East Asia. Traditional alcoholic beverages vary from crystal-clear products to turbid liquid or thick gruels and pastes. Clear products which are generally called *shaosingjiu* in China, *chongju* in Korea and *sake* in Japan contain around 15% alcohol and are designated as rice wine, while turbid beverages, *takju* in Korea and *tapuy* in the Philippines, contain less than 8% alcohol along with suspended insoluble solids and live yeasts, and are referred to as rice beer. Examples of alcoholic beverages prepared from cereals in the Asia-Pacific region are listed in Table 5.3 (Lee, 2001).

The process of cereal alcohol fermentation using *nuruk* involves two-step fermentation: a solid state fermentation growing mold on raw or cooked cereals which is called *nuruk*, and mashing the *nuruk* with

additional cereals to produce alcohol using yeast. The dried and powdered *nuruk* is mixed with water and stored in a cool place for several days to make a 'mother' brew. During this period the microbial amylases and proteases are activated and convert starches into sugars. The acid-forming bacteria in *nuruk* produce organic acids to bring the pH below 4.5. About 2–3 volumes of cooked grains and water are added to the mother brew to prepare a first fermentation mash. By the addition of new cooked grains and water to the mash the volume of production increases and the alcohol concentration and quality of the final product enhances. Multiple brews prepared by adding newly cooked grains to the fermenting mash two, three, four and up to nine times have been described in the old literature (Yoon, 1993).

Newly cooked cereals are added at the end of each step of the fermentation process. The incubation period for each step of the brewing process varies from 2 days to 1 month depending on the fermentation temperature. Low temperatures (ca. 10°C) are better for improving the taste and keeping quality of rice wine. Traditionally rice wines are prepared in late autumn or early spring, when ambient temperatures are below 10° in the Far East. The volume of wine produced is approximately the same as that of the raw grain used (Rhee *et al.*, 2003).

The traditional method of rice-wine brewing was industrialized by Japanese brewers in the early twentieth century, who adopted pure starter culture, rice *koji* and manufacturing technology from Europe and transferred it to Korea and China. Industrial production of rice wine uses pure cultured starter, *koji*, by the steaming of polished rice, inoculation of mold, *Aspergillus oryzae* or *kawachii*, and incubation at 25–30°C for 2–3 days. The mother brew is made by mixing *koji*, yeast seed mash and water followed by incubation for 3–4 days at 20°C. The main brew is made by adding ca. 10 times the volume of cooked rice and water to the mother brew and fermenting for 2–3 weeks. The fermented mash is filtered to obtain clear liquid and aged in a cool place for 1–2 weeks. It is filtered again and then bottled and pasteurized (Rhee *et al.*, 2003).

Rice beers are produced at a higher temperature of fermentation (ca. 20°C). The fermentation starter powder is mixed with cooked cereals (rice, wheat, barley or corn) and water, and then incubated at approximately 20°C for 2–3 days, following which it is filtered through a fine mesh sieve or cloth. These beers are usually prepared by either single or double

Table 5.3 Examples of cereal alcoholic beverages in the Asia-Pacific region (Lee, 2001).

Product	Country	Major ingredients	Microorganisms	Appearance and usage
Rice wine				
Shaosingjiu	China	Rice	*Sac. cerevisiae*	Clear liquid
Chongju	Korea	Rice	*Sac. cerevisiae*	Clear liquid
Sake	Japan	Rice	*Sac. sake*	Clear liquid
Rice beer				
Takju	Korea	Rice, wheat	Lactic acid bacteria *Sac. cerevisiae*	Turbid liquid
Tapuy	Philippines	Rice, glutinous rice	*Saccharomyces Mucor Rhizopus Aspergillus Leuconostoc Lb. plantarum*	Sour, sweet liquid, paste
Brem bali	Indonesia	Glutinous rice	*Mucor indicus Candida*	Dark brown liquid, alcoholic
Alcoholic rice paste				
Khaomak	Thailand	Glutinous rice	*Rhizopus Mucor Saccharomyces*	Semi-solid, sweet, alcoholic
Tapai pulut	Malaysia	Glutinous rice	*Chlamydomucor Hansenula*	Semi-solid, sweet, alcoholic
Tape-ketan	Indonesia	Glutinous rice	*Asp. rouxii Saccharomycopsis burtonii*	Sweet/sour, alcoholic paste
Lao-chao	China	Rice	*Rhizopus Asp. rouxii*	Paste
Alcoholic rice seasoning				
Mirin	Japan	Rice, alcohol	*Asp. oryzae Asp. usamii*	Clear liquid seasoning

brew. Cereal beers are abundant in micronutrients, such as vitamin B groups formed during the fermentation, and provide rapid energy supplements with the ethyl alcohol and partially hydrolyzed polysaccharides (Lee, 1998).

5.2.2 Acid fermentation

Lactic acid fermentation is probably one of the first biological processes from which human beings discovered the benefits of fermentation (Lee, 1998). The sour ferments of flour dough, milk, cereals and vegetables have been used for the enhancement of keeping quality and palatability of food from prehistoric times. The fermentation of dairy products in Europe has been widely studied over the past century, and the processes have been highly standardized and industrialized to ensure efficient production of safe and nutritious food products. Caucasian yogurt and Middle-Eastern cheese have become every-day food for people in Europe, America and Oceania and they are considered as gourmet foods for wealthy people in Asia and Africa. However, little scientific research has been carried out on other types of fermented foods, which have contributed greatly to diets in East-Asia and Africa (Lee *et al.*, 1994).

The most important microorganisms for acid fermented food are the lactic acid bacteria and these are differentiated into four genera: *Streptococcus, Pediococcus, Lactobacillus* and *Leuconostoc*. In addition, *Bifidobacterium*, belonging to the order *Actinomycetales*, is also important for dairy products. *Streptococcus,*

Pediococcus and some of *Lactobacillus* are homolactic, while *Leuconostoc* and *Bifidobacterium* are heterolactic. The metabolic pathways of glucose in lactic acid bacteria vary: glycolysis, the bifidus pathway and the 6-P-gluconate pathway.

Glycolysis:

$$C_6H_{12}O_6 \xrightarrow{\text{Homolactic bacteria}} 2\ CH_3\text{-CHOH-COOH}$$

Bifidus pathway:

$$\text{Glucose} \xrightarrow{\text{Bifidobacteria}} \text{Lactic acid} + \text{Acetic acid}$$

6-P-Gluconate pathway:

$$\text{Glucose} \xrightarrow{\text{Heterolactic bacteria}} \text{Lactic acid} \\ + \text{Acetic acid (ethanol)} + \text{Carbon dioxide}$$

5.2.2.1 Lactic acid fermented milk products

Cultured milk products are produced by the lactic acid fermentation of milk using various bacterial cultures. Fermented milk products originated in the Near East and then spread to parts of southern and eastern Europe. Today cultured milk products in various forms have been introduced throughout the world even in regions where milk is not a traditional food like Korea and Japan. There are great differences in cultured products depending on the variations in the starter cultures used and manufacturing principles. However, most culture products use the following basic manufacturing steps:

1 culture of starter preparation;
2 treatment of product, such as pasteurization, separation and homogenization;
3 inoculation with bacterial cultures;
4 incubation;
5 agitation and cooling;
6 packaging.

Table 5.4 lists the world's principal cultured milk products including type, location and bacterial culture used (McGregor, 1992).

5.2.2.2 Lactic acid fermented cereals and starchy tubers

Lactic acid fermentation of bread dough improves the keeping quality and flavor of the baked products. It also enhances the palatability of bread made from low-grade flours and underutilized cereals. Acid-fermented breads and pancakes are an important staple food for people in Africa and some parts of Europe and Asia (Lee, 1994). Sour bread is a typical German food and Scandinavian rye bread is highly favored by the Nordic people. The Indian *idli* bread types (*idli, dosa, dhokla, khaman*) are important staple foods of the Indian and Sri Lankan people and they are consumed three or four times a week at breakfast and supper. *Idli* is a small, white, acid leavened, steamed cake made by bacterial fermentation of a thick batter prepared from rice and dehulled black gram. Similar products are made from rice in the Philippines (*puto*) and in Korea (*kichudok*). *Puto* is made using year-old rice and the batter is neutralized in the course of fermentation. In Sri Lanka *hopper* is prepared from acid fermented dough made with rice or wheat and coconut water. In hopper fermentation, a very large inoculum of baker's yeast or coconut toddy, which includes acid-producing bacteria, is added. Table 5.5 lists various types of acid fermented bread, pancakes, porridges and starch materials used in different regions (Lee, 1994).

Acid porridges prepared from cereals are eaten in various regions of the world, particularly in Africa, where porridges may represent the basic diet (Table 5.6). Nigerian *ogi*, Kenyan *uji* and Ghanaian *kenkey* are examples of porridges prepared by the acid fermentation of maize, sorghum, millet or cassava, followed by wet-milling, wet-sieving and boiling.

Acid fermentation is also used to produce food starches with extended shelf-life, resistance to infectious microorganisms and palatable flavor in different regions of the world. Nigerian *Gari*, Ethiopian *Kocho*, Chinese mungbean starch and Mexican *pozol* are important acid fermented starch ingredients used for the preparation of porridges, steamed cakes, pastes, noodles, soups and drinks (Table 5.7).

Most countries in Asia produce mungbean starch, and mungbean starch noodles are a staple of the Chinese diet. The manufacturing process for mungbean starch involves acidic bacterial fermentation. The mungbeans are hydrated by soaking in water inoculated with 12-hour steeped water from a previous fermentation to ensure acidification of the beans. The

Table 5.4 Examples of cultured milk products (McGregor, 1992).

Product	Location	Bacteria
Acidophilus	Europe, North America	*Lactobacillus acidophilus, Bifidobacterium bifidum*
Bulgarian buttermilk	Europe	*Lactobacillus bulgaricus*
Buttermilk	North America, Europe, Middle East, North Africa, Indian subcontinent, Oceania	*Lac. lactis* subsp. *cremoris, Lac. lactis* subsp. *diacetylactis, Leuconostoc cremoris*
Filmjolk	Europe	*Lactococcus lactis* subsp. *cremoris, Lac. lactis, Lac. lactis* subsp. *diacetylactis, Leuc. cremoris, Alcaligenes viscosus, Geotrichum candidum*
Flummery	Europe, South Africa	Naturally present lactic bacteria
Ghee	Indian subcontinent, Middle East, South Africa, Southeast Asia	*Streptococcus, Lactobacillus* and *Leuconostoc* sp.
Junket	Europe	*Lactococcus* and *Lactobacillus* sp.
Kefir	Middle East, Europe, North Africa	*Streptococcus, Lactobacillus* and *Leuconostoc* sp., *Candida kefyr, Kluyveromyces fragilis*
Kishk	North Africa, Middle East, Europe, Indian subcontinent, East Asia	*Streptococcus, Lactobacillus* and *Leuconostoc* sp.
Kolatchen	Middle East, Europe	*Lac. lactis, Lac. lactis* subsp. *diacetylactis, Lactococcus lactis* subsp. *cremoris, Saccharomyces cerevisiae*
Koumiss	Europe, Middle East, East Asia	*Lac. lactis, Lb. bulgaricus, Candida kefyr, Torulopsis*
Kurut	North Africa, Middle East, Indian subcontinent, East Asia	*Lactobacillus* and *Lactococcus* sp., *Saccharomyces lactis, Penicillium*
Lassi	Indian subcontinent, East Asia, Middle East, North Africa, South Africa, Europe	*S. thermophilus, Lb. bulgaricus,* sometimes yeast
Prokllada	Europe	*Streptococcus* and *Lactobacillus* sp.
Sour cream	Europe, North America, Indian subcontinent, Middle East	*Lactococcus lactis* subsp. *cremoris, Lac. lactis* subsp. *diacetylactis*
Yakult	East Asia	*Lactobacillus casei*
Yogurt	Worldwide	*S. thermophilus, Lb. bulgaricus*

principal microorganisms found in the steep-water are *Leuconostoc mesenteroides, Lactobacillus casei, Lb. cellobiosus* and *Lb. fermentum*. The lactic fermentation, which reduces the pH to about 4.0, protects the beans from spoilage and putrefaction which would otherwise occur in ground bean slurries (Steinkraus, 1983).

Thai rice-noodle, *Khanom Jeen*, is also made from acid fermented raw rice. Soaked rice is drained and fermented for at least 3 days before grinding, and *Lactobacillus* sp. and *Streptococcus* sp. are involved in the acid fermentation.

5.2.2.3 Acid fermented vegetables

Acid fermented vegetables are important sources of vitamins and minerals. *Leuconostoc mesenteroides* has been found to be important in the initiation of the fermentation of many vegetables, i.e. cabbages, beets, turnips, cauliflower, green beans, sliced green tomatoes, cucumber, olives and sugar beet silages. In vegetables, *Leuc. mesenteroides* grows more rapidly and over a wider range of temperatures and salt concentrations than any other lactic acid bacteria. *Leuconostoc mesenteroides* produces carbon dioxide and acids, which quickly lower the pH, thereby inhibiting the development of undesirable microorganisms and the activity of their enzymes, which may soften the vegetables. The carbon dioxide produced replaces air and provides anaerobic conditions favorable for the stabilization of ascorbic acid and the natural color of the vegetables. The growth of this species modifies the environment, making it favorable for the

Table 5.5 Examples of acid-leavened bread and pancakes

Product	Country	Major ingredients	Microorganisms	Usage
Sourbread	Germany	Wheat	Lactic acid bacteria Yeast	Sandwich bread
Ryebread	Denmark	Rye	Lactic acid bacteria	Sandwich bread
Idli	India Sri Lanka	Rice Black gram	*Leuc. mesenteroides* *Enterococcus faecalis*	Steamed cake
Puto	Philippines	Rice	*Leuc. mesenteroides* *E. faecalis*	Steamed cake
Kichudok	Korea	Rice	Yeast	Steamed cake
Enjera	Ethiopia	Tef or other cereals	*Leuc. mesenteroides* *P. cerevisiae* *Lb. plantarum* *Sac. cerevisiae*	Pancake
Kisra	Sudan	Sorghum Millet	*Lactobacillus* sp. *Acetobacter* sp. *Sac. cerevisiae*	Pancake
Kishk	Egypt	Wheat + milk	*Lb. casei* *Lb. brevis* *Lb. plantarum* *Sac. cerevisiae*	
Hopper	Sri Lanka	Rice + coconut water	Yeast Lactic acid bacteria	Steam-baked pancake

growth of other lactic acid bacteria. The high acidity produced by the species and other subsequent lactic acid bacteria inhibits the growth of *Leuc. mesenteroides*. *Leuconostoc mesenteroides* converts glucose to approximately 45% levorotatory D-lactic acid, 25% carbon dioxide and 25% acetic acid and ethyl alcohol. Fructose is partially reduced to mannitol and is then readily fermented to yield equimolar quantities of lactic acid and acetic acid. The combination of acids and alcohol is conducive to the formation of esters that impart desirable flavors.

Table 5.8 shows examples of acid fermented vegetables produced in different regions of the world. The difference between sauerkraut and *kimchi* is the preferred end point of fermentation. The best tasting *kimchi* is attained before overgrowth of *Lb. brevis* and *Lb. plantarum* with an optimal product pH of 4.5. The overgrowth of *Lb. brevis* and *Lb. plantarum* diminishes

Table 5.6 Examples of acid fermented cereal gruels and non-alcoholic beverages.

Product	Country	Major ingredients	Microorganisms	Usage
Ogi	Nigeria	Maize, sorghum, or millet	*Lb. plantarum* *Corynebacterium* sp. *Acetobacter* Yeast	Sour porridge Baby food Main meal
Uji	Kenya Uganda Tanzania	Maize, sorghum, millet, or cassava flour	*Leuc. mesenteroides* *Lb. plantarum*	Sour porridge Main meal
Mahewu	South Africa	Malze + wheat flour	*Lac. lactis* *Lactobacillus* sp.	Sour drink 8–10% DM
Hulumur	Sudan	Red sorghum	*Lactobacillus* sp.	Clear drink
Busa	Turkey	Rice, millet	*Lactobacillus* sp.	

Table 5.7 Examples of acid fermented starch ingredients.

Product	Country	Major ingredients	Microorganisms	Usage
Gari	Nigeria	Cassava	*Leuconostoc* *Alcaligenes* *Corynebacterium* *Lactobacillus*	Staple Cake Porridge
Mungbean starch	China Thailand Korea Japan	Mungbean	*Leuc. mesenteroides* *Lb. casei* *Lb. cellobiosus* *Lb. fermenti*	Noodle
Khanom-jeen	Thailand	Rice	*Lactobacillus* sp. *Streptococcus* sp.	Noodle
Pozol	Mexico	Maize	Lactic acid bacteria *Candida*	Porridge
Me	Vietnam	Rice	Lactic acid bacteria	Sour food ingredient

product quality, but sauerkraut production depends on these organisms. The fermentation is manipulated by the salt concentration and temperature. The optimal range of salt concentration of sauerkraut is 0.7, approximately 3.0%, while that of *kimchi* is 3.0, approximately 5.0% (Lee, 1994).

5.2.2.4 Acid fermented fish and meat

The storage life of perishable fish and meats can be extended by acid fermentation with added carbohydrates and salt. In Scandinavian countries most traditional low-salt fermented fish products are transformed into pickled products in vinegar. These products generally require low-temperature storage. On the other hand, most Asian products are lactic fermented with added cereals, as shown in Table 5.9.

Rice, either cooked or roasted, is the most frequently used carbohydrate source, but other sources such as millet in *sikhae* are also used. In some cases fruits and vegetables, for example tamarind in *Bekasam* for the reduction of pH, and garlic and pepper in *sikhae*, are added. The antimicrobial effect of garlic to some putrefactive microorganisms such as *Bacillus* in lactic fermented fish products has been demonstrated (Souane *et al.*, 1987).

Fermented sausages, salami in Europe, *nham* in Thailand and *nem-chua* in Vietnam, are also made by a process involving lactic acid bacteria. Starter cultures for salami fermentation are isolated from fermented

Table 5.8 Examples of acid fermented vegetables produced in different regions of the world.

Product	Country	Major ingredients	Microorganisms	Usage
Sauerkraut	Germany	Cabbage, salt	*Leuc. mesenteroides* *Lb. brevis*	Salad Side dish
Kimchi	Korea	Korean cabbage, radish, various vegetables, salt	*Leuc. mesenteroides* *Lb. brevis* *Lb. plantarum*	Salad Side dish
Dhamuoi	Vietnam	Cabbage, various vegetables	*Leuc. mesenteroides* *Lb. plantarum*	Salad Side dish
Dakguadong	Thailand	Mustard leaf	*Lb. plantarum*	Salad Salt side dish
Burong mustala	Philippines	Mustard	*Lb. brevis* *P. cerevisiae*	Salad Side dish

Table 5.9 Examples of acid fermented seafood and meat products (Lee, 1994).

Product	Country	Major ingredients	Microorganisms	Usage
Sikhae	Korea	Sea water fish, cooked millet, salt	*Leuc. mesenteroides* *Lb. plantarum*	Side dish
Narezushi	Japan	Sea water fish, cooked millet, salt	*Leuc. mesenteroides* *Lb. plantarum*	Side dish
Burong-isda	Philippines	Freshwater fish, rice, salt	*Lb. brevis* *Streptococcus* sp.	Side dish
Pla-ra	Thailand	Freshwater fish, salt, roasted rice	*Pediococcus* sp.	Side dish
Balao-balao	Philippines	Shrimp, rice, salt	*Leuc. mesenteroides* *P. cerevisiae*	Condiment
Kungchao	Thailand	Shrimp, salt, sweetened rice	*P. cerevisiae*	Side dish
Nham	Thailand	Pork, garlic, salt, rice	*P. cerevisiae* *Lb. plantarum* *Lb. brevis*	Pork meat in banana leaves
Sai-krok-prieo	Thailand	Pork, rice, garlic, salt	*Lb. plantarum* *Lb. salivarius* *P. pentosaccus*	Sausage
Nem-chua	Vietnam	Pork, salt, cooked rice	*Pediococcus* sp. *Lactobacillus* sp.	Sausage

fish products in Korea as well as other Asian countries (Lee, 2001).

5.2.2.5 Vinegar

Vinegar fermentation is as ancient as alcoholic fermentation, since acetic acid is produced in any natural alcoholic fermentation upon exposure to the air.

$$\text{Ethanol in fruits, wine, toddy, rice-wine} \xrightarrow{\textit{Acetobacter aceti}} \text{Acetic acid}$$

Vinegars are produced from fruits in Europe, from tropical fruits, such as coconut, sugar cane and pineapple in the Asia Pacific region and from cereals in Northeast Asia. Cereal vinegars may be divided into three classes: rice vinegar, rice-wine filter-cake vinegar and malt vinegar. The indigenous processes are natural or spontaneous fermentation brought about by the growth of *Acetobacter aceti* on alcoholic substrates under aerobic conditions. Traditionally, degraded or poor quality wines were used for the production of low-grade vinegars at the household level. Today, vinegars of high quality standards are produced by industry.

Commercial vinegar is prepared from rice-wine filter-cake in Far-Eastern countries. Filter cakes from rice-wine factories are collected and packed tightly into a storage tank for 1–2 years. The filter-cake contains large amounts of unused carbohydrate and proteins, which are further hydrolyzed by inherent microorganisms and enzymes during storage, converting them into alcohol and other nutrients and flavor substances. The cake is slurried in 2–3 volumes of water prior to filtration. The filtrate is heated to 70°C, and cooled by mixing with fresh vinegar mash to a temperature of 36–38°C, and then fermented with *Acetobacter* for 1–3 months. It is further aged at the ambient temperature for 3–6 months and filtered to obtain clear vinegar (Lee, 2001).

5.2.3 Bread fermentation

Baking, brewing and enology all depend on the ability of yeasts to carry out anaerobic fermentation of sugars, yielding CO_2 and ethanol. In brewing and wine-making, alcohol is the prime product of interest, while in baking the leavening effect of CO_2 is more important. Breads are divided into two groups, leavened loaf bread and unleavened flat bread. Traditionally leavening is due to the products of fermentation, carbon dioxide and ethanol produced by yeast. Although the CO_2 generating chemical leavening agents such as food acid and soda (sodium bicarbonate) can replace yeast, biological leavening imparts physicochemical modification of dough constituents and flavor development.

Originally sour doughs were used for the production of all type of breads because commercial baker's yeast was not available. Baker's yeast was introduced in the market at the beginning of the twentieth century. Industrially produced yeasts are strains of the top-fermenting species *Saccharomyces cerevisiae* grown on molasses in an aerobic fed-batch fermentation. The optimum temperature for the growth and fermentation of baker's yeast is between 28 and 32°C, and the optimum pH is 4–5. Leavening of dough requires the addition of 1–6% yeast based on the weight of flour. The exact percentage depends on the recipe, the process and the quality of the flour and yeast as well as the operation conditions (Spicher and Brummer, 1995).

The production of baked goods consists of preparation of raw materials, dough formation (kneading, maturing), dough processing (fermentation and leavening, dividing, molding and shaping), baking in the oven and final preparation (slicing, packaging, etc.). Bread dough is fermented for a sufficiently long time to permit the yeast to act on the assimilable carbohydrates and convert them into alcohol and carbon dioxide as the principal end product. By the end of the proofing period the aqueous phase of the bread is saturated with CO_2 and the volume has roughly doubled owing to the pressure of CO_2 that has diffused to air cells. At the beginning of baking the loaf further expands (oven spring) by the expansion of air and steam during heating, and then at some temperature the matrix sets, expansion stops, and starch gelatinization, crust coloring and flavor development take place. The magnitude of oven spring depends on two factors: (1) the generation and expansion of gases, and (2) the amount of time available for loaf expansion before the structure sets. The first factor is primarily a function of yeast fermentation; the second factor is affected by dough components such as shortening, surfactants, gluten protein and flour lipids (Stauffer, 1992).

5.2.4 Amino acid/peptide fermentation

Fermented protein foods are used mainly for flavor-enhancing condiments and gourmet food ingredients due to the meaty and appetite-stimulating flavor of protein hydrolysate, which is formed during the fermentation. The type of indigenous fermented protein food is decided primarily by the availability of the raw material in the specific climatic and geographical conditions. Cheese is made in the Middle East and Europe where animals are the main food source. Fermented soybean products, for example soybean sauce and paste, are used in Northeastern Asian countries, and fermented fish products in the Asia-Pacific region.

5.2.4.1 Cheese

The worldwide number of cheese varieties has been estimated at 500, and there are several methods of classification. Cheeses can be divided by their texture: very hard (Parmesan, Romano), hard (Cheddar, Swiss), semisoft (Brick, Muenster, blue, Harvarti), soft (Brie, Camembert, feta) and acid (cottage, cream, Ricotta). A broad look at cheeses might divide them into two large categories, ripened and fresh. More technical classifications are also used, for example, those based on coagulating agent: rennet cheese (Cheddar, Brick, Muenster), acid cheese (cottage, Quarg, cream), heat-acid (Ricotta, Sapsago) and concentration-crystallization (Mysost) (Nuath *et al.*, 1992).

Cheese is manufactured by coagulating or curdling milk, stirring and heating the curd, draining off the whey, and collecting or pressing the curd. Characteristic flavor and texture are formed during the ripening of cheese depending on the type of starter culture and microorganisms involved as well as the coagulating agent and salting methods. Depending on the variety, the milk is pasteurized (generally at about 72°C for 16 sec), and a bacterial starter culture is added to the milk, which is at 30–36°C. The inoculated milk is generally ripened at the temperature for 30–60 min to allow the lactic acid bacteria to multiply sufficiently for their enzyme system to convert lactose to lactic acid. After ripening a milk-coagulating agent is added. For blue cheese, the mold (*Penicillium* sp.) is added to the starting milk or to the drained curds.

The starter cultures are organisms that ferment lactose to lactic acid and other products. These include *Streptococci, Leuconostocs, Lactobacilli* and *Streptococcus thermophilus*. Starter cultures also include *Propionibacteria*, brevibacteria and mold species of *Penicillium*. The latter organisms are used in conjunction with lactic acid bacteria for a particular characteristic of cheese, for example, the holes in Swiss cheese are due to *Propionibacteria*, and the yellowish color and typical flavor of brick cheese is due to *Brevibacterium linens*.

Coagulation of milk is essential to cheese making. Most proteolytic enzymes can cause milk to coagulate. Rennet (chymosin, EC 3.4.23.4) is widely used for milk coagulation for cheese making. However, due to the shortage of calf's chymosin, commercial rennet may include blends of chymosin and pepsin extracted from the stomach of other animals such as pig. Microbial rennets with similar functionality are also prepared from *Mucor miehei*, *Mucor pusillus* and *Endothia parasiticus*. Proteases from plants are known to coagulate milk but are not used in commercial cheese making.

5.2.4.2 Fish sauce and paste

Fish fermentation is an old technology used for the preservation of freshwater and marine animals, which are highly perishable and localized in production and seasonally fluctuating in catch (Ruddle, 1993). The technology appears to have evolved with the availability of salt and a non-pastoral way of life. There is a strong correlation throughout the world between the use of fermented fish products and the use of cereals, especially rice, and vegetables (Ishige, 1993). Although the use of fermented fish products is nowadays mainly confined to East and Southeast Asia, traces of this technology can be found throughout old human civilizations.

Aging salted (cured) fish in a container or earthen jar for longer period produces fish sauce. The enzymes in the gut and from the halophilic microorganisms grown in the system decompose fish meats, and the exuded liquid (protein hydrolysate) is fish sauce. The hydrolysate mainly consists of amino acids and peptides, which form the characteristic meaty flavor of fish sauce. In the case of Korean *Jeotkal* fermentation containing 20% salt, the total number of viable cells increases for the first 40 days, mainly attributed to the growth of *Pediococcus* and *Halobacterium*. The concentrations of soluble-N and amino-N increase steadily during the first 60 days and it coincides with the development of optimum taste. The volatile basic-N content increases in two steps, and the second step increase causes the taste deterioration, which is re-lated to the maximum growth of yeast (Lee *et al.*, 1993).

Depending on the amount of salt added, the products are classified as high-salt (> 20% salt of total weight), low-salt (6–18% salt) and no-salt products, as shown in Fig. 5.2. When the salt concentration is higher than 20% of total weight, pathogenic and putrifactive microorganisms cannot grow and the product does not need other preservative means. The first criterion for the subdivision of this group is the degree of hydrolysis, which is influenced by fermentation time and temperature, added enzyme sources and water content. The fully hydrolyzed liquid is fish sauce. The name cured fish is confined to represent the partially hydrolyzed fish products which retain the original shape of fish immersed in the exuded liquid, and this form as such is frequently used as a side dish for rice meals. Fish paste is characterized by the salted fish being partially dried in order to restrict the degree of hydrolysis and comminuted to produce the homogeneous, solid condiment. Each class can be further subdivided by the kind of raw materials, such as fish species and portion of fish, and thus there are numerous kinds of products (Lee, 1989).

Many Asian countries produce salt cured and dried fish products, for example, *plakem* in Thailand, *jambalroti* in Indonesia, Maldive fish in Sri Lanka and *gulbi* in Korea, but the role of fermentation in these products is not fully understood. Fish fermentation without added salt is not a common practice. In some local specialties, half-spoiled fish or alkaline fermentation in leafy plant ash is used. The propagation of mold in dried bonito (katsuobushi) processing in Japan is another example of non-salt fish fermentation .

Most countries in East and Southeast Asia have fish sauce, but the flavor, physical properties and raw materials used vary. Depending on the degree of hydrolysis or fermentation time and the separation method, two types of sauce, namely, clear and turbid, are produced. *Ngan-pya-ye, nuoc-man, nampla, shottsuru* and *yu-lu* are clear fish sauces, while *budu, patis, ketjap-ikan* and *jeot-kuk* are turbid. Some turbid sauces are

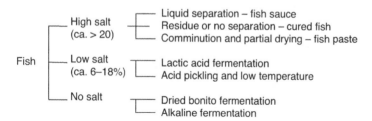

Figure 5.2 Classification of fermented fish products.

obtained from the exuded liquid of cured fish, for example, *patis* from *bagoong* production in the Philippines and *jeotkuk* from *jeotkal* production in Korea. In Northeast Asia cured fish products are more important than fish pastes. Fish pastes, especially those made from shrimp and planktonic animals such as *Seinsa ngapy, belacan, trassi, prahoc* and *kapi*, are important in Southeast Asian diets.

5.2.4.3 Fermented soybean products

At the early stage of soybean utilization the Northeastern *Dong-yi* probably first invented *shi*, the old Chinese term for Korean *meju*, by keeping cooked soybeans in a pottery jar. Cooked soybeans grown with mold and bacteria, which is called *meju*, are immersed in brine to leach out the protein hydrolysate, and the liquid part is soybean sauce (*kanjang*) and the residue is soybean paste (*doenjang*).

The traditional fermented soybean products are divided into three groups based on the type of fermentation starter used: *Shi* made from loose type soybean *meju, Maljang* from cake type soybean *meju* and *Jang* from soybean mixed with other cereals. The propa-

gation of these products in the Northeastern region, namely China, Korea and Japan, and their variations are shown in Fig. 5.3. According to S.W. Lee (1990), fermented soybean products were first introduced to China in the first century BC and to Japan in the sixth century AD. Varieties of products have been developed and have disappeared throughout history.

Korean kanjang and doenjang

Meju is prepared from cooked soybean. Soybeans are soaked in water overnight, cooked for 2–3 hours and mashed by pounding. The mash is then shaped like a brick or a ball, dried in the sun and kept in a stack covered during the night for several days. During this period, the surface is grown with mold, especially *Aspergillus oryzae*, and the inside with bacteria, typically *Bacillus subtilis*. The enzymes from mold and bacteria hydrolyze the soybean proteins into amino acids, and the carbohydrates into sugars and organic acids. The amino acids and sugars interact with each other during browning reaction, resulting in the characteristic dark brown color and meaty flavor. Properly fermented *meju* is immersed in brine in an earthen jar and ripened for several months. The brown color and

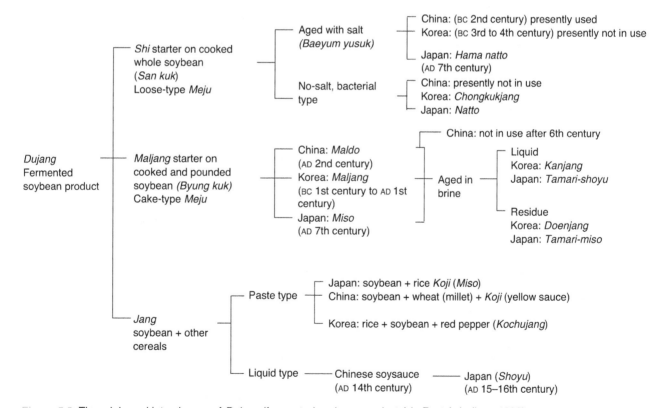

Figure 5.3 The origin and interchange of *Dujang* (fermented soybean products) in East Asia (Lee, 1990).

meaty flavor leach out into the brine. During this period, salt-tolerant yeasts grow in the mash, especially *Saccharomyces rouxii*, which produces the aroma of soy sauce. The liquid part is soy sauce and the precipitates are soybean paste. Soy sauce so produced is boiled once and stored in an earthen jar for years. The flavor of soy sauce gets richer as the storage time increases, just as the flavor of wine becomes smoother as it get old. It has been said in Korea that the taste of food in a household is decided by the taste of the household's fermented soybean products.

Japanese *shoyu* and *miso*

Japanese people modified the *meju* preparation method in the early twentieth century by controlled fermentation technology using a pure culture of mold isolated from the traditional starter (Shettleff and Aoyaki, 1976). The mold, normally *Aspergillus oryzae*, is grown on cooked rice or cooked wheat grits to make *koji*. It is mixed with cooked soybean for further fermentation, and then ripened in the brine. Soybean paste (*miso*) and soy sauce (*shoyu*) are made separately; for *shoyu*, *koji* is made with cooked defatted soybean flake and wheat grits and then mixed in brine for aging. After 4–6 months of aging, it is filtered to obtain shoyu, the liquid part, and the solid part is discarded. Miso is prepared by using koji made from cooked rice or other cereals mixed with cooked soybean and salt, and then mashed into a paste and ripened. These processes are easy for industrialization of the products. The flavor of Japanese *shoyu* and *miso* is mild and sweet compared to Korean counterparts. Korean people prefer the strong flavor of traditional soy sauce and soybean paste, the same as European people distinguish Roquefort from processed Cheddar cheese.

Korean *chongkukjang* and Japanese *natto*

Soybean is cooked and covered with straw mat or cloth, and placed on the warm stone floor, *ondol*, for 3–4 days until the mucous string is formed. It is mixed with chopped ginger, chopped garlic and salt, and pounded slightly until the bean kernels are separated into halves, and then stored in an earthen jar. The strong smell of fermented soybean is partially masked by the ginger and garlic smell, and creates the characteristic *chongkukjang* flavor. The spicy seasoning is thus prepared in 3–4 days, while ordinary soybean paste, *doenjang*, which uses *meju* as fermentation starter, takes over 6 months for complete ripening. In this respect, chongkukjang is a rapid fermentation method. The mucous substance in *chongkukjang* is peptido-polysaccharide produced by *Bacillus subtilis*.

Japanese *natto* is a modified form of *chongkukjang*. *Natto* is fermented soybean grown with *Bacillus subtilis* on cooked soybean. The fermented soybean with mucous string is consumed directly without further processing, so it is a non-salt fermented product. However, natto is not generally accepted by Korean people. It is always mixed with spices and used for the cooking of vegetable stew as a meaty flavored condiment. The amount of *chongkukjang* added to the stew is large enough to supplement protein in the diet significantly. *Chongkukjang* was also called *Jeonkukjang* in the old days. 'Chongkuk' means the Chinese kingdom 'Qing', while 'Jeonkuk' means 'a country at war' or 'a combat zone'. What all these names imply is that this product was made in extraordinary situations, for example during war time or famine conditions, for the urgent supply of a nutritious savory food ingredient.

Korean *kochujang*

The basic tastes of European people are sweet, sour, bitter and salty and Japanese people add umami, the meaty taste. Korean people add another: hot or pungent taste. The most remarkable difference of Korean food compared to food of neighboring Japan and China is the strong pungent taste of red pepper in most Korean dishes.

Kochujang is a unique hot bean paste seasoning popular in Korea. It is made from fermented soybean starter, *meju* and malt made from barley. Malt powder is mixed with cereal porridge made from rice, glutinous rice or barley. The enzymes in malt hydrolyze the starch into sugars and reduce the consistency of the mixture. *Meju* powder, red pepper powder and salt are added to the partly saccharified porridge with thorough mixing to make a paste, and put in an earthen jar. The top is covered by salt in order to prevent mold growth. The jar is placed in a sunny place for further fermentation. The proteins in soybean and cereals degrade into amino acids to produce the meaty flavor. During fermentation a wonderful harmony of the meaty flavor from hydrolyzed proteins and the sweet taste of hydrolyzed starches with the pungent taste of red pepper and salty taste is achieved, and a new characteristic flavor stimulating the appetite of Koreans is formed.

Tempe

Tempe is found in all parts of Indonesia but is particularly important in Java and Bali. It is also produced in some Malaysian villages and in Singapore. *Tempe* is a white, mold-covered cake produced by fungal fermentation of dehulled, soaked in water and partially cooked soybean cotyledons (Steinkraus, 1983). It is packed in wilted banana leaves and sold in the market. Essential steps in the preparation of tempe include cleaning the beans, soaking in water, dehulling and partial cooking of the dehulled beans. Dehulling is important for the growth of mold on the surface of cotyledons. Soybean is not necessarily cooked fully, because subsequent mold growth is able to softening the texture. Under the natural conditions in the tropics, *tempe* production involves two distinct fermentations: bacterial acidification of the beans during soaking and fungal overgrowth of the cooked bean cotyledon by the mold mycelium. A previous batch of sporulated *tempe* or sun-dried pulverized *tempe* powder (1–3 g) is sprinkled over the cooked and drained soybean cotyledon (1 kg) and thoroughly mixed to distribute the mold spores over the surface of all the beans. *Rhizopus oligosporus* is known as *tempe* mold, and the pure culture of strain NRRL2710 or CBS 338.62 can be used for the inoculum.

A handful of inoculated beans are placed on wilted banana leaves or other large leaves and packed. The leaf keeps the soybean cotyledon moist during the fermentation and allows for gaseous exchange. Incubation can be at temperatures from 25 to 37°C. The higher the incubation temperature, the more rapidly the *tempe* molds will grow. For example, 80 hours of incubation is required at 25°C, 26 hours at 28°C and 22 hours at 37°C. The *tempe* should be harvested as soon as the bean cotyledons have been completely overgrown and knitted into a compact cake. The cotyledons should be soft and pasty (not rubbery), and the pH should have risen to about 6.5.

Tempe should be consumed immediately after harvest. It can be stored for 1 or 2 days without refrigeration. If the *tempe* is not going to be consumed immediately, it should be deep-fried, in which form it remains stable for a considerable time, or it should be blanched by steaming and refrigerated. It can be stored after dehydration, either by sun-drying or by hot-air drying and keeping in plastic bags. Subsequent keeping quality is excellent because *tempe* contains a strong antioxidant produced by the mold and is resistant to the development of rancidity. *Tempe* is consumed fresh or in deep-fried form.

Chinese sufu

Chinese *sufu* (*tosufu*, *toufuru*, *fuyu* or *tauhuyi*) is a highly flavored, creamy bean paste made by overgrowing soybean curd with a mold belonging to the genus *Actinomucor*, *Rhizopus* or *Mucor* and fermenting the curd in a salt brine/rice wine mixture (Lee and Lee, 2002). In the West, *sufu* has been referred to as Chinese cheese. *Sufu* is usually sold in red or white blocks 2–4 cm^2 by 1–2 cm thick, and the white *sufu* is untreated, while the red variety is colored with Chinese red rice, *hung chu*. The procedure for making *sufu* consists of five steps: preparation of soybean curd (tofu), preparation of molded *tofu* (*pehtze*), salting, fermenting in salt brine/rice wine, and processing and packaging.

Soybeans are cleaned, soaked in water and ground to make soybean milk slurry. The slurry is heated to boiling and filtered through cloth, and the residue is discarded. To the filtrated soymilk are added coagulants (calcium chloride/calcium sulfate mixture or sea salt brine) to make soybean curd. The amount of coagulant used to produce *tofu* for sufu manufacture is 20% higher than that used for regular *tofu*. Moreover, after the coagulates are mixed with soybean milk, the mixture needs to be agitated vigorously in order to break the coagulated protein into small pieces, after which it is set aside for 10 min to complete the process of coagulation. This process reduces the water content of the curd and makes the texture harder. If the water content is more than 60%, the inoculation of fungi is deferred until the water remaining on the curd surface is reduced by drying.

Pehtze is the soybean curd overgrown with the grayish hair-like mycelium of molds belonging to genera *Actinomucor*, *Rhizopus* or *Mucor*. These fungi are normal contaminants in rice straw. Traditionally inoculation was performed by placing the *tofu* on the rice straw, but this method does not always yield a high quality because of undesirable contaminating microorganisms. In spring or autumn when the ambient temperature is 10–20°C, white fungal mycelium is visible on the surface of the cubes after 3–7 days, at which point the cubes are taken out and immediately salted in large earthenware jars. Each layer of *pehtze* is sprinkled with a layer of salt, and after 3–4 days when the salt is absorbed, the *pehtze* is removed, washed with water and put into another jar for processing.

For processing, a dressing mixture, which varies for each type of *sufu*, is placed in the jar. To make red *sufu*, *anka koji* and soy mash are added; to make fermented rice (*tsao*) *sufu*, fermented rice mash is added; to make Kwantung *sufu*, red pepper and anise are added in addition to salt and red *koji*. Alternate layers of *pehtze* and dressing mixture are packed into the jar until it is filled to 80% of its volume, then a brine with a concentration of approximately 20% NaCl is added. Finally, the mouth of the jar is wrapped with the sheath leaves of bamboo shoots and sealed with clay. After 3–6 months fermenting and aging, the *sufu* is ready for consumption (Steinkraus, 1983).

5.2.5 Other fermented products

5.2.5.1 Chinese red rice (Anka)

Anka, also known as *ang-kak*, *beni-koji* and red rice, is used in China, Taiwan, the Philippines, Thailand and Indonesia to color foods which include fish, rice wine, red soybean cheese, pickled vegetables and salted meats (Lee and Lee, 2002). It is a product of fermentation of rice with various strains of *Monascus purpureus* Went. A number of countries are gradually adopting this natural pigment to replace coal-tar dyes, as the latter have been implicated as carcinogens. The advantages of using *Anka* are that its raw materials are readily available, the yield is good, the color of pigment produced is consistent and stable, the pigment is water soluble, and there is no evidence of any toxicity or carcinogenicity.

Anka is produced at an industrial scale in Taiwan. Non-glutinous rice (1450 kg) is washed and steamed for 60 min. Water (1.8 hL) is sprayed on the rice, which is again steamed for 30 min. The steamed rice is mixed with 32 L of *chu chong tsaw*, a special variety of red rice inoculum in Taiwan, after cooling to 36°C, and heaped in a bamboo chamber. When the temperature of the rice rises to 42°C, it is spread on plates and shelved. *Anka* is produced by moistening the rice three times during incubation, followed by final drying; 700 kg of *Anka* is produced from 1450 kg of rice.

An uncommon phenomenon of the mold *Monascus purpureus* is the exudation of granular fluid through the tips of the hyphae. When the culture is still young, the freshly extruded fluid is colorless, but gradually changes to reddish-yellow and purple-red. The production of the red coloring matter is observed not only on the exuded granular fluid but also on the hyphal content within. The red coloring matter diffuses throughout the substrate. The dark red color consists of two pigments, the red monascorubrin ($C_{22}H_{24}O_5$) and the yellow monascoflavin ($C_{17}H_{22}O_4$). The strains of *Monascus purpureus* that are adopted in the production of red rice are only those that are capable of impregnating rice with a dark red color in the presence of water concentrations sufficiently low that no distortion is produced in the hydrated grains.

5.3 Enzyme technology

The application of enzymes in food processing is an important branch of biotechnology. Enzymes are the proteins that catalyze virtually all chemical reactions occurring in biological systems, and thousands of enzymes have been identified and characterized. Some of importance to the food industry include the production of high fructose corn syrups by using glucose isomerase, saccharification of starch by amylases in baking and brewing, juice clarification by using cellulases and pectinases, production of low lactose milk by using lactase, cheese making by rennin, and meat tenderization by proteases such as papain, bromelain and ficin. The cell wall degrading enzymes (hemicellulase and cellulase) are used to extract vegetable oil (olive and canola/rape seed) in an aqueous process by liquefying the structural cell wall components of the oil-containing crop. Table 5.10 summarizes some important food enzymes, their source, reaction specificity and applications in food processing.

In this chapter, enzymatic starch modification and protein hydrolysis will be discussed in detail.

5.3.1 Enzymatic starch modification

The major steps in the conversion of starch are liquefaction, saccharification and isomerization. During liquefaction the α-1,4 linkages of amylose and amylopectin are hydrolyzed at random by endo-α-amylase. This reduces the viscosity of the gelatinized starch and increases the dextrose equivalent (DE), a measure of the degree of hydrolysis of the starch. For saccharification to dextrose, a DE of 8–12 is commonly used, and the maximum DE obtainable is about 40 (Olsen, 1995).

β-Amylases are exo-enzymes which attack amylase chains, resulting in production of maltose from the non-reducing end. In the case of amylopectin the cleavage stops 2–3 glucose units from the α-1,6 branch points. Isoamylase and pullulanase hydrolyze α-1,6 glucosidic bonds of starch. When amylopectin is treated with pullulanase, linear amylose fragments are obtained.

Table 5.10 Some important food enzymes and their usage.

Name	Source	Action mode	Applications
α-Amylase	Malt, *Aspergillus*, *Bacillus* spp.	α-1,4 glycosidic linkage of amylase, amylopectin	Starch modification Brewing aid Reduce dough viscosity Prevent staling
β-Amylase	Malt, molds, bacteria	Split off β-maltose from non-reducing end of starch	Production of maltose syrup Brewing and baking aid
Glucoamylase	*Aspergillus*, *Rhizopus* spp.	Stepwise hydrolysis of α-1,4 linkages in starch	Production of glucose Analysis of starch content in food
Glucose isomerase	*Streptomyces*, *Actinoplanes*, *Bacillus* spp.	Conversion of glucose to fructose	Production of high-fructose corn syrup Immobilized form is used
Pullulanase	*Klebsiella pneumoniae*	α-1,6 bond of amylopectin	Production of maltose and malto-trios Removal of limit dextrins to produce high alcohol beer
Invertase (β-fructo-franosidase)	*Saccharomyces*, *Candida* spp.	Hydrolyze sucrose to glucose and fructose	Invert sugar syrup Chocolate coated sucrose candy Recovery of scrap candy Artificial honey Humectants
β-Glucanase	*Bacillus subtilis*, *Asp. niger*	β-1,3 or β-1,4 bonds of β-D-glucans	Solubilize barley gums in brewing Reduce viscosity of wort
Cellulase	*Trichoderma reesei*	Endo-cellulases to split β-1,4 linkage, exo-cellulases, cellobiases	Turn cellulosic waste into glucose to make ethanol Hydrolyze cellulose into β-dextrins and glucose
Pectinases (PG, PL, PE)	*Aspergillus* spp.	Split glycosidic bonds of pectin, endo/exo	Extraction and clarification of fruit juice
Lactase (β-galactosidase)	*Kluyveromyces marxianus*, *Asp. niger*	Hydrolyze lactose into glucose and galactose	Low-lactose dairy products Prevents crystallization of lactose in milk concentrate
Rennet (chymosin, pepsin)	Calf stomach, *Endothia parasitica*, *Mucor meihei*	Catalyze k-casein, destabilize casein micelle	Milk clotting in cheese making
Proteases	Plants (papain, ficin) Animal (trypsin) *Aspergillus niger*, *Bacillus* spp.	Hydrolyze peptide bond esterase activity	Meat tenderizer Chill-proof beer Recovery of scrap meat and fish Gluten modifier Decolorization of red blood cells
Lipases	*Mucor, Rhizopus*, *Aspergillus* spp.	Hydrolyze ester linkages of triglycerides	Accelerate cheese ripening Cheese flavor production

Maltodextrin (DE 15–25) produced from liquefied starch is commercially valuable for its rheological properties. Maltodextrins are used in the food industry as fillers, stabilizers, thickeners, pastes and glues. When saccharified by further hydrolysis using amyloglucosidase or fungal α-amylase, a variety of sweeteners can be produced having DE in the range 40–45 (maltose, 50–55 (high maltose) and 55–70 (high conversion syrup)).

5.3.2 Enzymic modification of proteins

The enzymic modification of proteins is an attractive means of obtaining better functional and nutritional properties of food proteins. The conversion of milk to cheese is an effect of the action of the protease of the microorganisms inhabiting the system. Enzymatic hydrolysis of milk proteins is used to produce non- and low allergic cow's milk products for baby food and highly digestible protein foods for hospital patients.

Protein structure is modified to improve solubility, emulsification and foaming properties, gelation and textural properties. Enzymatic processes provide several advantages compared to chemical treatments, including fast reaction rates in mild conditions with high specificity.

Proteases are classified according to their source of origin (animal, plant or microbial), their catalytic action (endo-peptidase or exo-peptidase) and the nature of the catalytic site. Based on a comparison of active sites, catalytic residues and three-dimensional structures, four major protease families have been recognized to date: the serine, the thiol, the aspartic and the metallo-proteases. The serin protease family contains two subgroups: the chymotrypsin-like and the subtilisin-like proteases. Many industrially important proteases are mixtures of the different types of proteases. This is especially the case for pancreatin, papain (crude), and some proteases from *Bacillus amyloliquefaciens*, *Aspergillus oryzae*, *Streptomyces* and *Penicillium duponti* (Olsen, 1995)

The degree of hydrolysis (DH) of enzyme-treated proteins determines the properties of relevance to food application. DH is measured by the pH-stat technique (Adler-Nissen, 1986). It is based on the principle that pH is kept constant during hydrolysis by means of automatic titration with a base, when the hydrolyzes are carried out under neutral to alkaline conditions. DH is calculated on the basis of the titration equations as follows:

$$DH = (h/h_{tot}) \times 100\%$$
$$DH = B \times Nb \times 1/a \times 1/MP \times 1/h_{tot} \times 100\%$$

where

B	is base consumption,
Nb	is the normality of the base,
a	is the average of the dissociation of the α-NH$_2$ groups,
MP	is the mass of protein,
h	is the equivalent peptide bonds cleaved per kilogram of protein (or milliequivalents per gram of protein),
h_{tot}	is the total number of peptide bonds in the protein substrate.

Proteases catalyze the hydrolytic degradation of the peptide chain. When a protease acts on a protein substrate, the catalytic reaction consists of three consecutive reactions:

1 formation of the Michaelis complex between the original peptide chain (the substrate) and the enzyme;
2 cleavage of the peptide bond to titrate one of the two resulting peptides;
3 a nucleophilic attack on the remains of the complex to split off the other peptide and to reconstitute the free enzyme.

The rate-determining step is the acylation step characterized by the reaction velocity constant k_{+2}.

Enzymatic hydrolysis of milk proteins and soybean proteins often produces bitter peptides (Kim *et al.*, 2003). The unpleasant bitterness can be managed by proper selection of the reaction parameters and the enzymes used. Bitterness is a complex problem, which can be influenced by many variables, for example, the hydrophobicity of the substrate, since the amino acid side chains containing hydrophobic groups become exposed due to the hydrolysis. DH is closely related to the bitterness of the protein hydrolysates.

5.3.3 Enzyme reaction kinetics

The activity of an enzyme is determined by many factors, including enzyme, substrate, and cofactor concentrations, ionic strength, pH and temperature. For conversion of substrate (S) to product (P) by an enzyme (E), the reaction scheme can be simply represented as:

$$E + S \xrightarrow{\ ks\ } ES \xrightarrow{\ kcat\ } E + P$$

The reaction velocity (V) is then given by the Michaelis–Menten equation:

$$V = \frac{k_{cat}[E][S]}{K_m + [S]}$$

where K_m is the substrate concentration at which V equals one-half the maximum velocity (V_{max}), as shown in Fig. 5.4.

Integration of this equation with respect to time gives:

$$V_{max} = K_m \ln\left(\frac{[S]}{[St]}\right) + ([S] - [S_t])$$

where S and S_t are the substrate concentrations at zero time and time t, respectively. This equation is particularly useful in industrial situations where a reaction is allowed to proceed to near completion or equilibrium. K_m and V_{max} can be determined by linearizing the above equation as follows, resulting in a Lineweaver–Burk plot (Karel and Lund, 2003):

$$\frac{1}{V} = \frac{1}{V_{max}} + \frac{K_m}{V_{max}[S]}$$

5.4 Modern biotechnology

The development of traditional fermentation technology and enzyme technology paved the way for modern biotechnology to achieve industrial mass production of amino acids, nucleic acids, organic acids and antibiotics in the twentieth century. The production of fine chemicals by fermentation methods has many advantages over chemical synthetic processes. It requires milder, safer and environmentally friendly reaction conditions, and provides higher productivity and wider varieties of physiologically active substances. Modern biotechnology was initially applied for the mass production of glutamate, the raw material of flavor enhancer, monosodium glutamate (MSG), in the 1950s in Japan, followed by the commercial production of other amino acids, nucleic acid-related substances and antibiotics. The modern fermentation process is carried out aseptically in the closed fermenter to form useful substances using microbial strains, and converts these to the higher-value-added chemicals either by enzymatic or chemical modification by necessity (Fig. 5.5). This so-called hybrid technology, combined biotechnology between fermentation and chemical/enzymatic modification, is recognized in both the food and fine chemical industries (Lim, 1999).

5.4.1 Amino acid production

Amino acids have many useful functions not only as nutrients but also as preventive pharmaceuticals. The world market for natural L-amino acids is led by glutamate (MSG), lysine, phenylalanine, methionine and glycine. MSG and feed additive amino acids (lys, met, thr, trp) account for 98% of the market. Phenylalanine is an important raw material for the production of aspartam, the synthetic sweetener.

Glutamate is produced from glucose (industrially raw sugar) through the Embden Meyerhoff Pathway (EMP) and the TCA cycle in the cells of *Corynebacterium glutamicum* or *Brevibacterium lactofermentum* (or *flavum* or *thiogenitalis*), as shown in Fig. 5.6.

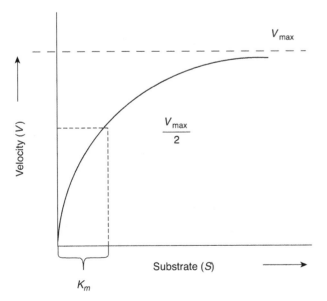

Figure 5.4 Enzyme-catalyzed reaction.

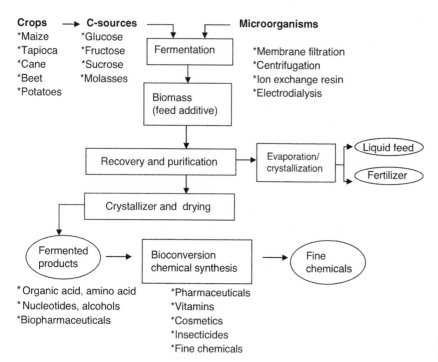

Figure 5.5 Schematic flow diagram of the fermentation bioindustry.

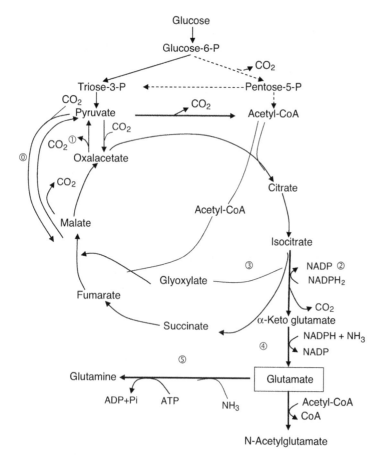

Figure 5.6 Biosynthetic pathways of glutamate production.

In order to convert glucose into amino acid nitrogen source (NH_4^+), energy (NADP) and oxygen are needed:

$$C_{12}H_{12}O_6 + NH_3 + O_2 \rightarrow C_5H_9O_4N + CO_2 + 3H_2O$$

One mole glucose produces 1 mol glutamate, but in the large fermenter the yield of glutamate is 60–67%.

MSG production involves fermentation, recovery, purification, crystallization, drying and packaging (Fig. 5.7).

Lysine is produced by *Corynebacterium glutamicum* or *Brevibacterium flavum* and their mutants. Aromatic amino acids, mainly phenylalanine, are produced by *Escherichia coli*, *Bacillus subtilis* and *B. flavum*.

5.4.2 Nucleic acid production

Nucleic acid is polynucleotide composed of a pentose (ribose or deoxyribose), phosphorus and a base unit. Base units are purines (adenine, guanine and hypoxanthine) and pyrimidines (thiamine, cytosine and uracil). Among the nucleic acid related substances, 5′-IMP (inosine monophosphate) and 5′-GMP (guanine monophosphate) are important for their flavor-enhancing properties, especially for their synergenic effect with MSG. Nucleic acids for flavor enhancement can be produced in many different ways. RNA is extracted from yeast cell mass and decomposed either chemically or by enzymatic method and then by deamination or phosphorylation to produce IMP, GMP and AMP. Another method is to produce inosine and guanine from carbohydrate substrate by *Bacillus subtilis* and then phosphorylation to make IMP and GMP. Direct fermentation process involves *Brevibacterium aminogenesis* fermentation on carbohydrate sources. Figure 5.8 shows the different methods of industrial nucleic acid production.

5.4.3 Organic acids production

Organic acids are also produced by industrial fermentation process or chemical synthetic methods. Over 70 different organic acids are produced by

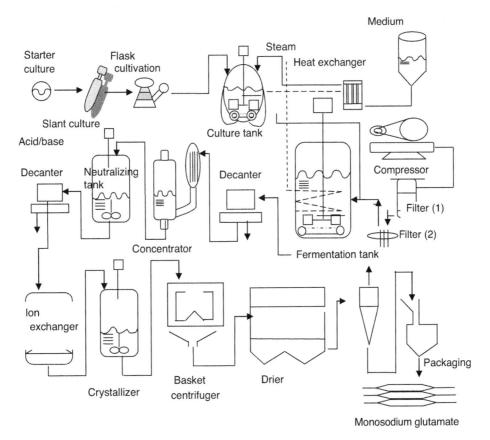

Figure 5.7 Flow diagram of the industrial production of glutamate.

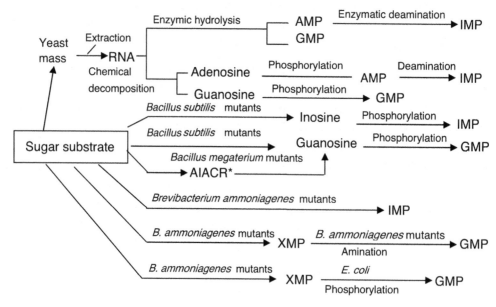

*5-amino-4-imidazole-carboxamide riboside

Figure 5.8 Production methods of flavor enhancer nucleic acids.

fermentation process. Table 5.11 summarizes the microbial strains used for the production of organic acids and their production yield from carbohydrate source.

Enzyme technology is a field aiming at production and improvement of enzymes of interest to the food industry. Modern biotechnology provides the means to select useful enzymes efficiently from conventional fermentation processes.

By using PCR techniques the newly developed enzymes are easily identified. The physiological functions of novel food ingredients are screened and confirmed rapidly by cell culture techniques.

Table 5.11 Microbial strains and yields of industrial organic acid production.

Acids	Microorganisms	Yield (%) (C source)
Acetic acid	Acetobacter aceti	95 (Ethanol)
Propionic acid	Propionibacterium shermanii	69 (Glucose)
Pyruvic acid	Pseudomonas aeruginosa	50 (Glucose)
Lactic acid	Lactobacillus delbrueckii	90 (Glucose)
Succinic acid	Cytophaga succinicans	57 (Malic acid)
Tartaric acid	Gluconobacter suboxydans	27 (Glucose)
Fumaric acid	Rhizopus delemar	58 (Glucose)
Malic acid	Lactobacillus brevis	100 (Glucose)
Itaconic acid	Aspergillus terreus	60 (Glucose)
α-Ketoglutaric acid	Candida hydrocarbofumarica	84 (N-paraffin)
Citric acid	Aspergillus niger	85 (Glucose)
	Candida lipolytica	140 (N-paraffin)
L(+)-Isocitric acid	Candida brumptii	28 (Glucose)
L(−)-Alloisocitric acid	Penicillium purpurogenum	40 (Glucose)
Gluconic acid	Aspergillus niger	95 (Glucose)
2-Ketogluconic acid	Pseudomonas fluorescens	90 (Glucose)
D-Araboascorbic acid	Penicillium notatum	45 (Glucose)
Kojic acid	Aspergillus oryzae	50 (Glucose)

5.5 Genetic engineering

Modern biotechnology is represented by the production of genetically modified organisms and their use in bioindustry. Since GM foods were first introduced in the 1980s, a quiet revolution in the food supply system has been going on. In 2001 46% of the world's soybean cultivated land and 7% of the world's corn fields were sown with transgenic crops (International Service for the Acquisition of Agri-biotech Applications, 2002). Among the 150 kinds of microbial enzymes in use in food production, over 40 food enzymes are now produced from GM microorganisms.

5.5.1 DNA transcription

Each protein is coded for by a piece of deoxyribonucleotide (DNA), commonly known as a gene. In most instances, DNA is located in the chromosome, although in some bacteria important DNA may be found on extrachromosomal elements called plasmids. In plants, mitochondrial and chloroplast DNA as well as nuclear DNA are important. DNA is composed of linear chains of nucleotide bases; adenine pairs with thymine (A-T) and guanine with cytosine (G-C). In DNA double helix, two strands of nucleotides twist round each other and the strands are linked by bonds between bases in each strand composed of alternating sugar and phosphate sections.

DNA copying is made by breaking the bonds between the bases on each strand and picking up to the free bases new nucleotides from a pool of such molecules provided by the cell. This process occurs with both parent strands and thus two identical DNA molecules are created from one original. The specific paring of the bases guarantees faithful reproduction. The strands of DNA separate and the messenger RNA molecule is built up according to the instructions contained in the DNA strand. As in DNA, the C and G bases pair together, but in mRNA a different base, uracil (U), replaces T as the partner of adenine (A). When the mRNA molecule is complete, it peels off the DNA template and moves to the protein assembly unit 'ribosome'. Ribosomes are built up from several sorts of protein plus a form of RNA called ribosomal RNA. Once the ribosomes have grasped the mRNA, the third type of RNA, transfer RNA (tRNA), comes into action.

There are many sorts of tRNA and each is able to recognize certain codons on the mRNA. Furthermore, each type of tRNA carries with it just one specific type of amino acid. The translation of genetic code depends on the fact that one end of a tRNA molecule recognizes specific codons, while the other end of the same tRNA molecule carries a particular amino acid. The tRNA deposits the amino acid it has brought on the last amino acid in the growing protein chain. The mRNA then moves another notch along the ribosome, exposing the next codon, and so the process continues. This process is known as transcription and translation.

Figure 5.9 shows mRNA translation for protein synthesis (Prentis, 1984).

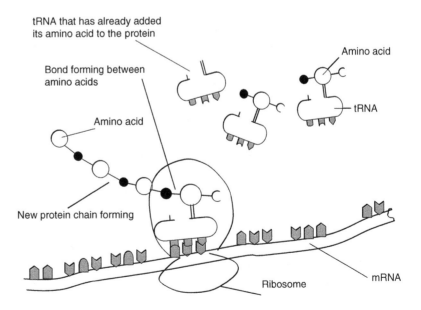

Figure 5.9 Schematic diagram of DNA transcription for protein synthesis.

5.5.2 Recombinant DNA technique

Recombinant DNA is made by excising a specific gene from one organism and inserting it into another organism. The development of gene transfer techniques requires a surgically precise means to cut and rejoin pieces of DNA. The discovery of restriction enzymes and ligases, both derived from bacteria in the early 1970s, enabled this. Restriction enzymes are able to cut DNA at specific sites, and ligases are enzymes which rejoin DNA fragments.

Figure 5.10 shows gene cloning in a simple bacterial system (Harlander, 1987). The process of introducing a gene into a host cell is known as transformation. The proper expressions or functions of transferred genes are tested for conformation. Procedures for transformation and expression in some simple bacteria and unicellular fungi such as *Escherichia coli* and yeast are now well established. Plants and filamentous fungi are more difficult to transform than bacteria or yeast. This is due in part to an increase in the number of chromosomes and the amount of DNA, and to more highly regulated mechanisms of transcription and translation.

Genetic engineering has been applied widely, among others, to crop improvement for high yield, disease resistance, herbicide tolerance and storage quality enhancement, as shown in Table 5.12.

Table 5.12 Application of genetic engineering in the food supply industry.

Agronomic application	Food technology application
Insect protection	Microbial strain improvement for enzyme production
Disease resistance	Ripening modification of fruits and vegetables
Herbicide tolerance	Increased levels of and modification of starch
Virus resistance	Increased levels of and modification of oils
Fungal disease resistance	Improved protein content and quality
Resistance to storage pests	Higher vitamin and mineral contents
Resistance to cold and draught	Reduced cyanogenic glucosides
Nitrogen fixation capability	Improved quality/ processing traits

The most immediate applications of genetic engineering in food technology have been in the dairy, baking and brewing industries. The gene for calf rennin has been isolated and cloned into yeast and fungi

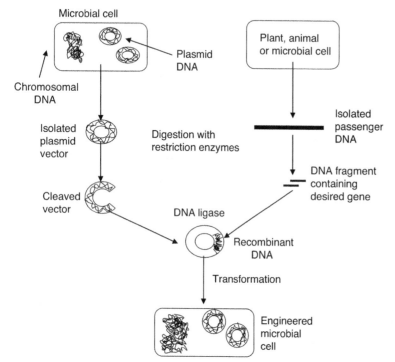

Figure 5.10 Gene cloning in a simple bacterial system.

Table 5.13 Examples of commercial food enzymes from GM microorganisms.

Enzyme	Production organism	Food application
Alpha-acetolactate decarboxylase	*Bacillus amyloliquefaciens* or *subtilis*	Beverages
Alpha-amylase	*Bacillus amyloliquefaciens* or *subtilis*	Baking, beverages
Aminopeptidase	*Trichoderma reesei* or *longibrachiatum*	Cheese, dairy
Arabinofuranosidase	*Aspergillus niger*	Beverages
Beta-glucanase	*Bacillus amyloliquefaciens* or *subtilis*	Beverages
Catalase	*Aspergillus niger*	Egg-based products
Chymosin	*Aspergillus niger*	Cheese
Cyclodextrin-glucosyl transferase	*Bacillus licheniformis*	Starch
Glucoamylase	*Aspergillus niger*	Beverages, baking
Glucose isomerase	*Streptomyces lividans*	Starch
Glucose oxidase	*Aspergillus niger*	Baking
Hemicellulase	*Bacillus amyloliquefaciens* or *subtilis*	Baking, starch
Lipase, triacylglycerol	*Aspergillus oryzae*	Fats
Maltogenic amylase	*Bacillus amyloliquefaciens* or *subtilis*	Baking, starch
Pectin lyase	*Aspergillus niger*	Beverages
Pectinesterase	*Trichoderma reesei* or *longibrachiatum*	Beverages
Phospholipase A	*Trichoderma reesei* or *longibrachiatum*	Baking, fats
Phospholipase B	*Trichoderma reesei* or *longibrachiatum*	Baking, starch
Polygalacturonase	*Trichoderma reesei* or *longibrachiatum*	Beverages
Protease	*Aspergillus oryzae*	Cheese
Pullulanase	*Bacillus licheniformis*	Starch
Xylanase	*Aspergillus niger*	Baking, beverages

to produce calf rennin from microorganisms. The lactose utilization genes of *E. coli* have been cloned into *Sac. cerevisiae*, *Xanthomonas campestris* and other microorganisms so that the lactose in whey permeate can be converted into ethanol, single-cell protein or xanthan gum. Interspecies gene transfer and protein engineering have been applied to microbial enzyme modification, such as in the production of thermotolerant amylase.

Table 5.13 lists the enzymes produced by GM microorganisms on the market (Robinson, 2001). However, no GM microorganisms have been utilized directly in food production yet.

5.6 Tissue culture

Plant tissue culture, the propagation of plant tissue in aseptic nutrient media, has widespread application in food and ingredient production (Wasserman *et al.*, 1988). Figure 5.11 shows the tissue culture process (Harlander, 1987). Plant tissue culture plays an important role in plant research, particularly genetic engineering because of time saving. The introduction of new genes into plant tissue requires the generation of single plant cells, or protoplasts. After genetic material is introduced, the protoplasts are grown in tissue culture, and, ultimately, whole plants are regenerated from single cells. Fusion of two genetically dissimilar protoplasts followed by regeneration often results in plants with desirable properties.

Highly valued natural products such as flavors, colors, preservatives and nutritional supplements are biosynthesized efficiently using this technology. For example, the world's supply of vanilla beans does not satisfy the demand for vanilla flavor. The in-vitro biosynthesis of natural vanilla flavor would alleviate this problem (Moshy, 1986). Product yields in plant cell culture are often many-fold higher than those found in the native plant. Technologies used to accomplish this include mutation, adjustment of nutrient and hormone levels, addition of appropriate metabolic precursors and plant cell immobilization (Knorr and Sinskey, 1985).

5.7 Future prospects

The biotechnological revolution we have witnessed in the last century has opened up the possibility of processing food through specific reaction and producing food having specific function. The initial application of genetic engineering for agronomic benefits, such as resistance to herbicides and pesticides,

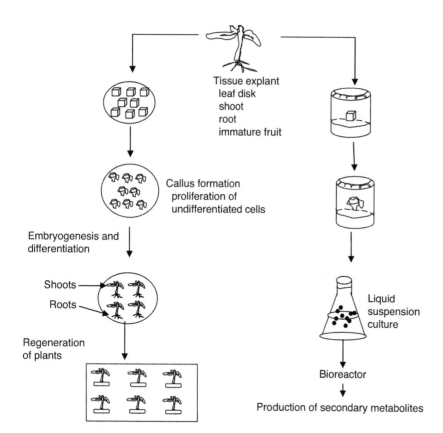

Tissue explant
leaf disk
shoot
root
immature fruit

Callus formation
proliferation of
undifferentiated cells

Embryogenesis and
differentiation

Shoots

Roots

Regeneration
of plants

Liquid
suspension
culture

Bioreactor

Production of secondary metabolites

Figure 5.11 Schematic diagram of tissue culture.

shorter time between sowing and harvesting and increased yields, is now gradually changing focus to quality traits for producing better nutritional value and keeping quality. Although the direct use of GM microorganisms in food has not yet occurred in the market place, new and powerful enzymes derived from GM microorganisms are widely used in food processing. By knowing the genetic map of major food crops and the amino acid sequence of proteins, the tailor-making of food ingredients having specific function is now possible through biotransformation and biocatalysis (Lee, 2003).

Modern biotechnology provides the means to select useful enzymes efficiently from conventional fermentation processes. By using PCR technique the newly developed enzymes are easily identified. The physiological functions of novel food ingredients are screened and confirmed rapidly by cell culture technique.

Further reading and references

Adler-Nissen, J. (1986) *Enzymic Hydrolysis of Food Proteins*. Applied Science, London.

Amerine, M.A., Kunkee, R.E. and Singleton, V.L. (1980) *The Technology of Wine Making*. AVI Publishing, Westport, Connecticut.

Barnes, G.L. (1993) *China, Korea and Japan. The Rise of Civilization in East Asia*. Thames and Hudson, London.

Harlander, S.K. (1987) Biotechnology; emerging and expanding opportunities for the food industry. *Nutrition Today*, **22**(4), 21.

Huang, H.T. (2001) *Science and Civilization in China, Vol. VI:5, Fermentations and Food Science*. Cambridge University Press, Cambridge.

International Service for the Acquisition of Agri-biotech Applications (2002) *Global Review of Commercialized Transgenic Crops: 2001*. ISAAA, Metro Manila.

Ishige, N. (1993) Cultural aspect of fermented fish products in Asia. In: *Fish Fermentation Technology* (eds C.H. Lee, K.H. Steinkraus and P.J.A. Reilly). UNU Press, Tokyo.

James, C. (2007) *Global Status of Commercialized Biotech/GM Crops*. ISAAA Brief No.37. ISAAA, Ithaca, New York.

Karel, M. and Lund, D.B. (2003) *Physical Principles of Food Preservation*. Marcel Dekker, New York.

Kim, M.-R., Kawamura, Y. and Lee, C.H. (2003) Isolation and identification of bitter peptides of tryptic hydrolysate of soybean 11S glycinin by reverse-phase HPLC. *Journal of Food Science*, **68**(8), 2416–22.

Knochel, S. (1993) Processing and properties of North European pickled fish products. In: *Fish Fermentation Technology* (eds C.H. Lee, K.H. Steinkraus and P.J.A. Reilly). UNU Press, Tokyo.

Knorr, D. and Sinskey, A.J. (1985) Biotechnology in food production and processing. *Science*, **229**, 1224.

Lee, C.H. (1989) Fish fermentation technology. *Korean Journal of Applied Microbiology and Bioengineering*, **17**, 645.

Lee, C.H. (1994) Importance of lactic acid bacteria in non-dairy food fermentation. In: *Lactic Acid Fermentation of Non-dairy Food and Beverages* (eds C.H. Lee, J. Adler-Nissen and G. Barwald). Harnlimwon, Seoul, pp. 8–25.

Lee, C.H. (1997) Lactic acid fermented foods and their benefits in Asia. *Food Control*, **9**(5/6), 259–69.

Lee, C.H. (1998) Cereal fermentations in the countries of the Asia-Pacific region. In: *Fermented Cereals – A Global Perspective* (eds N.F. Haard, S.A. Odunfa, C.H. Lee and R. Quintero-Ramirez). *FAO Agricultural Service Bulletin*, **138**, 63–97.

Lee, C.H. (2001) *Fermentation Technology in Korea*. Korea University Press, Seoul.

Lee, C.H. (2003) *The role of biotechnology in modern food production*. Proceedings 12th IUFoST World Congress, 16–20 July, Chicago.

Lee, C.H. and Kim, K.M. (1993) Korean rice-wine, the types and processing methods in old Korean literatures. *Bioindustry*, **6**(4), 6–23.

Lee, C.H. and Lee, S.S. (2002) Cereal fermentation by fungi. *Applied Mycology and Biotechnology*, **2**, 151–70.

Lee, C.H., Adler-Nissen, J. and Barwald, G. (1994) *Lactic Acid Fermentation of Non-Dairy Food and Beverages*. Harnlimwon, Seoul.

Lee, S.W. (1990) A study on the origin and interchange of Dujang (also known as soybean sauce) in ancient Asia. *Korean Journal of Dietary Culture*, **5**(3), 313.

Lim, B.S. (1999) *Present status and prospect of Korean bioindustry*. Symposium of 3rd Inauguration Anniversary for Institute of Bioscience and Biotechnology, Korea University.

McGregor, J.A. (1992) Cultured milk products. In: *Encyclopedia of Food Science and Technology* (ed. Y.H. Hui). John Wiley, New York.

Moshy, R. (1986) Biotechnology; its potential impact on traditional food processing. In: *Biotechnology in Food Processing* (eds S.K. Harlander and T.P. Labuza). Noyes Publications, Park Ridge, New Jersey.

Nuath, K.R., Hynes, J.T. and Harris, R.D. (1992) Cheese. In: *Encyclopedia of Food Science and Technology* (ed. Y.H. Hui). John Wiley, New York.

Olsen, H.S. (1995) Enzymes in food processing. In: *Biotechnology, Vol. 9* (eds H.-J. Rehm and G. Reed). VCM, Weinheim.

Owades, J.L. (1992) Beer. In: *Encyclopedia of Food Science and Technology, Vol. 1* (ed. Y.H. Hui). John Wiley, New York.

Prentis, S. (1984) *Biotechnology: A New Industrial Revolution*. G. Braziller, New York.

Rhee, S.J., Lee, C.Y.J., Kim, K.K. and Lee, C.H. (2003) Comparison of the traditional (Samhaeju) and industrial (Chongju) rice-wine brewing in Korea. *Food Science Biotechnology*, **12**(3), 242–7.

Robinson, C. (2001) *Genetic Modification Technology and Food*. ILSI Europe, Brussels.

Ruddle, K. (1993) The availability and supply of fish for fermentation in Southeast Asia. In: *Fish Fermentation Technology* (eds C.H. Lee, K.H. Steinkraus and P.J.A. Reilly). UNU Press, Tokyo.

Russell, J. and Stewart, G.G. (1995) Brewing. In: *Biotechnology, Vol. 9* (eds H.-J. Rehm and G. Reed). VCM, Weinheim.

Shettleff, W. and Aoyaki, A. (1976) *The Book of Miso*. Autumn Press, Berkeley, California.

Souane, M., Kim, Y.B. and Lee, C.H. (1987) Microbial characterization of *gajami sikhae* fermentation. *Korean Journal of Applied Microbiology Bioengineering*, **15**(3), 150.

Spicher, G. and Brummer, J.-M. (1995) Baked goods. In: *Biotechnology, Vol. 9* (eds H.-J. Rehm and G. Reed). VCM, Weinheim.

Stauffer, C.E. (1992) Bakery leavening agents. In: *Encyclopedia of Food Science and Technology* (ed. Y.H. Hui). John Wiley, New York.

Steinkraus, K.H. (1983) *Handbook of Indigenous Fermented Foods*. Marcel Dekker, New York.

Steinkraus, K.H. (1993) Comparison of fermented foods of the East and West. In: *Fish Fermentation Technology* (eds C.H. Lee, K.H. Steinkrause and P.J.A. Reilly). UNU Press, Tokyo, pp. 1–12.

Wasserman, B.P., Montville, T.J. and Korwek, E.L. (1988) Food biotechnology, a scientific status summary by IFT. *Food Technology*, January.

Yoon, S.S. (1993) *Cheminyosul*. A translation of *Chiminyaosu* in Korean. Mineumsa, Seoul.

Tim Aldsworth, Christine E.R. Dodd and Will Waites

Key points

- This chapter discusses the different microorganisms important to the food industry, including viruses, bacteria, yeasts, protozoa and worms.
- Microbial growth is examined and methods of measurement are discussed.
- The bacterial agents of foodborne illness are described, including *Clostridium botulinum*, *Staphylococcus aureus*, *Bacillus cereus*, *Vibrio* species, *Yersinia enterocolitica*, *Clostridium perfringens*, *Salmonella*, *Shigella*, *Escherichia coli*, *Campylobacter*, *Mycobacterium* and *Listeria monocytogenes*.
- Non-bacterial agents of foodborne illness are examined, including mycotoxins, protozoa, helminths, *Taenia*, *Trichinella*, *Toxoplasma*, scrapie, bovine spongiform encephalopathy, kuru and Creutzfeldt-Jacob disease.
- Specific outbreaks of foodborne illness are discussed, together with their incidence, in order to lead to a better understanding of the causes.
- Water as a source of illness is discussed.
- Traditional and novel methods of microbial detection are compared, together with sampling plans. The Hazard Analysis Critical Control Point system is also introduced.
- Production of foods by microbial fermentation including tempeh, beer, wine, saké, bread, cheese, yoghurt, kefir, fish, vegetables, soy sauce, miso and natto is discussed.

6.1 Introduction

Microorganisms are generally considered to include bacteria, yeasts, fungi and some protozoa. Bacteria especially, but also yeasts and fungi, are ubiquitous. Different species of bacteria are found growing in all natural and man-made environments – from the Antarctic, through refrigeration at 1–5°C, hot springs at the boiling point of water (100°C) to hypothermal vents where high pressure allows growth at 160°C and even higher temperatures found deep in the ocean floor. In addition, some bacteria form resting bodies called endospores or, more simply, spores (Colour Plate 20). Spores have no measurable metabolism and those of some species can survive heat at 132°C at a pressure of 1 atmosphere. They also resist ultraviolet light (Warriner *et al.*, 2000; Waites and Warriner, 2005), chemicals including disinfectants and enzymes (Setlow and Johnson, 2007). Spores have variously been claimed to survive for 25 million years (Cano and Borucki, 1995) or even 250 million years (Wreeland *et al.*, 2000). Yet when put into contact with some specific but simple chemicals, such as L- but not D-alanine (which is a competitive inhibitor for L-alanine), spores will germinate and begin to metabolise within 1 or 2 min.

Since a number of microorganisms including some of the spore formers produce toxins, and other microorganisms are able to cause illness when ingested, it is clear that their presence in food is a major problem. Others cause spoilage, while some microorganisms, bacteria especially, often members of the genus *Lactobacillus*, can prevent the growth of organisms able to cause foodborne illness and food spoilage by a process of fermentation, for example when milk is fermented to produce cheese (see section 6.18.4.1). This results in major changes in the food, and can also improve flavour, texture and digestibility. In the same way the concentration of toxic chemicals, such as cyanide in the very important plant crop cassava, can be reduced to non-toxic levels (Campbell-Platt, 1987).

6.2 Microorganisms important to the food industry

Those microorganisms of interest to the food scientist include mostly bacteria, yeasts and fungi but also some protozoans, and even worms and viruses. It is obvious from the heading that these are all very small organisms (Table 6.1) and that this leads to problems with detection and enumeration. As with so much else in life, size matters here.

Bacteria consist of a cell wall, membrane and cytoplasm and do not have any organelles within the cells, whereas yeasts and fungi and animal cells including protozoa contain mitochondria which have been described as the cell's energy producing units. Yeasts and fungi, like bacteria, have rigid cell walls. Viruses are much smaller, have a protein outer layer

(a capsid) and can contain either DNA or RNA as the means of passing on their characters. Interestingly bacteria can also be infected by viruses which are termed bacteriophages (phages). Like those that attack animals and plants, these phages are very specific, often infecting only some strains within a species. This allows their use in differentiating between strains as well as species (phage typing). In addition, there have been efforts to develop them as a novel rapid method of detection (see Rees and Loessner, 2005). One of the problems that microorganisms present is as a result of their ability to grow extremely rapidly. One species of bacterium, *Clostridium perfringens*, which is able to form spores and is also able to produce foodborne illness, has been detected in the absence of oxygen, that is anaerobically, doubling every 7.1 min. Even worse for its competitors, in the presence of adequate nutrients this rate of doubling will continue until the metabolism of the cells produces chemicals able to inhibit further growth. Although one bacterium weighs only 1×10^{-12} g, doubling every 30 min with unlimited nutrients would result in the total bacterial weight equalling that of the earth in less than 48 h. Obviously this cannot happen, but it does indicate the speed at which microorganisms can develop if not controlled.

6.3 Microscopic appearance of microorganisms

It is clear from microscopy that cells of one bacterial species may appear dissimilar to that of others. Their growth may be different as well. For example, those of the family Enterobacteriaceae, like *Escherichia coli* (Colour Plate 21) and *Salmonella enterica*, grow as short rods, *Bacillus megaterium* (Colour Plate 22), *Clostridium perfringens* (Colour Plate 25) and *Clostridium botulinum* as large rods, cells of *Streptococcus* in chains of spherical cells termed cocci, whilst *Staphylococcus aureus* grows as clumps of spherical cells (Colour Plate 26). Cells of the meat spoilage bacterium *Brochothrix thermosphacta* can best be described as a ball of wool after the cat has played with it.

The easiest method of visualising the microorganisms is to stain cells with methylene blue. This enables the cells to be seen under an ordinary light microscope. An alternative is to use the Gram stain (Table 6.2) on cells heat fixed to a glass slide; unlike with methylene blue staining, there is no addition of a cover slip. Often a better method is to use a phase contrast microscope. This allows cell mobility

Table 6.1 Size of microorganisms.

Microorganism	Size
Virus	0.065 μm (or 65 nm)
Rod-shaped bacterium	1–5 μm
Yeast	8–8 μm
Protozoan	5–20 μm
Trichinella	0.1–4 mm

Table 6.2 Gram staining of bacterial cells.

Gram positive	Gram negative
Bacillus	E. coli
Clostridium	Salmonella
Staphylococcus	Pseudomonas
Lactococcus	Campylobacter
Streptococcus	Actinobacter
Listeria	Shigella
Lactobacillus	Vibrio
Enterococcus	Yersinia

to be determined since the cells can be seen without staining and therefore there is no requirement to fix them to the slide by heating. However, one advantage of the Gram's stain is that it allows bacterial cells to be divided into Gram positive (dark blue/purple; genera such as *Bacillus* (Colour Plate 24), *Clostridium* (Colour Plate 25) and *Staphylococcus* (Colour Plate 26)) or Gram negative (red; genera such as *Escherichia, Salmonella* and *Pseudomonas*). This distinction is important, since it is often the first step in traditional methods of bacterial identification.

6.4 Culturing microorganisms

6.4.1 Culturing bacteria

Microorganisms such as bacteria can be grown on nutrient-rich agar plates or in liquid media, which must have oxygen removed for anaerobes or be oxygenated for aerobes. There are two forms of growth in liquid media, either batch or continuous culture. Batch culture most commonly involves growth in liquid media, either in a sterilised flask or in a test tube which is shaken under optimum conditions of temperature, pH and nutrients, not only to encourage oxygen intake, but also to equalise the cell growth in different parts of the vessel. Growth rates can be extremely high with *C. perfringens*, as mentioned previously (see section 6.2), the record breaker, having been found to be able to double every 7.1 min. With batch culture, no further nutrients are added and although samples may be removed for testing, it is intended that any sample removed should be exactly the same as the remainder of the culture. However, as soon as growth begins, the medium will change, since nutrients will be utilised and metabolites produced. Some of the metabolites will be secreted into

the growth medium, so that, for example, the pH may drop due to acid production.

If the samples are to be the same it is therefore particularly important for the tests carried out to be reproducible and that the time and conditions of culture and sampling be exactly the same from one day to the next. Continuous culture provides these conditions and will be discussed in detail later.

In order to predict future growth, growth parameters, such as culture turbidity, can be plotted against time. However, the exponential nature of bacterial growth means that instead of an arithmetic plot of the results, an exponential or semi-log plot is used. For a microbial culture doubling at the same time, this gives a straight line. Such a plot gives a further advantage that, provided exponential growth continues, it is possible to calculate the growth rate as well as to forecast when a particular growth point will be reached. Perhaps most important to the laboratory-bound researcher, it allows calculation of the time available for a lunch break. An arithmetic plot allows none of these possibilities to occur.

The use of a semi-logarithmic plot and batch culture also shows the stage of growth where there is an initial lag followed by growth, which may become exponential at least in axenic (one species growing alone) growth in the laboratory before nutrients or other conditions such as pH changes begin to limit growth. Eventually, stationary phase is reached where any new growth and division is matched by cell death. It is, however, particularly important to allow these comparisons, to make sure that measurements are taken by the same method for each experiment.

This may seem obvious, but it is worth thinking about the details. If a component of the cell is used as an indicator of growth, then more of this may be synthesised or lost at different times during the growth cycle. Equally, at the end of the growth, chains of cells may separate into single cells, giving an apparent increase in colony forming units, whereas measurements of a compound like ATP will show a fall, although turbidity measurements using a spectrophotometer will not decrease until cell breakdown (lysis) begins. At the point of lysis, it is very likely that the cells will have been non-viable for some time. It is usual to draw growth curves with an exponential fall in cell numbers after the end of stationary phase, but it is most important to recognise that the stationary phase may continue for some time (months, for example, with *Salmonella* in chocolate which has a water activity (a_w) suitable for survival but not for growth or division) and that after a fall in cell numbers another growth phase may occur as some cells lose viability, lyse and release nutrients which allow the small number of viable cells to grow and divide (Vulic and Kolter, 2001). Expressing the changes at the end of an exponential stationary phase as a semi-log fall in cell numbers is therefore a gross oversimplification. As a final point in batch culture, it is important to recognise that in natural environments, such as soil, it is possible as a result of lack of nutrients and/or low a_w for doubling to occur only four times in a year.

A semi-log plot is also important in producing kill curves. This allows the decimal reduction time to be determined. The decimal reduction time is the time required to reduce the number of viable microorganisms by 90% and is given in minutes with the temperature in degrees Centigrade indicated by a superscript. As a result the D_{121} of spores of *C. botulinum* is 0.21 min, but *Clostridium sporogenes* which is a spoilage organism whose spores are more resistant to wet heat than those of *C. botulinum* are used as an indicator. Hence, while the spoilage rate is less than 1 in 10^6 cans the extent of survival of the *C. botulinum* spores is about 24 or even more decimal reductions. As discussed previously, bacterial spores are generally the most resistant entities so far discovered. So resistant are they that some suggestions have been made that they were the first organisms to populate the planet. However, spores seem a widespread and evolved life form which would have developed considerably since they arrived; they are also especially heat resistant, with spores of thermophiles like *B. stearothermophilus* having a D_{121} value of 5.0 and *C. botulinum* types A and B of 0.2, whereas vegetative cells of the non-spore former *E. coli* have a D_{65} value of 0.1 min, and *Campylobacter jejuni* has a D_{55} value of 1.1 min. However, the vegetative cells of some species of *Micrococcus* which can act as spoilage organisms are able to survive temperatures of 75°C, with D values of 1–2 h at 65°C. Table 6.3 shows the heat resistance of some vegetative cells and spores.

Interestingly the heat resistance of spores of *B. stearothermophilus* is so high that most canned foods contain low levels of viable thermophilic spores which will germinate and grow at temperatures of 45°C or so. Canned food is therefore stored refrigerated in many countries where the ambient temperatures are 40°C or more. Would canning have been successful if the initial development of the process had been in those countries with high ambient tempera-

Table 6.3 Heat resistance of vegetative cells and spores.

Spores	D_{121} min
B. stearothermophilus	5.0
C. sporogenes	1.5
C. botulinum type A	0.2
B. coagulans	0.1
Vegetative cells	D_{65} min
Staphylococcus aureus	2.0
Salmonella Seftenberg	1.0
Escherichia coli	0.1
Campylobacter jejuni	D_{55} 1.1 min

tures such as parts of Australia, Saudi Arabia or even southern Spain?

As mentioned earlier, an alternative method of growing microorganisms is continuous culture which uses a chemostat. Industrially this can be extremely complex, but in its simplest form it involves a growth vessel with an inlet and an outlet tube. Liquid growth medium with one nutrient in limiting supply enters the vessel from a reservoir and, because of the arrangement of the liquid and the exit tube, an equivalent volume of liquid flows out of the fermentation flask and can be collected and used for testing. If no contaminating organism enters the vessel or the growth reservoir, then a continuous culture should be exactly that – continuous – and will deliver cells at the same stage of growth and the same physiology and biochemistry as yesterday, last week or even last year. It might be expected that a continuous culture system, which is stirred to make all parts of the culture the same, has its pH controlled and often has chemicals added to prevent foaming, is used frequently in industry for fermentations. The reverse is true, however, and batch culture is used much more frequently. It also takes a few days for a continuous culture to reach steady state. In addition, if it is set up too close to starvation, then growth will be too slow to maintain a microbial presence in the culture vessel. At the other end of the system, if the flow from the medium reservoir is too great, then the organism will not be able to divide fast enough and will be washed out of the vessel.

6.4.2 Culturing yeasts and fungi

Yeast cells are much (about 10 ×) bigger than the bacterial cells described so far. In addition, although some yeasts divide by binary fission and hence produce two daughter cells of equal size in the same way that bacteria do, others divide by budding. This last group can be differentiated easily when in axenic culture because cells of very different sizes appear in the same culture. In comparison, fungi generally grow as long hyphae and may produce large fruiting bodies. Mushrooms are an obvious example. At the other extreme, plant and animal viruses, as well as phages, are so small that they can be visualised only by electron microscopy and hence are always inactivated when seen. It is important to remember that viruses will only grow in their specific host.

6.5 Microbial growth

6.5.1 Microbial growth and its effect on foods

In natural environments, microorganisms often grow as mixed cultures in biofilms (Lappin-Scott and Costerton, 2003). This means that they attach to a surface and are not easily moved by liquid flow or, in the case of man-made environments, by liquid or mechanical cleaning systems, such as brushes. In laboratories, although the study of biofilms is advancing rapidly, the growth of a planktonic axenic culture has been used traditionally with the cells in a liquid growth medium. This allows the growth requirements, biochemistry and the physiology and growth rate to be determined in what is, despite our lack of knowledge, a simple system. Often the liquid medium is solidified by the addition of agar, an approach that allows individual cells to be enumerated by counting the number of colonies that are produced by growth. However, remembering the clumps of cells produced, for example, by *Staph. aureus* (Colour Plate 26), in many ways this is an approximation and alternative methods of measuring growth might give different but just as accurate results.

One of the difficulties that *Homo sapiens* has in competing with microorganisms is that many of them utilise the same nutrients as we do. Hence, food spoilage as well as foodborne illness can be a major problem. However, if microbes were less able to utilise such materials then the remains of dead plants and animals would not be broken down and the carbon, sulphur and nitrogen and other natural recycling systems would stop. In addition, spoilage and sometimes toxin production can be detected directly by changes in physical appearance, taste and/or smell or where new volatile chemicals are produced. Unfortunately, some organisms able to produce illness, such as *Clostridium botulinum*, fail to produce spoilage compounds and cannot be detected by changes in taste or smell. In some illnesses such as the progressive neurological disease referred to as Parkinson's disease, sufferers cannot smell or taste changes in food and it would be of interest to compare the risks of foodborne illness in Parkinson's disease patients with those of controls. Occasionally, spoilage of foods has been found to produce changes which the consumer accepts and then seeks out. Yogurt, cheese, beer, wine and fermented sausages would be examples (see fermented foods, section 6.18). A different occurrence (section 6.8.1.3) involved some minor cases of foodborne illness caused by *Bacillus subtilis* contaminated bread where the consumer's comments were along the lines of "Well, I knew it tasted different to normal bread but I thought it was a new type of loaf – it tasted of grapefruit!"

6.6 Methods of measuring growth

6.6.1 Microscopy

The most obvious method of measuring microbial growth is by means of a light microscope. For bacteria, the overall magnification required is at least $200 \times$ and most frequently $1000–1200 \times$ is used in order to differentiate easily between the different shapes of organisms. A haemocytometer with a grid pattern allows the number of organisms in a measured volume to be counted. Although some dyes have been used to differentiate between living and dead cells there are some doubts about their ability to differentiate to the same extent as the viable count. Other problems with microscopic counts include the length of time required to reach a statistically accurate number, the tediousness of the process for the operator (about 20 samples can be processed in an 8-h working day) and the poor sensitivity, with 5×10^6 to 1×10^7 ml^{-1} in a liquid suspension required before enough cells can be seen under the microscope to provide a statistically meaningful count.

6.6.2 Viable counting

Since it is generally the number of viable cells (that is, those that can grow and multiply) or, in some cases produce toxin, that is important, it is apparent that counting viable cells is likely to be a favoured method of understanding growth. Often in the past, and occasionally still, the term total viable count (TVC) has been used to describe the results from colonies on duplicate agar plates. A moment's thought will indicate that this is inaccurate. For example, if the plates have been incubated in air, then those organisms that will only grow in the absence of oxygen will not grow. Total aerobic count (TAC) is used here. However, those organisms that will grow best at 70°C will not grow at 37°C and those bacteria able to grow at 1°C will not grow at 45°C. In addition, some microorganisms have requirements for specific nutrients which might not be found in some media. The best system is to use the term viable count (VC) and indicate the conditions of the incubation. The best way to describe the results is as colony forming units (CFU g^{-1}, ml^{-1}). Finally, there are some bacteria, such as E. coli O157:H7, where only ten cells are sufficient to initiate illness and they may be present with 10^6 cells ml^{-1} of other microorganisms. As a result it is clear that it will not be possible to detect the E. coli cells with a simple viable count. There are far too many cells of other organisms that would grow and hide the presence of the E. coli, even if the colonies looked different.

One way of solving this problem is by the use of a selective enrichment. In this approach, chemicals that will prevent the growth of the majority of other species, but not those of E. coli, are added to the medium. This involves incubation in a liquid medium before plating out onto a selective agar medium.

There is usually a preliminary incubation in a non-selective medium, because the presence of the selective agents might prevent the growth of (or even kill) damaged cells of the organism that they are intended to enumerate. Since damage includes the effect of chemical disinfectants, heat treatment and drying it is clear that many cells found in the food industry, either in food or food processing units, will be damaged and can therefore be sensitive to selective agents.

One other important point about viable counts is the number of colonies on a plate required to produce a statistically meaningful result. In the older literature, this is usually quoted as between 30 and 300 colonies. More recently it has been suggested that 25–250 colonies is more realistic. In any case, some microorganisms, such as Bacillus cereus (Colour Plate 23) and especially fungi, will spread across the agar

surface and grow into each other so that it is difficult to find a plate with countable colonies. Where it is intended to count moulds, chemicals that prevent spreading can be used. Otherwise, the use of an incubation temperature of 37°C will reduce the growth of most moulds and help to encourage that of bacteria. The lower temperature of 30°C will often be used to allow rapid growth of human pathogens as well as reduce the effect of selective agents against the organism it is hoped to count. For spoilage organisms a temperature as low as 15°C might be an alternative, especially in temperate countries or where sampling is from a refrigerated environment.

For counting anaerobes, various combinations of gases, such as nitrogen, carbon dioxide and hydrogen, are used to exclude oxygen. For example, for Campylobacter jejuni and Lactobacillus species carbon dioxide is present together with reduced oxygen levels. Finally, it is usually assumed that one cell gives rise to one colony. Of course given the long chains and clumps of cells discussed earlier, this is nonsense, but it will give an approximation which will be different to that obtained by alternative methods which measure the biochemistry of the cells (for example, ATP or DNA based methods).

6.7 Microbial biochemistry and metabolism

6.7.1 Cell walls

Unlike animal cells, which have flexible surfaces, bacteria, fungi, yeasts and algae have rigid cell walls around the cytoplasm and the cytoplasmic membrane. The walls protect the cytoplasm from mechanical damage, but different organisms have very different cell wall structures. Gram positive cell walls are thicker (30–50 nm) than Gram negative cell walls (20–25 nm). Gram positive cell walls are mostly peptidoglycan which accounts for 20–40 nm of the thickness of the wall. The peptidoglycan chains run parallel to the cell surface. Gram positive cell walls consist of peptidoglycan and other polymers, such as teichoic acid, whereas Gram negative bacterial cell walls are largely lipopolysaccharide. Gram negative walls have a layered structure, with peptidoglycan making up 6% of the cell wall and the innermost part of the cell wall with an outer membrane covalently linked to the peptidoglycan. Fungal cell walls consist mainly of polysaccharides, such as chitin and cellulose. Different microorganisms have such a wide range of abilities to utilise nutrients that even disinfectants can be degraded if they are stored at the use dilution. It is clearly beyond the scope of this chapter to cover all the abilities of microorganisms to grow and metabolise, and only those important to foodborne illness, to food spoilage and to production of foods by microbial fermentation will be considered.

6.8 Agents of foodborne illness

The following microorganisms are those that are most common as agents of foodborne illness (see also Zoonoses Report UK, 2007). It is, however, not an exhaustive list.

It is worth noting that even in a highly developed country such as the United Kingdom (UK) only about 10% of samples thought to contain foodborne illness organisms are found to be positive. Hence, either detection methods are still poor or agents other than those searched for are responsible.

6.8.1 Bacteria

Bacteria are involved in two forms of foodborne illness. These are, first, as a result of the consumption of pre-formed toxin, as with *Clostridium botulinum*, *Staph. aureus* and *B. cereus* (which is a special case, because it also has the ability to produce infection as well as intoxication), and, second, those that produce infection. The latter group include *Campylobacter jejuni*, *Salmonella* spp., *E. coli* (especially O157:H7), *Listeria monocytogenes*, *C. perfringens*, *Vibrio* spp., *Yersinia enterocolitica* and *B. cereus*. Other important organisms, including protozoa and worms, are foodborne and are considered subsequently. The first five non-intoxicating bacteria are currently the most important five as far as the UK Food Standards Agency is concerned. This is based partly on the severity of

the illness produced, but also on the number of outbreaks and cases reported.

The intoxicating organisms will be discussed first since in some ways their effects are easier to understand.

6.8.1.1 Clostridium botulinum

Clostridium botulinum produces eight different types of neurotoxin: A, B, C_1, C_2, D, E, F and G. Although it is a rare cause of illness, the toxin is usually considered the most potent known since a lethal dose for a human is about 10^{-8} g. It causes an illness characterised by cranial neuropathy and descending flaccid paralysis. The bacterium is motile with peritrichous flagella, forms spores and is a strict anaerobe with straight or slightly curved rods 210 μm long. The eight toxins are differentiated by serology. Toxin C_2 is unusual in not being a neurotoxin. The toxins have been weaponised and are characterised as a category A biological agent by the Centers for Disease Control and Prevention (CCDC) in the USA. Types A, B and E cause most cases of botulism (Gupta *et al.*, 2005). In comparison, little is known about type F. During the years 1922–2005 only 62 cases were recognised in the UK (McLauchlin *et al.*, 2006) whereas between 1981 and 2002, 1269 cases of botulism in the USA were reported to the CCDC. Of these, only 13 (1%) were F, none of which were part of outbreaks. The situation is complicated since a toxigenic *Clostridium barati* was identified in nine cases and a toxigenic *Clostridium butyricum* has also been detected.

C. botulinum is divided into four groups and these differences have been confirmed by DNA homology and ribosomal RNA sequencing.

Physiological differences of C. botulinum

Some physiological differences of *C. botulinum* are shown in Table 6.4.

Table 6.4 Physiological differences of *C. botulinum*.

Group	Toxin	Pathogenicity	Heat resistance	Minimum growth temperatures (°C)	Proteolytic	Saccharolytic and lipolytic	Minimum a_W for growth
I	A, B or F	Humans	D_{121} 0.1–0.24 min	10–12	+	+	0.94
II	B, E or F	Humans	D_{80} 0.6–3.3 min	3–5	−	+	0.975
III	C_1, C_2 or D	Birds and other animals	High or low	15	−	+	3%
IV	G	Humans	?	12	+	−	>3%

Group I strains are similar to *Clostridium sporogenes*, an organism which does not produce toxin, and, although they are very proteolytic and can produce a slightly rancid odour, the toxin is so active that tasting the food and then spitting it out can be sufficient to cause illness. Group I strains produce more heat-resistant spores so that all heat processing is designed to kill them (see thermal death times, section 6.4.1), but such strains are not able to grow at refrigeration temperatures. Group II strains can grow in chilled foods and hence can grow (and produce toxin) at temperatures as low as 3°C. However, their spores are much less heat resistant than those of Group I. As indicated in Table 6.4 group III strains are usually responsible for illness in birds and other animals, while group IV strains have not been found in foods. The temperature, water activity and the acid present (or added to lower the pH) determine the minimum pH for growth. Research has shown that growth and toxin production can occur at pH 4.0, but this is in high protein media and most authorities would accept that pH 4.5 is low enough to prevent both growth and toxin formation.

The initial symptoms of botulism occur most frequently between 12 and 48 h after ingestion, but can occur as quickly as 8 h after or take as long as 8 days to show. The symptoms include double vision, vomiting, constipation, dry mouth and difficulty in speaking. After 1–7 days death can result from heart or respiratory failure and those patients who survive can take as long as 12 months to recover completely.

The toxin acts by binding to the nerve endings at the nerve–muscle junctions. This blocks release of the acetylcholine which is responsible for transmission of stimuli and produces paralysis.

Even with the current rapid treatment available in a modern, well funded hospital, mortality rates can be as high as 20%, although the particular food, the dose and the toxin type are important, since type A seems to produce a higher percentage of deaths than the other toxins. The toxins can be inactivated by heating at 80°C for 10 min.

In addition to the typical intoxication, there is also a form of the disease known as infant botulism where the toxin is produced by cells growing in the infant's gut (Johnson *et al.*, 2005). In this case, the infant fails to thrive and has difficulty feeding or holding his or her head up. This form of the illness seems to be more common in the USA and most frequently affects small infants between 2 weeks and 6 months after birth. It is thought that the infant's gut flora is not fully developed and is not able to exclude *C. botulinum* and prevent it from growing. The ingestion of spores, perhaps from honey (which may also be a problem in some parts of Europe outside the UK) or non-food sources such as contact with soil, may also be responsible for the illness. This form of botulism has a low mortality rate especially if treated quickly, although in the USA it has been suggested that it may be responsible for a low percentage (that is about 4%) of cases of sudden infant death syndrome.

While botulism would be expected to be a major problem because of the high activity of the toxin, the food industry is remarkably efficient in dealing with the organism and most cases are as a result of products prepared non-commercially, but examples of commercially produced outbreaks are given in the Outbreaks section (section 6.9). In most years since 1988, there have not been any cases of botulism in England and Wales (McLauchlin *et al.*, 2006). An exception to this was 1989, when 27 people became ill and one died as a result of consumption of yoghurt to which inadequately heated hazelnut puree had been added, allowing *C. botulinum* type B toxin to be produced (O'Mahony *et al.*, 1990). Unusually in this case, sufficient gas was present, presumably as a result of the metabolism of other species of clostridia, to allow spoilage to be detected. Other countries have a more significant problem, with Italy having more than 12 cases every year since 1987 and 50 cases in 1988, 1989 and 1996. In addition Germany had between four and 39 cases over the 11 years from 1988 to 1998 (Therre, 1999), with the USA reporting on average about 59 cases/year over the 22 years from 1981 to 2002.

Just to demonstrate that *H. sapiens* is able to make use of most things, the toxin has also been developed as a beauty aid. In this case, very small amounts of toxin are injected into the face and wrinkles are removed. Although this may seem trivial, the same procedure can be used to assist the movement of children suffering from such diseases as muscular dystrophy. Unfortunately, wound botulism has occurred as a result of the injection of drugs especially in the USA (Hunter and Poxton, 2002). In addition, in the USA in 2004, four cases of severe botulism were associated with unlicensed cosmetic treatment of facial wrinkles.

The food most commonly the cause of this form of intoxication depends very much on the country. As the organism occurs in the soil and forms heat-resistant spores it is no surprise that home canning in the USA is a problem, particularly since in the early part of the twentieth century, incorrect heating guidelines were published so that between 1899 and 1981 there were 522 outbreaks associated with products canned at home and 432 of them showed a problem with vegetables, whereas in Europe, meat products have been a problem in the past. In addition, in the USA in 1985 and 1989, there were two outbreaks caused by fermented Alaskan Native food, in 1994 by baked potatoes and in 2001 by chilli sauce, increasing the numbers to 49, 28, 49 and 38 respectively.

Isolation and identification is fraught with difficulty, partly because of the metabolic diversity of bacteria carrying the toxin genes (Hatheway, 1990), but also because it usually represents a small proportion of the flora. One approach is to enrich in a cooked meat broth at 30°C for 7 days before streaking out onto egg yolk or fresh horse blood agar and incubating in the absence of oxygen for 3 days. The colonies are smooth, 2–3 mm in diameter, with an irregular edge and, apart from type G, show lipolytic activity on the egg-yolk agar. Suspect colonies are transferred into broth to examine for toxin, by either immunoassay or a mouse test (with a lethal dose of picograms rather than micrograms) which must include a test with a specific antibody shown to neutralise the toxin.

6.8.1.2 Staphylococcus aureus

Staphylococcus aureus is an important human pathogen and can be a significant form of human disease and death, especially in its antibiotic resistant form of methicillin resistant *Staphylococcus aureus* or MRSA. However, it is more usually as an infectious agent that it is a major pathogen and in foods it causes a much less serious condition.

The genus *Staphylococcus* contains more than 20 species. All are flora of the skin and upper respiratory tract of warm blooded animals including man. They are Gram positive, facultatively anaerobic cocci that are catalase positive, a characteristic that distinguishes them from other significant Gram positive coccus genera such as *Streptococcus*, *Enterococcus* and *Lactococcus*. From a food standpoint, *S. aureus* is the only significant pathogen and is usually distinguished by its ability to coagulate blood plasma and is termed coagulase positive, with the others such as *S. epidermidis*, *S. chromogenes* and *S. xylosus* collectively termed coagulase negative species (CNS). *S. aureus* is not the only species with plasma coagulation ability; *S. intermedius* also is coagulase positive, but

this is an animal species that is rarely isolated from foods. However, when examining environmental isolates the possibility must always be kept in mind, especially where the presence of vermin is suspected.

Isolation of *S. aureus* is usually done on Baird–Parker agar without the need for enrichment. This contains glycine, lithium and tellurite as selective agents, and tellurite together with egg yolk emulsion are the diagnostic agents allowing the production of characteristic colonies. Staphylococci give black colonies due to a reduction of tellurite, but *S. aureus* produces a zone of clearing in the egg yolk around the colonies within 24 h due to proteolysis; an inner opaque halo also forms due to the action of lipase. The timing is important as other species may produce halos if incubated for longer and confirmation of any isolates as *S. aureus* using a coagulase test may be advisable where precise identification is needed.

As already indicated *S. aureus* is a significant pathogen and much of this pathogenic ability is through infection. A battery of virulence factors such as proteases and lipases mediate this disease ability, but certain syndromes are toxin-mediated. Toxic shock syndrome is mediated by toxic shock syndrome toxin 1 (TSST-1) previously described as enterotoxin F. In the 1980s there was a major problem particularly in the USA with this syndrome associated with healthy menstruating women using a particular type of tampon. However, it is also seen in people following surgery (especially with wound packing) or as a result of cutaneous infection. The symptoms include a high fever, headache, sunburn-like rash, sore throat, diarrhoea and vomiting, and these are followed by hypotensive shock and organ failure which can be fatal. Desquamation of affected areas, especially the palms of the hands and soles of the feet, is typical 1–2 weeks after onset. Around 40 cases occur a year in the UK, with 2–3 (5%) deaths.

Scalded skin syndrome is caused by the exfoliative toxin. This results in blistering lesions of the skin and is especially seen in babies and young children. In extreme cases sheets of epidermis (stratum corneum) exfoliate leaving a moist glistening surface. The mortality rate in children is 1–5% but much higher (20% to more than 50%) in adults.

Food poisoning, as already indicated, is due to a toxin preformed in food during growth of the organism. A structurally related family of enterotoxins has been described (types A, B, C1, C2, C3, D, E, G–I) and these can be distinguished serologically, although most kits have only been developed to distinguish between the better described toxins A–D; E cross-reacts with the antiserum for A. PCR-based identification using specific primers tends to be used for the more recently described enterotoxins and traditional and molecular methods of typing have been compared (Tenover *et al.*, 1994, 1997). The enterotoxins are globular single-chain proteins which are very heat stable (destroyed at 126°C in 90 min), pH stable and resistant to the proteolytic enzymes rennin, trypsin and pepsin. This means that when present in food the toxin is not destroyed by normal cooking procedures and will resist the pH and digestive enzymes, allowing it to pass through the intestinal defences. A minimal toxic dose of >90 ng (1 μg for toxin A) is usually quoted as causing disease which equates to around 10^5 cells. Symptoms are the rapid onset (within 2–6 h of consumption) of nausea, vomiting, abdominal cramps and diarrhoea which last for around 24 h. The illness is self-limiting and rarely fatal. Although described as an enterotoxin, the mode of action is not that of a classical enterotoxin (see *Vibrio*) (section 6.8.1.4) as the toxins do not act on intestinal cells directly. In fact they are neurotoxins which activate receptors on the abdominal viscera. The stimulus reaches the vomiting centre via the vagus nerve and sympathetic nervous system producing the response.

The minimum temperature for growth of the bacterium is 6°C and a water activity of 0.85 – the lowest for a non-specialised bacterium. However, toxin production does not occur over the whole growth range, with a minimum of 10°C and a_w 0.9 needed for toxin production. Thus toxin production is readily prevented by good refrigeration. Food vehicles causing disease are typically foods of animal origin and cooked foods contaminated by food handlers and then inadequately refrigerated.

A classic example of food poisoning caused by *S. aureus* occurred in June 1988 and involved particularly a group of Japanese businessmen. This occurred in the business area of London and involved a number of hospitals. Forty two people attended hospital with diarrhoea, vomiting and abdominal pain which onset 2–6 h after lunch; the duration of the symptoms was 3–6 h, but for some patients they lasted 12–24 h. All persons involved had eaten a pre-packed Japanese meal including scrambled egg, mange tout, salted herring roe and boiled rice. Of the 61 people eating the food 44 became ill – an attack rate of 72%, suggesting a close association of the disease with this food source. Twenty two samples were collected

and tested (either mixed or individual ingredients). All were found to contain staphylococcal enterotoxin A, an enterotoxin A producing strain of *S. aureus* at 7×10^6 to 5×10^9/g together with *B. cereus* at 1×10^6 to 1.5×10^8/g. Investigations into the production of the food revealed that the production premises were the basement of a hostel rented nightly and were described as having a "very low standard of cleanliness". There were five food handlers, all of whom were untrained and an *S. aureus* enterotoxin A producer was isolated from the nose of one of the handlers. The foods such as rice, eggs and meat had been cooked in bulk and allowed to stand at room temperature until packed. The foods were packed in metal trays with lids and stored at room temperature until delivered ~5 h later. Delivery was in an uninsulated sports bag with the first delivery at 9.30 a.m., around 6 h after production, and the latest at ~12.30 p.m., around 9 h after production. All the elements needed for cross contamination from an infected handler onto cooked foodstuffs which were not refrigerated were present, thus allowing the organism to grow and produce toxin leading to the intoxication events.

6.8.1.3 Bacillus cereus

Table 6.5 describes the characteristics of *Bacillus cereus* and related species. *B. cereus* is a Gram positive and spore-forming rod (cells are between 1 and 5 μm long). The bacterium grows in chains and is a facultative anaerobe with a temperature range for growth of 8–55°C and an optimum between 28 and 35°C. It grows down to a water activity of about 0.95 and a pH of 5.0. The organism is perhaps the most common found in natural environments with its spores having $D_{95°C}$ values varying between 1 and 36 min. As discussed earlier, it is able to cause two types of foodborne illness. Originally it was suggested that the two types of illness were produced by different strains, but it appears that at least some strains can produce both forms of the illness. There is a rapid onset emetic intoxication with an incubation period of between 0.5 and 5 h where the infectious dose is 10^5–10^8 cells g^{-1} with nausea and vomiting and an illness lasting between 6 and 24 h. It shows close similarity to *Staphylococcus aureus* intoxication. The organism can also cause diarrhoea with an incubation period of 8–16 h and an illness lasting between 12 and 24 h with abdominal pain, profuse watery diarrhoea and rectal tenesmus. This disease is described as

Table 6.5 Characteristics of *Bacillus cereus* and related species.

Character	B. cereus	B. thuringiensis	B. mycoides	B. anthracis	B. megaterium
Gram stain	1	1	1	1	1
Catalase	1	1	1	1	1
Motility	2	2	4	4	2
Reduction of nitrite	4	1	1	1	4
Tyrosine decomposition	1	1	2	4	2
Lyozyme resistant	1	1	1	1	4
Egg yolk reaction	1	1	1	1	4
Anaerobic utilisation of glucose	1	1	1	1	4
VP reaction	1	1	1	1	4
Acid from mannitol	4	4	4	4	4
Haemolysis of sheep red blood cells	—	—	—	—	—
Distinguishing features	Enterotoxins	Endotoxin crystals. Pathogenic to insects	Rhizoidal growth	Pathogenic to animals including *H. sapiens*	— —

1 90–100% of cells positive
2 50% of cells positive
3 90–100% of cells negative
4 Most strains are negative
From Rhodelamel and Harmon (1998)

C. perfringens-like. The emetic toxin is a 1.2-kDa cyclic peptide, cereulide, which is produced in the food in the late exponential and early stationary phase of growth and may act by stimulating the vagus nerve. This toxin is only inactivated by heating at 126°C for 90 min and so is effectively heat resistant in cooking procedures and survives pH values between 2 and 11; it is also protease resistant.

The diarrhoeal illness is as a result of toxin production in the small intestine after consumption of 10^5–10^6 cells. There are at least two enterotoxins (one of which is haemolytic) which bind to epithelial cells and disrupt the membrane after production in the late exponential/early stationary phases of bacterial growth and are sensitive to proteolytic enzymes, as well as being inactivated by heating at 56°C for 5 min.

The emetic illness is often related to consumption of cooked rice and pasta. In the UK it is referred to as the "Chinese restaurant syndrome" – where rice is heated sufficiently to activate the spores which germinate into vegetative cells. These grow if the rice is not cooled quickly. Subsequently, the rice is warmed through but often does not reach the 126°C required to inactivate the toxin. In the case of the diarrhoeal illness a variety of foods including meats, vegetables, soups, milk and sauces have been implicated.

Given the number of cells involved in the illness, an enrichment technique is not generally required and a blood agar medium with polymyxin to prevent the growth of Gram negative bacteria is sufficient for detection. After 24 h at 37°C, *B. cereus* forms large (3–7 mm diameter) flattish, grey-green colonies with a ground glass texture which are surrounded by haemolysis. A selective medium which contains polymyxin/pyruvate/egg yolk/mannitol/bromothymol blue agar (PEMBA) is often used, sometimes with actidione to reduce mould and yeast growth. *B cereus* has turquoise blue (the colour of the bromothymol blue) crenated colonies which cannot ferment mannitol to produce a yellow colour, but are surrounded by egg yolk precipitation caused by lecithinase. Pyruvate improves this precipitation while a low concentration of peptone increases spore formation. Confirmation is carried out by testing for growth on media containing glucose, mannitol, xylose and arabinose since only *B. cereus* can produce acid from glucose but not mannitol, xylose or arabinose. A diagnostic real-time PCR assay has been developed for the detection of emetic strains (Fricker *et al.*, 2007).

Although *B. cereus* can spoil milk, producing "bitty cream" when stored at too high a temperature, no cases of food poisoning from consuming bitty cream have been reported in developed countries (Mabbit *et al.*, 1987), possibly because milk is not suitable for toxin formation, or more simply because the product is obviously spoilt and no one will consume it.

Other species of *Bacillus*, including *B. subtilis*, have been known to cause foodborne illness, where *B. subtilis* is able to spoil bread by producing "ropey bread" which has a sticky, slimy interior. If this is eaten, a mild illness can occur.

6.8.1.4 Vibrio

The genus *Vibrio* (see Kaysner and DePaola (2004) for a review) includes a number of species pathogenic for man, especially where wound infection occurs. In addition, *V. cholerae* (which causes cholera) and *V. parahaemolyticus* are of concern to the water and food industries. Described as Gram negative, pleomorphic (curved/straight), short motile rods with the polar (end of cell) mostly single flagellum, they are facultatively anaerobic and oxidase and catalase positive. NaCl stimulates growth and *V. parahaemolyticus* grows best at 3% (w/v) NaCl but has a range of 0.5–8% and a minimum a_w between 0.937 and 0.986. Growth occurs at a temperature as low as 5°C and as high as 43°C with an optimum at 37°C. *V. parahaemolyticus* grows optimally at pH values of 7.5–8.5 but is able to grow at up to pH 11.0 and down to 4.5. *V. cholerae* is a marine organism, while *V parahaemolyticus* is generally associated with coastal waters but apparently is able to overwinter in the mud when temperatures fall below 15°C. Most isolates are non-pathogenic.

Cholera has an incubation period of between 1 and 3 days and can be life threatening. The infectious dose is about 10^3–10^4 cells. The cells grow in the lumen of the intestine to produce an enterotoxin which causes the hypersecretion of Na^+, K^+, Cl^- and bicarbonate. This, in turn, results in a profuse, pale, watery diarrhoea containing flakes of mucus described as rice water stools, and which can be up to 20 litres a day and contain as many as 10^8 bacterial cells ml^{-1} with vomiting but no fever or nausea. Without treatment (replacement of fluid and electrolytes) there is a fall in the amount of blood, leading to increased blood viscosity, renal failure, circulatory collapse and, within several days, death. Rapid treatment with an electrolyte/glucose solution will reduce the death rate from 50% to less than 1%. In the Indian subcontinent, cholera killed more than 20 million people in the twentieth century and, during the nineteenth century, moving at roughly 8 km each day, it spread across Europe, reaching England in 1831, with the imminent arrival of a second outbreak in 1848, and reaching New York in 1866. As a result the authorities were persuaded to begin improving the water and sewage systems.

In comparison, *V. parahaemolyticus* has had a less important place in man's history, although when conditions allow it will double in as little as 11 min. The illness generally has an incubation period of 9–20 h but can have a range from 2 to 4 days lasting for up to 8 days, again with a profuse watery diarrhoea but without blood or mucus, abdominal pain, vomiting or fever. This organism is, however, more enteroinvasive than *V. cholerae*. *V. parahaemolyticus* produces a heat-resistant haemolysin which acts as cardiotoxin, cytotoxin and enterotoxin.

The preferred selective agar is TCBS (thiosulphate/citrate/bile salts/sucrose agar) when *V. cholerae* ferments sucrose and produces yellow colonies, while other *Vibrio* including *V. parahaemolyticus* are unable to ferment sucrose and produce green colonies. Realtime PCR methods are being validated for detection and the cholera toxin gene can also be detected by PCR (Koch *et al.*, 1995).

V. cholerae originates from water, so that food washed with contaminated water is an obvious problem. With *V. parahaemolyticus* foodborne illness is as a result of consumption of shellfish and other fish and because of the high consumption of raw fish, Japan has had a major problem, with 45–70% of outbreaks of foodborne gastroenteritis or food cross-contaminated with *V. parahaemolyticus* due to eating fish. In addition, there have been outbreaks in the USA caused by the consumption of fish.

6.8.1.5 Yersinia enterocolitica

Yersinia enterocolitica is a member of the Enterobacteriaceae and is a short (0.5–1.0 × 1–2 μm) Gram negative rod. It is facultatively anaerobic and catalase positive but oxidase negative and is a psychrotroph able to grow slowly from a temperature at −1°C to above 40°C, although its optimum is about 29°C. Below 30°C it is motile with peritrichous flagella, but it is non-motile at 37°C. D values at 62.8°C vary from

0.7 to 57.6 s. Its pH range for growth is between 4.1 and 8 with an optimum between 7 and 8.

The organism can be isolated from fresh water, soil and the intestine of animals and surveys in the UK have shown that it has colonised many cattle, sheep and pigs. From there, it has been isolated from foods including meat (particularly pork), poultry, fish, shellfish, milk, fruits and vegetables.

Most isolates from food are non-pathogenic and, where illness occurs, it is generally in children under seven. The incubation period is between 1 and 11 days and the illness usually lasts 5–14 days, with abdominal pain, diarrhoea and a mild fever. The pain can be localised and leads to a misdiagnosis of appendicitis. In adults, especially women, complications (arthritis and red skin) can occur but are more often found from serotype isolates in Europe. The bacterial cells adhere to mucosal cells of the gut's lymphoid tissue. Severe disease is rare, but in immunocompromised patients septicaemia and death can result.

All pathogenic strains have a 40- to 48-MDa plasmid which encodes for bacterial outer membrane proteins, although invasion is controlled by chromosomal genes. Selective isolation is generally by the use of CIN (cefsulodin/irgasan/novobiocin) agar. This contains mannitol as a carbon source and deoxycholate and crystal violet (selective agents) as well as the antibiotics. Incubation is for 24 h at 28°C and typical colonies have a dark red centre with a transparent edge. About 300 cases are reported in England and Wales each year, but under-reporting is likely because of poor diagnosis. Only pigs carried the major pathogenic biotypes (McNally et al., 2004).

The organism is isolated most often from the tongue and tonsils of a healthy pig, but outbreaks of illness are not frequently related to pigs. Nevertheless if you keep a pet pig and you do want to show affection it is better to avoid kissing, especially with tongues!

Interestingly, a related organism, Yersinia pestis, causes bubonic plague which killed 25% of the population of Europe in the fourteenth century and, for a time, resulted in the working classes having the upper hand in the peasant/landlord relationship because labour was so scarce.

6.8.1.6 Clostridium perfringens

Clostridium perfringens not only causes foodborne illness and is the third most significant cause of foodborne illness in the USA, but also is responsible for gas gangrene and, during the First World War, trench foot as well as pulpy kidney in sheep. It is Gram positive, anaerobic, rod-shaped with subterminal oval spores. The rods are large (1 µm to 3–9 µm long) and are invariably said to be non-motile. However, recent research has shown that the cells are able to send out threads with which they can pull themselves along surfaces in the same way as cells of several Gram negative species. Interestingly the use of molecular biology has shown that other clostridia, such as C. botulinum, C. difficile and C. tetani, also carry these genes, suggesting that cells of these species may also be able to glide on an agar surface (Varga et al., 2006). This would provide cells on surfaces with a clear evolutionary advantage. Although catalase-negative, cells of C. perfringens are able to grow sufficiently to produce visible colonies on suitable agar medium aerobically. Growth will occur at temperatures as low as 12°C (but very slowly below 20°C) and as high as 50°C, with a temperature optimum between 43°C and 47°C. Amazingly, growth at 41°C has been measured at a scarcely believable generation time (doubling time) of 7.1 min. The cells will grow at pH 5.0 but with an optimum of pH 6.0–7.5 and have an a_w for growth of 0.95–0.97. Growth is, however, prevented by 6% sodium chloride.

The vegetative cells are relatively heat resistant and have D values at 60°C of several minutes. The spores of some strains are unexpectedly not especially heat resistant with D_{100} values of 0.31–38 min. The organism produces four major exotoxins as well as eight others and these are used to divide the species into five types (A to E). Type A causes food poisoning and gas gangrene and produces a lecithinase (the α major toxin) which is involved in gas gangrene but not foodborne illness.

Type C, which produces α and β toxins, causes a severe disease (enteritis necroticans) with the β-toxin causing necrosis of the intestinal mucosa. This problem, known as pig-bel, occurred especially in Papua New Guinea. Teenagers were allowed to take part in a celebration for the first time and ate pig which had been raised in muddy sties and then was cooked over an open fire. Spores of C. perfringens reached high numbers and the teenagers were exposed to large doses of the organism, which had not happened previously. This difficulty was made worse because of the mainly vegetable diet, which contained high levels of protease inhibitors, since protease enzymes would have broken down the toxins if they had been present in sufficiently large quantities. Under these conditions the organism produces abdominal pain and bloody diarrhoea and can result in the digestion of the infected intestine and death. The genes encoding the α-toxin or phospholipase C and Ø-toxin or perfringenolysin, are located on the chromosome, but many other genes encoding for other extracellular toxins are located on large plasmids (Rood, 1998).

Types B, C, D and E are usually believed to be strains that cause illness in animals, but type A is widespread and is generally regarded as the cause of both foodborne illness and gas gangrene and it can be present in soil at 10^3–10^4 g^{-1}, as well as being isolated from raw and processed foods, together with mud in streams and rivers. It can also be found in the faeces of healthy humans at 10^3–10^4 g^{-1}.

Generally the foodborne illness is characterised by nausea, abdominal pain and diarrhoea 8–24 h after consumption of food with usually 7×10^5 g^{-1}. The illness generally takes 1–2 days to run its course in immunocompetent people. It is apparent that those vegetative cells that gain entry to the small intestine grow and sporulate, producing the enterotoxin which is released by lysis of the mother cell. The protein enterotoxin is inactivated by heating at 60°C for 10 min, as well as being destroyed by proteases. It acts at the human cell membrane, reversing the flow of water, Na$^+$ and Cl$^-$ across the gut epithelium, resulting in secretion rather than absorption and killing the cell by producing pores in the membrane.

Detection uses plating media with antibiotics as selective agents (e.g. tryptose/sulphite/cycloserine-TSC or oleandomycin/polymyxin/sulphadiazine/perfringens (OPSP)). The plates are incubated at 37°C for 24 h to produce black colonies in a pour plate as a result of sulphite reduction. On spread plates the colonies can be white. Confirmation is by lactose fermentation, gelatine liquefaction, reduction of nitrate to nitrite, molecular techniques (Keto-Timonen et al., 2006) and the absence of flagella producing motility (but see earlier for a discussion of motility in C. perfringens, section 6.8.1.6). Problems usually occur after consumption of meat, generally where a dish has been prepared in advance of eating and has not been refrigerated adequately. Cured meats are not often a problem because of the salt and nitrite together with the heating process. Many outbreaks occur in large-scale catering establishments including, unfortunately, hospitals and residential care homes, where the residents are more likely to be immuno-deficient and susceptible. Recent work funded by the Food Standards Agency (FSA) in the UK has shown that growth during the cooling of meats can be predicted under different temperatures, pH values, a$_w$ and salt and nitrite levels (Peck and Baranyi et al., 2007); this is available as The Perfringens Predictor from the Institute of Food Research, Colney Lane, Norwich, UK (http://www.ifr.ac.uk/safety/growth predictor/perfringens/predictor.zip).

6.8.1.7 Salmonella

Salmonella is a member of the family Enterobacteriaceae like *Escherichia* and *Yersinia* and is a Gram negative rod which is facultatively anaerobic, has a growth optimum of 37°C and is bile salt tolerant. It is found primarily in the gastrointestinal system of infected humans and other animals and is therefore in their faeces, which is a route of transmission to food products, particularly those of animal origin. Unlike *E. coli*, the organism does not ferment lactose and is therefore not a coliform. Bile salt tolerance and lactose utilisation are the key features used for selective media which distinguish Enterobacteriaceae members (see MacConkey agar below). An unusual feature of *Salmonella* taxonomy has been the large number of species named (~2200). This has arisen from the Kauffman–White scheme which used serotyping to define each species.

A serotype or serovar is defined as an antigenically distinguishable member of a bacterial species. In the Enterobacteriaceae a number of antigens are used to define the serotype: the major antigen is the lipopolysaccharide of the cell wall (the O or somatic antigen); variation in the polysaccharide component gives rise to different O antigens. A second antigenic component is the flagella antigen (H antigen); most *Salmonella* isolates have two alternatives which are termed phases 1 and 2. A third antigen, found only in *Salmonella*, is the capsular or Vi antigen. Although O and H antigens are used for serotyping in *E. coli* and other Enterobacteriaceae, these have not been given specific names but a serovar designation: so *E. coli* O157: H7 is a specific serovar of *E. coli* with O antigen 157, H antigen 7. In the Kauffman–White scheme of classification for *Salmonella*, the combination of these antigens distinguished a particular serotype which was given a species name. This led to very fine distinctions being made between *Salmonella* species thus:

Salmonella typhimurium	O1, 4, (5), 12:Hi;1,2
Salmonella lagos	O1, 4, (5), 12:Hi;1,5

This was inconsistent with modern taxonomic ideas and a reconsideration of the *Salmonella* taxonomy was made. In this re-evaluation most *Salmonella* causing foodborne diseases are serovars of the species *Sal. enterica* and members of one specific subspecies *enterica*. Thus the full nomenclature for *Salmonella typhimurium* should be *Sal. enterica* subspecies *enterica* serovar Typhimurium. This is acceptably shortened to *Sal.* Typhimurium.

Pathogenicity of Salmonella

Salmonella serovars show important variations in their pathogenicity to man and other animals. In particular, a number show specific adaptations which allow them to cause serious disease in a particular host. Independent horizontal gene transmission events are thought to have allowed acquisition of all the virulence factors needed for disease and the acquisition of certain virulence genes explains the host-adaptive variations seen.

Human-adapted serovars are *S.* Typhi and *S.* Paratyphi A, B and C; it is this group that express the Vi antigen. These serovars cause typhoid and paratyphoid fever, a systemic enteric fever typified by headache, loss of appetite, abdominal pain, diarrhoea (but not with *S.* Typhi where constipation is more characteristic) and continued fever. Disease is caused by ingested bacterial cells invading from the intestine into the lymph nodes; from here they are released from macrophages into the blood and disseminated to organs (including the liver, especially the gall bladder, spleen, kidneys and bone marrow) where further multiplication and shedding occurs. After 24–72 h there is reinfection of the intestine via bile and ulceration may appear 8–15 days after infection. A high mortality of 10–15% can result for untreated patients.

Host-adapted species also exist for animals, for example *S.* Gallinarum, *S.* Pullorum (poultry); *S.* Abortus-ovis (sheep); *S.* Choleraesuis (pigs); *S.* Dublin (cattle). These last two can also be human pathogens and in the case of *S.* Choleraesuis can cause symptoms of septicaemia, pneumonia, osteomyelitis and meningitis in a human host. Other *Salmonella* serovars are considered unadapted and these comprise the majority of serovars causing disease in man including the common types *S.* Typhimurium and *S.* Enteritidis.

Gastroenteritis is caused by the ingested bacteria attaching to and invading the epithelium of the small intestine, causing disease by killing epithelial cells and inducing fluid accumulation (diarrhoea). However, even with these non-host adapted serovars, systemic infection can sometimes occur. The non-systemic gastrointestinal disease caused by these serovars is termed salmonellosis and consists of diarrhoea, vomiting and a low-grade fever; incubation is typically 12–48 h but can be as short as 8 h or as long as 4 days. Duration is typically from 1 to 7 days and the disease is generally self limiting with a low mortality rate of ~0.1%. The only treatment needed is fluid and electrolyte replacement, with antibiotics normally only given when systemic disease occurs.

The infectious dose needed to cause gastrointestinal disease is generally high, $>10^5$ cells, but examination of outbreaks has shown that this depends on a number of factors. The host is important, with fewer cells needed in susceptible individuals (such as the old and the very young). The type of food vehicle also has an influence, with foods like chocolate causing disease with an estimated 1 cell/g and cheese with estimates of 1.5–9.1 cells/100 g. Some strains of particular serotypes are also more pathogenic: *S.* Typhimurium DT (definitive phage type) 104 is a multi-antibiotic-resistant isolate typically resistant to ampicillin, chloramphenicol, streptomycin, sulphonamides and tetracycline. Whilst normally this is not a problem as treatment does not require antibiotics, this strain is highly virulent, becomes systemic more readily and hence needs antibiotic treatment. The mortality rate associated with this strain is consequently much higher at ~5%.

Live bacteria are shed in the faeces whilst symptoms persist, but even when these cease, asymptomatic carriage of non-host adapted species can follow infection for around 5 weeks; rarely this may persist for up to a year and is more prolonged in children under 5 years. Three successive negative faecal samples are required for the person to be considered free of the organism; this is clearly important for food handlers. This asymptomatic carriage is also important in livestock as animals may be carrying and transmitting the organism with no symptoms.

Carriage of *S.* Typhi can be prolonged, the longest known being 52 years. Typically the organism is carried in the gall bladder and is shed intermittently into the intestine in the bile. Treatment is problematic as antibiotic treatment is understood to exacerbate the situation.

Salmonella *food sources*

Salmonella cells are heat sensitive (but see Barrile and Cone, 1970) and so contamination of food is often a failure of undercooking or contamination of raw to cooked product. Food vehicles for salmonellosis are undercooked poultry and poultry products such as eggs, meat and meat products. Raw milk and its products such as raw milk cheeses may contain the organism, but pasteurised milk and cheeses should be safe unless a failure in production has occurred. Salad vegetables including salad leaf products and bean sprouts have increasingly been a cause of salmonellosis; this is because the products are eaten raw and often ready prepared and hence any organisms present will not be removed. The problem of internalisation of the organism into such products has been reviewed and will not be discussed further here. Another product that may present a problem is chocolate.

Salmonella *isolation*

Salmonella is normally present in foods in low numbers, but there is a need in ready-prepared foods to be able to detect 1 cell in 25 g of food product. Consequently enrichment procedures are used for isolation. This involves pre-enrichment of the sample in a low nutrient medium such as buffered peptone water, followed by selective enrichment and then selective plating and confirmation of isolates by serotyping. The whole process takes up to 4 days. Selective media may include bile salts and lactose as two important selective and diagnostic features. The classical medium is MacConkey agar in which bile salts and crystal violet are the selective agents and the diagnostic agents are neutral red (pH indicator) and lactose. Those Enterobacteriaceae that have the ability to ferment lactose with the production of acid (coliforms) appear red on this medium whilst non-lactose utilisers (e.g. *Salmonella, Shigella, Yersinia*) appear colourless. Other media more specific for salmonella have been devised (e.g. xylose lysine deoxycholate (XLD), Brilliant Green agar), but these still rely partly on bile salts and lactose utilisation in the composition. DNA hybridisation has also been developed (D'Aoust, 1998).

6.8.1.8 Escherichia

Escherichia, like *Salmonella*, is a member of the Enterobacteriaceae and therefore is a Gram negative facultatively anaerobic rod associated with the faeces of warm blooded animals. Unlike *Salmonella, Escherichia* is lactose fermenting and therefore a coliform. *E. coli* is the key member of the genus; this organism forms part of the normal flora of the alimentary tract of man, farm and domestic animals and consequently its presence in food and water is used as an indicator of faecal contamination.

Although considered as a gut commensal, *E. coli* has been recognised as a pathogen in recent years. The strains causing disease (diarrhoeagenic strains) have a number of characteristics which make them more similar to the genus most closely related to *Escherichia*, the genus *Shigella*. Like *Shigella*, the diarrhoeagenic *E. coli* strains have a high proportion of lactose negative strains or are late lactose fermenters, showing lactose utilisation after 14 days. Some such as the enteroinvasive (EIEC) strains are often non-motile and sometimes do not produce gas from carbohydrate fermentation (non-gas producing) and thus resemble *Shigella* isolates even more closely. There are also similarities in mechanisms of pathogenicity and routes of transmission.

6.8.1.9 Shigella

All members of the genus *Shigella* cause bacillary dysentery (shigellosis) in man. Severity varies from mild diarrhoea usually associated with the only species endemic in the UK, *S. sonnei*, to classical dysentery; symptoms of the latter occur 12–50 h after infection, with fever, abdominal cramps and frequent liquid stools which are often bloody. This lasts for 3–4 days or 10–14 days in severe cases but is rarely fatal as long as fluid and electrolyte replacement are available. Transmission is via water or person to person spread as there are no animal reservoirs; food outbreaks are known, but the original source is always a contaminated handler or contaminated water source. The infective dose is 10–10^4 cells depending on species. The mechanism of disease caused by *Shigella* is through invasion of the epithelial cells of the colon and this ability is a plasmid-determined character (120–140 MDa). Entry into epithelial cells and intracellular spread are by mechanisms similar to those used by *Listeria monocytogenes*, although the genetic determinants are quite different.

Some *Shigella* strains also produce a toxin, the Shiga toxin, which is a potent protein exotoxin produced in the gut. This has cytotoxic abilities and a key method of action is to bind to and inactivate 60S ribosomes of the host cell, thereby stopping protein synthesis and killing the cells. The Shiga toxin is found in high levels in strains of the species *S. dysenteriae* type 1, with other serotypes and species producing a Shiga toxin-like cytotoxin at lower levels. Other toxins may also be involved, but their roles are not well characterised.

6.8.1.10 Pathogenic E. coli

Pathogenic *E. coli* cause disease by a number of different mechanisms and they are given designations on this basis: enteroinvasive *E. coli* (EIEC), enterotoxigenic *E. coli* (ETEC), enteroaggregative *E. coli* (EAEC), diffusely adherent *E. coli* (DAEC), enteropathogenic *E. coli* (EPEC) and enterohaemorrhagic *E. coli* (EHEC). These abilities have been gained by the acquisition of a series of genes, either plasmid, phage or pathogenicity island determined. With the exception of EHEC, all are associated primarily with transmission through water and person to person spread; however, EHEC have become major food-borne pathogens.

6.8.1.11 Enteroinvasive E. coli

A close similarity exists between the pathogenicity mechanism of enteroinvasive *E. coli* (EIEC) and that of *Shigella* and, as indicated above, the two resemble each other in particular phenotypic characteristics. EIEC invade and destroy epithelial cells lining the colon by the same mechanism of invasion as *Shigella*. They also produce the same symptoms as shigellosis, from a mild form to severe, classical dysentery-like symptoms, and they do this by the presence of plasmid determined genes on a 140-MDa plasmid related to that of the shigellas. A major difference, however, is the number of cells needed to cause disease, with 10^9 cells needed for EIEC to be infective.

6.8.1.12 Enterotoxigenic E. coli

Enterotoxigenic *E. coli* (ETEC) cause disease by colonisation of the surface of the small bowel and the elaboration in situ of toxins. This results in diarrhoea which varies from mild to severe; typically this causes watery stools lacking blood or mucus. Sudden explosive onset is common and may be accompanied by abdominal cramps and vomiting. This is a typical cause of infant and traveller's diarrhoea. Again a high infective dose is needed. ETEC produce either a heat stable (ST) or a heat labile (LT) enterotoxin or both together; again this is a plasmid determined

characteristic. The LT toxin is related to cholera toxin with the cellular target being adenylate cyclase. Over-stimulation leads to a net secretion of Cl^- ions and inhibition of NaCl absorption; water moves into the lumen resulting in diarrhoea. The ST toxin mechanism is similar but targets a different site, that of the guanylate cyclase system. In addition, strains may also produce fimbrial adhesins which are encoded on the same plasmids as the toxins. These adhesins are host specific: K88 for pigs, K99 for calves and lambs and CFAI and II for humans. They allow adhesion of the bacteria to the host cell and encourage retention and secretion of the toxin close to the host cell; presence of the adhesins produces more severe disease, and in young animals like piglets which are susceptible to ETEC at the weaning stage, scours can result and spread rapidly, leading to death if untreated.

6.8.1.13 Enteropathogenic E. coli

Enteropathogenic *E. coli* (EPEC) cause a watery diarrhoea often with vomiting and low-grade fever; disease may be mild to severe, prolonged and life-threatening and is characteristic of infantile diarrhoea. Again 10^8–10^9 cells are needed in adults to cause disease. The mechanism of pathogenicity is complex with a characteristic attaching-and-effacing ability. Initial adherence is plasmid controlled and a result of the production of adhesive fimbriae, the bundle-forming pilus. Subsequently intimate adherence and the formation of pedestals occurs which is determined by a chromosomal pathogenicity island (*eae*) which codes amongst other things for the adhesin intimin and signalling pathways. These produce cytoskeletal changes to the cell resulting in the accumulation of polymerised actin beneath attached bacteria and the formation of pedestals. Characteristic lesions of the brush border develop.

6.8.1.14 Enterohaemorrhagic E. coli

Enterohaemorrhagic *E. coli* (EHEC) are part of the group known as verotoxin-producing *E. coli* (VTEC) and are associated with certain serotypes of *E. coli*, especially O157:H7. *E. coli* O157:H7 causes a number of syndromes that are generally more severe in their nature than those caused by the other pathogenic *E. coli*. Haemorrhagic colitis is a gastrointestinal disease typified by severe abdominal pain and the production of watery diarrhoea followed by grossly bloody diarrhoea. There is little or no fever and the syndrome is usually self limiting, resolving in 8 days, but it can be fatal in adults, principally the elderly where stroke and heart attack are common causes of death. The organism is shed for ~29 days after infection, but this can vary from 11 to 57 days in different individuals and often persists for longer in young children. Given the low infectious dose of the organism (see below) person to person spread is a serious risk.

Haemolytic uraemic syndrome (HUS) follows haemolytic colitis in around 7% of cases. Typical symptoms include acute renal failure, thrombocytopaenia (reduction in blood platelet number) and haemolytic anaemia (reduction in red blood cell number) and the syndrome is principally associated with children. Unexpectedly levels may increase through antibiotic treatment. Fatality levels, associated with particular outbreaks, have ranged from 6 to 31%, with long-term kidney damage a possible outcome. Interestingly, *Shigella dysenteriae* 1 can also produce HUS and this gives a clue to the cause of the disease.

A further syndrome is thrombotic thrombocytopaenic purpura; this is related to HUS but also induces fever and neurological symptoms such as agitation, headache and disorientation which may progress rapidly to hemiparesis, seizures, coma and death. A key feature is platelet agglutination, and mortality rates of around 90% were evident until plasma exchange therapy became available.

EHEC have a similar ability to EPEC strains of producing attaching and effacing lesions. In addition they (and all VTEC strains) produce a cytotoxin termed a verotoxin. Verotoxin 1 (VTI) and verotoxin II (VTII) show a strong relatedness to the Shiga toxin produced by *Shigella* and so are also known as Shiga-like toxin I and II (SLTI and SLTII). VTI is almost identical to Shiga toxin with one amino acid different in its sequence; VTII is 60% similar in amino acid sequence to the Shiga toxin. Both verotoxins are encoded by a chromosomally inserted lysogenic phage and it is the circulating toxin that is responsible for HUS, targeting and destroying cells in the kidney. The phage is believed to be induced by antibiotic therapy increasing the expression of the toxin genes; hence the poorer outcome of antibiotic-treated patients. There is a very low infectious dose for the organisms suggested as ~10 cells and the likelihood that it may be aerosol transmitted resulted in it being classed as an ACDP 3 pathogen from 1998.

The emergence of EHEC O157:H7 as a major pathogen in the 1980s has led people to look for the origin of the strain in EPEC strains, as effectively it is an EPEC which has gained a lysogenic phage. The EPEC strain O55:H7 and the *E. coli* O157:H7 have many intestinal adherence and other virulence factors in common and this appears to be the ancestral form of O157:H7.

Initially *E. coli* O157:H7 was mainly associated with cattle and cattle products (beef, milk), but it has now become more widespread and is associated with other animals and food types. Salad vegetables in particular have become a major vehicle, with a number of outbreaks caused by spinach and lettuce. However, food accounts for approximately 50% of cases, with direct zoonotic transmission from animal handling and contact with animal faeces a major cause of other cases.

E. coli O157:H7 is non-sorbitol fermenting and this is used as a diagnostic characteristic in the medium Sorbitol MacConkey agar (SMAC). This variation of MacConkey agar still contains bile salts and crystal violet to inhibit Gram positive flora but with a cefixime potassium tellurite (CT) supplement to increase the selectivity for *E. coli* O157:H7. Sorbitol is used as the carbohydrate source and this with the neutral red pH indicator turns sorbitol-positive colonies red. Sorbitol-negative strains form colourless colonies. Molecular methods are available (e.g. see Zhao *et al.*, 2000).

6.8.1.15 Non-O157 verocytotoxin-producing E. coli

Although O157:H7 has been the EHEC of most significance in Europe and the USA, the importance of particular serogroups varies with country and in other countries this serotype is unknown and other serotypes predominate; these other serogroups are in some cases also causes of disease in Europe and the USA. The main serogroups of concern are O26, O103 and O111; O26 and O111 also produce attaching and effacing lesions like O157:H7 and so are considered EHECs, although the site of *eae* insertion is not the same as in O157 and has come about by an independent transfer event. The prime source of isolation of these serogroups is cattle faeces, but sheep and goat have also been shown to carry them. The frequency of isolation of non-O157 VTEC from these sources is much greater than that of O157 itself. Studies in Canada showed the presence of non-O157 VTEC in 17% and 45% of cattle faeces respectively, but O157 was present in <1% of both cases. Similar studies in Germany showed that non-O157 VTEC were isolated at ten times the frequency of O157. However, despite this, the frequency of disease causation by non-O157 VTEC (which can cause bloody and non-bloody diarrhoea and HUS) is very much lower than that of O157. This may reflect a true difference in virulence between O157:H7 and these VTEC serogroups or it may be that the detection methods used are less well developed and so disease is attributed less frequently. The Shiga toxin is present on a phage in non-O157:H7 VTEC, but the toxin alone appears insufficient to cause significant disease without the background of other virulence factors.

6.8.1.16 Enteroaggregative and diffusely adherent E. coli

The other pathogenic strains are rather less well characterised. Enteroaggregative *E. coli* (EAEC) cause disease by the production of a plasmid-associated bundle-forming pilus which results in autoagglutination of the bacterial cells in a stacked-brick configuration on the surface of the host epithelial cells where they produce a cytotoxin; enhancement of mucus formation traps the bacteria which may aid their persistent retention. These bacteria may be associated with persistent diarrhoea which lasts for more than 14 days. The diffusely adherent *E. coli* (DAEC) are associated with the induction of finger-like projections from host cells mediated by the production of fimbriae.

6.8.1.17 Campylobacter

Although currently the most important cause of bacterial foodborne disease in industrialised countries and a major public health problem and economic burden, *Campylobacter* is a relatively recently identified foodborne pathogen. It was only in the 1970s that this organism was recognised as a cause of acute human enterocolitis, but it is now considered the most common cause of diarrhoea in the UK as well as the rest of Europe and the USA.

C. jejuni is the species considered most prevalent, causing around 90% of the reported foodborne disease. However, this may be influenced by the fact that characterisation of isolates is focused on differentiating *C. jejuni* from the rest. Other species important in human disease include *C. coli, C. lari (C. laridis), C. upsaliensis* and *C. hyointestinalis*. *Arcobacter* species can be isolated by the same methods as campylobacters

and are difficult to distinguish from them, but they are much more aero-tolerant. Whilst less common than campylobacters, they have also been shown to cause diarrhoea. A PCR assay has been developed to allow differentiation of the two.

Campylobacter is a slim Gram negative, oxidase positive spiral rod 0.5–0.8 µm long and 0.2–0.5 µm wide. However, under stress conditions it may become coccoid, e.g. if placed in water. It has an unsheathed polar flagellum and is highly motile. An unusual feature of its cell wall structure is that it has a lipo-oligosaccharide (LOS) attached to lipid A with a more loosely bound LPS capsular layer. Another feature that distinguishes it from other foodborne pathogens is that it is microaerophilic: this means it requires an oxygen level below atmospheric, with 5% O_2 being optimum. It is also capnophilic and grows better with elevated levels of CO_2. An atmosphere of 5% O_2, 10% CO_2 and 85% N_2 is therefore typically used for growth. Whilst all species grow at 37°C, some species (*C. jejuni, C. coli, C. lari*) grow optimally at 42°C and are thus termed thermophilic, although these are not thermophiles by the strict definition of the term. The lower limit for growth is considered to be 30°C, but certain species do grow at 25°C. The genus also has the characteristics of being pH sensitive below pH 5.9 and above pH 9, desiccation sensitive and temperature sensitive with D values for 50°C of 7.3 min and for 55°C of 1.1 min. Hippurate hydrolysis is a distinguishing feature for *C. jejuni* and is the only test routinely used to distinguish *Campylobacter* species.

Enterocolitis caused by campylobacter in industrialised countries typically involves acute abdominal cramps, fever, headache, dizziness and nausea followed by profuse watery diarrhoea with or without blood; less acute infections consist of a mild attack of diarrhoea. In developing countries disease generally takes the less acute form or results in asymptomatic excretion. The incubation period is typically 2–7 days with a mean of 3.2 days and lasts for approximately 7–10 days although relapse is common (~25% cases). Infection is usually self-limiting but can be prolonged and severe and in that situation will require antibiotic treatment, with erythromycin currently the antibiotic of use. Bacteraemia is rarely reported. The infectious dose has been reported as being as little as 500 organisms, but doubt has been expressed about the accuracy of this experiment.

Understanding of the way in which *Campylobacter* causes diarrhoea is progressing. Initially a cholera-like enterotoxin was suspected, but this was never shown and since sequencing of the *Campylobacter* genome no gene homologue has been found. The organism is known to cause infection and colonises the mucosal surface of the lower intestinal tract. The motility of the organism is a key factor in this. Flagella have been shown to be important and flagella rotation is adapted for penetration through the intestinal mucus to the epithelial cells. Serine, mucin and fucose are chemo-attractants, while bile acids are chemo-repellents. Hence flagella and chemotaxis mutants have an impaired ability to colonise. Multiple adhesins are also thought to be involved. CadF is an outer membrane binding protein which has been shown to bind to fibronectin and promotes binding to intestinal cells. The gene for CadF is conserved among all *C. jejuni* and *C. coli* isolates tested, indicating the importance of this gene for pathogenicity. It has been hypothesised that invasion of gut cells is contributory to causing inflammatory enteritis and a group of secreted proteins – the *Campylobacter* invasion antigens (Cia) – have been shown to be produced on cultivation with mammalian cells. Invasion of cultivated cells requires the protein CiaB, with strains lacking this protein being non-invasive (http://molecular.biosciences.wsu.edu/faculty/konkel.html).

A more significant outcome of campylobacter infection is Guillain–Barré syndrome (GBS), a serious autoimmune disease. This is a neurological condition which follows in 1/1000 cases of campylobacteriosis (Yuki *et al.*, 2004). Symptoms are an ascending paralysis which may lead to respiratory muscle paralysis and death; however, with intensive care patients may recover after several weeks. The link with *Campylobacter* to this syndrome has been that 8–50% of GBS cases are preceded by culture-positive *Campylobacter* infection and GBS patients are five times more likely to have serological evidence of *Campylobacter* infection. The syndrome comes about through the structure of the *Campylobacter* LOS showing molecular mimicry of human gangliosides. Antibodies produced in response to *Campylobacter* LOS also attack the nerves, resulting in this autoimmune disease.

The incidence of campylobacteriosis has continued to increase with many cases arising from sporadic incidences. There is a higher incidence in babies under 4 months of age and in young adults (with a greater incidence in males); in developing countries disease is usually restricted to children, with adults often asymptomatic when infected.

Virulence is not linked with particular serotypes and the same strain may cause different grades of illness in different people, suggesting the importance of host factors in the progress of the disease. Other animals can carry the organism in their intestinal flora and be asymptomatic, with a wide range of wild and domestic animals (pigs, cattle and companion animals particularly puppies and kittens) and birds of particular note. Chicken has been the major food vehicle implicated either from undercooked chicken or through cross contamination from chicken to other foods. A number of surveys have shown that a high proportion of carcasses on retail sale carry the organism. Barbecues have been highlighted in this regard. Occasional large outbreaks tend to be less important than sporadic incidences affecting only one or a few people. Large outbreaks have been associated with untreated water or milk. Raw milk is a potential source of the organisms, although the organism is killed by pasteurisation. However, door-step milk has been implicated in incidences and this has been attributed to birds pecking the bottle tops during early summer, although with the reduction in the extent of the door-stop delivery this must be less significant than previously.

Isolation of Campylobacter

The microaerophilic nature of *Campylobacter* is central to its growth and isolation. Modified gas atmospheres are essential and can be generated either using gas jars with sachets designed to give a specific gas mixture or using variable atmosphere incubators. A typical atmosphere used is 10% CO_2, 6% O_2, 82% N_2 (v/v) and these conditions are part of the selective process for isolation.

Media have also been designed to generate a low oxygen level and contain either blood or charcoal for this purpose. Selectivity is produced through the use of antibiotics such as rifampicin, polymixin B and trimethroprim, but specific agars use particular combinations, e.g. Preston agar contains these three antibiotics plus cyclohexamide; Exeter medium contains the first three plus amphoteracin and cefoperozone. In addition further supplements are used to enhance the aerotolerance of the organism and these include ferrous sulphate, sodium pyruvate and sodium metabisulphite. The most widely used blood-free medium is modified CCDA (campylobacter blood-free selective agar base); this is based on Preston agar but contains charcoal instead of blood and the antibiotic cefoperazone.

Temperature is also used as a selective feature as the three thermophilic species, *C. jejuni*, *C. coli* and *C. lari*, grow optimally at 42°C (which is the body temperature of birds which they readily colonise); many isolation procedures for *Campylobacter* therefore use this temperature as a selective element, but it must be remembered that not all *Campylobacter* species will grow at this temperature.

An alternative strategy exploits the small size of the organism. A 0.45-μm filter is placed on the surface of a non-selective blood agar plate and a drop of sample is placed on the top. *Campylobacter* will swim through this pore size and then the filter is discarded before incubation.

6.8.1.18 Mycobacterium

Mycobacterium tuberculosis *and* Mycobacterium bovis

In the nineteenth century tuberculosis caused by *Mycobacterium tuberculosis* was responsible for 30% of deaths under the age of 50 in Europe. Although this markedly reduced in the twentieth century, by 1980 the appearance of drug-resistant strains and the increase in susceptible individuals such as AIDS patients added to the continuing high levels of the illness in developing countries, resulting in a detected increase. By 1990 it was estimated that as much as a third of the population of the world was infected, with 7–8 million new cases found every year.

M. tuberculosis is spread from one person to another by aerial transfer. *M. bovis* causes tuberculosis in cattle and man, as well as other animals, and is also spread by aerial transfer. However, it can also be spread by drinking infected milk and, less frequently, by eating contaminated meat. In the 1930s as many as 3000 children each year were infected in England and Wales as a result of drinking contaminated milk. The organisms are Gram positive, pleomorphic aerobes between 1 and 4 μm long. Their cell wall is very hydrophobic so that nutrient uptake and, hence, growth and division are very slow (but see Stanley *et al.*, 2001, and Rees and Loessner, 2005). As a result, 7 days may be required before growth can be seen on agar media. The organism is resistant to drying and can remain viable for a long time in the environment. The cells are acid fast, that is, they are slow to stain but once stained they are difficult to de-stain.

M. bovis is ingested often by consuming milk, and cows in developed countries are regularly tested for the organism, with those positive slaughtered. In the

1930s in England and Wales as much as 12% of milk was contaminated. As a result milk pasteurisation was introduced into developed countries and this has markedly reduced the number of cases. At least one microbiologist has been heard to say that he would not drink unpasteurised milk or eat cheese made from unpasteurised milk unless he knew both the farmer and the cow personally. Nevertheless, despite this and the Richmond report which recommended only pasteurised milk be consumed, in the UK many people still refuse to avoid unpasteurised dairy products, while in lesser developed countries the situation is less clear with unpasteurised milk and contaminated meat widely on sale.

6.8.1.19 Listeria monocytogenes

Listeria monocytogenes is a Gram positive, facultative anaerobe. It has a coccoidal-rod shape and is up to 2 μm long. It is catalase positive and oxidase negative. When growing at 20°C or less it has peritrichous flagella and swims with a characteristic tumbling motion, but when growing at 37°C it does not produce flagella. Colonies on tryptose agar viewed under oblique illumination have a blue-green sheen. The optimum temperature for growth is between 30 and 35°C but the organism will grow down to 0°C and up to 42°C. Its minimum pH for growth is 5.5. *L. monocytogenes* will survive for up to a year in 16% (w/v) NaCl and will grow in 10% (w/v) NaCl. It appears that the organism can be isolated from most environments, including silage and other vegetation, soil, sewage and even drains in food production units, and from both fresh and sea water and it has been detected after surviving for several months. *L. monocytogenes* produces a 58-kDa β haemolysin and a listerolysin O. This interacts synergistically with the *Staph. aureus* haemolysin to give a stronger haemolysis on blood agar. This can be used to separate *L. monocytogenes* from *L. innocua* (the CAMP test).

The incubation period before symptoms develop can be 90 days. Symptoms can be a mild influenza-like illness, but in the very young, the pregnant, the elderly and the immunocompromised the infection can lead to meningitis. In pregnant women abortion, premature labour or stillbirth can result. In small babies abscesses and pneumonia can occur, but several days after birth meningitis is more likely.

During infection when flagella are not synthesised the bacterium uses the host cells' actin to move around the body. The system works by the actin of the host cell polymerising onto the bacterial cell surface. The bacterial cell then passes around the cell before moving into a new host cell. This allows the *L. monocytogenes* to avoid the immune system of the host. As the next step the bacterial cells come to the mesenteric lymph nodes and then pass into the blood, allowing them to move around the whole body and eventually into the central nervous system or the placenta in pregnant women. This can lead to a mortality rate of up to 34% in adults, unless the infection is stopped. The liver is significant here and an inflammatory reaction results from the infection of hepatocytes with the result that the bacteria are set free and then killed. The methods of isolation utilise selective agars with lithium chloride, glycine anhydride, phenylethanol and antibiotics. Sugar fermentation is then used to separate *L. monocytogenes* from other *Listeria* species.

Research has shown that there are 13 serotypes of *L. monocytogenes* but that most of the human cases are caused by only three (1/2a, 1/2b and 4b) (Lyytikäinen, 2000). Food sources include cheese, especially soft cheese, paté, milk, cook-chill chicken and coleslaw, as well as pig's tongues in aspic in France caused by serotype 4b where 279 cases were reported in 1992 with 63 deaths and 22 abortions. In addition, in 1985 Mexican-style soft cheese in California was responsible for 142 cases with a mortality rate of 34% (see Outbreaks (section 6.9)). In Europe Swiss cheese caused 122 cases and 31 deaths between 1983 and 1987.

Vatanyoopaisarn *et al.* (2000) found that cells of a flagella-less mutant did not attach well to stainless steel surfaces when grown at 25°C or 37°C whereas the parent flagellated strain attached well at 30°C but not at 37°C. This study showed that flagella were important in bringing cells into contact with surfaces and allowing the preliminary stages of attachment to occur.

In the UK recently there has been a significant increase in the number of cases of *L. monocytogenes* reported. This change has been in the elderly, but, as yet, it is not clear what this increase has been driven by.

6.8.2 Foodborne viruses

As discussed earlier, viruses, including those that infect plants, animals and bacteria, are extremely small (25–30 nm). They are made up of a protein capsid which is the outer layer surrounding either DNA or RNA. In general they cannot grow or divide outside their host and bacteriophages (which attack bacteria) are very specific, in some cases only infecting particular strains of a species. Hence, they have been used in typing systems to differentiate strains of bacteria. Nevertheless a number of different viruses are able to cause gastrointestinal illness in man. Typical of this is gastroenteritis caused by Noroviruses. There is an incubation period of between 15 and 50 h before vomiting and diarrhoea occurs. This can last for 1–2 days. As rapid onset projectile vomiting can result there is a likelihood of a quick spread, particularly amongst children under five without any resistance and with poor hygiene and in situations where people are mixing close together, such as cruise ships and care homes for the elderly. The biggest outbreak of foodborne illness recorded was in Shanghai in 1988 where cockles contaminated with hepatitis A made 300,000 people ill. Large outbreaks have also occurred in Australia, the UK and the USA. The shellfish are grown in water close to the shore which can easily be contaminated with sewage. Even worse, the shellfish feed by filtering the water so that they concentrate both bacteria and viruses. Shellfish can be decontaminated by putting them into uncontaminated water. Bacteria are removed by 2 days but virus removal appears to take longer and is somewhat variable. In the UK internal temperatures of at least 85°C for 1.5 min are supposed to be needed for cooking cockles, but oysters are eaten raw.

Hepatitis A has also been found to be a cause of illness as a result of eating some fruits, including strawberries, and salads such as lettuce, as well as milk, but these outbreaks, unlike those from shellfish, usually result from contamination during food preparation by food handlers as a result of infected food processors. The incubation period can be as long as 6 weeks before symptoms of liver damage including dark urine and a jaundice appearance become obvious.

6.8.3 Mycotoxins

Mycotoxins are produced by fungi and have been shown to be mutagenic and/or carcinogenic both in vitro and in animals other than man. Their effects on humans are more difficult to determine, but it would appear that consumption of 0.5–2.0 mg aflatoxin kg^{-1} of food is enough to cause death. All mycotoxins are described as secondary metabolites, with primary metabolites defined as those compounds essential for growth.

Mycotoxins are small molecular weight compounds, usually produced towards the end of logarithmic growth. The first well-researched outbreak of illness was in the UK where at least 100,000 turkeys died in 1959 after being fed contaminated peanuts. The causative agents were aflatoxins produced by *Aspergillus flavus*. *A. parasiticus* is also able to synthesise aflatoxin, with 25°C being the optimum temperature for production. Aflatoxins have been found in cassava, maize, wheat, cotton seed oil, rice, raisins, cocoa, soya bean meal, milk, cocoa and a number of other foods. Although not all animals show as great a sensitivity in comparison, in 1974 in India nearly 1000 people became ill after eating mouldy maize and more than 100 people died (Moss, 1987) generally as a result of severe liver disease, while in 2004, 317 people became ill and 125 died in Kenya after consumption of aflatoxin in maize (Lewis *et al.*, 2005).

Another mycotoxin, which is a potent nephrotoxin, is ochratoxin *A* (Jørgensen, 2005; Leong *et al.*, 2007). This is made by *A. ochraceus* and *Penicillium verrucosum* in temperate countries in barley and is associated with kidney damage, producing a high incidence of renal tumours in rodents (Rached *et al.*, 2006). Another Aspergillus toxin is *Sterigmatocystin* (produced by *A. versicolar*), but mycotoxins have also been isolated from Sterigmatocystin. *Penicillium italicum* and *P. digitatum* are usually produced by blue and green mould on oranges, lemons and grapefruits. In addition, human disease produced by *Claviceps purpurea* causes ergotism or St. Anthony's fire in humans after consumption of mouldy bread. Patients hallucinate and can feel as if they are burning. The alkaloids produced cause constriction of the peripheral capillaries, leading to fingers and toes becoming gangrenous. In addition, this and other related mould metabolites can affect the central nervous system, causing smooth muscle stimulation. Over 300 years from AD 1500 there were more than 65 epidemics in Europe and it has been suggested that the Witch Trials in Salem in the USA in the late 1600s were induced by ergot on rye leading to contaminated bread, the consumption of which led to hallucinations.

6.8.4 Foodborne animal parasites – protozoa, flatworms (helminths), roundworms (nematodes), liver flukes (*Fasciola*) and tapeworms (*Taenia*)

Unlike bacteria, animal parasites are generally unable to grow or divide in food or in the environment but only in their host. Equally they are not able to grow in enrichment or selective media and are usually detected by specific antibodies, staining or by assays in their animal hosts.

6.8.4.1 Protozoa

The protozoa belong to the Protista (as do Algae and fungi with flagella). They are generally thought to be the most primitive of animals, as well as the smallest. Those important in food and water include *Cryptosporidium, Giardia intestinalis, Entamoeba histolytica* and *Toxoplasma* (Georgiev, 1994).

Cryptosporidium

Worldwide *Cryptosporidium parvum* represents about 1–4% of patients with diarrhoea (Tzipori, 1988) and is increasing, with AIDS patients having infections of up to 38% in certain hospitals. However, in immunocompetent humans the disease is self limiting. Initial infection is usually as a result of drinking contaminated water, but washing food with such water can lead to contamination and, as a result, human infection. However, in many countries the faecal oral route appears to be the most important form of transmission. The cells (oocysts) of *C. parvum* are oval to spherical and about 5 μm in size. Each sporulated oocyst contains four sporozoites and can remain viable for several months under cold and moist conditions (Current, 1988). Oocysts are also resistant to disinfectants including hypochlorite and ozone. In immunocompetent individuals the protozoan attaches to the intestinal epithelium, producing diarrhoea after an incubation period of 6–14 days with symptoms lasting for 9–23 days. The organism has a life cycle that occurs in one host. The thick-walled oocytes excyst in the small intestine. The sporozoites are released and then penetrate the microvillous region of host enterocysts. Sexual reproduction results in zygotes which form spores within the cells of the host. The oocytes then sporulate within host cells and are shed in faeces which are eaten by other hosts.

Where antibodies are used to detect the oocysts, both viable and non-viable cells are detected. Of course, as discussed under the section on detecting bacteria (section 6.14), usually only viable organisms are of interest and vital stains can help here. In addition, although the problem with *C. parvum* has been described, it is necessary to distinguish between this organism and related species, since other species of *Cryptosporidium* do not attack the human host.

Giardia intestinalis

Giardia intestinalis is a flagellate protozoan which produces cysts up to 20 μm long and 12 μm wide. After ingestion the organisms excyst in the upper small intestine. Water is a common source of the disease and in the USA muskrats and beavers are the major source of infection, with 70% of muskrats excreting the cysts in one study in New Jersey (Kirkpatrick and Benson, 1987). Perhaps as many as 15% of the population of the USA are infected by this protozoan (Osterholm *et al.*, 1981). The incubation period can be as short as 7 days or as long as 13 days. The cysts appear in faeces after 3–4 weeks (Piekarski, 1989) and may persist for a year or so. As many as 9×10^8 can be passed each day, with an infectious dose of less than 10 (Rendtorff, 1954) and they can survive for 3 months in sewage (Barnard and Jackson, 1984). The disease is very contagious with an infection rate up to 67.5% (Chen, 1986) and mostly causes diarrhoea, cramps, fever, vomiting and weight loss. Unusually for a protozoan, *G. lamblia* can be grown in axenic culture and detection by ELISA is commercially available. The cysts are chlorine resistant but are killed by heating at cooking temperatures.

Entamoeba histolytica

Entamoeba histolytica causes amoebic dysentery and can be passed on by the faecal–oral route and in water as well as foods washed with contaminated water. The organism has no mitochondria and is an aero-tolerant anaerobe. The cells can be as big as 60 μm while the cysts are smaller (maximum 20 μm). The cells (but not the cysts) are motile. Up to 5.0×10^7 cysts can be passed each day (Barnard and Jackson, 1984) and they can survive in sewage for 3 months (Barnard and Jackson, 1984). In parts of the USA as much as 36.4% of the population may be infected (Chen, 1986) and worldwide perhaps 100 million cases occur each year (Walsh, 1986). The infection is endemic in lesser developed countries in many parts of the world, but there has been a decline in developed countries such as the UK. The disease may persist for a number of years.

6.8.4.2 Liver flukes

The mature liver fluke is 2.5×1 cm. After entering the host the organism feeds off the liver before coming to rest in the bile duct. Here it matures before producing eggs (which are 90×150 μm). These have a lid at one end and are secreted in the faeces. In water the eggs hatch and produce motile cells which are unable to infect the main host but need to infect a water snail, where after passing through a number of stages, they escape, encyst and then survive in the environment for as long as a year. Main hosts are sheep or cattle, or man where it is associated with raw or undercooked watercress.

6.8.4.3 Tapeworms (Taenia)

Tapeworms are associated with pork (or humans) and beef. The larval stages can cause a spotted appearance in the muscle tissue of pigs or cows, but the mature tapeworm can only develop in the human intestine, where it produces severe symptoms in the weak or young. The infection produces a mechanical irritation of the gut and general symptoms including anaemia, nausea and abdominal pain. The irritation can be severe enough to produce reversed peristalsis, resulting in segments of the tapeworm entering the stomach where eggs are released and subsequent invasion of the tissues can occur. Invasion of the central nervous system may result in death.

6.8.4.4 Roundworms (nematodes)

The only significant roundworm for man is *Trichinella spiralis*. This is passed from host to host, as it has no free-living stage. Its hosts, unfortunately, are a number of mammals including humans and pigs, and raw or inadequately cooked pork can be a problem. This may be one of the reasons why a number of religions have taboos on eating pigs. The active larval stages cause nausea, diarrhoea and sometimes death after consumption of the larval cyst. The cysts are able to survive for years in a living host, but after consumption of the larvae they are set free as a result of contact with the digestive enzymes in the stomach. After this, they elongate to 3–4 mm. No symptoms are produced by the adults, but one female can produce over a thousand larvae, which burrow into the gut wall to reach the muscles where they grow to 1 mm, causing muscle pain or fever before encysting.

To reduce the problem, the United States Department of Agriculture (USDA) recommends that all pork should be heated to 76.7°C, although fermentation, smoking and curing will kill the cysts. Several other species including *Trichinella nativa* occur in walrus and polar bears and may have led to the disease being present in the Inuit. It is not clear if consumption of polar bear or walrus meat are avoided by any religious groups. What is certain is that early man must have been able to live much longer once he started to cook the meat of large animals. Subsequently specific methods for the control of *Trichinella* have been published (Gamble *et al.*, 2000).

6.8.4.5 Toxoplasma gondii

Toxoplasma is a protozoan which is capable of infecting virtually any nucleated cell (Miller *et al.*, 1972). Its primary host is the cat, but humans and other animals can be infected by oocytes passed in the faeces. As a result, animals other than carnivores can become infected, for example, by eating faecally contaminated grass. After infection, tissues remain infectious for life and estimates suggest that 50% of Americans have circulating antibodies to *T. gondii* by the time they become adults (Plorde, 1984). Toxoplasmosis is usually without symptoms or is the cause of an illness similar to influenza. Management by drugs is especially important (Georgiev, 1994). Unfortunately, in those who are immunocompromised the infection can be much more significant and this organism is the most common cause of cerebral infection in AIDS victims.

6.8.5 Scrapie, bovine spongiform encephalopathy (BSE), kuru and Creutzfeldt-Jacob disease (CJD)

CJD is a worldwide human disease which is invariably fatal and is very rare, having an incidence of about 1 in 2 million of the human population. A variant of CJD began to be detected in Britain in 1985. By June 2007 this had killed 165 in the UK and six more in other countries. All these transmissible spongiform encephalopathies (TSE) produce changes in the brain, where microscopic vacuoles appear so that the grey matter looks sponge-like. Any animal with a spongiform encephalopathy has neurological symptoms which can make it aggressive as well as nervous and anxious and eventually it has difficulty standing.

6.8.5.1 Scrapie

Scrapie was detected in the eighteenth century and, apart from New Zealand and Australia where it is absent, it has been found in sheep and goats throughout the world. However, CJD is present at roughly the same level in the human population in New Zealand and Australia as it is in all other countries. In addition, in those nations with scrapie, there is the same

incidence of CJD even in countries where sheep's brains are eaten as a delicacy. There would appear, therefore, to be no relationship between consumption of sheep and goats and CJD in mankind.

As these diseases take a long time to develop they have been described as slow viruses (e.g. Collinge *et al.*, 2006). Although Manuelidis (2007) has found viruses and no prions in neural cells infected with scrapie and CJD, there is little or no evidence that any of these diseases are either viruses or bacteria. Instead, the infectious agents have been named prions. A prion cannot be cultured and does not cause the host to produce antibodies. Even worse, it cannot be detected by electron microscopy and is extremely resistant to chemicals, heat and irradiation. Scrapie which has been most carefully studied can be transmitted both horizontally (sibling to sibling) and vertically (parent to child). From sheep flocks it can also be transmitted by injection into the cerebellum, where a number of other species, including monkeys and cats, have been shown to be susceptible.

6.8.5.2 Kuru

Kuru was for many years the best understood spongiform encephalothopy and was a major cause of death amongst the women of the Fore tribe in Papua New Guinea. Kuru is probably the same disease as CJD and if not considered food-borne, at least it is caused by ingestion of animal protein. The tribe had a tradition that involved the women and children eating the brains of the dead as a sign of respect. This activity was eventually suppressed and the incidence of kuru declined over a 50-year period.

6.8.5.3 BSE

BSE or mad cow disease appeared in cattle in the UK in 1985 and was virtually a British disease. About 482,000 BSE-infected animals entered the UK human food chain before controls were introduced. It is thought to occur because of scrapie crossing the species barrier as a result of scrapie-infected material from sheep being fed to cattle in high concentrations. In addition, because tallow was no longer required, it was not removed from the sheep meat, whilst the temperature of the heating process was also decreased. It has been suggested that the tallow further protected the infectious agent from the lowered temperatures, allowing the agent to survive the heating process.

Some scientists thought the number of British dead would reach more than 100,000 and it is likely that the numbers of dead may still continue to rise somewhat, given the long incubation time. As a result of these concerns and the rest of the world refusing to import British beef a number of changes to British law were introduced in 1989. These included:

- The sale of bovine offal from animals over 6 months old was banned.
- The feeding of protein from ruminants to other ruminants was banned.
- The slaughter for meat of cattle over 30 months old was banned.
- In addition, in 1996, the use of all mammalian-derived protein in farm animal feed was banned.

Very briefly, the prion contains a protein called PrP^{sc}. This is a modified version of a protein (PrP^c) which normally occurs within neurons. The protein has almost the same structure as the "normal" protein. It has, however, a different tertiary structure which makes it resistant to protease enzymes. These would remove it from the brain at a suitable time. When the PrP^{sc} particle comes into contact with a PrP^c it converts it into a PrP^{sc}, starting an irreversible reaction which eventually results in the clinical symptoms described above. Ultimately these changes lead to death of the host.

6.9 Outbreaks

A number of well-known outbreaks of foodborne illness are discussed below with the aim of developing a greater understanding of the problems that lead to illness and, in some cases, death. For further examples see Pawsey (2002).

6.9.1 Meat pie in Wishaw, Scotland

In 1996 as many as 496 cases of illness with 21 deaths occurred as a result of eating meat contaminated with *E. coli* O157:H7 after a meal for churchgoers aged over 70 in Wishaw, Scotland. The meat pie was consumed by 81 over 70s on 17 November. By 20 November gastric illness was apparent, in some cases with bloody diarrhoea. Of these orig-

inal 81, 45 became ill and eight eventually died. On 22 November the Public Health Department became aware of events and an outbreak control team was set up. Very quickly it became obvious that the church meal was the source of the outbreak and the meat pie served at the dinner was known to have come from John Barr and Sons butcher's shop in Wishaw. This was a shop that had recently been voted "Butcher of the Year" in Scotland. On 23 November a local nursing home collected cooked meat and meat sandwiches from Barr's and on the 24 these were served to the mostly elderly residents, who soon developed symptoms. Barr's shop was closed voluntarily on 27 November.

On investigation the outbreak control team found that Barr's shop employed 40 people, although most of them were part time, and also had a bakery on the premises. From these businesses 85 other shops across central Scotland were supplied. Microbiological investigation found that *E. coli* phage type 2, which was verocytotoxigenic, was isolated from the patients and the shop. The outbreak control team found that on 16 November Barr's had delivered to the church hall two bags of cooked stew, the pastry tops for the pies and raw meat for the soup. These were all left unrefrigerated overnight until the 17th when they were heated up before serving at lunch. Surplus gravy divided among the helpers was found positive for *E. coli* O157 a week later.

During the investigation it became apparent that the shop:

- had no defined procedures for cooking meat;
- had little that was written down;
- failed to keep records of all the 85 customers;
- took no account of the possibility of the varying sizes of meat during the heating process;
- did not measure the temperature of the water bath used for cooking meat and the temperature of the meat during cooking.

The same organism was found on the surfaces in John Barr's shop, in the gravy and in those who became ill.

The outcome was that on 27 October 1997 John Barr was found not guilty of culpably, wilfully and recklessly supplying cooked meat, but on 21 January 1998 John Barr and Sons was fined £2,250 for breaches in the hygiene laws. However, Mr Barr himself was cleared of any blame.

6.9.2 North American salmon

This is an example of international trade causing a problem in a country distant from the site of production. Salmon was canned in Canada and consumed by four veterans/senior citizens in Birmingham, England, in 1978. The salmon was eaten without further heating and two of the people died as a result of intoxication with *C. botulinum*. On careful investigation it was discovered that the canning factory was designed as a "horse shoe" shape (rather than a straight line) and the workers who cleaned and skinned the fish were prone to walking across to the area where the cans were cooling subsequent to the canning operation. They then put their wet and dirty overalls over the cans in order to dry them. Unfortunately on one day a fault developed in the canning line so that there was a hole in some of the cans. This defect was then covered over by the label.

Clearly the faults were:

- the original design of the plant;
- the management of the operatives;
- checking the line efficiency.

6.9.3 Mexican-style soft cheese in California, USA

In California 142 cases of listeriosis were reported between 1 January and 15 August 1985; 93 were pregnant women or offspring including 20 fetuses, 10 neonates became ill and of those who became ill, 48 died as a result of consuming Mexican-style soft cheese. The milk used was pasteurised, but a second pipe line allowed non-pasteurised milk through to the cheese production area apparently in order to improve the flavour.

The faults were:

- poor design of the plant;
- failure to test the cheese adequately.

6.9.4 Radish sprouts in Osaka, Japan

More than 1000 people, many of them children, became ill allegedly as a result of consumption of white radish sprouts contaminated with *E. coli* O157:H7. This was apparently due to contamination of the sprouts with animal faeces.

The fault was:

- failure to keep the sprouts uncontaminated.

6.9.5 Corned beef in Aberdeen, Scotland

In 1964, 487 people in Aberdeen, Scotland, were admitted to hospital suffering from typhoid as a result of consumption of corned beef from one tin, giving further support to the joke that only an Aberdonian could make so many ill with meat from just one tin. In fact it was a tin of catering size corned beef and the slicing of the meat in the butcher's shop spread the bacteria through the slices.

The fault was:

- cooling the cans in faecally contaminated water in the River Plate in Argentina, where, as the cans lost heat, the water would be drawn into the cans through the seams.

6.9.6 Botulism in yoghurt, North West England

In 1989 in the North West of England 27 people became ill with botulism as a result of consuming hazelnut yoghurt. Of these 27, there was one fatality and 24 people had their future health reduced. Some, but not all, of those affected had noticed an unusual taste. An investigation showed that a can of hazelnut puree added to the yoghurt contained *Clostridium botulinum* type B toxin (O'Mahoney *et al.*, 1990). In addition 15 unopened cartons of yoghurt and two opened cartons from patients' homes all contained type B toxin. Furthermore some viable cells of type B toxin were discovered in cartons with the same sell by date of 13 June 1989.

Seventy six cans of the hazelnut conserve were sweetened, not with sugar, but with aspartame. These cans were produced in July 1988 and were subsequently stored at room temperature. These cans were reported as blown to the manufacturer and in October 1988 potassium sorbate was added to prevent yeast growth. It is clear that the manufacturers of the hazelnut preserve relied on a combination of treatments to provide safe products. These included:

- a low pH;
- a low A_w;
- heat treatment.

By changing from sucrose to aspartame the a_w was higher and allowed *C. botulinum* and yeasts to grow. Thus the spores of *C. botulinum* would survive the heat processing, germinate and grow.

6.9.7 C. botulinum and wild duck paste
in Scotland

Eight people died after eating wild duck paste at a hotel near Loch Maree in Scotland in August 1922. The cause of death was unclear for a number of days, since botulism was unknown in the UK at the time. Events occurred in the following manner.

Packed lunches consisting of sandwiches were provided on 14 August for 35 people including:

- 17 ghillies;
- 13 fishermen;
- 2 wives;
- 3 mountain climbers.

Eight became ill during the next 2 days and all of those who became ill subsequently died. The first signs of illness were at 3.00 a.m. on the morning of the 15th. Later several others also became ill and the local doctor came to check on them. As the 15th progressed the condition of at least one patient became serious and Professor Munro, a local expert, was called out, arriving in the late evening to find one patient already dead. One of the ghillies became ill during Professor Munro's visit and while he was visiting him, a second patient died. On the 16th a third death occurred. As a result the legal authorities were called. Despite this the number of deaths rapidly reached eight.

In the search for the cause, foodborne illness was rapidly suspected and the only meal shared by the ghillies with the hotel guests was the luncheon on the 14th. Further, since the patients were lucid and able to talk, it quickly became apparent that the problem was as a result of eating the sandwiches made with potted duck paste. It became clear that the two containers of potted meat had been used to make the sandwiches. Given the size of the containers there must have been no more than 12 sandwiches involved. The patients had very similar symptoms but there was no headache, pain, facial paralysis, deafness, fever, disturbance of the sphincters, diarrhoea and no interference with mental activity. The symptoms included dizziness and double vision, with indistinct speech and difficulty in swallowing, followed by paralysis and death. Careful examination showed that the wild duck paste contained a toxin which was neutralised by type A antitoxin. The amount of paste on the head of a pin would have been enough to kill 2000 mice.

It is apparent that the source of the botulism occurring in most countries differs from that found in other countries. For example:

- in Japan 99% of cases are associated with fish;
- in France 84% of cases are associated with meats;
- in Germany more that 75% of cases are associated with meat;
- in the USA 60% of cases are associated with vegetables, mostly home bottled.

This is partly due to differences in diet but also to differences in methods of food preparation (for example, the high number of cases associated with home-bottled vegetables in the USA).

The mistakes in the duck paste production are not totally clear at this distance in time. However, the subsequent enquiry into the Scottish outbreak found that the materials were cooked in milk and were then "sterilised" in open containers before heating again at boiling point and storing at room temperature. The problems would seem to be as follows:

- heating in open containers would allow contamination to occur and the temperature would not reach above 100°C;
- type A C. botulinum belongs to group I and requires heating to temperatures above 100°C in order to kill the spores. A kill process of 121°C for 3 min would be used today and a temperature of 100°C would have required 25 min of heating.
- storage at ambient temperature would allow any surviving spores to germinate, grow and produce toxin, since minimum growth temperature would be 10–12°C.

6.10 An outbreak that wasn't!

In 1995 the producer of a semi-soft mould-ripened cheese (Lanark Blue) made from raw sheep's milk in Lanarkshire, Scotland, was taken to court after *Listeria monocytogenes* was allegedly found in cheese samples. It is important to recognise that no one was reported as ill, despite the fact that 1000 CFU g^{-1} was allegedly found. It is also worth noting that in 1995 in the USA if the bacterium was present in 25 g of food, then the food would have been removed from sale. At this time most human cases of listeriosis were produced from just three of the 13 serotypes detected.

Thus serotype 4b was by far the most common cause of listeriosis, although 1/2a and 1/2b had also been shown to be responsible. The strain isolated from this cheese, 3a, had only once been associated with listeriosis (Lyytikäinen, 2000). However, some microbiologists believe that the presence of *Listeria* spp. of *whatever strain* is an indication of the possibility that a pathogenic strain of *L. monocytogenes* would be able to grow and cause listeriosis.

Foods such as sliced meat, soft ripened cheese and vacuum-packed paté have, at various times, been associated with outbreaks of listeriosis and all can be stored for long times at refrigeration temperatures. Since *L. monocytogenes* is able to grow at such temperatures its presence *at any* level is of significance and enrichment should be used to detect it in addition to simple plating out (Oravcová *et al.*, 2006).

6.11 Incidence of foodborne illness

Any comparison of foodborne illness between countries, causative agent or food source depends on the diet, the accuracy of testing and the organisms involved. An obvious example would be vegetarians who have completely different food sources to omnivores. Further, *S. aureus*, where the preformed toxin will result in illness within 6 h, is more likely to be detected than *L. monocytogenes*, where several weeks can occur before symptoms appear. In addition, an outbreak involving two people is less likely to be reported than that involving a hundred, unless, of course, the two both die.

Against this background Hughes *et al.* (2007) examined the incidence of foodborne illness between 1992 and 2003 in the 50 million people living in England and Wales. They estimated that there were 1,729 foodborne outbreaks of illness, with 39,625 affected, 1,573 admitted to hospital and 68 deaths. *Salmonella* was responsible for over half of the outbreaks (57%), those affected (57%) and the hospital admissions (53%), as well as the majority of deaths (82%). The next most significant organism was *C. perfringens* which was responsible for 12% of the outbreaks, with viruses (7%), *Campylobacter* (4%) and VTEC O157 (3%), but it is worth noting that the group described as mixed/other/unknown accounted for 14% of the outbreaks. The reason for VTEC O157 being considered as significant was that it was responsible for 11% of the hospitalisations and 7% of the deaths. The sites of the illness were most frequently commercial catering premises (55%), residential (armed services, holiday camps and residential homes) (13%) and private addresses (12%). It is, however, disappointing that residential homes were not separated from other groups, since it would seem to be very unlikely that an army camp has much relationship with a home for the elderly and immunocompromised.

With regard to the foods most commonly cited as responsible, poultry (24%), red meat (20%), fish and shellfish (14%) and salad, fruit and vegetables (8%) were the largest groups, although outbreaks caused by poultry and red meat in particular decreased over the period. Breaking the information down further showed that desserts, poultry, red meat and eggs were responsible for more than 75% of the *Salmonella* outbreaks, while the majority of *C. perfringens* outbreaks were linked to the consumption of red meat and poultry. Poultry and milk and its products accounted for most of the *Campylobacter* spp. outbreaks. The majority of infections caused by VTEC O157 were linked to red meat and milk.

Adak *et al.* (2002, 2005, 2007) examined indigenous foodborne diseases and deaths in England and Wales from 1992 to 2000. In 1995 they estimated that there were 2,365,909 cases, 21,138 admissions and 718 deaths, but by 2000 this figure had fallen to 1,338,772 cases, 20,759 hospital admissions and 408 deaths.

Although some authors have shown that the levels of foodborne illness in the UK post 1950 increased at the same rate as the consumption of poultry meat, there was a similar increase in the number of television licences, suggesting that such changes, while they appear to be meaningful, might instead be extremely misleading.

A study was carried out in Australia for 1 year in about 2000. Here pathogenic *E. coli*, followed by Noroviruses, *Campylobacter* and non-typhoidal *Salmonella* were found to be more significant in a country with a population of 20 million, with 1.48 million affected, 15,000 hospitalised and 45 deaths. Work in the USA (Mead *et al.*, 1999) has shown that a population of 300 million had approximately 76 million illnesses, 325,000 hospitalisations and 5,000 deaths each year, with three known pathogens (*Salmonella*, *Listeria* and *Toxoplasma*) responsible for 1,500 deaths, more than 75% of those caused by known pathogens.

Adjusted per million it seems clear that even with deaths there are considerable differences between highly developed countries and that the organisms responsible for the majority of foodborne illness also differ between countries.

6.12 The Richmond Report on the microbiological safety of food

Although Parts I and II of this report were submitted to the UK Government in June and November 1990, much of their comments and recommendations are still relevant today. The Committee concluded that foodborne illness was a common problem in the USA and Western European countries. The factors responsible for the increases were:

- levels of reporting and enhanced publicity;
- technology and its control;
- changes in lifestyle and habits;
- hygiene awareness;
- international travel.

The Committee saw poultry and poultry products as the most important source of gastrointestinal infections arising from food. It noted that a high proportion of chicken carcasses were contaminated with *Salmonella* and *Campylobacter* and that shell eggs could be a source of *Salmonella*. In addition, it observed that traditionally the food manufacturer has relied on microbiological testing. The Committee recognised, however, that the Hazard Analysis and Critical Control Point (HACCP) approach to safety assurance should be applied to all sections of the food industry. With regard to changes in lifestyles the Committee recognised a decrease in the formal meal and an increase in snacking. In addition, there has been an increase in the amount of poultry meat consumption, which rose by 41% from 744,000 tonnes in 1978–80 to 1,488,000 tonnes in 1988. It also noted a lack of understanding of the basic principles of microbiology and food hygiene and believed that there was a lack of effective training.

Subsequently, there has been a wide take-up of HACCP, although the actual practice particularly in small sales outlets may be less than ideal. In addition, there has been a reduction in the carriage rate for *Salmonella* in poultry but not for *Campylobacter*. A final change has been a number of outbreaks of foodborne illness associated with sprouting seeds and leafy herbs and vegetables. In this instance the USDA has acted to reduce the risk by advising those who are immunocompromised to avoid such products.

6.13 Water-borne diseases

Each and every one of us is 85% water. It follows that water is particularly important to us and the absence of liquid water on other planets is one of the most significant problems faced by carbon-based life forms in colonising new "worlds". Clearly drinking water must be free of pathogenic microorganisms. Despite this the World Health Organisation (WHO) has calculated that a child dies every 8 seconds as a result of waterborne disease (WHO, 2002).

Cholera (caused by *Vibrio cholerae*) killed more than 20 million people in the twentieth century alone and typhoid fever (caused by *Salmonella* Typhi) is generally transmitted through water, although it can also be passed from person to person or through food. In addition, *Shigella dysenteriae* (the cause of bacillary dysentery) has no known animal reservoir and transmission is either as a result of contaminated water or by spreading from person to person. It is important to remember that water is not just for drinking (and for producing ice); it is also used for washing clothes and for bathing as well as for washing and preparing food. Nevertheless, 80% of mortality is estimated to be caused by faecal contamination of drinking water.

In the past, beer and to a lesser extent other alcoholic drinks were developed at least in part because the level of microbial contamination was lower than it was in untreated water, and in England and Wales most households brewed their own beer while other countries concentrated on producing whisky or its equivalent. In addition to life-threatening waterborne illnesses, milder gastrointestinal infections can also be passed on through contaminated water. So-called traveller's diarrhoea can result from strains of *E. coli* generally considered as non-pathogenic in their home country, but when found in a foreign visitor who has not previously interacted with them, illness can result. Hence a European travelling to South America can suffer from what has been nick-named Montezuma's revenge. Equally South Americans coming to Northern Europe also find their resistance lower to, what are to them, novel strains and are likely to suffer similar symptoms. One of the authors of this chapter became so ill on a visit to Mexico City that he failed to notice an earthquake.

Nevertheless it would seem to be excessive to take a case full of chocolate bars to the Indian subcontinent in order to avoid foodborne and/or waterborne illness, as one well-known microbiologist has claimed to have done.

In addition to bacteria, protozoa including *Cryptosporidium parvum*, *Giardia intestinalis* and *Entamoeba histolytica* can be a problem, particularly when they form resting bodies (cysts) which are long lived and resistant to extremes of temperature. In some developed countries such as the UK the pipes of the water delivery systems were put in place more than 100 years ago and subsequently wear and tear have resulted in breakdowns and loss of water so that as much as a quarter or a third of water is lost before it reaches the taps of the consumer. Where such leaks occur contamination can be introduced into the system and attempts to repair the pipe work can result in further contamination, especially where heavy precipitation occurs. One of the biggest outbreaks of gastrointestinal illness was as a result of water contamination with *Cryptosporidium parvum* when 300,000 people became ill in the twin cities of Minneapolis and St Paul in the USA.

Testing the faecal contamination of water is of obvious importance. Currently faecal coliforms, *E. coli*, thermo-tolerant and other coliform bacteria, faecal streptococci, coliphages and spores of sulphate-reducing bacteria are used for testing, *E. coli* being more specific than the rest, while the coliphages and the spores are thought to be able to survive for longer and hence to give an idea of the past history of the contamination of the sample.

A carefully researched survey of the microbiological quality of water leaving water treatment works in Scotland from 1995 to 1999 found that the percentage containing coliforms fell from 1.8 to 0.71 while the percentage positive for faecal coliforms fell from 1.0 to 0.37. Detailed studies showed that the majority of failures occurred at small treatment works. A similar survey published in Ireland in 2004 found that only 84.6% of public water supplies were completely free of faecal contamination.

In addition *Salmonella* Typhi, as indicated above, is generally transmitted through water. However, typhoid fever has been almost removed from drinking water in developed countries by chlorination, which is an effective water treatment at a concentration of 0.6 ppm free chlorine, which also leads to the death of animal viruses.

The WHO states that any level of faecal coliforms in a water supply is a breach of drinking water regulations. Contaminated irrigation water can also be a problem, especially since recent studies have shown that internal contamination of leafy herbs and vegetables by *E. coli* can be produced. The outbreaks described in this chapter have demonstrated the difficulties resulting from contaminated water allowing transmission to food products. For example, water was directly responsible for the contamination of the tins of corned beef cooled in the River Plate, while the radish sprouts in Japan cannot have been washed carefully enough with potable water, and in the North American salmon drips were allowed to contaminate the cooling cans.

Most developed countries rely on chlorination of their water supply. Where this does not happen even *E. coli* O157:H7 (which is not often found in water) can cause significant problems. With chlorination at 0.6 ppm free chlorine, vegetative bacterial cells and viruses have been almost removed from drinking water. One alternative to chlorination is bottled water, which can cost up to 1000 times the price of tap water. The USA consumes the most bottles of water while the Italians drink more per head of the population, bottled water consumption doubled in China and tripled in India between 1999 and 2004. Unfortunately bottled water can support microbial growth and is not free of viable microorganisms. It has been found that carbonated water can carry as much as 10^4 organisms ml^{-1}, while still water can contain a hundred-fold more. Even worse, in 1974 there was an outbreak of cholera in Portugal, when water supplies became contaminated with cholera from faeces. This led to contamination of shellfish, other foods and bottled water.

Evidence for the success of this approach comes from the incidence of typhoid fever in Philadelphia in the USA, where there were about 2,000 cases every year from 1880 until 1890; then the number began to increase so that there were almost 10,000 in 1906. At this point water filtration was introduced and there was a rapid fall back to about 2,000 by 1914 when chlorination was begun, and the number of cases fell swiftly to 200 by 1926. Other waterborne diseases followed the same pattern.

In many ways it is not yet clear what the effects of global warning will be on our planet. It is very likely, however, that sea levels will rise, resulting in the mixing of salt water with fresh water. As there is already a shortage of drinking water, added to the problems in the water supply produced by any natural disaster (volcanic eruptions and earthquakes, for example), it seems clear that the supply of drinking water is certain to reduce more quickly and to a greater extent. There are already disputes between farmers and now countries over available water. This will undoubtedly get worse and it is entirely possible in the future for water to approach oil as a scarce commodity.

6.14 Traditional and novel methods of microbial detection

6.14.1 Plate counts

In the food industry the most frequently used method of counting bacteria, yeasts and moulds is still the spread plate, where 0.1 ml is spread over an agar surface with the agar containing suitable nutrients. Samples are spread on at least two plates. The growth medium is usually contained in sterile Petri dishes. The plate is then incubated at a suitable temperature until the cells multiply sufficiently to be visible as a colony. This can take 16 h for some fast growing organisms or weeks in the case of those that grow very slowly. There is an urgent need to detect microorganisms more rapidly and sensitively, given, for example, the need to detect one *Salmonella* cell in 25 g of food. Theoretically with spread plates, 1 CFU/0.1 ml can be detected. More realistically it is likely that 10^2 CFU/ml is the minimum possible. This compares favourably, however, with microscopy where 1×10^7 CFU/ml is the minimum easily countable even in a clear liquid. An advance on spread plates is the Spiral Plate Maker which automatically spreads across a plate in an Archimedes' spiral, with the result that the same volume is spread over a bigger area, hence resulting in a dilution effect. This method, although thought to be difficult to use routinely, can save growth medium and staff time.

6.14.2 Microscopic counts

Microscopic counts use a haemocytometer which requires a slide marked in squares. The number of cells per square are then counted to produce a statistically meaningful result. Most counts of this type fail to differentiate between viable and non-viable cells, although staining techniques which show live cells in a different colour from dead cells are now available (see direct epifluoresence technique, section 6.14.7). It is, however, important to recognise that cells able to grow and divide may be different from those able to take up particular stains. Cells that have been irradiated with ultra-violet light or X-rays, for example, may have lost the ability to divide but can still

stain as viable and will also contain ATP (see ATP as a method of measuring growth, section 6.14.6).

6.14.3 Impedance microbiology

Impedance microbiology has been used for a number of years and several laboratory equipment manufacturers produce impedance systems. The first commercially available system was produced by Don Whitley Scientific (Silley, 1991). All of the systems available rely on changes to the conductivity of a medium, brought about by microbial growth, to determine whether microorganisms are present in a sample and to estimate viable counts. Each of the systems is semi-automated and relies upon a computer to collect the data, although manual loading is still generally required. However, the systems differ in the types of growth vessels and incubation chambers used. This section will not describe the individual systems in detail, but will give an overview of the operating principles.

6.14.3.1 Direct impedance technique

All of the impedance microbiology systems measure changes to the conductivity of a medium brought about by microbial growth. The flow of current between an anode and a cathode can occur when some form of bridge is made between the two. Normally, for a microbial growth medium that bridging medium will be water, with a variety of nutrients added. The conductivity of water relies upon the number of ions present in it and the mobility of those ions. Pure water has a relatively low conductivity as it has a relatively small number of ions. The addition of molecules such as sugars to water, which do not dissociate into charged ions, does not increase the conductivity. However, the addition of molecules such as salts, acids or alkalis to water, which do dissociate into charged ions, does increase the conductivity.

Therefore, a microbial growth medium of given composition will have a constant conductivity, so long as the chemical composition of the medium remains constant. If viable microorganisms are added to the culture medium and they metabolise the nutrients present, they may either ferment, releasing a variety of molecules including acids, CO_2 and alcohols, or respire, releasing CO_2 and water. Organisms that produce acids can be detected by direct impedance since they will affect the conductivity of the medium as the acid molecules dissociate into acid anions and protons. Consequently, a growth medium of given composition will have a given conductivity until a population of microorganisms, capable of using its constituents, is added. As the organisms grow, ferment and excrete acids, so the conductivity of the medium will increase steadily as the concentration of acid anions present increases. On the other hand, alcohol and CO_2 do not dissociate into charged ions and so will not affect the conductivity of the medium. Consequently organisms that respire only, or that produce alcohol only, cannot be detected by direct impedance, although there is an alternative technique called indirect impedance (presented below) that can prove useful.

The conductivity of an aqueous solution is dependent upon the mobility of the ions present. The mobility of the ions present is directly related to the temperature of the solution – increasing the temperature increases the mobility of the ions and vice versa. The consequence of this is that the temperature of an impedance microbiology unit must be controlled very exactly, since each $1°C$ change in temperature will lead to a 1.8% change in conductance. Impedance microbiology systems strive for thermal regulation of better than $0.1°C$. The different systems take a different approach to thermal regulation, for example the Don Whitley RABIT and SyLab BacTrac systems use large aluminium incubator blocks to provide thermal buffering, whilst the BioMérieux Bactometer system uses a thermostatically controlled oven-type incubator, and the discontinued Malthus system used a thermostatically controlled water bath. The sample containers used by each of the systems are also physically different, although all are effectively a non-conducting chamber into which two electrodes protrude and make contact with the growth medium.

Figure 6.1 gives a schematic representation of a direct impedance sample chamber. The RABIT and BacTrac systems utilise plastic cylinders with the electrodes protruding through the base and a cap covering the top; in contrast the Bactometer system uses cards with a series of wells that each have the electrodes at the bottom.

6.14.3.2 Indirect impedance technique

Organisms that excrete molecules such as CO_2 or alcohol present something of a problem to impedance microbiology, since these molecules do not dissociate and alter the impedance of the medium. However, it is possible to estimate cell numbers indirectly,

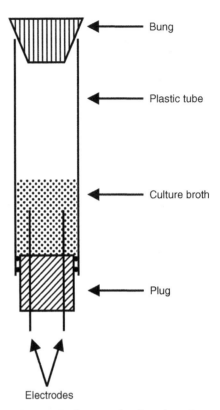

Figure 6.1 Schematic diagram of a direct impedance technique cell.

since all respiring organisms, along with those that produce alcohols, will also excrete CO_2. The CO_2 can be measured impedimetrically by its interaction with KOH. A solution of KOH is suspended in agar and cast over the electrodes of the sample cell. The KOH dissociates into highly mobile K^+ and OH^- ions, leading to low impedance. Meanwhile, CO_2 can diffuse into the agar, initially forming carbonic acid (H_2CO_3) which dissociates into protons (H^+) and bicarbonate anions (HCO_3^-). The HCO_3^- reacts with the K^+ to form $KHCO_3$, which is relatively stable and less likely to dissociate into charged ions than KOH, increasing the impedance of the medium. The H^+ reacts with the OH^- to form H_2O, which is also less mobile than the component ions, again increasing the impedance of the medium.

In the indirect impedance technique, the microbial culture is inoculated into a second smaller chamber inserted into the main sample chamber and is suspended above the electrodes and the KOH_{aq} bridge. The top of the sample chamber is sealed tightly to prevent gas escape. During incubation, the cells metabolise and excrete CO_2 which diffuses into the

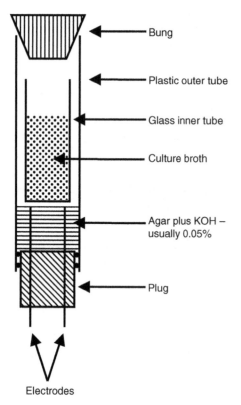

Bung

Plastic outer tube

Glass inner tube

Culture broth

Agar plus KOH – usually 0.05%

Plug

Electrodes

Figure 6.2 Schematic diagram of an indirect impedance technique cell.

agar containing KOH and increases the impedance. Figure 6.2 gives a schematic representation of an indirect impedance sample chamber.

6.14.3.3 Applications of impedance microbiology

Impedance microbiology has been used to determine overall population densities (i.e. microbial hygiene) in, for example, the dairy industry. The technique can yield good estimations of population density in a sample very rapidly, whilst at the same time permitting large numbers of samples to be analysed automatically. Historically, impedance microbiology has struggled in the rapid enumeration of particular (e.g. pathogenic) species; this is because selective media (e.g. Baird–Parker and XLD agars) can contain salts and ions which mask any ionised molecules (acids) produced by the bacteria. However, impedance microbiology system manufacturers have been working hard to address this problem and now selective media are becoming available to permit enumeration of particular organisms (pathogens) within samples.

6.14.4 Flow cytometry

Cytometry literally means the "measurement of cells". Measurement of cells – numbers, sizes or particular features – can be performed by eye under the microscope. However, this is a relatively slow and laborious process that is prone to operator error as fatigue sets in. Flow cytometry (FCM) is performed automatically, making it much faster and less prone to operator error than manual approaches. The basic techniques for FCM were developed in the 1940s, but for many years it was used purely as a theoretical technique. The first commercially available FCM system was developed by Becton Dickinson in the 1970s under the trade name of Fluorescence-Assisted Cell Sorting (FACS), which is now synonymous with FCM.

In essence, FCM involves taking a volume of microbial cells in liquid suspension, creating a stream of the liquid in which the microbial cells are separated and then passing the stream of individual cells past some form of detection and measurement device. To create the stream of individual cells from the volume, the technology of "hydrodynamic focusing" is essential. Hydrodynamic focusing relies upon the fact that water is incompressible at sea level. The inner of two concentric nozzles is fed with a cell suspension, whilst the outer of the two nozzles is fed with a stream of water. The outer stream of water, the *sheath*, is accelerated through a fine aperture in the nozzle and in doing so draws out an even finer stream of the cell suspension, the *core*. Figure 6.3 gives a schematic representation of hydrodynamic focusing.

This focused stream of cells is then passed through the detector unit which comprises a laser light source, to provide sufficient resolution for single microbial cells, and a series of optical detectors. To determine cell count and size, low-angle and high-angle scatter of light from the cells is measured. Cells can also be labelled with a variety of fluorescent dyes to indicate various features (e.g. species) and so a series of dichroic mirrors and photodetectors for specific wavelengths may also be utilised.

Finally, a number of techniques may be used to select and separate individual cells from the stream, dependent upon their labelling. One technology uses a piezoelectric vibration to thrust a collector tube into the stream of cells when a particular signal (e.g. a green fluorescent cell) is detected. Another technology gives the cells different electrical charges depending on the fluorescence characteristic of interest.

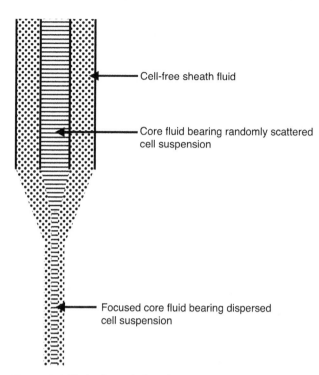

Cell-free sheath fluid

Core fluid bearing randomly scattered
cell suspension

Focused core fluid bearing dispersed
cell suspension

Figure 6.3 Hydrodynamic focusing.

The stream of cells is then passed between an anode and cathode, so that the cells are deflected dependent upon the charge they carry, and collected in separate containers.

With this type of fluorescence-assisted sorting, it is possible both to generate counts for particular features of cells and to collect populations of cells based upon the feature(s) of interest. Flow cytometers can operate at a rate from 100 cells s^{-1} up to 10,000 cells s^{-1} for the most sophisticated, analysing as many as five to ten different parameters associated with each cell.

6.14.5 Antibodies

Antibodies form one line of the body's defences against microbial invaders. They can have exquisite sensitivity to particular structures – antigens – associated with microorganisms, which makes them an ideal tool to detect pathogens within a sample. However, it is technically quite challenging to obtain a supply of suitable antibodies for the reliable detection of individual species. The reason for this is that when a mammalian system is challenged with a microorganism, it will respond with an array of antibodies to different antigens associated with the microorganism.

This array of antibodies is termed a *polyclonal* mix of antibodies and only some of those antibodies may target antigenic structures unique to the pathogen in question. However, other antibodies within that mix may target antigens that are common to several different species or even a whole genus.

To make use of antibody specificity to its full potential, *monoclonal* antibodies – i.e. antibodies that only target one antigenic structure – must be selected and propagated. This is time consuming since a host mammal such as a mouse or a rabbit needs to be challenged with the chosen pathogen and an immune response to that pathogen raised. Individual B-cells – the cells that produce antibodies – are then immortalised by fusion with myeloma cells and finally the antibodies are harvested in tissue culture. The supply of monoclonal antibodies produced then needs to be assessed for specificity to the chosen pathogen – in other words to make sure that it does not cross-react with other organisms to give a false-positive. Moreover, the sensitivity of the antibody to the chosen pathogen needs to be determined as the antibody may have been raised against a surface antigen that is not stably present (e.g. parts of capsular polysaccharide or inducible structures such as flagellae); in other words, to ensure that it will detect the chosen pathogen in as wide a set of circumstances as possible so that it does not give false-negatives.

Having raised, selected, propagated and purified a monoclonal antibody with suitable specificity and sensitivity, it is still necessary for the interaction of the antibody with the chosen pathogen to be detectable in some way. Historically, a variety of approaches have been taken such as simple *immunoprecipitation* in agarose (Ouchterlony's assay) and radiolabelling in the radioimmunoassay (RIA). Currently, two broad approaches seem to be more popular in antibody-based detection systems for food-borne pathogens. One approach involves labelling the antibodies with a reporter molecule (e.g. enzymes or fluorescent molecules) that makes their presence easy to detect. The other approach is to attach the antibodies to small beads that either precipitate out of colloidal suspension due to cross-linking of beads by antigens (e.g. reverse passive latex agglutination or RPLA) or trap the antigen and facilitate its extraction from the wider mix (e.g. immunomagnetic separation or IMS, described in more detail in section 6.14.5.1).

Antibodies labelled with a reporter molecule are most commonly used in various forms of the enzyme-linked immunosorbent assay (ELISA). The

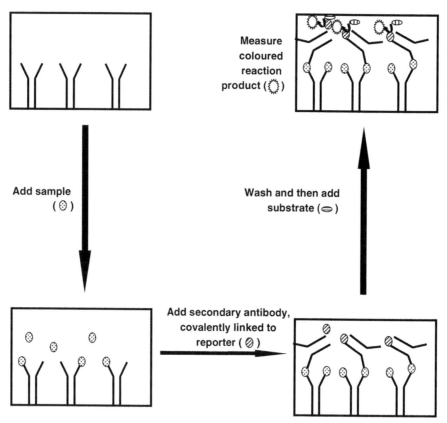

Figure 6.4 Sandwich ELISA.

simplest form of ELISA assay is the *sandwich ELISA*, so called because the target antigen is sandwiched between a capture antibody (immobilised onto a solid substratum) and a reporter antibody that carries the labelling molecule (Fig. 6.4). Sandwich ELISA is used for the direct detection of a target antigen within a mixed population, for instance to detect *Salmonella* in a sample of food product.

A slightly more complex form of ELISA test, and one that is not commonly used within the food industry, is *indirect ELISA*. For indirect ELISA, it is the target antigen that is immobilised to the solid substratum. A sample of fluid believed to contain antibodies to the target antigen is allowed to adsorb to the antigen and then a labelled antibody raised against the first one is added and the signal measured. This type of ELISA is more often used in medical microbiology laboratories.

The big advantages of ELISA over traditional culture methods are that a pathogen can be detected within a sample much more rapidly and that the assay can be miniaturised to a greater extent. In addition, an ELISA assay can be made semi-quantitative by performing, and testing, a serial dilution of the sample to determine at what dilution the signal ceases to be detectable; this is the *titre* of the sample. A sample with a high titre requires considerable dilution before the signal disappears, whilst a sample with a low titre can only be diluted a little before the signal goes. The sandwich ELISA can be performed in microtitre plate wells – the microwell assay – but the capture antibody can also be attached to a moveable surface – the dipstick assay. This makes it ideal for use outside of the laboratory and on the production line in the factory, giving very rapid and specific information on the hygienic standard of a food product as it is made. A wide range of ELISA kits, both microwell and dipstick-based, is available from a variety of manufacturers.

One problem with ELISA for use in the food industry is that it has a relatively high minimum detection limit – perhaps in the order of 10^4–10^6 CFU – in other words the sample of food product must contain substantial amounts of the target antigen for a positive response to occur in the ELISA assay. Since some pathogens such as *E. coli* O157 have infectious doses estimated in the hundreds of cells, this could prove problematic! One possible approach to partially overcome the problems of minimum detection limit is to simply perform a pre-enrichment step to boost the amount of target antigen present, if possible. Another problem with ELISA in the food industry is that some components of foods can interfere with the adsorption of antibody to antigen.

Antibodies can be raised against a wide range of chemical structures – so protein, carbohydrate and lipid molecules can all be targeted by individual antibodies. This means that not only can physical structures on bacterial cells be detected by antibodies, but also excreted/released materials such as toxins can also be detected. It is possible to detect toxins by ELISA, but another approach is to use RPLA. In RPLA, latex beads with a sufficiently small diameter to remain in colloidal suspension are coated with antibodies to a particular toxin. A sample thought to contain the toxin is mixed with the RPLA test suspension and mixed gently; if the toxin is present the beads agglutinate and precipitate out of suspension, giving it a clearly granular appearance. If the target toxin is not present then the beads do not agglutinate, but remain as a smooth homogeneous suspension. Mixing must be gentle; too vigorous and the precipitate will be broken up. As with ELISA, components of the food can interfere with adsorption of antibody to antigen. Assays using RPLA are available against, for example, staphylococcal enterotoxins and also streptococcal surface antigens, amongst others.

6.14.5.1 Immunomagnetic separation

Immunomagnetic separation, or IMS, uses the specificity of antibodies to target and capture organisms of interest and is becoming increasingly popular as the technology is developed. In the technique of IMS, monoclonal antibodies are raised against specific target antigens and then covalently linked to paramagnetic beads via the F_c portion of the antibody, leaving the F_{Ab} portion free in the medium. The paramagnetic beads comprise a material such as Fe_2O_3 and Fe_3O_4 in a ceramic, glass or polystyrene shell, and for bacteria are approximately 2.8 μm in diameter. The uniform size, surface area and chemical composition of the beads means that they display uniform chemical and physical characteristics. Finally, and most importantly, the beads are *paramagnetic* which means that they are not of themselves magnetic but that they will

be influenced by a magnetic field. This means that the beads will remain in colloidal suspension until a magnetic field is applied, at which point they will be attracted to the pole of the magnet.

Immunomagnetic separation of a target organism from a mixture of organisms in a food product involves first homogenising the food sample such that it is in a free-flowing liquid form. A pre-enrichment step may be incorporated at this stage if the organism is thought to be present in only low numbers. The antibody-coated paramagnetic beads are then added to the suspension, mixed to disperse and incubated for a short period to allow adsorption of target cells. A magnetic field is applied to the tube of sample and the beads collected at the magnetic pole. The sample can be washed carefully with sterile diluent to remove food matrix and unadsorbed (non-target) cells. Novel methods using antibodies have been developed (e.g. Rao *et al.*, 2006). Finally, the beads must be analysed to determine whether any cells are present; plate counting, ATP-bioluminescence, ELISA and direct epifluorescence technique (DEFT) are all suitable confirmatory techniques.

Immunomagnetic separation is a very rapid technique, potentially yielding qualitative presence/absence data within about an hour, although pre-enrichment and/or quantitative plate counting steps may slow it down somewhat. Selection of an appropriate monoclonal antibody can make the IMS technique specific for individual pathogen strains. However, development and selection of an appropriate monoclonal antibody can prove time-consuming initially. There are a number of commercial IMS kit manufacturers, meaning that a range of foodborne pathogens can be isolated by this technique. In addition, automated systems are now coming onto the market, which means that a sample of food product can be added to a tube and then the unit left to add the beads, perform the washes and determine whether target organisms have been captured by the beads. Problems can occur with IMS if the food sample contains significant amounts of particulate solids as these can interfere with pelleting of the beads at the magnetic pole, meaning that beads can be lost during aspiration/washing and the signal weakened significantly. Moreover, excessively vigorous washing may also remove beads and weaken the signal. Finally, chemical components within the food product can reduce the efficiency of antibody–antigen adsorption and so weaken the signal (as with ELISA).

6.14.6 ATP-bioluminescence

Adenosine triphosphate (ATP) is the energy currency of cells and is produced continuously by viable ones to power metabolic reactions. Dead cells, on the other hand, do not produce any new ATP, whilst the highly energetic and unstable ATP that has already been produced is rapidly hydrolysed. Consequently, the amount of ATP present in a sample of cells gives a good estimate of the number of viable ones in the sample, since only viable cells will yield ATP and the amount of ATP in each cell is roughly equal (with certain caveats described below).

The enzyme firefly luciferase catalyses the production of light from the substrate luciferin when supplied with energy from ATP and also oxygen. Consequently, a mixture of firefly luciferase and luciferin when mixed with a fresh lysate of viable cells (containing ATP) will produce a burst of light in the presence of oxygen. Provided that the amount of luciferin, oxygen and luciferase is in excess of requirements, then the amount of ATP released from the sample of cells will be the limiting factor and so the intensity of light produced will be in direct proportion to the amount of ATP. One photon of light will be emitted per ATP molecule in the sample. Since each cell in the sample will contain roughly the same amount of ATP, it follows that the intensity of light can be calibrated to give an estimate of the number of viable cells present in the original sample.

This technique is excellent as a rapid indicator of surface hygiene (Davidson et al., 1999). It is very quick and simple to take a swab sample from a surface, emulsify any cells in a lysing agent to release the ATP, add luciferase/luciferin, incubate for a few minutes and then measure the bioluminescence. Indeed, the technology has been simplified sufficiently that a hand-held luminometer with a traffic-light system (green for safe, etc.) has been produced so that cleaning staff do not need any sophisticated training to assess the quality of cleaning as it is performed. The technology can also give semi-quantitative data on the levels of microbial contamination present in a sample.

However, ATP-bioluminescence technology does have a number of drawbacks. First, the minimum detection limit is relatively high (perhaps 10^4 CFU/ 100 cm^2), so that small numbers of cells cannot be detected; this is a serious problem if those cells are virulent pathogens. Second, the system is not able to differentiate between pathogenic and non-pathogenic cells – all contain roughly the same amount of ATP. Finally, the presence of a mixed population of eukaryotic and prokaryotic cells can confound the test. Eukaryotic cells contain approximately 100-fold more ATP than prokaryotic cells. Consequently, for a given light level the question has to be asked: "Is this indicative of a 'large' population of bacteria or a 'small' population of yeast cells?"

6.14.7 Direct epifluorescence technique

Direct epifluorescence technique, or DEFT for short, is a technique that employs direct observation of samples under a microscope. Samples to be observed are stained with one of a variety of fluorescent dyes, examples of which include fluorescein and propidium iodide. A culture of cells stained with fluorescein will emit light of a green wavelength (520 nm) when irradiated with light of a blue wavelength (488 nm). In contrast, a culture of cells stained with propidium iodide will emit light of an orange-red wavelength (620 nm) when irradiated with light of a green wavelength (536 nm). Other fluorescent stains with other excitation and emission spectra are also available.

For DEFT to work, the microscope used needs not only a light source that can supply the appropriate excitation wavelength for the stain in use, but also suitable optical filters that will allow transmission of the emission wavelength of the stain used but prevent transmission of unwanted wavelengths (such as that of the excitation light) as these could swamp the light emitted from the stained samples. Dichroic mirrors are normally used to filter out the unwanted wavelengths as these usually only permit passage of a narrow range of wavelengths, whilst reflecting all other wavelengths. Figure 6.5 gives a schematic representation of how the light source, fluorescent sample and dichroic mirror would work together in a microscope used for DEFT.

Strictly speaking DEFT is simply the direct observation of cells, within a sample, that have been stained with a fluorescent dye. However, normally a laboratory using the DEFT technique would use aqueous samples and introduce a filtration step to concentrate the cells present. Thus the aqueous sample of cells would normally be filtered through a black hydrophobic membrane filter, the filter plus cells mounted on a microscope slide, fluorescent stain added to the cells and then the cells would be observed. The major advantage of DEFT, for an

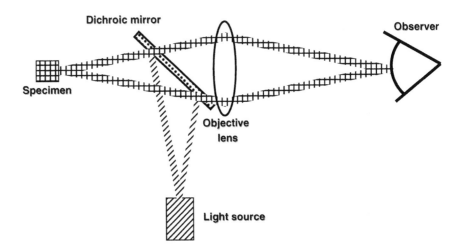

Figure 6.5 Direct epifluorescence technique (DEFT).

operator, is that it is much easier on the eye to observe, and count, brightly fluorescing cells against a dark background than it is to observe dully coloured cells against a bright background (as might be observed in a Gram's stain).

DEFT lends itself ideally to determining overall cell counts, particularly in diffuse aqueous samples such as drinking water, since large volumes of water can be passed through a single membrane to achieve a significant concentration of cells. However, the filters can easily become blocked by particulates present in the sample. Additionally, the basic technique does not give an indication of either the viability or the identity of any cells seen. A number of stain kits have become available that address the problems of viability and identity to some extent. For example, the BacLight LIVE/DEAD kit supplied by Molecular Probes does permit viable cells to be differentiated from non-viable. It is also possible to label specific antibodies with particular fluorescent reporter molecules so as to give an indication of identity, due to the specificity of suitably chosen antibody molecules. Finally, 4-methylumbelliferyl-β-D-glucuronide (MUG) is cleaved to the fluorescent molecule 4-methylumbelliferone by the enzyme glucuronidase found in *E. coli* but not other Enterobacteriaceae (Feng and Hartman, 1982). This means that a solution of MUG added to a mixture of cells containing *E. coli* will diffuse into all cells but only be cleaved to the fluorescent form in *E. coli*, meaning that only *E. coli* cells will fluoresce.

Lastly, some manufacturers are now developing the technology of solid-phase cytometry, which is some-thing of a hybrid of flow cytometry and DEFT. Samples of cells are immobilised onto a surface (such as a membrane filter), stained with fluorescent dyes and then the surface interrogated microscopically by a laser which rasters across it and the fluoresced light detected by a CCD camera linked to a computer. The data can be stored on the computer and can also be analysed automatically for features such as total cell number, or number of cells fluorescing a particular colour within a greater number.

6.15 Microbiological sampling plans

6.15.1 Introduction

It is essential for a food manufacturer to ensure the microbiological safety and quality of a product. This requires destructive sampling of a proportion of the product. To be absolutely sure of the complete absence of pathogens or toxins from a batch of product would entail testing the whole batch. Obviously, this would have a significant impact upon profitability and so a proportion of the whole batch – a *sample* – is selected for testing and the results from this extrapolated to the whole batch. The success of this extrapolation from sample to entire batch relies upon statistical probability:

- the way that a contaminant is distributed throughout the batch of food product;
- the likelihood that a sample taken from the batch will harbour some of the contaminant.

The greater the number of samples taken from the whole batch, the greater the likelihood that one of those samples will harbour the contaminant.

Microorganisms or toxins could, in theory, be distributed throughout a food product in three patterns:

1 There is the regular distribution in which every contaminant unit (cell or toxin molecule) within a batch of product is equidistant in space from every other one. For example, each 1-ml sample withdrawn from a hypothetical 1-litre batch of product containing a population of 1000 cells would harbour exactly one cell; a statistical summary of the data would demonstrate a mean of 1 and a variance of nil (i.e. $s^2 < x$, where s is the variance and x the mean).

2 There is the random distribution in which each contaminant unit is distributed throughout the batch of product in no particular pattern. Some may be spatially close to one another, whilst others may be more distant. Each 1-ml sample withdrawn from this hypothetical 1-litre batch could contain any number of units between 0 and 1000, although the greater likelihood is for only moderate numbers in each sample. A statistical summary of the data would still demonstrate a mean of 1 but now the variance would be greater than that of the regular distribution, such that $s^2 = x$.

3 There is the contagious distribution. In this distribution, contaminant units are present in discrete clumps with large (uncontaminated) distances between them. Most 1-ml samples withdrawn from a hypothetical 1-litre batch, if the contaminant units demonstrate a contagious distribution, will harbour few if any contaminant units. However, a small number of samples will contain a very large number of contaminant units. A statistical summary of the hypothetical batch would again yield a mean of 1, but now the variance would be greater even than that of a random distribution (i.e. $s^2 > x$).

In practice most food products are not truly homogeneous at a microscopic level and so microorganisms are likely to be distributed throughout a batch at best in a random fashion or, more likely, in a contagious fashion. Some regions of a food product may contain levels of preservative, salt or water activity that are lethal to microbial cells. Other regions may contain levels that are merely bacteriostatic, whilst other regions may contain little preservative and sufficient water to permit microbial growth. This would

lead to a contagious distribution. Since the aim of microbiological sampling is to estimate accurately the presence and levels of organisms and toxins within a batch of product, whilst at the same time sacrificing the minimum amount of product, it is necessary to collect the most representative samples possible to allow reasonable estimation. It is clear that the further a distribution departs from regularity the greater the proportion of the whole batch that requires to be tested to ensure a reasonable probability of selecting a sample that actually harbours some of the contaminant.

The International Commission on Microbiological Specifications for Foods (ICMSF) has developed a series of standardised terms relating to food safety and microbiological content. These terms are as follows: a food safety objective, which is a target set by government to achieve a given level of safety but leaves the manufacturer free to determine how it is achieved; a performance objective, which is a target for levels of a contaminant in a food that a manufacturer seeks to achieve when final preparation is carried out by the consumer not the manufacturer (e.g. a joint of meat); and a microbiological criterion. Microbiological criteria may be microbiological standards (a legal or regulatory requirement); microbiological specifications (a contractual obligation); or microbiological guidelines (a manufacturing target).

There are a number of components to a microbiological criterion and these need to be taken into account when designing a sampling plan:

1 a statement of the type of food to be assessed, since different ones will tend to harbour different microbial species;
2 a statement of the microorganisms or their toxins most reasonably expected to be a health or spoilage hazard in a given product;
3 whether there are any appropriate microbiological indicators for the expected microorganisms;
4 the size and number of samples to be collected from the batch or particular points along the processing line from which to sample;
5 a detailed statement of the methods (preferably validated) to be used to detect the organisms or toxins;
6 the appropriate microbiological limits for the food product (these can be described by the parameters m, M, n and c, which are discussed below).

The size and number of samples to be collected and the appropriate microbiological limits selected will be affected by the statistical distribution of the contaminant within the food product, along with the nature of the contaminant. Is the contaminant pathogenic? Is the contaminant capable of producing toxins under some, but not all, circumstances? Is the contaminant capable of spoiling the product? As a consequence, three types of sampling plans prove practicable: two-class attributes plans; three-class attributes plans; and variables acceptance sampling – each of which is more suitable than the others in certain circumstances.

The two-class attributes sampling plan has a simple pass/fail limit; the two attributes are that the product has either "passed" or "failed" the inspection. A two-class attributes plan may simply be a presence/absence test, e.g. the presence of *Salmonella* will lead the batch to fail whilst its absence will permit it to pass. On the other hand, a two-class attributes plan may specify a microbial count limit which must not be exceeded. For example it might stipulate a limit of 10^5 CFU per sample and any sample that exceeds this limit will fail. The ICMSF denotes this pass/fail limit with the lower case letter "m". It should be noted that a two-class attributes plan lacks some sophistication in that for m $= 1 \times 10^5$ CFU per sample, 1×10^2 and 9.5×10^4 CFU per sample would both pass, whilst 1.1×10^5 would fail. The ICMSF provides guidelines as to the food products and contaminant organisms or toxins for which a two-class attributes sampling plan is appropriate.

Three-class attributes sampling plans bring a greater degree of sophistication into the process since, in addition to the "unacceptable" limit which automatically leads a batch to fail the test, there is a further, lower, "marginally acceptable" limit. The "marginally acceptable" limit does not automatically lead to failure, but it does indicate that the process is beginning to lose control of the contaminant in question and that corrective action needs to be taken urgently. There would be a limit to the number of "marginally acceptable" samples tolerated before the batch is automatically rejected. The ICMSF denotes the "unacceptable" limit with the upper case letter "M" and the "marginally acceptable" limit with the lower case letter "m". As an example, the "unaccept-

able" limit may be 10^7 CFU per sample, whilst the "marginally acceptable" limit might be 10^5 CFU per sample, if this can be achieved normally under Good Manufacturing Practice (GMP).

For both two-class and three-class attributes plans, the number of samples collected and the tolerance of faulty samples will influence the likelihood of a contaminated batch being rejected – the *stringency* of the test. Again the ICMSF uses letters to denote these terms; the number of samples collected is denoted by the lower case letter "n" and the tolerance of fails is denoted by the lower case letter "c". The number of samples collected bears a direct relationship with the stringency of the test since the more samples tested, the more stringent it is, as the likelihood of finding contamination is greater. Note that it is the number of samples tested that is important and not merely the number of samples collected. On the other hand, the tolerance of contaminated samples bears an inverse relationship with the stringency of the test – the more tolerant of these the tester is, the lower the likelihood of finding sufficient faulty samples to trigger rejection of the whole batch.

Sampling plan applications are discussed in Roberts *et al.* (1996). Two-class and three-class attributes sampling plans are particularly useful in situations where the past processing and handling history of the food product are unknown, for instance shipments arriving from overseas at an airport or port. The batch of product should be subdivided into notional blocs that represent potential samples; e.g. individual packs on a pallet or three-dimensional coordinates within a bulk liquid container. Samples should then be selected and collected according to a predetermined randomised pattern until the appropriate number of samples has been collected. The requisite number of samples should be collected from each individual batch as this will influence the statistical likelihood of finding a contaminant. The samples should then be tested for the predetermined contaminant organisms and/or toxins as specified in the sampling plan.

The nature of the product and its intended use will have an impact upon the most appropriate sampling plan for it. A product that will not undergo any further processing (such as cooking by the consumer) and which is stored under conditions that might permit microbial growth (for example high ambient temperatures) will require a much more stringent sampling plan than a product that will undergo further processing and/or be stored under sub-optimal conditions for microbial growth (e.g. frozen, or low water activity). In practical terms, for example, pork paté would require greater stringency of testing than chicken breasts, since paté generally receives no further processing.

The probable final users of the product will also affect the stringency of the sampling plan required; infants and the elderly/immunocompromised would be more severely affected by an enteric pathogen such as *Salmonella* than a healthy adult would. Consequently, powdered baby milk would require greater stringency than joints of beef. Finally, the realistic ability of processors to remove a particular pathogen, particularly when further processing by the consumer will occur, will influence the stringency required. For example, a high stringency sampling plan would not be appropriate for chicken carcasses, since a large proportion of these are likely to harbour both salmonellae and campylobacter but adequate handling and cooking by the consumer should minimise the risks. The ICMSF has proposed a series of different sampling plans contingent upon the product, its likely processing and its likely use.

6.15.6 Operating characteristics curves

In a two-class attributes sampling plan, the statistical distribution of samples proving to be unacceptable or acceptable (i.e. exceeding "m" or otherwise) when taken from a batch of product will follow a binomial distribution. In a two-class plan, however, a sample can only be acceptable or unacceptable and so the probability of a selected sample being either acceptable or otherwise must add up to 1:

$$a + u = 1$$

where

a is the proportion of acceptable samples within a batch,
u is the proportion of unacceptable samples within a batch.

In a simple pass/fail test, for example for the presence of botulinum toxin in a canned food product when there would be no tolerance of unacceptable samples, or fails (c = 0), this means that:

$$P\% = (1 - u)^n$$

where

P% is the probability of accepting the batch,
n   is the number of samples taken.

Conversely:

$$R\% = 1 - (1 - u)^n$$

where

R% is the probability of rejecting a batch.

In other words, when a specified number of samples is taken from a batch of product with varying levels of contamination, the probability of accepting that batch will be inversely proportional to the level of contamination, as illustrated in Table 6.6.

If the probabilities of accepting or rejecting a batch for an increasing percentage of contamination are plotted on a graph, then a sine curve will become apparent (Fig. 6.6). This sine curve is described as an operating characteristics curve. Increasing the number of samples "n" tested makes it more likely that a given level of contamination will be detected and the batch rejected. Decreasing the tolerance of faulty samples "c", when a proportion of faulty samples is permitted, will have the same effect of increasing the

Table 6.6 The probability of accepting/rejecting a batch of product, for a given level of contamination.

Level of contamination	For 50 samples tested, probability of:	
	Acceptance (%)	Rejection (%)
0.01	99.5	0.5
0.025	98.75	0.25
0.1	95.1	4.9
0.25	88.2	11.8
0.5	77.8	22.2
1	60.5	39.5
2	36.4	63.6
5	7.7	92.3
10	0.5	99.5

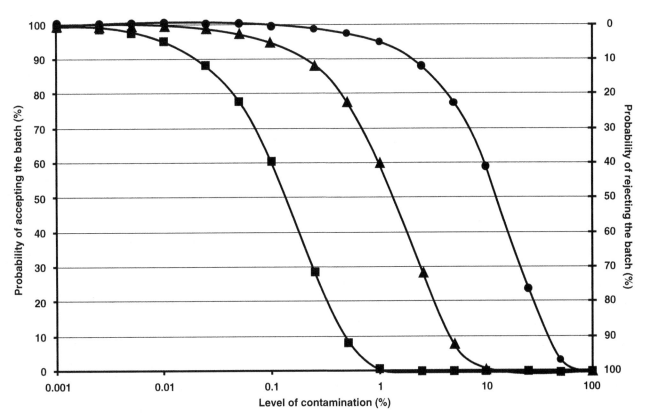

Figure 6.6 An operating characteristics curve, with decreasing stringency based upon a decreasing number of samples tested (n). Squares, n = 500; triangles, n = 50; circles, n = 5.

likelihood of rejecting a batch with a given level of contamination. The likelihood of finding contamination and rejecting a contaminated batch is known as the *stringency* of the sampling plan and so increasing "n" and decreasing "c" makes the sampling plan more stringent. The effect on the operating characteristics curve is to push the sine curve to the left. Conversely, the curve will be pushed to the right of the graph if the sampling plan is made less stringent, by decreasing "n" or increasing "c", and for a given level of contamination there is a greater probability of accepting a batch, and a lesser probability of rejecting it. Altering "n" and "c" in a three-class attributes plan will have the same effect as in a two-class attributes plan. The operating characteristics curve is important to a manufacturer as, ideally, the manufacturer would always like to reject contaminated batches and accept uncontaminated batches. However, in practice this is not possible and so an amount of health risk to the consumer of accepting a contaminated batch and economic risk to the manufacturer of rejecting an uncontaminated batch has to be tolerated but minimised.

The third potential sampling plan is a variables acceptance sampling plan, which is more appropriate to a product of known history. To use variables acceptance sampling, the microorganisms in the food product must be log-normally distributed, since this approach seeks to determine that an acceptable proportion of the batch harbours counts below a predetermined limit. For a hypothetical batch of food product of infinite size, a log-normally distributed microbial population within it could be summarised by the parameters μ, the mean, and σ, the standard deviation. If a sample were to be collected from this batch of product and the viable counts V determined, then a proportion of the possible samples remaining in the batch would harbour fewer counts than this value and a further proportion would harbour higher counts than this value. Moreover, if the parameters μ and σ are known for the batch, then it is possible to determine what proportion of the batch remaining actually harbours counts of greater than V. The formula used to determine the number of those samples that would harbour counts greater than the found counts V in a batch is:

$$V = \mu + K\sigma$$

where

μ is the overall population mean,
K is a constant – the standardised normal deviate,
σ is the overall population standard deviation.

This can then be used to determine the acceptability of a batch of food product if V is set as a safety limit, similar to M in a three-class attributes sampling plan, and K is set as the number of samples with counts in excess of V that will be tolerated, similar to c in an attributes sampling plan.

μ and σ are hypothetical values for a batch of infinite size. In reality, the values x and s would be determined from the $\log_{10}$ counts of microorganisms within a real batch of food product, whilst K is estimated from published tables as k_1. The value k_1 represents the minimum probability of rejecting a batch with unacceptably high counts, in excess of the safety limit V. Consequently, when samples from a batch of product are tested the conditions for acceptance of the batch would be:

$$V \geq x + k_1 s$$

whilst rejection would be triggered by:

$$V < x + k_1 s$$

Since k_1 represents the minimum probability of rejecting a batch with unacceptable counts, it follows that a reduction in k_1 will reduce the stringency of a variables acceptance sampling plan since the probability of rejecting a batch with unacceptable counts will be decreased.

In addition, to the value k_1 which determines the safety limits for a batch of food product, the value k_2 can be used to determine limits that would normally be achievable under GMP. In this case the value v is set as the count limit and k_2 gives the minimum probability of accepting a batch with counts lower than v, which acts similarly to m in a three-class attributes sampling plan. The formula to express this relationship is:

$$v > x + k_2 s$$

As with three-class attributes sampling plans, the failure of a batch using this second formula does not automatically mean that the batch should be rejected, but that GMP is failing and that in the absence of corrective action unacceptable contamination may occur.

6.15.8 Summary

Sampling plans allow the manufacturers of food products, or regulatory agencies in receiving countries, to assess a batch of food product for its microbiological safety according to standardised and agreed procedures without destroying the whole batch in the process. Attributes sampling plans are more appropriate for use when the distribution of microorganisms within a batch of product is not clearly understood. In contrast, variables acceptance plans are more appropriate for use in situations where the distribution of organisms within a batch is understood. Each type of sampling plan sets out to estimate the likelihood that a batch will harbour a population of organisms, or a dose of toxin, in excess of predetermined safety limits. Consequently, each type of sampling plan exhibits what is known as an operating characteristics curve which shows the probability of rejecting or accepting the batch for a given level of contamination. It has to be said that no sampling plan will offer an absolute guarantee of always finding and only rejecting contaminated batches, which is why HACCP programmes for individual products are becoming so important in an attempt to design out the possibility of contamination.

6.16 Hazard Analysis and Critical Control Points

6.16.1 Introduction

The Hazard Analysis and Critical Control Points (HACCP) system was originally introduced in the 1970s by NASA and the US Army in collaboration with the Pillsbury Company. The intention was to design a quality management programme for foods that removed any potential health hazards from the beginning; in other words, it was a "nil defects" philosophy. The system was originally developed for the space programme, since long-duration space travel requires food to be well preserved and the impact of foodborne illness in space is probably best not thought about. The system was originally implemented by forward-thinking food manufacturers, but more recently it became a mandatory requirement both in the USA (from 1997 onwards) and the EU (Regulations 852-854/2004, but see Untermann, 1999).

Devising a HACCP programme for a food product requires systematic analysis of the product; its formulation, its manufacture and the hazards that might arise at any intermediate step as well as in the finished product. The HACCP programme itself is devised and implemented by a HACCP team, which works according to the seven guiding principles laid out by the CODEX Alimentarius Commission, an advisory body formed jointly by the Food and Agriculture Organisation (FAO) and the World Health Organisation (WHO) of the United Nations. However, to be fully successful a HACCP programme requires total commitment from every member of the factory staff, since it is a holistic manufacturing philosophy. A brief guide to the composition of the HACCP team and the processes followed to develop a HACCP programme is given below (Mortimore and Wallace, 1994).

6.16.2 The HACCP team

The first step in HACCP is to assemble an expert team that covers the major disciplines involved in the production process. The team will comprise individuals such as microbiologists, engineers, production managers and production personnel, quality assurance managers and a chair. The key point is that all members should be suitably experienced and willing to express an opinion; the job of the chair is to ensure that all opinions are aired fully (Dillon and Griffith, 1996).

6.16.3 The CODEX principles

The implementation of a HACCP programme for a particular food product follows the seven stages set out in the CODEX Alimentarius:

1. Hazard analysis.
2. Identification of critical control points (CCPs).
3. Establishment of CCP criteria.
4. Monitoring procedures for CCPs.
5. Protocols for CCP deviations.
6. Record keeping.
7. Verification.

6.16.4 Hazard analysis

The hazard analysis starts with a flow diagram describing the whole process in detail from raw materials arriving at the plant to the finished product arriving on the consumer's plate. The hazard analysis will determine all of the potential hazards associated with a food: microbiological (e.g. pathogenic organisms or their toxins), physical (e.g. bone shards or glass) or chemical (e.g. sanitiser residues). The hazard analysis will also assess the severity and level of risk associated with each hazard. For example, botulinum toxin is a more severe hazard than *Bacillus cereus* enterotoxin when ingested since it is considerably more lethal, but the likelihood or risk of finding botulinum toxin in a product stored in air is much less than that of finding *Bacillus* enterotoxin since *Clostridium botulinum* is a strict anaerobe whilst *Bacillus cereus* is an aerobe. There are, therefore, three important terms associated with a hazard analysis, and these are: a *hazard* – an agent that could cause harm to a consumer; the *severity* of a hazard – in other words how much harm it might cause; and *risk* – the likelihood of a particular hazard actually arising in a specified food product.

6.16.5 Identification of critical control points

Next the critical control points – the CCPs – within the process must be identified. This is performed using a CCP decision tree, which is a logical series of questions that is asked at each stage of the process and ensures a consistent approach at each. It also promotes structured thinking and encourages discussion amongst all the members of the team. The CODEX Alimentarius Commission has published an example of a decision tree to assist in the identification of CCPs, although it recognises that this may not be most appropriate in all situations and permits other approaches.

6.16.6 Establishment of CCP criteria

Next the critical control point criteria must be established (Khandke and Mayes, 1998). CCP criteria may define features such as the temperature and time exposure profile of a product to exert control over a microbiological hazard (e.g. 121°C for 15 min) or they may define the maximum size of particles permitted in the product (e.g. particles of no greater than 3 mm diameter by X-ray). However, CCP criteria can also define features such as depth of fill in cooling trays (as this can influence the rate of cooling and therefore the possibility of microbial growth), or the concentration of free chlorine in can cooling water (as insufficient chlorine may permit build-up of microbial contamination and its potential entry into cans via seam failures). Ideally, raw materials supplied to a manufacturer should be provided "supplier assured"; the supplier should certify that the raw materials meet specified criteria, although it is wise to perform checks also. Finally, the HACCP team will also define any post-purchase handling that must be specified by the manufacturer to instruct both retailers and consumers in the safe storage and preparation of the product (for example, "keep refrigerated" or "use by date"). Some intermediate steps in the process may actually be more prone to hazards than either the raw material or the finished product (e.g. liquid whole egg after removal from the shell but before pasteurisation) and the manufacturer must specify suitable criteria for safe handling of these intermediates.

6.16.7 CCP monitoring and protocols for CCP deviations

Having identified the CCPs and defined the criteria necessary to ensure that control is exerted, the manufacturer must introduce robust monitoring protocols. For instance, in the time/temperature example above, a thermocouple with a data-logger attached would be inserted in an appropriate location within the batch of product and the temperature achieved and the duration it was held for recorded by the data logger. The manufacturer must also introduce standard protocols to be followed by staff in the event of a failure at a CCP – for instance, if the required temperature is not achieved, or not held for long enough, can the product be reworked safely, could it be incorporated safely into a subsequent batch, or should it be destroyed. This decision is for the manufacturer, but it must have a sound scientific rationale.

6.16.8 Record-keeping and verification

Under EU HACCP regulations, it is a requirement for manufacturers to keep the records from all CCP monitoring for "an appropriate period" (852/2005 Article 4c). The data records can also help to form the basis of a defence of "due diligence" should a

manufacturer be sued in the event of problems. Finally, a manufacturer should verify routinely that the HACCP programme is operating effectively for the product in question and that all of the monitoring equipment is functioning correctly and within tolerances (see also Anonymous, 2006).

6.16.9 When changing the product formulation

A HACCP programme should be developed for each individual product that a manufacturer makes, since differences in ingredients and processing stages/methods between products can influence both the risk and the severity of potential hazards associated with each product. Moreover, should the formulation of a product be altered after a HACCP programme has been implemented for it, then the HACCP team should reassess the product and its process as even small changes in ingredients can have a major influence on the risk and severity of possible hazards (see comments on *Cl. botulinum* in yoghurt in section 6.9.6).

6.16.10 Summary

HACCP is as much a production philosophy as it is a manufacturing process required in an increasingly large number of countries around the world. Seven basic principles have been laid down in the CODEX Alimentarius (see above). A team of expert individuals will be gathered together and, following these seven principles, will assess the process used to manufacture individual products and will design a HACCP programme for each. Should the formulation of the product be altered in any way, then the HACCP team would reassess both the product and the process to ensure that safety standards are maintained. Ultimately it may be important to differentiate between strains of the same species, often by using molecular methods (Zabeau and Vos, 1993).

6.17 Hygienic factory design

Hygienic factory design is essential to the safe production of food. The topic is a large one and could easily occupy a textbook in itself. However, the principles underlying it are relatively straightforward and basically encompass the choice of appropriate materials for the construction of both plant and factory buildings and the construction of the plant and buildings in such a fashion as to minimise the accretion of contamination, exclude pests and facilitate access for cleaning (see also Anonymous, 2003, 2004a–c, 2006; Cole 2004).

6.17.1 Plant construction materials

To select suitable plant construction materials it is necessary to understand the chemical and physical properties of the food product and thus how they will interact with the construction materials. For example, although stainless steel does not corrode readily, it is actually relatively sensitive to chloride ions and so products high in salt (e.g. brine) may well promote corrosion of the stainless steel surface if in contact for too long. Corrosion pitting may generate difficult-to-clean foci for microbial growth in biofilms that can then contaminate the product. Equally, products high in fat or oil may leach plasticiser from plastic contact surfaces if the wrong formulation is used.

6.17.2 Factory construction materials

To select suitable factory building construction materials it is necessary to consider the likely wear-and-tear that they may be subjected to in addition to exposure to dust or moisture and proximity to exposed food product. For example, bare concrete, "breeze block" or brick would be inappropriate in the main processing hall as they all have relatively rough surfaces that can collect dust that can then contaminate the product. These materials, left unprotected, would also be inappropriate in locations where there is considerable movement, for instance in passageways (either as floor or wall), since there is a risk of a trolley or vehicle colliding with and damaging the wall, leading both to a risk of walling material dropping into product and a focus for dirt collection. Finally, a material like bare concrete is unsuitable as a flooring material as liquids may drip onto it, soak in and act as a focus for microbial growth.

6.17.3 Plant design

Plant design should ensure that there are no locations where food material may collect and permit microbial growth that can go on to contaminate future batches of product. There should be no "dead-spaces": tanks should drain from the very lowest point, fluid flow through a pump should scour the whole unit, valves should be of the pinch-cock design and pipework should be joined in such a way as to minimise gaps (e.g. using IDF-type ring joints or using ring welds). Moreover, if T-pieces are required in a pipeline, their depth should be minimised and their location designed to promote scouring by fluid flow, bends should be of smooth and gentle radius, and any reductions in pipe diameter should occur gradually, rather than leaving a shoulder. All of these pipeline "dead-spaces" can collect fluid and so act as foci for microbial growth that can then contaminate the product. Exterior surfaces of the plant should be sloping or curved, to minimise the collection of dust and contamination.

The plant should also be located so as to promote easy cleaning, for example with sufficient space underneath to allow a brush access or with *safely* removable covers to permit access to enclosed spaces. Ideally, plant should be positioned in a logical flow progressing from raw materials inwards to finished product storehouse or dispatch to prevent raw materials from crossing the path of, and potentially contaminating, finished product (see *Cl. botulinum* in tinned salmon, section 6.9.2). Many factories working with liquid products have moved over to a cleaning-in-place design for their plant, which incorporates secondary lines next to the main pipelines that permit circulation of cleaning agents and rinses.

6.17.4 Building design

Factory building design should also permit easy cleaning, where smooth non-porous surfaces have important uses. In addition, windowsills and unavoidable overhead service lines should not be out of easy reach for cleaning and should have sloped or curved surfaces to minimise dust collection. Ideally, though, overhead services should be enclosed in the roof space in order to allow technical staff access for repairs (and cleaning) without entering food processing or handling areas. Light fittings should also be enclosed in non-breakable plastic housings and only accessible from the overhead service space, to prevent broken glass from dropping onto the production line. Finally, factory buildings should be proofed against vermin (such as insects and rodents). This should include an absence of opening windows in areas where product, intermediates or raw materials are exposed and all doors should be tight-fitting.

6.17.5 Summary

This is not intended to be an exhaustive description of hygienic factory design, but indicates that there is a wide range of factors that interact to make factory design hygienic. Effectively, though, the aim of hygienic factory design is to use the most cost-effective materials in such a fashion as to minimise the likelihood of contamination collecting, entering the food product and causing a hazard to consumers. Arvanitoyannis *et al.* (2005) has more details.

6.18 Microbial fermentation

6.18.1 Food products

6.18.1.1 General introduction

Fermentation has been used for many centuries both to preserve food products and to make favourable alterations to their sensory qualities. Fermentation has been known perhaps since the time of the Sumerians, whilst the Egyptians brewed a fermented grain product called "boozah" during the time of the Pharaohs. The Greeks and Romans both fermented grapes to produce wine, whilst there is controversy over who first developed distillation, with the Chinese being the first in the eighth century with freeze-distillation and Muslim alchemists probably the first to use heat-distillation in the ninth century. Distillation for drinking purposes was probably developed in Europe in the fourteenth century. Production of "beer" and wine provided a source of safe drinkable liquids at a time when water quality was poor, as the water for beer is heated prior to brewing and the alcohol present in both beer and wine would discourage growth of microbial pathogens. The mood-altering effects of alcoholic products may also have been a factor in their adoption.

Cheese may first have been made around 8000 years ago, with the Egyptians recording its production 4000 years ago. Fermentation of milk to make a cultured, yoghurt-like product is believed to have been discovered in the plains of Central Asia by the Bulgars around 5000 years ago, probably as a result of storing milk at ambient temperature in skins. The preparation of cheese, yoghurt and other fermented milk products would have provided a means to extend the shelf-life of milk when refrigeration was unknown and the milk would have become unpalatable

or unsafe within a few days. Fermentation of milk also alters its sensory qualities significantly, often producing solids or gels with significantly increased levels of acidity, which can make it more palatable to some and also more digestible to those who are, for example, lactose intolerant.

Fermented meat products, particularly sausages, also seem to have an ancient history and were certainly made by the Greeks and Romans. Lactic acid fermentation of meats can have a substantial impact upon the safety of products made from it as many meat-borne pathogens are relatively sensitive to the effects of organic acids. One notable exception to this is *Listeria monocytogenes* which is becoming more common in outbreaks from fermented meat products due to its greater resistance to organic acids. In a historical context, though, fermentation of meat for storage would have yielded enhanced safety throughout the products' shelf-life at a time when refrigeration was unknown. In addition, and just as with milk products, the altered sensory qualities of fermented meat products would have been an attraction too. Fish may also be fermented, although this is usually to produce a strongly flavoured condiment to enliven a bland diet and introduce essential amino acids that might otherwise be lacking.

A wide range of vegetables are fermented in many different countries as a means of both enhancing their shelf-life, allowing consumption out of season, and introducing some variety into the diet. Along with fermented vegetable products can also be classed what are described as "traditional" fermentations, which can include, for example, the fermentation of soy beans to make soy sauce, tempeh and miso, and also cocoa, coffee and tea. This section gives an overview of each of these types of products.

See also Doyle *et al.* (2001).

6.18.2 Alcoholic products

In simple terms "beer" is produced through the fermentation of grain sugars by strains of the yeast *Saccharomyces cerevisiae*. In practical terms, the process is a little more complicated than this since the grain sugar is a polymer, starch. Brewing strains of *Saccharomyces* lack amylase and so cannot utilise starch. Traditionally, the grain is "malted" to convert the starch to more simple mono- and disaccharides that *Saccharomyces* can ferment. In the malting process, the grain is allowed to initiate germination, so that amylases are produced in the aleurone layer of the grain.

Germination is then terminated by roasting the grains in a kiln, the duration of the roast affecting the colour and flavour of the malt.

A short roast yields a malt of light golden colour and mild flavour, used to brew light beers such as lagers, whilst a longer roast leads to a much darker grain with a much stronger toasted or burnt flavour used to brew stronger beers such as ale, bitter (heavy to the Scots) and stout. The malt is coarse milled to form "grist" that can be transported to the brewery and/or stored. Fermentable sugar is extracted from the grist during "mashing", when the grist is steeped in water at 65°C for about 1 h to yield "sweet wort". The "wort" is extracted from the spent grains by filtration, using the barley husk of the grist as the filter-bed. At this stage hops, or processed hops, are added to the wort which is boiled for several hours to extract the hop resins and to concentrate the wort, reducing its volume by around 5–15%. In addition, boiling helps to kill contaminant bacteria in the wort. The hop resins extracted during boiling have an antibacterial effect as well as imparting a balancing bitterness to the otherwise potentially cloying sweet flavour of the wort.

Finally the wort is cooled to between 8 and 18°C, depending on the type of beer, and "pitched" (inoculated) with *Saccharomyces*, traditionally collected from the previous fermentation. Fermentation proceeds for up to 10 days, again depending on the beer variety, towards the end of which period the yeast will flocculate. The quality of the water used to make the wort has a significant impact on the quality of the finished beer; water with a particular calcium content, found in the traditional brewing areas of Burton, Edinburgh and London, causes precipitation of phosphate and a reduction in the pH to around that most suitable for barley amylases and to limit dextrinase.

Lager-type beers are fermented at lower temperatures – traditionally 8–12°C – for about 8–10 days and at the end of this period the lager strain of *Saccharomyces* will flocculate and settle to the bottom of the fermentation vessel, leading to the term "bottom fermenter". In contrast, ale-type beers are traditionally fermented at slightly higher temperatures – of 12–18°C – for a shorter period of 5–7 days. At the end of the fermentation the ale strain of *Saccharomyces* will flocculate and float to the top of the vessel, leading to the term "top-fermenter". At this stage of the fermentation, the freshly brewed "green" beer will not be particularly palatable since there will be a range of *Saccharomyces* secondary metabolites present in addi-

tion to the alcohol. The beer will undergo a period of "maturation" during which chemical interactions between the various secondary metabolites can occur and yield the full flavour of a good beer.

Historically, a number of species of *Saccharomyces* were considered to be involved in the various beer fermentations. Brewers' yeast, *Saccharomyces cerevisiae*, would have been used to produce ales and bitters, whilst lager yeast was named *Saccharomyces carlsbergensis*, and is believed to be a relative of the wine yeast *Saccharomyces uvarum*. Recently, the taxonomy of *Saccharomyces* has undergone some revision, such that *S. uvarum* was stripped of species status at one point, but it has now been reinstated.

The major biochemical alteration wrought by *Saccharomyces* during beer fermentation is the anaerobic fermentation of 1 molecule of glucose to 2 molecules of ethanol and 2 of CO_2. *Saccharomyces cerevisiae* is not capable of growth and cell division anaerobically, since it requires O_2 for sterol synthesis, but fortunately for brewers it is capable of fermentation to meet a proportion of its energy needs. Consequently, the initial stage of brewing, immediately after pitching, is undertaken aerobically to permit multiplication of the *Saccharomyces* inoculum to a working density. Subsequently, the culture is incubated anaerobically to encourage maximal ethanol production.

It is possible for brewers to use a process called "high-gravity brewing". In this process concentrated wort is fermented to yield a higher concentration of alcohol and then diluted to the required concentration of alcohol (alcohol by volume or ABV) with water. This is a technically difficult process, but, if mastered, it can lead to process efficiencies as larger volumes of beer can be produced for a given unit of plant size.

Boiling the wort, and the presence of hop acids and alcohol, mean that the finished beer is not particularly hospitable to many microorganisms, particularly the foodborne pathogens.

A greater risk to beers is post-process contamination by spoilage organisms, particularly *Acetobacter* and *Gluconobacter* species, which convert the ethanol to acetic acid (vinegar). Other spoilage organisms of beer include *Lactobacillus* and *Pediococcus* which can produce extracellular polysaccharide leading to "ropy" beer. Occasionally, Enterobacteriaceae may contaminate the early stages of the fermentation, leading to off-flavours and odours, but this is relatively rare. Yeasts and moulds may cause problems during brewing, both by spoiling the barley grains, for instance growth of *Fusarium* species on the grain can lead to "gushing" beer, and by contaminating the pitching yeast, leading to spoilage of the fermenting wort. Species such as *Candida*, *Saccharomyces*, *Zygosaccharomyces*, *Pichia*, *Brettanomyces*, *Torulospora* and *Debaryomyces* can cause problems in the fermenting wort. Another, although unlikely, health hazard could arise from the presence of mycotoxins, and particularly aflatoxins, in the original grains if these were harvested and stored in damp conditions permitting growth of moulds such as *Aspergillus* or *Penicillium*.

A major hazard for brewers is contamination with what are termed "wild yeasts". A wild yeast is any strain of yeast – *Saccharomyces* or otherwise – that was not added intentionally by the brewer. Contamination with wild yeasts can cause problems as these may outgrow the strain used by the brewer and lead to unpredictable rates of fermentation and potentially to undesirable flavours. This can affect the customer's perception of product quality as major brands are normally built upon predicable product characteristics; in other words does "brand x" beer taste the same today as it did previously? It should be noted that for a small number of "artisanal" brews, wild yeasts are encouraged. "Lambic" beer, produced in Belgium, is the result of a "natural" fermentation in which the wort is left open to the atmosphere and yeasts allowed to settle into it to initiate the fermentation. This method of brewing beer yields a product that has a much more fruity and complex flavour than other beers.

Another potential problem that brewers can encounter is "killer yeast", which secretes a lethal factor that is toxic to other strains of yeast. If a batch of beer becomes contaminated with a killer strain, it will kill the brewing strain and take over the fermentation vessel, leading to an unpredictable fermentation and possibly to undesirable flavours. There are more than 90 species of killer yeast, in addition to strains that do not produce lethal factors.

Some brewers have attempted to overcome the problem of wild yeasts and killer yeasts by selecting strains of killer yeast (*Saccharomyces cerevisiae*) with favourable fermentation and flavour profiles to use in brewing such that these will exclude the wild yeasts and other killer yeast strains (Buzzini *et al.*, 2007).

6.18.2.1 Wine

The basic principle of brewing wine is as for brewing beer in that a sugar source is converted to ethanol through the action of a culture of *Saccharomyces cerevisiae*. However, the source of sugar is the fruit of *Vitis vinifera*, the grape vine. One molecule of glucose or of the fruit sugar fructose is converted to 2 molecules of ethanol and 2 molecules of CO_2 by the wine yeast. Wine-makers would traditionally allow the natural yeast flora of the grapes to ferment the batch of grape juice, or "must", and some still do this. A variety of yeasts colonise grape skins, with *Kloeckera* and *Hanseniospora* being the most prevalent and *Saccharomyces* present in very low numbers. In contrast, the surfaces of winery equipment generally harbour a yeast flora in which *Saccharomyces* predominates and these may be the major source of this species. Alternatively, the must to be fermented can also be inoculated with a starter culture of freeze-dried wine yeast to initiate the fermentation. Again, as for beer, the initial stage of the wine-making process is allowed to proceed aerobically in order to develop a sufficient yeast cell population and then the main fermentation is performed anaerobically to maximise ethanol yield.

The strains of yeast used to produce wine commercially are most often *Saccharomyces uvarum* and *S. pastorianus* (Nguyen and Gaillardin, 2005). Both *S. uvarum* and *S. pastorianus* have a greater ethanol tolerance than the beer strains, meaning that the fermentation can proceed for longer and yield a higher concentration of ethanol. This can be up to 15%, compared with around 8% for beer strains.

To produce red wine, the grapes are crushed and the skins and must are fermented together to extract the anthocyanins that give red wine its characteristic colour. Red wine is usually fermented for 4–10 days at 25–30°C. In contrast, to produce white wine the skins and the must are separated immediately and then the pale-coloured must is fermented for 4–10 days at 10–15°C. To produce a rosé wine the must and skins are simply left in contact for a shorter period than for a red wine. At the end of the fermentation, the flocculated yeast and any particulate matter are allowed to settle out of the wine before the clarified wine is "racked off" into bottles (or other final containers) for consumption.

The production of a sparkling wine such as champagne goes through a very interesting second fermentation process. The wine is initially fermented as for any other wine. However, after the primary fermentation, the wine is transferred to thick-walled pressure-resistant bottles and dosed with a small amount of a yeast and sugar suspension and then given a temporary cap. The dosed wine is incubated inverted at an angle and the bottles are turned or "riddled" at frequent intervals so that the yeast cells settle towards the cap. When the champagne is judged to be ready, the neck of the bottle is frozen to freeze the plug of cells into it. The cap is then gently tapped to release it and the pressure of CO_2 within the bottle ejects the plug. After this the bottle is corked and a wire cage applied over the cork to prevent its subsequent unintended ejection.

6.18.2.2 Saké

The production of saké relies on the fermentation of grain sugars to ethanol by *Saccharomyces cerevisiae* just as for beer and wine. However, saké production takes a different approach to provide a source of fermentable carbohydrate for the yeast. Rice grains are used to produce saké and the complex carbohydrate, starch, of which the grains are composed is hydrolysed by amylase from the mould *Aspergillus oryzae*. The rice is inoculated with *Aspergillus oryzae* mycelia in several steps. First a batch of rice is steamed to hydrate it, and a portion mixed with *Aspergillus* spores and incubated aerobically at 35°C for 7 days to allow germination and amylase production by the new *Aspergillus* mycelia. This "koji" is then mixed with more rice and with saké strains of *S. cerevisiae* to yield the "moto", which is mixed with the remainder of the rice and water to form the "moromi" and incubated a further 21 days at 13–18°C to yield saké.

6.18.3 Bread products

The production of leavened bread uses a strain of *Saccharomyces cerevisiae* selected for enhanced CO_2 production rather than ethanol production. The process involves fermentation of simple sugars to ethanol and CO_2, but the majority of the ethanol evaporates during the baking process. However, unlike beer, the grain is not encouraged to produce endogenous amylase and so yield fermentable sugar from itself. Instead, a small amount of sucrose is added either to the dry mix or to the yeast inoculum, and this provides sufficient sugar for CO_2 production to raise the bread.

Not all bread is raised using the CO_2 produced by yeast. For some breads, "sourdough" fermentation is used. In the sourdough process a small portion of the previous batch is held back and used to inoculate the subsequent batch of bread. The major microbial species in sourdough fermentations are the lactic acid bacteria (LAB). Fermentation of sugars by the LAB yields CO_2, which raises the bread, and also a quantity of lactic acid which gives the bread its distinctive tangy flavour. In addition, some breads are manufactured using chemical raising agents, such as baking powder, but this is beyond the scope of this chapter.

6.18.4 Fermented Milk Products

6.18.4.1 Cheese

Types of cheese

There is a huge variety of different cheese types made around the world. The variety of milk types, starter cultures, fermentation and ripening conditions and the secondary inoculum is enormous and all contribute to the variety of cheeses available. Nevertheless, a classification scheme of sorts has been devised, and has gained some degree of acceptance; cheeses can be broadly divided into hard, semi-soft and soft varieties, which is adequate for general description.

Hard cheeses

Hard cheeses are prepared over the longest period of time and generally have the highest solids and lowest moisture content. To prepare a hard cheese, milk is collected, often heat-treated and then cooled to around 30°C (e.g. in the case of English cheddar cheese). A starter culture comprising mesophilic (*Lactococcus lactis*) or thermophilic (*Lactobacillus delbrueckii* ssp. *bulgaricus*, *Lb. casei* or *Str. thermophilus*) lactic acid bacterium strains, either as a mix or as a pure culture, is mixed with the milk and incubated for approximately 45 min at 30°C to initiate acidification and casein precipitation. The enzyme rennet, from a variety of sources (calf abomasum, a number of fungi – although this may not yield such high quality cheese – or from recombinant *E. coli*), is next added to the milk and cleaves the macropeptide moeity from κ-casein, yielding para-κ-casein and causing the casein micelles to coagulate. Continued incubation for 30–45 min at around 30°C leads to further lactic acid production and curd formation as the solids start to precipitate out of the liquid whey. The curds are then cut into small cubes which permits contraction of the curd gel and expulsion of the whey fluid. The curds may then be scalded at around 40°C which causes further shrinkage of the curds and expulsion of more whey from the gel. Thermophilic starter strains can continue to ferment during scalding, even

at the higher temperatures used for the hardest varieties of cheese such as parmeggiano regiano and pec-
corino romano.

For hard cheese varieties, salt is mixed into the
curd crumbs and then the curds are pressed in a
mould to help expel whey and produce the characteristic wheels or cylinders of cheese. Finally, most
hard cheese varieties undergo an extended maturation or ripening stage, during which production
of secondary metabolites from endogenous enzymatic activity and also non-starter lactic acid bacterial metabolism can alter the flavour further. In some
varieties of cheese, some metabolism of the lactate
present by *Propionibacterium freudenreichii* ssp. *sher-
manii* to yield propionate and CO_2 is encouraged as
this can impart a desirable nutty flavour and cause
the typical holes present in Emmenthal and Gruyere
cheeses.

Semi-soft cheeses

Semi-soft cheese production follows the same basic
protocol and uses the same strains of bacteria in the
starter culture as hard cheese production. However,
some specifics may differ, such that the scald temperature may be lower, expelling less whey. Equally, the
curd crumb may not be pressed to mechanically expel
whey, but may simply be moulded into wheels and
allowed to dry naturally. Semi-soft cheeses may finally undergo a shorter maturation process than hard
cheeses, but that maturation may be assisted by organisms other than bacteria. For instance, the blue
cheeses – Stilton, Roquefort and Gorgonzola – are matured with the aid of the mould strains *Penicillium
roqueforti* and *P. glabrum*. Mould-ripened cheeses typically have a higher pH than bacterial-ripened cheese,
due to the proteolytic activity of the moulds and the
release of ammonium ion which buffers the lactate.
In addition, the moulds initiate lipolysis, leading to
a complex mixture of fatty acids which contribute
to the pungency of blue cheeses. Of the bacteria-
ripened semi-soft cheeses, some such as Gouda and
Edam are actually salted by soaking in brine, rather
than mixing salt with the curd crumb. Other bacteria-
ripened semi-soft cheeses (e.g. Limburger, Brick and
Port Salut) may be matured by smearing the surface
of the cheese with a mixture of bacteria and moulds
(Deetae *et al.*, 2007). The range of species in the surface smear can be wide (e.g. *Brevibacterium linens, Pro-
teus vulgaris, Staphylococcus equorum* and *Geotrichum
candidum*) and can lead to some extremely pungent
products (Arfi *et al.*, 2003).

Soft cheeses

On the whole, soft cheeses are manufactured using
the same basic protocols and starter cultures as hard
and semi-soft cheeses. However, some soft cheeses
are not made using a starter culture, but the natural
microflora of raw milk is simply permitted to perform
the fermentation; an example of this type of cheese
would be Mexican queso blanco. However, serious
problems with listeriosis in inadequately prepared
cheese during the 1980s in southern California have
demonstrated the potential problems in making such
a high water activity, low salt product from raw milk.
The most important feature of the soft cheeses is that
the whey is only partially separated from the curds,
and in the case of cottage cheese not at all separated.
As a result, the water activity for these types is the
highest of all the cheese varieties. This results in a reduced shelf-life for such cheeses compared with those
that are harder and have a lower water activity.

Some soft cheeses may undergo a ripening period.
For example both Camembert and Brie are ripened
cheeses and both also make use of mould ripening
using species such as *Penicillium camemberti, P. case-
icolum* and *G. candidum*. For both Camembert and
Brie, the ripening mould grows over the surface of the
cheese, rather than through it as in Stilton, although
the effect of the mould is similar. In both cases, proteolysis occurs leading to increased pH through ammonium release and lipolysis occurs leading to fatty
acid release and a concomitant increase in aroma and
flavour intensity. The increased pH brought about by
mould ripening can lead to problems with spoilage
and pathogenic organism survival due to the higher
water activity of these cheeses. Some soft cheeses
are pickled in brine, for example Feta cheese, whilst
some cheeses such as Mozzarella are not ripened at
all. Mozzarella is formed by heating the curd and
whey mixture until the curds form strings. These are
then stretched and formed into balls which are held
in the whey until use. Similarly, paneer is made by
mixing lemon juice with milk, which causes the casein micelles to destabilise, and then removing the
curdled solids. This cheese is not matured, but eaten
immediately.

6.18.4.2 Yoghurt

Yoghurt is manufactured from milk using *Lactobacillus delbrueckii* ssp. *delbrueckii* and *Streptococcus thermophilus*. Both of these species ferment lactose in the

milk to lactic acid as the major end-product. Small amounts of minor components, which are important for flavour, are also evolved with acetaldehyde, diacetyl and acetone being the most important.

Although both *Lactobacillus* and *Streptococcus* are capable of fermenting lactose to lactic acid, milk is quite limiting in the levels of other nutrients that are required by these species. Consequently, a pure culture of either *Lactobacillus* or *Streptococcus* in milk will not yield such a significant, nor rapid, pH change as occurs with a mixed fermentation of both species. This has implications for both the quality and the safety of the yoghurt made, since gelation will not be so great and the pH may not be reduced enough to inhibit the growth of enteric pathogens such as salmonellae.

When inoculated together into milk, streptococci initiate fermentation producing some lactic acid from lactose and also formic acid and CO_2. However, *Streptococcus thermophilus* lacks proteases and so cannot acquire a sufficient supply of nitrogen for vigorous growth, although it does produce peptidases that help it to hydrolyse the limiting concentrations of peptides present in the milk. The lactobacilli do not grow so vigorously in the very early stages of the fermentation, but are stimulated to grow more rapidly by the formate and CO_2 produced by the streptococci. *Lactobacillus delbrueckii* ssp. *delbrueckii* also produces proteases that can hydrolyse milk protein to peptides that *Streptococcus* can then fully hydrolyse, thus yielding amino acids that can be utilised by both species. In the later stages of fermentation, *Lactobacillus* will predominate and perform most of the fermentation as the pH will decline below that which *Streptococcus* can tolerate. The optimum growth temperature for *Streptococcus* is 39°C, whilst that for *Lactobacillus* is 45°C. Consequently, yoghurt fermentation is normally performed at 40–42°C to accommodate both organisms.

6.18.4.3 Kefir

Kefir is a soured, effervescent and slightly alcoholic milk drink originating from the eastern steppes of Russia (Lopitz-Otsoa *et al.*, 2006). The fermentation is performed by a mixture of microorganisms, both yeasts and bacteria bound together by a polysaccharide called kefiran, that are added to the milk in the form of small pellets or granules called "kefir grains". A wide range of organisms is present in the kefir grains, but the most common are various lactobacilli

(e.g. *Lb. kefirofaciens*, *Lb. kefiri*, *Lb. acidophilus* and *Lb. delbrueckii* ssp. *bulgaricum*) and various yeasts (e.g. *Candida kefir* and *Saccharomyces cerevisiae*). Commercially, kefir is produced by heating milk to 85–95°C for 3–10 min, cooling to 22°C and adding the grains to about 5% w/v. Fermentation proceeds for 8–12 h at 22°C, after which the mix is cooled to 8°C and incubated for another 10–12 h.

6.18.5 Fermented fish products

Fermented fish products (Thapa *et al.*, 2004) tend to be used as strongly flavoured additives to recipes, so as to make an otherwise relatively bland diet a little more interesting. Fermented fish can also serve to introduce nutrients such as vitamins and essential amino acids that can be missing from solely vegetable-based diets.

Fermented fish can be broadly classified into two forms: fish paste and fish sauce. The majority of fermented fish products are not strictly microbial fermentations, but are the result of autolysis resulting from enzymes endogenous to the fish. In these fermentations, whole fish are packed into containers along with salt at high concentrations (up to 25% salt and 75% fish) which inhibits virtually all microbial growth. The material in the containers does not have enough carbohydrate to permit significant microbial growth either. True fermented fish products require the addition of exogenous carbohydrate to promote fermentation by the naturally occurring microflora. The organisms that have been found in "fermented" fish products include *Lactobacillus*, *Lactococcus*, *Micrococcus*, *Staphylococcus*, *Moraxella* and halotolerant *Bacillus* species.

6.18.6 Fermented vegetable products

Very many different vegetables can be fermented, but perhaps the most widely known fermented vegetable product is cabbage (sauerkraut in Germany or kimchi in Korea). Although the details of the fermentation method differ between the products, the basic process is the same, being a lactic fermentation of the plant sugars. To make sauerkraut, the washed and shredded cabbage is stacked in large containers and layered with salt to between 2 and 3% w/w, before being covered with a weight that helps to force liquid from the cabbage. The salt acts as an inhibitor for many bacterial species, particularly the enteric pathogens. In addition, the salt causes the excretion of water from the plant cells, so creating a brine environment. Some sugar is also released from the cut surfaces of the cabbage leaves and this is sufficient for lactobacilli, which are more halotolerant than many other species, to ferment to lactic acid. The reduced pH of the sauerkraut, along with the high salt concentration, acts to extend the shelf-life of the cabbage significantly.

Sauerkraut is not usually inoculated with a starter culture, but instead relies on the natural flora of the cabbage leaves and the fermentation vessels. The species that predominate in sauerkraut production include *Leuconostoc mesenteroides* and *Lactobacillus plantarum*. Due to the low pH and high salt content of sauerkraut, it is relatively free from microbiological problems if made correctly. One pathogen that has occasionally been found to contaminate sauerkraut is *Listeria monocytogenes*, which has greater acid and halotolerance than most enteric pathogens. In addition, yeasts and moulds can occasionally cause problems with spoilage, again due to their greater acid and halotolerance.

Kimchi is produced in a similar way to sauerkraut, using shredded cabbage and salt. However, the fermentation does not proceed for as long with kimchi and is completed in 2–3 days at 20°C. The shorter incubation period means that the microflora does not mature in quite the same way as in sauerkraut. The predominant species in kimchi production is *Leuconostoc mesenteroides*.

Examples of other vegetables that are also fermented to improve their keeping qualities and/or their palatability include olives, cucumbers, cauliflower, peppers and swedes. Generally, the principle in most of these fermentations is as for sauerkraut production. In other words, the vegetable material is mixed with either salt or brine, which excludes many pathogenic and spoilage organisms, but not halotolerant lactic acid bacteria. The LAB population ferments sugars released from the plant cells into the brine to produce lactic acid, and the combined effects of low pH and high salt concentration help to exclude pathogens and spoilers, making the product stable for longer than the raw material.

Another interesting example of a fermented product made using vegetable matter is cocoa (Ardhana and Fleet, 2003). This is interesting from both a food technology and a microbiological standpoint. It is not currently possible to replicate accurately the flavour of cocoa synthetically. A complex mixture of species is involved in the development of the full cocoa flavour, and this mixture must undergo population succession to achieve a full flavour. The ripe cocoa pods are harvested from the *Theobroma cacao* tree, of which there are two main varieties (forastero and criollo). Cocoa made from criollo seeds is considered to be of finer quality than that made from forastero, but the criollo trees are much less hardy and more prone to disease than the forastero trees.

After harvesting, the cocoa pods are split manually and the seeds, along with their mucilagenous sheath, are scooped out of the pods into fermentation containers by the harvesters. These are probably the key primary inoculation stages in the process, with important species coming from harvesters, their implements and the fermentation containers. The raw cocoa seeds, or more precisely their surrounding mucilage, are then left to ferment: either spread in thin layers on racks in the sun; placed in covered heaps on the ground; placed in covered baskets; or placed into wooden containers of approximately 1 m³ volume.

Fermentation of the cocoa beans and their attached mucilage then goes through a number of phases which have distinct microbiological and biochemical profiles. During the primary, aerobic, stage of the fermentation (usually up to about 24–48 h) many different genera of yeasts, including *Candida*, *Hanseniospora*, *Saccharomyces*, *Kloeckera*, *Pichia* and *Kluyveromyces*, tend to predominate, forming around half the microflora initially and rising to about 90% at their peak. The yeast species degrade the carbohydrates in the mucilage and the major metabolic products of this are ethanol and CO_2. The yeasts also metabolise citrate in the pulp, reducing its levels and therefore increasing the pH. Since the mass is initially aerobic, there will be a small amount of acetic acid produced by acetic acid bacteria from the ethanol. As the fermentation proceeds, the mass becomes anaerobic and the pH is reduced by acetate production. At this stage lactic acid bacteria such as *Lactobacillus*, *Pediococcus*, *Lactococcus* and *Leuconostoc* increase in numbers and predominate, whilst the early colonisers decline.

The major metabolic products during the secondary phase of fermentation are lactic acid, acetic acid, CO_2 and acetylmethylcarbinol. As the secondary phase of fermentation proceeds, the lactate and acetate begin to kill the LAB and periodic turning of the heap introduces oxygen which permits the final colonisers to increase at the expense of the secondary ones. This middle, anaerobic, stage of the fermentation is followed by a final, aerobic, stage. Cocoa from high-quality criollo beans can be fully fermented within a few days, whereas the fermentation of forastero beans can take up to 7 days to complete all the stages.

The predominant organisms in the final stages of the cocoa fermentation are *Acetobacter* and *Gluconobacter*. The numbers of bacilli can also increase significantly in the final stages of the fermentation although there is some controversy over the role of this in the overall cocoa quality. The major metabolic product of the organisms in the final stages of the fermentation is acetic acid, which is produced from the ethanol present in the heap. Oxidation of ethanol to acetate is highly exothermic and the internal temperature of the heap will climb sufficiently to kill the beans and a proportion of the microflora. The beans, by now mucilage-free, are finally roasted and milled.

Some workers have tried, with varying degrees of success, to select the most important organisms from the flora found in cocoa fermentations and to develop these selections as starter inocula to improve the reproducibility of the fermentation. However, the quality of the cocoa produced by these starter culture initiated fermentations does not yet fully compete with that of cocoa produced using a wild flora. Moreover, although the major flavour components of the cocoa beans have now been determined, a synthetic cocktail does not seem to yield as rich a flavour as that of cocoa beans fermented using a wild microflora.

6.18.7 Traditional fermented products

Traditional fermented products is a "catch-all" term for products that are not necessarily produced on a global scale, but are nonetheless important locally as either dietary adjuncts or a means of food preservation. The variety of traditional fermented products is too great to recount within the confines of this book, and so a small selection of the better-known traditional products will be presented.

Perhaps the best known of the traditional fermented products are those made with soya beans. Examples of fermented soy products are tempeh, miso, natto and soy sauce. The great benefit of the fermented soy products is that the nutritional value of the soy beans is increased, since they do not contain the full range of essential amino acids for humans when raw.

Tempeh is produced from soya beans that have been soaked in water to approximately double their volume, which softens the beans and loosens the husk. The soaked beans are inoculated either with a sample of tempeh held back from a previous batch or with a commercial starter, and incubated at around 30–38°C for 1–2 days. Traditionally, tempeh would have been made wrapped in banana leaves, although it can be made just as well in perforated plastic bags. Incubation in a semi-enclosed environment slows water loss and controls the exchange of O_2 and CO_2. It is important that the mix does not become anaerobic or this will select for putrefying organisms, nor does the mix want to be incubated at too high a temperature or this will inhibit or kill some of the necessary microorganisms. A complex mixture of bacteria and yeasts is involved in the fermentation of tempeh: *Lactobacillus casei*, *Lactococcus* spp., *Pichia burtonii*, *Candida diddensiae*, *Rhodotorula mucilagenosa* and also *Rhizopus oligosporus*, whose mycelia bind the finished product together, have all been isolated. Tempeh has a shelf-life of around 1 day and the nutritional quality of the beans is altered; the glucose concentration is reduced, the fibre content is increased and, although the range of amino acids is unchanged, the free amino acid concentration is increased markedly.

Miso is also prepared from fermented soya beans, but has a paste-like consistency, rather than the firmer and more cake-like consistency of tempeh (Onda *et al.*, 2002). There are three main recipes for making miso: rice miso is prepared using rice, soya beans and salt; barley miso is prepared using barley, soya beans and salt; and soya miso is prepared using just soya beans and salt. For each of the recipes, an initial koji, or inoculum, is prepared as described for saké; seed koji, which contains *Aspergillus oryzae* spores, is added to a batch of rehydrated rice, barley or soya beans and incubated for 40–50 h at 30–35°C. After incubation, overgrowth of the *Aspergillus* mycelia is prevented by the addition of salt. Whilst the koji is developing, a batch of soya beans is prepared by an initial soak in water for 18–22 h and then the soya beans are cooked at 0.5 atm (115°C) for 20 min. After the cooked soya beans have cooled, the koji is mixed with them and a small batch of miso held back from a previous fermentation; this mixture is packed into anaerobic tanks and allowed to ferment at 25–30°C for differing periods depending on the type of miso being prepared. Soya bean miso, the darkest and strongest, is incubated for periods in excess of 1 year. Salty miso can be incubated for between 1 and 3 months, whilst white miso, the lightest and most delicate, is incubated for about 1 week.

The *Aspergillus* in the original koji produces proteases that break down the protein structures of the soya beans. However, other microbial species, in particular *Zygomyces rouxii*, *Torulopsis* spp. and *Pediococcus halophilus*, are also important in miso fermentation. Due to the high salt concentration and the initial steaming of the soya beans, the range of organisms likely to cause problems is limited, but occasionally include *Pediococcus acidilacti*, *Bacillus subtilis*, *Lactobacillus plantarum* and *fructivorans*, micrococci and clostridia.

Natto is also produced using soya beans, but a glutinous mass surrounds the beans. The most common form of natto is prepared from beans that have been soaked and steamed for about 15 min, inoculated with the bacterium *Bacillus natto*, and incubated for 18–20 h at between 40 and 45°C. The glutinous mass surrounding the soya beans is an extracellular product of the bacilli, poly-DL-glutamic acid. Other forms of natto can be made by mixing the basic natto with either a rice koji or a koji made using a mixture of wheat and barley, along with ginger and brine, and then incubating for longer. When made with rice koji, the natto is incubated for 2 weeks at 25–30°C. When made with the mixed koji, ginger and brine, the koji and soya bean mixture is incubated for around 20 h, dried and then stored for a further 6–12 months in brine and mixed with ginger.

6.18.8 Summary

An extensive range of food and drink products is prepared with the help of microorganisms and the origins of many of these fermentations are lost in the mists of time. Examples include milk products such as cheese and yoghurt, grain products such as bread and beer, fruit products such as wine, and vegetable products such as sauerkraut, cocoa, miso and natto. It seems likely that the original fermentations were discovered by accident and that they enhanced the shelf-life of otherwise perishable materials at a time when their long-term storage would have proven difficult. In addition, the flavour changes wrought by the fermentation might have proved attractive, whilst the mood-altering effects of some would probably also have appealed. However, the flavour changes occurring in some products may have taken a little more effort to appreciate, at least to modern tastes! Finally, fermentations can also yield nutritional benefits, either by enhancing the levels of otherwise scarce nutrients (such as essential amino acids or vitamins) or by

reducing the levels of potentially problematic components (such as reduced lactose levels in cheese reducing its impact upon lactose-intolerant individuals). An excellent review of fermented foods has been published by Campbell-Platt (1987).

Acknowledgements

The authors would like to thank Judith Arris and Dave Fowler and especially Babs Perkins, without whose hard work and efforts this chapter would not have been completed.

Further reading and references

Adak, G.K., Long, S.M. and O'Brien, S.J. (2002) Trends in indigenous foodborne disease and deaths, England and Wales: 1992 to 2000. *Gut*, **51**, 832–41.

Adak, G.K., Meakins, S.M., Yip, H., Lopman, B.A. and O'Brien, S. (2005) Disease risks from foods, England and Wales: 1996 to 2000. *Emerging Infectious Disease*, **11**, 365–72.

Adak, G.K., Long, S.M. and O'Brien, S.J. (2007) Food-borne transmission of infectious intestinal disease in England and Wales 1992–2003. *Food Control*, **18**, 766–72.

Adams, M.R. and Moss, M.O. (1997) *Food Microbiology*, pp. 252–302, 323–36. The Royal Society of Chemistry, Cambridge.

Anonymous (1994) Commission on Tropical Diseases of the International League Against Epilepsy. Relationship between epilepsy and tropical diseases. *Epilepsia*, **35**, 89–93.

Anonymous (2003) *Recommended International Code of Practice – General Principles of Food Hygiene. CAC/RCP 1-1969, Revision 4–2003.* Food and Agriculture Organisation of the United Nations, Rome.

Anonymous (2004a) Regulation (EC) No. 852/2004 of the European Parliament and of the Council of 29 April 2004 on the hygiene of foodstuffs. *Official Journal of the European Union*, L226 **47**, 3–21.

Anonymous (2004b) Regulation (EC) No. 853/2004 of the European Parliament and of the Council of 29 April 2004 on the hygiene of foodstuffs. *Official Journal of the European Union*, L226 **47**, 22–82.

Anonymous (2004c) Regulation (EC) No. 854/2004 of the European Parliament and of the Council of 29 April 2004 on the hygiene of foodstuffs. *Official Journal of the European Union*, L226 **47**, 83–127.

Anonymous (2006) *A Simplified Guide to Understanding and Using Food Safety Objectives and Performance Objectives*. International Commission on Microbiological Specifications for Foods. Kluwer Academic, Dordrecht/Plenum Press, New York.

Arvanitoyannis, I.S., Choreftaki, S. and Tserkezou, P. (2005) An update of EU legislation (Directives and Regulations) on food-related issues (Safety, Hygiene, Packaging, Technology, GMOs, Additives, Radiation, Labelling): presentation and comments. *International Journal of Food Science and Technology*, **40**, 1021–112.

Ardhana, M.M. and Fleet, G.H. (2003) The microbial ecology of cocoa bean fermentations in Indonesia. *International Journal of Food Microbiology*, **86**, 87–99.

Arfi, K., Amarita, F., Spinnler, H.E. and Bonnarme, P. (2003) Catabolism of volatile sulfur compound precursors by *Brevibacterium linens* and *Geotrichum candidum*, two microorganisms of the cheese ecosystem. *Journal of Biotechnology*, **105**, 245–53.

Barnard, R.J. and Jackson, G.J. (1984) *Giardia lamblia*: the transfer of human infections by foods. In: *Giardia and Giardiasis Diseases: Biology, Pathogenesis and Epidemiology* (eds S.L. Erlandsen and E.A. Meyer), pp. 365–78. Plenum Press, New York.

Barrile, J.C. and Cone, J.F. (1970) Effect of added moisture on the heat resistance of *Salmonella anatum* in milk chocolate. *Applied Microbiology*, **19**, 177–8.

Buzzini, P., Turchetti, B. and Vaughan-Martini, A.E. (2007) The use of killer sensitivity patterns for biotyping yeast strains: the state of the art, potentialities and limitations. *FEMS Yeast Research*, **7**, 749–60.

Campbell-Platt, G. (1987) *Fermented Foods of the World: A Dictionary and Guide*. Butterworths, London.

Cano, R.J. and Borucki, M.K. (1995) Revival and identification of bacterial spores in 25–40 million year old Dominican amber. *Science*, **268**, 1060–64.

Chen, T.C. (1986) *General Parasitology*, 2nd edn. Academic Press, New York.

Cole, M. (2004) Food safety objectives – concept and current status. *Mitteilungen aus Lebensmitteluntersuchung und Hygiene*, **95**, 13–20.

Collinge, J., Whitfield, J., Mckintosh, E., *et al.* (2006) Kuru in the 21st century – an acquired human prion disease with very long incubation periods. *Lancet*, **367**, 2068–74.

Current, W.L. (1988) The biology of *Cryptosporidium*. *American Society of Microbiology News*, **54**, 605–611.

D'Aoust, J.Y. (1998) *Detection of Salmonella spp. in Food and Agricultural Products by the Gene-trak® DNA Hybridisation Method*. Health Protection Agency, Government of Canada, MFLP-5.

Davidson, C.A., Griffith, C.J., Peters, A.C. and Fielding L.M. (1999) Evaluation of two methods for monitoring surface cleanliness – ATP bioluminescence and traditional hygiene swabbing. *Journal of Bioluminescence and Chemiluminescence*, **14**, 33–8.

Deetae, P., Bonnarme, P., Spinnler, H.E. and Helinck, S. (2007) Production of volatile aroma compounds by bacterial strains isolated from different surface-ripened French cheeses. *Applied Microbiology and Biotechnology*, **76**(5), 1161–71.

Dillon, M. and Griffith, L. (1996) *How to HACCP*, 2nd edn. M.D. Associates, Cleethorpes.

Doyle, M.P., Beuchat, L.R. and Montville, T.J. (2001) Food fermentations. In: *Food Microbiology: Fundamentals and Frontiers*, 2nd edn, pp. 651–772. ASM Press, Washington.

Feng, P.C.S. and Hartman, P.A. (1982) Fluorogenic Assays for immediate confirmation of *Escherichia coli*. *Applied and Environmental Microbiology*, **43**, 1320–1329.

Fricker, M., Messelhäußer, U., Busch, U., Scherer, S. and Ehling-Schulz, M. (2007) Diagnostic real-time PCR assays for the detection of emetic *Bacillus cereus* strains in foods and recent foodborne outbreaks. *Applied Environmental Microbiology*, **73**, 3092–8.

Gamble, H.R., Bessonov, A.S., Cuperlovic, K., *et al.* (2000) Recommendations on methods for the control of *Trichinella* in domestic and wild animals intended for human consumption. *Veterinary Parasitology*, **93**, 393–408.

Georgiev, V.S. (1994) Management of toxoplasmosis. *Drugs*, **48**, 179–88.

Gupta, A., Sumner, C.J., Castor, M., Maslanka, S. and Sobel, J. (2005) Adult botulism type F in the United States, 1981–2002. *Neurology*, **13**(65), 1694–700.

Hatheway, C.L. (1990) Toxigenic clostridia. *Clinical Microbiology Review*, **3**, 66–98.

Hughes, C., Gillespie, I.A., O'Brien, S.J., *et al.* (2007) Foodborne transmission of infectious intestinal disease in England and Wales, 1992–2003. *Food Control*, **18**, 766–72.

Hunter, L.C. and Poxton, I.R. (2002) *Clostridium botulinum* types C and D and the closely related *Clostridium novyi*. *Reviews in Medical Microbiology*, **13**, 75–90.

Johnson, E.A., Tepp, W.H., Bradshaw, M., Gilbert, R.J., Cook, P.E. and McKintosh, E.D.G. (2005) Characterization of *Clostridium botulinum* strains associated with an infant botulism case in the United Kingdom. *Journal of Clinical Microbiology*, **43**, 2602–7.

Jørgensen, K. (2005) Occurrence of ochratoxin in commodities and processed food – a review of EU occurrence data. *Food Additive and Contaminants*, **22**,26–30.

Kaysner, C.A. and DePaola, A.J. (2004) *Vibrio cholerae, V. parahaemolyticus, V. vulnificus and other Vibrio spp*. In: *FDA Bacteriological Analytical Manual 2004*, 8th edn, Chapter 9. AOAC International, Gaithersburg.

Keto-Timonen, R., Heikinheimo, A., Eerola, E. and Korkeala, H. (2006) Identification of *Clostridium* species and DNA fingerprinting of *Clostridium perfringens* by amplified fragment length polymorphism analysis. *Journal of Clinical Microbiology*, **44**, 4057–65.

Khandke, S.S. and Mayes, T. (1998) HACCP implementation: a practical guide to the implementation of the HACCP plan. *Food Control*, **9**, 103–109.

Kirkpatrick, C.E. and Benson, C.E. (1987) Presence of *Giardia* spp and absence of *Salmonella* spp in New Jersey muskrats (*Ondatra zibethicus*). *Applied and Environmental Microbiology*, **53**, 1790–92.

Koch, W.H., Payne, W.L. and Cebula, T.A. (1995) Detection of enterotoxigenic *Vibrio cholerae* in foods by the polymerase chain method. In: *FDA Bacteriological Analytical Manual 1995*, 8th edn, pp. 28.01–28.09. AOAC International, Gaithersburg.

Lappin-Scott, H.M. and Costerton, J.W. (2003) *Microbial Biofilms*. Cambridge University Press, Cambridge.

Leighton, G. (1923) *Botulism and Food Preservation (The Loch Maree Tragedy)*. Collins, London.

Leong, S.C., Hien, L.T., An, T. V., Trang, N.T., Hocking, A.D. and Scott, E.S. (2007) Ochratoxin A-producing Aspergilli in Vietnamese green coffee beans. *Letters in Applied Microbiology*, **45**, 301–306.

Lewis, L., Onsongo, M., Njapau, H., *et al.* (2005) Aflatoxin contamination of commercial maize products during an outbreak of acute aflatoxicosis in Eastern and Central Kenya. *Environmental Health Perspectives*, **113**, 1763–7.

Lopitz-Otsoa, F., Rementeria, A., Elguezebal, N. and Garaizar, J. (2006) Kefir: a symbiotic yeast–bacteria community with alleged healthy capabilities. *Revista Iberoamericana de Micrologia*, **23**, 67–74.

Lyytikäinen, O. (2000) An outbreak of *Listeria monocytogenes* serotype 3a infections from butter in Finland. *Journal of Infectious Diseases*, **181**, 1838–41.

Mabbit, L.A., Davies, F.L., Law, B.A. and Marshall, V.M. (1987) Microbiology of milk and milk products. In: *Essays in Agricultural and Food Microbiology* (eds J.R. Norris and G.L. Petiffer), pp. 135–66. Wiley, Chichester.

Manuelidis, L. (2007) Viruses in the frame for prion diseases. *New Scientist*, 12 February 2007. Accessed 25 September 2007.

McLauchlin, J., Grant, K.A. and Little, C.L. (2006) Foodborne botulism in the United Kingdom. *Journal of Public Health*, **28**, 337–42.

McNally, A., Cheasty, T., Fearnley, C., *et al.* (2004) Comparison of the biotype of *Yersinia enterocolitica* isolated from pigs, cattle and sheep and slaughter and from humans with yersiniosis in Great Britain during 1999–2000. *Letters in Applied Microbiology*, **399**, 103–108.

Mead, P.S., Slutsker, L., Dietz, V., *et al.* (1999) Food related illness and death in the United States. *Emerging Infectious Disease*, **5**, 605–27.

Miller, N.L., Frenkel, J.K. and Dubey, J.B. (1972) Oral infections with *Toxoplasma* cysts and oocysts in felines, other mammals and in birds. *Journal of Parasitology*, **58**, 928–37.

Mortimore, S. and Wallace, C. (1994) *HACCP – A Practical Approach*. Chapman and Hall, London.

Moss, M.O. (1987) Microbial food poisoning. In: *Essays in Agricultural and Food Microbiology* (eds J.R. Norris and G.L. Pettifer), pp. 369–400. Wiley, Chichester.

Nguyen, H.-V. and Gaillardin, C. (2005) Evolutionary relationships between the former species *Saccharomyces uvarum* and the hybrids *Saccharomyces bayanus* and *Saccharomyces pastorianus*; reinstatement of *Saccharomyces uvarum* (Beijerinck) as a distinct species. *FEMS Yeast Research*, **5**, 471–83.

O'Mahony M., Mitchell E., Gilbert R.J., *et al.* (1990) An outbreak of foodborne botulism associated with contaminated hazelnut yoghurt. *Epidemiology Infection*, **104**, 389–95.

Onda, T., Yanagida, F., Uchimura, T., *et al.* (2002) Widespread distribution of the bacteriocin-producing lactic acid cocci in Miso-paste products. *Journal of Applied Microbiology*, **92**, 695–705.

Oravcová, K., Kaclíková, E., Krascsenicsova, K., *et al.* (2006) Detection and quantification of *Listeria monocytogenes* by 5′-nuclease polymerase chain reaction targeting the *Act A* gene. *Letters in Applied Microbiology*, **42**, 15–18.

Osterholm, M.T., Forfang, J.C., Ristinen, T.L., *et al.* (1981) An outbreak of foodborne giardiasis. *New England Journal of Medicine*, **304**, 24–8.

Pawsey, R.K. (2002) *Case Studies in Food Microbiology for Food Safety and Quality*. Royal Society of Chemistry, Cambridge.

Peck, M. and Baranyi, J. (2007) *Perfringens Predictor*. Accessed 11 September 2007. Institute of Food Research, Colney Lane, Norwich. http://www.ifr.ac.uk/safety/growthpredictor/perfringens/predictor.zip.

Piekarski, G. (1989) *Medical Parasitology*. Springer, New York.

Plorde, J.J. (1984) Sporozoan infections. In: *Medical Microbiology: An Introduction to Infectious Diseases* (eds J.C. Sherris, *et al.*), pp. 469–83. Elsevier, New York.

Rached, E., Pfeiffer, E., Dekant, W. and Mally, A. (2006) Ochratoxin A: apoptosis and aberrant exit from mitosis due to perturbation of microtubule dynamics. *Toxicological Sciences*, **92**, 78–86.

Rao, V.K., Sharma, M.K., Goel, A.K., Singh, L. and Sekhar, K. (2006) Amperometric immunosensor for the detection of *Vibrio cholerae* O1 using disposable screen-printed electrodes. *Analytical Sciences*, **22**, 1207–11.

Rees, C.E.D. and Loessner, M. (2005) Phage for the detection of pathogenic bacteria. In: *Bacteriophages Biology and Applications* (eds E. Kutter and A. Sulakvelidze), pp. 267–84. CRC Press, Boca Raton.

Rendtorff, R.C. (1954) The experimental transmission of human intestinal protozoan parasites II *Giardia lamblia* cysts given in capsules. *American Journal of Hygiene*, **58**,209–220.

Rhodelamel, E.J. and Harmon, S.M. (1998) *FDA Bacteriological Analytical Manual*, 8th edn, Chapter 14. Revision A. AOAC International, Gaithersburg.

Richmond, M. (1990) *The Microbiological Safety of Food. Parts I and Part II.* HMSO, London.

Roberts, T.A. (ed.) (1996) *Microorganisms in Foods 5. Microbiological Specifications of Food Pathogens.* ICMSF, Blackie, London.

Rood, J.I. (1998) Virulence genes of *Clostridium perfringens. Annual Reviews of Microbiology*, **52**, 333–60.

Setlow, P. and Johnson, E.A. (2007) Spores and their significance. In: *Food Microbiology: Fundamentals and Frontiers* (eds M.P. Doyle and L.R. Beuchat), pp. 35–68. ASM Press, Washington, DC.

Silley, P. (1991) Rapid automated bacterial impedance technique (RABIT). *SGM Quarterly*, **18**, 48–52.

Stanley, E.C., Mole, R.J., Smith, R.J., *et al.* (2001) Development of a new, combined rapid method using phage and PCR for detection and identification of viable *Mycobacterium paratuberculosis* bacteria within 48 hours. *Applied and Environmental Microbiology*, **73**, 1851–7.

Tenover, F.C., Abeit, R.D., Archer, G., Biddles, J., *et al.* (1994) Comparison of traditional and molecular methods of typing isolates of *Staphylococcus aureus. Journal of Clinical Microbiology*, **32**, 407–15.

Tenover, F.C., Abeit, R.D. and Goering R.N. (1997) How to select and interpret molecular strain typing methods for epidemiological studies of bacterial infections. A review for healthcare epidemiologists. *Infectious Control in Hospital*, **18**, 426–39.

Thapa, N., Pal, J. and Tamang, J.P. (2004) Microbial diversity in ngari, hentak and tungtap, fermented fish products of North-East India. *World Journal of Microbiology and Biotechnology*, **20**, 599–607.

Therre, H. (1999) Botulism in the European Union. *Eurosurveillance Monthly*, **4**, 2–7.

Tzipori, S. (1988) Cryptosporidiosis in perspective. *Advances in Parasitology*, **27**, 63–129.

Untermann, F. (1999) Food safety management and misinterpretation of HACCP. *Food Control*, **10**, 161–7.

Varga, J.J., Nguyen, V., O'Brien, D.K., Rodgers, K., Walker, R.A. and Melville, S.B. (2006) Type IV pili-dependent gliding motility in the Gram-positive pathogen *Clostridium perfringens* and other clostridia. *Molecular Microbiology*, **62**, 680–94.

Vatanyoopaisarn, S., Nazli, A., Dodd, C.E.R., Rees, C.E.D. and Waites, W.M. (2000) Effect of flagella on initial attachment of *Listeria monocytogenes* to stainless steel. *Applied and Environmental Microbiology*, **66**(2), 860–63.

Vulic, M. and Kolter, R. (2001) Evolutionary cheating in *Escherichia coli* stationary phase cultures. *Genetics*, **158**, 519–26.

Waites, W.M. and Warriner, K. (2005) Ultraviolet sterilisation of food packaging. *Culture*, **26**, 1–4.

Walsh, J.A. (1986) Problems in recognition and diagnosis of amebiasis. Estimation of the global magnitude of morbidity and mortality. *Review of Infectious Diseases*, **8**, 228–38.

Warriner, K., Rysstad, G., Murden, A., Rumsby, P., Thomas, D. and Waites, W.M. (2000) Inactivation of *Bacillus subtilis* spores on aluminium and polyethylene preformed cartons by uv-excimer laser irradiation. *Journal of Food Protection*, **63**, 753–7.

WHO (2002) *Fact Sheet No. 237 Revised January 2002. Food Safety and Foodborne Illness.* http://www.who.int/mediacentre/factsheets/Fs237/en/. Accessed 18 December 2006.

Wreeland, R.H., Rosenzweig, W.D. and Powers, D.W. (2000) Isolation of a 250 million-year-old halotolerant bacterium from a primary salt crystal. *Nature*, **407**, 897–900.

Yuki, N., Susuki, K., Koga, M., *et al.* (2004) Carbohydrate mimicry between human ganglioside GM1 and *Campylobacter jejuni* lipopolysaccharide causes Guillain–Barré syndrome. *PNAS*, **101**, 11404–409.

Zabeau, M. and Vos, P. (1993) *Selective Restriction Fragment Amplification: A General Method for DNA Fingerprinting.* European Patent Office Publication 534 858 A1, Bulletin 93/13.

Zhao, S., Mitchell, S.E. Meng, J., *et al.* (2000) Genomic typing of *Escherichia coli* O157:H7 by semi-automated fluorescent AFLP analysis. *Microbes and Infection*, **2**, 107–113.

Zoonoses Report UK (2007) http://defraweb/animal/diseases/zoonoses/zoonosesreports/zoonoses2005.patt. Accessed July 2007.

Supplementary material is available at www.wiley.com/go/campbellplatt

Numerical procedures

R. Paul Singh

Key points

- Solving problems in food processing requires a thorough knowledge of numerical procedures. Frequently, we need to express given quantities with their respective units. For this purpose, standard rules developed to employ an international system of units must be followed.
- The fundamental expression used in numerical calculations is an equation. The different forms of an equation, such as linear and nonlinear, are solved using appropriate techniques.
- The frequently encountered time dependent processes in food processing require application of methods learnt in calculus.
- These topics are reviewed in this chapter so that appropriate numerical procedures are selected in solving a given problem.

7.1 SI system of units

In scientific measurements, physical quantities are determined using a wide variety of unit systems. The most common systems include the Imperial (English) system: the centimeter, gram, second (cgs) system; and the meter, kilogram, second (mks) system. Considerable confusion occurs when one uses a variety of symbols to express units of measurement. To address these issues in a systematic manner, international organizations have made attempts to standardize unit systems, symbols, and the quantities. International agreements have resulted in the "Système International d'Unités," or the SI units.

The SI units consist of seven base units, two supplementary units, and a series of derived units, as described below (Singh and Heldman, 2009).

7.1.1 Base units

The SI system is based on seven well-defined units. The base units, along with their symbols, are summarized in Table 7.1. By convention, base units are regarded as dimensionally independent. These seven base units are defined as follows:

1. Unit of length (meter): the *meter* (m) is the length equal to 1,650,763.73 wavelengths in vacuum of the radiation corresponding to the transition between the levels $2p_{10}$ and $5d_5$ of the krypton-86 atom.
2. Unit of mass (kilogram): the *kilogram* (kg) is equal to the mass of the international prototype of the kilogram. The international prototype of the kilogram is a particular cylinder of platinum-iridium alloy, which is kept by the International Bureau of Weights and Measures in Sèvres, France.

Table 7.1 SI base units.

Measurable attribute of phenomena or matter	Name	Symbol
Length	meter	m
Mass	kilogram	kg
Time	second	s
Electric current	ampere	A
Thermodynamic temperature	kelvin	K
Amount of substance	mole	mol
Luminous intensity	candela	cd

3 Unit of time (second): the *second* (s) is the duration of 9,192,631,770 periods of radiation corresponding to the transition between the two hyperfine levels of the ground state of the cesium-133 atom.

4 Unit of electric current (ampere): the *ampere* (A) is the constant current that, if maintained in two straight parallel conductors of infinite length, of negligible circular cross section, and placed 1 m apart in vacuum, would produce between those conductors a force equal to 2×10^{-7} newton per meter length.

5 Unit of thermodynamic temperature (kelvin): the *kelvin* (K) is the fraction $1/273.16$ of the thermodynamic temperature of the triple point of water. Most temperatures are measured in degrees Centigrade ($^\circ$C). The kelvin temperature is used in scientific literature.

6 Unit of amount of substance (mole): the *mole* (mol) is the amount of substance of a system that contains as many elementary entities (atoms, molecules, electrons, and other particles) as there are atoms in 0.012 kg of carbon 12.

7 Unit of luminous intensity (candela): the *candela* (cd) is the luminous intensity, in the perpendicular direction, of a surface of $1/600,000$ m^2 of a blackbody at the temperature of freezing platinum under a pressure of 101,325 newton/m^2.

7.1.2 Derived units

The preceding base units are combined by means of multiplication and division to obtain derived units. Derived units often carry special names and symbols, mainly for the sake of simplicity, and they may be used to obtain other derived units. Some commonly used derived units are defined as follows:

1 *Force.* Newton (N): the *newton* is the force that gives to a mass of 1 kg an acceleration of 1 m/s^2.

2 *Energy and work.* Joule (J): the *joule* is the work done when due to force of 1 N the point of application is displaced by a distance of 1 m in the direction of the force.

3 *Power.* Watt (W): the *watt* is the power that gives rise to the production of energy at the rate of 1 J/s.

4 *Electrical potential difference.* Volt (V): the *volt* is the difference of electric potential between two points of a conducting wire carrying a constant current of 1 A, when the power dissipated between these points is equal to 1 W.

5 *Electrical resistance.* Ohm (Ω): the *ohm* is the electric resistance between two points of a conductor when a constant difference of potential of 1 V, applied between theses two points, produces in this conductor a current of 1 A, when this conductor is not being the source of any electromotive force.

6 *Quantity of electricity.* Coulomb (C): the *coulomb* is the quantity of electricity transported in 1 s by a current of 1 A.

7 *Electrical capacitance.* Farad (F): the *farad* is the capacitance of a capacitor, between the plates of which there appears a difference of potential of 1 V when it is charged by a quantity of electricity equal to 1 C.

8 *Electrical inductance.* Henry (H): the *henry* is the inductance of a closed circuit in which an electromotive force of 1 V is produced when the electric current in the circuit varies uniformly at a rate of 1 A/s.

9 *Electrical conductance.* Siemens (s): the Siemens is the electrical conductance of a conductor in which a current of 1 ampere is produced by an electrical potential difference of 1 volt.

10 *Magnetic flux.* Weber (Wb): the *weber* is the magnetic flux that, linking a circuit of one turn, produces in it an electromotive force of 1 V as it is reduced to zero at a uniform rate in 1 s.

11 *Luminous flux.* Lumen (lm): the lumen is the luminous flux emitted in a point solid angle of 1 steradian by a uniform point source having an intensity of 1 cd.

Some examples of SI-derived units expressed in terms of base units, SI-derived units with special names, and SI-derived units expressed by means of special names are given in Tables 7.2–7.4, respectively.

Table 7.2 Selected SI-derived units expressed in terms of base units.

Quantity	Name	Symbol
Area	square meter	m^2
Volume	cubic meter	m^3
Speed, velocity	meter per second	m/s
Acceleration	meter per second squared	m/s^2
Density, mass density	kilogram per cubic meter	kg/m^3
Current density	ampere per square meter	A/m^2
Magnetic field strength	ampere per meter	A/m
Concentration (of amount of substance)	mole per cubic meter	mol/m^3
Specific volume	cubic meter per kilogram	m^3/kg
Luminance	candela per square meter	cd/m^2

Table 7.3 Selected SI-derived units with special names.

Quantity	Name	Symbol	Expression in terms of other units	Expression in terms of SI base units
Frequency	hertz	Hz		s^{-1}
Force	newton	N		$m\ kg\ s^{-2}$
Pressure, stress	pascal	Pa	N/m^2	$m^{-1}\ kg\ s^{-2}$
Energy, work, quantity of heat	joule	J	N m	$m^2\ kg\ s^{-2}$
Power, radiant flux	watt	W	J/s	$m^2\ kg\ s^{-3}$
Quantity of electricity, electric charge	coulomb	C		s A
Electric potential, potential difference, electromotive force	volt	V	W/A	$m^2\ kg\ s^{-3}\ A^{-1}$
Capacitance	farad	F	C/V	$m^{-2}\ kg^{-1}\ s^4\ A^2$
Electric resistance	ohm	O	V/A	$m^2\ kg\ s^{-3}\ A^{-2}$
Conductance	siemens	S	A/V	$m^{-2}\ kg^{-1}\ s^3\ A^2$
Celsius temperature degree	Celsius	°C		K
Luminous flux	lumen	lm		cd sr
Illuminance	lux	lx	lm/m^2	$m^{-2}\ cd\ sr$

Table 7.4 Selected SI-derived units expressed by means of special names.

Quantity	Name	Symbol	Expression in terms of SI base units
Dynamic viscosity	pascal second	Pa s	$m^{-1}\ kg\ s^{-1}$
Moment of force	newton meter	N m	$m^2\ kg\ s^{-2}$
Surface tension	newton per meter	N/m	$kg\ s^{-2}$
Power density, heat flux density, irradiance	watt per square meter	W/m^2	$kg\ s^{-3}$
Heat capacity, entropy	joule per kelvin	J/K	$m^2\ kg\ s^{-2}\ K^{-1}$
Specific heat capacity	joule per kilogram kelvin	J/(kg K)	$m^2\ s^{-2}\ K^{-1}$
Specific energy	joule per kilogram	J/kg	$m^2\ s^{-2}$
Thermal conductivity	watt per meter kelvin	W/(m K)	$m\ kg\ s^{-3}\ K^{-1}$
Energy density	joule per cubic meter	J/m^3	$m^{-1}\ kg\ s^{-2}$
Electric field strength	volt per meter	V/m	$m\ kg\ s^{-3}A^{-1}$
Electric charge density	coulomb per cubic meter	C/m^3	$m^{-3}\ s\ A$
Electric flux density	coulomb per square meter	C/m^2	$m^{-2}\ s\ A$

Table 7.5 Supplementary units.

Quantity	Name	Symbol
Plane angle	radian	rad
Solid angle	steradian	sr

7.1.3 Supplementary units

Another class of SI units is purely geometric, which may be regarded either as base units or as derived units:

1 Unit of plane angle (radian): the *radian* (rad) is the plane angle between two radii of a circle that cut off on the circumference an arc equal in length to the radius.
2 Unit of solid angle (steradian): the *steradian* (sr) is the solid angle that, having its vertex in the center of a sphere, cuts off an area of the surface of the sphere equal to that of a square with sides of length equal to the radius of the sphere.

The supplementary units are summarized in Table 7.5.

Example 7.1 Determine the following unit conversions to SI units:

a a density value of 70 lb_m/ft^3 to kg/m^3
b an energy value of 3.4×10^4 Btu to kJ
c an enthalpy value of 3482 Btu/lb_m to kJ/kg
d a pressure value of 0 psig to kPa
e a viscosity value of 51 cp to Pa s

Solution
To determine the required conversions, we will first use the conversion factors separately for each quantity.

a From conversion tables,

$$1 \, lb_m = 0.45359 \, kg$$
$$1 \, ft = 0.3048 \, m$$

Thus,

$$(70 \, lb_m/ft^3)(0.45359 \, kg/lb_m)\left(\frac{1}{0.3048 \, m/ft}\right)^3$$
$$1121.3 \, kg/m^3$$

An alternative solution involves the direct use of the conversion factor for density,

$$\frac{(70 \, lb_m/ft^3)(16.0185 \, kg/m^3)}{(1 \, lb_m/ft^3)} = 1121.3 \, kg/m^3$$

b For energy

$$1 \, Btu = 1.055 \, kJ$$

Therefore,

$$\frac{(3.4 \times 10^4 \, Btu)(1.055 \, kJ)}{(1 \, Btu)} = 3.6 \times 10^4 \, kJ$$

c For enthalpy, the conversion units for each dimension are

$$1 \, Btu = 1.055 \, kJ$$
$$1 \, lb_m = 0.45359 \, kg$$

Thus,

$$(3482 \, Btu/lb_m)(1.055 \, kJ/Btu)\left(\frac{1}{0.45359 \, kg/lb_m}\right)$$

$$= 8099 \, kJ/kg$$

Alternatively, using the composite conversion factor for enthalpy of

$$1 \, Btu / lb_m = 2.3258 \, kJ/kg$$
$$\frac{(3482 \, Btu/lb_m)(2.3258 \, kj/kg)}{1 \, Btu/lb_m} = 8098 \, KJ/Kg$$

d For pressure

$$psia = psig + 14.69$$

The gauge pressure, 0 psig, is first converted to the absolute pressure, psia

$$0 \, psig + 14.69 = 14.69 \, psia$$

The unit conversions for each dimension are

$$1 \, lb = 4.4483 \, N$$
$$1 \, in = 2.54 \times 10^{-2} \, m$$
$$1 \, Pa = 1 \, N/m^2$$

Thus,

$$(14.69\,\text{lb/in}^2)(4.4482\,\text{N/lb})$$

$$\times \left(\frac{1}{2.54 \times 10^{-2}\,\text{m/in}}\right)^2 \left(\frac{1\,\text{Pa}}{1\,\text{N/m}^2}\right)$$

$$= 101{,}283\,\text{Pa}$$

$$= 101.3\,\text{kPa}$$

Alternatively, since

$$1\,\text{psia} = 6.895\,\text{kPa}$$

$$\frac{(14.69\,\text{psia})(6.895\,\text{kPa})}{(1\,\text{psia})} = 101.28\,\text{kPa}$$

For viscosity

$$1\,\text{cp} = 10^{-3}\,\text{Pa s}$$

Thus,

$$\frac{(51\,\text{cp})(10^{-3}\,\text{Pa s})}{(1\,\text{cp})} = 5.1 \times 10^{-2}\,\text{Pa s}$$

7.2 Rules for using SI units

The use of the SI system of units requires that certain rules be followed. These rules have been developed by several organizations with a focus on their unique areas. The following rules are based on the recommendations of the International Organization for Standardization, and the American Society of Agricultural and Biological Engineers (Singh and Heldman, 2009).

7.2.1 Symbols

Symbols are the short forms of SI unit names and prefixes. Symbols are not acronyms or abbreviations. All symbols are written in upright roman letters except two (Greek omega Ω for ohm and mu μ for micro). SI symbols should not be confused with symbols used for physical quantities, e.g. m is often used as a symbol for mass of an object and written in italics.

7.2.2 Prefixes

When expressing numerical quantities there is often a need to eliminate insignificant digits, and this is accomplished by providing an order of magnitude with the use of prefixes. For example,

$$27{,}300\,\text{m or } 27.3 \times 10^3\,\text{m becomes } 27.3\,\text{km}$$

In this case, the prefix "k" for kilo is used. The prefixes along with the SI symbols are given in Table 7.6. The prefix symbols are printed in roman (upright) type without spacing between the prefix symbol and the unit symbol. Note that the first five prefixes in the table are written in capital letters and the remaining eleven are written in lowercase letters. The precise use of upper and lowercase letters is important to avoid confusion, such as,

K is used for kelvin, and k for kilo

N is used for newton, and n for nano

M is used for Mega, and m for milli

When a symbol has an exponent attached to it, then a prefix implies that the multiple or submultiple of the unit is also raised to the power expressed by the exponent. For example,

$$1\,\text{cm}^3 = (10^{-2}\,\text{m})^3 = 10^{-6}\,\text{m}^3$$
$$1\,\text{mm}^{-1} = (10^{-3}\,\text{m})^{-1} = 10^3\,\text{m}^{-1}$$

Table 7.6 SI prefixes.

Factor	Prefix	Symbol	Factor	Prefix	Symbol
10^{18}	exa	E	10^{-1}	deci	d
10^{15}	peta	P	10^{-2}	centi	c
10^{12}	tera	T	10^{-3}	milli	m
10^{9}	giga	G	10^{-6}	micro	μ
10^{6}	mega	M	10^{-9}	nano	n
10^{3}	kilo	k	10^{-12}	pico	p
10^{2}	hecto	h	10^{-15}	femto	f
10^{1}	deka	da	10^{-18}	atto	a

Juxtaposing of two or more SI prefixes to create a compound prefix is not allowed. For example,

$$1 \text{ nm is allowed} \quad \text{but not } 1 \text{ m } \mu\text{m}$$

The unit of mass is the only one in the base units whose name, for historical reasons, contains a prefix. Therefore to obtain names of decimal multiples and submultiples of the unit mass, attach an appropriate prefix to the word "gram."

When two units are combined as a compound unit, then attach prefixes to the numerator of compound units, except when using "kilogram" in the denominator. For example, use

$$3.5 \text{ kJ/s} \quad \text{not} \quad 3.5 \text{ J/ms}$$

$$420 \text{ J/kg} \quad \text{not} \quad 4.2 \text{ dJ/g}$$

One has a choice in selecting prefixes. It is advisable to choose a prefix so that the numerical value preferably lies between 0.1 and 1000. However, double prefixes and hyphenated prefixes should not be used. For example, use

$$\text{GJ} \quad \text{not} \quad \text{kMJ}$$

7.2.3 Capitalization

When writing unit symbols it is important to adhere to the following rules: Roman (upright) type is used, in lower case, for symbols of units; however, if the symbols are derived from the name of a person, then capital roman type is used (for the first letter), for example, K and N. Furthermore, these symbols are not followed by a full stop (period). When writing in a sentence, SI unit names are not capitalized except the first letter of a symbol when used as a first word in a sentence. In case the units are written in an unabbreviated form, the first letter is not capitalized (even for those derived from proper nouns): for example, kelvin and newton. The exception to this rule is the capital letter L used as a symbol for liter. Because there is no discernable difference between number "1" and lower case "l" in typing, many technical organizations including the US National Bureau of Standards recommend using 'L' for liter.

The numerical prefixes are not capitalized except for symbols E (exa), P (peta), T (tera), G (giga), and M (mega).

Do not use a capital letter for a symbol in the title; to avoid confusion, use the unit name.

7.2.4 Plurals

The unit symbols remain the same in the plural or singular form. In unabbreviated form the plural units are written in the usual manner. For example:

$$23 \text{ newtons} \quad \text{or} \quad 23 \text{ N}$$
$$14 \text{ centimeters} \quad \text{or} \quad 14 \text{ cm}$$

The following units are both singular and plural: lux, hertz, and siemens.

7.2.5 Punctuation

When the numerical value is less than one, a zero should precede the decimal point. The SI symbols should not be followed by a period, except at the end of a sentence. English-speaking countries use a centered dot for a decimal point; others use a comma. When the numbers are large, then they should be grouped into threes (thousands) by using spaces instead of commas. For example,

$$2\,743\,637.236\,77$$

not

$$2,743,637.236,77$$

7.2.6 Derived units

Derived units are the largest class of SI units. They are obtained by combining base, supplementary, and other derived units. When the units involve a product of two or more units, then they may be written in either of the following ways:

$$\text{N} \cdot \text{m} \quad \text{N m}$$

When a derived unit is formed from two or more units by division, then a solidus (oblique stroke, /), a horizontal line, or negative powers may be used. For example:

$$\text{m/s} \quad \frac{\text{m}}{\text{s}} \quad \text{ms}^{-1}$$

However, a solidus must not be repeated on the same line. When there is a complex combination of units, parentheses or negative powers should be used. For example:

$$\text{m/s}^2 \text{ or m s}^{-2} \quad \text{but not} \quad \text{m/s/s}$$
$$\text{J/(s m}^2\text{ K)} \quad \text{or} \quad \text{J s}^{-1}\text{ m}^{-2}\text{ K}^{-1} \quad \text{but not} \quad \text{J/s/m}^2\text{/K}$$

Note that many derived units are given special names and symbols, e.g. a joule is a product of newton and meter.

7.3 Equation

Equation is a mathematical expression that allows us to create a balance between different quantities. These quantities may be variable or constant. Constants in the equations are items that never change, e.g. 35 is a constant, and its value is always 35. On the other hand, a variable can take a value that may change; we express a variable with a letter, e.g. x may be called a variable where the value of x may change. In addition to the variables and constants used in an equation, different mathematical operatives are used such as multiply, divide, add, or subtract (Hartel *et al.*, 1997). Thus, we may consider a simple equation such as

$$4x = 48 \qquad (7.1)$$

In this equation x is a variable, and 48 is a constant, and 4 is a coefficient of x. In order to obtain a balance between the left and right hand sides of the equation, the value of the variable x must be equal to 12, otherwise the equation will be unbalanced.

Any modification of an equation should be done so that it is the same action on both sides of the equation. For example, we may add 10 on both sides of Equation (7.1):

$$4x + 10 = 48 + 10$$

Similarly, we could subtract, multiply, or divide both sides with 10, and the result of the equation will remain the same.

Variables are further categorized as dependent or independent. Consider an equation of the form:

$$y = 5t + 4 \qquad (7.2)$$

It is a normal practice that we write the dependent variable on the left-hand side and the independent variables on the right-hand side of Equation (7.2). In this equation, y is a dependent variable and t is an independent variable. We note from Equation (7.2) that the value of y depends upon the value of t. For example, if t is 1, then y = 9. We designate t to be

an independent variable, since it can have any value, and it does not depend upon y. We could rearrange the terms in Equation (7.2) by subtracting 4 from both sides and dividing by 5 to obtain

$$t = \frac{y}{5} - 0.8 \qquad (7.3)$$

Following the general convention in Equation (7.3), t is the dependent variable and y is the independent variable.

Constant terms used in the equation imply that these values do not change. Obviously, the value of a number (48 in Equation (7.1)) does not change and it is a constant. In addition, we have certain other constants that are frequently used in engineering calculations. For example, π is a constant used in calculating area, circumference, and cross-sectional area of rounded objects, and its value is 3.14159; Avogadro's number represents the number of atoms in a mole of a substance, its value is 6.0221×10^{23} atoms/mole; another commonly used constant is the acceleration due to gravity, g = 9.8 m/s^2.

Equations are sometimes written to describe a relationship as a function. For example, we may state that y is a function of t, where the function of t is a mathematical expression. In the notation of equations, we write

$$y = f(t) \qquad (7.4)$$

This means that y is a function of t, where $f(t) = 5t - 72$. This implies that the function of t is equal to $5t - 72$.

A shorter notation of the preceding is

$$y(t) = 5t - 72 \qquad (7.5)$$

The use of parentheses in Equation (7.5) implies a function. This should not be confused with a situation where parentheses are used to indicate multiplication, for example,

$$42(t - 10) = 33 \qquad (7.6)$$

In this equation, 42 is multiplied with $t - 10$ on the left-hand side.

7.3.1 Solving equations

In solving equations, both sides of the equation must be operated in the same manner to ensure that the equation remains valid. For example, as we observed earlier, we may add, subtract, multiply, or divide the same quantity on both sides of an equation. Thus:

$$33x = 44t - 20 \qquad (7.7)$$

We may add 10 to both sides of Equation (7.7) thus: $33x + 10 = 44t - 20 + 10$.

Or we may subtract 10 from both sides of Equation (7.7) thus: $33x - 10 = 44t - 20 - 10$.

Or we may divide both sides of Equation (7.7) by 10 thus: $\dfrac{33x}{10} = \dfrac{(44t - 20)}{10}$.

Or we may multiply both sides of Equation (7.7) by 10 thus: $33x \times 10 = (44t - 20) \times 10$.

All the preceding four equations are the same as Equation (7.7). Such manipulations are important in solving equations. Consider, the ideal gas law,

$$PV = nRT \qquad (7.8)$$

Let us say that we want to determine P and we know values of n, T, and V. Note that R is a constant, and we can determine its value from an appropriate source.

To solve Equation (7.8) for P, we divide both sides by V and get

$$P = \frac{nRT}{V}$$

Now substituting the values of n, T, V, and the constant R, we can obtain a value of P.

7.3.2 Linear equations

In a linear equation, each term of the equation is either a constant or a multiple of a constant with a variable of power one. The most common form of a linear equation is

$$y = mx + b \qquad (7.9)$$

In this equation, y and x are variables, whereas m and b are constants. This equation meets our definition as the variables, y and x, are only of power one. Note that Equation (7.9) is also used to express a straight line, where y is a dependent variable, x is the independent variable, m is the slope of the straight line,

and b is the y-intercept. The dependent variable y is the value on the y-ordinate and x is the value shown on the abscissa.

7.3.3 Non-linear equations

In a non-linear equation there are terms that contain variables to the power other than one. Thus, the following equations are considered non-linear:

$$y = x^3 + 4$$
$$y^3 = x + 67$$
$$y^{0.2} = x$$

7.3.4 Converting non-linear to linear equations

In solving problems, often we convert a non-linear equation into a linear form to simplify the solution. For example, consider a non-linear equation such as

$$x^3 + 4x = c \qquad (7.10)$$

This equation may be linearized by defining a new variable u as

$$u = x^3$$

Then the linear form of equation (7.10) is

$$u + 4x = c$$

7.3.5 Factorization

In a mathematical description, it is common to use a concise way to represent relationships. One way to accomplish this is the use of parentheses. In arithmetic, distributive law is used to expand an expression containing parentheses. For example,

$$(a + b)c + d = ac + bc + d$$

where $a, b, c,$ and $d,$ are real numbers.

Similarly laws of signs (from arithmetic) allow us to incorporate signs in expressions containing parentheses, for example,

$$-(a + b)c - 3(4 - d) = -ac - bc - 12 + 3d$$

The process of factorization works backwards to the preceding examples.

As simple examples, let us first work with common factors:

$$ac + bc + d$$

is modified by taking the common c as a factor out of the first two terms and rewriting it as

$$(a + b)c + d$$

or, in the case of

$$-ac - bc - 12 + 3d$$

we take $-c$ as the factor out of the first two terms and also factor out -3 out of the second two terms to obtain

$$-c(a + b) - 3(4 - d)$$

7.3.6 Factorization of quadratic equations

A common quadratic expression is

$$ax^2 + bx + c \qquad (7.11)$$

In this equation a, b, and c are real numbers, a and b are coefficients of x^2 and x, respectively, and c is a constant.

Examples of quadratic expressions are as follows:

$$5x^2 + 3x + 4$$
$$x^2 + 2x$$
$$x^2 + 4$$

In the first expression, $a = 5$, $b = 3$, and $c = 4$; in the second expression, $a = 1$, $b = 2$, and $c = 0$; in the third expression, $a = 1, b = 0$, and $c = 4$.

For a quadratic expression, the roots of the equation are determined by using the following formula:

$$x = \frac{-b \pm \sqrt{b^2 - 4ac}}{2a} \qquad (7.12)$$

The roots may be positive, negative, or indeterminate.

7.3.7 Solving simultaneous equations

In this section we will use matrix algebra and Cramer's rule to solve simultaneous equations. We will work with a system of two equations in two unknowns, for example:

$$a_1x + b_1y = c_1$$
$$a_2x + b_2y = c_2 \qquad (7.13)$$

In this set of equations, a_1, a_2 are coefficients of x, b_1 and b_2 are coefficients of y, and c_1 and c_2 are constants. Then in matrix format the coefficients are written as

$$\begin{vmatrix} a_1 & b_1 \\ a_2 & b_2 \end{vmatrix} \qquad (7.14)$$

For this matrix, the number obtained by calculating $a_1b_2 - a_2b_1$ is called the determinant of the matrix, and is abbreviated as det. Thus,

$$det \begin{vmatrix} a_1 & b_1 \\ a_2 & b_2 \end{vmatrix} = a_1b_2 - a_2b_1 = D \qquad (7.15)$$

In the matrix given in Equation (7.14), if we replace the coefficient of x with the constant c, we obtain,

$$\begin{vmatrix} c_1 & b_1 \\ c_2 & b_2 \end{vmatrix} \qquad (7.16)$$

Then, the determinant of this matrix is

$$det \begin{vmatrix} c_1 & b_1 \\ c_2 & b_2 \end{vmatrix} = c_1b_2 - c_2b_1 = D_x \qquad (7.17)$$

Again, in the matrix given in Equation (7.14), if we replace the coefficient of y with the constant c, we obtain,

$$\begin{vmatrix} a_1 & c_1 \\ a_2 & c_2 \end{vmatrix} \qquad (7.18)$$

The determinant of this matrix is,

$$det \begin{vmatrix} a_1 & c_1 \\ a_2 & c_2 \end{vmatrix} = a_1c_2 - a_2c_1 = D_y \qquad (7.19)$$

The preceding steps give us the definition of the Cramer's rule. It states that in every system of two equations with two unknowns in which determinant D is not equal to 0,

$$x = \frac{D_x}{D} \qquad (7.20)$$

and

$$y = \frac{D_y}{D} \qquad (7.21)$$

We will use Cramer's rule to solve two simultaneous equations in the following example.

Example 7.2 Solve the following simultaneous equations:

$$x - y = 450$$
$$0.1x - 0.05y = 90$$

To solve these two equations we will first calculate the determinate of a matrix containing the coefficients of x and y.

$$D = det \begin{vmatrix} 1 & -1 \\ 0.1 & -0.05 \end{vmatrix} = 1 \times (-0.05) - (0.1) \times (-1)$$
$$= 0.05$$

Then

$$D_x = det \begin{vmatrix} 450 & -1 \\ 90 & -0.05 \end{vmatrix} = 450 \times (-0.05) - (90) \times (-1)$$
$$= 67.5$$

and

$$D_y = det \begin{vmatrix} 1 & 450 \\ 0.1 & 90 \end{vmatrix} = 1 \times (90) - (0.1) \times (450) = 45$$

then

$$x = \frac{D_x}{D} = \frac{67.5}{0.05} = 1350$$

and

$$y = \frac{D_y}{D} = \frac{45}{0.05} = 900$$

Similar procedure is used for solving any number of simultaneous equations.

7.3.8 Transposition

It is often necessary to rearrange the variables in an equation to rewrite the equation in a manner that will allow the unknown quantity to become the subject of the equation. For this purpose, transposition of terms is useful in solving for an unknown when the unknown resides are in the right-hand side of an equation. For example, we know that the circumference of a circle, C, is given by

$$C = 2\pi r \qquad (7.22)$$

where r is the radius. Thus for a circle of radius 5 cm, the circumference is equal to 31.42 cm.

On the other hand, if the circumference was given and the radius is to be determined then we will need to rewrite Equation (7.22) to make radius the subject of the equation, thus:

$$r = \frac{C}{2\pi} \qquad (7.23)$$

Therefore, if the circumference is 50 cm, using the above equation we obtain the radius 7.96 cm.

7.3.9 Ratio and proportions

A ratio or a fraction is when a number or a variable is divided by another number or variable. Examples of ratios are

$$\frac{x}{2}; \frac{3}{y}; \frac{x}{y}$$

Let us consider an equation

$$2x - y = 0 \qquad (7.24)$$

We can rewrite this as $2x = y$, or,

$$\frac{x}{y} = \frac{1}{2}$$

Therefore the ratio of x to y is 1 to 2.

Similarly, if we were to consider the wall of a room that has a length = 5 m and height = 4 m, the ratio of length to height is 5 to 4.

Proportion is expressed with an equation containing two ratios as

$$\frac{x}{y} = \frac{a}{b} \qquad (7.25)$$

This may also be written as

$$x : y = a : b \qquad (7.26)$$

In other words, the Equation (7.26) is expressed as "x is to y as a is to b." This may also be written as

$$xb = ya$$

Using this method, we can solve an equation as shown below.

Consider an equation

$$\frac{(4a + 1)}{6} = \frac{(a - 1)}{2}$$

This may be written as

$$2(4a + 1) = 6(a - 1)$$
$$8a + 2 = 6a - 6$$
$$2a = -8$$
$$a = -4$$

Example 7.3 Let us consider a word problem: a 10-hour process produces 1000 kg of product; how much product is produced in 44 hours of operation?

Let the amount of product produced in 44 hours of operation be x.

Then, we can write the above problem statement in proportions as:

$$10 \text{ hours}: 1000 \text{ kg} = 44 \text{ hours}: x \text{ kg}$$
$$\text{then } 10x = 1000 \times 44$$
$$x = \frac{1000 \times 44}{10}$$
$$x = 4400 \text{ kg}$$

Thus, 4400 kg of product will be produced in 44 hours.

7.4 Graphs – linear and exponential plots

In analyzing experimental data, it is common to plot the results and then seek an equation that best describes the data.

An equation of a straight line is given as

$$y = mx + c \qquad (7.27)$$

where m is the slope and c is the intercept.

Example 7.4 Consider the following data for a linear plot:

x	y
1	5.5
2	6.5
3	7.5
4	8.5
5	9.5

Since the above points represent a straight line, we may determine slope of the line by selecting any two points; e.g. (2, 6.5) and (4, 8.5). Then the slope of the line is

$$m = \frac{y_2 - y_1}{x_2 - x_1}$$
$$m = \frac{8.5 - 6.5}{4 - 2} = 1.0$$

and the intercept is obtained as

$$c = y_1 - mx_1 = 5.5 - 1.0(1) = 4.5$$

Therefore the equation of the straight line is

$$y = x + 4.5$$

7.4.1 Logarithmic transformation to obtain a linear equation

In mathematical calculations we encounter two types of logarithms, logarithm of base e and logarithm of base 10. Logarithm of base e is commonly called natural log. The equation for log of base 10 is written as

$$y = 10^x \qquad (7.28)$$

and we may take log of both sides to obtain

$$\log_{10} y = x \qquad (7.29)$$

Similarly, for the following equation

$$y = e^x \qquad (7.30)$$

we may take log of base e of both sides

$$\log_e y = x \qquad (7.31)$$

Commonly, $\log_e$ is abbreviated as ln, and $\log_{10}$ is simply written as log.

Below are some of the common rules used in simplifying equations involving logarithm functions. Note that these apply to both log of base 10 and log of base e:

$$\log(xy) = \log x + \log y \quad (7.32)$$

$$\log\left(\frac{x}{y}\right) = \log x - \log y \quad (7.33)$$

$$\log(x^y) = y\log x \quad (7.34)$$

The relationship between log of base 10 and log of base e may be determined as follows:

$$\ln x = 2.303\log x \quad (7.35)$$

Note that $e^1 = 2.7183$, and $\ln(e) = 1$.

Logarithmic transformations are useful in analyzing data that exhibit exponential relationships. For example, in food sterilization processes, the microbial destruction is observed to be exponential. The following equation describes microbial inactivation due to heat:

$$N = N_o e^{-kt} \quad (7.36)$$

where N is the number of microorganisms at some time t. N_o is the initial population of microorganisms prior to heating, k is the rate constant, and t is the time.

We may transform this equation into a linear equation by taking the natural logarithm of both sides, and noting that

$$\ln(e^{-kt}) = -kt\ln(e) = -kt \quad (7.37)$$

we get

$$\ln N = \ln N_0 - kt \quad (7.38)$$

The preceding equation is an equation of a straight line, if we consider $\ln N$ to be y ordinate and t as x axis. We may note the resemblance to Equation (7.27), or

$$y = mx + c$$

where x is t, the slope is $-k$, and the y-intercept, c, is $\ln N_0$.

Table 7.7 Common derivatives.

Function $f(x)$	Derivative $\frac{df}{dx}$
constant	0
$\sin x$	$\cos x$
$\cos x$	$-\sin x$
x^n	nx^{n-1}
e^x	e^x
$\ln x$	$1/x$

7.5 Calculus

In food science and engineering problems we are often concerned with change in a variable quantity occurring over time. For example, the color of a food may change during storage. Calculus helps us in describing change (Browne and Mukhopadhyay, 2004). In studying calculus, both differentiation and integration are essential in understanding how to analyze change.

7.5.1 Differentiation

If a function, f(x), describes a certain change, then the derivative of the function (df/dx), obtained by differentiation, gives us the rate of change. Differentiation of a function is carried out by using derivatives as given in Table 7.7.

Thus to obtain a derivative of x^5, we will use Table 7.7 to obtain a derivative of x^n as nx^{n-1}, and for x^5, the derivative is $5x^{5-1}$ or $5x^4$. If the function is plotted then the slope of the graph gives the derivative.

For a combination of functions, certain rules of differentiation apply. These rules, for two functions, f and h, are as follows:

Addition Rule:

$$\frac{d}{dx}(f + h) = \frac{df}{dx} + \frac{dh}{dx} \quad (7.39)$$

Product Rule:

$$\frac{d}{dx}(f \cdot h) = f\frac{dh}{dx} + h\frac{df}{dx} \quad (7.40)$$

Quotient Rule:

$$\frac{d}{dx}\left(\frac{f}{h}\right) = \frac{h\dfrac{df}{dx} - f\dfrac{dh}{dx}}{h^2} \quad (7.41)$$

Function-of-a-Function Rule:

$$\frac{d}{dx}(f(h)) = \frac{df}{dh} \cdot \frac{dh}{dx} \qquad (7.42)$$

The following examples illustrate differentiation using these rules.

Example 7.5   Determine the derivative of $f = (4x - 7)(x - 8)$.

We may first expand the right-hand side to obtain:

$$f = 4x^2 - 39x + 56$$

Then, the derivative is obtained using the rule for differentiation of x^n in Table 7.7:

$$\frac{df}{dx} = 8x - 39$$

Example 7.6   Differentiate $f = \dfrac{7x - 2}{3x + 5}$

Using the rule for derivative of fractions we obtain

$$\frac{df}{dx} = \frac{7(3x + 5) - (7x - 2)3}{(3x + 5)^2}$$

or

$$\frac{df}{dx} = \frac{41}{(3x + 5)^2}$$

Example 7.7   Determine the derivative of $f = \sin(x^2 + 5)$.

Using the chain rule and the derivatives of trigonometric functions from Table 7.7 we have

$$\frac{df}{dx} = 2x \cos(x^2 + 5)$$

Example 7.8   Differentiate $f = (2x^3 + 3)^5$.

This is a case of a function of a function; therefore we apply the rule given in Equation (7.42).

To calculate $\dfrac{df}{dx}$, we first let

$$u = 2x^3 + 3$$

then

$$f = u^5$$

Thus, we have

$$\frac{df}{du} = 5u^4$$

and

$$\frac{du}{dx} = 6x^2$$
$$\frac{df}{dx} = 5u^4 \cdot 6x^2$$
$$\frac{df}{dx} = 5(2x^3 + 3)^4 6x^2$$
$$\frac{df}{dx} = 30x^2(2x^3 + 3)^4$$

7.5.2 Logarithmic differentiation

There are cases where the ordinary rules of differentiation do not apply. In many such situations, logarithmic differentiation is helpful. We will examine logarithmic differentiation using the following example:

Example 7.9   Determine $\dfrac{df}{dx}$ if $f = x^x$.

First we take the natural logarithm of both sides:

$$\ln f = \ln x^x$$
$$= x \ln x$$

Now we apply the chain rule to the left-hand side, since f is a function of x, and we will use the product rule on the right-hand side. Thus,

$$\frac{1}{f}\frac{df}{dx} = x\frac{1}{x} + \ln x$$
$$= 1 + \ln x$$

or

$$\frac{df}{dx} = f(1 + \ln x) = x^x(1 + \ln x)$$

7.5.3 Partial derivatives

In many situations, a function may involve more than one variable, and the differentiation may be required for only one variable. In such cases, we use partial derivatives, where the function is differentiated with respect to one variable while the second variable is treated as constant. We use ∂ in place of d. This procedure can be best understood from the following example:

Example 7.10   Calculate the partial derivative of the function f(x,z) with respect to x, where f(x,z) = 2x²z³+4xz².

We will obtain the derivative of the x variable while keeping the term containing z constant, thus

$$\frac{\partial f}{\partial x} = 4xz^3 + 4z^2$$

If we wanted to find $\frac{\partial f}{\partial z}$, then we will carry out the differentiation, by keeping terms of x as constant

or

$$\frac{\partial f}{\partial z} = 6x^2z^2 + 8zx$$

7.5.4 Second derivative

There are occasions when we require differentiation of a function carried out twice. In mathematical terms, this is referred to as a second derivative. For a function $f(x)$, a second derivative is determined as:

$$\frac{d^2 f}{dx^2} = \frac{d}{dx}\left(\frac{df}{dx}\right) \tag{7.43}$$

We will consider the use of a second derivative in determining maximum or minimum value of a function.

Let us consider a continuous function f(x), defined in a domain a<x<b. The function describes a curve with maximum and minimum points. These points are called maxima or minima. At these points, the tangent to the curve is horizontal. In other words, the gradient of the curve is zero. Therefore, we can use differentiation of the function to determine the local maxima or minima:

$$\frac{df}{dx} = 0 \tag{7.44}$$

For a point with a local maximum, the second derivative will be less than zero. Similarly, for a point with a local minimum, the second derivative will be more than zero. Thus, we may write a condition for a local maximum as:

$$\frac{df}{dx} = 0, \quad \frac{d^2 f}{dx^2} \leq 0 \tag{7.45}$$

And for a local minimum:

$$\frac{df}{dx} = 0, \quad \frac{d^2 f}{dx^2} \geq 0 \tag{7.46}$$

Example 7.11   A cylindrical food package is being designed that must have a volume of 0.1 m³ . To minimize heat transfer, it will be advantageous to have a package radius that results in minimum surface area. Calculate the radius.

Consider a package of radius r m, height h m, surface area A m², and volume V m³.
For a cylindrical package,

$$\text{Area} = 2\pi r^2 + 2\pi rh \tag{7.47}$$
$$\text{Volume} = \pi r^2 h = 0.1 \tag{7.48}$$

Therefore,

$$h = \frac{0.1}{\pi r^2} \tag{7.49}$$

Substituting h in the equation for area we obtain

$$\text{Area} = 2\pi r^2 + \frac{0.2\pi r}{\pi r^2} \tag{7.50}$$

or

$$\text{or, } A = 2\pi r^2 + \frac{0.2}{r} \tag{7.51}$$

Since we want to determine minimum area by determining radius, we will differentiate the above equation with respect to r and set the equation equal to zero:

$$\frac{dA}{dr} = 4\pi r - \frac{0.2}{r^2} = 0 \tag{7.52}$$

Thus, we get

$$r = \sqrt[3]{\frac{1}{20\pi}} = 0.251 \text{ m} \tag{7.53}$$

Next, to ensure that we have a minimum area, we take the second derivative of area with respect to r:

$$\frac{d^2 A}{dr^2} = 4\pi + \frac{0.4}{r^3} \tag{7.54}$$

The right-hand side of Equation (7.54) is a positive number for $r = 0.251$. Therefore, we are assured that the selected radius gives the minimum area.

Thus the package dimensions should be: radius = 0.251 m and height = 0.503 m.

7.5.5 Integration

Integration is the reverse of differentiation. The procedural steps are similar to those carried out for differentiation, but in a reverse order. Thus, we may consider taking an integral to be an anti-derivative. For example, if a derivative of a function is a constant A, then we know that since $\frac{dy}{dx} = A$ then the original expression must be $y = Ax$. However, it is possible that there may be additional constants present in the original equation; then $y = Ax + B$ is the complete expression that results in the derivative $\frac{dy}{dx} = A$. We can use the same logic in determining the anti-derivative of another expression,

$$\frac{dy}{dx} = 4x - 3$$

The original expression must be:

$$y = 2x^2 - 3x + c$$

In integration, we use the notation of an integral. Thus,

$$\int y\, dx$$

means that a function $y(x)$ will be integrated with respect to x. Some of the key rules of integration are shown in Table 7.8.

Furthermore, if f and g are functions of x. then

$$\int (df + dg) = \int df + \int dg \qquad (7.55)$$

Table 7.8 Selected rules of integration.

$f(x)$	$\int f(x)dx$
constant, K	$Kx + C$
x^n, where $n \neq -1$	$\frac{x^{n+1}}{n+1} + C$
$\frac{1}{x}$	$\ln x + C$
e^{Ax}	$\frac{e^{Ax}}{A} + C$
$\sin x$	$-\cos x + C$
$\cos x$	$\sin x + C$

7.5.6 Definite integrals

Definite integrals are similar to the indefinite integrals discussed in the preceding section. In definite integrals, there is an upper and lower limit noted on the integral sign. The solution of these integrals involves evaluating the integral at the upper and lower limits. This procedure is explained in the following example:

Example 7.12 Evaluate the following integral:

$$y = \int_3^6 2x^4 dx$$

The integral is evaluated in the following steps:

$$y = \frac{2}{4+1}x^{4+1}\Big|_3^6$$

$$y = \frac{2x^5}{5}\Big|_3^6$$

$$y = \frac{2(6^5 - 3^5)}{5}$$

$$y = \frac{2(7776 - 243)}{5}$$

$$y = 3013.2$$

7.5.7 Trapezoidal rule

An important and useful application of definite integrals is in determining area under a curve. We can approximate the area under the curve by assuming that the area is divided into several small trapezoids. Then we determine the total area by summing the area of individual trapezoids.

Consider Fig. 7.1, where a curve is described by a function $f(x)$, and the curve is bounded by two points a and b on either ends of the curve. The area

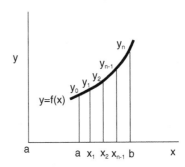

Figure 7.1 A plot of function $f(x)$.

of each individual trapezoid is calculated from the height and width. As seen in Fig. 7.1, the height of the trapezoid is $\dfrac{y_n + y_{n+1}}{2}$ and the width is $x_{n+1} - x_n$.

If we select n number of increments between a and b, then the width of the trapezoid will be

$$width = \frac{b - a}{n}$$

Therefore, the area of the trapezoid is

$$A = \left(\frac{b - a}{n}\right)\left(\frac{y_n + y_{n+1}}{2}\right)$$

If we select a large number of increments, n, then the accuracy in calculating the area improves.

In a more general way, we can write the trapezoidal rule to determine the total area under the curve as

$$A = \left(\frac{b - a}{2n}\right)(y_0 + 2y_1 + 2y_2 + \cdots + 2y_{n-1} + y_n)$$

Example 7.13   Find area under the curve $y = x^3 + 2$ evaluated between the limits 2 and 4.

For this example we will select $n = 5$.

The width of the trapezoid is determined as $x = \dfrac{4 - 2}{5} = 0.4$.

Next we need to determine values of y for the incremental increases in x; this may be set up in a table as follows:

x	y
2	10
2.4	15.824
2.8	23.952
3.2	34.768
3.6	48.656
4.0	66

Using the trapezoidal rule,

$$A = \left(\frac{b - a}{2n}\right)(y_0 + 2y_1 + 2y_2 + \cdots + 2y_{n-1} + y_n)$$

we obtain the total area as follows:

$$A = \left(\frac{4 - 2}{2 \times 5}\right)(10 + 2 \times 15.824 + 2 \times 23.952$$
$$+ 2 \times 34.768 + 2 \times 48.656 + 66)$$
$$A = 64.48$$

This same problem may be solved by using definite integrals, as follows:

$$A = \int_2^4 (x^3 + 2)dx$$

$$A = \left(\frac{x^4}{4} + 2x\right)\Big|_2^4$$

$$A = \frac{4^4 - 2^4}{4} + 2 \times 4 - 2 \times 2$$

$$A = 64$$

Thus, the exact solution obtained from using definite integrals gives the area as 64, whereas using the trapezoidal rule with five increments the estimate is 64.48. The accuracy of the trapezoidal rule may be further improved by using more increments.

7.5.8 Simpson's rule

As we saw in Fig. 7.1, in case of the trapezoidal rule, the curve on the top of each trapezoid is approximated as a straight line. Simpson's rule is similar to the trapezoidal rule, except the top curve is approximated as a parabola.

According to Simpson's rule:

$$Area = \int_a^b f(x)dx \approx$$

$$\frac{\Delta x}{3}(y_0 + 4y_1 + 2y_2 + 4y_3 + 2y_4 + \cdots + 4y_{n-1} + y_n) \tag{7.56}$$

In Simpson's rule, n must be even.

Let us solve the previous example using Simpson's rule:

$$\Delta x = \frac{4 - 2}{4} = 0.5$$

$$y_0 = f(a) = f(2) = 2^3 + 2 = 10$$

$$y_1 = f(a + \Delta x) = f(2.5) = 2.5^3 + 2 = 17.625$$

$$y_2 = f(a + 2\Delta x) = f(3) = 3^3 + 2 = 29$$

$$y_3 = f(a + 3\Delta x) = f(3.5) = 3.5^3 + 2 = 44.875$$

$$y_4 = f(b) = f(4) = 4^3 + 2 = 66$$

$$Area = \frac{0.5}{3}(10 + 4(17.625) + 2(29) + 4(44.875) + 66)$$

$$Area = 64.0$$

The area calculated using Simpson's rule is the same as obtained from solving the definite integral.

7.5.9 Differential equations

Differential equations are used to model quantities that change continuously with respect to an independent variable. In a differential equation, we have terms such as $\frac{dy}{dx}$, where y is a dependent variable and x is an independent variable. Similarly, we may have a term $\frac{dy}{dt}$ describing the change of a dependent variable y with respect to time t. If in the same equation we have two independent variables, then we use partial sign, for example, if y varies with x and t, then we have terms $\frac{\partial y}{\partial x}$ and $\frac{\partial y}{\partial t}$. If the differential term in the equation is of the first order, then the equation is also of the first order; on the other hand, if we have higher order differential terms such as $\frac{\partial^2 y}{\partial x^2}$ then the equation is also of the higher order.

To solve differential equations, we need to know the value of the dependent variable at some value of the independent variable. For example, the value of y may be 5 when $x = 0$. These values are called boundary values. The number of boundary values required to solve a differential equation depends upon the order of the differential equation. First order differential equation requires one boundary condition. When the differential term involves time as the independent variable, then we must know the value of the dependent variable at some time, usually at time $= 0$, and it is called initial value.

We will consider a solution of a first order differential equation used to describe heat conduction in a solid:

$$q = -kA\frac{dT}{dx} \tag{7.57}$$

This equation is also known as the Fourier equation. In this equation, q is the rate of heat transfer, k is the thermal conductivity, A is the area perpendicular to heat flow, T is the temperature, and x is the location along x axis.

Since $\frac{dT}{dx}$ is a first order differential, the equation is a first order differential equation. To solve this equation we need to know one boundary value.

The solution method involves first separating the variables. In this equation, q, k, A are constant terms.

Temperature, T, and location, x, are variables. Therefore, we move the variables to either side of the equation thus:

$$q\,dx = -kA\,dT \tag{7.58}$$

Next, we perform integration of both sides of the equation:

$$q \int dx = -kA \int dT \tag{7.59}$$

Note that we keep q, k, and A outside the integrals because they are constant terms.

To solve the above integral, we need boundary values. For this equation,

$$T = T_1 \text{ when } x = x_1$$

and

$$T = T_2 \text{ when } x = x_2$$

Therefore, we incorporate the boundary conditions into the integrals:

$$q \int_{x_1}^{x_2} dx = -kA \int_{T_1}^{T_2} dT \tag{7.60}$$

Next, we evaluate the integrals

$$q\,x \Big|_{x_1}^{x_2} = -kAT \Big|_{T_1}^{T_2} \tag{7.61}$$

or

$$q(x_2 - x_1) = -kA(T_2 - T_1) \tag{7.62}$$

We may rearrange the terms to obtain a value for the rate of heat transfer as

$$q = \frac{-kA(T_2 - T1)}{(x_2 - x_1)} \tag{7.63}$$

This example is a brief introduction to the topic of differential equations. Separation of variables is a common method employed to solve first order differential equations. More complex equations, such as higher order or partial differential equations, require special methods for solving them.

Further reading and references

Browne, R. and Mukhopadhyay, S. (2004) *Mathematics for Engineers and Technologists*, 2nd edn. Pearson Education New Zealand, Auckland.

Hartel, R.W., Howell, T.A. Jr. and Hyslop, D.B. (1997) *Math Concepts for Food Engineering*. Technomic, Lancaster, Pennsylvania.

Singh, R.P. and Heldman, D.R. (2009) *Introduction to Food Engineering*, 4th edn. Elsevier, London.

Supplementary material is available at www.wiley.com/go/campbellplatt

Keshavan Niranjan and Gustavo Fidel Gutiérrez-López

Key points

- This chapter introduces the basic concepts of general physics as applied to food materials and will serve food science students and professionals engaged in dealing with food materials.
- This chapter highlights the interrelations between physical, material and interfacial properties.
- A detailed understanding of the basic principles of physics and the physical properties of food is needed for product development, manufacturing, quality control, and also to understand how food feels in the mouth and how it disintegrates within the gastrointestinal tract.
- The key physical properties are covered: however, it should be borne in mind that other properties are also relevant for particular foods or processes.

8.1 Physical principles

8.1.1 Physical dimensions and units

In order to have systems of measurement which are consistent and understood by all, the SI system of units (*Système Internationale d'Unités*) has been adopted since 1960 as the set of standard measurement units. It is imperative to work within this system of units, and if data from any other system is encountered, it is essential to convert the data to the SI system before embarking on further manipulation. In general, a *unit* (e.g. kilogram, meter, second) is a basic division of a measured quantity (e.g. mass, length, time) and the quantity is also known as *dimension*. The SI system divides dimensions into two classes: *fundamental* dimensions and *derived* dimensions, i.e.

those that can be derived from, and expressed in terms of, the fundamental ones. Although the SI system recognizes a total of seven fundamental dimensions (see Table 8.1), the most commonly encountered physical quantities in food science and technology are related to just three dimensions and units: mass (kilogram), length (meter) and time (second). Each of the fundamental units has been precisely defined and regularly reviewed to improve upon precision. For example, the meter is the length of the path travelled by light in a vacuum during the time interval of 2,997,924,581 of a second (Bird *et al.*, 2002). Thus, the meter is defined in terms of both the second and a universal constant – the speed of light in a vacuum. The second, on the other hand, is the duration of 9,192,631,770 periods of the radiation corresponding to the transition between the two hyperfine levels

Table 8.1 SI base units for the seven fundamental dimensions.

Physical dimension	Name of SI unit	Symbol
Length	Meter	m
Mass	Kilogram	kg
Time	Second	s
Electric current intensity	Ampere	A
Temperature	Kelvin	K
Amount of substance	Mole	mol
Luminous intensity	Candela	cd

of the ground state of cesium-133 atoms. This definition references time to radiation from a stable atomic structure. The kilogram, on the other hand, is arbitrarily taken to be the mass of a unique lump of metal stored in Paris (Fishbane *et al.*, 1996).

For the sake of completeness, the remaining four fundamental quantities listed in Table 8.1 will now be defined (Bloomfield and Stephens, 1996; Sandler, 1999; Chang, 2005; Spencer *et al.*, 2006).

■ The *mole* (mol) is the amount of substance that contains as many elementary entities as there are atoms in 0.012 kg of carbon-12. The elementary entities may be atoms, molecules, ions, or specified groups of such particles. The number of elementary entities in a mole is the Avogadro number. Hence, a mole is the mass of substance (in grams) containing 6.02×10^{23} particles of the specified entity.

■ The *Kelvin* (K) is the unit of temperature, and it is defined as 1/273.16 of the temperature of the triple point of water. On the Kelvin scale, the triple point of water (effectively its freezing point at standard temperature and pressure (STP)) is 273.16 K. It is noteworthy that the magnitude of 1 K is effectively the same as 1 degree Celsius; the difference being the setting of the zero point. Hence the Celsius temperature is converted to Kelvin by adding (approximately) 273.

■ The *ampere* (A) is the unit of electric current and the value of current that, if maintained in two straight parallel conductors of negligible cross section, infinite length, and placed 1 m apart in a vacuum, would produce a force equal to 2×10^{-7} Newtons per meter of length. Note that the unit of electrostatic charge – the Coulomb (C) – is not fundamental but is linked to the ampere as follows: 1A = $1 \, C \, s^{-1}$.

■ The *Candela* (cd) is the luminous intensity of a light source which emits monochromatic radiation of frequency 540×10^{12} Hz that has a radiant intensity of 1/683 watt per steradian, which is the standard international (SI) unit of solid angular measure (Bloomfield and Stephens, 1996) in a specified direction.

The abbreviation or symbol of each unit should be carefully noted in Table 8.1 (including the fact whether the letters are in upper or lower case), and it would be incorrect to use any other symbol (e.g. sec for seconds is incorrect).

Table 8.2 lists commonly encountered derived quantities and their respective SI units (Sandler, 1999; Chang, 2005; Spencer *et al.*, 2006). It may be noted that the relationship between fundamental and derived units is based on the definitions of the physical quantities under consideration, and there is no need to memorise the units.

The SI system also allows the sizes of units to be made bigger or smaller by the use of appropriate prefixes. For example, the unit of power stated in Table 8.2, i.e. the watt, is not a big unit even in terms of ordinary household use, so it is commonly expressed in terms of kilowatts (i.e. 10^3 W). Likewise, the unit of length, i.e. the meter, is too large a value to express the size of a microorganism. Therefore, microbial size is expressed in terms of micrometers, more commonly known as microns, which represent 10^{-6} m. Each order of magnitude is known by a prefix, and the range of prefixes normally used in food science and technology, together with their symbols or abbreviations and their multiplying factors are given in Table 8.3.

8.1.2 Explanations of basic physical quantities

For the understanding of physical principles, it is necessary to know the definition and explanations of the key physical concepts. Some commonly encountered terms and concepts are described below.

Area (m^2) is a physical quantity expressing the size of a surface. *Surface area* is the net area of the exposed sides of any object.

The *volume* (m^3) of an object is a measure of how much space it occupies. One-dimensional objects (such as lines) and two-dimensional objects (such as squares) are assigned zero volume in the three-dimensional space. Formulae are available to estimate the volumes of regularly shaped objects (e.g. those that have straight edges). Volumes of curved

Table 8.2 A selection of derived dimensions and their units commonly encountered in food science.

Physical quantity	Units based on fundamental units	Name of SI unit (if specified)	Symbol of SI unit
Area	m^2		
Volume	m^3		
Speed/velocity	$m\ s^{-1}$		
Acceleration	$m\ s^{-2}$		
Force	$kg\ m\ s^2$	Newton	N
Pressure/shear stress	$kg\ m^{-1}\ s^{-2}$	Pascal	Pa
Energy/work	$kg\ m^{-2}\ s^{-2}$	Joule	J
Power	$kg\ m^{-2}\ s^{-3}$	Watt	W
Density	$kg\ m^{-3}$		
Interfacial tension	$kg\ s^{-2}$		$N\ m^{-1}$
Viscosity	$kg\ m^{-1}\ s^{-1}$		Pa s
Thermal conductivity	$kg\ m\ s^{-3}\ K^{-1}$		$W\ m^{-1}\ K^{-1}$
Heat transfer coefficient	$kg\ s^{-3}\ K^{-1}$		$W\ m^{-2}\ K^{-1}$
Electric charge	A s	Coulomb	C
Frequency	s^{-1}	Hertz	Hz

shapes can be calculated by using the principles of integral calculus.

Bulk density is the mass of the object divided by its volume; it is expressed in kilograms per cubic meter (kg m^{-3}). The volume includes the space between particles as well as the space inside the pores of individual particles which make up the object. Bulk density can change depending on how the material is packed in a container. For example, grain poured into a cylinder will have a particular bulk density. If the cylinder is tapped, the grain particles will move and settle closer together, resulting in a higher bulk density. Therefore, the bulk density of powders is usually reported both as "freely settled" and "tap" density (where the tap density refers to the bulk density after a specified compaction process, usually involving vibration of the container). The term *specific gravity* is the ratio of the density of a material to that of water.

The term *volumetric flow rate* appears frequently in fluid mechanics and it plays an important role in the study of the processing of liquid and particulate food materials. This represents also the volume of fluid that passes through a given volume of space per unit time, and it has the unit cubic meters per second (m^3 s^{-1}) in basic SI units.

In general, the rate of change of any quantity will have units of that quantity multiplied by second (s^{-1}). The rate of change of distance is known as *speed* if the direction in which the distance is measured is not specified and *velocity* if the direction is specified. Since both these terms are rates, their units will be in meters per second (m s^{-1}). Likewise, the rate of change of velocity is known as *acceleration*, and its unit will be that of velocity multiplied by s^{-1} or meters per square second (m s^{-2}). A freely falling particle moves under a uniform acceleration of approximately 9.8 m s^{-2}. On the other hand, if a particle has already acquired a velocity in any direction other than the vertical, its motion will still be subjected to gravitational acceleration (which is vertical), and the

Table 8.3 SI unit prefixes.

Multiple	Prefix	Symbol	Multiple	Prefix	Symbol
10^{-1}	deci	d	10	deca	da
10^{-2}	centi	c	10^2	hecto	h
10^{-3}	milli	m	10^3	kilo	k
10^{-6}	micro	μ	10^6	Mega	M
10^{-9}	nano	n	10^9	giga	G
10^{-12}	pico	p	10^{12}	tera	T
10^{-15}	femto	f	10^{15}	peta	P
10^{-18}	atto	a	10^{18}	exa	E

resulting trajectory, known as that of a *projectile*, will be parabolic. A particle thrown in any direction other than vertical will observe projectile motion.

Force is the interaction between two similar entities. The entities may be two masses, or two electrostatic charges, or two nuclear particles, or two magnetic poles. The interaction between like entities is itself the force. Functionally, force results in acceleration, and it follows from Newton's second law of motion that *force is the product of mass and acceleration*. The unit of force can thus be deduced from this definition as the product of the units of mass and acceleration, i.e. kilograms per meter per square second ($kg\ m\ s^{-2}$). This unit is also known as Newton, abbreviated to the letter N (note that only units named after scientists are abbreviated using upper case; i.e. kilogram will not be abbreviated as Kg but as kg).

Newton's second law of motion also defines the term *momentum* as the quantity of motion or movement associated with a mass. This term is quantified as the product of the mass and velocity ($kg\ m\ s^{-1}$). If force is the product of mass and acceleration, and momentum is the product of mass and velocity, it follows from the definition of acceleration that *force is the rate of change of momentum* (this, indeed, is the enunciation of Newton's second law). The main characteristic of momentum worth noting is that it is a physical quantity which is *universally conserved*. In events such as collision or flow, the momentum is transferred from one particle to another, but, overall, it is unambiguously conserved.

Momentum transfer underpins the principles governing the flow of fluids. When a force acts perpendicular to a given area, it is said to exert a *pressure*, which is given by the force divided by the area. In other words, pressure will have the unit $N\ m^{-2}$, also known as Pascal (Pa). If the force acts *along* the area, instead of acting normally, then it is said to exert a *shear stress*, which is also given by the force divided by the area. In other words, shear stress and pressure will have the same units. Food materials are often processed either under positive pressure or under vacuum. In such processes it is common to express the pressure in terms of the *gauge pressure*, which represents the value of the pressure above or below normal atmospheric pressure.

The *bar* is a practical unit of pressure and it is taken to be approximately equal to 10^5 Pa. When a material is processed under positive pressure, the value of the pressure can be expressed either in terms of the absolute pressure, say in *bars*, or in terms of *barg*

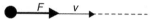

Figure 8.1a Constant force, *F* and velocity, *v* acting along the same line; the particle trajectory (- - - - -) will continue to be linear.

(bar gauge) which represents the value of the pressure above 1 bar.

As explained above, force and acceleration are intimately related. If a mass accelerates (i.e. changes its velocity – be it in terms of magnitude, direction, or both), it must be acted upon by a net force. If no net force acts on the particle, then its velocity remains unchanged; and it moves in a straight line at a constant speed (or velocity). If this mass is acted upon by a net force *along the line of its motion* (Fig. 8.1a), then it will experience an acceleration that will change the magnitude of the velocity; the mass will, however, continue to be in motion along the same line. If the net force were to act uniformly in a direction *perpendicular to the line*, the mass will deviate from the line and follow a parabolic trajectory as explained above (see Fig. 8.1b).

Now consider a third case where the same mass, moving uniformly along the line initially, is acted upon by a net force which is perpendicular to the line *but always towards a fixed point not on its path* (see Fig. 8.1c); the magnitude of its velocity, i.e. its speed, will not change. On the other hand, its direction will progressively change and the mass will be set into a *circular motion*. The force setting the mass into a circular motion is known as the *centripetal force*.

An important consequence of Newton's third law of motion is that the mass itself will experience an equal and opposite force (i.e. acting radially away from the centre) which is known as the *centrifugal force*. It is under the influence of this force field – which, in principle, can be raised at will to several multiples of the gravitational force – that particles tend to get thrown outwards when set into rapid circular motion; this forms the basis of *centrifugation* or centrifugal separation. The angle swept by the radius of a particle performing circular motion, per second, is known as the angular velocity, ω (radians s^{-1}),

Figure 8.1b Constant vertical force *F* acting perpendicular to velocity *v*; the particle trajectory (- - - - -) will be parabolic.

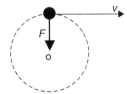

Figure 8.1c Constant force *F* acting towards a fixed point 'o' and perpendicular to velocity *v*; the particle trajectory (- - - - -) will be a uniform circular path. *F* is known as centrifugal force.

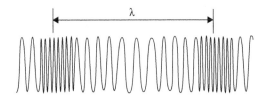

Figure 8.2a Longitudinal wave illustrated by the horizontal oscillation of a spring (direction of wave propagation is left to right).

and the number of revolutions completed per second is known as the frequency, v (s^{-1}). It can easily be shown that $T = 1/v = 2\pi/\omega$. The angular velocity is related to the linear velocity, v, by the expression $v = r\omega$. Further, the centrifugal force acting on a particle of mass m is given by:

$$F = \frac{mv^2}{r} = mr\omega^2 \tag{8.1}$$

The high accelerations required for separating, say, small drops from emulsions or small particles from suspensions, are generated in centrifuges, which essentially generate centrifugal force fields which can be several orders of magnitude greater than the gravitational force field.

Finally, if the mass moving uniformly was acted upon by a force that was always directed towards a fixed point in the path of motion, with its magnitude proportional to the distance from the fixed point, the mass will be set into what is commonly known as a *simple harmonic motion* (Fig. 8.1d). In other words, the mass will end up oscillating on either side of the fixed point, with the maximum displacement from the fixed point being known as the amplitude of motion. Both circular and simple harmonic motions are *periodic*, i.e. these are repetitive motions in which the particle appears at the same position with the same velocity after regular time intervals known as the *period* (*T*). It may be noted that the relation between *T* and *v* for oscillatory motion is the same as that for circular motion.

8.1.3 Wave motion

The oscillation of particles forms the basis of *mechanical wave motion* where oscillating particles can transfer their movement to adjacent particles either partly or wholly. Mechanical waves propagate through a material medium (solid, liquid, or gas) at speeds that depend on the elastic and inertial properties of that medium (Bodner and Pardue, 1995).

There are two basic types of wave motion for mechanical waves: *longitudinal* waves and *transverse* waves. In a longitudinal wave, the particle displacement is parallel to the direction of wave propagation, and particles simply oscillate back and forth about their individual equilibrium positions, creating compressions and rarefactions (Fig. 8.2a). The air motion which accompanies the passage of a sound wave will be back and forth in the direction of the propagation of the sound, and this is a typical example of a longitudinal wave.

Transverse waves, on the other hand, are characterised by particle displacement that is perpendicular to the direction of wave propagation, resulting in the formation of peaks and troughs (Fig. 8.2b). A wave on the string attached to a stringed instrument is a typical transverse wave. Further, such waves cannot propagate in a gas or in the bulk of a liquid because

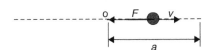

Figure 8.1d Variable force proportional to its distance from a fixed point 'o'; the particle trajectory (- - - - -) will be a simple harmonic motion with amplitude '*a*'.

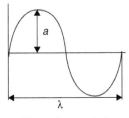

Time taken: period

Figure 8.2b One complete cycle of a transverse wave (direction of wave propagation is left to right, and direction of particle oscillation is vertical).

there is no mechanism for driving oscillatory motion perpendicular to the propagation of the wave.

The *amplitude* of a wave is defined as the maximum displacement experienced by any particle from the mean position. The amplitude is a measure of the energy of a wave, and, in general, *the wave energy is proportional to the square of the amplitude*. In the case of sound waves, the amplitude indicates the loudness of the sound, whereas in the case of visible light, it represents the intensity (Bonder and Pardue, 1995). The wavelength is the distance between two peaks (or two troughs) or two compressions (or rarefactions); see Figs 8.2a and b. The wavelength can be defined as the distance the wave has travelled during one complete cycle. Wavelength is denoted by the symbol λ. The frequency of a wave is the number of complete cycles per second and is measured in Hertz (Hz), which represents one cycle per second. Just as in the case of simple harmonic motion or circular motion, the period – which represents the time taken for a complete cycle – is the reciprocal of the frequency. Further, the frequency and wavelength are related to each other through the wave velocity by the following equation which is a pivotal relation in the science of waves:

$$v = \nu\lambda \qquad (8.2)$$

While mechanical waves require a medium for propagation, *electromagnetic waves* are self-propagating waves having electric and magnetic components. Electromagnetic radiations differ in wavelength (and hence in frequency), but all radiations travel at the speed of light in free space ($\approx 3 \times 10^8$ m s^{-1}). Visible light varies in frequency between red (4×10^{14} Hz) and violet (8.5×10^{14} Hz) colours. Radiations in the infrared region of the spectrum (see Fig. 8.3) can be directly absorbed by matter in the form of sensible heat. Radiation in other components of the spectrum shown in Fig. 8.3, notably in the ra-

dio and microwave region, when absorbed by matter, can be converted by the absorbing material into heat energy depending on the dielectric properties of the material. This is indeed the principle underpinning dielectric and microwave heating of foods. Radiation is much more sensitive to temperature and it is of dominating importance in the instrumental analysis of foods and in food processing.

8.1.4 Conservation of mass

Mass is the property of a physical object that quantifies the amount of matter it is equivalent to. The *law of conservation of mass* states that the mass of a closed system of substances will remain constant, regardless of the processes occurring inside it. This implies that for any process in a closed system, the total mass of the substances at the beginning must equal the total mass of the substances present at, say, the end of the process. The conservation of mass is a fundamental concept of classical or Newtonian physics.

The mass entering a system must either leave the system or accumulate within the system. A *mass balance* (also called a material balance) is an audit of material entering and leaving a system. Mass balances are used, for example, to design and evaluate the performance of process equipment such as cookers, dryers and mixers. Mass balances are often developed for total mass crossing the boundaries of a system, but they can also focus on one component within the system. When mass balances are written for specific components rather than for the entire system, a production term is introduced such that the production term may then describe chemical reaction rates, the difference between formation and destruction (this term might be positive or negative, just as the term involving accumulation in the system).

Consider the control volume defined in Fig. 8.4 and examine the inputs and outputs. Two types of mass balances can be written around this control volume:

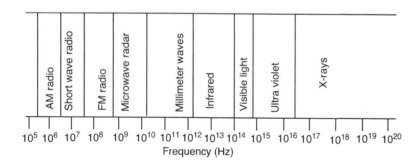

10^5 10^6 10^7 10^8 10^9 10^{10} 10^{11} 10^{12} 10^{13} 10^{14} 10^{15} 10^{16} 10^{17} 10^{18} 10^{19} 10^{20}

Frequency (Hz)

Figure 8.3 Spectrum of electromagnetic radiation.

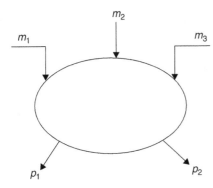

Figure 8.4 Mass balance over a control volume.

(1) mass balance over a period of time and (2) mass balance at any given instant. If m_1, m_2 and m_3 are the masses entering in kg, and p_1 and p_2 are the masses of products leaving over the same period of time, the mass conservation principle says that the sum total of mass entering must be equal to the sum total of mass leaving and mass accumulated within this control volume. In other words,

$$m_1 + m_2 + m_3 = p_1 + p_2 + \textit{mass accumulated} \qquad (8.3)$$

If, on the other hand, the masses are continuously entering and leaving the control volume, it would be more appropriate to write down an *instantaneous* mass balance equation, which is essentially Equation 8.3, except that m_1, m_2, m_3, p_1 and p_2 are *mass flow rates* (kg s^{-1}), and the accumulation term essentially reflects the *rate of accumulation*. If the mass flow rates are all *uniform*, there will be no net accumulation within the control volume, and the system is said to operate under *steady state*.

8.1.5 Energy

Energy can be defined as the work done by a given force, and it is quantified as the force multiplied by the displacement that it brings about. It follows that the unit of work or energy will be the product of the units of force and displacement, i.e. N m, which is also given the name Joule (J). The rate at which work is done or energy is spent (or gained) is known as *power* and, once again, following the same rationale, its units will be J s^{-1} which is also known as Watt (W).

The *principle of conservation of energy* states that energy cannot be created or destroyed, but it can be altered from one form into another. Thus, in any isolated system, the sum of all forms of energy remains constant even though there may be interconversion between the many different forms (mechanical, electrical, magnetic, thermal, chemical and nuclear).

The two common forms of mechanical energy are *potential* and *kinetic*. The former is the energy associated with position in the gravitational field, and is given by *mgh*, where *m* is the mass of the object, *g* is the acceleration due to gravity and *h* is its position in the field with reference to a base position. The kinetic energy is the energy associated with movement and if a mass *m* moves at a velocity *v*, its kinetic energy is given by $^{1}/_{2}$ mv^2. In the context of the fluid flow, there is a third form of mechanical energy that is associated with the pressure which is quantified as the product of pressure and volume terms. Note that dimensionally Pa $\times$ m^3 is the same as J. In a number of processes, it is possible that mechanical energy itself may largely be conserved. This principle is found to hold good in the case of the flow of fluids such as air and water. Its mathematical formulation is known as Bernoulli's equation, which represents a mechanical energy balance between any two system conditions 1 and 2 of a moving fluid (Fig. 8.5) and can be expressed as follows (Landau and Lifshitz, 1987; Acheson, 1990):

$$\frac{v_1^2}{2} + gh_1 + \frac{p_1}{\rho} = \frac{v_2^2}{2} + gh_2 + \frac{p_2}{\rho} \qquad (8.4)$$

Equation 8.4 represents an energy balance with no energy loss or input from (or to) the system, where

v is velocity of the fluid,
g is acceleration due to gravity,
h is height measured from a defined base level,
p is pressure of the system,
ρ is the density of the fluid.

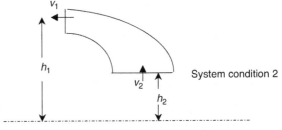

Figure 8.5 Flow through an arbitrary conduit between system conditions 1 and 2 (Equation 8.4).

If the fluid received energy, say, from a pump, just after it passed system condition 1, a term must be added to the left-hand side of the above equation. Also, energy losses due to friction (F) in pipelines and fittings must be taken away from the left-hand side (or added to the right-hand side). In general, Equation 8.4 is useful for evaluating pumping energy for a given fluid motion system.

8.1.6 Thermal or heat energy

Thermal energy or heat energy is essentially the mechanical energy of particles at the atomic or molecular level. Heat, as we all know it, is the internal thermal energy that *flows from one body to another*. Strictly speaking, it is not the same as the total internal energy or enthalpy of the system. It is only the "energy-in-transit". However, in practice, thermal energy and enthalpy are used synonymously and no attempt will be made to distinguish between the two in this chapter.

The principal unit of thermal energy is indeed the Joule (J), as explained above. However, the unit *calorie* is also used, especially to represent the energy generated by the food we consume. A calorie is equal to the heat that must be added to or removed from 1 gram of water to change its temperature by 1°C. This is, however, a very small unit, and a dietary *Calorie* (note the use of capital C), with which most people are familiar, is the same as the kilocalorie. Incidentally, the *temperature* essentially indicates the direction of internal energy flow between bodies and the average molecular kinetic energy in transit between the bodies (Fishbane *et al.*, 1996; Sandler, 1999).

There are two types of thermal energy which can be transferred. One relates to the heat added (or removed) from a substance in order to raise (or lower) its temperature; this heat is known as *sensible heat*. In general, the heat needed to change the temperature of 1 kg of a substance by 1 Kelvin (K) is known as its *specific heat* (C_p) which is commonly expressed in kJ kg^{-1}K^{-1}; and the sensible heat needed to change the temperature of m kg of the substance by ΔT is $mC_p \Delta T$. The second form of thermal energy is *latent heat*, which is defined as the heat necessary to change the physical state of 1 kg of any substance, i.e. from solid to liquid or liquid to vapour, or vice versa. The net latent heat for changing the state of m kg is therefore given by mL where L is the latent heat normally expressed in kJ kg^{-1}.

It is apparent from the above discussion that the thermal energy level of any substance is not an absolute property; it has to be defined relative to a base temperature. If T_0 is the base temperature and the substance is in a solid state at this temperature, its enthalpy at any other temperature T, say, in the vapour state, is obtained by calculating the energy needed to heat 1 kg of substance from T_0 to T, going through melting at T_f and boiling at T_b. Thus, the change in *enthalpy* (ΔH) is given by (Beiser, 1991; Brown, 1994):

$$\Delta H = C_{pS}(T_f - T_0) + L_f + C_{PL}(T_b - T_f) + L_b \\ + C_{PV}(T_b - T) \tag{8.5}$$

where

C_{pS}, C_{pL} and C_{pV} are the specific heats of the solid, liquid and vapour,

L_f and L_b represent latent heats of fusion and vaporisation at the respective temperatures.

The *first law of thermodynamics* is widely used as a means of stating the energy conservation principle which accounts for thermal energy as well as its interconversion to mechanical energy (American Society of Heating, Refrigeration and Air-Conditioning Engineers, 1997; Sandler, 1999; Moran, 2001; Fleisher, 2002). According to this law, the external heat supplied to the system (Q) is partly utilised to change the internal energy (ΔE) (which is enthalpy, for all practical purposes), and the remaining fraction is utilised for doing mechanical work (W). In other words,

$$Q = \Delta E + W \tag{8.6}$$

If work is done *on* the system, instead of the *system doing work*, W is assumed to be negative while if the system does work, W would be assumed to be positive.

The *second law of thermodynamics* states that the entropy of a given system tends to a maximum value:

$$\int \frac{\partial Q}{T} = S \geq 0 \tag{8.7}$$

The above statement and the equation may seem very abstract, but entropy essentially measures the spontaneous dispersal of energy. In other words, it reflects how much energy is spread out in a process, or how widely spread out it becomes *at a given temperature*.

Hence entropy change (ΔS) is given by q/T. In Equation 8.6, W can also be expressed as the product of the pressure of the system and the change in its volume ($P \Delta V$), and considering that $Q = T \Delta S$, Equation 8.6 can be rewritten as:

$$\Delta E = T \Delta S - P \Delta V \tag{8.8}$$

This is the equation that combines the first and second laws of thermodynamics and expresses limitations to the first law since it clearly suggests that not all the energy of the system can be used to produce work; there will be some spontaneous dispersal of energy which is normally reflected in the configurational changes accompanying the process.

8.1.7 Transfer of heat energy

Most food processing operations involve either addition or removal of heat. Ovens, driers, cookers, thermal baths and blanching equipment are among those equipments that transfer heat to the product, while chilling devices and freezers remove heat from the product and lower its temperature.

Heat transfer occurs when particles having *different temperatures* are either directly in contact or brought into direct contact, or simply happen to be exposed to each other. *Conduction, convection* and *radiation* are recognised as the three modes by which heat can be transferred (Beiser, 1991; Brown, 1994). Conduction refers to the transfer of heat from particle to particle within a material, without bulk movement of the particles themselves. This type of heat transfer is common in solids; it may also dominate heat transfer in highly viscous liquids where bulk movement is not significant. For example, when meat is roasted in an oven, heat penetrates from the surface inwards by conduction.

Convection, on the other hand, involves bulk material movement, which brings together particles having different temperatures. Bulk material movement (or convection) itself can be induced by temperature difference. For example, when a kettle containing water is heated by an electric heating element, the water in contact with the element becomes hotter than the water further away from the element, as a consequence of which a density difference is developed causing the higher density water from farther regions to displace lower density water closer to the element, thereby establishing circulation. When convection occurs as a consequence of density difference,

it is known as *natural convection*, e.g. atmospheric air convection is natural. In food and other materials that are in the fluid state, it is also possible to *force* movement, by agitating or pumping the materials. This is common industrial practice, and such movements are known as *forced convection*. When materials having different temperatures get mixed, either "naturally" or "forcibly", heat transfer occurs by convection and, accordingly, the process is known as *natural convective* or *forced convective* heat transfer (Beiser, 1991; Brown, 1994).

Oven baking is another example of a process using convection. Ovens, typically, contain two heating elements, one at the top and the other at the bottom. During baking, the bottom element heats up the air inside the oven, which rises, circulates and distributes heat throughout the oven. Such natural convection currents are easily blocked by large pans, creating non-uniform temperatures within the oven. Convection ovens improve temperature distribution by using a fan located inside the oven, which forces convection currents. The increased airflow reduces cooking time.

Unlike the above two modes of heat transfer, both of which involve direct contact between the particle exchanging heat, radiative heat transfer occurs when particles having different temperatures are simply exposed to one another, regardless of whether there is a medium separating them or not. For example, heat from the sun reaches the planets by radiation. Even in food processes, radiation can play a significant role: for example, the heat reaching a piece of meat in a hot oven can largely be transmitted by radiation from the heating element.

Two points may be noted from the above discussion on modes of heat transfer. First, the existence of a temperature difference is a prerequisite for *net* heat transfer to occur. Second, in any practical process, no single mode of transfer operates in isolation. For example, a piece of meat gets cooked in an oven because heat from the heating element reaches its surface by a combination of radiation and air convection, and traverses inside it by conduction. Thus all three modes are simultaneously involved. However, in the analysis of the *rate* at which the temperature at the center of the meat rises, only one mode may be critical. If the rate of temperature rise is regarded as the net result of two processes occurring *in series*, i.e. heat transfer from the element to the surface by radiation and air convection, followed by internal transmission by conduction, the conductive step is far slower in relation to the former, and tends to control the overall rate. *It may be noted that the slowest of a series of steps involved in any rate process controls the overall rate.*

8.2 Material properties

8.2.1 Elasticity

Elasticity is essentially the behavior of materials which deform or strain when stresses (defined in section 8.1.2) are applied, but relax completely to their original form when the stresses cease to act. The stress–strain relationships are explained by elasticity principles, which, in general terms, state that achieved extension of an elastic object (x) is proportional to the applied force (F). If the relationship between stress and strain is linear, then Hooke's law applies (Kuhn and Försterling, 1999):

$$F = kx \qquad (8.9)$$

In the above equation, k is called Young's modulus and is a measure of the stiffness of the material. Hooke's law is generally valid for small values of x. For many materials, the linearity between F and x does not apply, particularly when F and x are high. In other words, k is not constant and changes during deformation. However, if F is sufficiently high, the stress–strain relationships will reach a critical value beyond which permanent deformation may result.

8.2.2 Rheological properties

In the simplest terms, the study of the response of a material to applied *shear stress* is known as *rheology*. The rheological characteristics of food materials play a key role in processing operations such as transport of fluids through pipelines and mixing operations, as well as in determining texture, mouth-feel and sensory response. It is important to note that the term rheology applies to materials in all the states of matter: solids, liquids or gases. In fact, the rheological response of a material can help to identify the state of a material under any given set of conditions.

The effect of applied shear stress on a *fluid* (commonly taken to mean a liquid or a gas) is to produce a *shear rate*, and many common fluids, such as air and water, obey Newton's law which states that the shear

stress and shear rate are proportional. The former is commonly denoted by the symbol τ and has units of Pa, while the latter is denoted by γ and has units of s^{-1}. Thus according to Newton's law:

$$\tau \alpha \gamma, \text{ or } \tau = \mu\gamma, \qquad (8.10)$$

where μ is the constant of proportionality, known as *viscosity*. Thus Newton's law formally defines viscosity, and it has the units Pas. It may be noted that water has an approximate viscosity of 10^{-3} Pas or 1 mPas under ambient conditions (around 18°C).

It is worth pointing out at this stage that not all fluids obey Newton's law. Indeed, a significant number of "biological" fluids do not strictly comply with this law, and are collectively known to exhibit non-Newtonian behavior. In general, for non-Newtonian flows, the viscosity is not constant but depends on the shear level (de Nerves, 2005). Such a varying viscosity is known as apparent or effective viscosity, and mathematically can be represented as:

$$\mu_a(\gamma) = \frac{\tau}{\gamma} \qquad (8.11)$$

It is important to note that, in addition to shear levels prevailing, the viscosity can also be dependent on time. Many fluids are endowed with some "structure" which shows an elastic behavior like a solid when stress is applied, *in addition* to viscous behavior. Such fluids are known as viscoelastic. A number of polymeric solutions exhibit this form of behavior. As a consequence of viscoelasticity, a number of materials exhibit what is known as "die-swell" in extrusion.

A large variety of non-Newtonian behavior is exhibited by materials, but for the sake of simplicity these can be classified as: (1) *time independent* behavior, (2) *time dependent* behavior without elastic effects, and *linear viscoelastic* behavior. In the case of time independent behavior, the apparent viscosity (μ_a) only depends on shear rate (γ). When μ_a decreases with γ, the flow is described as *shear thinning* or *pseudoplastic*. On the other hand, when μ_a increases with γ, the flow is described as *shear thickening* or *dilatent*. A number of concentrated suspensions exhibit shear thickening behavior (e.g. corn flour pastes). Mathematically, the power law model can be used to describe the time independent behaviors described above:

$$\tau = k\gamma^n \qquad (8.12)$$

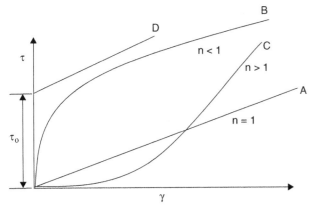

Figure 8.6 Commonly observed relationships between shear stress and shear rate. A, B and C represent flows illustrated by the power law model: $\tau = k\gamma^n$; A represents Newtonian flow where $n = 1$; B represents shear thinning or pseudoplastic flow where $n < 1$; and C represents shear thickening or dilatent flow where $n > 1$. Line D represents what is commonly known as Bingham plastic flow, where the fluid must experience a minimum shear stress (or yield stress) τ_0 in order to commence flowing; a behavior similar to Newtonian flow is observed at shear stress values greater than τ_0.

where

k is the *consistency index*,
n is the *power law coefficient* (or flow behavior index).

It is obvious that when $n = 1$, this equation reduces to indicate Newtonian flow. When $n < 1$, the flow is shear thinning, whereas when $n > 1$, flow is shear thickening. The relation between shear stress and shear rate for Newtonian and Power law flows can be graphically represented, as shown in Fig. 8.6.

In addition to the time independent flow behavior described above, some materials are known to exhibit yield stress, which means that such materials have to be subjected to a minimum shear stress below which they will not flow. Chocolate is known to exhibit such behavior under processing conditions. If the flow behavior resembles Newtonian flow when shear stress levels exceed the yield stress (as shown in Fig. 8.6, line D), the flow is described as being Bingham plastic.

In time dependent non-Newtonian flows, the rheological behavior depends on the duration of the shear applied, in addition to its magnitude. Two common types of time dependent flows are known as *thixotropic* and *rheopectic* behavior. Mayonnaise and gelatins are known to demonstrate thixotropicity, where, due to continuous structural breakdown occurring under the application of a progressively in-

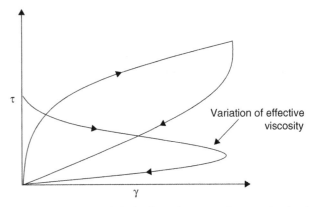

Figure 8.7 Thixotropic behavior – due to continuous structural breakdown during the ramp test, viscosity progressively decreases, e.g. mayonnaise, gelatins, etc.

creasing shear rate, the viscosity decreases (see Fig. 8.7). Rheopectic behavior is the corresponding behavior with shear thickening fluids. It occurs generally at low shear rates. Generally, the structure of time independent fluids is recoverable; and the breakdown of structural interactions is reversible.

As mentioned earlier, viscoelastic fluids are endowed with "structure" and show elastic behavior like a Hookean solid when stress is applied, in addition to viscous behavior. Linear viscoelasticity means that the mathematical principle of superposition can be applied to viscous and elastic properties. The mathematical formulation of linear viscoelasticity is based on what is commonly known as Maxwell's spring and dashpot model. With reference to Fig. 8.8, if $\dot{\gamma}_1$ and $\dot{\gamma}_2$ are the elastic and viscous strain rates generated in a material that is subjected to a shear stress τ at any instant, then, according to the superposition principle, the net strain rate is:

$$\dot{\gamma} = \dot{\gamma}_1 + \dot{\gamma}_2 \qquad (8.12)$$

As explained above, Newton's law states: $\dot{\gamma}_2 = \tau/\mu$, where μ is the material viscosity, and Hooke's

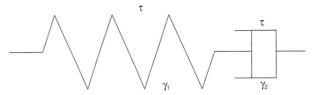

Figure 8.8 Maxwell's spring and dashpot model for viscoelasticity.

law states $\gamma_1 = \tau/G$, whence,

$$\dot{\gamma}_1 = \frac{1}{G}\frac{d\tau}{dt} \qquad (8.13)$$

Here, G is the elastic modulus of the material. From the above equations, $\dot{\gamma}_1$ and $\dot{\gamma}_2$ can be eliminated to give:

$$\frac{d\tau}{dt} + \frac{\tau}{\lambda} = G\dot{\gamma} \qquad (8.14)$$

where $\lambda = \mu/G$ is known as the relaxation time. The above differential equation, also known as the Maxwell equation, can be solved to yield an expression which relates the shear stress to the net shear rate:

$$\tau(t) = G\int_{-\infty}^{t}\exp\left(-\frac{t-t'}{\lambda}\right)\dot{\gamma}(t')dt' \qquad (8.15)$$

This is the constitutive equation for viscoelasticity just like the equation for Newtonian or Power law flow. It is evident that the shear stress at any time t depends on previous history, i.e. all times t' from $-\infty$ to the present, i.e. t. Thus if the shear rates can be independently varied with time, the corresponding shear stress can be determined, provided the material relaxation time and viscosity are known. Alternatively, if shear rates can be varied according to a predetermined function and the corresponding shear stress can be experimentally measured, then the material properties, relaxation time and viscosity can be determined. This forms the principle of rheometry, where, normally, the shear rate is varied sinusoidally as follows:

$$\dot{\gamma}(t') = \lambda_0 \sin\omega t' \qquad (8.16)$$

Substituting this expression for $\dot{\gamma}(t')$ in Equation 8.16, it can be shown that:

$$\tau(t) = \frac{G\lambda^2\omega^2}{(1+\lambda^2\omega^2)}\gamma_0\sin\omega t + \frac{G\lambda\omega}{(1+\lambda^2\omega^2)}\gamma_0\cos\omega t$$
$$= G'\gamma_0\sin\omega t + G''\gamma_0\cos\omega t \qquad (8.17)$$

where

G' is the storage modulus,
G'' is the loss modulus.

Thus if the applied shear rate is sinusoidal, it is evident from Equation 8.17 that the stress response will also be sinusoidal, having the same frequency, but shifted out of phase by an angle, δ, where $\tan\delta = G''/G'$. The storage and loss modulus are very important in the description of food rheology and these parameters are used to characterise the texture and mouth-feel of foods.

8.2.3 Interfacial properties

Most foods are complex multiphase systems, and their production and properties depend on interfacial properties. It is therefore critical to understand the properties as well as composition of interfaces, which, as one would expect, are very different to those prevailing in the bulk. Food processing applications of interfacial properties play a key role in the manufacture of foams and emulsions, particulate suspensions, and even in operations such as rinsing and cleaning in place (CIP). The most fundamental interfacial property is *surface tension* or *interfacial tension*. The term surface tension normally refers to the boundary between any liquid and air, whereas interfacial tension is more generic and refers to any boundary formed between two phases (e.g. gas–liquid, solid–liquid or liquid–liquid).

As mentioned above, the surface or interfacial tension may be defined by considering that the molecules present at the physical boundary with another phase have different energy-related properties than when they are in the bulk. The difference in energy arises because the molecules present in any bulk are experiencing *the same intermolecular forces in all spatial directions* which maintain the molecule in a state of true equilibrium, whereas the molecules present at the interface are only experiencing forces in spatial directions that lead into the bulk. Thus the molecules present at the interface or surface possess a higher energy than those present in the bulk, and are in a state of tension. The surface tension (σ) is then defined as the force needed to transport molecules from the bulk of the fluid to its surface and is generated by the difference in energy fields between molecules in the bulk and those at the interface. Surface tension is measured in units of force per unit length ($\mathrm{N\,m^{-1}}$) and is normally denoted by the letter σ (Kuhn and Försterling, 1999). It is interesting to note that the unit $\mathrm{N\,m^{-1}}$ may also be written as $\mathrm{N\,m/m^{-2}}$ (this has been obtained by multiplying the unit by $\mathrm{m\,m^{-1}}$ which essentially does not change the unit in any way). Since

N m is J, the unit of energy, surface tension may alternatively looked upon as the energy of the surface (or interface) per unit area. Thus, surface tension can be treated as either the force acting per unit length of an interface, or the energy associated with unit area of the interface.

Several experimental methods have been developed to evaluate the interfacial or surface tension, and these methods generally depend on generating a well-characterised interface and measuring the force or energy required to do so. The most common experimental methods are (see, for instance, Kuhn and Försterling, 1999):

- *Du Noüy ring*: measures surface or interfacial tension by evaluating the maximum force exerted on a ring to pull a sheet of liquid away from a liquid.
- *Wilhelmy plate*: evaluates surface tension by measuring the force exerted by a sheet of fluid formed when it is allowed to wet a vertical plate suspended from a balance.
- *Capillary rise*: the tip of a capillary is immersed in a liquid and surface tension is evaluated by measuring the height to which the liquid rises in the capillary.
- *Drop formation methods*: these methods are based on measuring the force required to form a single drop of precisely known geometry, without any significant interference of gravity.
- *Bubble pressure*: surface tension is evaluated by measuring the excess pressure inside bubbles over and above the pressure outside (also known as the Laplace pressure), given that the excess pressure is related to the bubble radius by the equation:

$$\Delta P = \frac{4\sigma}{r} \qquad (8.18)$$

The Wilhelmy plate and capillary rise methods rely on the liquid wetting solid surfaces. *Wetting* of a surface by a liquid essentially occurs when the solid molecules attract the liquid molecules by a force that is greater than the intermolecular force prevailing within the liquid. In other words, the *adhesive* force is greater than the *cohesive* force which is a measure of the surface tension. If cohesive forces turn out to be greater (i.e. the liquid has a very high surface tension), the liquid will simply bead up and not wet the solid surface. This is very clearly evident when mercury accidentally drops out of a thermometer bulb.

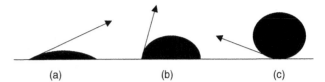

Figure 8.9 Wetting of liquids on a solid surface. The angle made by the tangent with the surface, measured anticlockwise, is known as the *contact angle*: (a) the liquid significantly wets the solid and the contact angle is very low (approaches zero in case the liquid is highly wetting), (b) the liquid is only partially wetting, and (c) the liquid does not wet the surface.

The cohesive forces within mercury are so strong that the drops formed are incapable of wetting the floor. Figure 8.9 illustrates how three different liquids could possibly wet a solid surface. The *contact angle* and the capillary rise together are used to determine surface tension by capillary rise methods.

Surface activity is exhibited by molecules which contain both affinity as well as repulsion causing groups. For instance, molecules that are surface active in water will contain both hydrophilic and hydrophobic groups. In other words, the interface becomes the most hospitable place for these molecules because both groups can coexist satisfactorily with the hydrophilic group pointing towards water and the hydrophobic group pointing away from it. Surfactants clearly lower the interfacial tension of the liquid and can be used to stabilise interfaces. For example, *foaming agents* (e.g. egg white) stabilise foams by occupying the gas–liquid interface, whereas *emulsifying agents* (e.g. lecithin) stabilise emulsions (dispersion of a liquid in another insoluble liquid) by occupying the interface formed by droplets. Very fine multiphase dispersions are known by the generic name *colloids*, and they include a variety of complex dispersions of solid in liquid (sol or suspension), liquid in gas (aerosol), solid in gas (solid aerosol) or gas in solid (solid foam).

Interfacial science is currently an exciting area with a number of novel applications, none more significant than *nanotechnology*, which involves the manipulation of individual or small groups of atoms and molecules (i.e. over length scales involving a nanometer or 10^{-9} m). Nature is known to perform such assemblies very well in the creation of molecular machinery that supports life on earth. It is only recently that we have acquired the capability to "see" (i.e. create images) of individual molecules and move them to create novel assemblies and, thus, novel materials. Developments in classical interfacial science have contributed significantly to developments in nanotechnologies.

Further reading and references

Acheson, D.J. (1990) *Elementary Fluid Dynamics*. Clarendon Press, Oxford.

American Society of Heating, Refrigeration and Air-Conditioning Engineers (1997) *Handbook of Fundamentals*. ASHRAE, Atlanta.

Beiser, A. (1991) *Physics*, 5th edn. Addison-Wesley, Upper Saddle River, New Jersey.

Bird, R.B., Warren, E.S. and Lightfoot, E.N. (2002) *Transport Phenomena*, 2nd edn. John Wiley, Chichester, pp. 488, 867–71.

Bloomfield, M. and Stephens, L.J. (1996) *Chemistry and the Living Organism*, 6th edn. John Wiley, Chichester, pp. 12–14.

Bodner, G.M. and Pardue, H.L. (1995) *Chemistry*, 2nd edn. John Wiley, Chichester, pp. 216–217.

Brown, W. (1994) *Alternative Sources of Energy*. Chelsea House, New York.

Chang, R. (2005) Chemistry. In: *Thermochemistry*, 8th edn. McGraw Hill, Maidenhead.

de Nerves, N. (2005) *Fluid Mechanics for Chemical Engineers*. McGraw Hill, Maidenhead, pp. 428–31.

Fishbane, P.M., Gasiorowicz, S. and Thornton, S.T. (1996) *Physics*, 2nd edn. Prentice Hall, Upper Saddle River, New Jersey.

Fleisher, P. (2002) *Matter and Energy: Principles of Matter and Thermodynamics*. Lerner Publications, Minneapolis.

Kuhn, H. and Försterling, H. (1999) *Principles of Physical Chemistry*. John Wiley, Chichester, pp. 133–6, 732–9.

Landau, L.D. and Lifshitz, E.M. (1987) *Fluid Mechanics*. Pergamon Press, Oxford.

Moran, J.B. (2001) *How Do We Know the Laws of Thermodynamics?* Rosen Publishing, New York.

Sandler, S.I. (1999) *Chemical and Engineering Thermodynamics*, 3rd edn. John Wiley, Chichester.

Spencer, J.N. Bodner, G.M. and Rickard, L.H. (2006) *Chemistry: Structure and Dynamics*, 3rd edn. John Wiley, Chichester, pp. 17–21.

Jianshe Chen and Andrew Rosenthal

Key points

- Food processing transforms raw food materials through a variety of cleaning, separation, size reduction, mixing, heating, cooling and packaging operations into high quality, nutritious products.
- Without food processing we become reliant on indigenous, in-season foods. Food processing extends shelf life while introducing variety and an enhanced sensory eating experience.
- The physical properties of foods dictate their behaviour during processing. In general foods are poor conductors of heat, which introduces the challenge of achieving homogeneous temperatures.
- Generally, untreated foods are hospitable environments for microbes to thrive; food preservation techniques employ food processing operations which extend the shelf life by either eliminating the microbes or by making the food less hospitable.

Most of our food originates from living organisms, both animal and plant. Food scientists and technologists are primarily concerned with the post harvest quality, processing and preservation of such materials, thus controlling the natural processes of decay and manipulating the food from its original form to a variety of high quality, safe and nutritious products, to suit the consumer. While food scientists and technologists are not directly involved in pre-harvest production (such as animal husbandry, arable crop production or fisheries), these areas are of concern as the quality of raw materials has consequences on processed food quality.

Raw materials for food processing frequently contain non-food contaminants from the environment in which they have been produced. Such contaminants need to be removed from the food prior to further processing.

- Inorganic soils and stones are frequently present on root crops (e.g. potatoes).
- Stalks, leaves, seeds from other plants and other plant parts are often present in harvested grains (e.g. wheat).
- Animal droppings, rodent hairs, insects and their parts can be present in raw materials from bulk stores.
- Damaged items of food are of inferior quality and unsuitable for consumption.

Sorting is based on a physical property such as weight, size, shape, density, colour and magnetism.

Machines exist which separate on the basis of such phenomena. Grading is based on quality criteria. While quality is sometimes based on a physical property, more often grading requires some human intervention such as visual inspection. Cleaning processes to remove contaminants may involve both sorting and grading operations. It is worth noting that these operations may be used both for cleaning and for other separation processes. Such techniques are dealt with in more depth in section 9.3.1.

9.1 Fundamentals of fluid flow

9.1.1 Properties of fluids

A fluid is by nature a substance that deforms continuously under the action of a shearing force, however small it may be. Therefore, a fluid is unable to retain any unsupported shape but takes the shape of any solid with which it comes into contact. The transportation or flow behaviour of a fluid food is directly related to the properties of the fluid, primarily viscosity, density and compressibility. These properties are critically important in influencing the power requirement of fluid transportation as well as its flow characteristics within a pipeline.

9.1.1.1 Density

Density is the quantity of matter in a unit volume of the substance and is expressed as kilograms per cubic meter (kg.m^{-3}) in the SI unit system. The density of a fluid can be expressed in different forms: mass density, relative density and specific volume. Mass density is the mass of the substance per unit volume. Relative density (or specific gravity) is the ratio of the mass density of a substance to a standard mass density (such as water) and is a dimensionless quantity. Specific volume is the volume per unit mass and is expressed as cubic meters per kilogram (m^3·kg^{-1}) in the SI unit system.

The density of a fluid is influenced by the temperature and the concentration of solute or dispersed particles. It is common to observe a density decrease of a fluid at an elevated temperature. For example, water has a density of 1000 kg.m^{-3} at 4°C but only 988 kg.m^{-3} at 50°C and 958 kg.m^{-3} at 100°C.

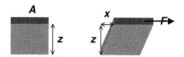

Figure 9.1 Shear deformation and fluid flow.

9.1.1.2 Viscosity

Viscosity is a measure of the resistance of a fluid to shear and hence to flow. Internal friction between fluid elements is the cause of viscosity. Imagine a fluid body of height z and surface area A within two parallel plates and the fluid is divided into a number of imaginary layers. Once a force F is applied to the top plate and carries it forward at a certain speed, the fluid layers will move forward at different speeds. The top layer moves at the same speed as the top plate but the boundary layer next to the bottom plate will be stationary (Fig. 9.1). The relationship between the force applied to the unit area of surface (defined as the shear stress $\sigma = F/A$) and the rate of deformation (defined as the shear rate $\dot{\gamma} = v/z$) is a characteristic property of a fluid and the ratio of the two is defined as the viscosity of a fluid:

$$\eta = \frac{\sigma}{\dot{\gamma}} = \frac{\sigma z}{v} \qquad (9.1)$$

The viscosity has a unit of Pa.s or mPa.s in the SI unit system. The viscosity of a fluid can be influenced by a number of factors, of which temperature, concentration of solute, molecular weight of solute and the presence of suspended matter are the most prominent. Table 9.1 gives the viscosities of some typical food samples.

9.1.1.3 Compressibility

Strictly speaking, all fluids are compressible under pressure. Air and gases are the most obvious examples. The densities of these compressible fluids change as they are compressed. However, the compressibility of liquids is very low and, for the convenience of calculation, liquid fluids are often treated as being incompressible.

9.1.2 Types of flow and Reynolds number

A fluid flow within a pipe could be streamlined and smooth or violently turbulent, depending on the properties of the fluid, the dimension of the pipe and

Table 9.1 Viscosities of some typical substances.

Substance	Temperature (°C)	Viscosity (mPa.s)
Air	27	0.0186
Water	0	1.793
	20	1.002
	100	0.2818
Milk	0	3.4
	20	2.0
	50	1.0
	80	0.6
20% sucrose solution	20	1.967
40% sucrose solution	20	6.223
60% sucrose solution	20	56.7

the rate of flow. A turbulent flow is favoured for promoting mixing and heat and mass transfer, whilst a streamlined flow has a lower power requirement for transportation.

Reynolds first demonstrated the two types of flow with a simple experimental set-up (Fig. 9.2). A narrow tube enters through a main glass tube containing a flow of water. The mouth of the narrow tube is positioned along the centre line of the larger one. A dye is fed through the narrow tube and its flow is negligible compared to the flow of the main tube. At "low" flow rates, the dye moves in a straight-line manner in the axial direction. However, once the flow rate in the main tube exceeds a certain value, the dye spreads in a random manner along both radial and axial directions and becomes blurred. The former is called laminar flow and the latter is called turbulent flow. In a laminar flow, fluid flows without any mixture of adjacent layers (except on the molecular scale), whilst in a turbulent flow particle motion at any point varies rapidly both in magnitude and in direction.

Whether a flow will be laminar or turbulent depends on the balance of the two forces acting on the fluid particles: the inertial force and the viscous force. The viscous force originates from the surrounding fluid and makes the particle conform to the motion of the rest of the stream. However, the particle could also move into a new direction due to the disturbance and its inertial force will carry it on in the new direction. The balance between the two is conveniently expressed as the Reynolds number (Re):

$$\text{Re} = \frac{\rho \bar{v} d}{\eta} \qquad (9.2)$$

where

ρ is the density of the fluid (kg.m^{-3}),
$\bar{v}$ is the mean velocity (m.s^{-1}),
d is the internal diameter of the pipe (m),
η is the viscosity of the fluid.

Reynolds number is a dimensionless number and is most useful in quantitatively describing the flow characteristics of a fluid. As a simple rule, a flow will be laminar if the Reynolds number is less than

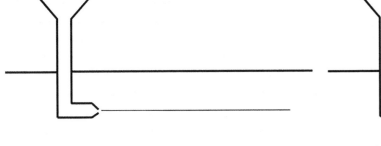

A

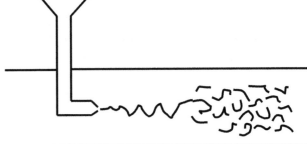

B

Figure 9.2 Types of flow: **A** laminar flow; **B** turbulent flow.

2100 but will be turbulent once the Reynolds number exceeds 4000. For a Reynolds number between 2100 and 4000, the flow pattern is unstable or called transitional. If there is no disturbance, a streamline flow can be maintained in this region, but any slight disturbance tends to upset the pattern.

9.1.3 Velocity profile

The rate of a fluid flow can be expressed as the volumetric rate, Q, (volume per unit of time, such as $m^3.s^{-1}$) or mean velocity, $\bar{v}$. The relationship between the two can be seen from the following equation:

$$\bar{v} = \frac{Q}{A} = \frac{Q}{\pi r_0^2} \quad (9.3)$$

where

A is the cross-section area of the pipe,
r_0 is the radius of the pipe.

The concept of mean velocity is practically very important due to the fact that fluid travels at different speeds inside a pipe, depending on the position of the flow element. For a Newtonian fluid, the overall flow profile within a pipe of length L and radius r_0 can be described by the following equation:

$$v = \frac{\Delta P}{4L\eta} \left(r_0^2 - r^2 \right), \quad (9.4)$$

where

ΔP is the pressure drop (Pa) along the pipe,
η is the viscosity of the fluid.

Equation (9.4) shows that fluid at the centre of the pipe travels at the maximum speed but the fluid next to the wall of the pipe is stationary (Fig. 9.3). This stationary layer is also known as the boundary layer and

the thickness of this layer is critically important in influencing the efficiency of heat and mass transfer in many food processing operations. The integration of Equation (9.4) across from the centre of the pipe to the wall of the pipe gives the volumetric flow rate:

$$Q = 2\pi \int_0^{r_0} vr\,dr = \frac{\Delta P r_0^4}{16L\eta} \quad (9.5)$$

The mean velocity can then be expressed as

$$\bar{v} = \frac{Q}{A} = \frac{Q}{\pi r_0^2} = \frac{\Delta P r_0^2}{16\pi L\eta} \quad (9.6)$$

9.1.4 Mass balance

Mass balance is employed to establish quantities of fluids within a processing stream, based on the principle that matter neither is created nor vanishes. For example, if an inflow of fluids is discharged through two different exit pipes after process (Fig. 9.4), the mass flow at location 1 should be the sum of the mass flows at locations 2 and 3:

$$\rho_1 A_1 v_1 = \rho_2 A_2 v_2 + \rho_3 A_3 v_3 \quad (9.7)$$

and for incompressible fluids ($\rho_1 = \rho_2 = \rho_3$)

$$A_1 v_1 = A_2 v_2 + A_3 v_3 \quad (9.8)$$

9.1.5 Energy conservation for steady fluid flows

The cause for a fluid flowing from one place to another is the energy difference at the two locations. A fluid always tends to flow from a place of high energy to one of lower energy. The energy of a flowing element can be stored or released in different forms. Assuming no heat exchange is involved, forms of energy stored within a fluid body are potential energy (E_p),

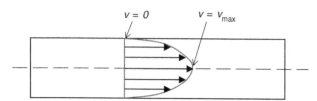

Figure 9.3 The velocity profile of a fluid flow.

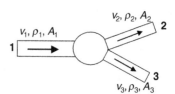

Figure 9.4 The mass flow at location 1 equals the sum of mass flows at locations 2 and 3.

Box 9.1 Definition of energy terms

Potential energy E_p is the energy due to the relative height of the fluid and is defined as

$$E_p = zg$$

where

E_p is the potential energy (J) of 1 kg fluid,
z is the height (m),
g is the acceleration due to gravity (9.81 m.s^{-2})

Kinetic energy E_k is the energy stored within a moving fluid body and its amount equals the work required to bring a body from rest to the same velocity:

$$E_k = \frac{v^2}{2}$$

where

E_k is the kinetic energy (J) of 1 kg fluid.
v is the velocity.

Pressure energy E_r is the energy released or required due to pressure change from one location to another.

$$E_r = \frac{\Delta P}{\rho}$$

where

E_r is the pressure energy (J) of 1 kg fluid,
ΔP is the pressure difference,
ρ is the density of the fluid.

kinetic energy (E_k) and pressure energy (E_r), as related to the changes of the relative height, the travelling speed and the pressure, respectively. Definitions of these energy terms are given in Box 9.1.

The energy conservation for a flowing fluid at locations 1 and 2 can then be expressed as:

$$E_{P1} + E_{K1} + E_{r1} = E_{P2} + E_{k2} + E_{r2} - E_c + E_f \tag{9.9}$$

or

$$Z_1 g + \frac{v_1^2}{2} + \frac{\Delta P_1}{\rho_1} = z_2 g + \frac{v_2^2}{2} + \frac{\Delta P_2}{\rho_2} - E_c + E_f \tag{9.10}$$

where

E_c is the mechanical energy input (such as pumps),
E_f is the energy loss due to friction.

If there is no mechanical energy input and friction energy loss is negligible, then the above equation can be simplified as:

$$z_1 g + \frac{v_1^2}{2} + \frac{\Delta P_1}{\rho_1} = z_2 g + \frac{v_2^2}{2} + \frac{\Delta P_2}{\rho_2} \tag{9.11}$$

This is called Bernoulli's equation and is very useful for calculating the flow properties of a fluid and estimating the mechanical energy input for a transportation system.

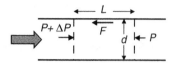

Figure 9.5 Friction of fluid flow within a pipe of length L and internal diameter d.

9.1.6 Friction energy loss

Friction is the major energy loss during fluid transportation. The magnitude of friction force is proportional to the velocity pressure and the internal surface area of a pipe wall:

$$F = f\frac{\rho v^2}{2}A_w = f\frac{\rho v^2}{2}\pi d L \qquad (9.12)$$

where

F is the friction force,
f is the friction factor,
A_w is the surface area of pipe wall,
d is the internal diameter of the pipe,
L is the pipe length.

The friction on the wall must be overcome by a pressure force which gives rise to a pressure drop of ΔP along the pipe (Fig. 9.5):

$$F = \Delta P A_c = \Delta P \frac{\pi d^2}{4} \qquad (9.13)$$

where A_c is the cross-sectional area of pipe.

Combine Equations (9.12) and (9.13) and we have

$$\Delta P_f = 2f\rho v^2\left(\frac{L}{d}\right) \qquad (9.14)$$

or

$$E_f = 2fv^2\left(\frac{L}{d}\right) \qquad (9.15)$$

Equation (9.14) is known as the Fanning equation which effectively predicts the pressure drop of a flowing fluid within a pipe. Friction factor, f, in this Fanning equation depends on the nature of the flow (Reynolds number) and the surface properties of the pipe (roughness). For a laminar flow, the friction factor can be calculated based on the Reynolds number, thus:

$$f = \frac{16}{Re} \qquad (9.16)$$

However, for a turbulent flow, the estimation of the friction factor is not as straightforward as for a laminar flow. Various models have been proposed and Fig. 9.6 is a friction factor chart which shows the dependence of friction factor on the Reynolds number and the pipe's relative roughness (the ratio of roughness to the internal diameter of the pipe).

Energy losses could also occur when the direction of a flow is altered or distorted, as when a fluid flows around pipe bends or through fittings of varying cross sections. Such energy losses can be estimated

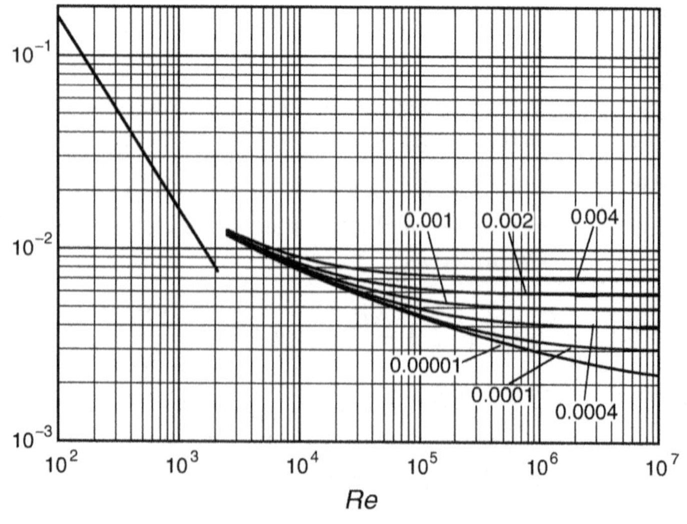

Figure 9.6 Friction factor chart (numbers shown in the graph are the relative roughness of the pipe wall, e/d).

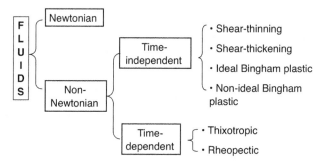

Figure 9.7 Classification of fluids.

using a friction factor of fittings k:

$$E_f = k\frac{v^2}{2} \tag{9.17}$$

9.1.7 Flow of non-Newtonian fluids

The way a fluid responds to an applied stress can vary. For a Newtonian fluid, a constant viscosity is observed for varied shear stresses and rates. For a non-Newtonian fluid, its viscosity varies depending on the rate of shear or the history of shear. Figure 9.7 summaries the types of fluids according to their flow behaviour. Correlations between shear stress and shear rate for various types of time-independent fluids are shown in Fig. 9.8. A dilatant fluid distinguishes itself with increased viscosity at higher shear rates (shear thickening), but a pseudoplastic fluid distinguishes itself with decreased viscosity (shear thinning). Because of the changing viscosity for non-Newtonian fluids, caution must be taken in comparing fluids' viscosities. It is almost meaningless and sometimes could be misleading to make such comparison unless certain testing conditions (shear stress or shear rates) are specified.

Various models have been developed to describe the characteristics of non-Newtonian fluids. The Bingham equation is most useful for Bingham plastic fluids, where σ_0 expresses the yield stress and η_p is the plastic viscosity:

$$\sigma = \sigma_0 + \eta_p\dot{\gamma} \tag{9.18}$$

Many fluid food materials have flow behaviour of the power law model, where a consistency index, K, and a flow behaviour index, n, are used to characterize the relationship between shear stress and shear rate:

$$\sigma = K\dot{\gamma}^n \tag{9.19}$$

Because of the changing viscosity for non-Newtonian fluids, a generalized Reynolds number (Re') should be used to calculate the balance between inertial force and viscous force, where an apparent viscosity (η_a) is used for a non-Newtonian fluid.

$$Re' = \frac{\rho\bar{v}d}{\eta_a} \tag{9.20}$$

A fluid's response to an applied stress is extremely important in determining its behaviour and, therefore, the energy consumption requirement during transportation. A typical example is the huge variation of the velocity profile of non-Newtonian fluids. Fig. 9.9 shows the flow profiles of power law fluids. Unlike a parabolic flow profile for a Newtonian fluid ($n = 1$), velocity profiles of power law fluids strongly depend on the rheological properties of fluids. For an infinite shear thickening fluid ($n = \infty$), the velocity has a linear relationship with the position; but for an infinite shear thinning fluid ($n = 0$), the velocity has no dependence on the location and a flat front line is expected for this type of fluid.

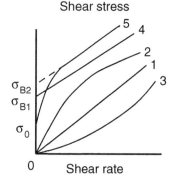
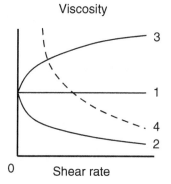

Figure 9.8 Correlations between shear stress and shear rate and the viscosity of Newtonian and non-Newtonian fluids: 1 Newtonian; 2 shear-thinning (pseudoplastic); 3 shear-thickening (dilatant); 4 ideal Bingham plastic; 5 non-ideal Bingham plastic.

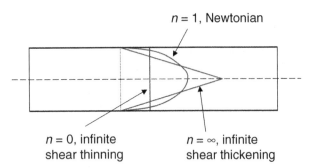

Figure 9.9 Velocity profile for power law fluids.

9.1.8 Pumps for fluid transportation

Pumps are devices that provide mechanical energy to flowing fluids during transportation. Two types of pumps are commonly used in food industry: centrifugal pumps and positive displacement pumps. Centrifugal pumps convert rotational energy into velocity and pressure energy. The typical structure of a centrifugal pump can be seen in Fig. 9.10. A motor drives the impeller (B) within the enclosed case (C). The rotation of the impeller sucks fluid into the centre of the impeller rotation (D) and moves it to the periphery. The fluid achieves the maximum pressure and then moves through the exit (E). Centrifugal pumps are most efficient for low viscosity liquids when high flow rates and moderate pressure are required; they are not suitable for high viscosity and shear-thickening fluids.

Positive displacement pumps draw fluid into the pump and force it through the outlet as a result of the high pressure head developed within the system. This type of pump is very suitable for the transportation of high viscosity fluids. Flow rates are accurately

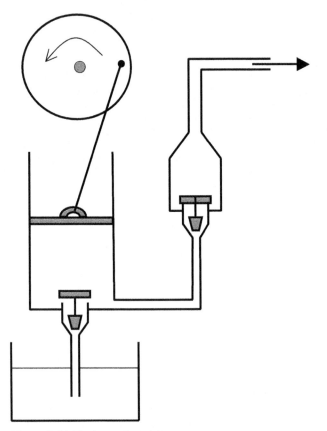

Figure 9.11 Reciprocating piston pump.

controlled by drive speed to the pump. Because of the high pressure head, positive displacement pumps cannot tolerate blockages in the discharge. Figure 9.11 shows a typical type of positive displacement pump: a reciprocating piston pump, where an upward movement of the piston closes the exit pipe but opens the inlet pipe and draws fluid into the chamber, while the downward movement of the piston shuts the inlet pipe but opens the outlet pipe and forces fluid out of the chamber. Other types of positive displacement pumps include gear pumps and rotary pumps.

The overall performance of a pumping system depends on the characteristics of the pump and the transportation system (including the pump and pipes). For example, the energy provided by a pump to a fluid has a reverse relationship with the increase of the volumetric flow rate, but the energy required by the transportation system increases with the increase of flow rate. That is, at a higher volumetric flow rate, the pump provides lower kinetic energy to the flowing body, but to transport such a larger

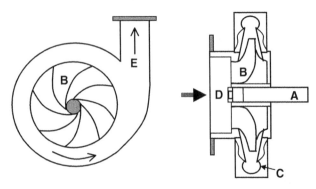

Figure 9.10 Centrifugal pump: A shaft; B impeller; C external shell; D suction side; E discharge point.

volume of fluid requires higher energy input. Therefore, for every transporting system there is an optimum flow rate where the energy obtained by the fluid meets the energy needed for transportation.

Another important feature of a centrifugal pump is its net positive suction head (NPSH). This is a characteristic parameter for a pump and the location of the pump must be carefully chosen to satisfy this requirement. If a minimum net positive suction head cannot be obtained, cavitation will occur at the suction side, leading to greatly reduced transportation efficiency and possible damage to the transportation system.

9.2 Principles of heat transfer

9.2.1 Resistances to heat flow

Foods are generally poor conductors of heat. However, the rate of heat flow through foods is not solely dependent on their thermal conductivity. In practice heat flow also depends on the ease with which heat can pass through the surface of the food. The resistance to heat flow at the surface is due to a stationary layer of fluid referred to as the boundary layer (see section 9.1.3). In the case of food processing this layer, which is normally either air or water, acts as insulation, preventing heat flow. In addition to reducing the heat flow, the boundary layer acts to retard other processes such as mass transfer during drying.

The relative rates of heat flow through the surface of the food and from the surface to the centre of the food are sometimes expressed by the Biot number, where:

$$Biot\ number = \frac{xh}{K} \qquad (9.21)$$

where

x   is the distance from the surface to the centre (m),
h is the surface heat transfer coefficient ($W.m^{-2}.K^{-1}$),
K is the thermal conductivity of the food ($W.m^{-1}.K^{-1}$).

Biot numbers in excess of 40 suggest negligible resistance to heat transfer at the surface, as occurs with condensing steam.

9.2.2 Fourier equation

In food processing the driving force to all heat transfer is the temperature difference, $\Delta\theta$ (K or °C) between the food and the heating/cooling medium. Other factors that affect the rate of heat flux, q (W), include the area of contact with the heating medium, A (m^2), and all the thermal properties of the food, and of the surfaces and/or fluids in contact with the food. Thus:

$$q = UA\Delta\theta \qquad (9.22)$$

U is the overall heat transfer coefficient ($W.m^{-9.}K^{-1}$) and takes account of the thermal conductivity of the food along with resistances to heat flow such as the insulating effects of the boundary layer. While U is a measure of how easily heat will flow, it is normally calculated by looking at the resisting forces:

$$\frac{1}{U} = \frac{x}{K} + \frac{1}{h} \qquad (9.23)$$

The Fourier equation explains the factors involved in steady state heat transfer and can be used to predict the rates of heating and cooling of liquid foods flowing through heat exchangers. However, as solid foods heat or cool, their own temperature changes and thus the magnitude of the driving force ($\Delta\theta$) is not constant. Such situations are referred to as unsteady state and need another solution to predict rates of heating.

9.2.3 Heat exchangers for food processing

Consider two fluids, A and B, separated by a heat conducting surface. If $\theta_A > \theta_B$ then the zeroth law of thermodynamics tells us that heat will flow from A to B and the rate of heat flux (q) is given by the Fourier equation. If A and B are flowing at set velocities over their sides of the surface, then a temperature equilibrium develops across the surface. Such a system is effectively in a steady state, with heat being constantly provided from A and continuously moving through the surface to B where it is taken away in the flowing liquid. This kind of apparatus is a description of a heat exchanger. Commercial heat exchangers are designed to maximize heat flux from one liquid to another, with minimal losses to the environment. In addition to heat transfer efficiency, heat exchangers used for food processing need to be hygienic and easy to clean.

Heat exchangers can be configured in different ways. In parallel flow the two fluids enter at the same end of the heat exchanger. As the hottest end of the warm liquid is adjacent to the coldest end of the cool liquid, there is very rapid initial heat flux, though the temperatures along the length of the heat exchanger converge on each other. Consequently the temperature of the cool liquid can never exceed the coldest temperature of the warm liquid (i.e. at the exit of the heat exchanger). In contrast it is possible to configure the heat exchanger in counter-current flow, whereby the warm and the cool liquids enter the heat exchanger at opposite ends. In this situation there is a more moderate temperature difference along the length of the heat exchanger; however, where the cooler liquid exits the heat exchanger it is adjacent to the hottest temperature of the warm liquid and the temperature of the cool liquid at its exit can be greater than the temperature of the warm liquid at its exit. Clearly both flow patterns have a variety of temperature differences along the length of the heat exchanger. While the Fourier equation considers $\Delta\theta$ to be constant, in a heat exchanger $\Delta\theta$ is changing along its length. By integrating the temperature differences we obtain the log mean temperature difference $\Delta\theta_{lm}$, and this log mean temperature difference can be found by measuring the temperature difference at the two ends the heat exchanger and designating them $\Delta\theta_1$ and $\Delta\theta_9$.

$$\Delta\theta_{lm} = \frac{\Delta\theta_1 - \Delta\theta_2}{\ln\left(\frac{\Delta\theta_1}{\Delta\theta_2}\right)} \quad (9.24)$$

The heat flux can also be represented in terms of the heat flow "into" or "out of" either of the two liquids. In such a situation,

$$q = c_p G(\theta_{warm} - \theta_{cool}) \quad (9.25)$$

where

c_p is the specific heat of that liquid (J.kg^{-1}.K^{-1}),
G is the mass flow rate (kg.s^{-1}),

θ_{warm} and θ_{cool} are the temperatures of the liquid at the warm and cool ends of the heat exchanger. Effectively ($\theta_{warm} - \theta_{cool}$) is the temperature gain or loss of the liquid in the heat exchanger.

Measuring data for the various phenomena involved in heat exchangers is relatively straightforward with the exception of the overall heat transfer

coefficient (U). However, as the heat flux (q) occurs in both Equations 9.22 and 9.25, we can see that

$$U = \frac{c_p G(\Delta\theta_{warm} - \Delta\theta_{cool})}{A\Delta\theta_{lm}} \quad (9.26)$$

where the numerator refers to one of the two liquids.

We know from Equation 9.23 that U is made up of more than one component. In practice Equation 9.23 is a simplification and we can add additional resistances to heat transfer caused by changes to the heat transfer surfaces with time. Many foods are affected by high temperatures, and protein denaturation as well as mineral precipitation can occur during heating. Deposition of such substances causes further resistances to heat transfer (just as foods are poor conductors of heat, so are precipitated foods!). Additional terms can thus be added to Equation 9.23 such as $\frac{1}{h_{fouling}}$, which is a resistance due to the development of a fouling layer with time. There are no real solutions to fouling other than intermittent shut down and cleaning of the equipment. In contrast the influence of the boundary layers can be minimized by creating turbulence within the liquids on both sides of the heat exchanger.

The following sections provide examples of heat exchangers used for foods.

9.2.3.1 Plate heat exchanger

A common type of heat exchanger used for foods consists of a series of steel plates separated from each other by sealing gaskets on the edges of their surfaces (Fig. 9.12). The gap between the plates is filled with either the food or the heat transfer fluid. At each end of every plate are two holes and on the face of the plates around each of these holes a gasket can be fitted. If a series of plates are stacked next to each other so that the holes line up, then holes with a gasket around them form a "tube" through which one of the liquids can flow. If a hole is not surrounded by a gasket then the liquid in that "tube" will flow through the gap between the two adjacent plates. In this way we can organize the plates so that the four holes correspond to the supply of fluid A, the supply of fluid B, the exit of fluid A and the exit of fluid B.

The plates have a ribbed pattern pressed into their surface. These ribs have two functions: (1) to increase the surface area of the plate; and (2) to promote turbulence within the gap between the plates, thus

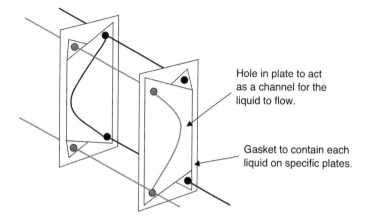

Hole in plate to act
as a channel for the
liquid to flow.

Gasket to contain each
liquid on specific plates.

Figure 9.12 Plate heat exchanger.

minimizing the boundary layer and enhancing heat transfer.

Plate heat exchangers are ideal to heat or cool low viscosity, particle free liquids such as milk. They can be disassembled easily for thorough cleaning.

9.2.3.2 Scraped surface heat exchanger

Scraped surface heat exchangers normally consist of a wide bore tube surrounded by a jacket through which the heating/cooling fluid flows. Scraper blades attached to a central rotor within the tube wipe the surface and keep the food agitated as it is pumped through the heat exchanger. Thus viscous materials or those containing particles can be heated or cooled. Additionally, materials that thicken during heating or cooling can be dealt with, for example margarine can be cooled while agitated, allowing fat crystals to form. Similar equipment is used to make ice cream (of course the jacket contains a refrigerant).

9.2.4 Prediction of temperature during unsteady state heat transfer

Unlike steady state heat transfer, in unsteady state the driving force for heat transfer gradually decreases with time. A typical example is the heat transfer within a jacketed pan consisting of a vessel surrounded by a heating/cooling jacket in which the heat transfer medium is held (Fig. 9.13). Such pans are used extensively in food processing to prepare batch recipes. Initial temperature difference is relatively large, and as the temperature of the food in the pan approaches that of the surrounding jacket, the temperature difference gradually becomes less. Such a situation is more difficult to model than steady state heat transfer, and a number of predictive equations have been developed. However, these equations are unwieldy and only made practicable by high speed computers which allow iterative solutions to be undertaken. Prior to cheap and powerful computers the prediction of temperature was achieved using graphical solutions to the equations (e.g. Gurney–Lurie charts) and this approach will be used here.

Since the variety of initial temperature differences is large the charts make use of a dimensionless temperature, which is defined as:

$$\theta_{lm} = \frac{\theta_\infty - \theta_t}{\theta_\infty - \theta_i} \qquad (9.27)$$

Thus as the dimensionless temperature of the food tends towards zero it approaches the temperature of the heat transfer medium.

We have already considered the Biot number which looks at the relative rate of heat flux at the surface compared with that through the bulk (Equation 9.21). In addition to heat flux, we would expect

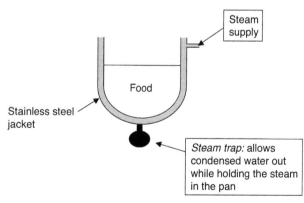

Steam
supply

Food

Stainless steel
jacket

Steam trap: allows
condensed water out
while holding the steam
in the pan

Figure 9.13 Jacketed pan.

a change in temperature when a food is placed in a heating/cooling medium. Another useful relationship compares the rate of thermal energy flux with the rate of thermal energy absorption resulting in a rise in the temperature of the food. This relationship can be expressed as another dimensionless number and is termed the Fourier number:

$$Fourier\ number = \frac{kt}{\rho c_p x^2} \qquad (9.28)$$

Clearly energy absorption is time dependent and any relationship that compares it to the amount carried away is also subject to time. Consequently the Fourier number progressively rises.

In real situations heat flows in and out of foods in three dimensions. For simplicity Gurney–Lurie charts consider heat flow in one dimension only; this is achieved by treating solids as infinite in all but one direction (Fig. 9.14). Thus an infinite cylinder has radial heat flow with negligible heat flow along its length and an infinite slab has one directional heat flow through its thickness with negligible heat flow along its length and width.

Figure 9.15 is a simplified Gurney–Lurie chart which allows prediction of a centre temperature for either an infinite slab (thickness $2x$) or an infinite cylinder (radius x). This particular chart is drawn for an infinite Biot number. For any particular food, we

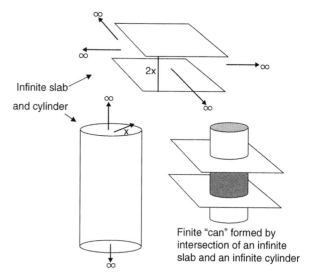

Figure 9.14 Infinite and finite bodies.

can find values for the thermal conductivity, the density, the specific heat and the shortest distance (x) from the centre to the surface. We can thus calculate the Fourier number at different times. By interpolating the chart we are able to identify the dimensionless temperature at any particular time.

It is usually appropriate to convert predicted temperatures for infinite objects to actual temperatures in finite shapes, like a "can" of food. In terms of heat flux, such an item consists of an infinite cylinder

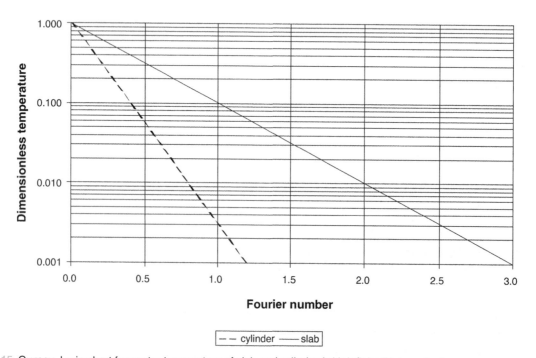

Figure 9.15 Gurney–Lurie chart for centre temperature of slab and cylinder (with infinite Biot number).

whose radius is that of the can and an infinite slab whose thickness is the length of the can. The centre temperature of a can of food may be determined by multiplying the dimensionless temperature of such a cylinder by that of such a slab. Finally the real temperature can be determined by rearranging Equation 9.27.

Most published unsteady state predictive temperature charts are more comprehensive, allowing the user to work with other Biot numbers, as well as predicting the temperature in places other than the centre.

9.2.5 Electromagnetic radiation

Infrared and microwave energy are both examples of electromagnetic radiation used to process foods. The essential difference between the classification of electromagnetic radiation is in their wavelengths and consequently the depth of penetration into the food. Infrared radiation has wavelengths between 400 nm and 400 μm, and generally the depth of penetration into foods is slight. In contrast microwaves have a longer wavelength at around 300 mm which penetrate much deeper into the food.

9.2.5.1 Infrared heating

All objects radiate energy to varying extents, and the amount of energy flowing from one body to another is given by:

$$q = 5.7 \times 10^{-8} \varepsilon A \left(\theta_1^4 - \theta_2^4\right) \qquad (9.29)$$

where

the constant (5.7×10^{-8} W m^{-2} K^{-4}) is the Stefan Boltzmann constant,
ε is the emissivity of the object (black body = 1, highly polished surface → 0),
A is the area,
$\theta_{1\&2}$ are the absolute temperatures of the two bodies.

Since the depth of penetration of infrared energy is slight, conductive heat transfer is a significant factor in infrared heating of foods.

9.2.5.2 Microwave heating

Only specified frequencies may be used (such as 2450 MHz in Europe). Within the food, microwaves cause molecules with dipoles (such as water) to oscillate, the energy being dissipated in the form of heat. While microwave energy will penetrate into foods, as with visible electromagnetic radiation, it can be diffracted when it comes into contact with a solid object. Consequently the shape of the object can cause an unequal distribution of microwave energy such that spherical objects can focus the radiation at its centre, causing excessive heating and even burning.

9.3 Unit operations

The processing of food products often needs the employment of different forms of operation procedures. The devices and equipments employed for these operations could vary to a great extent in terms of design, scale, size, etc., but the principles of many of these individual operations are often the same. For example, the separation of fat droplets from the milk for skimmed milk can be achieved using the same principle as the separation of pulp from the fruit juice for smooth juice (e.g. centrifugal separation). This same principle can also be used in the manufacturing of many other food products. Therefore, various industrial operations can be grouped into different categories according to their operation principles and/or purposes and are called unit operations. Examples of unit operations commonly applied in the food industry include cleaning, material handling, coating, concentrating, evaporation, drying, heating/cooling, freezing, fermentation and forming. This section will discuss unit operations of separation processes, mixing processes, size reduction processes and extrusion processes. Other unit operations, such as heating treatment, drying and freezing, will be discussed in section 9.4.

9.3.1 Separation processes

Separation operations are designed for the separation or physical removal of food components based on physical/mechanical mechanisms or chemical equilibrium principles. Separations based on physical/mechanical mechanisms involve the application of physical forces to the fluid of concern and the separation of product components is achieved due to the different reactions of components to the applied force. Typical examples include centrifugation, filtration and expression. Solvent extraction is an example of separation based on the equilibrium of components in solid and fluid phases. The separation of a component is achieved as a result of the composition change of the two contacting phases. This section will discuss centrifugation, filtration and solvent extraction operations. Operations designed to sort foods by separating them into classes based on size, colour or shape, to clean them by separating contaminating materials are discussed in section 9.1. Selective removal of water from foods by evaporation or dehydration is discussed in section 9.4.3.2.

9.3.1.1 Centrifugation

Centrifugation can be applied for the separations of two immiscible liquids or of solid particles from liquids. The separation factor is the centrifugal force that arises when particles/droplets and fluid elements move in a circle path. The magnitude of centrifugal force equals the acceleration force needed for such circular movement, but in an opposite direction. The acceleration force always acts towards the centre of the circle (perpendicular to the instantaneous velocity), while the centrifugal force acts outwards from the centre. The magnitude of the acceleration a_{rad} depends on the radius r and speed of rotation v:

$$\alpha_{rad} = \frac{v^2}{r} \tag{9.30}$$

The force acting on a particle of mass m is also inward towards the centre of rotation (Fig. 9.16) and its magnitude is given by the following equation:

$$F = ma_{rad} = m\frac{v^2}{r} \tag{9.31}$$

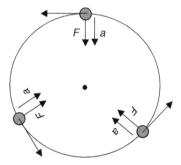

Figure 9.16 A centrifugal force acting on a particle in circular movement.

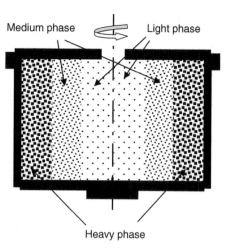

Figure 9.17 Zone distribution of centrifugation separation, with light phase at the annular core, heavy phase to the wall and a medium phase in between.

Once a dispersion system is exposed to a centrifugal field, dispersed particles/droplets will displace themselves according to the magnitude of the centrifugal force acting on them. The heavier phase (such as water in milk) moves to the wall of the centrifuge and the lighter phase (such as fat in milk) will be displaced to an inner annulus. A phase with a medium density could also exist and be separated (Fig. 9.17). A boundary region between heavy and light phases at a given centrifuge speed forms at a radius where the hydrostatic pressure of the two layers is equal. The radius position of this boundary region depends on the geometry size of the centrifuge and the densities of the two phases.

While a high rotation speed is essential for efficient centrifugal separation, it is also worth noting that the density difference between the two separating phases is critically important for centrifugal separation. The higher the density difference, the more efficient the centrifugal separation. For the separation of a dispersion system whose two phases have very close densities, centrifugation should never be a choice.

Centrifugal devices are normally categorized according to the purpose of separation: liquid–liquid centrifuges, such as a tubular bowl centrifuge or a disk centrifuge (Fig. 9.18A and B); centrifugal clarifiers, such as a nozzle-disk centrifuge (Fig. 9.18C); and centrifugal decanters, such as a conveyor bowl centrifuge (Fig. 9.18D). Table 9.2 lists various centrifuges and their possible applications in food processing.

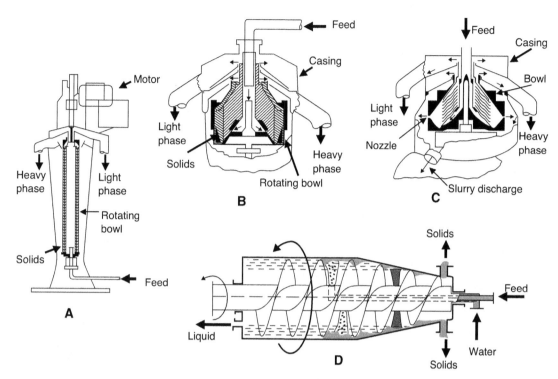

Figure 9.18 Examples of centrifuges: **A** a tubular centrifuge; **B** a disc centrifuge; **C** a nozzle-discharge centrifuge (McCabe *et al.*, 2001); **D** a conveyor bowl centrifuge. (From Leniger and Beverloo (1975), *Food Process Engineering*, Edidel Publishing, Holland.)

9.3.1.2 Filtration

Filtration separations are the removal of solid particles from a fluid by passing the fluid through a filtering medium. The fluid could be a gas or a liquid. The size of the separated particles is the most important factor that determines the choice of filtering medium and is often used as a parameter for the categorization of filtration techniques (see Table 9.3).

Hyperfiltration is applied for the separation of nano-size particles or small molecules, such as ions, sucrose and flavour molecules. Ultrafiltration is used for the separation of sub-micron particles (1–100 nm), such as the separation of enzymes, viruses, gelatine and egg albumin. Microfiltration is most useful for the separation of particles above a micron-size, such as fat droplets in an emulsion, yeast and bacteria.

Table 9.2 Applications of centrifuges in food processing.

Centrifuge type	Range of particle sizes (μm)	Feed solid content (% of w/w)	Applications							
			A	B	C	D	E	F	G	H
Disc bowl										
Clarifier	0.5–500	<5	★	★	★					
Self-cleaning	0.5–500	2–10	★	★	★	★	★			★
Nozzle bowl	0.5–500	5–25		★	★	★	★	★		★
Decanter	5–50,000	3–60		★	★	★	★	★	★	★
Basket	7.5–10,000	5–60						★	★	
Reciprocating conveyor	100–80,000	20–75						★	★	

A, liquid–liquid extraction; B, separation of liquid mixtures; C clarification of liquids; D, concentration of slurries; E, liquid–solid–liquid extraction; F, dehydration of amorphous materials; G, de-watering of crystalline foods; H, wet classification. Source: Fellows (2000).

Table 9.3 Classification of filtration techniques.

Filtration	Particle size	Applied pressure	Application
Microfiltration	0.1 mm–10 mm	200–500 kPa	Suspended particles, fat globules
Ultrafiltration	1–100 nm	350–1000 kPa	Colloids, macromolecules
Hyperfiltration	<1 nm	1–10 MPa	Small molecules

Particular filter mediums (membranes) are required for the separations at molecular and/or colloidal level filtration. The smaller the particles, the higher the technical demand on the filtering membrane.

Effective filtration always requires a pressure to be applied across the filter medium acting as the driving force for the fluid flow. Typical ranges of applied pressures are shown in Table 9.3. Acting against the driving force is the resistance from the filter and the filter cake. The resistance from the filter cake increases when more particles are trapped by the filter. The overall rate of filtration, dV/dt (m³.s⁻¹), would depend on the ratio of the driving force and the resistance:

$$\frac{dV}{dt} = \frac{A\Delta P}{\eta R(x_c + x)} = \frac{A\Delta P}{\eta R\left(\dfrac{SV}{A} + x\right)} \qquad (9.32)$$

where

ΔP (N.m⁻²) is the pressure difference across the filter medium,
A (m²) is the area of filter medium,
η (Pa.s) is the viscosity of filtrate,
R is the specific resistance of filter cake (m⁻²),
x_c (m) is the thickness of filter cake,
x (m) is the thickness of filter,

V (m³) is the volume of feed fluid,
S is the solid fraction of feed fluid.

Filtration can be carried out in two different ways: constant-pressure filtration or constant-rate filtration. The former applies a constant pressure drop across the filter medium and allows the flow rate to fall with time. The latter keeps the flow rate constant by progressively increasing the pressure drop across the filter medium.

In constant-pressure filtration, the rate decrease of filtration depends on the resistance increase of the cake layer. Integration of Equation 9.32 gives a relationship between the volume of filtration and the time of filtration thus:

$$\frac{t}{V/A} = \frac{\eta RSV}{2\Delta PA} + \frac{\eta Rx}{\Delta P} \qquad (9.33)$$

This indicates that for a constant-pressure filtration, the time required for a unit volume of filtration increases linearly with the filtrate volume. That is, the efficiency of filtration decreases linearly with the filtrate volume. If $t/(V/A)$ is plotted against V/A, a straight line can be obtained (Fig. 9.19). This relationship can be used to estimate the specific resistance of the cake (R) and equivalent cake thickness of the filter medium (x). Similarly, for a constant-rate filtration, a

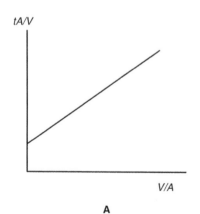

A

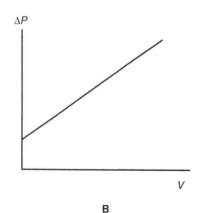

B

Figure 9.19 Filtration profile for **A** a constant-pressure filtration and **B** a constant-rate filtration.

linear relationship exists between the pressure drop across the filter medium and the volume of filtration:

$$\Delta P = \frac{\eta RSQ}{A^2}V + \frac{\eta RQL}{A} \qquad (9.34)$$

where Q represents the volumetric rate of filtration.

This relationship indicates that in order to keep a constant rate of filtration, the pressure across the filter medium must increase linearly with the filtrate volume (Fig. 9.19). Equation 9.34 can also be used to calculate specific resistance of the cake (R) and equivalent cake thickness of the filter medium (x).

Filtration devices can be categorized as pressure filters or vacuum filters. For the former, a pressure is applied to force the fluid through the filter, while for the latter a vacuum is created beyond the filter medium.

Pressure filters apply a large pressure on the upstream side of the filter medium to give economically rapid filtration of viscous suspensions. The filter press is an example of a pressure filter (Fig 9.20A). It contains a set of plates designed to provide a series of chambers or compartments for solid collection. The plates are covered with a filter medium such as canvas or paper filters. Feed liquor is pumped into each compartment at a pressure of 3–10 atm and liquid passes through the filter cloths and out a discharge

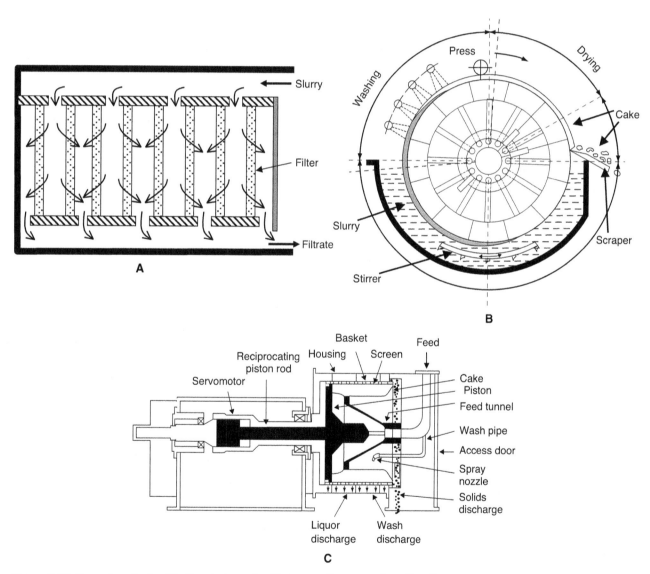

Figure 9.20 Examples of microfiltration devices. **A** a filter press; **B** a rotary drum filter (from Leniger and Beverloo, 1975); **C** a reciprocating conveyor centrifugal filter (from McCabe *et al.*, 2001).

pipe, leaving a layer of wet cake of solids behind. The filtration process would need to be regularly stopped for clean water to be pumped through in the opposite direction to wash cake off the plate. Vacuum filters are often designed for continuous operation. The pressure difference across the filter medium is created by a vacuum pump attached to the downstream of the filter. Figure 9.20B illustrates a rotary drum filter. A cylinder covered with filter cloth and connected to a vacuum pump is horizontally arranged. As the drum rotates, it dips into a bath of liquor and filtrate flows through the filter and out through channels in the drum. The filter cake is sucked free of liquor and washed with water once it moves out of the slurry bath. A scraper removes the cake and releases the vacuum.

Enhanced microfiltration can be achieved by combining centrifugal action with filtration. Pressure resulting from the centrifugal action forces the liquor through the filter medium and leaves the solid behind. Figure 9.20C shows a reciprocating conveyor continuous centrifuge filter, where slurry is introduced into a rotating basket through a feed pipe. Filtrate flows through the basket screen and leaves a layer of cake. The cake is pushed forward by the reciprocating piston rod and dried further before being removed. Compared to normal filtration techniques, centrifugal filtration gives a much drier cake. This could lead to a considerable saving if the filtered material must subsequently be dried by thermal means.

Ultrafiltration is also called membrane filtration or membrane concentration. Ultrafiltration membranes have a high porosity and a narrow pore size distribution in the selective layer and are often characterized by a molecular weight cut-off. Molecules larger than the cut-off size would be rejected, but partial rejection for a wide range of sizes is very common. The most common application of ultrafiltration is for whey concentration in the dairy industry. Other applications include the concentration of sucrose and tomato paste; the treatment of still effluents in the brewing and distilling industries; the separation and concentration of enzymes, proteins or polysaccharides; and the removal of hazes from honey and syrups.

Hyperfiltration is commonly known as reverse osmosis in the food industry. This technique uses a semipermeable membrane which selectively allows water and some small solutes in a solution to pass through, but nothing else. The operation factor of reverse osmosis is the osmotic pressure of a solution. Once a solution is separated by a membrane

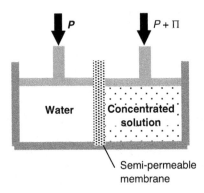

Figure 9.21 Reverse osmosis (P is the osmotic pressure of the solution).

from pure water, water molecules will pass through the membrane to make the solution diluted until the height difference between the solution and the water reaches a typical value. This typical potential difference is called osmotic pressure of the solute solution.

Reverse osmosis works by applying a pressure higher than the osmotic pressure on the solution side so that water molecules migrate from the solution side to pure water and make the solution concentrated (Fig. 9.21). Typical osmotic pressures of selected food solutions are given in Table 9.4, but reverse osmosis generally operates at a much higher pressure (4000–8000 kPa) than osmotic pressure. Main applications of reverse osmosis are: concentration and purification of fruit juices, enzymes, fermentation liquors and vegetable oils; concentration of wheat starch, egg white, milk, coffee, syrups, natural extracts and flavours; clarification of wine

Table 9.4 Osmotic pressure of selected foods and food constituents at room temperature.

Food	Concentration	Osmotic pressure (kPa)
Milk	9% non-fat-solids	690
Whey	6% total solids	690
Orange juice	11% total solids	1587
Apple juice	15% total solids	2070
Grape juice	16% total solids	2070
Coffee extract	28% total solids	3450
Lactose	5% w/v	380
Sodium chloride	1% w/v	862
Lactic acid	1% w/v	552

Source: M. Cheryan (1998), *Ultrafiltration and Microfiltration Handbook*. Technomic Publishing Co., Lancaster, Pennsylvania.

and brew products; dehydration of fruits and vegetables; and desalination of sea water.

9.3.1.3 Solvent extraction

Solvent extraction is a separation operation which applies a solvent to extract/separate a desired component (the solute) from solid food. The separation factor for solvent extraction is the chemical equilibrium of the component between solid and solvent phases and the driving force for solvent extraction is the concentration difference of the component between the two phases. Once a solid is in contact with a solvent, the concentration difference drives a net flow of solutes from the solid phase to the solvent in an attempt to reach equilibrium. The bigger the concentration difference, the larger the driving force and the more efficient the extraction.

Solvent extraction is not a single stage operation but involves mixing the food with solvent, holding for a period of time, and then separating the solvent. Further separation of solute from the extracting solvent, such as concentration and/or dehydration, is often needed after the extraction separation. Applications in the food industry include the extraction of cooking oils from nuts and oilseeds; flavours, spices and essential oils from fruits and vegetables; coffee; tea; and removal of caffeine from coffee and tea.

It is highly desirable to use a solvent at an optimum condition which gives a high capacity of dissolving the target component. Common types of solvents used for extraction are water, organic solvents and supercritical fluids (Table 9.5). Water is by far the most convenient, cost-effective and environmentally friendly solvent and is widely used for the production of sugar, instant coffee and tea. Oils and fats are water-insoluble and would need organic solvents for extraction. In principle, an ideal solvent should have the following desirable features:

- It should have a high capacity for the solute being separated into it.
- It should be selective, dissolving the specific component to a large extent while having a minimum capacity for the other components.
- It should be chemically stable; no irreversible reactions with contacting components.
- It should be regenerable, so that the extracted species can be separated from it readily and it can be reused again and again.
- It should be non-toxic and non-corrosive and be environmentally friendly.
- It should have low viscosity for easy pumping and transportation.

Equipments for solvent extraction can be in the form of stationary solid beds, moving beds or dispersed-solid. Stationary solid-bed extraction is done in a tank with a perforated bottom to support the solids and permit drainage of the solvent. Solids are loaded into the tank, sprayed with solvent until their solute content is reduced to the economical minimum, and excavated. This technique is commonly used for the production of instant coffee, where a series of extraction tanks are linked together to form an extraction battery. Hot water is fed into the tank containing roasted coffee granules that are almost extracted and then flows through the various tanks in series before it is withdrawn from the freshly charged tank (Fig. 9.22). Moving-bed extraction design often

Table 9.5 Solvents used for the extraction of food components.

Food	Solvent	Temperature (°C)
Decaffeinated coffee	Supercritical CO_2, water or methylene chloride	30–50 (CO_2)
Fish livers, meat byproducts	Acetone or ethyl ether	30–50
Hop extract	Supercritical CO_2	N/A
Instant coffee	Water	70–90
Instant tea	Water	N/A
Olive oil	Carbon disulphide	
Seed, bean and nut oils	Hexane	60–70
	Heptane	90–99
	Cyclohexane	71–85
Sugar beet	Water	55–85

Source: Fellows (2000).

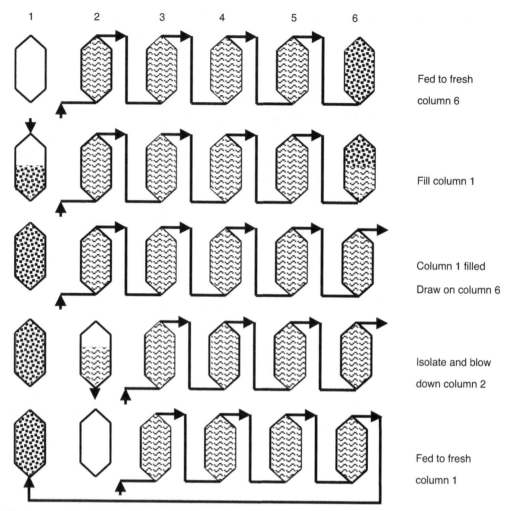

Figure 9.22 Stationary extraction battery for instant coffee production.

uses a series of solid-containing compartments (such as baskets) moved around within a larger container and passed through a number of solvent sprays, a drainage section, a discharge point and a feeding point. Very large quantities of solvent are often used to maximize extraction, but this will have increased operation costs for the following concentration/dehydration processes. High temperature and high flow rate of solvent are desirable for enhanced efficiency of solvent extraction. Higher temperature leads to a lower viscosity of solvent but higher component solubility in the solvent. High solvent flow rate minimizes the stationary layer between the solid phase and flowing solvent and increases the rate of mass transfer between the two phases.

Supercritical fluids (SCF) extraction technique has become increasingly popular in biomaterial processing. Unlike normal solvent extraction, supercritical fluid extractions use fluids in their supercritical status. The term supercritical comes from the fact that a gas (such as CO_2), when compressed isothermally to a pressure higher than its critical pressure, exhibits an enhanced solvent power in the vicinity of its critical temperature (Fig. 9.23). Supercritical fluids exhibit desirable transport properties that enhance their adaptability as solvents for extraction processes. They have densities close to that of liquids but viscosities comparable to that of gases. High density means a high diffusivity of supercritical fluid and hence faster dissolution of solute particles. Compared to normal solvent extraction techniques, supercritical fluid extraction has the advantages of non-toxic operation and causing no damage to the environment.

Common examples of supercritical fluids applied in the food industry include supercritical carbon dioxide, nitrogen and ethylene. Critical points

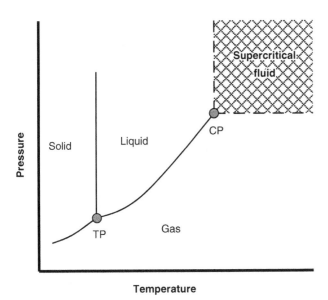

Figure 9.23 Phase diagram of a fluid. TP represents the triple point and CP represents the critical point.

(temperature and pressure) of these gases are listed in Table 9.6. A typical application of supercritical fluid extraction in food processing could be the decaffeination of coffee. Coffee beans are first soaked in water to make the extraction more selective and then are loaded into an extraction vessel. Near-critical CO_2 from a storage condenser is pumped in under high pressure and through a heat exchanger to the extraction vessel. The state of CO_2 in the extractor is carefully tuned by pressure and temperature regulation. The solution is then passed to the separation vessel where it is mixed with fresh water under high pressure and the caffeine is transferred from the CO_2 to water. The caffeine-rich water is drained for further treatment and the CO_2 from the separation vessel is returned to the cooled condenser for re-use (Fig. 9.24).

Table 9.6 Critical points of some gases.

Fluid	Critical temperature (K)	Critical pressure (MPa)
Hydrogen	32.97	1.293
Neon	44.40	2.76
Nitrogen	126.21	3.39
Oxygen	154.59	5.043
Carbon dioxide	304.13	7.375
Methane	190.56	4.599
Ethylene	282.34	5.041
Propane	369.83	4.248

9.3.2 Mixing

Mixing is by nature an intermingling process of two or more dissimilar portions of a material, resulting in the attainment of a desired level of uniformity (either physical or chemical) in the final product. Mixing applications in the food industry are designed to blend ingredients and to reduce the non-uniformities or gradients in composition, properties or temperature of bulk materials or products. Mixing has no preservative effect and is applied solely as a processing aid in order to achieve consistency of properties and eating quality of food products. Mixing different forms of materials, such as the mixing of liquids, solids or solid–liquid, involves different principles and mechanisms and requires different equipment and methods.

The primary concern in mixing is to achieve a satisfactory mixture which has a uniform distribution of the ingredients. The performance of a mixing operation or a mixer is assessed by the time required, the power consumption and the uniformity of the product. Assessing criteria could vary widely from one case to another. For example, sometimes a very high degree of uniformity is required, while at other times a rapid mixing is essential or a minimum amount of power is critical.

The degree of mixing or uniformity of the mixing product is measured by the concentration variance, where spot samples are analysed and the standard deviation is calculated at various mixing lengths. By plotting the concentration variance against the time of mixing (Fig. 9.25), we can estimate the minimum time required for a particular mixer or a particular mixing design. Other considerations of mixing operations are (1) rate of mixing, (2) power requirement, (3) hygienic design, and (4) possible effects of mixing operation on the properties of food components. Care should also be taken in designing a mixing operation to avoid unnecessary over-mixing which may cause waste of energy and damage to the food materials.

9.3.2.1 Mixing of liquid materials

The mixing of liquids is achieved through diffusion, deformation flow and redistribution mechanisms. The rheology properties and the flow behaviour of the liquids are extremely important in selecting a mixing device and in setting up operation conditions. For low viscosity Newtonian fluids, the mixing is relatively easy and straightforward. Turbulence and

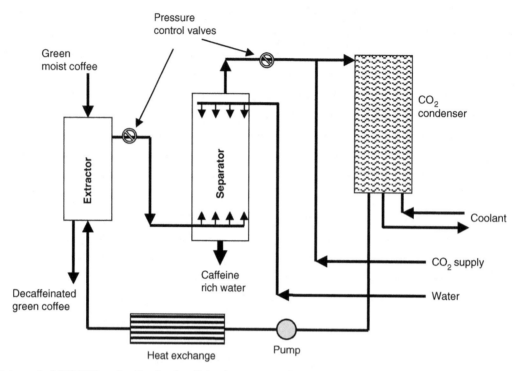

Figure 9.24 Layout of SCF CO_2 extraction for decaffeination process. (Redrawn from McHugh and Krukonis (1994), *Supercritical Fluid Extraction: Principles and Practice*, 2nd edn. Butterworth-Heinemann, Boston.)

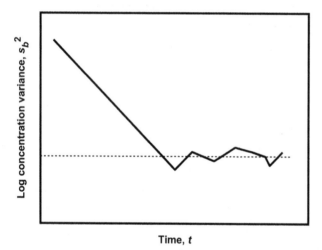

Figure 9.25 Uniformity assessing of a mixing operation, where s_b^2 is the standard deviation between samples.

mixing of low viscosity systems. Concave-shaped blades or angled blades may also be used in turbine design for better mixing.

Mixing of highly viscous fluids (such as dough mixture) is much more difficult. These viscous fluids

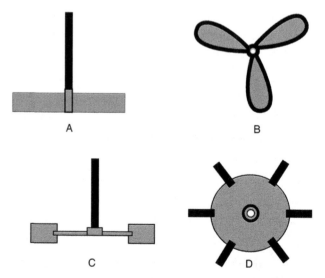

Figure 9.26 Devices for the mixing of low viscosity liquids: **A** a paddle; **B** a three-blade propeller; **C** (side view) and **D** (top view) a six-blade turbine.

diffusion can lead to enhanced redistribution of fluid elements and results in successful mixing. A vortex would have liquid rotating in a circle where the adjoining layers of rotating liquid travel at a similar speed and, therefore, have relatively small effect on the mixing. Devices such as simple paddles, propellers and turbines (Fig. 9.26) can be used for the

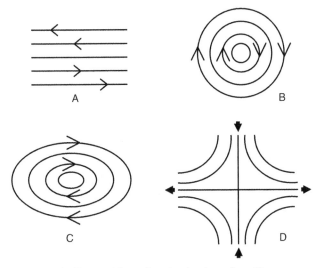

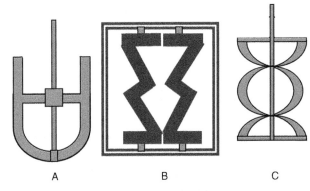

Figure 9.28 Devices for the mixing of high-viscosity fluids: **A** an anchor impeller; **B** a double z-blade mixer; **C** a double-flight helical-ribbon impeller.

Figure 9.27 Types of flow: **A** a simple shear flow; **B** a pure rotational flow; **C** an elliptical flow; **D** a mixed extensional and shear flow.

are highly resistant to deformation and very slow in molecular motion. Therefore, neither diffusion nor turbulence is of much use in mixing these fluids. Instead, deformational flow must be introduced in order to have effective mixing of viscous fluids. Deformational flow could be in the form of a simple shear deformation flow, a pure rotational flow, an elliptical flow, an extensional flow or a mixture of these flows (Fig. 9.27). The net effect of these flows is designed to have particles or fluid elements moving at different speeds and also possibly in different directions. For example, within a simple shear flow, fluid layers will move in the same direction but at different speeds as they slide over each other (Fig. 9.27A). The mixed flow in Fig. 9.27D will have particles and fluid elements moving in different direction as well as at different speeds.

Different forms of flows could be created by the actions of pumping the fluid through the narrow throat of a pipe, squeezing the fluid through a narrow exit, kneading the material against the vessel wall, folding the unmixed part into the mixed part, stretching the material, etc. High rate mixing of viscous fluids could cause stress on the machinery and high power requirements and, therefore, is not recommended. Over-mixing of viscous materials may cause changes of the physical and rheological properties of the material and lead to undesired texture and microstructure of the final product.

A wide variety of devices and equipment are available for the mixing of highly viscous fluids, of which

an anchor impeller, a Z-blade mixer and a helical ribbon are probably the most common choices for the mixing of such systems (Fig. 9.28).

9.3.2.2 Mixing of solid materials

The mixing of solid materials differs significantly from the mixing of liquids in terms of mixing mechanisms, extent of homogeneity, optimum mixing length, energy consumption, equipment design, etc. Liquid mixing depends on the creation of flow currents, which transport unmixed material to the mixing zone adjacent to the impeller, but there is no such current for the mixing of solid materials. For liquid materials the homogeneity normally increases with extended length of mixing, but for the mixing of solids there is often a maximum homogeneity at a certain length of mixing and segregation will occur after an extended length of mixing. Furthermore, the meanings of "well-mixed" products are different for liquid and solid materials. For the mixing of liquid materials, "well-mixed" usually means a truly homogeneous liquid phase, from which random samples should have the same composition. However, for "well-mixed" solid products (pastes or powders) small random samples could differ markedly in composition; samples from any given such mixture must be larger than a certain critical size (several times the size of the largest individual particle in the mix) if the results are to be significant.

Dry food materials that require a mixing operation include flour, sugar, salt, flavouring materials, flaked cereals, dried milk, dried vegetables and fruits. Primary mechanisms for the mixing of cohesionless solid materials include convection mixing, surface mixing and inter-particle percolation. Convection mixing is the mass movement of a group of particles forced by the impeller from one location to another. Surface mixing occurs as a result of the rolling of particles down a free surface. This random movement of particles on the free surface results in the redistribution of solid components. Inter-particle percolation is a process where the smaller particles under the influence of gravity tend to drain through an array of deforming larger ones, and this often occurs when the particle mass is dilated under the applied strain. The above processes will help in the mixing of solid materials, but could also lead to segregation if over-processed. Major factors that affect the mixing process include particle sizes, particle shapes and density differences between solid phases. As a general rule, the larger the size difference and density difference between solid phases, the more difficult the mixing and the greater tendency for segregation.

The mixing performance or mixing extent of particulate materials is harder to assess than it is with liquids. A common method is to analyse spot samples taken from the mix at various times. Either the number fraction (for non-cohesive solids) or mass fraction (for cohesive solids) of components is calculated for each sample and used as a quantitative measure of mixing. A mixture in which one component is randomly distributed through another is said to be completely mixed. The effectiveness or efficiency of mixing operation can be assessed by the time required for the maximum mixing, the power load and the properties of the product.

The design of mixers for solid materials can be roughly categorized as follows:

- mixers with rotating shell: double cone, rotocube, Y-blender;
- mixers with a fixed shell with either a rotating horizontal impeller (such as ribbon blender, Z-blade mixer) or a rotating vertical impeller (such as Kenwood mixer, Nauta mixer);
- fluidizing mixers: air mixer, Young gravity blender.

9.3.3 Size reduction

The term size reduction is applied to all the operations where particles of materials are cut or broken down into smaller pieces. Size reduction has little or no preservation benefit to foods but is essential for the consistency and eating quality of food products.

The main benefits of size reduction for food processing are: (1) an increased surface-area-to-volume ratio for enhanced rates of mass and/or heat transfer; and (2) reduced particle size for more complete and easier mixing of ingredients. The major effects of size reduction on foods are on sensory quality and possible nutrition loss. With reduced size of food components, the texture of foods can be substantially altered. For solid foods, size reduction can increase the smoothness of food and quick release of hydrolytic enzymes. For liquid foods, size reduction of dispersed droplets will affect the viscosity and mouth feeling. However, the increase of surface area of foods during size reduction will inevitably increase surface contact of foods with the surrounding environment (liquid or air) and could cause losses of nutritional compounds and increased oxidation of fatty acids.

Size reduction operation can be divided into two major categories depending on whether the material is a solid or a liquid. For solid materials, the operations are called grinding or cutting, and for liquid materials, emulsification or atomization are the terms.

9.3.3.1 Size reduction of solid foods

Size reduction of solid foods requires the application of mechanical stresses (or forces). Once the applied stress exceeds its yield value, the material would become deformed and finally break up. The relationship between the magnitudes of the applied stress and the resultant strain would depend on the mechanical properties of the material. As shown in Fig. 9.29, strong materials (curves 1 and 2) would need high stresses to break up, but weak materials (curves 3, 4 and 5) break under much smaller stresses. Hard materials (curves 1, 2 and 3) break under small

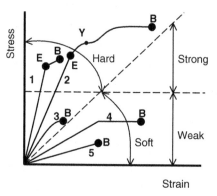

Figure 9.29 Stress–strain diagram for various solid materials: *E* elastic limit; *Y* yield point; *B* breaking point.

strains, but soft materials (curves 4 and 5) are able to stand higher strains before they break up.

Forces used for breaking solid foods can be in the form of a compression, an impact or a shearing (attrition) force. A compression force is applied to fracture friable or brittle foods (e.g. nuts, sugar crystals, roasted coffee beans), while a shearing force is used to cut fibrous foods (e.g. meats, fruits, vegetables). Impact force is often used to break up highly brittle materials (e.g. sugar crystals) by using a moving object (such as hammers in hammer mills) in short-time contact with the material. Devices for size reduction of solid foods apply either a single type of force or more often a combination of different types of forces.

The amount of energy required for the size reduction of solid foods can be theoretically calculated based on the extent of size reduction:

$$\frac{dE}{dD} = cD^{-n_e} \qquad (9.35)$$

where

E is the energy required in breaking a unit mass of diameter D;
c and n_e are constants.

This relationship has been classically interpreted in three ways, referred to as Rittinger, Kick and Bond's laws (Table 9.7) to estimate energy consumption for different extents of size reduction.

However, the total energy applied for size reduction is always much higher than the theoretical predication. This is because a very large amount of energy is wasted as a result of heat dissipation. It was estimated that only about 25–60% of energy consumption was delivered to the solid material and the percentage of energy actually used for the creation of new surface could be as low as 1%. The calculated energy consumption based on the equations shown in Table 9.7 can be used to estimate the energy efficiency of the size reduction operation which is the ratio of the energy used for surface creation against the total energy consumption.

The design of the size reduction operation should aim for minimum heat dissipation. Heat dissipation not only causes waste of energy, but also, more importantly, may cause loss of heat sensitive components (such as flavour and aroma compounds). Factors needed to be considered when making a choice of size reduction device to use include:

- the mechanical properties of the material (hard, brittle, soft, strong, weak, etc);
- the extent of size reduction (large, medium or fine);
- the energy efficiency of the device.

For strong but soft materials (e.g. meat, fruits, vegetables and many other fibrous materials), devices that apply shearing and compression forces should be considered, such as choppers, slicers, dicers, shredders and pulpers. For hard but brittle materials (e.g. sugar, dried starches, roasted nuts, roasted coffee beans), devices that apply impact and compression forces should be used, such as hammer mills, ball mills, disc mills and roller mills.

9.3.3.2 Size reduction of liquid foods

Size reduction of liquid foods is more commonly referred to as homogenization or emulsification, where two immiscible liquids (such as water and oil) are mixed up by dispersing one liquid in the form of very small droplets (called the dispersed phase) in another

Table 9.7 Energy consumption of size reduction of solid foods.

Law	Assumption	n_e	ΔE	Application
Rittinger	Required work is proportional to the newly created surface area	2	$c_R \left(\dfrac{1}{D_2} - \dfrac{1}{D_1} \right)$	Suitable for cases of large surface increase, such as fine grinding
Kick	Required work is constant for the same reduction ratio of the particle size	1	$c_K \ln \left(\dfrac{D_1}{D_2} \right)$	Suitable for cases of small surface increase, such as coarse grinding
Bond	Required work is proportional to the square root of the surface-to-volume ratio of the product	1.5	$c_B \left(D_2^{-0.5} - D_1^{-0.5} \right)$	Suitable for intermediate surface increase

(called the continuous phase). Emulsion is the general term for these dispersed systems. Examples of foods that require such a size reduction operation are margarines, salad creams, mayonnaises, ice cream, soft drinks and homogenized milks. The purpose of size reduction is not for preservation but largely for the texture and sensory alteration and for long-term stability against phase separation.

There are two types of emulsions: oil-in-water (O/W) emulsions (e.g. milk, ice cream) and water-in-oil (W/O) emulsions (e.g. margarine, low-fat spreads). Multiple emulsions, such as water-in-oil-in-water (W/O/W) or oil-in-water-in-oil (O/W/O), are now used increasingly in the food industry for better control of flavour and nutrient release. Emulsification requires an application of an intense shear to break liquid down into droplets of micron or sub-micron sizes. This shearing effect can be achieved by applying a high pressure (hundreds of atmospheric pressure) to force liquids through a narrow opening. Centrifugal forces may also be used to obtain the shearing action. A typical example would be the atomization using a spinning disc. High speed spinning of the disc gives rise to high shearing forces to the liquid flowing over it. Shearing action to a liquid can also be obtained by using ultrasonic vibration energy, such as in the ultrasonic emulsification process.

One consequence of liquid size reduction is the creation of huge contacting surface areas between the two phases and surface energy. For example, if 1 m^3 of oil is dispersed into 1 m^3 of water with an average size of 1 μm, the total surface area within the emulsion will be 3,000,000 m^2. Because the surface energy is unfavourably high, dispersed droplets would have a natural tendency to emerge together via a coalescence process for a lower energy status. Therefore, surface active agent(s) (or emulsifiers) would be needed to reduce surface tension in order to keep the emulsion stable for a specified period of time. Emulsifiers suitable for food uses include proteins (e.g. milk proteins, soy proteins, egg white), phospholipids (e.g. lecithins) and various esters of fatty acids (e.g. glycerol monostearate).

Polysaccharides and other hydrocolloids (such as pectin, gums) are often added to food emulsions. These large molecules have low surface activity and have little preference for surface adsorption. The main function or benefit of the addition of polysaccharides is to increase the viscosity of the continuous phase and, therefore, to slow down the emulsion's destabilization process.

Equipment used for size reduction of liquids includes high-speed mixers, pressure homogenizers, colloid mills and ultrasonic homogenizers.

9.3.4 Extrusion cooking

Extrusion is one of the most versatile operations available to the food industry for transforming ingredients into intermediate or finished products. An extruder is a device that shapes materials by the process of extrusion. Extruders can work at room temperature or at an elevated temperature. If food inside the extruder is heated above 100°C the process would be called extrusion cooking (or hot extrusion). Most extrusion processes in the food industry are operated at high temperatures in order to achieve better gelatinization and/or denaturation of starches and proteins.

The extrusion cooking technique has become increasingly popular in the food industry. The main advantages of extrusion cooking are as follows:

- Versatility. By changing formulation ingredients, processing conditions and the shape and size of the dies, extrusion cooking can be applied for producing a wide variety of food products, such as cereal-based products (e.g. pasta products, expanded snacks), sugar-based products (e.g. fruit gums, toffee) and protein-based products (e.g. sausage products, hot dogs).
- High productivity. An extruder is a continuous processing system and has much greater production capability than other cooking/forming systems.
- Low costs. Labour/operation costs and space requirements of extrusion processing are lower than for other cooking/forming systems.
- High product quality. High temperatures during short time extrusion cooking help retain heat sensitive components.
- No process effluents. Extrusion is a low-moisture operation and does not produce process effluent.

Extrusion is primarily an operation of the continuous forming of plastic or soft materials. Extrusion cooking can be roughly divided into four different stages: feeding, kneading, final cooking and expansion (Fig. 9.30):

- Feeding. The feeding zone is designed at the far end of the extruder where various ingredients are fed into the extruder barrel. Premix of ingredients

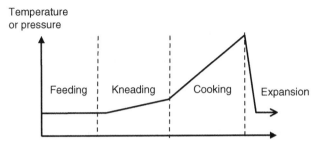

Figure 9.30 Stages of extrusion cooking.

Expansion. Food material is forced through the die and experiences a sudden release of pressure and a decrease of temperature. Sudden expansion leads to the creation of the characteristic microstructure and texture of the final product. This is the stage of shaping and forming of food. The size and geometric properties of the die play an important role in the shaping and forming of foods and in textural properties of the final product.

is not essential. The overall feeding rate depends on the transporting capability of the screws and the overall production capability. Water or other liquid ingredients can be either added together or injected downstream of the feeding zone. Initial mixing occurs at this stage.

Kneading. At this stage, food ingredients are further transported forward and compressed. The screw pitch becomes smaller in this part of the extruder. The raw materials lose their granular identity texture. The density begins to increase as well as the pressure and temperature. After this stage raw particulate materials will become homogeneous viscoelastic and ready for final cooking.

Final cooking. Temperature and pressure increase most rapidly during this stage due to the combined effect of further reduced screw pitch, the presence of the die and the heating jacket. This stage gives food material final transformation and has the most important influence on the density, texture, colour and other functional properties of the final product.

The material flow within a single-screw extruder is the combination of the drag flow and the pressure flow. Drag flow is a forward viscous flow caused by the rotating screw and is proportional to screw speed. However, the pressure flow is in a reverse direction and is caused by the higher pressure at the die end of the extruder. The rheological properties of food material and the operation conditions (temperature, pressure, the diameter of the die aperture, screw speed, etc.) are the main factors affecting the flow within the extruder. For a chosen die, compromise has to be made in overall output rate (productivity) and the pressure within the extruder for an optimum operating point.

There are two types of screw extruders: single-screw extruders and twin-screw extruders. A single-screw extruder (Fig. 9.31) is the simplest food manufacturing device and is very economic to operate. The performance of a single-screw extruder can be adjusted for different purposes by using different dies (either different shapes or different diameters) and operating at different temperatures. However,

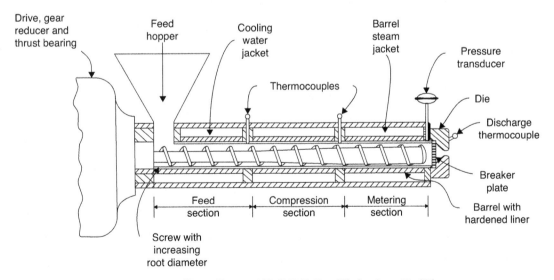

Figure 9.31 A typical single-screw extruder. (From Harper, J.M. (1978) *Food Technology*, **32**, 67.)

single-screw extruders are only suitable for manufacturing of foods that contain less than 4% fat, 10% sugar and 30% water. The presence of high contents of fat, sugar and moisture will significantly reduce the friction between food material and the inner barrel surface and, therefore, impair the mixing and flow of food. Twin-screw extruders consist of two intermeshing screws either co-rotating or counter-rotating against each other. They have much higher mixing capability than single-screw extruders. One significant advantage of twin-screw extruders is the much extended product range. Food may contain as much as 20% fat, 40% sugar and 65% moisture and can be comfortably handled by a twin-screw extruder.

9.4 Food preservation

Foods originate as biological material which has been growing and developing prior to harvest; post-harvest, foods continue to change. These changes can include:

- Enzymic developments such as ripening which in some circumstances occurs off the plant. While ripening itself is desirable, sometimes undesirable enzymic changes may occur in foods.
- Drying out of the food.
- Physical damage through poor transportation and handling, and the ensuing enzymic changes that may result.
- Microbial attack resulting in spoilage.
- Physical damage by pests such as insects or rodents and the subsequent microbial damage that will likely follow.
- Chemical changes such as oxidation.

Food preservation aims at controlling such natural processes of decay in order to maintain high quality, safe and nutritious foods with an extended shelf life than if the food was left at ambient conditions. Thus foods are available out of season and in regions of the world where they are not indigenous, both adding to consumer choice. Some processes of food preservation maintain the food in a native state. Other processes transform the food into a different product which may be desirable in its own right, such as drying grapes to produce sultanas; both have distinctive uses, and there is no pretence that sultanas can be rehydrated to produce grapes.

9.4.1 High temperature treatment

High temperatures are used to destroy enzymes and micro-organisms which might otherwise cause organoleptic changes to the food or result in intoxication. Once a critical temperature is reached, the rate of destruction of both enzymes and micro-organisms is exponential, with an increase in rate of destruction as the temperature rises. Just as foods can be made microbiologically safe, high temperature treatments can give rise to quality changes in foods as they are cooked.

Of primary importance to food manufacturers is the safety of their products and in the case of canning the greatest concern has been over the presence of *Clostridium botulinum* as the most dangerous spore-forming organism potentially present in foods. While thermal sterilization has its origins with Tyndall and Appert in the nineteenth century, much of the theory used today was developed during the twentieth century.

Clostridium botulinum will not produce its toxin below pH 4.5, and consequently high acid products with a pH below 4.5 do not need as severe a heat treatment. The following section examines the rationale behind the thermal processing of low acid foods (i.e. pH > 4.5).

As destruction of micro-organisms (including their spores) is exponential at any given temperature, plotting logs (base 10) of the numbers of survivors vs. time gives a straight line (see Fig. 9.32). The time taken to pass through one log cycle is denoted as the *decimal reduction time* (or D value). Effectively the population of microbes is reduced by 90%, for example it may fall from 10^6 to 10^5, representing one log cycle. A dilemma exists in that exponential destruction never reaches zero, suggesting that there may be a surviving spore which could give rise to poisoning. The resolution of this predicament is the 12D cook. Consider the worst possible scenario – a can full of solid packed *Clostridium botulinum* spores; in such a situation we have about 10^{12} spores per gram. Thus putting the food through a cooking process which achieves 12 decimal reductions should destroy all the spores of *Clostridium botulinum* in a gram in the worst possible case. Most of the theory for canning was developed prior to the introduction of the SI system of units, and a reference temperature of 250°F was used (the units of time used was minutes). These conditions have remained in general use, though when converted the unusual temperature

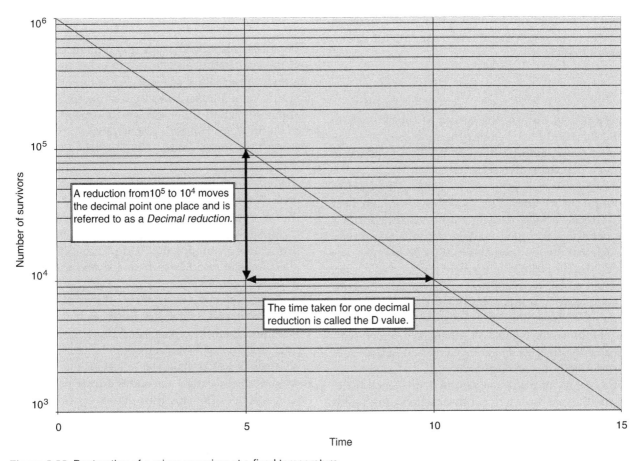

Figure 9.32 Destruction of a micro-organism at a fixed temperature.

of 121°C is used as a reference temperature. At this temperature the D value for *Clostridium botulinum* is 0.21 minute; thus a 12D cook is equivalent to 2.52 minutes (which is normally rounded up to 3). Effectively if we can hold the food for 3 minutes at 121°C we should have a safe and commercially sterile product.

As pointed out in section 9.2.4 cans take time to reach temperatures like 121°C, yet we know that there is a destruction of spores at temperatures below this. In order that we can integrate the destruction at other temperatures and still make use of this 12D concept, we can examine the relationship between D values at various temperatures. We find that D values change exponentially with temperature and plotting the log (base 10) of D values vs. temperature gives us a straight line. The temperature range that achieves a 10-fold reduction in D values is termed the z value. It turns out that this temperature range for *Clostridium botulinum* is 10°C. With this information we can actually represent the destruction at any temperature in terms of equivalent destruction at the

reference temperature:

$$\frac{D_\theta}{D_{121}} = Lethality = 10^{\frac{(\theta-121)}{10}} \qquad (9.36)$$

This ratio of D values at any given temperature compared to the reference temperature is also referred to as the *Lethality* and can be envisaged as the equivalent destruction in minutes at 121°C achieved by 1 minute at the temperature in question. We can add up all the lethalities to get an equivalent time at 121°C; this sum of the lethalities is termed the F_0 value and as we desire 12 decimal reductions at this temperature an F_0 value of 3 should provide a safe food as far as *Clostridium botulinum* is concerned.

9.4.1.1 Blanching

Blanching involves a brief exposure of the food to hot water (or steam). Blanching does reduce the microbial load, but the main reasons for blanching foods are to destroy enzymes which would otherwise result in spoilage and to purge gas from the material, as gases like oxygen might otherwise result in chemical reactions in storage.

9.4.1.2 Pasteurization and HTST

Developed to prevent the spread of tuberculosis in milk, this mild heat treatment is now used on a variety of liquid food products as it kills a wide range of microbial vegetative cells.

Pasteurization is achieved by heating the food to 63°C for 30 minutes. The food flows through heat exchangers to raise its temperature and then into a temperature-controlled holding tube whose volume is such that the food remains in it for 30 minutes. After the holding time is complete the food is passed through a heat recovery heat exchanger (heating incoming food) and it is then cooled further. The temperature at the exit of the holding tube is monitored and if found to be below 63°C, then the product is automatically diverted to another line as it has not met the criteria for pasteurization.

The theory of microbial destruction (section 9.4.1) relates the rate of microbial destruction to different temperatures, and the conditions for pasteurization can be achieved by a high temperature short time process (HTST) employing a temperature of 72°C for 15 seconds.

Just as micro-organisms are killed by pasteurization the destruction of certain enzymes present in the food also occurs. This has led to the development of tests to check that products have been adequately pasteurized. In the case of milk the enzyme alkaline phosphatase is destroyed by similar conditions as exist in pasteurization; thus assaying for its presence gives an indication of the success of the process. Similarly in liquid egg α-amylase is used.

During heating of liquid foods a number of quality changes can occur such as development of cooked flavours and browning reactions. Like micro-organisms and enzymes, these quality changes depend on both temperature and time. Just as D and z values have been derived for micro-organisms, they can also be determined for the rates of change in quality, giving us a sense of how *cooked* the food has become. The z value for quality changes is higher than that for microbes and carrying out pasteurization at higher temperatures generally leads to a better quality product than at lower temperatures.

9.4.1.3 Canning

Chapter 11 considers the nature and filling of the various tin-cans, bottles and flexible pouches used for the thermal preservation of foods. After filling and sealing, the cans are sterilized by heating in a retort to temperatures and times known to achieve commercial sterility. Several types of retort are available to process the canned foods:

- Batch retorts are essentially cylindrical pressure vessels which can be filled with steam. A space efficient option is the *vertical retort* in which case the cylinder stands on end and the top is opened to allow crates containing the food packages to be lowered in. The retort is then sealed and the heating process commences.

  Horizontal retorts lie on their sides with the opening at the end, crates are filled with food packages and then the crates loaded into the retort using fork-lift trucks or conveyor systems. The advantage of horizontal batch retorts is that they can be designed to rotate the crates during heating. As crates rotate the cans turn end-over-end and the headspace in each can is displaced by the denser liquid food. Effectively the can is agitated during heating, resulting in forced convection and more rapid heating than achieved by conduction.

- Continuous retorts require an airlock system to continuously transfer the cans from atmospheric conditions to the high steam pressure used to process them. A second airlock is required to allow the cans to exit after processing . An elegant solution which allows such a continuous operation is the hydrostatic retort (Fig. 9.33). This consists of a tower with three interconnecting chambers. The

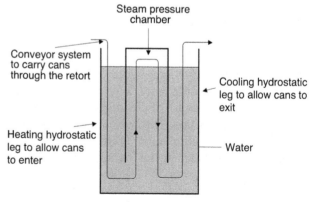

Figure 9.33 Hydrostatic retort.

base of the tower is filled with water and steam is injected into the central chamber forcing the water up the two outer legs. A conveyor system carries cans through the heating leg into the steam chamber and then out of the retort through the cooling leg.

9.4.2 Food additives

Food additives are covered in Chapters 2 and 16 of this book. From a food preservation perspective we are mainly concerned with antimicrobial agents which prevent the growth of micro-organisms in the food and antioxidants which prevent deterioration of fats and oils. Having said this, any additive that prevents spoilage in one form or another will preserve the quality of the product and in that respect stabilizers, anticaking agents and moisture barriers are all worth considering as additives that preserve foods.

9.4.3 Control of water activity

Control of water activity can be achieved in two ways: by raising the solute concentration or by physically removing water by either evaporation or sublimation.

9.4.3.1 Solute preservation

Solutes used to preserve foods must be water soluble and safe to consume. The commonest solutes used for solute preservation of foods are sucrose and sodium chloride, though other salts may be included such as sodium nitrite and sodium tripolyphosphate, each having its own specific functional contribution to the product. Usually concentrated or saturated solutions are used, though in some situations the food is placed in direct contact with the dry solute, in which case surface water causes some dissolution and a saturated solution is quickly formed at the surface. The concentration gradient which develops between the food and the solution results in two simultaneous phenomena:

- Osmotic dehydration, whereby water in the cell tissues is drawn out of the food into the surrounding solution. Such water loss results in dehydration of the tissues. As water is drawn out, so too are soluble substances such as sugars, salts, short chain organic acids and water soluble pigments. These soluble substances end up in the surrounding so-

lution which can over a period of time become discoloured and putrid.
- Diffusion of salts from the surrounding medium into the bulk of the food. Thus the solute concentration within the tissues increases and as the solute becomes associated with water, the water activity is depressed.

A number of traditional products have been produced by solute preservation, such as hams and bacon. In the case of Wiltshire cure, dry salts are rubbed on the surface and allowed to diffuse in over a period of time. As a saturated solution of sodium chloride is around 20%, it can be seen that this is actually quite a wasteful process. The speed of diffusion of solutes is explained by Fick's law. When dealing with large items of food such as a joint of meat, the time involved for the salt to penetrate through to the centre can be considerable. Some modern methods of solute preservation overcome this problem by injecting brine solutions through a bank of hollow needles into the bulk of the item being preserved. Thus the time taken for achieving a homogeneous salt concentration is greatly reduced.

Solute preservation can be used as a preliminary step prior to water removal through drying. For example when kippers are produced the fish are gutted and then soaked in a brine solution prior to being dried.

9.4.3.2 Drying

Drying is the physical removal of water. The controlled application of heat can be used to dry foods at atmospheric pressure through evaporation or at reduced pressures (below the triple point of water) through sublimation. Traditionally foods were dried as a form of preservation; more recently convenience foods which can be rehydrated to their original form have been produced. Generally speaking when liquids or slurries are dried, they are for convenience, e.g. products such as dried coffee, dried milk and instant dried mashed potato. In contrast solid foods are generally dried as a form of preservation. While this distinction is fairly arbitrary, the technology used to dry solid and liquid foods does tend to differ. However, two main principles are applied to both: air drying and freeze drying. Some foods are also dried by direct contact with a heated surface, though this technology is increasingly less common as the quality of the product is variable.

Air drying

For water to be lost from a solid particle of food or from a droplet of liquid food, water must be evaporated from the surface and then water within the bulk must diffuse to the surface for subsequent evaporation. As mentioned in section 9.2 there is a boundary layer surrounding the food which acts to inhibit both heat and mass transfer. Disrupting the boundary layer, by creating turbulent air flow, aids water loss from the surface to the air. When foods are dried by contact with hot air they tend to undergo an initial constant rate of water loss from their surface as free water at the surface is evaporated (Fig. 9.34). This phase of the drying process is referred to as the constant rate drying period and the actual rate and its duration depend on factors like the air flow rate of the dryer, the moisture content and temperature of the air, and the size of the food particles. Once the surface water has been removed, the so-called *critical moisture concentration*, the rate of evaporation is governed by the speed with which water can diffuse from the centre of the food to its surface. This rate of diffusion is normally slower than the evaporation during the constant rate period, in fact any water that reaches the surface is rapidly evaporated. The driving force for diffusion of water from the centre to the surface is governed by its concentration gradient, but as the food becomes drier, the concentration gradient declines; thus once the speed of drying becomes dominated by the ease of diffusion of water from the

bulk of the food to its surface, the rate of dehydration gradually falls and this latter period is often referred to as the falling rate period.

The saturation curve in Fig. 9.35 shows the relationship between the amount of water that air will hold at various temperatures (specific humidity being the mass of water that a unit mass of air holds). At any particular temperature, the 50% humidity curve is half that of the saturation curve. In order that air can take up water it must be less than saturated. As air is warmed its capacity to hold water increases, and thus saturated air at 40°C (Fig. 9.35, point A) can be used to dry foods if it is first of all heated (Fig. 9.35, point B). If less than saturated hot air is passed over the surface of a food, its heat causes some of the water to evaporate and as that water is taken up, the temperature of the air falls. If there is ample water at the surface, as arises during the constant rate drying period, then the air will continue to pick up moisture and lose heat until it is saturated (Fig. 9.35, point C). The temperature of the food when the air is saturated will be at the air's wet bulb temperature. Once the food enters the falling rate period, there is no longer enough water at the surface of the food to saturate the air and the surface temperature of the food starts to rise with subsequent quality changes such as the development of browning reactions and shrinkage. Figure 9.35 is in fact a simplified psychrometric chart. Such charts can be used to predict the amount of air needed to dry foods to a particular moisture content.

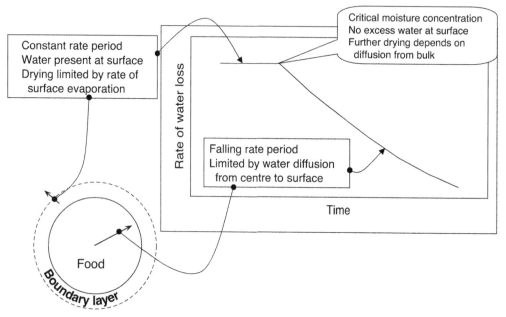

Figure 9.34 Rate of migration of water in foods during drying.

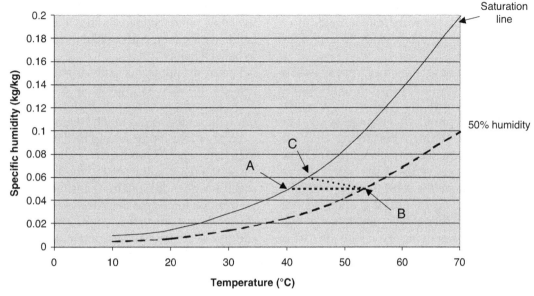

Figure 9.35 Water holding capacity of air at various temperatures (see text for key).

Traditional drying techniques such as sun drying or smoking are variants of air drying. The foods are often given a preliminary soak in a solute solution. They are then heated in direct sunlight or in the warm smoke from smouldering sawdust and the water present is carried away in the breeze or in the circulating smoke from the fire. The surrounding air/smoke must not be saturated for drying to take place and there must be some circulation of the air/smoke around the food. Generally these processes are relatively slow and poorly controlled.

Mechanical dryers employ fans to generate an air current which is heated before being passed over the food, thus achieving a controlled environment in which the food is dried. Differences between the different types of mechanical dryer arise from the orientation of the air flow in relation to the food:

- Fluid bed dryers employ a high velocity air flow directed upwards from below the food. Such dryers are used on small food items up to about 10 mm diameter. The particles of food become suspended in the air, forming a bed of about 100 mm deep which behaves as a fluid. The high velocity of the air causes a large degree of turbulence which strips away the boundary layer, leading to high levels of heat and mass transfer at the surface.
- Tunnel dryers employ trolleys or conveyor belts which carry the food through a tunnel with the air blowing horizontally through. The flow of the air

may be in the same direction as the food (parallel flow), in which case the wettest food is adjacent to the driest air, resulting in very high rates of drying at the start of the tunnel. However, the air quickly becomes saturated, losing its drying capacity, and the foods rarely achieve low levels of moisture.

In contrast the air flow may be in the opposite direction to the one in which the food is passing down the tunnel (counter current). While the initial rate of drying is lower than in the case of parallel flow, the hottest air is adjacent to the driest food. Drying frequently enters the falling rate period, and the surface temperature of the food is likely to exceed the wet bulb temperature as the moisture in the food is brought down. Browning reactions are likely to ensue as the water activity passes through the range of 0.8 to 0.6, a range known to be optimal for such reactions.

Hybrid dryers exist in which the food initially passes through a parallel section achieving rapid initial rates of drying, before passing into a countercurrent section in which very low moisture levels can be achieved.

- Spray dryers utilize hot air to dry liquid foods and slurries. A fine aerosol of the food is produced and sprayed into the top of a tall chamber which has hot air being blown through. As the droplets fall through the hot air, the water at the droplet surface evaporates. Since droplets are small there is only a short distance for any water in the bulk to travel before it reaches the surface. When operated correctly, the aerosol drops are transformed into a dusty powder by the time they reach the exit of the dryer.

Spray forming devices are either a spray nozzle which produces a very fine spray, but can be prone to blocking, or a centrifugal type atomiser which does not block but produces a coarser spray. The air is either blown through the drying chamber or sucked through by negative pressure. The flow configuration of dryers varies. A separating device is necessary to remove the solid particles from the suspending air at the end of drying; often cyclone separators are used, though very fine material will be lost and needs removing by means of filters. While the temperature of the air can be in the region of 250°C the contact time of the food with the air is very short and relatively little damage to proteins occurs. However, water is not the only material to vaporize under these conditions and volatile flavour compounds can be lost as well.

Freeze drying

In freeze drying water is frozen and then sublimed by controlled heating below the triple point (640 Pa and 0.01°C). The freezing step usually is carried out in separate freezers. While a food becomes fully solid at its eutectic temperature, it is normally chilled well below that before being transferred to the freeze dryer. Once in the dryer a vacuum is drawn so that the pressure over the food is less than the triple point. Maintaining low pressures is expensive and normally no attempt to reach an absolute vacuum is made; instead the dryer operates at about 500 Pa and the temperature in the food is gradually raised. Heating is normally achieved by conduction through a heating plate on which the trays of food are placed, and as the heat reaches the frozen water there is a gradual sublimation.

Unlike air drying, where the water loss is from the surface of the food, in freeze drying as water sublimes it leaves a porous sponge-like structure, and further sublimation is from the frozen water further inside the bulk of the food. This raises the obvious problem of how heat is able to reach the frozen layer through a porous structure in a vacuum, to which the answer is "slowly". When most of the water is lost, the pressure can be reduced further to lose the final traces. Another problem with freeze dryers is that as water vapour is produced by sublimation, it fills the vacuum, raising the pressure. Rather than run the vacuum pump excessively, the vapour can be removed

with a condenser, though as the pressure is below the triple point, condensation will be from vapour to solid and the temperature of the condenser will need to be well below 0.01°C.

Freeze drying is normally carried out as a batch process, though continuous systems with sophisticated air locks to allow raw materials in and product out without losing too much vacuum exist.

Freeze dried foods tend to be a better quality than their counterparts produced by air drying. This is because the temperature of the food during heating is much lower. Additionally the sample food does not go through a gradual dehydration, resulting in a period of time at which the a_w is around 0.6–0.8 (optimal conditions for browning reactions); instead any part of the food is either fully hydrated (though frozen) or dehydrated because the subliming front moves gradually through the food. Furthermore the low temperature means that volatile flavour components will tend not to be lost during drying.

Drying by contact with a heated surface

A variety of designs of roller dryer exist. Essentially they all consist of a rotating metal drum which is heated from the inside by steam. A food slurry is brought into contact with the surface (some designs dip a part of the drum in a vat, others trickle and spread the liquid). The slurry dries on the surface as the drum rotates and the dried food is removed from the surface by a spring-loaded scraper.

Drum dryers are difficult to control and thus only used with foods that can withstand exposure to high temperatures, such as starches.

9.4.4 Low temperature preservation

Lowering the temperature of a food reduces the rates of enzymic, chemical and microbial activity. Often the foods are blanched prior to cooling to reduce enzyme activity which, while retarded, may not be halted. The change of state that occurs during freezing and subsequent thawing can cause damage to the structure of the food with knock-on effects to the texture of the food when consumed. Sub-ambient storage, particularly in conjunction with other preservation techniques (such as controlled atmosphere, or preservatives), can give rise to excellent quality products.

9.4.4.1 Freezing

To appreciate the changes that take place during freezing we need to understand the process of ice crystal formation and the factors that influence it.

Ice crystals either grow on existing inhomogeneities within the food or must be formed through homogeneous nucleation. In pure water without suspended particles present, surface tension acts as a resisting force to the formation of ice crystals. The Laplace equation (see Chapter 8 of this book) shows how very small ice crystals have enormously high pressures exerted on them, so much so that for new ice crystals to form an activation energy must be overcome. In practice this is achieved when the liquid is supercooled – it remains liquid below the equilibrium freezing point and then rapidly solidifies (returning to the freezing point), the activation energy coming from the spontaneous temperature rise to the freezing point. If this is achieved then small ice crystals form throughout the liquid.

Homogeneous nucleation depends on a high degree of supercooling and this requires very rapid cooling rates. Rapid cooling is achieved by a large temperature difference between the food and the freezing medium. If only limited supercooling is achieved, then freezing still occurs, though the ice crystals form on existing inhomogeneities such as cell components (endoplasmic reticulum, cell walls, mitochondria, etc.); as these are not spread throughout the food, the crystals are more isolated from each other.

In both cases, freezing progresses by crystal growth, whereby further ice is formed by deposition of solid water onto existing crystals. Obviously when homogeneous nucleation has occurred the nuclei are numerous and spread throughout the tissues so further crystal growth is also throughout the tissues, whereas when limited supercooling has occurred the number of nuclei are less numerous and fewer larger crystals grow. When large crystals develop they tend to physically disrupt the tissues in which they are growing, rupturing the delicate cell walls. Furthermore as isolated crystals grow they draw water from the surrounding tissues, resulting in dehydration. When such foods thaw, the dehydrated tissues do not rehydrate, the melt water from the large ice crystals is lost as *drip* and the texture of the food is generally inferior to similar products frozen with a large degree of supercooling.

During early stages of freezing the surface freezes and subsequent heat loss is through a frozen layer at the surface. As the latent heat of freezing of water is relatively high the time taken for foods to freeze can be protracted. Factors that influence the freezing time include the actual latent heat λ ($J.kg^{-1}$) of the food material, its density, size and shape, as well as being inversely proportional to the temperature difference between the food and the freezing medium, the thermal conductivity of the frozen food and the surface heat transfer coefficient.

A number of attempts have been made to model freezing time, and an example predictive equation is that developed by Plank which makes a number of assumptions:

- the food is at its freezing point but not frozen (i.e. only loss of latent heat is considered);
- the food freezes under steady state conditions (i.e. the temperature difference during freezing remains constant);
- only one freezing temperature exists and the food has a defined shape (e.g. a sphere).

In Plank's equation the freezing time, t (s) is calculated as follows:

$$t = \frac{\lambda \rho}{\Delta \theta} \left(\frac{0.167x}{h} + \frac{0.042x^2}{k} \right) \qquad (9.37)$$

where

$\Delta \theta$ is the temperature difference between the freezing point of the food and the freezing medium, the constants 0.167 and 0.042 are shape specific for a sphere; other shapes for which constants exist include an infinite cylinder and an infinite slab.

While such predictive equations provide an estimate of freezing time they are limited by the validity of the assumptions they make. In the case of Plank's equation, foods rarely enter freezers at their freezing point and while the specific heat of foods is generally small compared with the latent heat, in practice this results in an underestimate. More of a problem is the idea that there is a single freezing point, for while pure materials have defined temperature–pressure relationships; when solutes are present the freezing point is depressed, yet the composition of the mixture changes as water freezes out of solution, leaving more concentrated solutions which have even lower freezing points.

Technology of freezing

Refrigeration is discussed in depth in Chapter 10 of this book, so this brief introduction is to contextualize the terms used. When a substance evaporates there is an absorption of heat from the surroundings. Such materials are referred to as *primary refrigerant*, and in the case of refrigeration they are normally contained inside a sealed system where they are able to be re-compressed to a liquid. The surface of the equipment where the evaporation occurs becomes cold as heat is taken up from the surroundings. Foods are chilled either by bringing them into direct contact with such a surface or by passing another material such as air over the cold surface and driving that substance over the food – such materials are referred to as *secondary refrigerants*. Foods can be frozen in the following ways:

- In the case of plate freezers the food is separated from the primary refrigerant by a flat metal plate, and heat is lost from the food through the heat conducting plate into the refrigerant. As the food is sandwiched between such plates there is good contact and very low resistances to heat transfer. Plate freezers are very space efficient and can be used to freeze fish into blocks onboard ships; they can also be used to freeze packaged foods. Clearly the geometry of the plates is a limitation on the types of food that are frozen.
- Fluid bed freezers blow cold air vertically upward through a shallow bed of the food. The food particles must be smaller than about 15 mm to become fluidized. The high air velocity strips away the boundary layer, and this and the relatively small size of the particles leads to rapid freezing.
- Blast freezing employs cold air being blown through a cold store which contains the food. The food is not limited by size or shape, though these do affect the freezing time. Lengthy exposure to a cold air blast can result in dehydration of the surface which is referred to as freezer burn. This can be minimized by adding a small amount of water to the surface of the food in the form of a glaze.
- Cryogenic freezing involves bringing the food into direct contact with a material which absorbs heat while changing state. For example, liquid nitrogen can be sprayed over the food as it is passes through a tunnel. The very low temperature of the liquid nitrogen ensures a rapid rate of freezing, resulting in little damage of the frozen food and an excellent quality of the food after thawing. Of course such processes are not without cost and consequently can only be used to freeze high value products.

9.4.4.2 Sub-ambient temperatures

Chilled distribution extends the shelf life of fresh food while not changing the apparent properties of the products. The low temperature retards but does not arrest microbial and enzymic spoilage. Chilled storage is often undertaken in conjunction with modified atmosphere packaging (see Chapter 11 of this book).

While chill storage effectively distributes fresh produce, it should be noted that not all foods are suitable and some tropical fruits such as avocado suffer enzymic damage through chill injury.

9.4.4.3 Hurdle technology

It is recognized that some of the preservation techniques mentioned in this chapter can be detrimental to food quality. Severe processes both kill microbes and reduce organoleptic excellence. Hurdle technology is the synergistic application of several sub-lethal preservation techniques. This might comprise a mild heat treatment followed by chill storage in a modified atmosphere pack which also contains low levels of a food preservative.

9.4.5 Irradiation

Irradiation is potentially a valuable way of preserving foods; however, its use has been impeded by public perception, misunderstanding and some well publicized abuses of the technology.

The depth of penetration of particles and rays from radioactive decay result in only two practical entities to be used:

- β-particles (electrons) can be used to sterilize thin film packaging material, but will not achieve any appreciable penetration into foods.
- γ-rays normally generated from the decay of ^{60}Co can be used to expose foods for sterilization. The cobalt source is housed in a specially constructed maze-type building to prevent the rays (which travel in straight lines) from escaping. A conveyor passes in front of the source carrying the packaged food. As processed and raw foods are almost indistinguishable it is necessary to separate the processed portion of the irradiation plant from the raw materials section.

Once seen as a panacea for preserving foods, in practice irradiation is not suitable for fatty foods as it causes radiolysis, resulting in rancidity. Another problem is the effect it has on many packaging materials – resulting in darkening and opacity. The high cost of the process obviously adds to the product price and limits its use to relatively high value items.

9.5 Food processes and flowcharts

The main aims and purposes of modern food processing are:

- to produce foods that remain wholesome for a satisfactory length of time (shelf life);
- to increase variety in the diet and to make foods consumer friendly (convenient to consume);
- to maintain consistency of the quality of foods (eating and sensory quality and nutritional quality);
- to develop methods/techniques that can facilitate various steps of manufacturing;
- to increase time economy and profitability of the manufacturing company.

To achieve these goals would require great care from every aspect of the manufacturing procedure, from raw material handling to final packaging and distribution.

In developing new products, food technologists would need to investigate various formulations and choices of ingredients. A manufacturing procedure that transforms raw materials into a high quality product also needs to be developed and a flowchart diagram is often used in industry to serve such purposes.

The main purpose of flowchart design is to define unit operations required in the manufacturing of a product and to put these operations in proper order. A flowchart should be comprehensive, informative but also brief. Key technical information (devices, equipment, operation conditions, etc.) should be added to the flowchart either at the side of each unit operation box or as footnotes.

Figure 9.36 is a flowchart for the production of dehydrated milk, where unit operations such as mixing, heating, concentration, size reduction, dehydration and packaging are needed. Key technical information (including principles, operation conditions) is shown to the side of each operation step.

Figure 9.37 is a very simple flowchart for the manufacturing of instant coffee, where only a list of applied unit operations is given. Technical details for such a flowchart would be described in detail as separate notes, as follows:

- *Raw materials and blending*. Green coffee beans from different sources (locations and seasons) are selected. The blend could give a particular taste of

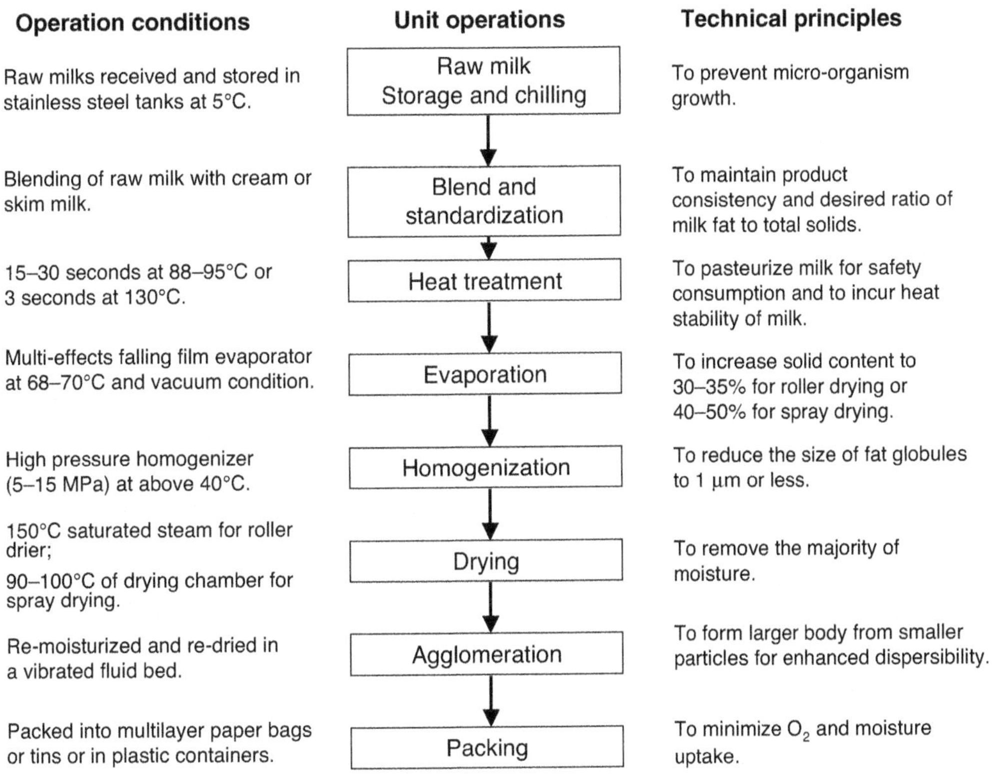

Figure 9.36 Flowchart for the production of dehydrated milk.

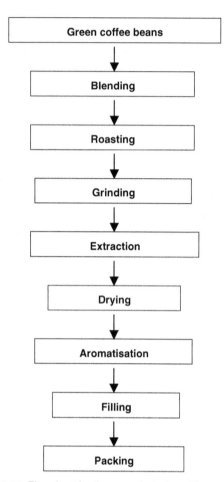

```
┌─────────────────────────┐
│    Green coffee beans    │
└─────────────────────────┘
            │
            ▼
┌─────────────────────────┐
│        Blending          │
└─────────────────────────┘
            │
            ▼
┌─────────────────────────┐
│        Roasting          │
└─────────────────────────┘
            │
            ▼
┌─────────────────────────┐
│        Grinding          │
└─────────────────────────┘
            │
            ▼
┌─────────────────────────┐
│        Extraction        │
└─────────────────────────┘
            │
            ▼
┌─────────────────────────┐
│         Drying           │
└─────────────────────────┘
            │
            ▼
┌─────────────────────────┐
│      Aromatisation       │
└─────────────────────────┘
            │
            ▼
┌─────────────────────────┐
│         Filling          │
└─────────────────────────┘
            │
            ▼
┌─────────────────────────┐
│         Packing          │
└─────────────────────────┘
```

Figure 9.37 Flowchart for the manufacturing of instant coffee.

the brand and also helps to reduce the risk of shortage or price fluctuation from a single raw material supplier. Beans are mixed by weight portion and no special arrangement of mixing is required. The following roasting operation will provide the necessary blending for beans.

- *Roasting*. This is a key operation, in which characteristic flavour and headspace aroma develops. There are roughly two stages of transformation: driving off the 12% of free moisture and pyrolysis with swelling of the beans. The first stage takes up about 80% of the roasting time and green beans gradually change to straw colour then to pale brown. Rapid darkening, accompanied by the emission of oily smoke and crackling sounds, occurs during the second stage of roasting. Chemical composition of the beans also changes rapidly during this stage. Porous microstructure is formed and the density of coffee bean is almost halved after roasting (from ca. 1.3 g.ml^{-1} to ca. 0.7 g.ml^{-1}).

Roasting degree is the key parameter for quality consistency of the final product. External colour or density of roasted beans can be used as a measure of roasting degree. For a well-controlled roasting operation (temperature, hot air speed, etc.), roasting time can be set for predetermined roasting degree. Roasters available for roasting operation include vertical rotating bowl roasters, vertical static drum roasters, horizontal rotating drum roasters, fluidized bed roasters and pressure roasters. Horizontal rotating drum roasters are probably the most popular ones with either a perforated wall or a solid wall.

- *Grinding*. This is a size reduction operation, where roasted coffee beans are ground to small particles. A multi-roller mill is normally used, where coffee beans pass through as many as four stages of size reduction. The gap between rollers decreases with each stage of size reduction.

- *Extraction*. This is a separation operation, where soluble solids and volatile aroma/flavour compounds are extracted from ground coffee granules. Hot water is used as the solvent. A percolation battery extraction device (Fig. 9.22) is a typical example of an extractor. A counter-current continuous screw extractor is another possible choice, where a pressurized water feeding system can be used to enhance extraction efficiency.

- *Drying*. Both spray drying and freeze drying (see section 9.4.3.2) are commonly used for instant coffee manufacturing. Operating at a high temperature, spray drying provides an efficient and economic method for the dehydration of coffee solution. Freeze drying has much better retention of flavour/aroma compounds, but at a relatively higher cost.

- *Aromatisation*. Dried coffee has little or no aroma. Manufacturers usually recover aromatic volatiles by various means during the bean grinding or extraction processes and spray them back onto the product just before the final filling operation. This will provide a coffee-like fragrance when the pack is opened. Coffee oil is usually used as a carrier for these volatiles and it is necessary to fill the packs under a blanket of inert gas such as CO_2 to reduce the risk of oxidation.

- *Filling and packing*. Sealed packing is necessary for volatile compound retention and to prevent moisture pick-up. At 7% moisture content, instant coffee may start to "cake". Glass bottles or metal tin are often used as containers for instant coffee.

Appendix – symbols used

Roman characters

A	Area (m^2)
c	Energy consumption constant of particle size reduction
c_p	Specific heat at constant pressure ($J.kg^{-1}.K^{-1}$)
D	Particle diameter (m)
D value	Decimal reduction time – time (normally in minutes) at a fixed temperature to achieve a 10-fold reduction in the numbers of an organism or spore.
	In the case of D_{121}, the subscript refers to the temperature in °C
d	Internal pipe diameter (m)
e	Surface roughness
E	Energy

      Subscripts:

E_p	Potential energy of a unit fluid ($J.kg^{-1}$)
E_k	Kinetic energy of a unit fluid ($J.kg^{-1}$)
E_r	Pressure energy of a unit fluid ($J.kg^{-1}$)

F	Force (N)
G	Mass flow rate ($kg.s^{-1}$)
g	Acceleration due to gravity ($9.81\ m.s^{-2}$)
h	Surface heat transfer coefficient ($W.m^{-2}.K^{-1}$)
K	Thermal conductivity ($W.m^{-1}.K^{-1}$)
k	Friction factor of pipe fittings
L	Pipe length (m)
m	Mass (kg)
n	Flow behaviour index of power law fluids
n_e	Energy consumption index of particle size reduction
P	Pressure (Pa)
Q	Volumetric flow rate ($m^3.s^{-1}$)
q	Heat flux (W)
R	Specific resistance of filter cake and filter medium (m^{-2})
r_0	Internal radius of pipe (m)
S	Solid fraction of filtrate
t	Time (s)
U	Overall heat transfer coefficient ($W.m^{-2}.K^{-1}$)
V	Volume of filtrate (m^3)
v	Velocity ($m.s^{-1}$)
$\bar{v}$	Mean velocity ($m.s^{-1}$)
x	Distance or thickness of filter (m)
x_c	Thickness of filter cake (m)
z	Height (m)
z value	Temperature range to achieve a 10-fold reduction in D values

Greek characters

Δ	As a prefix denotes a difference, e.g. ΔP being a pressure difference, or $\Delta\theta$ being a temperature difference
ε	Emissivity
$\dot{\gamma}$	Shear rate (s^{-1})
η	Viscosity (Pa.s)
K	Consistency index of power law fluids
λ	Latent heat ($J.kg^{-1}$)
Π	Osmotic pressure (Pa)
θ	Temperature K (or °C)

      Subscripts:

θ_∞	Infinite, i.e. if left for an infinite period
θ_t	At a given time t
θ_i	Initial
$\Delta\theta_{lm}$	Log mean temperature difference

ρ	Density ($kg.m^{-3}$)
σ	Shear stress (Pa)
σ_0	Yield stress (Pa)

Further reading and references

Fellows, P.J. (2000) *Food Processing Technology: Principles and Practice*, 2nd edn. Woodhead, Cambridge.

Lindley, J.A. (1991) Mixing process for agricultural and food materials: 1. Fundamentals of mixing; 2. Highly viscous liquids and cohesive materials; 3. Powders and particulates. *Journal of Agricultural Engineering Research*, **48**, 153–70; **48**, 229–47; **49**, 1–19.

McCabe, W.L., Smith, J.C. and Harriott, P. (2001) *Unit Operations of Chemical Engineering*, 6th edn. McGraw-Hill, Boston.

McHugh, M. and Krukonis, V. (1994) *Supercritical Fluid Extraction: Principles and Practice*, 2nd edn. Butterworth-Heinemann, Boston

Supplementary material is available at www.wiley.com/go/campbellplatt

R. Paul Singh

Key points

- Food engineering involves a detailed study of numerous unit operations and a fundamental understanding of momentum, heat and mass transfer relevant to food processing.
- This chapter introduces concepts relevant to the hygienic design of equipment used for processing, handling and storage of foods, application of common process control systems, and approaches used in handling wastewater generated in a food processing plant.
- These topics are presented in sufficient detail to gain an appreciation of the important role of engineering in selected aspects of food processing.

10.1 Engineering aspects of hygienic design and operation

In a food processing plant, raw food is converted and processed into desired products using a variety of equipment. In designing food processing equipment for any given purpose, an engineer must consider numerous criteria that are inherent to a process. For example, in designing a heat exchanger, one must consider the heat transfer, fluid flow, and various physical, chemical, and biological changes occurring in a food. Furthermore, an underlying criterion in designing food processing equipment is the sanitary design. Each equipment and product in a food processing plant must adhere to some unique requirements to ensure sanitary operation. There are several general sanitary requirements that are common to most equipment design. In this section, we will consider many of these considerations that a food engineer must carefully address whenever designing food processing equipment. More details on topics presented in this section are available in Jowitt (1980) and Ogrydziak (2004).

10.1.1 Food process equipment design

Hygienic design of equipment is essential in a modern food processing plant. A major concern in processing foods is to prevent microbial contamination that may be facilitated in poorly designed equipment. Such equipment is difficult to clean or may require longer cleaning times and more use of chemicals. The key principles of hygienic design are as follows:

- Use materials of construction that are suitable for hygienic processing of food.
- Product contact surfaces must be easily accessible for inspection and cleaning.
- Incorporate design features that prevent harboring microbial accumulation and their growth.

To ensure clean food processing equipment, the product contact surface has an important role. If the surface is rough and/or porous it will most likely allow food particles to build up and it will be more difficult to clean when compared with a smooth, polished surface. The contact surface must not chemically interact with the food and its ingredients. It should be free of corrosion and it must be inert to the chemicals used during cleaning.

If the contact surface is hidden during visual inspection, it would be impossible to know whether it is properly cleaned. Therefore, all contact surfaces must be accessible during inspection and cleaning. In some cases, it may require complete disassembly; in other circumstances ports for inspection may be strategically located. Often, access doors are provided to inspect the contact surface. The door fasteners should be easy to open without the use of tools; quick release type fasteners are preferred. Many of the small equipment such as pumps should be installed 15 cm or more from the floor, whereas larger pieces of equipment should be elevated 30 cm from the floor so that the floor areas under them are easily cleaned. Sealing process equipment to the floor should be avoided as sealants (such as caulking) crack over time.

The noncontact surfaces should be designed to prevent accumulation of any solid materials. These surfaces should prevent any absorption of liquids or water.

Motors used to power equipment should be placed where any lubricant used in the motor does not contaminate the product. Direct drive systems are generally preferable. While drip pans may be used, they should be avoided as much as possible by using a direct drive system. Another source of contamination in drive systems is bearings. Use of food grade material such as nylon, sealed or self-lubricating bearings are preferable. Additionally, seals should be nontoxic and nonabsorbent. It should be easy to remove seals for inspection and sanitation purposes. Hoods used to exhaust steam or dust collection should be easy to clean. For processing liquid foods, kettles should be of the self-draining type. For kettles equipped with mixers, the mixer lubricant should not enter the product.

If steam is to be directly injected, then any additives used in the boiling water must be food grade. Compressed air coming into contact with food or food contact surfaces must be free of dust, pollens, and lubricant oils. Many lubricants used in compressors are toxic. Filters placed on the discharge may be necessary to prevent any dust or pollen. Desiccants and filters are useful in removing undesirable materials.

10.1.2 Construction materials

Numerous materials are available to fabricate food processing equipment. However, each material has its advantages and limitations, and these should be carefully assessed prior to their selection. For product contact surfaces, the interaction between the product and contact surface must be carefully evaluated.

10.1.2.1 Stainless steel

This is the most common material used for fabricating food processing equipment. Stainless steel is an alloy of iron and chromium. When chromium is added to iron in excess of 10%, it imparts resistance to corrosion. Additional elements are added for specific purposes. In reference to food processing equipment, an 18-8 grade (18% chromium and 8% nickel) is ideal in the fabrication of processing equipment. Within the 18-8 grade, there are different types that impart special properties:

- Type 302 is used for outside surfaces mainly for appearance.
- Type 303 has additives such as S and Se. It is mostly used for fabricating shafts, and castings. Its precipitates are harmful.
- Type 304 is less susceptible to corrosion. It is used in tubing and in situations where mild corrosion is anticipated.
- Type 316 is highly heat resistant and it has superior corrosion resistance. When severe corrosion is anticipated, type 316 is most suitable. When processing equipment is to be used for high temperature processing, then type 316 is more durable.

There are different types of finishes available for stainless steel. A flat finish, referred to as 2 B finish, is available as standard from steel mills, whereas numbers 6 to 8 refer to highly polished finish.

10.1.2.2 Titanium

Titanium is a light weight (approximately 44% less than stainless steel) but very strong metal. It is resistant to corrosion.

10.1.2.3 Inconel

Inconel, an Ni-Cr alloy (containing 77% nickel and 18% chromium), is more ductile than steel and it is resistant to corrosion.

10.1.2.4 Mild steel/iron

This is used for non-contact surfaces or products such as dry ingredients and syrups. Steel and iron corrode in such applications.

10.1.2.5 Aluminum

Aluminum reacts with hydrochloric acid and caustic solution. It is a soft material and easily subject to gouging and scratching. Aluminum is used for certain butter and dry product applications. It should not be used when cleaning requires strong caustic solutions or corrosive action of dissimilar metals.

10.1.2.6 Brass/copper/bronze

These may corrode when cleaning chemicals are present. Also, they may impart undesirable taste and flavor to the food, which may be problematic. As a result, brass or bronze are not acceptable for product contact surfaces or surfaces that may come into contact with cleaning solutions. However, in nonfood contact areas they are acceptable.

10.1.2.7 Plating materials

The use of plated metals should be carefully evaluated. For example, galvanized iron (iron coated with zinc) is unsuitable for juice, as the fruit acids can dissolve zinc. However, for framework applications it may be quite suitable.

10.1.2.8 Tin

Tin is extremely resistant to corrosion, but it is soft and easily scratched.

10.1.2.9 Cadmium

Cadmium is toxic and any cadmium-plated surfaces including fasteners should not be used.

10.1.2.10 Glass

Glass should be avoided and replaced with polymeric materials. Glass is used only when there is a demonstrated functional need. In that case clear, heat-resistant and shatter-resistant type of glass is used.

10.1.2.11 Wood

Wood should be avoided as splinters and slivers can cause problems.

10.1.2.12 Wire

Wire should be of materials that have magnetic properties so that metal detectors can effectively remove it along the processing line.

10.1.3 Construction features

During equipment fabrication, several construction features require careful attention to avoid locations for insect infestation or product soiling that is difficult to remove during cleaning.

- Lap seam: insects may harbor in tiny cracks, crevices, and where one piece of metal is spot welded onto another. Weld material should be ground. Metal ends should butt together.
- Ledges: contact zone ledges must be avoided where product may accumulate.
- Void areas: spaces that are difficult to access for cleaning, where insects may harbor, must be eliminated and sealed.
- Dead ends: dead ends in pipes and screw conveyors may trap product and therefore they must be avoided.
- Rolled edges: rolled edges are necessary to strengthen the edge of sheet metal. If the rolled edges are not properly sealed then they may harbor bacteria.
- Rounded corners: corners should be rounded to be easily cleaned.
- Coves: corner welds should be ground smooth.
- Seams: it is preferable to have continuously welded joints to avoid seams.
- Cracks: continuous welding should be used. Any caulking may be used in nonproduct contact surface areas.
- Frames: use of tubular shapes is preferred to avoid excessive dust collection. Horizontal framing members should be at least 12 inches from the ground.

- Product zone welds: these should be continuous. For milk, and egg processing equipment, a ground flush finish is necessary.
- Caulking materials: silicone for sealing crevices or exterior nonproduct contact surface only may be used. Caulking is not acceptable for product contact surface.
- Paint: only nonproduct contact surfaces may be painted. Those parts that have both product contact and nonproduct contact are subject to washing and should not be painted.
- Lubricants: where a lubricant may come into contact with food, only those given in 21 CFR Part 178-375 (US FDA) are permitted. For example, a light coat of mineral oil on rolls may be used to prevent the sticking of cheese.
- Finish: smoother finish results in easier cleaning of the surface. The product contact surfaces are milled or polished to a high degree of smoothness that prevents microbial adherence. The most recommended surface finish is number 4. A number 4 finish has a maximum Ra of 32 micro-inches or 0.8 microns. Ra number is the average height of roughness expressed in microns or micro-inches. The welded junctions are also ground and polished to number 4 finish. A 150 grit silicon carbide, when properly applied to stainless steel, is equivalent to number 4 finish.
- Gaskets: for junctions containing gaskets, there should be no tight recesses or protruding unsupported gasket material that may harbor microorganisms.
- Fasteners: wing nuts, "T" nuts, or Palm nuts are preferable over hex or dome nuts. The fasteners must facilitate easy cleaning and dismantling. Exceptions are made where vacuum, pressure, or safety issues are involved.

10.2 Cleaning and sanitizing

Cleaning and sanitizing in a food processing plant involves a number of steps. The first step is to remove any gross soil. Then a chemical agent is used to remove any visible soil residues. Next, a rinse of a cleaning agent is used. The rinse is followed by the use of a sanitizer that assists in killing, removal, or inhibiting any microorganisms. If necessary, a final rinse cycle may be used to remove the sanitizer.

Cleaning is influenced by temperature, time of cleaning, concentration of chemicals, and the mechanical action used in cleaning. Use of higher temperatures during cleaning of fat and grease is beneficial; however, it should not be excessively high to cause protein adhesion to a surface.

The various types of soils from food are shown in Table 10.1 based on their solubility in water. Many properties of soil are important in determining the ease or difficulty in removing it, for example, particle size, viscosity, surface tension, wettability, solubility of a liquid soil in a solid soil, the chemical reactivity with the substrate, the attachment of soil to a surface or entrapped in voids, and any forces such as cohesion, wetting, or chemical bonds that influence the attachment.

The selection of a cleaning compound depends upon:

- the type of soil on the surface;
- the type of surface to be cleaned;
- the amount of soil on the surface;
- the method of cleaning (such as soaking, use of a foam, or clean-in-place);
- the type of cleaning agent – liquid or powder;
- the quality of the water;

Table 10.1 Various types of soil from foods and the detergents used to remove them (adapted from Katsuyama, 1993).

	Type of soil (from food)	Detergents used
Water soluble	Sugars, salt, organic acids	Alkaline (mild)
	High protein foods (meat, fish and poultry)	Chlorinated alkaline
Partly water soluble	Starchy foods, tomatoes, fruits and vegetables	Alkaline (mild)
Water insoluble	Fatty foods (fatty meats, butter, margarine, oils)	Mild or strong alkaline
	Stone forming foods: mineral scale from milk, beer and spinach	Chlorinated or mildly alkaline, alternated with acid cleaner each 5th day
	Heat-precipitated water hardness	Acid

Table 10.2 Various types of detergents used in cleaning food processing equipment (Ogrydziak, 2004).

Detergents	Advantages	Disadvantages
Water	Water effectively dissolves sugars and salt Use of high pressure (600–1200 psi) water is effective in removal of many soluble and insoluble solids	Limited use in cleaning
Alkaline	In presence of fats, it produces soap In presence of denatured proteins, it produces soluble peptides	Corrodes aluminum, galvanized metal and tin Rinses poorly Causes precipitates to form in hard water
Strong alkalis (NaOH)	Strongest detergent Low cost Good germicidal value	Very corrosive to nearly all surfaces including metal, glass and skin Rinses poorly No buffering capacity Poor deflocculating and emulsifying power
Mild alkalis (carbonates, borates, silicates, phosphates)	Moderate dissolving power Less corrosive than strong alkalis	
Soaps (Na$^+$ or K$^+$ of fatty acids)	Effective in washing hands in soft water	Not very soluble in cold water Limited use as cleaner in food plants
Acids	Dissolve mineral deposits, hard water stone, beer stone, milk stone and calcium oxalate	Not effective against fats, oils and proteins
Inorganic acids (hydrochloric, sulfuric, nitric, phosphoric acids)	Some acids are sequestering agents Used at pH 2.5 or lower (0.5% acid) Used with corrosion inhibitors Phosphoric acid is used for hard water films on tile	Hydrogen ion is corrosive to metals especially stainless steel and galvanized iron
Organic acids (acetic, lactic, citric)	Not as corrosive as inorganic acids Causes less irritation of skin Citric, tartaric and gluconic acids have chelating properties Used with corrosive inhibitors	

- the amount of time available for the cleaning cycle;
- the cost of the compound.

Cleaning involves first the separation of the soil from the surface, followed by dispersion of the soil in the detergent medium. It is important that the soil must not redeposit on the surface.

The effectiveness of a detergent is evaluated based on its:

- penetration and wetting ability;
- control of water hardness;
- efficient removal of soil;
- ease of rinsing;
- noncorrosiveness to the surface.

The desired properties of detergents are obtained by proper mixing of selected chemicals. Various types of detergents and their key characteristics are noted in Table 10.2.

10.2.1 Sanitizer

A sanitizer used in the food industry must produce a 99.999% (or 5 log) reduction in populations of 75–125 million *Escherichia coli* and *Staphylococcus aureus* within 30 s at 20°C (70°F). The purpose of using a sanitizer is to destroy pathogens or other organisms on a clean surface. Furthermore, the sanitizer should not adversely affect the equipment or the health of the consumer.

Sanitizers may be classified as physical and chemical. Some of the commonly used physical

Table 10.3 Physical sanitizers used in food processing equipment (Ogrydziak, 2004).

Physical sanitizers	Typical treatment	Comments
Steam	15 min at $>=$ 170°F or 5 min at $>=$ 200°F	
Hot water	Immersion for 5 min at 170°F	In practice, use $>$ 180°F for $>$ 15 min for large pieces of equipment
Hot air	$>$ 356°F for $>$ 20 min in hot-air cabinet	Measure temperature at the coldest zone
UV light	Limited to translucent fluid streams	Used for treating bottled water

sanitizers used in food processing are shown in Table 10.3.

A large number of chemical sanitizers have been developed specifically for the food industry. Some of the more common chemical sanitizers are as follows:

- Hypochlorites: one of the most widely used sanitizer in the food industry is the liquid form of sodium hypochlorite. Microbes are destroyed by hypochlorous acid (HOCl).

- Chlorine gas: when chlorine gas is injected in water, hypochlorous acid is formed. Since the solubility of gases decreases with increasing temperature, its effectiveness in high temperature applications should be carefully examined.

- Chlorine dioxide: chlorine dioxide gas is widely used in treating process water for fruits and vegetables at concentrations up to 1 ppm. It is more effective than hypochlorous acid under alkaline conditions up to about pH 10. Often, chlorine dioxide is produced on site and it is an expensive method of sanitizing.

- Organic chlorides: organic chlorides form hypochlorous acid at a slow rate. Their rate of microbial kill is also slow.

- Iodophors: an iodophor contains iodine and a surfactant that acts as a solubilizing agent. When mixed with water, iodophor releases free iodine at a slow rate. Often, iodophors are blended with phosphoric or citric acid to ensure optimal pH (4.0–4.5). Iodophors are effective sanitizers for a large range of microorganisms but less effective against spores and bacteriophages. They may stain plastic and uncleaned stainless steel surfaces. In diluted form they are nontoxic.

- Acid–anionic surfactant compounds: these compounds include anionic surfactants and acids such as phosphoric or citric. They are stable at high temperature but are ineffective at pH above 3.5. In dairy processing equipment, they are effective in controlling milkstone (a carbonate formed during the processing of dairy-based products), and they are noncorrosive to stainless steel.

- Fatty acid–anionic surfactant compounds: fatty acids such as octanoic and decanoic are used along with surfactants. They are notable for causing reduced foam.

- Peroxyacetic acid sanitizers: these include an equilibrium mixture of hydrogen peroxide, acetic acid, and peroxyacetic acid. They break down into oxygen, water, and acetic acid. They are effective against a broad range of microorganisms including spores, viruses, and molds. They are effective at cold temperature.

- Quaternary ammonium sanitizers: also referred to as "quats," these sanitizers involve a cationic surfactant molecule combined with a chlorine anion. They are stable at high temperatures and effective over a broad range of pH and in the presence of organic matter. They are not compatible with chlorine sanitizers. If the concentration of quats is less than 200 ppm then no water rinse is required after treatment.

- Sequestering agents: these compounds form soluble complexes with metal ions. Their primary function is to prevent film formation on equipment and utensils. Common chemicals used for this purpose include tetrasodium pyrophosphate, sodium tripolyphosphate, and sodium hexametaphosphate. Phosphates are unstable in acid solutions.

- Wetting agents: wetting agents are used to wet surfaces. They can penetrate crevices and woven fabrics. Anionic wetting agents act as emulsifiers for oils, fats, waxes, and pigments. Examples are soaps, sulphonated amides, and alkyl-aryl sulphonates. Some of these agents foam excessively. Nonionic wetting agents such as ethylene oxide-fatty acid condensates are excellent detergents for oil. They may be sensitive to acids.

Table 10.4 Cleaning related items in and around a food processing plant (Ogrydziak, 2004).

Zone	Specific items	Items that require attention during cleaning
Outside the factory	Grounds	Weeds are harbors of pests such as rodents
	Parking areas	Dirt and weeds introduce contamination into plant
	Waste disposal	Odor, food source for pests, rodents
Receiving area	Containers	Any residual food product that will attract pests
	Silos, tanks, bins	Any residual organic matter and encrusted product that may serve as food for pests
	Floors, gutters, walls	Any residual raw material, cracks, or crevices
	Construction materials	Peeling paint, rust, corroded parts
	Loading docks	Dirt, filth, broken wood, plastic, splinters
	Freezers and coolers	Dirty drains, cracks in walls
Preparation	Flumes	Any product residue, biofilms
	Belts, conveyors, elevators	Any food residues or organics
	Washers	Grime and residual food, dirt, organics
	Peelers	Product residues, grime, organics
	Slicers	Product residues, fats, oil and grease
	Floor, gutters, walkways	Dirt, crevices, cracks
	Insect and rodent control	Cracks, access areas, gaps
Processing	Conveyors	Spaces between interlocking belts, underneath belts
	Tanks and pipes	Welds, CIP equipment maintenance
	Fryers	Oil filters, deposits in steam hoods
	Floors, drains, gutters, walkways	Bacterial infestation
	Hoods, filters, screens	Dirt
Packaging	Conveyors, packaging machinery	Dust, dirt
	Filters	Dust
	Tubing, piping, pumps	Surface dirt
Warehousing	Pallets	Rodent droppings, insects, splinters
	Floors and walls	Rodent droppings, spilled product
	Docks	Dirt, spilled products
	Trucks	Maggots under trailer floor, dirt

Cationic wetting agents such as quaternary ammonium compounds have an antibacterial effect. They are not compatible with anionic wetting agents.

In a food processing plant, cleaning extends both indoor and outdoors. Different areas inside the plant and surrounding areas require special considerations to keep them clean. Table 10.4 describes various items that should be considered in maintaining a clean environment in food processing plants.

Since food plants require frequent cleaning, dismantling equipment can be time consuming. To avoid long shutdown periods, it has become common to use the clean-in-place (CIP) technique. In a CIP system, pipework is cleaned by pumping water with appropriate detergent in a turbulent mode. Chambers, tanks, and other larger vessels are cleaned by installing spray balls or rotating jets. These devices ensure that every nook and corner inside the vessels is cleaned. After cleaning with detergent, sanitizers are used to disinfect the interior surfaces. With appropriate process controls, the CIP system can be operated in a completely automated mode after each shift.

10.3 Process controls

Process control may be defined as manipulation of process variables so that the desired product attributes are obtained. The variables employed in a process have a marked influence on the final attributes of the product. Therefore, an appropriate control of these variables is an important objective in food processing operations.

Advances made in computer technology since the 1970s have made it possible to automate process operations. By controlling process variables, the desired consistency of operation is achieved along with reduced costs of production and improvements in safety. When equipment begins to deviate from what it is designed for, automatic controls provide a higher level of consistency than if human intervention is allowed that often leads to higher levels of variability. The production productivity is enhanced as less out-of-specification product is produced. Automatic controls provide a higher level of screening of unsafe conditions to improve overall safety of the equipment.

10.3.1 A feedback process model

Consider a manual temperature control installed on juice being pumped through a steam heat exchanger. The temperature of the juice is the control parameter. The measurement device is a thermometer used to measure temperature. An operator decides whether the temperature is too hot or too cold. A steam valve is used to make adjustments. If the juice temperature is too high, then the operator adjusts the valve towards the closed position. This is an example of a negative feedback control, because a positive error requires a negative response from the operator. Initially the operator adjustment may be too large. When the desired set point is approached, the operator is able to make finer adjustments.

A simple feedback process model is shown in Fig. 10.1. The process variable is measured and compared with a set point, and this generates an error signal. For the given error signal, an algorithm is used to determine the type of control response. The control response manipulates the control element. Thus the control variable is modified and the loop is repeated. With decreasing error, the control response becomes smaller.

As shown in Fig. 10.1, the information flows through various elements of the control loop. The key elements of a control loop are described below.

10.3.1.1 Transducer

A transducer is a sensing element that detects the process variable. It converts the signal into some measurable quantity. Often the measurable quantity is an electrical signal. For example, a thermocouple receives information about temperature and converts it into a millivolt signal.

The output signal of a transducer may or may not be convenient to transmit for long distance. The modern control systems can accommodate a variety of signals, such as millivolt, frequency, and variation in current. The output signal may or may not be linear with respect to the measured quantity.

10.3.1.2 Transmitter

Transmitters help in converting the measured variable into a standardized signal. Often the variable is linearized to the measured signal. Typical output of

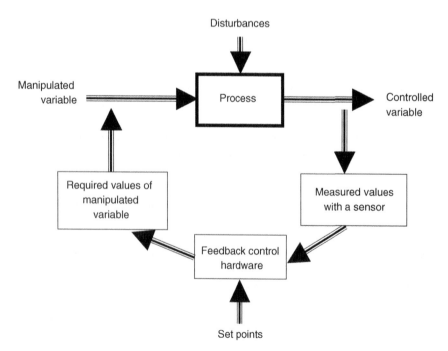

Figure 10.1 A feedback control system.

Table 10.5 Common sensors used in food processing operations.

Sensing parameter	Sensor	Range of application
Temperature	Thermocouple	
	Type J	−320 to 1400°F
	Type T	−310 to 750°F
	Type K	−310 to 2500°F
	Resistance temperature detector (RTD)	430–1200°F
Volumetric flow	Magnetic	Down to 0.01 gal/min
	Vortex-shedding	Down to 0.1 lb/min
Mass flow	Coriolis	Down to 0.1 lb/min
	Heat loss	Down to 0.5 cc/min
Density	Vibration	Down to 0.2 g/cc
	Nuclear	Down to 0.1 g/cc
Pressure	Strain gage	Down to 2 psi
Level	Differential pressure	Down to < 1 inch of water column
	Capacitance	Point level to > 20 ft
	RF impedence	Point level to > 20 ft
	Ultrasonic	Several inches to 100 ft
Moisture	Infrared	1–100%
	Microwave	0 to > 35%
Viscosity	Vibration	0.1–106 cP

a transmitter is 4–20 mA. Often a power source of around 24 V is used as a direct current source, and any other voltage generated due to electrical noise is usually not a problem as only the change in the current is measured. Within limits of the power source, devices may be driven by including them within the 4- to 20-mA loop.

10.3.1.3 Controller

A controller reads the transmitted signal and relates it to the set point. The controllers are able to handle a variety of electrical signals such as current, voltage, or frequency. In special situations where electrical circuits may cause hazardous conditions such as explosion, pneumatic controllers are used.

Digital controllers are used to convert analog to digital signals. The digital signal is read by a digital computer that processes the data and calculates the deviation of the transmitted signal from the set point. The digital controller then converts the digital signal received from the computer into an analog signal in the form of 4–20 mA; however, other output signals are also possible. The output signal is used to adjust the control element. In case of a current to pneumatic converter, the 4- to 20-mA signal is converted into a 3- to 15-psig output signal. In addition to controlling valves, other food processing devices may also be controlled such as a variable speed motor used to drive a pump.

10.3.1.4 Sensors

There are a wide variety of sensors used in the industry, as shown in Table 10.5. Several considerations are necessary when seeking a sensor for food processing application, including type of material contacting with food, range, accuracy, and cost.

In the case of CIP applications, it is important to make sure that no dead volumes are created around the sensor. Furthermore, sensor housing may require wash down, and therefore specifications call for an appropriate National Electrical Manufacturers Association (NEMA) rating of 3 (weatherproof), 4 (watertight), or 5 (dust-tight).

10.3.2 Process dynamics

Whenever a process is to be adjusted, certain factors may cause a delaying time before the system responds. These factors causing a delay in response time include the inertia, lag, or dead time.

Inertia is often associated with mechanical systems such as those involving fluid control. In case of liquids, their incompressible nature minimizes the delay due to inertia.

Lag is quite common in many applications. For example, consider a liquid food being heated in a steam-jacketed vessel. After the steam is turned on, it takes time before the product begins to heat due to inherent resistances of the vessel and the product.

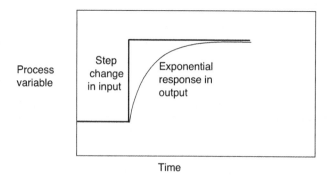

Figure 10.2 Exponential response to a step change in input.

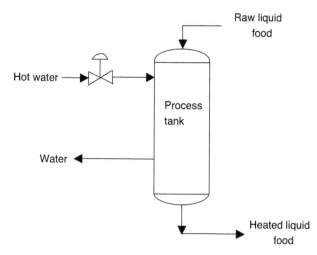

Figure 10.3 A process tank to heat liquid food using hot water.

Typically, a first order equation is used to describe the lag, as shown in the following equation:

$$\tau \frac{dy}{dt} + y = Kx \qquad (10.1)$$

where

y is the output as a function of time,
x is the input as a function of time,
τ is the system time constant,
K is a constant.

First order lag is the most common type of response in process control. Figure 10.2 shows how a process responds when there is a sudden change in the input. The response curve is exponential; it approaches the new steady state value in an asymptotic manner. The response behavior of such a system is characterized by calculating the time constant. After one time constant the system responds to 63.2% of the step change in the input.

Dead time is associated with how the equipment is designed. For example, in measuring the temperature of a liquid in a tank, if the liquid is being pumped to a pipe where the temperature detector is placed then it will take time before the liquid reaches the temperature detector and this will result in a delay.

10.3.3 Modes of process control

To understand different modes of process control, consider a vessel containing a submerged heating coil (Fig. 10.3). A liquid food enters the vessel from the top and exits at the bottom. To heat the food, hot water circulates through the heating coil. A valve is installed on the pipe feeding hot water into the coil. A temperature sensor is used to measure the tempera-

ture of the food inside the vessel. It is desired that the temperature of the food inside the vessel is maintained at some constant temperature (for this example, let us assume 50°C). We will consider different modes of control that may be used to achieve this objective.

10.3.3.1 On/off control

The on/off control is perhaps the simplest method of control. For our example, to maintain a juice temperature of 50°C, an operator would observe the temperature sensor; if the temperature falls below the set point of 50°C, the valve will be fully opened (on) to allow hot water to circulate through the coil. When the temperature of the juice goes above the set point of 50°C, the valve is fully closed (off).

The on/off control may be expressed mathematically as follows:

$$e = P_v - S_p \qquad (10.2)$$

where

e is the error,
P_v is the process variable,
S_p is the set point.

This control algorithm responds to a change in the sign of the error by turning the system off or on. Although on/off control will be able to maintain an average temperature, there could be large variations in

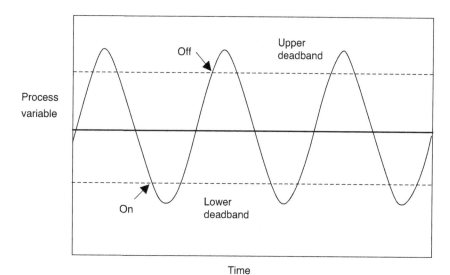

Figure 10.4 On/off control with dead bands. (Adapted from Bresnahan, 1997.)

temperature around the set point. In a batch heater, some reduction in variation in temperature may be realized by good agitation.

To prevent too quick oscillations, dead bands are used as shown in Fig. 10.4. The system is turned on prior to reaching the lower dead band and then turned off when it reaches the upper dead band.

10.3.3.2 Proportional control

In our example, if we conduct a simple energy balance on our system, we will find that there is a certain ideal steady flow rate of hot water that will maintain the juice temperature at 50°C. However, this ideal flow rate of water will be different for different flow rates of juice in and out of the vessel. Thus, to control the process we need to accomplish two tasks: (1) determine a flow rate of hot water that will maintain the juice at 50°C for some normal flow rate of juice in and out of the vessel, and (2) any increase or decrease in the error (difference in set point and temperature of the juice) must be allowed to cause a corresponding change in the flow rate of the hot water. This is the basis of proportional control. The proportional control may be thought in terms of the gain. Mathematically, in proportional control algorithm, the following calculations are used to determine the output based on the error between the set point and the measured variable:

$$C_o = Ge + m \qquad (10.3)$$

where

C_o is the controller output (such as position of a control valve),
G is the proportional gain,
e is the error,
m is the controller bias.

This equation suggests that there is a direct relationship between the error and the controller output (or position of the controller valve). For our example, the valve used to control hot water must be adjustable (such as an electrically or pneumatically operated diaphragm actuator). In the case of a reverse acting controller, if there is a larger positive error then the output will decrease, and vice versa for direct acting controllers.

In many industrial controllers, the gain adjusting mechanism is expressed in terms of the proportionality band. The proportionality band represents a percent change in the output based on the change in the input. The following equation is used to calculate the proportional band:

$$B_p = \frac{100}{G} \qquad (10.4)$$

where

B_p is the proportional band,
G is the proportional gain.

Thus, a gain of 1.0 corresponds to a proportional band of 100%, whereas a gain of 0.5 corresponds to a proportional band of 200%.

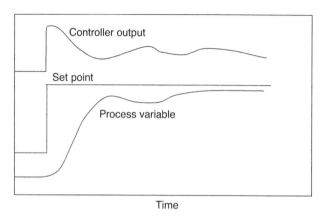

Figure 10.5 Process variable, set point and controller output in a proportional control.

Proportional-only control can rarely keep the process variable at the set point. As shown in Fig. 10.5 with a step change in the error, the controller output also experiences a step change. Due to this step change in the controller, the process variable begins to respond. This causes a decrease in the error, which results in a decrease in the controller output. After some time the error decreases and there is no change in the error. Similarly, the controller output also does not change because it is a product of the gain and change in error. No change in error results in no change in the output. This means that there will be a constant error or offset. In order to minimize it, the gain, G, may be made large. However, this can also cause excessive oscillations when there is a response lag in the system. In some applications, such as measurement of pressure, the control variable has a very fast response, and a high gain proportional controller is well suited. In case of flow measurements, due to considerable noise, these controllers are not well suited as they give erroneous response to the noise as well as to the real signal.

10.3.3.3 Proportional-plus-integral control

To eliminate the offset error, noted in the proportional control, one strategy would be to adjust the proportional controller. The manual reset may be automated by moving the valve at a rate that is proportional to the error. This means that if the deviation or the error is doubled, then the control element will move twice as fast to respond. On the other hand, if there is no deviation, the error is zero, and the control element remains stationary.

The integral action (or reset) is often combined with the proportional control and called proportional integral (PI) control.

Integral controls involve the use of a cumulative error in determining the controller output along with the instantaneous action of the proportional component of the controller. Thus only the bias term for each set point needs to be adjusted.

The following equation applies to the proportional and integral control:

$$C_o = Ge + \frac{G}{t_i} \int edt \qquad (10.5)$$

where t_i is the reset time tuning parameter.

The adjustable parameter for the integral mode is t_i or the reset time (its units are time per repeat, e.g. minutes per repeat). In some controllers, $1/t_i$ is used, which is called repeats per unit of time. When using the preceding equation, the terms and units should be carefully checked.

The significance of the reset time as used in the above equations may be understood as follows. For a given error, the reset time is the time for the integral action to provide a change in output signal equal to that provided by the proportional control mode. Consider an example when $t_i = 1$ minute; if the error at time zero undergoes a step change from 0 to 1, the output from the preceding equation will have an instantaneous magnitude of G for the first term of the right-hand side of the equation. After 1 minute if the error stays constant at 1, then the output will equal 2 G (contributions from both proportional and integral parts of the equation). After another minute, another G will be added and so on. This will continue until either the error disappears or the controller reaches saturation point of either 0 or 100%. The advantage of the PI control is the elimination of offset. However, some instability may be present due to the integral component.

10.3.3.4 Derivative control

The derivative of a controller adds to the controller output a term in proportion to the rate of change in error with respect to time. On a theoretical basis, one may consider a controller that is based only on the rate of change of the error, but in practical situations it will mean that in cases where there is a large but constant error there will be zero controller output. Therefore, one needs to incorporate proportional control to a derivative control. An equation for a proportional, integral, and derivative control (PID) may be written as

$$C_o = Ge + \frac{G}{t_i} \int edt + Gt_d \frac{de}{dt} \qquad (10.6)$$

where t_d is the derivative time tuning parameter.

An additional corrective action for the PID controller is obtained by determining the slope of the error vs. the time curve and multiplying with the derivative tuning parameter. Another approach to determining the derivative term is to modify the process variable by using the slope of its change with time to predict a new value in future. The predicted value is then used in error calculation instead of the actual process variable (Bresnahan, 1997).

The derivative action in PID controllers circumvents the problems encountered by other algorithms with significant lag or large variation from set points. Consider a case of heating a viscous liquid where there is slow dynamic response. If a controller does not contain derivative action, then error will change signs and there will be considerable overshoot past the process set point; this may result in excessive burning of the product or fouling of the heat transfer surface. A PID controller reduces the oscillations. Normally, a derivative control is not employed when the system response is too fast.

A more detailed discussion on process controls used in the food industry is available in Bresnahan (1997), Murrill (2000), and Hughes (2002).

10.4 Storage vessels

10.4.1 Tanks for liquid foods

In processing liquid foods, tanks are essential for short- and long-term storage. Consider milk processing in a dairy plant. In a modern dairy plant, the size of tanks may vary from 110 to 150,000 liters. The design of a tank must adhere to the specific requirements of the process and product being handled.

Raw milk received at a dairy plant is stored in large vertical tanks with capacities varying from 25,000 to 150,000 liters. Larger tanks, generally located outdoors, are usually double-wall construction, with the interior wall made of stainless steel and the outside wall of welded sheet metal. In between the walls,

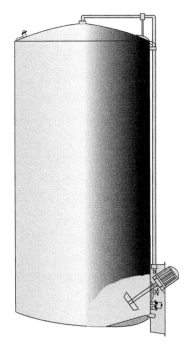

Figure 10.6 A tank with a propeller agitator. (Source: *Dairy Processing Handbook*, Tetra Pak.)

a minimum of 70 mm of mineral-wool insulation is used. In raw milk tanks, gentle agitation, using a propeller agitator, is provided to prevent gravity separation of cream (Fig. 10.6). Level indicators are used to provide low-level protection to ensure that the agitator is submerged before it is turned on, and overflow protection to prevent overfilling. An empty tank indication is used to ensure that the tank is completely empty before the rinse cycle. In modern facilities, the data obtained from these indicators are transmitted directly to a central location.

The bottom of the tank slopes downwards towards the outlet with an inclination of about 6% to provide easy drainage (Fig. 10.7). Appropriate sanitary connections and vents are used to prevent back-pressure buildup during filling and vacuum during emptying.

After milk is heat treated, it is often stored in intermediate storage tanks which are insulated to maintain a constant temperature. In these tanks, both inner and outer walls are made of stainless steel and the space between is filled with mineral wool to provide insulation. These intermediate tanks are also used as a buffer storage; typically a buffer capacity of a maximum of 1.5 hours of normal operation is used (Fig. 10.8).

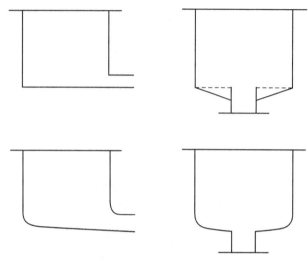

Figure 10.7 Different designs of tank floors with slope to aid drainage.

In addition to storage tanks, a dairy processing plant may also use several process tanks where milk or other dairy products are processed: for example, tanks for cultured products such as yogurt, ripening tanks for butter cream, and tanks for preparing starter cultures for fermented products.

When designing a transport system for liquid foods such as milk, certain potential problems must be considered. For example, the product being pumped must be free of air for a centrifugal pump to work properly, pressure at all points of the inlet must be higher than the vapor pressure of the liquid to prevent cavitation, there should be provision for redirecting the flow if the process has been inadequate, and the suction pressure at the pump must remain constant for uniform flow. In order to avoid these types of problems, balance tanks are located along the suction side of the pump (Fig. 10.9). The liquid level in a balance tank is always maintained at a certain minimum level, using a float, to provide a constant head on the suction side of the pump.

10.5 Handling solid foods in a processing plant

The design and layout of the plant and how materials are handled between various processing equipment have a major impact on the production efficiency. In the case of transporting solid foods in a food processing plant, it is typical to consider the movement of product in any direction, horizontal or vertical. A variety of conveyors and elevators are used for this purpose. These include belt, chain, screw, gravity, and pneumatic conveyors, and bucket elevators. In some cases fork-lifts and cranes are employed. Some salient features of these different conveyors are described below.

10.5.1 Belt conveyors

An endless belt operating between two or more pulleys is one of the most ubiquitous conveyors used in transporting solid foods in a processing plant (Fig. 10.10). Between the pulleys, idlers are used to support the weight of the belt. Some key advantages and disadvantages of belt conveyors are as follows:

- High mechanical efficiency as load is carried on antifriction bearings.

Figure 10.8 A tank used to provide buffer capacity in a processing line. (Source: *Dairy Processing Handbook*, Tetra Pak.)

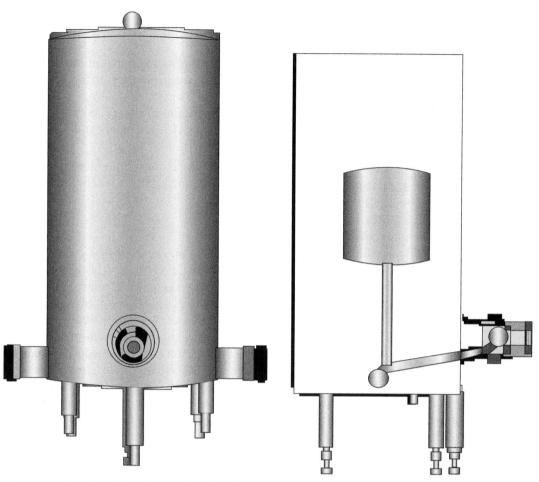

Figure 10.9 A balance tank located on the suction side of a pump. (Source: *Dairy Processing Handbook*, Tetra Pak.)

- Minimal damage to product, as no relative motion between the product and the belt.
- High carrying capacity.
- Ability to convey long distances.
- Long service life.
- High initial cost.
- Requires significant floor area.

In designing belt conveyors, the type of drive, belt, belt tension, idlers, and belt loading and discharge devices must be considered. A wide range of belt materials are used depending on the product requirements. In many cases, the belts must be washed after each shift to maintain sanitary conditions. The drive is located at the discharge end of the belt, and

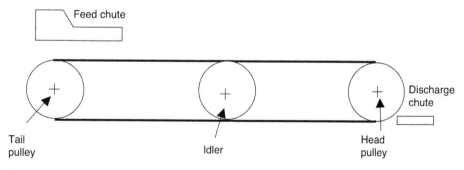

Figure 10.10 A belt conveyor.

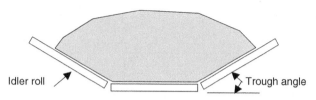

Figure 10.11 Cross-section of a troughed-belt conveyor.

sufficient contact area between the pulley and the belt is necessary to obtain a positive drive.

Belt conveyors may be either flat or troughed (Fig. 10.11). Troughed belts are suited for grains, flour, and other small particulate foods. The angle between the idler rolls and the horizontal is called the troughing angle. For conveying small particulates like grain, the troughing angle is from 20 to 45 degrees. The belt speed is kept below 2.5 m/s to minimize spillage and dust when conveying small particulates.

10.5.2 Screw conveyors

Screw conveyors consists of a helix turning inside a circular or U-shaped trough (Fig. 10.12). Screw conveyors are suited for handling powders, sticky and viscous products such as peanut butter, and granular materials. They are also useful for mixing in batch or continuous mode. Screw conveyors are useful to empty silos of flour and powder materials. Often they are used as metering devices.

The flights of screw conveyor are made of a variety of materials including stainless steel. Their operating power requirements are high, and they are used for distances less than 25 m. In a standard screw conveyor, the pitch of a screw is the same as the diameter. Screw conveyors are suitable for horizontal as well as incline up to 20 degrees. For horizontal screw conveyors an oval trough is used, whereas for a steep incline a cylindrical trough is necessary.

The power requirement of a screw conveyor depends upon a number of factors including:

- the length of the conveyor;

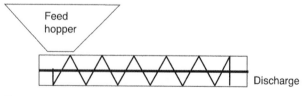

Figure 10.12 A screw conveyor.

- elevation;
- pitch;
- speed;
- type of flights and hanger brackets;
- weight and properties of the material being conveyed;
- the coefficient of friction between the product and the material of the flights and housing.

The startup power requirement of a screw conveyor is generally higher than for continuous operation.

10.5.3 Bucket elevators

The bucket elevator consists of an endless belt with buckets attached to it. The belt operates on two wheels; the top wheel is referred to as the head and the bottom is called the foot. Bucket elevators are highly efficient as there is absence of frictional loss between the product and the housing material. The bucket elevators are enclosed in a single housing referred to as a leg; in certain cases, the return is housed in a second leg. A chain or a belt is used to carry the buckets that are shaped with either rounded or sharp bottoms. The belt or chain operates between two wheels – the head and foot. For longer lengths, idlers are installed to prevent belt whip.

The product carried by the buckets is discharged at the top when the bucket turns around the head wheel, and the product is thrown out by the centrifugal force. The speed of the bucket as it goes around the head wheel must be maintained within limits to ensure that the product is discharged in a desired region (Fig. 10.13).

The conveying capacity of a bucket elevator depends on the product density, belt speed, bucket size,

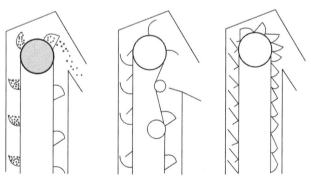

Figure 10.13 A bucket conveyor with buckets going around a head wheel.

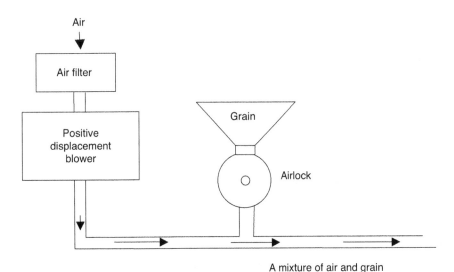

Air

Air filter

Positive displacement blower

Grain

Airlock

A mixture of air and grain

Figure 10.14 A pneumatic conveyor.

and spacing of buckets on the belt. Typical applications of bucket elevators are for handling cereal grain, animal feed, and meal. The energy requirements of bucket conveyors for cereal grains range from 0.1 to 0.2 kWh/m^3.

10.5.4 Pneumatic conveyors

A pneumatic conveyor consists of a blower, transport duct, and a device to introduce the product into the duct at the entrance and out of the duct at the exit (Fig. 10.14).

In a pneumatic conveyor, particulate food is conveyed in a closed duct by a high-velocity stream of air. The pneumatic conveyor may be operated as:

- a suction system, operating at pressures lower than atmospheric pressure;
- a low pressure suction system using high-velocity low-density air powered by a centrifugal fan;
- a high pressure system, using low-velocity high-density air powered by positive displacement blowers;
- a fluidized system that uses high-pressure, high-density air to move material with low conveying velocities.

Examples of pneumatic conveyors include the use of suction systems for unloading grain from trucks, freight cars, and pressure systems for loading freight cars or storage tanks.

Typical air velocities used in pneumatic conveying of some common products are:

- coffee beans, 3000–3500 fpm;
- corn, 500–7000 fpm;
- oats, 4500–6000 fpm;
- salt, 5500–7500 fpm;
- wheat, 5000–7000 fpm.

Empirical procedures are generally used to determine the energy requirements of pneumatic conveying. At the entrance to the conveyor, the velocity of the particle is zero, and the energy required to accelerate the product can be substantial. The mechanisms involved in moving the particulates in a pneumatic conveyor include horizontal force acting on the particulate due to the moving air and vertical force due to gravity. When particulates reach the bottom of the conduit, they slide and roll, and they are again lifted due to the action of the moving air; they may also clump with other particulates and move as a slug. These mechanisms bring complexities in developing theoretical description of the process. The power requirements of pneumatic conveyors for grains range from 0.6 to 0.7 kWh/m^3.

The key advantages and disadvantages of a pneumatic conveyor are:

- Low initial cost.
- Simple mechanical design with only one moving part (a fan).
- Random conveying path with many branches.

- Easy to change conveying path.
- Wide variety of materials can be conveyed.
- Self cleaning system.
- High power requirements.
- Possible damage to product.

More information on conveying systems used in the agricultural and food processing industry is available in Labiak and Hines (1999).

10.6 Storage of fruits and vegetables

Many fruits and vegetables are highly perishable. Therefore in post harvest management, proper techniques of handling and storage of fruits and vegetables are essential to minimize losses. The range of post harvest losses varies anywhere from 5 to 50% or even higher. In developing countries, post harvest losses are enormous often due to lack of adequate infrastructure and poor handling practices. As a result, the growers and those engaged in the food handling chain suffer major financial losses. Moreover, the shelf life of these products is severely reduced and poor quality product is delivered to the consumer. Unfortunately any gains made in increasing the production yields of fruits and vegetables are compromised by increased post harvest losses due to inadequate practices.

In industrialized countries, major progress has been made in developing proper systems for the handling of fruits and vegetables. Post harvest losses are reduced in a significant manner and the product is delivered to the consumer with minimal quality loss.

10.6.1 The respiration process

Fruits and vegetables continue to undergo physiological changes after harvest. These changes are largely the result of the respiration process. The metabolic pathways active in a respiration process are complex. As a result of the respiration process, the starch and sugars present in the plant tissue are converted into carbon dioxide and water. Oxygen plays an important role in the respiration process. The oxygen concentration within a product is very similar to that of the normal atmosphere. When sufficient amount of oxygen is available, the respiration process is called aerobic. If the surrounding atmosphere becomes deficient in oxygen then anaerobic respiration occurs. Anaerobic respiration results in the production of ketones, aldehydes, and alcohol. These products are often toxic to the plant tissue and hasten its death and decay. Therefore anaerobic respiration must be prevented. Furthermore, to extend the shelf life of fruits and vegetables, the oxygen concentration must be controlled in such a manner as to allow aerobic respiration at a reduced rate. Control of the respiration process has become one of the most important methods of storage in commercial practice.

Along with the final products of aerobic respiration, carbon dioxide and oxygen, there is also the evolution of heat. The amount of heat generated due

Table 10.6 Heat of respiration of selected fruits and vegetables.

Commodity	Watts per megagram (W/Mg)			
	0°C	5°C	10°C	15°C
Apples	10–12	15–21	41–61	41–92
Apricots	15–17	19–27	33–56	63–101
Beans, green or snap	—	101–103	161–172	251–276
Broccoli, sprouting	55–63	102–474	—	514–1000
Cabbage	12–40	28–63	36–86	66–169
Carrots, topped	46	58	93	117
Garlic	9–32	17–29	27–29	32–81
Peas, green (in pod)	90–138	163–226	—	529–599
Potatoes, mature	—	17–20	20–30	20–35
Radishes, topped	16–17	23–24	45–47	82–97
Spinach	—	136	327	529
Strawberries	36–52	48–98	145–280	210–273
Turnips, roots	26	28–30	—	63–71

Table 10.7 Classification of fruits and vegetables based on their respiration rates.

Respiration rates	Range of CO_2 production at 5°C (mg CO_2/kg h)	Commodities
Very low	< 5	Nuts, dates, dried fruits, vegetables
Low	5–10	Apple, citrus, grape, kiwifruit, garlic, onion, potato (mature), sweet potato
Moderate	10–20	Apricot, banana, cherry, peach, nectarine, pear, plum, fig (fresh), cabbage, carrot, lettuce, pepper, tomato, potato (immature)
High	20–40	Strawberry, blackberry, raspberry, cauliflower, lima bean, avocado
Very high	40–60	Artichoke, snap beans, green onion, Brussels sprouts
Extremely high	> 60	Asparagus, broccoli, mushroom, pea, spinach, sweet corn

to respiration varies with different commodities, as shown in Table 10.6. The growing parts of a plant such as a leafy vegetable have higher rates of heat generation than plant tissues where the growth has ceased such as a tuber crop. Reducing the storage temperature controls the rate of respiration process. The respiration rate is expressed in terms of the rate of carbon dioxide production per unit mass. A classification of commodities based on their respiration rates is given in Table 10.7.

An important physiological change during fruit storage is the production of ethylene gas. Based on their ethylene production, fruits are classified as either climacteric or nonclimacteric. Climacteric fruits exhibit a high production of ethylene and carbon dioxide at the ripening stage. Table 10.8 lists some of the fruits according to this classification. The rate of production of ethylene may be controlled by storage temperature, atmospheric oxygen, and carbon dioxide concentration. The results

of ethylene gas on the maturation process of fruits and vegetables include change in green color due to loss of chlorophyll, browning of tissues due to changes in anthocyanin and phenolic compounds, and development of yellow and red color due to the development of anthocyanin and carotenoids, respectively.

During storage, the loss of water from a commodity causes major deteriorative changes. Not only is there a loss of weight but also the textural quality is altered causing a commodity to lose its crispness and juiciness.

Fruits and vegetables may undergo three types of physiological breakdown that may be caused by poor post harvest practices. These include chilling injury, freezing injury, and heat injury. Chilling injury occurs mostly in commodities of tropical and subtropical regions when they are stored at temperatures above their freezing point and below 5–15°C. This type of injury causes uneven ripening, decay, growth of

Table 10.8 Classification of fruits and vegetables based on their respiration behavior during ripening.

Climacteric fruits		Nonclimacteric fruits	
Apple	Muskmelon	Blackberry	Olive
Apricot	Nectarine	Cacao	Orange
Avocado	Papaya	Cashew apple	Pepper
Banana	Passion fruit	Cherry	Pineapple
Blueberry	Peach	Cucumber	Pomegranate
Breadfruit	Pear	Eggplant	Raspberry
Cherimoya	Persimmon	Grape	Satsuma mandarin
Feijoa	Plantain	Grapefruit	Strawberry
Fig	Plum	Jujube	Summer squash
Guava	Sapote	Lemon	Tamarillo
Jackfruit	Soursop	Lime	Tangerine
Kiwifruit	Tomato	Loquat	
Mango	Watermelon	Lychee	

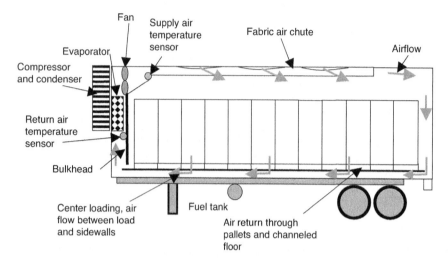

Figure 10.15 Airflow in a refrigerated trailer.

surface molds, development of off flavors, and both surface and internal discoloration.

Improper handling causes surface injuries and internal damage in fruits and vegetables. Similarly pathological breakdown caused by bacteria and fungi enhances product deterioration. Often physical damage makes it easier for the bacteria and fungi to infect plant tissues. Therefore, handling systems such as conveying and packing should be designed to minimize physical damage to the commodity being handled.

10.6.2 Controlled atmosphere storage

A large number of fruits and vegetables benefit from storage under controlled atmosphere conditions. A reduced level of oxygen and increased concentration of carbon dioxide in the immediate environment surrounding a fruit or vegetable retards its respiration rate. With a reduced respiration rate, the storage life of the product is enhanced. Considerable research has been done to determine the most suitable concentrations for oxygen and carbon dioxide that extend the storage life of fruits and vegetables. Table 10.9 is a compilation of recommended conditions for gas composition. The controlled atmosphere technology is well developed, and for certain products like apples it is used worldwide.

10.7 Refrigerated transport of fruits and vegetables

One of the most common ways to transport perishable foods such as fruits and vegetables is by using a refrigerated trailer that may be either pulled via a truck on a highway or placed in a ship for transocean shipment. Any perishable product stored in the trailer must be kept refrigerated for the entire duration. A typical refrigerated trailer, shown in Fig. 10.15, includes a refrigeration system to cool air and an air handling system to distribute air within the trailer. It is vital that there is uniform air distribution within the trailer; otherwise regions with no air circulation can lead to product heating and spoilage (Thompson et al., 2002).

Temperature control of the air circulating inside the trailer is vital to the shipment of perishable foods. Modern trailers are equipped with temperature sensors and controllers that automatically control the refrigeration unit based on the temperature of air exiting the refrigeration unit. The control of air temperature at the exit of the refrigeration unit is important in protecting fresh produce that is sensitive to chilling injury or freeze damage. The temperature of the thermostat in these systems is set within 0.5°C of the long-term storage temperature. In older trailers used for chilling/freeze-sensitive produce, the temperature control is typically based on the return air to the refrigeration unit, and the controllers should be set at least 1.5–2.5°C above the long-term storage temperature of the produce. For frozen products, the temperature is controlled based on the return air temperature. The trailer used for frozen products should be set at −18°C or colder. The frozen food industry generally requires that at the time frozen food is loaded into a trailer, the temperature of the product should be less than −12°C

When a trailer load contains more than one type of produce, care must be taken that these are compatible in terms of their storage temperature and ethylene

Table 10.9 Recommended conditions for controlled atmosphere (CA) storage of fruits and vegetables.

Common name	Scientific name	Storage temperature (°C)	Relative humidity (%)	Highest freezing temperature (°C)	Ethylene production	Ethylene sensitivity	Approximate shelf-life	Beneficial controlled atmosphere
Apple, nonchilling sensitive varieties		−1.1	90–95	−1.5	Vh	H	3–6 months	CA varies by cultivar
Apple, chilling sensitive	Yellow Newtown, Grimes Golden, McIntosh	4	90–95	−1.5	Vh	H	1–2 years	CA varies by cultivar
Apricot	*Prunus armeniaca*	−0.5–0.0	90–95	−1.1	M	H	1–3 weeks	2–3% O_2 + 2–3% CO_2
Artichoke, Globe	*Cynara acolymus*	0	95–100	−1.2	VL	L	2–3 weeks	2–3% O_2 + 3–5% CO_2
Asparagus, green, white	*Asparagus officinalis*	2.5	95–100	−0.6	VL	M	2–3 weeks	5–12% CO_2 in air
Avocado Cv Fuerta, Haas	*Persea americana*	3–7	85–90	−1.6	H	H	2–4 weeks	2–5% O_2 + 3–10% CO_2
Banana	*Musa paradisiaca var. sapientum*	13–15	90–95	−0.8	M	H	1–4 weeks	2–5% O_2 + 2–5% CO_2
Beans, snap, wax, green	*Phaseolus vulgaris*	4–7	95	−0.7	L	M	7–10 days	2–3% O_2 + 4–7% CO_2
Lima beans	*Phaseolus lunatus*	5–6	95	−0.6	L	M	5–7 days	
Strawberry	*Fragaria* spp.	0	90–95	−0.8	L	L	7–10 days	5–10% O_2 + 15–20% CO_2
Cabbage, Chinese, Napa	*Brassica campestris var. perkinensis*	0	95–100	−0.9	VL	H	2–3 months	1–2% O_2 + 0–5% CO_2
Carrots, topped	*Daucus carota*	0	98–100	−1.4	VL	H	6–8 months	No CA benefit
Carrots, bunched	*Daucus carota*	0	98100	−1.4	VL	H	10–14 days	Ethylene causes bitterness

(*Continued*)

Table 10.9 (Continued)

Common name	Scientific name	Storage temperature (°C)	Relative humidity (%)	Highest freezing temperature (°C)	Ethylene production	Ethylene sensitivity	Approximate shelf-life	Beneficial controlled atmosphere
Cauliflower	B. oleracea var. botrytis	0	95–98	−0.8	VL	H	3–4 weeks	2–5% O_2 + 2–5% CO_2
Cherimoya, Custard Apple	Annona cherimola	13	90–95	−2.2	H	H	2–4 weeks	3–5% O_2 + 5–10% CO_2
Citrus, lemon	Citrus limon	10–13	85–90	−1.4			1–6 months	5–10% O_2 + 0–10% CO_2
Citrus, orange	Citrus sinensis, California, dry	3–9	85–90	−0.8	VL	M	3–8 weeks	5–10% O_2 + 0–5% CO_2
Citrus, orange	Citrus sinensis, Florida, humid	0–2	85–90	−0.8	VL	M	8–12 weeks	5–10% O_2 + 0–5% CO_2
Cucumber	Cucumis sativus	10–12	85–90	−0.5	L	H	10–14 days	3–5% O_2 + 3–5% CO_2
Eggplant	Solanum melongena	10–12	90–95	−0.8	L	M	1–2 weeks	3–5% O_2 + 0% CO_2
Garlic	Allium sativum	0	65–70	−0.8	VL	L	6–7 months	0.5% O_2 + 5–10% CO_2
Ginger	Zingiber officinale	13	65		VL	L	6 months	No CA benefit
Grape	Vitis vinifera	−0.5–0	90–95	−2.7	VL	L	2–8 weeks	2–5% O_2 + 1–3% CO_2
Guava	Psidium guajava	5–10	90		L	M	2–3 weeks	
Lettuce	Lactuca sativa	0	98–100	−0.2	VL	H	2–3 weeks	2–5% O_2 + 0% CO_2
Loquat	Eriobotrya japonica	0	90	−1.9			3 weeks	
Lychee, litchi	Litchi chinensis	1–2	90–95		M	M	3–5 weeks	3–5% O_2 + 3–5% CO_2
Mango	Mangifera indica	13	85–90	−1.4	M	M	2–3 weeks	3–5% O_2 + 5–10% CO_2
Melon, honey dew, orange flesh	Cucurbita melo	5–10	85–90	−1.1	M	H	3–4 weeks	3–5% O_2 + 5–10% CO_2

Commodity	Scientific name	Temperature (°C)	RH (%)	Freezing point			Storage life	CA conditions
Mushroom	Agaricus	0	90	−0.9	VL	M	7–14 days	3–21% O_2 + 5–15% CO_2
Okra	Abelmoschus esculentus	7–10	90–95	−1.8	L	M	7–10 days	Air + 4–10% CO_2
Papaya	Carica papaya	7–13	85–90		H	H	1–3 weeks	2–5% O_2 + 5–8% CO_2
Peach	Prunus persica	−0.5–0	90–95	−0.9	H	L	2–4 weeks	1–2% O_2 + 3–5% CO_2
Pepper, Bell	Capsicum annuum	7–10	95–98	−0.7	L	L	2–3 weeks	2–5% O_2 + 2–5% CO_2
Persimmon, Fuyu	Dispyros kaki	7–10	95–98	−0.7	L	L	2–3 weeks	2–5% O_2 + 2–5% CO_2
Persimmon, Hachiya	Dispyros kaki	10	90–95	−2.2	L	H	1–3 months	
Pineapple	Ananas comosus	5	90–95	−2.2	L	H	2–3 months	
Pomegranate	Punica granatum	5	90–95	−3.0			2–3 months	3–5% O_2 + 5–10% CO_2
Potato, early crop	Solanum turbersom	10–15	90–95	−0.8	VL	M	10–14 days	
Potato, late crop	Solanum tubersom	4–12	95–98	−0.8	VI	M	5–10 months	
Spinach	Spinacia oleracea	0	95–100	−0.3	VL	H	10–14 days	5–10% O_2 + 5–10% CO_2
Tomato, mature-green	Lycopersicon esculentum	10–13	90–95	−0.5	VL	H	1–3 weeks	3–5% O_2 + 2–3% CO_2
Tomato, firm ripe	Lycopersicon esculentum	10	85–90	−0.5	H	L	7–10 days	3–5% O_2 + 3–5% CO_2
Watermelon	Citrullus vulgaris	10–15	90	−0.4	VL	H	2–3 weeks	No CA benefit

VL, very low; L, low; M, medium; H, high; VH, very high.

Table 10.10 Ethylene sensitivity of selected vegetables during storage.

Commodity	Symptoms of ethylene injury
Asparagus	Increased lignification (toughness) of spears
Beans, snap	Loss of green color
Broccoli	Yellowing, abscission of florets
Cabbage	Yellowing, abscission of leaves
Carrot	Development of bitter flavor
Cauliflower	Abscission and yellowing of leaves
Cucumber	Yellowing and softening
Eggplant	Calyx abscission, browning of pulp and seeds, accelerated decay
Leafy vegetables	Loss of green color
Lettuce	Russet spotting
Parsnip	Development of bitter flavor
Potato	Sprouting
Sweet potato	Brown flesh discoloration and off flavor detectable when cooked
Turnip	Increased lignification (toughness)
Watermelon	Reduced firmness, flesh tissue maceration resulting in thinner rind, poor flavor

sensitivity. Table 10.10 lists some of the deleterious effects on vegetables due to excessive ethylene exposure. Ethylene-sensitive vegetables should not be mixed with ethylene-producing fruits.

In a refrigerated trailer, it is usually difficult to force sufficient air through boxes. As a result, it is not generally possible to cool the stored product. In fact, in a poorly designed or managed trailer, the product may warm during transit.

Figure 10.16 shows a checklist for trailer conditions before loading food. The loading pattern of pallets inside a trailer has a marked influence on the airflow. Airflow in a typical refrigerated highway trailer is shown in Fig. 10.15. Air exiting the evaporator of the refrigeration unit is directed to the ceiling (usually through an air chute), between the walls, around the rear door, and it returns through the channeled floor to the front bulk head. For air to circulate in this manner, it is important that adequate spacing is provided between the load and the ceiling, walls, rear doors, and floor. Several types of loading arrangements for pallets are used; some

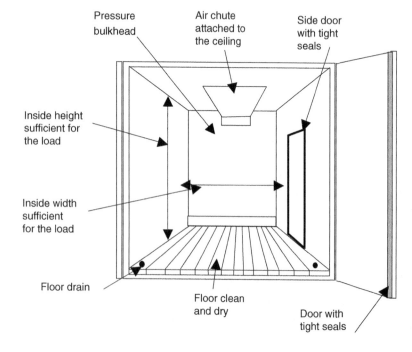

Pressure bulkhead

Air chute attached to the ceiling

Side door with tight seals

Inside height sufficient for the load

Inside width sufficient for the load

Floor drain

Floor clean and dry

Door with tight seals

Figure 10.16 Key features of a refrigerated trailer.

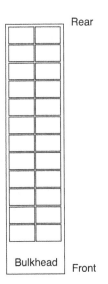

Figure 10.17 An arrangement of 24 pallet loads in a refrigerated truck.

common patterns for 24 standard pallets are shown in Fig. 10.17.

In long-distance shipping, vibration damage to the produce can be a significant factor. Vibration damage to product is more severe for trailers using steel-spring-suspended axles. Dramatic reductions in vibration damage are observed in trailers using air ride suspension. If the rear-axle of the trailer has steel springs, then vibration-sensitive products such as pears and berries should not be loaded in the rear section of the trailer.

10.8 Water quality and wastewater treatment in food processing

Water is a ubiquitous resource on our planet, yet the availability of a clean and reliable supply of water is becoming increasingly scarce. The food processing industry relies heavily on access to clean water. Water use in food processing has increased with the widespread mechanization of harvesting operations; raw agricultural products arriving at a food processing plant require large quantities of water for cleaning (Fig. 10.18). Inside the plant, water is used in a variety of processing and handling operations such as product conveying, peeling, blanching, cooling, generating steam, and washing equipment and floors (Figs 10.19 and 10.20).

The quality of water used in food processing depends on the function of water in the manufacturing process. For example, water quality for initial cleaning of raw produce has a different requirement than water required for the formulation of carbonated

Figure 10.18 Water sprays are used in washing spinach. Leafy vegetables require a considerable amount of water to remove any insects and other debris attached to the leaves.

Figure 10.19 Mechanically harvested tomatoes, brought in gondola containers in trucks to a processing plant, are removed by pumping water into gondolas and transferred into water flumes.

beverages, beer, and bottled drinking water. Water obtained from ground wells or surface areas (such as lakes, rivers, and springs) may require certain treatment before its use in the food processing plant. If the water has high levels of hardness due to the presence of dissolved solids, then processes such as precipitation, ion exchange, distillation, or reverse osmosis may be used. To remove turbidity, domestic water is

Figure 10.20 A water flume is used to convey tomatoes from the receiving area to the processing equipment.

often pretreated using coagulation, flocculation, sedimentation, or filtration. The presence of dissolved organics in water causes off-flavor, odor, or color that is often removed by activated carbon adsorption.

10.8.1 Characteristics of food processing wastewater

Foods processed in a processing plant influence the composition of the discharged wastewater. For example, in a fruit and vegetable canning plant, wastewater contains the product residues generated from operations such as peeling, blanching, cutting, washing, heating and cooling, and cooking. For sanitary requirements, equipment and floors are frequently washed, creating large volumes of wastewater. Any detergents and lubricants used in the processing equipment and chemicals such as caustic solution (NaOH) for lye-peeling vegetables are also mixed in the wastewater. Other typical constituents of wastewater include emulsified oil, organic colloids, dissolved inorganics, and suspended solids.

The quantity and composition of wastewater generated by different food processing plants are highly variable. Most of the pollutants in food processing wastewater are organic in nature. Up to 80% of total organic matter in the wastewater from food processing plants may be in a dissolved form.

The quality of wastewater is expressed by two commonly measured quantities, namely biological oxygen demand and chemical oxygen demand.

10.8.1.1 Biological oxygen demand (BOD)

BOD is a measure of oxygen required to oxidize the organic content in a water sample with the action of microorganisms. The biodegradable content present in the wastewater is expressed by the 5-day (120-hour), 20°C biological oxygen demand (BOD_5). The BOD value is commonly used to determine the efficiency of the wastewater treatment. The procedure for measuring BOD_5 involves the following steps (Schroeder, 1977):

- Procure samples of wastewater ensuring that there is minimal time delay before the test is started.
- Make dilution with a nutrient solution so that a maximum BOD of <6 mg/L will be obtained.
- Add a bacterial "seed" to the sample.
- Fill the standard (300 mL) bottles with diluted wastewater and seal. Also, prepare blank samples

with bottles filled with dilution water containing the "seed".
- Immediately determine oxygen content of at least two samples and two blank bottles.
- Incubate samples at 20°C for 5 days, determine the oxygen content of remaining samples and blanks.
- Calculate BOD_5 using the following equation:

$$BOD_5 = D_f[(DO_0-DO_5)]_{sample}-[(DO_0-DO_5)]_{blank}$$
(10.7)

where

DO_0 is the initial dissolved oxygen,
DO_5 is the dissolved oxygen at the end of 5 days,
D_f is dilution factor

Some typical BOD_5 values of wastewater measured in food processing plants are shown in Table 10.11.

10.8.1.2 Chemical oxygen demand (COD)

COD is a measure of oxygen (in parts per million) required to oxidize organic and inorganic matter in a sample of wastewater. A strong chemical oxidant is used to determine the COD value. Although no direct correlation exists between COD and BOD_5, COD is useful to estimate the oxygen demand because it is

Table 10.11 Typical values of BOD measured in food processing plants (Environmental Protection Service, 1979b).

Industry sector	BOD_5 (mg/L)
Dairy	
Cheese	790–5900
Fluid milk	1210–9150
Ice cream	330–230
Fruit and vegetables	
Apple products	660–3200
Carrots	640–2200
Corn	680–5300
Green beans	130–380
Peaches	750–1900
Peas	270–2400
Fish	
Herring filleting	3200–5800
Meat and poultry	
Red meat slaughtering	200–6000
Poultry processing	100–2400
Poultry slaughtering	400–600

a rapid test taking less than 2 hours compared to mea-surement of BOD_5 that takes 120 hours. Standard test-ing procedures to measure COD of wastewater are available (American Society of Testing and Materials, 2006).

10.8.2 Wastewater treatment

Wastewater treatment is often classified as primary, secondary, or tertiary. Primary treatment is often a physico-chemical process involving sedimentation; secondary treatment comprises biological treatment with sedimentation; and tertiary treatment involves removal of residual and nonbiodegradable materials.

Dissolved organic matter in the wastewater is re-moved using biological treatment and adsorption, whereas dissolved inorganic matter requires the use of ion exchange, reverse osmosis, evaporation, and/or distillation. Any suspended organic matter is typically removed using physico-chemical and bi-ological treatment methods. Other suspended inor-ganic or organic content is removed by screening, sedimentation, filtration, and coagulation.

10.8.3 Physico-chemical methods of wastewater treatment

A variety of physico-chemical methods are used in separating solids from wastewater. In this sec-tion some of the common methods are introduced. More details on these operations are presented in Schroeder (1977) and Liu (2007).

10.8.3.1 Screening

Screens are used to separate any debris or other suspended solid materials from wastewater. Typi-cal mesh size for coarse screens is 6 mm or larger, whereas for fine screens it is 1.5–6 mm. Screens are made of stainless steel, and they can be effective in re-ducing the suspended solids to levels similar to those obtained in sedimentation. To minimize clogging of screens, a scraping system is employed. Rotary drum screens are also used where the screen is formed into a cylindrical drum shape that is used to separate par-ticulate matter.

10.8.3.2 Flotation systems

For wastewater containing oil and grease, flotation systems are frequently used. In a flotation system, air is diffused into the wastewater causing the oil and grease to float to the top. Other slowly settling partic-ulates are also separated as they attach themselves to the air bubbles and rise to the surface, where they are skimmed off with the use of skimmers. In case there is emulsified fat present in the wastewater then the emulsion must be first destabilized by use of addi-tives to improve the efficiency of fat removal.

10.8.3.3 Sedimentation

Sedimentation is a widely used method for treating wastewater. The process is simple, as it involves fill-ing a tank with wastewater and letting gravity cause the settling of solid particulates with a specific grav-ity of >1 to the bottom of the tank. The wastewater enters from the bottom of the tank and moves either upward or in a radial direction in the tank. The solid matter that settles in a sedimentation tank is referred to as sludge. In food processing wastewater, sludge is organic in nature and it is periodically removed for further treatment.

If we conduct a force balance on a rigid sphere falling through a Newtonian liquid, we obtain the fol-lowing expression for the particle velocity, v_p:

$$v_p = \frac{d_p^2 g(\rho_p - \rho_L)}{18\mu} \qquad (10.8)$$

where

d_p is the solid particle diameter,
g is gravitational constant,
ρ_p is the density of particles,
ρ_L is the density of liquid,
μ is the viscosity of the liquid.

From Equation (10.8), the particle velocity, v_p, is a function of the diameter and density of the particles, and density and viscosity of the liquid. In the sed-imentation tank, the density and viscosity of liquid cannot be changed, but aggregating small particles into larger ones can increase the particle size and den-sity. According to Equation (10.8), larger particles will descend faster, and thus a coagulation step is com-monly employed.

Furthermore, particles present in the wastewater are often colloidal, and they carry the same charge. As a result, they repel each other. This creates a sta-ble suspension. However, to allow particles to grow we need to destabilize the suspension. This is accomplished by the use of coagulants such as alum ($Al_2(SO_4)$), ferric chloride ($FeCl_3$), and metal oxides or hydroxides (CaO or $Ca(OH)_2$).

10.8.3.4 Filtration

In nature, water is filtered as it moves through different layers of sand, soil, and granular materials. Analogous to the natural systems, filtration systems have been developed using materials such as sand, diatomaceous earth, powdered carbon, and perlite.

The complexity of the filtration process is evident when we consider that the movement of water through a granular bed may involve a variety of different interactions between the filter medium and the material being separated. For example, there is the influence of gravity, diffusion, and adsorption on the filter media. The variability in the composition of the feed stream adds to the complexity of the process. When a high concentration of suspended solids is present then frequent cleaning and backwashing of the filter medium is required. Any organic matter present in the wastewater results in a buildup of biological slime on the filter, causing problems with cleaning. For this reason, filtration is typically used only in the tertiary treatment of water.

Two commonly used filtration systems are pre-coat filters and depth filters.

Pre-coat filters

In a precoat filter, a coating of particulates is applied on a support medium made of cloth or finely woven wire. The support medium and the particulate coating act as the filter medium. In some cases, the solids present in the wastewater stream provide the pre-coating medium.

Depth filters

A depth filter is constructed using granular material with varying porosity supported on a gravel layer. The filtration medium is usually graded sand. Depth filters require backwashing to keep them clean. The flow of liquid in a filtration system is described using Darcy's law,

$$v = KS \tag{10.9}$$

where

v is the apparent velocity obtained by dividing flow rate by cross-sectional area,
K is the coefficient of permeability,
S is the pressure gradient.

Using Darcy's law, Kozeny proposed the following equation to calculate flow through a medium with uniform porosity (Schroeder, 1977):

$$\frac{h_L}{H} = \frac{k\mu v}{g\rho}\frac{(1-\phi)^2}{\phi^3}a_v \tag{10.10}$$

where

H is the bed depth,
h_L is the head loss through bed of depth H,
ϕ is the bed porosity,
ρ is the liquid density,
g is the gravitational constant,
a_v is the average grain surface area to volume ratio,
k is the dimensionless coefficient (typically its value is 5 for wastewater filtration).

10.8.4 Biological treatment of wastewater

Although physicochemical operations described in the preceding sections are useful in the separation and removal of suspended solids of varying dimensions, these operations are often unable or inefficient to remove dissolved and colloidal organic matter. For this purpose, biological treatment of wastewater is commonly employed. Any organic matter that does not settle or is dissolved is treated with microorganisms. In the presence of oxygen, aerobic microorganisms break down the organic matter. Anaerobic treatment involves microbial activity in the absence of oxygen.

Lagoons or ponds are widely used for the biological treatment of food processing wastewater. Lagoons cover a large surface area up to several acres. The base of these ponds is lined with impervious material such as plastic. Wastewater is held in these large ponds for a certain number of days and then pumped out. Lagoons or ponds are suitable when sufficient land is available. Two common types of lagoons used in treating food processing wastewater are anaerobic lagoons and aerobic lagoons.

In anaerobic lagoons, the breakdown of organic matter involves two steps. In the first step, acid-producing bacteria break down organic matter into compounds such as fatty acids, aldehydes, and alcohols. The second step involves bacteria that convert these compounds into methane, carbon dioxide, ammonia, and hydrogen. Anaerobic ponds are typically 3–5 m deep and mostly devoid of oxygen.

In aerobic lagoons, mechanical systems are used for complete mixing and aeration of the wastewater. With excess oxygen, the microbial growth is under aerobic conditions. High concentration of dissolved oxygen is maintained in the wastewater by aeration as well as the growth of algae. As bacteria break down the organic matter, nutrients become available for more algae to grow. The photosynthetic activity of algae helps maintain aerobic conditions. Aerobic lagoons are shallow in depth (around 1 m) so that sun rays can penetrate to the bottom of the lagoon to promote the growth of algae. Case studies involving use of lagoons in the treatment of different food processing wastewater are presented in Environmental Protection Service (1979b).

Microbial activity during biological treatment of wastewater generates solid material. Sedimentation procedures, as described in previous sections, are used to settle suspended solids. The solid material removed from the sedimentation tanks is commonly referred to as biological sludge. The role of bacteria in wastewater processing and the rate of kinetics of microbial activity are elaborated by Liu (2007).

Another commonly used method for biological treatment of food processing wastewater is a trickling filter. A trickling filter is a large tank containing the following components:

- an inert filter medium (such as gravel, stones, wood, or plastic particulates) with microorganisms attached to it forming a slime layer or biofilm;
- a water distribution system;
- a pipe that conveys incoming wastewater to a water distribution system so that water can uniformly trickle down through the filter medium;
- an under-drain system to support the filter medium and ensure that oxygen is uniformly available throughout the tank.

A trickling filter is an example of a film flow system whereby a thin film of wastewater flows over the biofilm attached to the filter medium. As the wastewater is allowed to trickle through the filter medium, the microorganisms attached to the filter medium in the biofilm utilize the organic matter. To avoid clogging of the filter medium, wastewater undergoes some primary treatment to remove suspended solids and other coarse material before it is fed to the trickling filter.

Around the world, the cost of water for food processing continues to increase due to competitive demands for water by urban areas as well as changing climatic conditions with periodic droughts. Regulations governing wastewater disposal have become stricter, adding to the cost of disposing wastewater generated by the processing plants. For both environmental and economic reasons, there is an increasing need to reduce water use in food processing plants. There are considerable opportunities to reuse water within a food processing plant. Recycling water from one operation to another, with proper treatment of water between the operations, would help achieve this goal (Maté and Singh (1993).

Further reading and references

American Society of Testing and Materials (2006) *Standard Test Methods for Chemical Oxygen Demand (Dichromate Oxygen Demand) of Water, Standard D1252 – 06.* ASTM International, West Conshohocken, Pennsylvania., www.astm.org.

Brennan, J.G., Butters, J.R., Cowell, N.D. and Lilley, A.E.V. (1990) *Food Engineering Operations*, 3rd edn. Elsevier Applied Science, London.

Bresnahan, D. (1997) Process control. In: *Handbook of Food Engineering Practice* (eds E. Rotstein, R.P. Singh and K. Valentas). CRC Press, Boca Raton, Florida.

Bylund, G. (1995) *Dairy Processing Handbook.* TetraPak, Lund.

Chakravarti, A, and Singh, R.P. (2002) *Postharvest Technology. Cereals, Pulses, Fruits and Vegetables.* Science Publishers, New York.

Environmental Protection Service (1979a) *Evaluation of Physical-Chemical Technologies for Water Reuse, Byproduct Recovery and Wastewater Treatment in the Food Processing Industry. Economic and Technical Review Report EPS-3-WP-79-3.*EPS, Environment Canada, Ottawa.

Environmental Protection Service (1979b) *Biological Treatment of Food Processing Wastewater Design and Operations Manual. Economic and Technical Review Report EPS-3-WP-79-7.* EPS, Environment Canada, Ottawa.

Hughes, T.A. (2002) *Measurement and Control Basics*, 3rd edn. ISI – The Instrumentation Systems and Automation Society, Research Triangle Park, North Carolina.

Jowitt, R.E. (1980) *Hygienic Design and Operation of Food Plant*. AVI Publishing Co., Westport, Connecticut.

Kader, A.A. (2002) *Postharvest Technology of Horticultural Crops*, 3rd edn. DANR Publication 3311, University of California, Davis.

Katsuyama, A.M. (1993) *Principles of Food Processing Sanitation*. Food Processors Institute, London.

Labiak, J.S. and Hines, R.E. (1999) Grain handling. In: *CIGR Handbook of Agricultural Engineering, Vol IV, Agro Processing Engineering* (eds F.W. Bakker-Arkema, J. DeBaerdemaker, P. Amirante, M. Ruiz-Altisent and C.J. Studman). American Society of Agricultural Engineers, St Joseph, Michigan.

Liu, S.X. (2007) *Food and Agricultural Wastewater Utilization and Treatment*. Blackwell Publishing, Ames, Indiana.

Maté, J.I. and Singh, R.P. (1993) Simulation of the water management system of a peach canning plant. *Computers and Electronics in Agriculture*, **9**, 301–317.

Murrill, P.W. (2000) *Fundamentals of Process Control Theory*, 3rd edn. Instrument Society of America, Research Triangle Park, North Carolina.

Ogrydziak, D. (2004) *Food Plant Sanitation*. Unpublished Class Notes. Department of Food Science, University of California, Davis, California.

Rotstein, E., Singh, R.P. and Valentas, K. (1997) *Handbook of Food Engineering Practice*. CRC Press, Boca Raton, Florida.

Schroeder, E.D. (1977) *Water and Wastewater Treatment*. McGraw Hill, New York.

Singh, R.P. and Erdogdu, F. (2009) *Virtual Experiments in Food Processing*, 2nd edn. RAR Press, Davis, California.

Singh, R.P. and Heldman, D.R. (2009) *Introduction to Food Engineering*, 4th edn. Academic Press, London.

Thompson, J.F., Brecht, P.E. and Hinsch, T. (2002) *Refrigerated Trailer Transport of Perishable Products*. *ANR Publication 21614*. University of California, Davis, California.

Gordon L. Robertson

- Requirements of packaging materials: containment; protection; convenience; communication.
- Classification of packaging materials: metals; glass; paper; plastics.
- Permeability characteristics of plastic packaging.
- Interactions between packaging materials and food: corrosion; migration.
- Packaging systems: modified atmosphere packaging; active packaging.
- Package closures and integrity.
- Environmental impacts of packaging: municipal solid waste; source reduction; recycling and composting; waste-to-energy incineration; landfill; life cycle assessment; packaging waste legislation.

Packaging is pervasive and essential in today's society as it surrounds, enhances and protects the goods we buy from the moment they are processed and manufactured through storage and retailing to the final consumer. The importance of packaging hardly needs stressing because it is almost impossible to think of more than a handful of foods that are sold in an unpackaged state. However, despite the importance and key role that packaging plays, it is often viewed negatively by society and regarded as an unnecessary cost. Such views arise because the functions that packaging performs are either unknown or misunderstood, and by the time most consumers come into contact with a package, its role is almost over.

11.1 Requirements of packaging materials

A primary package is one that is in direct contact with the contained product. It provides the initial and usually the major protective barrier. Examples of primary packages include metal cans, paperboard cartons, glass bottles and plastic pouches. Frequently it is only the primary package which the consumer purchases at retail outlets. A secondary package contains a number of primary packages, e.g. a corrugated case or box. It is the physical distribution carrier and is increasingly being designed so that it can be placed directly onto retail shelves for the display of primary packages (so-called shelf-ready packaging). A tertiary package is made up of a number of secondary packages, the most common example being a stretch-wrapped pallet of corrugated cases. This chapter will confine itself to a consideration of the primary package.

Packaging is a socio-scientific discipline which ensures delivery of goods to the ultimate consumer of those goods in the best condition appropriate for their use. Packing involves the enclosure of products in a package such as a pouch, bag, box, cup, tray, can, tube, bottle or other container form to perform one or more of the following functions: containment; protection; convenience; and communication.

11.1.1 Containment

All products must be contained before they can be stored or moved from one place to another; without containment, product loss and environmental pollution would be widespread.

11.1.2 Protection

The package must protect its contents from outside environmental effects, be they water, water vapour, gases, odours, micro-organisms, dust, shocks, vibrations, compressive forces, etc., and protect the environment from the product. For many food products, the protection afforded by the package is an essential part of the preservation process. In general, once the integrity of the package is breached, the product is no longer preserved.

11.1.3 Convenience

Modern industrialized societies have created a demand for greater convenience in foods; e.g. foods that are pre-prepared and can be consumed outside the home or cooked/reheated in a very short time, preferably without removing them from their primary package; and sauces, dressings and condiments that can be applied simply through pump-action packages which minimise mess, etc. Thus packaging plays an important role in allowing products to be used conveniently. The shape (relative proportions) of the primary package is related to convenience of use by consumers (e.g. easy to hold, open, pour, reclose) and efficiency in building into secondary and tertiary packages.

11.1.4 Communication

Modern methods of consumer marketing would fail were it not for the messages, graphics, distinctive shapes, branding and labelling communicated by the package, enabling consumers to instantly recognize and correctly use products. The package must function as a silent sales person.

11.1.5 Attributes

In addition to the above functions, there are also several attributes of packaging that are important. One (related to the convenience function) is that it should be efficient from a production or commercial viewpoint, i.e. in filling, closing, handling, transportation and storage. Another is that the package should have, throughout its life cycle from raw material extraction to final disposal after use, minimal environmental impacts. A third attribute is that the package should not impart to the food any undesirable contaminants. Although this last attribute may seem self-evident, there has been a long history of so-called food contact substances migrating from the packaging material into the food. Not surprisingly, food packaging materials are highly regulated in most countries to ensure consumer safety.

11.2 Classification of packaging materials

The protection offered by a package is determined by the nature of the packaging material and the format or type of package construction. A wide variety of materials is used in packaging and primary packaging materials consist of one or more of the following materials: metals; glass; paper; and plastic polymers. These are briefly described below.

11.2.1 Metals

Four metals are commonly used for the packaging of foods: steel, aluminium, tin and chromium. Tin and steel, and chromium and steel, are used as composite materials in the form of tinplate and electrolytically chromium-coated steel (ECCS), the latter being sometimes referred to as tin-free steel (TFS). Aluminium is used in the form of purified alloys containing small and carefully controlled amounts of magnesium and manganese.

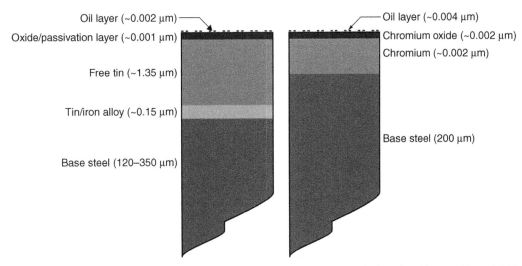

Oil layer (~0.002 μm)
Oxide/passivation layer (~0.001 μm)
Free tin (~1.35 μm)
Tin/iron alloy (~0.15 μm)
Base steel (120–350 μm)

Oil layer (~0.004 μm)
Chromium oxide (~0.002 μm)
Chromium (~0.002 μm)
Base steel (200 μm)

Figure 11.1 Schematic structure (not to scale) of tinplate and ECCS showing the main functional layers. (Copyright 2006. From *Food Packaging Principles & Practice* by G.L. Robertson. Reproduced by permission of Routledge/Taylor & Francis Group, LLC.)

The term tinplate refers to low carbon mild steel sheet varying in thickness from around 0.15 to 0.5 mm with a coating of tin between 2.8 and 17 gsm $(g\,m^{-2})$ (0.4–2.5 μm thick) on each surface of the material. The combination of tin and steel produces a material which has good strength combined with excellent fabrication qualities as well as a corrosion-resistant surface of bright appearance due to the unique properties of tin.

Tinplate is manufactured by electrolytically applying a thin coating of tin to a steel sheet which, depending on the final application, contains various levels of carbon, silicon, manganese, phosphorus, copper and sulphur. After plating, the coating is passivated by electrolytic treatment in sodium dichromate to render the surface more stable and resistant, and then lightly oiled. The final structure of the completed coating is shown in Fig. 11.1.

The production of ECCS is very similar to electrotinning. ECCS consists of a duplex coating of metallic chromium and chromium sesquioxide to give a total coating weight of approximately 0.15 gsm. This is much thinner than the lowest grade of electrolytic tinplate which has a tin thickness of 2.8 gsm. The surface of ECCS is more acceptable for protective lacquer coatings, printing inks and varnishes than tinplate. However, ECCS is less resistant to corrosion than tinplate and therefore must be lacquered on both sides.

Aluminium is used to manufacture both metal cans and thin foil in thicknesses ranging from 4 to 150 μm, foils thinner than 25 μm containing minute

pinholes that are permeable to gases and water vapour. In both applications alloying agents including silicon, iron, copper, manganese, magnesium, chromium, zinc and titanium are added to impart strength, and improve formability and corrosion resistance.

11.2.3 Glass

Glass is an amorphous, inorganic product of fusion that has been cooled to a rigid condition without crystallizing. Although rigid, glass is a highly viscous liquid that exists in a vitreous or glassy state. A typical formula for soda-lime glass is:

- silica, SiO_2 68–73%;
- calcia, CaO 10–13%;
- soda, Na_2O 12–15%;
- alumina, Al_2O_3 1.5–2%;
- iron oxide, FeO 0.05–0.25%.

The loss on ignition or fusion loss (generally the oxides of carbon and sulphur) can vary from 7 to 15%, depending on the quantity of cullet (used and scrap glass), there being less fusion loss the greater the quantity of cullet. Soda-lime glass accounts for nearly 90% of all glass produced and is used for the manufacture of containers where exceptional chemical durability and heat resistance are not required.

The two main types of glass containers used in food packaging are bottles (which have narrow necks) and jars (which have wide openings). About

75% of all glass food containers are bottles and approximately 85% of container glass is clear, the remainder being mainly amber or green. Today's glass containers are lighter but stronger than their predecessors, and through such developments the glass container has remained competitive and continues to play a significant role in the packaging of food products.

The container finish (so-called because in the early days of glass manufacturing it was the part of the container fabricated last) is the glass surrounding the opening in the container that holds the cap or closure. It must be compatible with the cap or closure and can be broadly classified by size (i.e. diameter) and sealing method (e.g. twist cap, cork).

11.2.3 Paper

Pulp is the fibrous raw material used for the production of paper, paperboard, corrugated board and similar manufactured products. It is obtained from plant fibre and is therefore a renewable resource. Almost all paper is converted by undergoing further treatment after manufacture such as embossing, coating, laminating and forming into special shapes and sizes such as bags and boxes. Further surface treatment involving the application of adhesives and printing inks are common, depending on the end use. While paper that has been laminated or coated with plastic polymers can provide a good barrier to gases and water vapour, other paper packaging provides little more than protection from light and minor mechanical damage.

Paper is generally termed board when its grammage exceeds 224 gsm. Multi-ply boards are produced by the consolidation of one or more web plies into a single sheet of paperboard which is then subsequently converted into rigid boxes, folding cartons, beverage cartons and similar products.

11.2.4 Plastics

Plastics are organic polymers with the unique characteristic that each molecule is either a long chain or a network of repeating units. The properties of plastics are determined by the chemical and physical nature of the polymers used in their manufacture, the properties of polymers being determined by their molecular structure, molecular weight, degree of crystallinity and chemical composition. These factors in turn affect the density of the polymers and the temperatures at which they undergo physical transitions.

Polymer chains can and do align themselves in ordered structures, and the thermodynamics of this ordered state determine such properties as melting point, glass transition temperature, and mechanical and electrical properties. However, it is the chemical nature of the polymer that determines its stability to temperature, light, water and solvents, and hence the degree of protection it will provide to food when used as a packaging material.

A wide range of polymers is used in food packaging and the major categories are briefly reviewed below.

11.2.4.1 Polyolefins

These form an important class of thermoplastics and include low, linear and high density polyethylenes (LDPE, LLDPE and HDPE) and polypropylene (PP). The polyethylenes have the nominal formula $-(CH_2-CH_2)_n-$ and are produced with a variable amount of branching, each branch containing a terminal ($-CH_3$) group. Branch chains prevent close packing of the main polymer chains, resulting in the production of low density polyethylenes (nominal density < 940 kg m^{-3}).

LDPE is a tough, flexible, slightly translucent material that provides a good barrier to water vapour but a poor barrier to gases. It is widely used to package foods and is easily heat sealed to itself.

LLDPE contains numerous short side chains and has improved chemical and puncture resistance and higher strength than LDPE.

HDPE has a much more linear structure than LDPE, is stiffer and harder and provides superior oil and grease resistance. It is used in both film form where it has a white, translucent appearance, and as rigid packs such as bottles.

PP is a linear polymer with lower density, higher softening point and better barrier properties than the polyethylenes. In film form it is commonly used in the biaxially oriented state (BOPP) where it has sparkling clarity; it can also be blow and injection moulded to produce closures and thin-walled containers.

11.2.4.2 Substituted olefins

Monomers in which each ethylene group has a single substituent are called vinyl compounds and the properties of the resultant polymers depend on the nature of the substituent, molecular weight, crystallinity and degree of orientation.

The simplest is polyvinyl chloride (PVC) with a repeating unit of $(-(CH_2-CHCl)_n-)$. A range of PVC films with widely varying properties can be obtained from the basic polymer. The two main variables are changes in formulation (principally plasticizer content) and orientation.

Thin, plasticized PVC film is widely used for the stretch wrapping of trays containing fresh red meat and produce. The relatively high water vapour transmission rate of PVC prevents condensation on the inside of the film. Oriented films are used for shrink wrapping of produce and fresh meat, but in recent years LLDPE films have increasingly replaced them in many applications.

Unplasticized PVC as a rigid sheet material is thermoformed to produce a wide range of inserts from chocolate boxes to biscuit trays. Unplasticized PVC bottles have better clarity, oil resistance and barrier properties than those made from HDPE. However, they are softened by certain solvents, notably ketones and chlorinated hydrocarbons. They have made extensive penetration into the market for a wide range of foods including fruit juices and edible oils, but in recent years they have been increasingly replaced by PET.

Polyvinylidene chloride (PVdC) has a repeating unit of $(-(CH_2-CCl_2)_n-)$ and the homopolymer yields a rather stiff film which is unsuitable for packaging purposes. When PVdC is copolymerized with 5–50% (but typically 20%) of vinyl chloride, a soft, tough and relatively impermeable film results. Although the films are copolymers of VdC and VC, they are usually referred to simply as PVdC copolymer and their specific properties vary according to the degree of polymerization and the properties and relative proportions of the copolymers present. Properties include a unique combination of low permeability to water vapour, gases and odours, as well as greases and alcohols. They also have the ability to withstand hot filling and retorting and so find use as a component in multilayer barrier containers. Although highly transparent, they have a yellowish tinge. As an important component of many laminates, PVdC copolymers can be sealed to themselves and to other materials. The copolymer is frequently used as a shrink film since orientation improves tensile strength, flexibility, clarity, transparency and impact strength. As well, gas and moisture permeabilities are lowered and tear initiation becomes difficult.

Polyvinyl alcohol (PVOH) has the general formula of $(-(CH_2-CHOH)_n-)$ and is made by alcoholysis of polyvinyl acetate (PVA) $(-(CH_2-CHOCOCH_3)_n-)$. PVOH films are poor barriers to water vapour but excellent barriers to O_2 (when dry) and greases. Wet film has little strength but the strength of dry film is high. Being water soluble, it is difficult to process.

Ethylene vinyl alcohol (EVOH) copolymers are produced by a controlled hydrolysis of ethylene vinyl acetate (EVA) copolymer, the hydrolytic process transforming the VA group into VOH; there is no VOH involved in the copolymerization. EVOH copolymers offer not only excellent processability, but also superior barriers to gases, odours, fragrances, solvents, etc. when dry. It is these characteristics that have allowed plastic containers incorporating EVOH barrier layers to replace many glass and metal containers for packaging food.

Polystyrene (PS) has the general formula $(-(CH_2-CHC_6H_5)_n-)$. Crystal grade PS can be made into film, but it is brittle unless the film is biaxially oriented. While a reasonably good barrier to gases, it is a poor barrier to water vapour. The oriented film can be thermoformed into a variety of shapes. To overcome the brittleness of PS, synthetic rubbers (typically 1,3-butadiene isomer $CH_2=CH-CH-CH_2$) can be added during polymerization at levels generally not exceeding 25% w/w for rigid plastics. The chemical properties of this toughened or high impact polystyrene (HIPS) are much the same as those for unmodified or general purpose polystyrene (GPPS); in addition, HIPS is an excellent material for thermoforming. It is injection moulded into tubs which find wide use in food packaging.

11.2.4.3 Polyesters

Polyethylene terephthalate (PET) is a condensation product of ethylene glycol (EG) and terephthalic acid and has the general formula $(-OOC-C_6H_5-COOCH_2-CH_2-)_n$. The outstanding properties of PET film as a food packaging material are its great tensile strength, excellent chemical resistance, light weight, elasticity and stability over a wide range of temperatures ($-60°$ to $220°C$). PET films are most widely used in the biaxially oriented, heat-stabilized form.

To improve the barrier properties of PET, coatings of LDPE and PVdC copolymer have been used. PET film extrusion-coated with LDPE is very easy to seal and very tough. Two-side, PVdC copolymer-coated grades provide a high barrier and a major special application is the single-slice cheese wrap. PET is also used to make 'ovenable' trays for frozen foods and prepared meals, where they are preferable to foil trays because of their ability to be microwaved without the necessity for an outer paperboard carton.

PET bottles are stretch blow moulded, the stretching or biaxial orientation being necessary to get maximum tensile strength and gas barrier, which in turn enables bottle weights to be low enough to be economical.

11.2.4.4 Polyamides

Polyamides (PA) are condensation, generally linear thermoplastics made from monomers with amine and carboxylic acid functional groups resulting in amide (–CONH–) linkages in the main polymer chain that provide mechanical strength and barrier properties. Nylon 6 refers to PA made from a polymer of ε-caprolactam. One material is used containing 6 carbon atoms. Nylon 11 refers to PA made from a polymer of ω-undecanolactam which has 11 carbon atoms. Nylon 6,6 is formed by reacting hexamethylenediamine with adipic acid. Both materials each contain 6 carbon atoms. Nylon 6,10 is made by reacting hexamethylene diamine with sebacic acid (HOOC–$(CH_2)_8$–COOH). The diamine has 6 carbon atoms and is numerically first, followed by the acid which contains 10 carbon atoms. Films from nylon 6 have higher temperature, grease and oil resistance than nylon 11 films.

A relatively new polyamide is MXD6 made from meta-xylylene diamine and adipic acid, the 6 indicating the number of carbon atoms in the acid. It has better gas barrier properties than nylon 6 and PET at all humidities, and is better than EVOH at 100% RH, due to the existence of the benzene ring in the MXD6 polymer chain. Biaxially oriented film produced from MXD6 is used in several packaging applications as it has significantly higher gas and water vapour barrier properties, and greater strength and stiffness, than other PAs. Together with its high clarity and good processability, the above properties make MXD6 film suitable as a base substrate for laminated film structures for use in lidding and pouches, especially when the film is exposed to retort conditions.

11.2.4.5 Regenerated cellulose

Regenerated cellulose film (RCF) is made from cellulose and is therefore a natural and renewable polymer. However, since it competes with synthetic polymers in food packaging applications it is discussed here. It is commonly referred to by the generic term cellophane which is still a registered trade name in some countries. RCF can be regarded as transparent paper and for food packaging applications it is plasticized (typically with ethylene glycol) and coated on one or both sides, the type of coating largely determining the protective properties of the film. The most common coatings are LDPE, PVC and PVdC copolymer.

11.3 Permeability characteristics of plastic packaging

11.3.1 Permeability

In contrast to packaging materials made from glass or metal, packages made from thermoplastic polymers are permeable to varying degrees to small molecules such as gases, water vapour, organic vapours and other low molecular weight compounds. A plastic polymer that is a good barrier has a low permeability.

Under steady-state conditions, a gas or vapour will diffuse through a polymer at a constant rate if a constant pressure difference is maintained across the polymer. The diffusive flux, J, of a permeant in a polymer can be defined as the amount passing through a plane (surface) of unit area normal to the direction of flow during unit time, i.e.:

$$J = Q / A \bullet t \qquad (11.1)$$

where Q is the total amount of permeant which has passed through area A during time t.

The relationship between the rate of permeation and the concentration gradient is one of direct proportionality and is embodied in Fick's first law:

$$J = -D\frac{\delta c}{\delta x} \qquad (11.2)$$

where

J   is the flux per unit area of permeant through the polymer,

D   is the diffusion coefficient (it reflects the speed at which the permeant diffuses through the polymer),

By substituting for J using Eq. 11.1, the quantity of permeant diffusing through a polymer of area A in time t can be calculated:

$$Q = \frac{D \bullet (c_1 - c_2) \bullet A \bullet t}{X} \qquad (11.5)$$

Rather than the actual concentration, it is more convenient when the permeant is a gas to measure the vapour pressure p which is at equilibrium with the polymer. Henry's law applies at low concentrations and c can be expressed as:

$$c = S \bullet p \qquad (11.6)$$

where S is the solubility coefficient of the permeant in the polymer (it reflects the amount of permeant in the polymer).

Combining Eqs. 11.5 and 11.6:

$$Q = \frac{D \bullet S \bullet (p_1 - p_2) \bullet A \bullet t}{X} \qquad (11.7)$$

The product D $\bullet$ S is referred to as the *permeability coefficient* (or *constant*) or simply just the *permeability* and is represented by the symbol P. Thus:

$$P = \frac{Q \bullet X}{A \bullet t \bullet (p_1 - p_2)} \qquad (11.8)$$

or

$$\frac{Q}{t} = \frac{P}{X} \bullet A \bullet (\Delta p) \qquad (11.9)$$

The term P/X is called the *permeance*.

Consideration of Eq. 11.8 suggests that the dimensions of P are:

$$P = \frac{(\text{quantity of permeant under stated conditions})(\text{thickness})}{(\text{area})(\text{time})(\text{pressure drop across polymer})} \qquad (11.10)$$

The quantity of permeant can be expressed in mass, mol or volume units. For gases, volume is preferred, expressed as the amount permeating under conditions of standard temperature and pressure (STP: 273.15 K and 1.01325×10^5 Pa). Although over 30 different units for P appear in the scientific literature, the

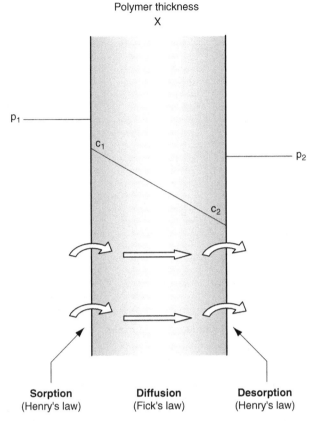

Polymer thickness
X

p_1

c_1

p_2

c_2

| Sorption | Diffusion | Desorption |
| (Henry's law) | (Fick's law) | (Henry's law) |

Figure 11.2 Permeability model for gas or vapour transfer through a polymer. (Copyright 2006. From *Food Packaging Principles & Practice* by G.L. Robertson. Reproduced by permission of Routledge/Taylor & Francis Group, LLC.)

c   is the concentration of the permeant,
$\delta c / \delta x$ is the concentration gradient of the permeant across a thickness δx.

Consider Fig. 11.2 which depicts a polymeric material X mm thick, of area A, exposed to a permeant at pressure p_1 on one side and at a lower pressure p_2 on the other. The concentration of permeant in the first layer of the polymer is c_1 and in the last layer c_2. When steady-state diffusion has been reached, J = constant and Eq. 11.2 can be integrated across the total thickness of the polymer X, and between the two concentrations, assuming D to be constant and independent of c:

$$J \bullet X = -D \bullet (c_2 - c_1) \qquad (11.3)$$

and

$$J = \frac{D \bullet (c_1 - c_2)}{X} \qquad (11.4)$$

Table 11.1 Representative permeability coefficients of various polymers and permeants at 25°C and 90% relative humidity.

Polymer	$P \times 10^{11}$ [mL(STP) cm cm^{-2} s^{-1} (cm Hg)$^{-1}$]			
	O_2	CO_2	N_2	H_2O 90% RH
Low density polyethylene	30–69	130–280	1.9–3.1	800
High density polyethylene	6–11	45	3.3	180
Polypropylene	9–15	92	4.4	680
Polyvinyl chloride film	0.05–1.2	10	0.4	1560
Polystyrene film (oriented)	15–27	105	7.8	12–18,000
Nylon 6 (0% RH)	0.12–0.18	0.4–0.8	0.95	7000
Nylon MXD6	0.01			
Polyethylene terephthlate				
(amorphous)	0.55–0.75	3.0	0.04–0.06	
(40% crystalline)	0.30	1.6	0.07	1300
Polycarbonate film	15	64		
PVdC copolymer	0.05	0.3	0.009	14
EVOH copolymer				
27 mol% ethylene	0.0018	0.024		
44 mol% ethylene	0.0042	0.012		

© 2006. Adapted from *Food Packaging Principles & Practice* by G.L. Robertson. Reproduced by permission of Routledge/Taylor & Francis Group, LLC.

following metric units are the most widely used:

$$\frac{10^{-11}(\text{mL at STP})\,\text{cm}}{\text{cm}^2\,\text{s}\,(\text{cm Hg})}$$

Broadly representative permeability coefficients for a number of polymers to several gases and water vapour are given in Table 11.1 and the interpretation of such data is shown below.

Example   What is the permeability coefficient of polypropylene to O_2 at 25°C expressed in standard units? Taking the upper range value from Table 11.1:

$$P \times 10^{11} = 15\ [\text{mL(STP)}\,\text{cm}\,\text{cm}^{-2}\,\text{s}^{-1}(\text{cm Hg})^{-1}]$$

and therefore

$$P = 15 \times 10^{-11}\ [\text{mL(STP)}\,\text{cm}\,\text{cm}^{-2}\,\text{s}^{-1}(\text{cm Hg})^{-1}]$$
$$= 1.5 \times 10^{-10}\ [\text{mL(STP)}\,\text{cm}\,\text{cm}^{-2}\,\text{s}^{-1}(\text{cm Hg})^{-1}]$$

The above treatment of steady-state diffusion assumes that both D and S are independent of concentration, but in practice deviations do occur. Equation 11.8 does not hold when there is interaction such as occurs between hydrophilic materials (e.g. EVOH copolymers and some of the PAs) and water vapour, or for heterogeneous materials such as coated or laminated films. The property is then defined as the

transmission rate (TR) of the material, where:

$$TR = \frac{Q}{A \bullet t} \qquad (11.11)$$

where

Q is the amount of permeant passing through the polymer,
A is the area,
t is the time.

Permeabilities of polymers to water and organic compounds are often presented in this way, and in the case of water and oxygen, the terms WVTR (water vapour transmission rate) and OTR (oxygen transmission rate) are in common usage. It is critical that the thickness of the film or laminate, temperature and the partial pressure difference of the gas or water vapour are specified for a particular TR. It has become common for the units for WVTR to include a thickness term in which case the WVTR should strictly speaking be referred to as the thickness normalized flux. Data on WVTRs at 38°C and 95% relative humidity are given in Table 11.2 where, for example, low density polyethylene has a WVTR of 0.315–0.59 g mm m^{-2} day^{-1}. To convert a measured WVTR or OTR to P, multiply by the thickness of the film and

Table 11.2 Water vapour transmission rates at 38°C and 95% relative humidity.

Polymer	Transmission rate $(\text{g mm m}^{-2}\text{ day}^{-1}) \times 10^{-2}$
PVdC copolymer	4.1–19.7
Polypropylene	7.8–15.7
High density polyethylene	0.1–0.2
Polyvinyl chloride	19.7–31.5
Low density polyethylene	31.5–59
Polyethylene terephthalate	1–10
Polystyrene	280–393
EVOH copolymer	546
Nylon 6	634–863

Adapted from Karel and Lund (2003).

divide by the partial pressure difference used to make the measurement.

Example Calculate the permeability coefficient of a low density polyethylene film to O_2 at 25°C given that the OTR through a 2.54×10^{-3} cm thick film with air on one side and inert gas on the other is 3.5×10^{-6} mL cm^{-2} s^{-1}.

O_2 partial pressure difference across the film is 0.21 atm = 16 (cm Hg)

$$P = \frac{OTR}{\Delta p} \times \text{thickness}$$
$$= \frac{3.5 \times 10^{-6} \text{ mL cm}^{-2}\text{s}^{-1}}{16 \text{(cm Hg)}} \times 2.54 \times 10^{-3} \text{ cm}$$
$$= 0.55 \times 10^{-9} [\text{mL (STP) cm cm}^{-2}\text{ s}^{-1}\text{ (cm Hg)}^{-1}]$$
$$= 55 \times 10^{-11} [\text{mL (STP) cm cm}^{-2}\text{ s}^{-1}\text{ (cm Hg)}^{-1}]$$

which is within the range given in Table 11.1.

In many food packaging applications the transmission rates of various organic compounds such as flavours, aromas, odours and solvents through polymers are of interest. The permeation of organic vapours through polymer films is much more complicated than that of gases, due to the pressure-dependent solubility coefficient and the concentration-dependent diffusion coefficient. Although extensive studies have been made on the permeability, solubility and diffusivity of various organic vapours in LDPE, much less data are available for other polymers. Since the solvent action of organic vapours varies from polymer to polymer, the perme-

ability cannot be compared in a similar fashion as it can for permanent gases and water vapour.

Many foods require more protection than a single material can provide to give the product its intended shelf life. Where increased barriers to gases and/or water vapour are necessary, it is more economical to incorporate a thin layer of barrier material than to simply increase the thickness of a monolayer. Multi-layer materials can be considered as a number of membranes in series. For the case of three layers in series (total thickness $X_T = X_1 + X_2 + X_3$) and assuming steady-state flux, the rate of permeation through each layer must be constant, i.e.

$$Q_T = Q_1 = Q_2 = Q_3 \qquad (11.12)$$

Likewise, the areas will also be constant so that:

$$A_T = A_1 = A_2 = A_3 \qquad (11.13)$$

If the individual thicknesses and permeability coefficients are known for each layer, and provided that the permeability coefficients are independent of pressure, then Eq. 11.14 can be used to calculate the permeability coefficient for any multi-layer material:

$$P_T = \frac{X_T}{(X_1/P_1) + (X_2/P_2) + (X_3/P_3)} \qquad (11.14)$$

11.3.2 Effect of temperature

The temperature dependence of the solubility coefficient over relatively small ranges of temperature can be represented by an Arrhenius-type relationship:

$$S = S_o \exp(-\Delta H_s/RT) \qquad (11.15)$$

where ΔH_s is the heat of solution. For the permanent gases, ΔH_s is small and positive and therefore S increases slightly with temperature. For easily condensable vapours ΔH_s is negative due to the contribution of the heat of condensation, and thus S decreases with increasing temperature.

The temperature dependence of the diffusion coefficient can also be represented by an Arrhenius-type relationship:

$$D = D_o \exp(-E_d/RT) \qquad (11.16)$$

where E_d is the activation energy for the diffusion process. E_d is always positive and the diffusion coefficient increases with increasing temperature. From the above two equations it follows that:

$$P = P_o \exp(-E_p/RT) \qquad (11.17)$$
$$= (D_o S_o) \exp[-(E_d + \Delta H_s)/RT] \qquad (11.18)$$

where

E_p $(= E_d + \Delta H_s)$ is the apparent activation energy for permeation,

E_p, E_d and ΔH_s are expressed in kJ mol^{-1},

R   = 8.3145 J mol^{-1} K^{-1},

T   is the absolute temperature in Kelvin.

Thus it follows that the permeability coefficient of a specific polymer-permeant system may increase or decrease with increases in temperature depending on the relative effect of temperature on the solubility and diffusion coefficients of the system. Generally, the solubility coefficient increases with increasing temperature for gases and decreases for vapours, and the diffusion coefficient increases with temperature for both gases and vapours. For these reasons, permeability coefficients of different polymers determined at one temperature may not be in the same relative order at other temperatures.

11.3.3 Moisture exchange and shelf life

When a food is placed in an environment at a constant temperature and relative humidity, it will eventually come to equilibrium with that environment. The corresponding moisture content at steady-state is referred to as the equilibrium moisture content. When this moisture content (expressed as mass of water per unit mass of dry matter) is plotted against the corresponding relative humidity or a_w at constant temperature, a moisture sorption isotherm results. Such plots are very useful in assessing the stability of foods and selecting effective packaging. Since a_w is temperature dependent, it follows that moisture sorption isotherms must also exhibit temperature dependence. Thus at any given moisture content, a_w increases with increasing temperature as shown in Fig. 11.3.

The shelf life of a food is controlled by product characteristics including:

- formulation and processing parameters (intrinsic factors);

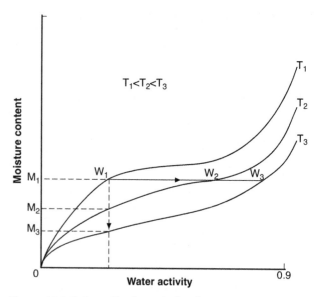

Figure 11.3 Schematic of a typical moisture sorption isotherm showing effect of temperature on water activity and moisture content. (Copyright 2006. From *Food Packaging Principles & Practice* by G.L. Robertson. Reproduced by permission of Routledge/Taylor & Francis Group, LLC.)

- the environment to which the product is exposed during distribution and storage (extrinsic factors);
- the properties of the package.

Examples of intrinsic factors include pH, water activity, enzymes, micro-organisms and concentration of reactive compounds. Many of these factors can be controlled by selection of raw materials and ingredients, as well as the choice of processing parameters.

Examples of extrinsic factors include temperature, relative humidity, light, total pressure and partial pressure of different gases, and mechanical stresses including consumer handling. Many of these factors can affect the rates of deteriorative reactions which occur during the shelf life of a product.

The properties of the package can have a significant effect on many of the extrinsic factors and thus indirectly on the rates of the deteriorative reactions. Thus the shelf life of a food can be altered by changing its composition and formulation, processing parameters, packaging system or the environment to which it is exposed.

Foods can be classified according to the degree of protection required (see Table 11.3), which focuses attention on the key requirements of the package such as maximum moisture gain or O_2 uptake and enables calculations to be made to determine whether or not

Table 11.3 Degree of protection required by various foods and beverages (assuming 1-year shelf life at 25°C).

Food/beverage	Maximum amount of O_2 gain (ppm)	Other gas protection needed	Maximum water gain or loss	Requires high oil resistance	Requires good barrier to volatile organics
Canned milk and flesh foods	1–5	no	3% loss	yes	no
Baby foods	1–5	no	3% loss	yes	yes
Beers and wine	1–5	<20% CO_2 (or SO_2) loss	3% loss	no	yes
Instant coffee	1–5	no	2% gain	yes	yes
Canned soups, vegetables, and sauces	1–5	no	3% loss	no	no
Canned fruits	5–15	no	3% loss	no	yes
Nuts, snacks	5–15	no	5% gain	yes	no
Dried foods	5–15	no	1% gain	no	no
Fruit juices and drinks	10–40	no	3% loss	no	yes
Carbonated soft drinks	10–40	<20% CO_2 loss	3% loss	no	yes
Oils and shortenings	50–200	no	10% gain	yes	no
Salad dressings	50–200	no	10% gain	yes	yes
Jams, jellies, syrups, pickles, olives, vinegars	50–200	no	10% gain	yes	no
Liquors	50–200	no	3% loss	no	yes
Condiments	50–200	no	1% gain	no	yes
Peanut butter	50–200	no	10% gain	yes	no

Adapted from Salame (1974) with kind permission of Springer Science and Business Media.

a particular packaging material would provide the necessary barrier required to give the desired product shelf life. Metal cans and glass containers with good closures can be regarded as essentially impermeable to the passage of gases, odours and water vapour, while paper-based packaging materials can be regarded as permeable. This leaves plastics-based packaging materials which provide varying degrees of protection, depending largely on the nature of the polymers used in their manufacture.

The expression for the steady-state permeation of a gas or vapour through a thermoplastic material was derived above (see Eq. 11.9) and can be rewritten as:

$$\frac{\delta w}{\delta t} = \frac{P}{X} \bullet A \bullet (p_1 - p_2) \qquad (11.19)$$

where $\delta w/\delta t$ is the rate of gas or vapour transport across the film, the latter term corresponding to Q/t in the integrated form of the expression (Eq. 11.9).

The prediction of moisture transfer either to or from a packaged food requires analysis of the above equation given certain boundary conditions. If it is assumed that P/X is constant, that the external environment is at constant temperature and humidity, and that p_2, the vapour pressure of the water in the food, follows some simple function of the moisture content, then a simple analysis can be made.

However, because external conditions will not remain constant during storage, distribution and retailing of a packaged food, P/X will not be constant. If the food is being sold in markets in temperate climates, then WVTRs determined at 25°C/75% RH can be used. A 'worst-case' analysis can be made using WVTRs determined at 38°C/90% RH. A further assumption is that the moisture gradient inside the package is negligible, i.e. the package should be the major resistance to water vapour transport. This is the case whenever P/X is less than about 10 g m^{-2} day^{-1} (cm Hg)$^{-1}$, which is the case for most films under high humidity conditions.

Consideration of Eq. 11.19 shows that the internal vapour pressure is not constant but varies with the moisture content of the food at any time. Consequently the rate of gain or loss of moisture is not constant but falls as Δp gets smaller. Thus to be able to make accurate predictions, some function of p_2, the internal vapour pressure, as a function of the moisture content, must be inserted into the equation. Assuming a constant rate results in the product being overprotected.

In low and intermediate moisture foods, the internal vapour pressure is determined solely by the moisture sorption isotherm of the food. In the simplest case when the isotherm is treated as a linear function:

$$m = b \bullet a_w + c \qquad (11.20)$$

where

m   is moisture content in g H_2O per g solids,
a_w is water activity,
b   is slope of curve,
c   is constant.

The moisture content can be substituted for water gain and after some mathematical manipulation the following expression is obtained:

$$\ln = \frac{m_e - m_i}{m_e - m} = \frac{P}{X} \bullet \frac{A}{W_s} \bullet \frac{p_o}{b} \bullet t \qquad (11.21)$$

where

m_e is equilibrium moisture content of the food if exposed to external package RH,
m_i is initial moisture content of the food,
m   is moisture content of the food at time t,
p_o is vapour pressure of pure water at the storage temperature (*not* the actual vapour pressure outside the package).

The end of product shelf life is reached when m = m_c, the critical moisture content, at which time t = θ_s, the shelf life.

Equations such as Eq. 11.21 have been extensively tested for foods and found to give excellent predictions of actual weight gain or loss. These equations can be used to calculate the effect on shelf life of changes in the permeability of the packaging film, external conditions such as temperature and humidity, the surface area:volume ratio of the package,

and variations in the initial moisture content of the product.

11.4 Interactions between packaging materials and food

11.4.1 Corrosion

The chemical structure which gives metals their valuable practical properties is also responsible for their main weakness: susceptibility to corrosion, the chemical reaction between a metal and its environment. All metals are affected to a greater or lesser extent, and because the reaction takes place at the metal surface, the rate of attack can be reduced and controlled by modifying the conditions at the surface.

When a metal corrodes, atoms of the metal are lost from the surface as cations, leaving behind the requisite number of electrons in the body of the metal. Thus, in the case of a metal M:

$$M \longrightarrow M^{n+} + n \text{ electrons}$$
$$(\text{reduced})\,(\text{oxidized}) \qquad (11.22)$$

Iron and tin will always tend to corrode in aqueous environments since the ionization or corrosion reaction can be balanced by hydrogen-ion reduction (i.e. evolution of hydrogen gas):

$$2Fe \longrightarrow Fe^{2+} + 2e \qquad (11.23)$$
$$2H^+ + 2e \longrightarrow H_2 \qquad (11.24)$$

If the concentration of hydrogen ions is increased (i.e. the aqueous environment has a lower pH), the rate of the reaction will tend to increase.

If free O_2 were available, the corrosion reaction would be balanced by O_2 absorption and, in the case of tin, the balanced reactions would be:

$$Sn \longrightarrow Sn^{2+} + 2e \qquad (11.25)$$
$$\tfrac{1}{2} O_2 + 2H^+ + 2e \longrightarrow H_2O \qquad (11.26)$$

Food products and beverages are extremely complex chemical systems covering a wide range of pH and buffering properties, as well as a variable content of corrosion inhibitors or accelerators. Factors that influence their corrosiveness can be divided into two groups: intensity and type of corrosive attack inherent in the food itself, and corrosiveness due to the

Table 11.4 Some corrosion promoting agents and their mode of reaction.

Corrosion accelerator	Reduction product	Equivalent in weight
Proton (H$^+$)	H$_2$	1 mL H$_2$ ≡ 5.3 mg Sn^{2+}
Oxygen (O$_2$)	H$_2$O	1 mL O$_2$ ≡ 10.6 mg Sn^{2+}
Sulphur dioxide (SO$_2$)	H$_2$S	1 mL SO$_2$ ≡ 5.5 mg Sn^{2+}
Sulphur (S)	H$_2$S	1 mL S ≡ 3.7 mg Sn^{2+}
Nitrate (NO$_3$)	NH$_3$	1 mg NO$_3$ ≡ 7.65 mg Sn^{2+}
Trimethylamine oxide (TMAO)	TMA	1 mg TMAO ≡ 1.57 mg Sn^{2+}

From Mannheim and Passy (1982). Reproduced by permission of Routledge/Taylor & Francis Group, LLC.

processing and storage conditions. All these factors are interrelated and may combine in a synergistic manner to accelerate corrosion.

The most important corrosion accelerators in foods include O$_2$, anthocyanins, nitrates, sulphur compounds and trimethylamines. Some typical corrosion reactions associated with these accelerators and their stoichiometric equivalents of dissolved tin are presented in Table 11.4. While high concentrations of tin in food may cause stomach upsets in some individuals, this is unlikely to be the case where tin concentrations remain below the legal limit of 200 mg kg^{-1} (100 mg kg^{-1} in canned beverages and 50 mg kg^{-1} in canned baby foods).

From a corrosiveness point of view, it is convenient to divide foods into five classes:

1 those that are highly corrosive such as apple and grape juices, berries, cherries, prunes, pickles and sauerkraut;
2 those that are moderately corrosive such as apples, peaches, pears, citrus fruits and tomato juice;
3 those that are mildly corrosive such as peas, corn, meat and fish;
4 strong detinners such as green beans, spinach, asparagus and tomato products;
5 beverages are conveniently considered as a fifth class.

The chromium/chromium oxide layer on ECCS cans is only approximately one-thirtieth to one-fiftieth the thickness of a typical tinplate coating. Thus ECCS cannot be used in food packaging unless it is first lacquered, because of its lack of resistance to corrosion.

Aluminium rapidly forms a protective oxide film when exposed to air or water:

$$4Al + 3O_2 \longrightarrow 2Al_2O_3 \qquad (11.27)$$

The film is extremely thin (about 10 nm) but renders the metal completely passive in the pH range 4–9.

11.4.2 Migration

In addition to permeability, there are two other mass transport phenomena in package systems: sorption and migration. Sorption (also called scalping) involves the take-up of molecules from the food by the package (e.g. flavour compounds from fruit juices by plastics). In addition, compounds present in the environment which surrounds the packaged food can be sorbed by the packaging and migrate into the food (e.g. perfumes from soaps can be picked up by fatty foods packaged in plastics under certain circumstances).

Migration is the transfer of molecules originally contained in the packaging material (e.g. plasticizer, residual monomer, antioxidants) into the product and possibly to the external environment. Overall migration (OM) is the sum of all (usually unknown) mobile packaging components released per unit area of packaging material under defined test conditions, whereas specific migration (SM) relates to an individual and identifiable compound only. Overall migration therefore is a measure of all compounds transferred into the food whether they are of toxicological interest or not, and will include substances that are physiologically harmless.

The migration of molecules from the packaging material into the food is a complex phenomenon, and most mathematical treatments of transport processes are derived initially from a consideration of gaseous diffusion as discussed above. It is worth noting that diffusion in liquids is approximately one million times slower than in gases, and diffusion in solids about one million times slower than in liquids.

11.5 Packaging systems

11.5.1 Modified atmosphere packaging (MAP)

MAP is the enclosure of food in a package in which the atmosphere inside the package is modified or altered to provide an optimum atmosphere for increasing shelf life while maintaining food quality. Modification of the atmosphere may be achieved either actively or passively. Active modification involves displacing the air with a controlled, desired mixture of gases, and is generally referred to as gas flushing. Passive modification (also known as commodity-generated MA) occurs as a consequence of the food's respiration and/or the metabolism of micro-organisms associated with the food; the package structure normally incorporates a polymeric film and so the permeation of gases through the film (which varies depending on the nature of the film and the storage temperature) also influences the composition of the atmosphere that develops.

In vacuum packaging, elevated levels of CO_2 can be produced by micro-organisms or by respiring fruits and vegetables. Thus vacuum packaging of respiring foods or foods containing viable micro-organisms such as flesh foods is a form of MAP since after initial modification of the atmosphere by removal of most of the air, biological action continues to alter or modify the atmosphere inside the package.

With the exception of baked goods, MAP is always used in association with chill temperatures (usually taken as $-1°$ to $+7°C$). The preservative effect of chilling can be greatly enhanced when it is combined with modification of the gas atmosphere because many deteriorative reactions involve aerobic respiration in which the food or micro-organism consumes O_2 and produces CO_2 and water. By reducing O_2 concentration, aerobic respiration can be slowed. By increasing CO_2 concentration, microbial growth can be slowed or inhibited.

The normal composition of air by volume is 78.08% nitrogen, 20.95% oxygen, 0.93% argon, 0.03% carbon dioxide, and traces of nine other gases in very low concentrations. The three main gases used in MAP are O_2, CO_2 and N_2, either singly or in combination. Although the literature on their application and benefits is limited, noble or 'inert' gases such as argon are being used commercially for a wide range of products. Argon is 1.43 times denser than N_2 and

can thus be made to flow like a liquid through air spaces, whereas N_2 cannot. Use of carbon monoxide (CO) and sulphur dioxide (SO_2) has also been reported.

The gas mixtures used for MAP of different foods depend on the nature of the food and the likely spoilage mechanisms. Where spoilage is mainly microbial, the CO_2 levels in the gas mix should be as high as possible, limited only by the negative effects of CO_2 (e.g. package collapse) on the specific food. Typical gas compositions for this situation are 30–60% CO_2 and 40–70% N_2. For oxygen-sensitive products where spoilage is mainly by oxidative rancidity, 100% N_2 or N_2/CO_2 mixtures (if microbial spoilage is also important) are used. For respiring products it is important to prevent anaerobic respiration by avoiding too high a CO_2 level or too low an O_2 level.

Equipment for MAP must generally be capable of removing air from the package and replacing it with a mixture of gases. Three types of packaging equipment are generally used for MAP: chamber machines using preformed pouches or trays; snorkel machines using preformed bags or pouches; and horizontal or vertical form-fill-seal (FFS) machines using pouches or trays.

The main characteristics to be considered when selecting packaging materials for MAP are the package permeability to gases and water vapour, mechanical properties, heat sealability and transparency. For non-respiring products, all the common high gas barrier structures have been used in MAP, including laminates and coextruded films containing PVdC copolymer, EVOH and PAs as a barrier layer. To provide a good heat seal and moisture vapour barrier, the inside layer is usually LDPE.

The choice of suitable packaging materials for the MAP of respiring produce such as fruits and vegetables is much more complex and no easy solutions are available due to the dynamic nature of the product. Ideally the packaging material should maintain a low O_2 concentration (3–5%) in the headspace and prevent CO_2 levels exceeding 10–20%. None of the plastic polymers discussed earlier is able to do this.

MAP is being successfully used around the world to extend the shelf life and retain the quality of a wide variety of foods. Table 11.5 gives examples of foods currently packaged in MAs, together with the gas mixtures typically used.

Table 11.5 Examples of gas mixtures for selected food products.

Product	Temp (°C)	O_2 (%)	CO_2 (%)	N_2 (%)
Meat products				
Fresh red meat	0–2	40–80	20	Balance
Cured meat	1–3	0	30	70
Pork	0–2	40–80	20	Balance
Offal	0–1	40	50	10
Poultry	0–2	0	20–100	Balance
Fish				
White fish	0–2	30	40	30
Oily fish	0–2	0	60	40
Salmon	0–2	20	60	20
Scampi	0–2	30	40	30
Shrimp	0–2	30	40	30
Plant products				
Apples	0–4	1–3	0–3	Balance
Broccoli	0–1	3–5	10–15	Balance
Celery	2–5	4–6	3–5	Balance
Lettuce	< 5	2–3	5–6	Balance
Tomatoes	7–12	4	4	Balance
Baked products				
Bread	RT*		60	40
Cakes	RT		60	40
Crumpets	RT		60	40
Crepes	RT		60	40
Fruit pies	RT		60	40
Pita bread	RT		60	40
Pasta and ready meals				
Pasta	4		80	20
Lasagna	2–4		70	30
Pizza	5		52	50
Quiche	5		50	50
Sausage rolls	4		80	20

* Room temperature; staling is accelerated at refrigerated temperatures. From Brody (2000). Reprinted with permission of John Wiley & Sons, Inc.

11.5.2 Active packaging

Active packaging is *packaging in which subsidiary constituents have been deliberately included in or on either the packaging material or the package headspace to enhance the performance of the package system.* The two key words are deliberately and enhance. Implicit in this definition is that performance of the package system includes maintaining the sensory, safety and quality aspects of the food.

A common type of active packaging involves the absorption or scavenging of O_2 inside the package. The most widely used O_2 scavengers consist of small sachets containing various iron-based powders together with an assortment of catalysts that scavenge O_2 within the food package and

irreversibly convert it to a stable oxide. Water is essential for O_2 absorbents to function and in some sachets the water required is added during manufacture, while in others moisture must be absorbed from the food before O_2 can be absorbed. The iron powder is separated from the food by keeping it in a small sachet (labelled *Do not eat*) that is highly permeable to O_2 and, in some cases, to water vapour.

O_2 absorbers have been used for a range of foods including sliced cooked and cured meat and poultry products, coffee, pizzas, specialty bakery goods, dried food ingredients, cakes, breads, biscuits, croissants, fresh pastas, cured fish, tea, powdered milk, dried egg, spices, herbs, confectionery and snack food.

Sachets that contain $Ca(OH)_2$ in addition to iron powder absorb CO_2 as well as O_2 (sachets which absorb only CO_2 are rare). They find a niche application inside packages of roasted or ground coffee as fresh roasted coffee releases considerable amounts of CO_2 (formed by the Maillard reaction during roasting) that can cause swelling or even bursting of the package unless removed.

During the ripening of fruits and vegetables the plant hormone ethylene (C_2H_4) is produced; it can have both positive and negative effects on fresh produce. Many C_2H_4-adsorbing substances have been described in the patent literature, but those that have been commercialized are based on potassium permanganate which oxidizes C_2H_4 in a series of reactions to acetaldehyde and then acetic acid which can be further oxidized to CO_2 and H_2O.

Ethanol exhibits antimicrobial effects even at low concentrations and sachets containing ethanol (55%) and water (10%) which are adsorbed onto SiO_2 powder (35%) and placed inside a paper/EVA copolymer sachet are available commercially. The sachet contents absorb moisture from the food and release ethanol vapour.

Liquid water can accumulate in packages as a result of transpiration of horticultural produce, temperature fluctuations in high moisture packages, and drip or tissue fluid from flesh foods. This water can lead to the growth of moulds and bacteria as well as fogging of films if allowed to build up in the package. Therefore drip-absorbent pads (consisting of granules of a superabsorbent polymer sandwiched between two layers of a microporous or non-woven polymer) have been used in the packaging of flesh foods to absorb liquid water. Polyacrylate salts and graft copolymers of starch are the most frequently used polymers and they can absorb 100–500 times their own weight of liquid water.

Antimicrobial agents incorporated into packaging materials can be used to prevent the growth of microorganisms on the food surface and thus lead to an extension in shelf life and/or improved microbial safety of the food. The burgeoning interest in antimicrobial food packaging is driven by increasing consumer demand for minimally processed, preservative-free foods. The use of antimicrobial films ensures that only low levels of preservative come into contact with the food compared to the direct addition of preservatives to the food.

Despite the large number of experimental studies on antimicrobials in packaging materials, there have been few commercial applications, the legislative status of antimicrobials being one limiting factor in their commercialization.

11.6 Package closures and integrity

While the choice of suitable packaging material is critically important to achieve the desired product shelf life, adequate closure or sealing of the package after filling is equally important. The quality of the resultant seal is of paramount importance to the ultimate integrity of the package.

For glass containers, a wide range of closures made from either metal or plastics is available. Metal closures are stamped out of sheets of tinplate, ECCS or aluminium and can take four forms: screw caps, crowns, lug caps and spin-on or roll-on closures. Plastic closures are generally compression or injection moulded, the former being based on urea-formaldehyde or phenolic-formaldehyde resins, and the latter on a variety of thermoplastic polymers including PS, LDPE, HDPE, PP and PVC.

The closure used to retain internal pressures of 200–800 kPa as found in carbonated drinks and beer has traditionally been the crown cork, a crimp-on/pry-off friction-fitting closure made from tinplate with a fluted skirt and a cork or plastisol liner. A roll-on tamper-evident aluminium or plastic closure is used where critical sealing requirements, such as carbonation retention, vacuum retention and hermetic sealing are to be met and is especially popular for soft drinks in large containers where reuse is common. The same closures are applied to glass and plastic bottles. The most common closure designed to contain and protect the contents with no internal pressure (e.g. wine in a bottle) has been the traditional bark cork obtained from the holm oak tree *Quercus suber*.

Three types of closures made from metal (either tinplate or ECCS) are used to maintain a vacuum inside a glass container which typically contains heat processed food: a lug-type or twist cap; a press-on twist-off cap held on mainly by vacuum with some assistance from the thread impressions in the gasket wall; and a pry-off (side seal) cap widely used on retorted products and consisting of a cut rubber gasket held in place by being crimped under the curl.

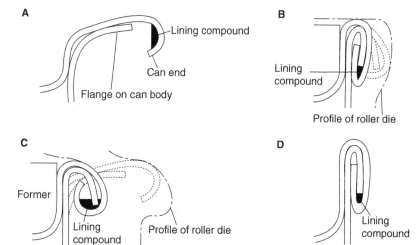

Figure 11.4 Double seaming of metal ends onto metal containers: **A**, end and body are brought together; **B**, first seaming operation; **C**, second seaming operation; **D**, section through final seam. (Copyright 2006. From *Food Packaging Principles & Practice* by G.L. Robertson. Reproduced by permission of Routledge/Taylor & Francis Group, LLC.)

Vacuum closures often have a safety button or flip panel consisting of a raised, circular area in the centre of the panel that provides a visual indicator to the consumer that the package is properly sealed.

For metal containers, the end is mechanically joined to the cylindrical can body by a double seaming operation as illustrated in Fig. 11.4. In the first operation, the end curl is gradually rolled inwards radially so that its flange is well tucked up underneath the body hook. In the second operation, the seam is tightened (closed up) by a shallower seaming roll. The final quality of the double seam is defined by its length, thickness and the extent of the overlap of the end hook with the body hook.

Heat sealable films are considered to be those films that can be bonded together by the normal application of heat. Non-heat-sealable films cannot be sealed this way, but they can often be made heat sealable by applying a heat-sealable coating. In this way the two facing coated surfaces become bonded to each other by application of heat and pressure for the required dwell time. Methods to heat-seal plastic films include conduction, impulse, induction, ultrasonic, dielectric and hot-wire.

Paper packages are typically sealed by the use of adhesives which can be made from either natural materials (e.g. starch, protein or rubber latex) or synthetic materials (e.g. PVA). The latter category can be either water- or solvent-borne; hot-melt and cold-seal type adhesives are also widely available. To confer gas and/or water vapour barrier properties, paper is coated with a continuous film of typically LDPE which also makes it possible to heat seal the coated layers.

11.7 Environmental impacts of packaging

Those involved in the design, development, production or use of packaging and packaging materials can no longer remain oblivious to the environmental demands now placed on them. These demands arise as a consequence of both the materials and processes that are used, and the packaging that is produced, utilized and discarded.

11.7.1 Municipal solid waste (MSW)

MSW – more commonly known as trash, garbage, refuse or rubbish – is simply what is left of the products that have been used or consumed by homes, offices, institutions and businesses, and includes packaging and food scraps.

In its 1989 report entitled *The Solid Waste Dilemma: Agenda for Action*, the EPA in the USA outlined what is referred to as a hierarchy of waste management options, with reuse, reduction and recycling at its apex and landfilling and incineration at its base. Several variations of the hierarchy are currently in circulation, but it is important to note that the hierarchy is not the result of any scientific study and makes no attempt to measure the impacts of individual options or of the overall system. Despite these short-comings, the hierarchy has become accepted as dogma in some countries and among some policymakers, politicians and environmentalists who insist, for example, that reuse is always preferable to recycling, despite the realities in a specific geographical location, e.g. the distance that refillable bottles might have to travel to be refilled.

Because of the wide variability in MSW composition, it follows that there can be no single, global solution to the issue of packaging recovery and recycling. Each waste management programme requires specific technical approaches which reflect geographic differences in both composition and the quantities of waste generated, as well as differences in the availability of some disposal options (e.g. MSW incinerators are rare in many countries). As well, the economic costs of using different waste management options show large variations between and within countries (e.g. the costs for sorting post-consumer packaging).

The concept of integrated waste management (IWM) has, since the mid-1990s, begun to replace the hierarchy as a more useful organizing framework for thinking holistically about waste management. IWM recognizes that all disposal options can have a role to play in waste management and stresses the interrelationships between the options. A mix of waste management options is generally employed depending on the specific local conditions, with the objective being to optimize the whole system rather than its parts, making it economically and environmentally sustainable.

Several MSW management practices such as source reduction, recycling and composting prevent or divert materials from the waste stream. Other practices such as incineration and landfill address those materials that require disposal.

11.7.2 Source reduction

Source reduction involves altering the design, manufacture or use of products and materials so that less total material is used and consequently there is less waste at the end. Lighter packages also require less energy for transportation, thus reducing the environmental impacts from energy production and use. There have been dramatic reductions of 20–46% in the weight of primary food packages over the past 40 years.

11.7.3 Recycling and composting

Recycling diverts items such as paper, glass, plastics and metals from the waste stream. Closed loop recycling refers to the recycling of a particular material back into a similar product, e.g. the recycling of glass bottles back into new glass bottles. Before any post-consumer packaging material can be recycled, it first has to be collected and sorted so that a clean stream of material can be delivered to the recycler. Although some package types are sorted by consumers prior to collection, most packaging materials are collected commingled and sorting takes place at a materials recovery facility (MRF – rhymes with 'surf'). The design and operation of MRFs vary widely within and between countries, with developed countries installing more automated sorting machinery in an effort to increase efficiency and reduce costs.

Composting involves microbial decomposition of organic waste such as food scraps and yard trimmings, as well as uncoated paper and other biodegradable packaging materials, to produce a humus-like substance.

11.7.4 Waste-to-energy incineration

Combustion or incineration is another MSW practice that helps reduce the amount of landfill space needed. Incinerators burn MSW at high temperatures, reducing waste volume and usually generating electricity from the waste heat. The energy content of MSW ranges from 6 to 8 MJ kg^{-1} depending on the proportion of food scraps and green waste. Of the common plastics packaging materials, LDPE has an energy content of 43.6, PS 38.3 and PVC 22.7 MJ kg^{-1}. Beverage cartons (laminates of paperboard, aluminium foil and LDPE) have an energy content of 21.3 MJ kg^{-1}. By way of comparison, wood chips have an energy content of 8.3, coal 26.0 and oil 41.0 MJ kg^{-1}.

The operation of incinerators results in the production of a variety of gaseous and particulate emissions, many of which are thought to have serious health impacts. Electrostatic precipitators and cyclones can be used to remove fly ash (particulates from the flue gases), and acid gases (HCl, SO_2 and HF) can be removed using scrubbers and CaO or $NaOH$ solutions.

11.7.5 Landfill

Landfills are the physical facilities used for the disposal of residual solid wastes in the surface soils of the earth. An engineered facility for the disposal of MSW designed and operated to minimize public health and environmental impacts is referred to as a sanitary landfill.

The anaerobic decomposition of organic materials in an MSW landfill will generate a combination of gases (roughly 50% CH_4 and 50% CO_2) at a rate of approximately $0.002 \, m^3 \, kg^{-1}$ of waste per year. Methane has 22 times more impact as a greenhouse gas than CO_2 and consequently recovering CH_4 and converting it to electricity reduces the potency of landfill gas as a greenhouse gas.

11.7.6 Life cycle assessment (LCA)

LCA is an environmental management tool which attempts to consider the resource and energy use and the resultant environmental impacts or burdens over the entire life cycle of a package, product or service – from extraction of the raw materials through manufacture/conversion, distribution and use to recovery or disposal. It is sometimes referred to as 'cradle to grave' analysis and typically compares two or more products that provide the same function or equivalent use. Thus an LCA makes it possible to isolate the stages in the life cycle of a process or product that make the most significant contribution to its environmental impacts. The first LCAs were performed on beverage containers in 1972 in the USA.

Despite the increasing popularity of LCAs by both industry and governments, the technique does have significant limitations which are often overlooked. One is that LCAs are not able to assess the actual environmental effects of emissions and wastes from the product or package because the actual effects will depend on when, where and how they are released into the environment. Another is that LCAs take no account of economic factors such as the costs of raw materials, manufacturing, transport and recovery or disposal. Furthermore the conclusions from LCAs are specific only for the precise system under study and cannot be extrapolated to provide universal generalizations that, for example, one particular package is always better than another in every situation.

Notwithstanding the limitations described above, LCA can be a very useful tool in two major areas. First, package design, development and improvement all benefit from having LCA results available which can help identify where significant resource use, wastes and emissions occur, and thus suggest where significant changes or improvements can be made. In situations where a 'quick and dirty' comparison is sufficient, only the weight of the competing packages is used since the lightest-weight package almost always has the lowest environmental impacts (PVC is one notable exception). Second, LCA is a useful tool for assessing waste management options and planning integrated solid waste management systems on a case-by-case and regional basis.

11.7.7 Packaging waste legislation

The EU Directive 94/62/EC on Packaging and Packaging Waste was the first product-specific regulation in the field of EU waste policy and laid down quantitative targets for recovery (i.e. material recycling, incineration with energy recovery and composting) and recycling of packaging waste. A substantial increase in the targets for recovery and recycling was contained in the amendment to the Directive (2004/12/EC) released in February 2004.

By 31 December 2008, 60% as a minimum by weight of packaging waste must be recovered and 55% as a minimum and 80% as a maximum by weight must be recycled, with the following minimum recycling targets for each material (by weight):

- 60% for glass, paper and board;
- 50% for metals;
- 22.5% for plastics (counting exclusively material that is recycled back into plastics);
- 15% for wood.

The Directive states that energy recovery shall be encouraged where it is preferable to material recycling for environmental and cost-benefit reasons, and recycling targets for each specific waste material should take account of life cycle assessments and cost-benefit analyses which have indicated clear differences both in the costs and in the benefits of recycling.

Further reading and references

Brody, A.L. (2000) Packaging: Part IV – controlled/modified atmosphere/vacuum food packaging. In: *The Wiley Encyclopedia of Food Science and Technology*, 2nd edn. (ed. F.J. Francis), Vol 3, pp. 1830–39. John Wiley, New York.

Brody, A.L. and Marsh, K.S. (eds) (1997) *The Wiley Encyclopedia of Packaging Technology*, 2nd edn. John Wiley, New York.

Chiellini, E. (ed.) (2008) *Environmentally Compatible Food Packaging*. Woodhead Publishing, Cambridge.

Han, J.H. (ed.) (2005) *Innovations in Food Packaging*. Elsevier Academic Press, San Diego.

Karel, M. and Lund, D.B. (2003) Protective packaging. In: *Physical Principles of Food Preservation*, 2nd edn, p. 551. Marcel Dekker, New York.

Krochta, J.M. (2007) Food packaging. In: *Handbook of Food Engineering,* 2nd edn (eds D.R. Heldman and D.B. Lund), pp. 847–927. CRC Press, Boca Raton.

Lee, D.S., Yam, K.L. and Piergiovanni. L. (2008) *Food Packaging Science and Technology*. CRC Press, Boca Raton.

Mannheim, C. and Passy, N. (1982) Internal corrosion and shelf-life of food cans and methods of evaluation. *CRC Critical Reviews in Food Science and Nutrition,* **17,** 371–407.

Piringer, O.-G. and Baner, A.L. (eds) (2008) *Plastic Packaging Interactions with Food and Pharmaceuticals*, 2nd edn. Wiley-VCH, Weinheim.

Robertson, G.L. (2006) *Food Packaging Principles & Practice*, 2nd edn. CRC Press, Boca Raton.

Robertson, G.L. (2009) Packaging of food. In: *The Wiley Encyclopedia of Packaging Technology*, 3rd edn. (ed. K.L. Yam). John Wiley, New York.

Robertson, G.L. (ed.) (2009) *Food Packaging and Shelf Life: A Practical Guide*. CRC Press, Boca Raton.

Salame, M. (1974) The use of low permeation thermoplastics in food and beverage packaging. In: *Permeability of Plastic Films and Coatings* (ed. H.B. Hopfenberg), p. 275. Plenum Press, New York.

Selke, S.E.M., Culter, J.D. and Hernandez, R.J. (2004) *Plastics Packaging Properties, Processing, Aplications and Regulations*, 2nd edn. Hanser Publishers, Munich.

C. Jeya Henry and Lis Ahlström

Key points

- Estimation of energy and protein requirements.
- Optimal nutrient needs for health and well-being.
- Nutrient needs during the life cycle.
- Special needs for growth, pregnancy and lactation.
- Food sources rich in nutrients.

12.1 Introduction

Nutrition may be defined as "The science of food, the nutrients and substances therein, their action, interaction, and balance in relation to health and disease, and the process by which the organisms ingest, digest, absorb, transport, utilize, and excrete food substances" (The Council on Food and Nutrition of the American Medical Association). The discovery of vitamins in the early part of the twentieth century (1906) may be viewed as a landmark in the inception of the science of nutrition. Nevertheless, it is only in the last 40 years that it has captured the interest and imagination of policy makers and the public alike. Today, not only is nutrition shaping the development of new food products, but technological innovations have helped create a range of nutritious and functional foods. This chapter provides basic information relevant to food scientists, technologists and food suppliers.

We require food for four main reasons:

1. as a source of energy;
2. as a source of raw materials for growth and development;
3. to supply minute chemicals that serve to regulate vital metabolic processes;
4. to supply food components (phytochemicals) that retard the development and progression of degenerative diseases.

12.2 Human energy requirements

It is no exaggeration to claim that the science of nutrition was founded on the study of energy metabolism. The largest contribution to energy expenditure is basal metabolic rate (BMR). BMR may be defined as the sum total of the minimum activity of all tissue cells of the body under steady-state conditions. It is also referred to as the minimal rate of energy expenditure compatible with life.

Although the BMR can be measured using direct calorimetry, its measurement is usually made indirectly. Indirect calorimetry measures the consumption of oxygen, the expiration of carbon dioxide, and the elimination of urinary nitrogen. These may then be used to measure the oxidation of fuel. From the respiratory quotient (RQ), which is the ratio of the carbon dioxide expired to the oxygen inhaled, the amount of heat being produced can be calculated.

12.2.1 Indirect calorimetry

Indirect calorimetry refers to the calculation of heat production using the measurement of gaseous exchange – notably, oxygen consumed and carbon dioxide expired. Whilst the heat equivalent of respiratory exchange is usually calculated from this, it is also dependent on the ratio of the moles of carbon dioxide produced to the moles of oxygen consumed, and this is called the respiratory quotient (RQ):

$$RQ = \frac{moles\ CO_2}{moles\ O_2}$$

The RQ varies when carbohydrate (carbohydrates), fat, and protein are oxidized. Differences in their composition determine the amount of oxygen required for complete oxidation. RQ for carbohydrates is 1.0. The amount of molecular oxygen required for oxidation is equal to the carbon dioxide produced during the combustion of carbohydrates. The oxidation of glucose may be illustrated as:

$$C_6H_{12}O6 + 6\ O_2 \rightarrow 6\ CO_2 + 6\ H_2O$$

Fats require more oxygen than carbohydrates for combustion as the fat molecule contains a lower ratio of oxygen to carbon and hydrogen. Thus, for fat, the RQ is represented as:

$$2\ C_{57}H_{110}O_6 + 163\ O_2 \rightarrow 114\ CO_2 + 110\ H_2O$$

$$\frac{CO_2}{O_2} = \frac{114}{163} = 0.70$$

The calculation of the RQ for protein is more complex than that for fat or carbohydrates as protein is not completely oxidized. Both carbon and oxygen are

Table 12.1 Non-protein RQ and calorific equivalent for oxygen and carbon dioxide.

Non-protein respiratory quotient	Oxygen		Carbon dioxide	
	kcal/l	kJ/l	kcal/l	kJ/l
0.70	4.686	19.60	6.694	28.01
0.72	4.702	19.67	6.531	27.32
0.74	4.727	19.78	6.388	26.73
0.76	4.732	19.80	6.253	26.16
0.78	4.776	19.98	6.123	25.62
0.80	4.801	20.09	6.001	25.11
0.82	4.825	20.19	5.884	24.62
0.84	4.850	20.29	5.774	24.16
0.86	4.875	20.40	5.669	23.72
0.88	4.900	20.50	5.568	23.30
0.90	4.928	20.62	5.471	22.89
0.92	4.948	20.70	5.378	22.50
0.94	4.973	20.81	5.290	22.13
0.96	4.997	20.91	5.205	21.78
0.98	5.022	21.01	5.124	21.44
1.00	5.047	21.12	5.047	21.12

excreted in the urine mainly as urea. When adjustment is made for urinary excretion, the ratio of carbon dioxide produced to oxygen consumed is approximately 1:1.2, producing an RQ of 0.80.

Table 12.1 shows the non-protein RQ and the calorific equivalent for oxygen and carbon dioxide. The calorie equivalence of oxygen alone is usually used for the estimation of energy expenditure as it varies little within an RQ range of 0.7 to 0.86 (compared to CO_2).

Table 12.2 represents the oxygen consumed, carbon dioxide produced and energy equivalence of oxygen when protein, fat and carbohydrates are metabolized.

12.2.2 Energy expenditure estimation: a shortcut method

Almost 60 years ago, Weir (1949) showed that energy expenditure (E) may be calculated without the need to obtain RQ, as shown in the equation below:

$$E\ (kJ^{-1}min) = \frac{20.58\,\dot{V}}{100}(20.93 - O_{2e})$$

where

$\dot{V}$ is the volume of expired air in liters per minute at standard temperature and pressure (STP),

O_{2e} is expressed as percentage oxygen content of expired air.

Table 12.2 Energy release from starch, fat, and protein.

Nutrient	O_2 consumed $(I^{-1}g)$	CO_2 produced $(I^{-1}g)$	RQ	Energy released $(kJ^{-1}g)$	Energy released $(kJ^{-1} I O_2)$
Starch	0.83	0.83	1.0	17.5	21.1
Fat	1.98	1.40	0.7	39.1	19.8
Protein	0.96	0.78	0.8	18.5	19.3

12.2.3 Method for the estimation of energy requirements

One practical use of BMR is in the estimation of energy requirements for population groups and subsequently their food needs. The Food and Agricultural Organization (FAO)/World Health Organization (WHO)/United Nations University (UNU) report Energy and Protein Requirements (1985) made clear for the first time the two main purposes of determining energy requirements. The first was for prescriptive purposes, i.e. for making recommendations about the level of consumption that ought to be maintained in a population; the second, for diagnostic purposes, i.e. the assessment of adequacy or inadequacy of the food situation of a population. The document recommended that daily requirements should be estimated using a measurement of energy expenditure. This recommendation dramatically enhanced the importance of measuring the BMR, which represents the largest component of total energy expenditure (TEE). Historically, BMR was measured using indirect calorimetry, i.e. assessment of oxygen consumption. By 1919, a series of predictive equations was developed, most notably the Harris and Benedict equations to predict BMR. The Harris and Benedict equations may be described as follows:

- Males: h = 66.4730 + 13.7516W − 5.0033S − 6.7750A
- Females: h = 665.0955 + 9.5634W + 1.8496S − 4.6756A

  (h = kcal day^{-1}, W = weight in kilograms, S = stature in centimeters, A = age in years)

The Harris–Benedict equations are still widely used in clinical nutrition. Quenouille analysis in 1951 represented the first comprehensive survey of all the BMR studies conducted worldwide and represented over 8600 subjects. Quenouille review of the early BMR literature formed the basis of the Schofield equations, now widely referred to as the FAO/WHO/UNU (1985) equations for the prediction of BMR. Table 12.3 shows the predictive equations for basal metabolic rate based on age, gender, and body size.

Whilst numerous equations for the estimation of BMR are available, a simple and easy-to-use equation is the Kleiber–Brody equation, as shown below. Its simplicity and ease of application make it an ideal equation to predict BMR in most mammals, including man:

$$BMR \ (kcal^{-1}day) = 70W^{.75}$$

where W is weight in kilograms.

12.2.4 Estimation of BMR

BMR represents approximately 50–75% of TEE in adults, and is influenced by gender, body size, body composition, and age. It is measured using direct or

Table 12.3 Equation to estimate BMR from body weight.

Age (years)	BMR (MJ/day)	BMR (kcal/day)
Males		
< 3	0.249 kg − 0.127	59.512 kg − 30.4
3–10	0.095 kg + 2.110	22.706 kg + 504.3
10–18	0.074 kg + 2.754	17.686 kg + 658.2
18–30	0.063 kg + 2.896	15.057 kg + 692.2
30–60	0.048 kg + 3.653	11.472 kg + 873.1
≥ 60	0.049 kg + 2.459	11.711 kg + 587.7
Females		
< 3	0.244 kg − 0.130	58.317 kg − 31.1
3–10	0.085 kg + 2.033	20.315 kg + 485.9
10–18	0.056 kg + 2.898	13.384 kg + 692.6
18–30	0.062 kg + 2.036	14.818 kg + 486.6
30–60	0.034 kg + 3.538	8.126 kg + 845.6
≥ 60	0.038 kg + 2.755	9.082 kg + 658.5

Adapted from FAO/WHO/UNU (1985).

indirect calorimetry under very strict metabolic conditions, including the following:

1 10–14 hours after a meal;
2 awake in a supine position;
3 following 8–10 hours of physical rest with no strenuous exercise the previous day;
4 in a thermoneutral environment that does not stimulate shivering or sweating.

Thus, the measurement of BMR is a highly technical and a relatively invasive process. It is therefore common to estimate BMR using predictive equations.

TEE is composed of:

- BMR;
- dietary-induced thermogenesis (DIT) or the thermic effect of food (TEF);
- physical activity (PA);
- growth.

In non-pregnant, non-lactating adults, growth makes no contribution to energy expenditure and therefore,

$$TEE = BMR + DIT + PA$$

DIT (TEF) is the second component of energy expenditure. The percentage increase in energy expenditure over BMR due to DIT ranges from 8% to 15% depending on the composition of the food consumed. It is recognized that a high protein diet elicits the greatest DIT (up to 15%). The most commonly used value for DIT is 10% of the caloric value of mixed meals consumed over a 24-h period.

Physical activity or exercise (PA) is the most variable of the components; it is also the only component that can be easily altered.

12.2.5 Factorial estimation of total energy expenditure and physical activity level (PAL)

Once BMR has been estimated using the predictive equations, it is necessary to increase the BMR by a "factor" to accommodate the energy cost of PA and DIT. This factorial approach is now widely recognized as the best method to estimate energy requirements in humans. Humans vary in body size, body composition, and levels of PA. In order to account for the differences in PA, TEE is estimated using factorial calculations, combining the time allocated to habitual activities and the energy cost of each activity. Some examples of these calculations are shown in Table 12.4. The energy cost of different activities is also expressed as a multiple of BMR per minute. This is referred to as the physical activity ratio (PAR). The 24-h energy requirements, when expressed as a multiple of BMR, are known as physical activity level (PAL).

The PAL values based on levels of PA are set the same for both genders. However, in order to recognize biological variability in levels of physical fitness and differences in body composition, a range of PAL values are presented in each category (Table 12.5).

Table 12.4 Factorial calculations of total energy expenditure for a population group that has an active or moderately active lifestyle.

Main daily activity	Time allocation (hours)	Energy cost (PAR)	Time × energy cost	Mean PAL (multiple of 24-h BMR)
Sleeping	8	1	8	
Personal care (dressing, showering)	1	2.3	2.3	
Eating	1	1.5	1.5	
Standing, carrying light loads	8	2.2	17.6	
Commuting to/from work on bus	1	1.2	1.2	
Walking at varying paces without load	1	3.2	3.2	
Light-intensity aerobic exercise	1	4.2	4.2	
Light leisure activities (TV, chatting)	3	1.4	4.2	
Total	**24**		**42.2**	**42.2 / 24 = 1.76**

PAR Physical activity ratio; PAL physical activity level.
Adapted from FAO/WHO/UNU (2001).

Table 12.5 Classification of lifestyle in relation to the intensity of habitual physical activity, or PAL.

Category	PAL value
Sedentary or light active lifestyle	1.40–1.69
Active or moderately active lifestyle	1.70–1.99
Vigorous or vigorously active lifestyle	2.00–2.40*

* A PAL value > 2.40 is difficult to maintain over a long period of time.
Adapted from FAO/WHO/UNU (2001).

12.2.6 Procedure for the calculation of human energy requirements

1 Use predictive equations (Table 12.3) to calculate BMR. Example: an 18—30 year old male, body weight 70 kg. BMR = 7.306 MJ/day.
2 PAL (moderately active lifestyle), midpoint value 1.85 (Table 12.5).
3 TEE = BMR (7.306) × PAL (1.85)
    = 13.5 MJ/day (3230 kcal/day) or 13.5 MJ/70 (3230/70)
    = 193 kJ/kg/day (46 kcal/kg/day).

12.2.7 Energy requirements during pregnancy

Energy requirements during pregnancy may be considered as an additional energy requirement over and above that calculated for normal adult females (see above procedure). The additional energy requirements for pregnancy are based on a mean gestational weight gain of approximately 12 kg. Since the energy cost of pregnancy is not evenly distributed during the gestation period, the total additional energy requirement for a successful pregnancy has been estimated at 321MJ (77,000 kcal). Additional energy requirements for pregnancy during the first, second, and third trimesters are shown in Table 12.6.

12.2.8 Energy requirements during lactation

The energy cost of lactation is very high for women who exclusively breastfeed their infants. This re-

Table 12.6 Additional energy requirements during pregnancy.

Trimester	MJ/day	kcal/day
1	0.4	85
2	1.2	285
3	2.0	475

quires additional energy requirements over and above their daily energy requirements. Whilst the fat stores accumulated during pregnancy may contribute, in part, to the energy demands of lactation, lactating women need to consume additional energy. It has been estimated that an additional energy requirement of 2.8 MJ/day (675 kcal/day) should meet the demands of lactation.

12.2.9 Units of energy

Energy may be measured in many different ways (calories, ergs, joules, watts) However, nutritionists normally use kilocalories (kcal) or the SI unit for energy, the joule (J), to express energy.

1 calorie = 1 kcal = 1000 calories
1 joule = 10 ergs = 0.2239 calories
1 kJ = 1000 joules = 0.239 calories
1 kcal = 4.184 kJ (kilo joules)
1000 kJ = 1 megajoule (MJ)

Specifically, 1 kcal is the amount of heat required to raise 1 kg of water from 14.5°C to 15.5°C. The energy needs of a human can therefore be described in any one of these units.

Conventionally in nutrition, energy requirements are expressed in kilocalories (usually called "calories" in the popular media) or megajoules. An energy requirement of 3000 kcal/day can also be expressed as 12.5 MJ/day.

12.2.10 Energy value of foods: estimation of the energy content of foods

Since Lavoisier's classical experiments on the origin of animal heat production, it has been recognized that foods burnt outside the body produce an identical amount of heat as foods oxidized by the slow metabolic process of the human body. This is the basis of Hess's law. When food is combusted in a calorimeter, the energy generated is defined as gross energy. The gross energy value of food, however, is not representative of the energy available to the human body as it does not take into account the digestibility of the foods. Some undigested food will be excreted in the faeces. Gross energy corrected for digestibility is called digestible energy. Digestibility of most foods is high; on average, 97% of ingested carbohydrates, 95%

of fats, and 92% of proteins are absorbed by the human gut.

Whilst carbohydrates and fats are completely oxidized to carbon dioxide and water in the human body, proteins are inefficiently oxidized. Thus, compounds such as urea, uric acid, creatinine, and other nitrogenous proteins are excreted in the urine. This metabolic loss must be subtracted from the digestible energy of protein. Digestible energy corrected for urine losses is called metabolizable energy (ME). Metabolizable energy represents the potential energy available to the human body when foods are ingested. Figure 12.1 illustrates the schematic flow of the energy value in food. The energy value of food is usually expressed in terms of either kilocalories or kilojoules (1 kcal = 4.184 kJ).

"How many calories does a pizza have"? is a common query today. Once the composition of pizza is known (the amount of protein, fat, and carbohydrates per 100 g) these values are converted to energy units using the "Atwater factors". The procedure used to calculate the energy value of foods can be traced to the work of Atwater in 1899. Today we use the Atwater factors of 4, 9, and 4 kcal as the ME of protein, fat, and carbohydrate respectively.

12.2.11 Preparation of food composition tables

The composition of food and the breakdown to its various constituents have been topics of interest since the time of Liebig (1803–1873). In 1843, Jonathan Pereira published his book *Treaties on Food and Diet*, which presented numerous analyses of different foods. As the nineteenth century closed, Atwater (1906) in America produced one of the most comprehensive food composition tables. From the 1930s, McCance and Widdowson started a partnership that lasted for nearly 70 years analyzing and presenting food composition tables in the UK.

The early determination of carbohydrates by "difference" meant that protein, fat, and water was estimated and the remainder assumed to be carbohydrates. As this included a considerable amount of unavailable cell wall structure in the carbohydrates fraction, it invariably overestimated the carbohydrate content. Today, chemical methods are adopted to determine separately the various kinds of carbohydrates including starch, sucrose, glucose, fructose, and maltose.

12.2.12 Energy density

The energy density is defined as the energy content per gram of food. The high fat and low moisture content of most western foods makes energy density an important contributor to energy regulation and overweight. Whilst the addition of fat and sugar increases palatability and taste, they both contribute to the energy density of foods. In many developing countries where the staples are cereal based or tubers, their energy density on cooking is very low. This is largely due to a considerable absorption of water during the gelatinization of starch. Table 12.7 presents the energy density of various foods.

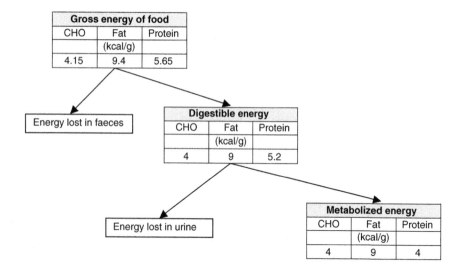

Figure 12.1 The metabolized energy value of foods.

Table 12.7 Fat content and energy density of some common foods.

Food	Fat content (g)	Energy density (kcal/g)	Energy density (kJ/g)
Soy flour	23.5	4.47	18.7
Wheat flour	2.0	3.24	13.6
Fresh pasta, raw	2.4	2.74	11.5
Fresh pasta, cooked	1.5	1.59	6.7
Easy cook rice, raw	3.6	3.83	16.0
Easy cook rice, cooked	1.3	1.38	5.8
Cheese and tomato pizza	10.3	2.77	11.6
Milk, full fat	3.9	0.66	2.8
Double cream	53.7	4.96	20.8
Cheddar cheese	34.9	4.16	17.4
Chicken eggs, raw	11.2	1.51	6.3
Bacon, back raw	16.5	2.15	9.0
Cornish pasty	16.3	2.67	11.2
Chicken curry	9.8	1.45	6.1
Cod in batter	15.4	2.47	10.3
Salmon, raw	11.0	1.80	7.5
Peanuts, roasted	53.0	6.02	25.2
Potato crisps	34.2	5.30	22.2
French fries	15.5	2.80	11.7

Source: McCance and Widdowson's The Composition of Foods (Food Standards Agency, 2002).

The energy density of foods can range from 0 (water) to 37 kJ/g (fat) (0—9 kcal/g). The average energy density of foods consumed in Europe and the USA ranges from 4.2 to 8 kJ/g (1—2 kcal/g). As energy density is essentially dependent on the moisture and fat content, numerous low calorie foods available today are manufactured by altering either the moisture and/or the fat content.

Low-energy dense foods are made with considerable attention being paid to their palatability. However, several studies have shown that they have little or no effect on suppressing appetite.

12.3 Protein

The determination of protein requirements for humans has generated considerable debate over the last century. For example, in the late nineteenth century, both Voit and Atwater suggested an intake of approximately 120 g protein/day. Their estimation was based on surveying the habitual food intake in a predominantly German population. In contrast, Chittenden working in the USA suggested a daily intake of 55 g of protein. Today, it is recognized that the mini-

mum protein requirements for humans is closer to the value suggested by Chittenden than that suggested by Voit and Atwater.

As proteins are known to be hydrolyzed into amino acids during digestion, the digestibility of proteins is dependent on the presence of fibre polyphenols, and other minor food constituents. As proteins are composed of amino acids (both essential and non-essential amino acids) the amino acid composition dictates the biological value of protein. Essential amino acids (indispensable) in humans are those that cannot be synthesized by the body in adequate amounts.

To estimate the protein content of a food, the total nitrogen content of that food is multiplied by 6.25. This value is derived from the knowledge that pure protein contains 16% nitrogen. Thus, 1 g of nitrogen corresponds to $100 \div 16 = 6.25$ g of protein. Protein content $= N \times 6.25$.

As 6.25 is an average factor, it is not appropriate for certain proteins. For example, the nitrogen value of cereals should be multiplied by 5.7, and for milk by 6.36.

The terms protein and nitrogen content will be used interchangeably in this chapter.

12.3.1 Protein content of food

The protein content of food varies markedly along with its amino acid composition and functional properties. Table 12.8 provides the protein content of some major food categories. To further illustrate the variability in protein content of cereals, Table 12.9 presents some examples.

12.3.2 Digestibility of proteins

The presence of fibre and polyphenols and the influence of high-temperature processing may affect

Table 12.8 Protein content of food groups.

Foods	Protein content range (g/100g)
Cereal	6–15
Legume	18–45
Oil seeds	17–28
Shell fish	11–23
Fish	18–22
Meat	18–24
Milk (fresh)	3.5–4.0

Table 12.9 Protein content of cereals.

Cereal (dry)	Protein content (g/100g)
Wheat	11.6
Rice	7.9
Corn	9.2
Barley	10.6
Oats	12.5
Rye	12.0
Sorghum	10.4
Millet	11.8

Table 12.10 Protein sources and true digestibility.

Protein source	True digestibility
Egg	97
Milk, cheese	95
Meat, fish	94
Maize	85
Rice, polished	88
Wheat, whole	86
Oatmeal	86
Peas, mature	88
Soyflour	86
Beans	78
Maize + beans	78
Maize + beans + milk	84
Indian rice diet	77
Chinese mixed diet	96
Filipino mixed diet	88
American mixed diet	96
Indian rice + beans diet	78

Adapted from *Energy Requirements* (FAO/WHO/UNU, 2001).

the digestibility of proteins. In its simplest form, digestibility may be assessed by measuring the difference between intake and faecal losses. "Apparent protein (N) digestibility" and "true protein (N) digestibility" are calculated as shown below. The difference between the two is that true digestibility takes account of the faecal loss when fed a non-protein diet.

$$\text{Apparent protein (N) digestibility (\%)} = \frac{I-F}{I} \times 100$$

$$\text{True protein (N) digestibility (\%)} = \frac{I-(F-F_k)}{I} \times 100$$

where

I is nitrogen intake,
F is faecal nitrogen output on the test diet,
F_k is faecal nitrogen output on a non-protein diet.

The digestibility of animal proteins is much higher than vegetable proteins. Moreover, the digestibility of diets from developing countries (due to the presence of a higher fibre and polyphenols content) is lower than that in western diets. Interestingly, the addition of a small amount of animal protein (for example, milk) enhances the digestibility of vegetable protein considerably.

Table 12.10 shows some protein sources and the true digestibility of some common foods.

12.3.3 Dietary sources

Foods provide all the 20 odd amino acids used for the synthesis of protein, peptides, and other nitrogen compounds. Most plant proteins are deficient in one or more essential amino acids. This deficiency may be overcome by consuming mixed meals containing different sources of plant proteins. For example, when bread or baked beans are consumed separately, the two protein sources are deficient in either lysine (bread) or methionine (baked beans). However, when the bread and baked beans are consumed together, the amino acids complement one another, leading to a "complete" protein source. Table 12.11 shows some protein sources and their limiting amino acids.

Whilst cooking tends to generate flavours and aroma, several amino acids are lost during the cooking process: notably the sulphur amino acids, threonine, and tryptophan. Whilst the overall bioavailability of protein improves with cooking,

Table 12.11 Protein sources and their limiting amino acids.

Protein source	Limiting amino acid
Wheat	Lysine
Corn	Tryptophan
Rice	Threonine
Legumes	Methionine
Soy bean	Methionine

Table 12.12 Amino acids.

Indispensable	Conditionally indispensable	Dispensable
Valine	Glycine	Glutamic acids (?)
Isoleucine	Arginine	Alanine
Leucine	Glutamine	Serine
Lysine	Proline	Aspartic acid
Methionine	Cystine	Asparagine
Phenylalanine	Tyrosine	
Threonine		
Tryptophan		
Histidine		

certain amino acids combined with sugars induce browning reactions. This produces the attractive golden colour of French fries and the brown crust of freshly made bread.

12.3.4 Influences of thermal processing on the nutrient values of foods

Heat treatment of foods affords the following advantages and disadvantages:

- inactivates or deactivates micro organisms;
- inactivates antinutritional factors (trypsin inhibitors, haemaglutinins);
- enhances the development of flavours and colours;
- produces flavour volatiles;
- denaturizes protein;
- gelatinizes starch;
- destroys certain vitamins;
- enhances the rancidity of oils.

12.3.5 Amino acid requirements

Using metabolic studies in adult males fed a purified diet, Rose in 1935 initially demonstrated that eight amino acids were required for the maintenance of nitrogen balance. These were isoleucine, leucine, lysine, methoinine, phenylaline, threonine, tryptophan, and valine. Historically, amino acids that could not be synthesized at an adequate level or speed were called essential amino acids (indispensable). Currently, the terms indispensable, conditionally indispensable, and dispensable amino acids are used (Table 12.12). The estimates of amino acid requirements at various ages is illustrated in Table 12.13.

12.3.6 Protein requirements

The protein requirements of humans are dependent on a number of factors. Some of these include: gender, age, body weight and composition, energy intake and micronutrient composition of the diet. The protein requirements of an individual represent the dietary needs necessary to prevent losses of body protein and to accommodate, as appropriate, rates of deposition for growth, pregnancy, and lactation.

When a diet is lacking in protein, the nitrogen lost in the urine and faeces amounts to approximately 49 mg N/kg body weight in adults. To this figure 5 mg N/kg body weight must be added to accommodate losses in sweat, hair, skin, etc. The total nitrogen loss is therefore 54 mg N/kg body weight when fed a protein-free diet. This is sometimes called obligatory nitrogen loss and represents the inevitable loss of nitrogen (protein) when the body is fed a protein-free diet.

Table 12.13 Estimates of amino acid requirements for infants and adults.

Amino acid	Infants (3–4 months) (mg/kg/day)	Pre-school children (2 years) (mg/kg/day)	Adults (mg/kg/day)
Histidine	28	?	8 –12
Isoleucine	70	31	10
Leucine	161	73	14
Lysine	103	64	12
Methionine + cystine	58	28	13
Phenylalanine + tyrosine	125	69	14
Threonine	87	37	7
Tryptophan	17	12.5	3.5
Valine	93	38	10
Total essential amino acids	714	352	84

Adapted from *Energy and Protein Requirements* (FAO/WHO/UNU, 1985).

Historically, obligatory nitrogen loss was used to estimate protein requirements. The amount of protein required was calculated as the amount required to replace this obligatory loss after adjustments for the amino acid pattern of the proteins and the inefficiency of protein utilization. This method is often called the "factorial method for the estimation of protein requirements". However, numerous studies have demonstrated that nitrogen balance could not be achieved even when high quality protein was fed to replace obligatory nitrogen loss. It was therefore apparent that a newer approach for the estimation of protein requirements was necessary.

12.3.7 Concept of nitrogen balance

The nitrogen balance method involves the subtraction of nitrogen intake from the amount excreted in the faeces, urine, sweat, and other minor routes of loss. Simplistically, N balance = N intake − N loss (in faeces + N loss in urine). To predict protein requirements, graded amounts of protein are fed and the requirement is estimated as the amount required to achieve zero nitrogen balance.

Table 12.14 summarizes the nitrogen balance studies conducted in healthy young men using single,

Table 12.14 Nitrogen balance studies using single, high quality sources of protein or mixed diets.

Single, high quality proteins

Protein source	Subjects (n)	Mean requirement (g protein/kg/day)
Egg	31	0.63
Egg-white	9	0.49
Beef	7	0.56
Fish	7	0.71
Average		0.62

Usual, mixed diets

Country	Subjects (n)	Requirement (g protein/kg/day)
China	10	0.99
India	6	0.54
Chile	7	0.82
Japan	8	0.73
Mexico	8	0.78

Adapted from *Energy and Protein Requirements* (FAO/WHO/UNU, 1985).

high quality sources of protein or mixed diets in various countries.

Based on a series of short-term and long-term nitrogen balance studies, it has been suggested that a value of 0.6g protein/kg/day represents the average requirements for high quality protein such as egg, milk, cheese, and fish. In order to accommodate population variation in requirements and variability in protein quality, the safe level of protein intake has been set at 0.75 g protein/kg/day for high quality proteins. Thus, for an adult weighing 70 kg the protein requirements per day would be 52.5 g, very similar to that recommended by Chittenden in 1901! An intake of 0.75 g protein/kg body weight/day is set as the safe level of intake of protein for adults. This amount refers to proteins with a full complement of essential amino acids and high digestibility. It is obvious that proteins with an inadequate complement of essential amino acids and /or poor digestibility will have to be consumed in larger amounts to achieve nitrogen balance.

12.3.8 Protein requirements for pregnancy

Assuming a weight gain of 12.5 kg during pregnancy and a birth weight of 3.5 kg, it is estimated that the total protein requirement during pregnancy is approximately 925 g. As the rate of protein storage during pregnancy is not a constant over the three trimesters, Table 12.15 shows the safe level of additional protein required during pregnancy.

12.3.9 Protein requirements during lactation

Assuming a protein content of breast milk of 1.15 g/ml from the second month of lactation and a breast milk volume of 700–800 ml/day during the first 6 months of lactation, Table 12.16 shows the additional protein requirements during lactation. For example,

Table 12.15 Additional protein requirements during pregnancy.

Trimester	Additional protein requirement (g/day)
1	1.2
2	6.1
3	10.7

Adapted from *Energy and Protein Requirements* (FAO/WHO/UNU, 1985).

Table 12.16 Additional protein requirements during lactation.

Month	Volume (ml/day)	Additional protein requirement (g/day) Average	+ 2 SD
0–1	719	13.3	16.6
1–2	795	13.0	16.3
2–3	848	13.9	17.3
3–6	822	13.5	16.9
6–12	600	9.9	12.3

Adapted from *Energy and Protein Requirements* (FAO/WHO/UNU, 1985).

the additional protein requirement during the first month of lactation is 16.6 g of protein.

The values reported in Table 12.17 are based on the consumption of high quality protein such as milk, meat, fish, or egg. It is therefore important to remember that the protein requirements per kilogram of body weight will be higher if consuming a wholly plant protein based diet. Whilst the protein requirements for children and adults are based on short- and long-term nitrogen studies, estimation of protein requirements in infants is based on estimating the protein intake in wholly breast fed infants.

12.3.10 Estimation of protein quality

The nutritive value of proteins has been of biological interest since the mid eighteenth century. Over the years several methods have been developed to estimate protein quality based on growth, nitrogen excretion, and nitrogen balance. The earliest method to estimate protein quality was developed by Osborne and Mendel in 1919 called the protein efficiency ratio (PER):

$$PER = \frac{\text{Gain in body weight}}{\text{Quantity of protein consumed (N} \times 6.25)}$$

PER is often assessed using laboratory rats and is one of the simplest methods to evaluate protein quality. As PER depends on assessing weight gain alone, it is not possible to assess the protein requirements for maintenance.

12.3.11 Biological value (BV)

Biological value is a simple measure of nitrogen retained for growth or maintenance divided by nitrogen absorbed. It is determined by nitrogen balance (the balance between intake and excretion) and is applicable in both humans and laboratory animals. BV may be written as:

$$BV = \frac{I - (F - F_0) - (U - U_0)}{I - (F - F_0)}$$

where

I   is nitrogen intake,
U   is urinary nitrogen,
F   is faecal nitrogen,
U_0 and F_0 are urinary and nitrogen excretion when subjects are fed a nitrogen-free diet.

The protein quality and the major source of protein from various foods are presented in Table 12.18.

12.3.12 Net protein utilization (NPU)

NPU is a single measurement that combines both the BV and D of a protein. Thus:

NPU = biological value (BV) × Digestibility (D).

Table 12.17 Summary of safe levels of protein requirements at different ages.

Group		Age (years)	Safe protein levels (g/kg/day)
Infants		0.3–0.5	1.47
		0.75–1.0	1.15
Children		3–4	1.09
		9–10	0.99
Young people	(boys)	13–14	0.94
	(girls)	13–14	0.97
		≥ 19	0.75
Elderly women		> 60	0.75

Adapted from *Energy and Protein Requirements* (FAO/WHO/UNU, 1985).

Table 12.18 Protein quality and protein source.

Protein	Source	Biological value
Lactalbumin	Dairy products	High
Casein	Dairy products	High
Ovalumin	Egg white	High
Myosin	Lean meat	High
Gelatin	Hydrolysis of animal tissue	Low
Gliadin	Wheat	Low
Glutenin	Wheat	High
Prolamin	Rye	Low
Glutelin	Corn	High
Zein	Corn	Low
Glycinin	Soybean	High
Legumelin	Soybean	Low
Legumin	Peas and beans	Low
Phaseolin	Navy beans	Low

When both equations as previously described are rearranged the following equation describes NPU:

$$NPU = \frac{I - (F - F_0) - (U - U_0)}{I - (F - F_0)} \times \frac{I - (F - F_0)}{I}$$

$$NPU = \frac{I - (F - F_0) - (U - U_0)}{I}$$

Thus:

$$NPU = \frac{N \text{ retained}}{N \text{ intake}}$$

NPU is commonly assessed using laboratory rats and an extensive literature exists on the protein quality assessment of various foods using this method. When human subjects are used, NPU is assessed by measuring nitrogen balance.

Table 12.19 NPU values of some foods.

Food	NPU
Barley	65
Maize	50
Oats	73
Rice	62
Wheat	40
Cow pea	47
Beef	74
Chicken	79
Milk (cows)	84
Fish	94
Egg	98
Soy bean	66

The NPU of animal proteins is much higher than vegetable proteins, as shown in Table 12.19. This is primarily because vegetable proteins have a lower digestibility and are deficient in one or more essential amino acids.

12.3.13 Net dietary protein energy percent (NDpE %) quantity

NDpE % is a ratio of protein quality and quantity to energy intake. This single factor may be used to evaluate the adequacy of human protein intake. NDpE % is obtained by the following equation:

$$NDpE \% = \frac{\text{protein energy}}{\text{total energy intake}} \times 100 \times NPUop$$

$$\text{Protein energy } \% = \frac{N \times 6.25 \times 4 \times 4.18 \text{ (kJ)}}{ME \text{ of food}} \times 100$$

where

ME   is metabolized energy,
NPUop is NPU operative.

NPUop is the NPU when a protein source is fed above that required to maintain nitrogen equilibrium. When the safe level of protein is expressed as a percentage of total energy requirements, this value can be used to predict the adequacy of any food to meet the protein requirements.

Example using the protein and energy requirement for a 30 year old male weighing 70 kg undergoing light physical activity:

52.5 g protein (protein requirements)
2514 kcal (energy requirements)
52.5 × 4 = 210/2514 × 100 = 8.3% provided by dietary protein.

Thus, any food that provides greater than 8% protein energy will meet the protein requirements for an individual.

Table 12.20 presents the protein energy percent, NPUop, and the NDpE % of some common staple foods. A comparison of the figures with the estimated protein energy percentage for sedentary adults suggests that most staples with the exception of barley

Table 12.20 Net protein utilization and net protein energy percent in selected staples.

Staple	Protein calories (%)	NPUop	NDpE %
Barley	14	60	8.4
Maize	11	48	5.3
Oats	12	66	8.0
Rice	9	57	5.1
Sorghum	11	56	6.2
Wheat	13	40	5.2
Cassava	2	50	1.0
Plantain	3	50	1.5

will not meet the protein requirements if consumed by themselves. This is merely an illustrative example of how the NDpE % of a food or diet may be assessed for its adequacy to meet protein requirements.

Whilst NPU is a useful measure of protein quality, its practical application is limited by the need to be conducted on animals. In 1991, the FAO proposed a method called protein digestibility-corrected amino acid score (PDCAAS), which may be calculated using international food composition tables. The method uses the digestibility of protein and compares the amino acid score of a diet or single protein with the amino acid requirements for a 2–5 year old child. Thus,

$$PDCAAS = \frac{\text{Concentration of most limiting, digestibility corrected amino acid in a test protein}}{\text{Concentration of that amino acid in the 1991 FAO/WHO amino acid scoring reference pattern}}$$

The PDCAAS is a simple and useful method to estimate the protein quality of diets worldwide. A worked example of how to determine PDCAAS is shown in Table 12.21.

12.4 Carbohydrates

Carbohydrates form a major source of energy for most humans around the world. For example, rice, wheat, or corn is consumed by approximately 70% of the global population. Other staple carbohydrate sources include potato, sweet potato, taro, arrowroot, teff, sorghum, millet, bread, fruit, and cassava.

Starch is composed of two forms of polysaccharides, amylose (linear) and amylo-pectin (branched). The amylose content of selected foods is listed in Table 12.22. The ratio influences gelatinization temperature, viscosity, retrogradation, and gel formation. From a nutritional perspective the amylose content significantly influences the glycaemic index (GI) of foods. Foods with greater amylose content have been shown to produce a lower glycaemic response and have a lower GI.

Several factors are known to influence the GI of starchy foods. These include:

- amylose:amylopectin ratio;
- inclusion of kernels of coarse flour;
- pH;
- presence of fat and protein;
- presence of polyphenols;
- presence of viscous fibres;
- extent of starch gelatinization;
- whether the starch is protected by encapsulation.

The glycaemic index (GI) was first introduced in 1981 by Jenkins. It is a classification of the blood glucose raising potential of carbohydrate foods. It is defined as the incremental area under the blood glucose curve (IAUC) of a 50-g carbohydrate portion of a test food expressed as a percentage of the response to 50 g carbohydrate of a reference food taken by the same subject, on a different day (FAO/WHO, 1998). GI may be defined as follows:

$$\text{GI of a food} = \frac{\text{Incremental area under the blood glucose response curve for the test food containing 50 g available carbohydrate}}{\text{Corresponding area after equi-carbohydrate portion of a standard food}} \times 100$$

Foods may be classified as low, medium, or high GI depending on their ability to raise blood glucose. The cut-off points used for classification are:

- Low ≤ 55
- Medium   56–69
- High ≥ 70

Table 12.21 A worked example of how to determine PDCAAS for a mixture of wheat, chickpea, and milk powder.

| | Chemical analysis | | | | | | | Quantities in mixture | | | | |
	Weight (g) A	Protein (g/100 g) B	Lys C	Sulphur amino acids (mg/g protein) D	Thr E	Trp F	Digestibility factor G	Protein (g) A × B/100 = P	Lys (mg) P×C	TSAA (mg) P×D	Thr (mg) P×E	Trp (mg) P×F
Wheat	350	13	25	35	30	11	0.85	45.5	1138	1593	1365	501
Chickpea	150	22	70	25	42	13	0.80	33	2310	825	1386	429
Milk powder	50	34	80	30	37	12	0.95	17	1360	510	629	204
Totals								95.5	4808	2928	3380	1134
Amino acids (mg/g)									50	31	35	12
Reference scoring pattern used			58	25	34	11						
Amino acids scoring for mixture									0.86	1.24	1.03	1.09
Amino acids/g protein divided by reference pattern												
Weighted average protein digestibility sum of							0.85					
[protein x factor (P×G)] divided by protein total												
Score adjusted for digestibility (PDCAAS) (0.85 x 0.86)									0.73 (or 73%) with lysine limiting			

Adapted from Protein Quality Evaluation. FAO Food and Nutrition Paper 51 (FAO, 1991).

Table 12.22 Amylose content of native starch sample (% dry weight of starch).

Food	Mean amylose content (% of starch)
Whole rice	23
Polished rice	24
Corn	23
White spaghetti	25
White bread	24
Potatoes	17
Peas	32
Beans	28
Lentils	28
Chick-peas	28

Adapted from Rosin *et al.* (2002).

Since the concept of GI was first introduced, many studies have investigated the potential health benefits of low GI foods. Today, there is an important body of evidence to support the therapeutic potential of low GI diets, not only in diabetes, but also in subjects with hyperlipidaemia. In addition, low GI foods have been associated with prolonged endurance during physical activity, improved insulin sensitivity, and appetite regulation.

The use of GI for the classification of carbohydrate-rich foods has been endorsed by the FAO and WHO who recommended that the GI of foods be considered together with food composition to guide food choices (FAO/WHO, 1998). GI values represent the glycaemic response of isoglucidic foods (usually 50 g available carbohydrate), and therefore are not always representative of the glycaemic effect of a typical serving of that food. To quantify the overall glycaemic effect of a standard portion of food, the concept of glycaemic load (GL) was introduced. The GL of a typical serving of food is the product of the amount of available carbohydrate in that serving and the GI of the food divided by 100. It is often necessary to consider the GL alongside GI values, especially when the carbohydrate content of the food is relatively small. For example, broad beans have been shown to have a high GI but because they contain very little carbohydrate they have a low GL. GL may be defined as GI multiplied by the carbohydrate content of the food portion consumed.

The GI of foods varies significantly due to factors such as particle size, cooking and food processing, other food components (e.g. fat, protein, and dietary fibre), and starch structure. Consequently there is of-ten considerable variation in the GI of the same food produced in different countries or by different manufacturers.

The largest table of GI and GL values lists over 750 different items across a range of food products and brands and was published by Foster-Powell and colleagues (Foster-Powell et al., 2002). A selection of values is reported in Table 12.23.

The blood glucose lowering effects of consuming a low GI bread in contrast to a high GI bread in ten subjects over a 24-h period are shown in Fig. 12.2. The blood glucose concentration at breakfast, lunch, and supper were considerably lower when subjects were

Table 12.23 Some GI and GL values.

Food	GI	GL
Kellog's All Bran®	30	4
Kellog's Corn Flakes(TM)	92	24
Kellog's Nutrigrain(TM)	66	10
Old-fashioned oatmeal	42	9
Kellog's Raisin Bran(tm)	61	12
Basmati rice	58	22
Pasta – egg fettucine	40	12
Spaghetti	38	18
Bagel	72	25
Croissant	67	17
"Grainy breads"	49	6
Pitta bread	57	10
Rye bread	58	8
White bread	70	10
Carrots	47	3
Baked potato	85	26
New potato	57	12
Sweet corn	60	11
Sweet potato	61	17
Baked beans	48	7
Broad beans	79	9
Butter beans	31	6
Chickpeas	28	8
Kidney beans	28	7
Soy beans	18	1
Apple	38	6
Banana	51	13
Cherries	22	3
Grapes	46	8
Kiwi fruit	53	6
Mango	51	8
Peach, fresh	42	5
Pear	38	4
Pineapple	59	7
Watermelon	72	4
Milk, full fat	27	3
Ice cream, regular	61	8
Yoghurt, low fat	33	10

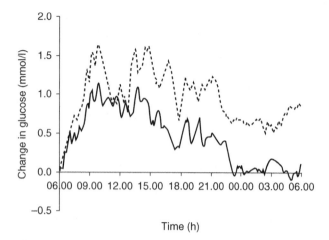

Figure 12.2 Change in blood glucose over 24 h for ten subjects in response to low GI bread (—) and high GI bread (---) (Henry *et al.*, 2006).

fed the low GI bread. The classification of carbohydrates into low, medium, or high GI has been one of the most important observations in carbohydrate nutrition. The consumption of a low GI diet has been shown to have positive health benefits in a variety of chronic diseases including insulin resistance, diabetes, cardiovascular disease, obesity, and cancers.

The monosaccharides in various plant materials and composition of various fruits are displayed in Tables 12.24 and 12.25 respectively.

12.5 Lipids and energy density

Amongst the macronutrients proteins, fats, and carbohydrates, fats are the richest source of energy, representing 9 kcal/g (37 kJ/g). Therefore fats supply almost twice as much metabolized energy per unit weight (9 kcal/g, 38 kJ/g) as protein (4 kcal/g, 17 kJ/g) and carbohydrates (4 kcal/g, 17 kJ/g). Fats act as a carrier of fat soluble vitamins, enhance the palatability of foods, and alter mouth feel and the texture of food. They are also an important source of flavour compound. In addition, fats contribute to the essential fatty acid requirements. With increasing levels of obesity and cardiovascular disease worldwide, the quality and quantity of fat consumed is of considerable public health interest.

Fats and oils are sometimes considered together as lipids. The term lipid is defined as: "A group of substances that are usually soluble in chloroform, ether or other solvents and are only sparingly soluble in water". It is customary to define oils as liquids at room temperature and fats as solid at room temperature. Globally, there is considerable variation in the quality and quantity of dietary fat consumed. North America and Europe consume the highest amounts of fat, with Asia and Africa consuming a modest amount. Simple classifications of fatty acids are:

Table 12.24 Types, compositions, and sources of some important food carbohydrates.

Carbohydrate type	Type	D-Fructose	L-Fructose	D-Glucose	D-Glucuronic acid	D-Galactose	D-Galacturonic acid	D-Mannose	L-Rhamnose	D-Xylose	Some common dietary sources
Lactose	D			x		x					Dairy products
Maltose, isomaltose	D			x							Malt
Sucrose	D	x		x							Most fruits and vegetables
α-Trehalose	D			x							Fungi
Raffinose, stachyose	O	x		x		x					Legumes
Cellulose	P			x							Plant cell walls and fibres
Glycogen	P			x							Animal tissues
Hemicellulose	P			x	x	x	x	x	x	x	Plant fibre, cell walls, cereal, and bran
Inulin (a fructan)	P	x									Tubers of Jerusalem artichoke
Pectic substances	P		x			x	x		x		Most fruits
Pentosans	P									x	Occurs with hemicellulose and pectic substances
Starch, dextrins	P			x							Cereals, legumes, roots, and tubers

D Disaccharide; O oligosaccharide; P polysaccharide.
Adapted from Zapsalis and Beck (1985).

Table 12.25 Type and amount of sugar in selected fruits.

Fruit	D-Fructose	Percentage D-Glucose	Sucrose
Apple	5.0	1.7	3.1
Cherries	7.2	4.7	0.1
Grapes (Concord)	4.3	4.8	0.2
Melon (Cantaloupe)	0.9	1.2	4.4
Oranges (Composite)	1.8	2.5	4.6
Peaches	1.6	1.5	6.6
Pear (Bartlett)	5.0	2.5	1.5
Pineapple (ripe)	1.4	2.3	7.9
Plum (sweet)	2.9	4.5	4.4
Raspberries	2.4	2.3	1.0

Adapted from Zapsalis and Beck (1985).

- saturated fatty acids
- *cis* monounsaturated fatty acids
- *cis* polyunsaturated fatty acids
    - *n*-6 fatty acids
    - *n*-3 fatty acids
- *trans* fatty acids

Tables 12.26a and 12.26b detail the common fatty acids composition of various oils.

Hydrogenation is an early example of the use of food technology to simulate a nutritious food. Hydrogenation is the direct induction of hydrogen to fatty acids with double bonds in the presence of a catalyst. The addition of hydrogen to these double bonds alters the physical chemical characteristics of fat. Hydrogenation therefore converts liquid oils to plastic fats enabling the manufacture of margarine. Hydrogenation is known to generate trans fatty acids (TFA). TFA are unsaturated fatty acids that contain double bonds in the trans position (Fig. 12.3). This is in contrast to the cis double bond (Fig. 12.4), unsaturated fatty acids that are commonly found in nature. The consumption of TFA has been shown to have considerable health risks, notably an increased risk of cardiovascular disease. Denmark was the first country in the world that set limits on the TFA content of foods. It proposed that TFA should not constitute more than 2% of the fat content of any food.

12.5.1 Conjugated linoleic acid (CLA)

CLA has generated considerable interest in nutrition as animal studies have shown that the consumption of these fatty acids reduces fat disposition and enhances lean tissue accretion. Common sources of CLA are meat and dairy products. The concentration of CLA ranges from 0.5 to 5.0 mg/g fat depending on the type of dairy product. Meat products may contain between 5 and 15 mg/g of fat.

Table 12.26a Classification of fatty acids found in foods.

Saturated	Monounsaturated	Polyunsaturated
Propionic (3:0)	Oleic (18:1n-9)	Linoleic (18:2n-6)
Butyric (4:0)	Elaidic (*trans*-18:1n-9)	γ-Linolenic (18:3n-6)
Valeric (5:0)	Vaccenic (18:1n-12)	Dihomo-γ-linolenic (20:3n-6)
Caproic (6:0)	Euricic (22:1n-9)	n-6 Docosapentaenoic (22:5n-6)
Caprylic (8:0)		α-Linolenic (18:3n-3)
Capric (10:0)		Eicosapentaenoic (20:5n-3)
Lauric (12:0)		n-3 Docosapentaenoic (22:5n-3)
Myristic (14:0)		Docosahexaenoic (22:6n-3)
Palmitic (16:0)		
Margeric (17:0)		
Stearic (18:0)		
Arachidic (20:0)		
Behenic (22:0)		
Lignoceric (24:0)		

Table 12.26b Fatty acid composition of various plant-based oils (g 100 g^{-1} total fatty acids).

Fatty acid	Coconut	Corn	Olive	Palm	Palm kernel	Peanut	Soyabean	Sunflower
8:0	8	0	0	0	4	0	0	0
10:0	7	0	0	0	4	0	0	0
12:0	48	0	0	tr	45	tr	tr	tr
14:0	16	1	tr	1	18	1	tr	tr
16:0	9	14	12	42	9	11	10	6
16:1	tr	tr	1	tr	0	tr	tr	tr
18:0	2	2	2	4	3	3	4	6
18:1	7	30	72	43	15	49	25	33
18:2	2	50	11	8	2	29	52	52
18:3	0	2	1	tr	0	1	7	tr
20:0	1	tr	tr	tr	0	1	tr	tr
20:1	0	0	0	0	0	0	2	3
22:0	0	tr	0	0	0	3	tr	tr
22:1	0	0	0	0	0	0	0	0
Others	0	1	1	2	0	2	2	3

tr Trace
Adapted from Gurr (1992).

12.5.2 Essential fatty acids (n-3 and n-6)

12.5.2.1 n-3 polyunsaturated fatty acids (PUFA)

These fatty acids are highly unsaturated (and therefore prone to rancidity), with one of the double bonds located at the carbon 3 position from the methyl end. The most important n-3 PUFA are:

- 18:3 α-linolenic acid
- 20:5 eicosapentaenoic acid
- 22:5 docosapentaenoic acid
- 22:6 docosahexaenoic acid

As 18:3 α-linolenic acid cannot be synthesized by the human body it is termed an essential fatty acid. It is also the precursor of 20:5 eicosapentaenoic acid and 22:6 docosahexaenoic acid. These are present in oily fish. Fish oils have been widely recommended for their health benefits in the prevention of coronary heart disease and thrombosis, and more recently for improving cognition.

12.5.2.2 n-6 Polyunsaturated fatty acids (PUFA)

In addition to being highly unsaturated, one of the double bonds in n-6 polyunsaturated fatty acids is located at the carbon 6 position from the methyl end. The most important n-6 PUFA are:

- 18:2 linoleic acid
- 18:3 γ-linolenic acid
- 20:3 dihomo-γ-linolenic acid
- 20:4 arachidonic acid
- 22:5 docosapentaeonic acid

As linoleic acid cannot be synthesized by the human body, it is also classified as an essential fatty acid. It is also a precursor of 20:4 arachidonic acid. It is recommended that 5–10% of daily energy is derived from linoleic acid and 0.5–1.3% of energy from linolenic acid.

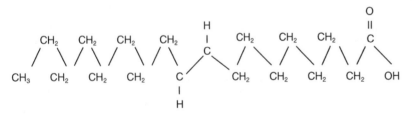

Figure 12.3 *trans* Acid.

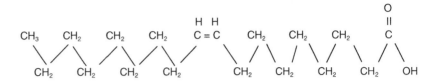

Figure 12.4 *cis* Acid.

12.6 Micronutrients – vitamins, minerals and trace minerals

There is a fundamental difference between micronutrients and macronutrients. Whilst micronutrients are only required in minute quantities each day (milligrams or micrograms), macronutrients are required in much larger quantities (from tens to hundreds of grams per day).

Whilst vitamins may not yield any energy, they are vital during energy transformation. Vitamins are also necessary for cell growth and repair. Both macro and micro nutrients are organic, available in food, and vital to life. In fact the term vitamin was derived from <u>vital</u> <u>amines</u>. The body synthesizes special protein carriers for certain vitamins to assist and promote absorption. Being organic compounds they are easily destroyed and can be oxidized or broken down (Tables 12.27 and 12.28). When vitamins are destroyed they cannot carry out their functions. Therefore special care must be taken when handling and preparing food in order to minimize loss.

12.6.1 Losses of nutrients during food processing

The term "food processing" covers a broad spectrum ranging from boiling to irradiation. Nutrients can be lost from the food in three different ways:

1 Intentional losses such as those occurring from milling cereals, peeling vegetables, or individual nutrients extracted from raw materials.
2 Inevitable losses due to the blanching, sterilizing, cooking, and drying of foods.
3 Avoidable or accidental losses as a result of either insufficient processing or poor storage systems.

Milling involves the mechanical separation of the endosperm from the germ, seed coat, and pericarp, resulting in changes to the micronutrient composition. Flours of high extraction rate retain many more micronutrients than those of lower extraction rate. As nutrients are unevenly distributed throughout the grain, the nutrient losses during processing

Table 12.27 Stability of nutrients as influenced by pH, oxygen, light, and heat.

| Nutrient | Effect of pH | | | Air or oxygen | Light | Heat | Maximum cooking losses (%) |
	Neutral pH 7	Acid < pH 7	Alkaline > pH 7				
Ascorbic acid (C)	U	S	U	U	U	U	90–100
Carotene (pro-A)	S	U	S	U	U	U	20–30
Choline	S	S	S	U	S	S	0–5
Cobalamin (B_{12})	S	S	S	U	U	S	0–10
Essential fatty acids	S	S	U	U	U	S	0–10
Folic acid	U	U	S	U	U	U	90–100
Niacin (PP)	S	S	S	S	S	S	65–75
Pyridoxine (B_6)	S	S	S	S	U	U	30–40
Riboflavin (B_2)	S	S	U	S	U	U	65–75
Thiamin (B_1)	U	S	U	U	S	U	70–80
Tocopheral (E)	S	S	S	U	U	U	45–55
Vitamin A	S	U	S	U	U	U	30–40
Vitamin D	S	S	U	U	U	U	30–40
Vitamin K	S	U	U	S	U	S	0–5

U Unstable; S stable.
Adapted from Harris (1988).

Table 12.28 Factors affecting the stability of vitamins and minerals.

Factor
☑ Oxygen
☑ Temperature
☑ Moisture
☑ Light
☑ pH
☑ Presence of ions such as Fe, Cu
☑ Reducing and oxidizing agents
☑ Other food additives such as sulphur dioxide

are non-linear and are characteristic for each nutrient. For example, thiamine is most concentrated in the scutellum and the aleurone layer of most cereals while riboflavin is more evenly spread throughout the grain, although it is predominantly concentrated in the germ. Commercial milling removes approximately 70% of thiamine, 60–65% of riboflavin, and 85% of pyridoxine from the whole wheat. Iron and zinc, which are located at the periphery of the kernel, are also considerably reduced by commercial extraction rates.

12.6.2 Bioavailability of micronutrients

Nutritionists now recognize that only a proportion of the nutrients ingested are biologically available. One of the most important factors that influence bioavailability is the presence of phytates and polyphenols.

12.6.3 Phytates

Most cereals and legumes contain high levels of phytate or phytic acid. Phytic acid is released when the grain germinates. It is the storage form of phosphorus in cereals and is usually located in the outer layer of cereals and legumes. Most of the phytate is removed during the milling of cereals and decortication of legumes.

12.6.4 Phenolic compounds (tannins)

Phenolics cover a range of compounds including flavenoids, phenolic acid, polyphenols, and tannins. They are widely distributed throughout the plant kingdom and are found in tea, vegetables (e.g. legumes and aubergines), cereals (sorghum), and many seeds. Phenolic compounds are important nu-

tritionally since they bind with nutrients such as iron and proteins, reducing their availability.

Vitamins have numerous roles within the human body. Vitamins are divided into two categories, water soluble (hydrophilic) and fat soluble (hydrophobic) vitamins. The water soluble vitamins consist of B vitamins and vitamin C. They can be absorbed directly into the blood stream where they are able to travel freely, diffusing into the water-filled compartments of the body. The fat soluble vitamins (A, D, E, and K) must enter the lymphatic system before they are absorbed into the blood stream. Fat soluble vitamins tend to be confined to cells related to fat. Any excess water soluble vitamins are excreted by the kidneys, whilst any excess fat soluble vitamins are stored in fat deposits in the body. This is why the fat soluble vitamins are more prone to reach toxic levels.

12.6.5 Water soluble vitamins (Table 12.29)

Thiamin was first discovered and isolated in 1937. Thiamin is important for energy metabolism and is part of the coenzyme thiamine diphosphate (TDP). The development and maintenance of the nervous system is also dependent on thiamin.

Riboflavin, like thiamine, plays a vital part during energy production from fat, carbohydrates, and protein. As the coenzymes flavin mononucleotide and flavin adenine dinonucleotide, riboflavin is part of the electron transport chain.

Niacin – NAD and NAFP – are the coenzymes of niacin, which play a central role in the metabolism of glucose, fat, and alcohol. Unlike the other B vitamins, the liver is able to synthesize niacin from tryptophan.

Vitamin B_6 has many vital functions within the body, which are facilitated via numerous enzyme systems. Vitamin B_6 is important for a healthy immune system through the production of white blood cells. It also converts tryptophan to niacin and is involved in transamination (the synthesis of amino acids). Vitamin B_6 plays important roles in energy production through glycogen breakdown and gluconeogenisis.

Folate, also known as folic acid or folacin, is vital during fetal development. Folate is essential for growth and development during cell division and DNA synthesis. It is also important in the maturation of red blood cells and tissue repair. Hence, folate is important in the prevention of anaemia.

Vitamin B_{12} is vital for folate metabolism. It is also essential for growth and development of tissues and for a healthy cardiovascular system. In the nervous

Table 12.29 Water soluble vitamins.

Standard name	Other name	Food sources	Deficiency	Toxicity
Thiamin	B_1	Meat, nuts, legumes, fortified cereals, wheat germ bran, yeast	Beriberi	> 3 g/day
Riboflavin	B_2	Liver, kidney, dairy products, fortified cereals, Marmite	Ariboflavinosis	No reports up to 120 mg/day
Niacin	B_3	Liver, kidney, rice, wheat, oatmeal, Marmite	Pellagra	> 200 mg/day
Vitamin B_6	Pyridoxine	Meat, fish, pulses, potatoes, nuts, seeds, bananas, avocados, milk	Anaemia (small cell)	2–7 g/day
Folate	Folic acid	Liver, kidneys, nuts and seeds, fortified breakfast cereals, fresh vegetables	Anaemia (large cell)	
Vitamin B_{12}	Cobalamin	Liver, sardines, oysters, meat or animal produce such as egg, cheese, milk	Pernicious anaemia	
Biotin		Liver, egg yolk, yeast, cereals, soy flour	Very rare	No reports up to 10 mg/day
Pantothenic acid		Whole grains, legumes, animal products	Very rare	No reports up to 10 g/day
Vitamin C	Ascorbic acid	Fresh fruit and vegetables, especially spinach, potatoes, broccoli, tomatoes, strawberries	Scurvy	5–10 g/day

system, vitamin B_{12} helps to maintain the myelin sheath that protects nerves. It also helps fatty acids to enter the Krebs cycle.

Biotin is a cofactor for many enzymes and is important in carbohydrate, fat, and protein metabolism. It is also involved in fatty acid breakdown and synthesis including gluconeogenesis.

Pantothenic acid stimulates growth and is part of coenzyme A (CoA). CoA helps to circulate acetate and other molecules through the glucose, fatty acid, and energy metabolism pathways.

Vitamin C is an antioxidant that prevents free radical damage and is important for a healthy immune system. It also helps promote iron absorption and is essential for collagen synthesis.

12.6.6 Fat soluble vitamins: A, D, E and K (Table 12.30)

Vitamin A: there are three active forms of vitamin A in the body known as retinoids – retinol (an alcohol), retinal (an aldehyde), and retinoic acid (an acid).

Table 12.30 Fat soluble vitamins.

Vitamin	Food sources	Deficiency	Toxicity
A	Liver, dairy products, oily fish β-carotene: carrots, apricots, dark green leafy vegetables		> 100 × RNI
D	Egg yolk, oily fish, fortified milk and butter; sunlight is the best source	Osteomalacia (adults) Rickets (children)	> 150 ng/ml (plasma)
E	Vegetable oils	Myopathies, neuropathies, liver necrosis	
K	Green vegetables such as kale, spinach, parsley, cabbage, broccoli	Vitamin K deficiency bleeding (VKDB)	

RNI Reference nutrient intake

Carotenoids (plant pigments) can also be converted to vitamin A by the liver. Vitamin A is required for many different body functions including the maintenance of cell nerve sheaths, red blood cell production, preservation of mucous membranes and the skin, immune function, promotion of good night vision, and stabilization of cell membranes.

Vitamin D can be synthesized by the body with the help of sunshine and is therefore not an essential vitamin. As long as the body is exposed to adequate sun, there is no dietary requirement for vitamin D.

Vitamin E is a powerful antioxidant. As long as there is sufficient vitamin E stored in the cell membranes, it will protect membranes and lipids from oxidative damage. Vitamin E is also effective in protecting the body from free radical formation derived from polyunsaturated fatty acids (PUFA) oxidation. It is sometimes called the anti-ageing vitamin as it protects deep connective tissues.

Vitamin K's primary function is in blood clotting, being essential in the synthesis of a quarter of the proteins required during blood clotting. It is also known to be required by one of the proteins during normal bone formation. Without the presence of vitamin K the protein cannot bind to the bone minerals.

12.6.7 Minerals and trace elements (Table 12.31)

Calcium is essential for the formation and maintenance of bone health. It is also necessary for blood clotting and for muscle and nerve functions. Calcium deficiency can lead to osteomalacia and osteoporosis in adults and rickets and retarded growth in children.

An excess intake can cause the formation of kidney stones and neural motor dysfunction.

Magnesium is important for teeth and bone structure. In addition, magnesium is required as a cofactor for various enzymes involved in energy metabolism and for RNA, DNA, and protein synthesis. Like calcium, magnesium is necessary for blood clotting.

Iron comes in two different types: haem (meat, offal) and non haem (pulses, vegetables, cereals, and dairy). Iron aids the transport of oxygen via haemoglobin present in red blood cells. Iron is important for immune function and participates in energy production via the various enzymes. Iron deficiency is probably the most common nutrient deficiency throughout the world and affects all populations.

Zinc is present in all body tissues and important for the immune system, protein synthesis, growth, and wound healing. Zinc is also vital for the synthesis of insulin.

Sodium and chloride help maintain the body's water balance and sodium is essential for both nerve and muscle functions. An excess of sodium chloride may result in high blood pressure whilst a lack of salt may cause muscle cramps.

Selenium is essential in the production of red blood cells and development of the immune system. It is also important in thyroid metabolism. Areas with low selenium content in the soil have higher prevalence of selenium deficiency.

Iodine is essential in the synthesis of thyroxin. The thyroid hormone controls the metabolic processes in the body and affects energy metabolism as well as mental function.

Table 12.31   Minerals and trace elements.

Name	Food sources	Deficiency	Toxicity
Selenium	Fish, offal, meat, cereals and dairy	Keshan disease, Kaschin–Beck disease	> 400 μ/day
Magnesium	Oysters, fish, shellfish, legumes, grains, vegetables	Hypertension, impaired CHO metabolism	Hypermagnesaemia > 350 mg/day
Zinc	Lean red meat, whole grain cereals, legumes,	Growth retardation, hypogonadism and delayed sexual maturity, impaired wound healing, immune deficiency	> 1 g/day
Iron	Meat, liver, breakfast cereals, bread	Anaemia	Organ damage
Iodine	Fish, shell fish, meat, milk, eggs, cereals	Impaired mental function, hypothyrodism, goitre, cretinism	Wolff–Chaikoff effect
Calcium	Meat, fish, dairy products	Adults: osteomalacia, osteoporosis	Kidney stones

Table 12.32 Haemoglobin cut-off values used to define anaemia.

Groups	Values (g/L)
Children	
0.5–5 years	< 110
5–11 years	< 115
12–13 years	< 120
Men	< 130
Non-pregnant women	< 120
Pregnant women	< 110

The three most important micronutrient deficiencies of global public health significance are iron deficiency anaemia, vitamin A deficiency, and iodine deficiency. Iron deficiency anaemia may be defined as a low concentration of haemoglobin (Hb) in the blood. In addition to iron, a deficiency of folic acid, riboflavin, and B_{12} can also lead to anaemia. Anaemia is diagnosed by measuring Hb in blood. The current cut-off points for anaemia at various ages are given in Table 12.32. Iron deficiency is of public health concern as it increases the morbidity and mortality of pregnant women, adversely affects physical capacity and work performance, and may impair cognitive performance at all ages.

12.6.8 Functional foods (Table 12.33)

An emerging interest in phytochemicals has spawned the development of functional foods. Functional foods are sometimes also called nutraceuticals, pharmafoods, or designer foods. Functional foods may be defined as any food that has a positive impact on an

individual's health, physical performance, or state of mind in addition to its nutrient value. Moreover, the Japanese have added three additional conditions that need to be met in order for a food to be classified as a functional food:

1 The ingredient is a food derived from a naturally occurring source.
2 The product can and should be consumed as part of a daily diet.
3 The product has specific function when eaten, promoting one or more of the following:
   a prevention or retardation of a specific disease;
   b improvement in cognition;
   c enhancement of immunological response;
   d retardation of the ageing process.

Food has been considered a source of fuel and energy since the time of early man. Today the need for food is not only for survival, but also a source of nutrients for extended longevity and good health. A newborn baby weighs approximately 3.5 kg at birth; by the time it attains adulthood its body weight would have increased to 70 kg. This 20-fold increase in body weight is entirely due to tissue accretion. This accretion is the outcome of the nutrients absorbed from food and retained by the human body. On average an individual adult consumes approximately one metric ton of food and drink per year. Thus, the often quoted statement "you are what you eat" should read "you are what you eat and retain"!

Advances in nutrition are likely to improve the quality of life of people worldwide. Moreover, they

Table 12.33 Some phytochemical food sources.

Phytochemical	Rich food sources
Polyphenols	Fruits, vegetables, garlic, onion, red wine, dark beer, tea (green tea in particular)
Indoles	Cruciferous vegetables
Isothiocyanates	Cruciferous vegetables (broccoli in particular)
Carotenoids	Fruits and vegetables (green, yellow, and orange)
Allyl sulphides	Garlic, onion, chives
Isoflavones	Legumes
Monoterpens	Oil from nuts, seeds, citrus fruit, and cherries
Phytic acid	Legumes, whole grains
Lignans	Fruits and vegetables, flax seeds
Phenolic acids	Fruits and vegetables, seeds in strawberries and banana
Chlorogenic acid	Fruits and vegetables
Saponins	Beans, legumes
Curcumin	Turmeric

will also provide nutrient requirements specially designed for individual needs.

Further reading and references

Brand-Miller, J., Wolever, T.M.S, Foster-Powell, K. and Colagiuri, S. (2003) *The New Glucose Revolution*, 2nd edn. Marlowe and Company, New York.

FAO (1991) *Protein Quality Evaluation. FAO Food and Nutrition Paper 51*. Food and Agriculture Organization of the United Nations, Rome.

FAO/WHO (1998) *Carbohydrates in Human Nutrition. Report of a Joint FAO/WHO Expert Consultation.* Food and Agriculture Organization of the United Nations, Rome.

FAO/WHO/UNU (1985) *Technical Report Series 724.* World Health Organization, Geneva.

FAO/WHO/UNU (2001) *Human Energy Requirements. Food and Nutrition Technical Report Series.* World Health Organization, Rome.

Food Standards Agency (2002) *McCance and Widdowson's The Composition of Foods*, 6th summary edn. Royal Society of Chemistry, Cambridge.

Foster-Powell, K., Holt, S. and Brand-Miller, J. (2002) International table of glycaemic index and glycaemic load values: 2002. *American Journal of Clinical Nutrition*, **76**, 5–56.

Gurr, M.I. (1992) *Role of Fats in Food and Nutrition*, 2nd edn. Elsevier, Oxford.

Harris, R.S. (1988) General discussion on the stability of nutrients. In: *Nutritional Evaluation of Food Processing* (eds E. Karmes and R.S. Harris), 3rd edn. AVl/Van Nostrand Reinhold, New York.

Henry, C.J.K., Lightowler, H.J., Tydeman, E.A. and Skeath, R. (2006) Use of low-glycaemic index bread to reduce 24-h blood glucose: implications for dietary advice to non-diabetic and diabetic subjects. *International Journal of Food Science and Nutrition*, **57**(3/4), 273–8.

Karmas, E. and Harris, R.S. (1988) *Nutritional Evaluation of Food Processing*, 3rd edn. AVI, New York.

Rosin, P.M., Lajolo, F.M. and Menezes, E.W. (2002) Measurement and characterization of dietary starches. *Journal of Food Composition and Analysis*, **15**(4), 367–77.

Weir, J.B. de V. (1949) New methods for calculating metabolic rate with special reference to protein metabolism. *Journal of Physiology*, **109**, 1–9.

Zapsalis, C. and Beck, R.A. (1985) *Food Chemistry and Nutritional Biochemistry*. John Wiley and Sons, Toronto.

Supplementary material is available at www.wiley.com/go/campbellplatt

Herbert Stone and Rebecca N. Bleibaum

Key points

- Sensory evaluation is a science that measures responses of people to products as perceived by the senses. Knowledge of human behavior and physiology of the senses is critical to obtain meaningful information.
- Sensory information is unique, not easily or directly obtained by other means, and it has value beyond the direct results of a specific test.
- An important reason for sensory tests is to measure the impact of physical properties on the sensory characteristics and, ultimately, how they affect consumer preference and purchase behavior.
- Important to the success of any sensory testing program is the laboratory setting (the facilities) where testing is conducted.
- There are two categories of methods, analytical and affective, each providing different kinds of information.
- Analytical methods consist of discrimination testing and descriptive analysis. Descriptive analysis is the most useful of sensory methods, enabling the researcher to identify perceptual product similarities and differences.
- Subjects should be qualified to participate in analytical methods based on sensory acuity with actual products from the category being tested.
- Target consumers should be qualified for affective methods based on usage and consumption criteria. Affective methods provide affect, liking, and preference information about products, concepts, benefits, uses, attitudes, etc.

13.1 Introduction

Sensory evaluation is a science that measures the responses of people to products as perceived by the senses. The products can be anything from a pure stimulus such as an aqueous solution of sodium chloride, a combination of fruit extracts used to flavor a beverage, or a finished product available for purchase such as a frozen prepared dinner, or it can be a nonfood such as running shoes or golf clubs. Results from a sensory test are used in a variety of applications in addition to evaluating theories of perception. Of

major importance are measuring responses to ingredient changes, the effects of process changes, cor-
relating ingredient changes with preferences, and
so forth. The benefits of these latter applications
are easy to understand and are responsible for the
growing interest in the use of sensory resources;
however, these opportunities present some unique
challenges.

First and foremost, sensory evaluation is a "people science" and knowledge of human behavior and
its measurement and the physiology of the senses
are critical to obtaining meaningful information. For
some, sensory evaluation is thought of in psychological and/or in physiological terms, while for others
it is thought of in statistical terms. For still others it
is not thought of as a scientific process; i.e., anyone
can do it, and therefore it has little scientific standing in the same way as one considers such disciplines
as chemistry, microbiology, or engineering. However,
to organize and field a test that yields actionable information requires a thorough understanding of the
aforementioned disciplines, as well as knowledge of
the technology of the products being tested. Failure
to do so compromises the potential value of the information, especially when that information is not consistent with expectations.

Part of the problem, as previously noted, is the
seemingly simplistic nature of the process. Given
a scorecard and a product, most consumers, when
asked, will respond regardless of whether the task
was understood or whether that person was qualified
to participate. Marks on a scorecard (a touch screen,
or any other means of capturing a response) represent nothing more than that, marks; it is only with
responses from a sufficient number of qualified subjects that one can expect to have confidence that the
information is more than a series of random events.
The issue of what constitutes a qualified consumer or
subject also needs explanation. As already noted, the
apparent ease with which a response is obtained is
often misunderstood, leading to situations in which
a technologist and/or a brand manager challenges
results that are inconsistent with his or her expectations. Since these individuals evaluate products regularly, it was/is easy to assume that one's judgments
are as good as the consumer's, if not better, or as good
as a qualified subject in a sensory test. In fact, some
will interpret results quite differently in the belief
that they know more than the consumers who were
tested. When combined with other kinds of product
information, this can have unintended and usually

negative business consequences (for more on this issue, see Stone and Sidel, 2007).

For sensory professionals these are some of the
challenges that must be addressed if accurate and actionable information is to be obtained. Other issues
that are equally important and part of the process
include deciding which consumers will be tested,
which test method and data analysis will be used,
and, finally, the interpretation and reporting of the
results. As noted at the outset, sensory information
is unique, not easily or directly obtained by other
means, and it has a value well beyond the direct
result of a specific test. Companies make substantial investments looking for and/or developing new
products, reformulating existing products, changing
processes, and protecting franchises from competition. Sensory information provides answers to many
of these objectives and achieves it quickly and at a
much lower cost than by other means. This chapter focuses on the principles and practices of sensory
testing and how to organize and field a test.

13.2 Background and definition

Before discussing resources and applications, it is
useful to provide a definition for sensory evaluation
(Anonymous, 1975). Such a definition is especially
important as it provides a context for the subsequent discussion and provides the reader with a
perspective:

> Sensory evaluation is a scientific discipline used to
> evoke, measure, analyze, and interpret reactions to
> those characteristics of foods and materials as per-
> ceived by the senses of sight, smell, taste, touch,
> and hearing.

This definition states it is a science that involves
measuring perceptions, and in this specific instance,
perceptions derived from foods and beverages as the
source of the stimulation. It also is relevant for products other than foods and beverages. The responses
reflect what is perceived in its entirety, what that
product looks like, how it smells, tastes, etc., and also
takes into account each person's past experience with
that product or related products.

This matter of "perceived in its entirety" is an
aspect of the sensory process that is often disre-
garded or not fully appreciated even by sensory
professionals. When planning a test and considering

the evaluation process, there is a tendency to think in narrow terms of "aroma or flavor"; however, this is not the entire picture of what is perceived. Even when testing non-foods, one must be careful about the extent to which a modality, e.g., appearance, is excluded from a questionnaire because the experimenter does not consider it an important property of the product or relevant to the objective. Yet, the interrelationships among the modalities cannot be ignored. The sensory receptors and their pathways to higher centers in the brain are anatomically unique; i.e., we perceive color and appearance with our eyes and the pathways to the functioning centers in the brain are well defined and independent of the pathways for taste, etc. However, as receptors are stimulated and the signals transmitted to the higher centers in the brain, modifications of those signals occur. These modifications reflect other sensory inputs (e.g., aroma) as well as stimulation from associated structures, leading to a response that contains more than that derived from the initial stimulus.

Memory (cognitive factors) also impacts the response. In practical terms, this means that, for example, the appearance of a product will be influenced by one's memory of prior consumption of that product (or at the least it will recall to that individual's conscious a variety of thoughts relevant to that product). It also will impact the consumer's expectations and judgment about that product's aroma, taste, and so forth.

These sensory interrelationships must be accounted for regardless of the type of test that is planned. If not accounted for, useful information will be lost and the overall value of the sensory resource is compromised. For example, it is inferred or suggested that through relevant training one can develop panels whose focus is a single modality; e.g., texture, using the Texture Profile Analysis© as developed by Brandt et al. (1963), Szczesniak, (1963), and Szczesniak et al. (1963). There is no question that one can teach subjects to limit their responses to specific characteristics and ignore others (modify behavior); however, it is naïve to assume that individuals respond in that way when evaluating a product, as if the appearance, aroma, etc. are not there. In the case of texture, it too is not a single perceptual experience.

The risk, as already noted, is high that perceptions not accounted for on a scorecard will be imbedded in those judgments subjects are able to make. Alternatively, some subjects will make a conscious effort to respond as requested but subconsciously will not, and that will add to the error and impact any conclusions reached. There can be situations, for example, where separating appearance from aroma is needed and a suitable protocol can be developed. For example, using a container with a clear plastic cover will allow for the visual evaluation without the aroma influencing the response. Such a test would be part of an evaluation in which all perceptions are measured, but there could be a business reason to isolate and measure a modality on its own. Another common trap is the request to instruct subjects to ignore a modality such as the appearance, and "only evaluate the flavor" because only a flavor change was made, the assumption being that flavor does not impact other modalities, an assumption that almost always is incorrect. This is an example of a practice that continues despite all efforts to eliminate such wishful thinking on the part of requestors (or sensory professionals). It reflects a lack of appreciation for the complexities and integrative nature of the sensory system and a lack of understanding of human behavior.

For most readers, awareness of these interactions should be intuitively obvious yet much research is done in which this is ignored with no appreciation of the consequences. In more practical terms it certainly helps explain many new product failures.

The definition of sensory evaluation emphasizes what is perceived as distinct from a product's physical properties. An important reason for many sensory tests is to measure the impact of those physical properties (derived from the ingredients and the process) on the sensory characteristics, and how those changes affect preference, purchase intent, etc. Another purpose could be to identify those physical and chemical measures that can explain specific sensory differences as well as identify the specific sensory differences that cannot be measured by existing physical and chemical measures. Such information has considerable practical application; for example, identifying the product characteristics that are important to the consumer's quality expectations. This provides quality control staff with the opportunity to provide weightings of importance to specific measures, physical, chemical, and sensory measures that are most important to consumer product quality. All of this is possible provided there are qualified subjects, the relevant methodologies have been used, and the analyses reflect the design parameters.

We now turn attention to the key elements that constitute a contemporary sensory capability – facilities, subjects, and methods.

13.3 Facilities

Important to the success of any sensory testing program is the laboratory setting (the facilities) in which tests are conducted. If facilities are not planned properly, then the ability to provide information quickly is compromised, test results are not available when expected and eventually the service will not be used. Therefore, having a facility means having a plan as to how subjects get to and from a test site, how products are prepared and tested, how results are analyzed, and the manner in which information is communicated to requesters. This section provides a kind of checklist for the sensory professional when planning the design of a facility.

A sensory program should have a dedicated area for its activities, in part as a demonstration of management support for this activity, and also demonstrating the extent to which care is taken when testing products. A contemporary facility has environmental controls, an area to greet and orient subjects, space to conduct screening, training, data collection, storage areas, and space allocated for staff. As a general guide, companies testing foods and beverages have to use construction materials that conform to domestic and, if appropriate, international food safety regulations.

Detailed descriptions of facilities and their basic requirements are described in Eggert and Zook (1986) Stone and Sidel (2004), and Kuesten and Kruse (2008). The interested reader should access these documents before any serious consideration is given to the actual design of facilities and initiated. When a new or existing facility is being considered, the sensory professional should review available information, such as the dimensions of the space, and location of all structural columns and utility lines. Once this is known, it makes it easier to identify how best to use the space. The following is a checklist covering the areas and activities within the space:

1 Preparation and serving
2 Storage
3 Climate controls
4 Reception, training, and booths
5 Satellite facilities
6 Administrative office

13.3.1 Preparation and serving

List all equipment needed in the course of preparing and serving products. Equipment such as freezers, refrigerators, ranges, ovens, microwaves, toasters, mixers, trays, and utensils are usually needed. Where necessary, the grade of equipment, commercial and/or retail, should be taken into account; i.e., if products are used by consumers, then the equipment would be typical of that available to consumers vs equipment used in restaurants.

There should be sufficient cabinets, counters, and central islands, space permitting, to enable products to be easily handled during a test and to be displayed prior to and between tests. All counters need to be high enough to minimize posture stress during serving, and wide enough to accommodate products, supplies, and scorecards or electronic equipment used for direct data entry. Booths should have sample pass-through openings, wide enough to accommodate serving trays, and space behind each booth to accommodate subject movement to/from a booth. On the serving side, the space above and below the counters can be used for additional storage.

A communication system connecting the reception area with the preparation/serving area and between this area and the subjects has to be built into the plan. This minimizes unnecessary communications during the testing activity. Telephones should be equipped with a visual alert such as a light, so that the auditory alert may be turned off during testing.

13.3.2 Storage

All facilities need to have sufficient storage capacity not only for utensils, cutlery, and serving containers (bowls, cups, etc.), but also for products awaiting a test. In some instances, the products could be in large case volumes. For some companies the building location has such space, but if not, then the sensory staff must give consideration to having sufficient space for storage. Another consideration is the number of stoves, refrigerators, freezers, and mixers that may be needed for some but not all tests. The available space may not be able to accommodate these units and the ability to store them when not being used needs consideration.

13.3.3 Climate controls

The air temperature and quality is important in the testing facility. Key areas for strict controls are the individual booths and in the training area/room. Climate controls include the ventilation system with a slight positive air pressure from the booths into the preparation area, and in the training area/room. The intent is to minimize odor transfer from the preparation or other adjacent areas to evaluation areas. These areas also should be as quiet as possible, minimizing the influence of external noises. Temperature in the area is recommended to be a comfortable room temperature (20–23°C), with a relative humidity of between 50 and 55%.

13.3.4 Reception, training, and booths

Product evaluation consists of three functional areas – reception, training, and testing. These areas should be adjacent to one another for ease of access by subjects but separated from the product preparation and serving areas. Ideally it should be in a quiet area; however, accessibility should take precedence over a location that takes time to get to.

13.3.4.1 Reception

On arrival at a facility, subjects identify themselves and are provided with information about the test. This area should be welcoming, well lit, and with some reading materials. There should be a counter and space for a staff person, communications with the preparation area, and a system for contacting subjects who may have forgotten to attend their scheduled session, etc. There should be sufficient chairs, about 10–15, to enable all subjects to be seated for orientation. A communication board (white or bulletin) can be located in the area to provide information on the product category and number of products to be evaluated in the session. A table also may be provided for subject treats, such as juice, water, cookies, candy, etc., when appropriate, immediately following test participation. The reception area should allow easy access to, yet be visually separate from, the booths. This will minimize distracting subjects in the midst of a test. Dimensions for a reception area will depend on the number of subjects it is required to accommodate and the amount of space available for all sensory activities.

13.3.4.2 Training

It is often necessary for subjects to receive some kind of instructions or training, especially when initiating a descriptive analysis panel. These discussions are typically held in a conference room, like a focus group facility, that can seat between 12 and 15 subjects and sensory staff. Ideally, this room is situated close to the product preparation area, is readily accessible by subjects and research staff, and free from distraction of noise and external odors. Sufficient whiteboard space is recommended to accommodate sensory attribute lists developed by a panel. A one-way mirror to an adjacent viewing room, equipped with video and audio taping capabilities, can accommodate observers (e.g., from marketing and the development team) without distracting subjects during training, similar to more traditional types of focus group facilities. This room can also be used for presentations, staff training, etc.

13.3.4.3 Booths

To minimize distractions and allow subjects to focus on the specific task, individual booths are required. Individual booths minimize visual distractions, allowing the subjects to concentrate on the task. They should be comfortable, large enough to hold trays, and screens if using a direct data entry system. The surroundings should be made of neutral surface colors and materials, provide adequate lighting, a sample pass-through structure, an expectoration system (disposable cups or sinks), and a means for communication with servers. Standards for colors and lighting in the booths exist (for example, ASTM International MNL 60 (Kuesten and Kruse, 2008) and ISO standards). In addition, the booth area should have positive pressure, that is, clean filtered air should flow into the booth area, and air should flow from the booths to the kitchen, and not the reverse. Electrical outlets for computers or electrical appliances may be needed in the evaluation area. Individual sinks, used for expectoration, are often a major source of odor. If sinks are chosen, it is advisable to choose one with an excellent rinsing system.

Most sensory booths are permanent structures; however, temporary booths can be constructed from sturdy coated art board or plywood, and these can be made collapsible or foldable with hinges or fitted slots for portability. Temporary booths can be moved to off-site testing locations.

13.3.5 Satellite facilities

While most tests are conducted in laboratories, as described above, there can be situations where evaluations are best done outside an identified location; i.e., a company office building. Tests also may be conducted in locations specific to product quality maintenance and production. Such satellite facilities are generally smaller, as they require less preparation equipment and space, and fewer booths. Regardless of where testing is done, the same general rules for maintaining these locations apply; that is, a clean, quiet, noise, and odor free test environment.

13.3.6 Administrative office

The sensory staff need an area/work station for managing the testing activities, preparing reports, meeting with requesters, and associated activities. Ideally it will be located within the sensory space. This area should include a library with relevant literature, journals, books, and access to online literature search services. In some companies this latter resource is in a centralized location for all staff to use.

As previously noted, there are several useful references on facilities design and the interested reader is urged to access this information, as well as to visit facilities, where permitted.

13.4 Subjects

A high priority in any discussion about sensory evaluation must be the choice of subjects. All individuals interested in participating must be voluntary, and qualified to participate based on relevant sensory test, not based on friendship, status, or position within a company, or if using non-employees it should not be one's immediate family, etc. For sensory analytical tests (discrimination and descriptive – defined in the 'Methods' section) the qualification is based on demonstrated sensory skill. For a preference test, the qualification is based on product usage and not sensory skill. Most sensory professionals will agree with the need for subjects to be qualified, yet the qualification specifics are different depending on one's perspective.

The literature reveals a range of approaches including a comprehensive program taking as long as 3 or more months, while others require as little as 2 weeks. In some industries, individuals spend sev-

eral years of apprenticeship with an experienced professional before being considered "qualified". However, these latter requirements are more reflective of the qualifications of a perfumer or a flavorist and not a subject for a sensory test. In other instances, the basis for qualifying an individual is described as having participated in sensory tests for several years and/or involvement in product quality judgments for an association (see, for example, Irigoyen *et al.*, 2002; Larráyoz *et al.*, 2002).

Various industry associations (such as ASTM International) and quasi-governmental bodies (such as the International Organization for Standardization) have organized committees and issued comprehensive directions for qualifying subjects. Guidelines have value, serving as a reminder that subjects must be qualified as well as providing directions as to how best to proceed with the qualifying process. However, they should not be transformed into a recipe, because each product category and each test type will have some unique requirements. Therefore, we believe that a basic set of guidelines is far more useful for the sensory professional when developing a pool of qualified subjects, to include the following:

- confirming to the volunteer that all information is confidential (privacy issue);
- confirming that the individual is not suffering from allergies nor has a medical condition that could be impacted by the testing;
- making sure the subject has no direct involvement in the technology of the products;
- selecting subjects who are average or above average users of the products of interest;
- selecting subjects who have supervisor approval to participate (if an employee);
- selecting subjects who demonstrate sensory skill for the products or product category that will be tested;
- confirming that the subject can stop participating at any time – understanding that it is strictly a volunteer effort.

This list is unique in that it does not identify products or "standard materials", nor does it identify a specific test method. Listing a product can be misleading since it may not be available in a specific market or it may be culturally inappropriate. Stating a specific method is equally risky because it, too, may not be relevant to the nature of the problem. For example, a product may have a lingering effect that interferes with the subject's ability to detect a difference compared with another product. The specified method may increase product exposure to such a degree that sensory fatigue occurs and sensitivity is significantly reduced, leading to selection of individuals who are not qualified.

Qualifying subjects, as noted above, should be based on sensory skill with the products that will be tested; i.e., the product(s) used should be the actual product or products from the category being tested. For example, if one's company is in the carbonated cola beverage business, then using colas is appropriate and using, for example, biscuits is not. The second requirement is that the subjects must demonstrate they can discriminate differences among those products at better than chance. The decision about qualifying will be made by the experimenter based on his/her knowledge about the individual's past performance and the objective(s) for the test, etc. This will be discussed later in this section.

Another interesting and related aspect of qualifying subjects is the use of threshold testing for specific kinds of stimuli, e.g., sweet, sour, bitter., the threshold being defined as the lowest concentration of a stimulus that can be detected 50% or more of the time. This is a practice that has been demonstrated to be ineffective in predicting a subject's performance when evaluating a product (Mackey and Jones, 1954). Periodically the authors are informed that the original results are correct, to wit, threshold testing is not a good predictor of future product evaluation performance. This should not be surprising. Absolute threshold testing makes use of pure stimuli such as a salt solution and the response is whether a difference was detected; i.e., given two samples, are they the same or different or some similar paradigm? Products are chemically more complex, no two products are the same, and the subject's perceptions will be equally complex. While it may be of interest to determine threshold sensitivity as an exercise, it is not useful in identifying who is best qualified to evaluate specific products.

A similar type of situation exists when it comes to training subjects for a descriptive analysis test. Use of references has been discussed for some time (see Rainey, 1986; Meilgaard et al., 1999) and there are numerous examples of their use reported in the more recent literature (see, for example, O'Sullivan et al., 2002 and Gambaro et al., 2003) or any current issue of journals that publish results from sensory descriptive tests. Some researchers spend as much as 3 or more months (reports from 60 to over 100 hours) on this

activity; however, no evidence for its effectiveness is reported, only that it was the procedure used prior to and in some instances during data collection.

Considering all the psychological issues associated with such an extended training effort, it is surprising that no evidence is presented for its effectiveness. An extended process runs the risk of modifying behavior to some greater or lesser degree for each of the participants. There is no question some individuals learn to use their senses more slowly than others; however, delaying the start of testing makes less sense from a business perspective. In addition, those already qualified will likely lose interest waiting for the others to reach the qualifying state.

The primary purposes for screening (not necessarily in order of importance) are to identify and eliminate those individuals who cannot follow instructions, are insensitive to differences among products being evaluated, or who exhibit little interest in the activity (volunteer for a reward), and, perhaps most important, to identify those individuals who can detect differences at better than chance among the products of interest. The latter individuals are the ones identified as qualified. This can easily be achieved in about 5–6 hours (over several days and sessions) starting with individuals with no prior experience, i.e., naïve subjects.

Training is a related activity, often as a part of the screening process. It is here where references and standards are most often used. Proponents will argue that the use of standards or references in training ensures that all subjects agree with each other, enabling comparisons of results from test to test, from one location or time to another, etc. Such arguments are surprising since statistical procedures have existed for many years that provide for such comparisons and are more meaningful than attempting to modify behavior and/or mask sensitivity differences or make claims to that effect. It is unrealistic to think that one can train subjects to always be in agreement with each other or to think that use of references can achieve this. Each individual is different from every other individual in terms of his or her sensitivity and ability to discriminate among products. If one could train people to be equally sensitive, why would one need a panel of more than one? If a purpose for such training is to demonstrate that a group of people can be trained to provide the same response each time a specific stimulus was presented (a learned association), then one might argue the merits of this approach, but of what practical value is this? The ability of a person to repeat a response when presented with the same stimulus on several occasions would represent evidence of reliability but hardly represents or constitutes anything else.

One can easily demonstrate this "skill" using an appropriate stimulus, one that is easily recognized, and an insensitive scale (preferably one with numbers that are easy to remember). This demonstration would not be worth much in relation to what people perceive, which is what sensory evaluation is all about; however, it may be effective for individuals with no appreciation for behavioral psychology or physiology of the senses. It would be far more productive if subjects demonstrated their sensory skills with the products they will be evaluating.

A primary goal of screening is to test products with consumers from the discriminating segment of the population. These consumers will be more likely to detect differences and, as a consequence, will be most likely to react to any perceived changes. Consumers who are poor discriminators may or may not detect a difference and therefore are less likely to respond to a change. This means that concluding there is no difference when there is has serious business consequences. Just as market research/consumer insights recruit specific population segments such as average or above average users of a specific brand, household size, or concept acceptor, so too must sensory abilities recruit those individuals most likely to detect differences. It is a matter of minimizing risk in product decisions.

Individuals can be recruited from within a company or from the local community. The decision as to which (or both) depends on cost, availability, the believability of results by management, and access to the testing location. The decision is always based on company-specific criteria. Regardless of the source, all individuals are contacted using a variety of techniques such as telephone, advertising, and/or the internet. Those who express interest, as already noted, must be users/likers of the product category or categories, and, most important, demonstrate the requisite sensory skills. Identifying users/likers as an "intellectual exercise" serves as an initial screening by identifying those who do not use a product at some previously specified frequency and those who do not follow instructions. Not surprisingly, those who do not follow instructions also tend to not follow instructions when testing and rarely demonstrate the required level of skill. Details about this are described by Stone and Sidel (2004).

Once individuals have been identified and meet the initial product usage criteria, actual sensory screening can be scheduled. We recommend using the discrimination model as the most useful method for qualifying subjects for sensory analytical tests. If a person cannot discriminate differences among products he/she normally consumes (at better than chance), it is unlikely that this same person will perform well in actual testing, and empirically this is what we observe. The discrimination model, a type of rank order, is fundamental to any other sensory task, such as scoring. The key to success with this screening model is to take into account the different rates at which individuals learn to use their senses, and the importance of motivation. By creating a paradigm in which the task becomes progressively more difficult, subjects have the opportunity to learn how to take a test and how to use their senses, and the experimenter can more easily identify those who are learning and qualifying. In addition, all modalities are tested, in case there are individuals who are color blind or anosmic, for example.

In practical terms this means that the sensory professional must bench screen the array of products (to be tested), identify observable differences, and prepare a series of product pairs that include visual, aroma, taste, texture, and aftertaste differences. The number of pairs should be at least 15, to as many as 20, and the differences should range from easy to difficult across modalities. With replication, this provides 30–40 trials/judgments, sufficient for subjects to demonstrate their abilities and for the sensory professional to be able to classify individuals based on their sensitivity and reliability. As a guide, one can select any individual who has achieved at least 51% correct matches across all trials; however, one can take a more conservative approach; for example, a 65% cutoff. The decision as to what percentage to use is the responsibility of the sensory professional. If there are a sufficient number of individuals who meet the 65% and higher criterion, than one can proceed with that cut-off for participation.

The effectiveness of this approach is based on the subsequent test performance of those individuals. Of course one must be mindful of the degree of difficulty of the product pairs. If the product pairs are easy to differentiate, then the screening process will not be effective as almost everyone will qualify. Only after this procedure has been used once or twice will the sensory professional know whether pairings represent the desired easy, moderate, and difficult options. It is reasonable to expect that some changes will be necessary; however, this will lead to a screening process that can be used with confidence. In the equivalent of about 6 test hours, one can identify qualified subjects and be ready to initiate testing. Interestingly enough, we observe that about 30% of those who volunteer do not qualify and this observation is independent of age, gender, employment, country, etc. More discussion about this can be found in Stone and Sidel (2004).

13.5 Methods

There are many methods available to the sensory professional and new methods are regularly described in the literature; most are extensions or modifications of existing methods, as opposed to being entirely new methods. Details about methods can be found in Lawless and Heymann (1999), Schutz and Cardello (2001), and Stone and Sidel (2004). The discussion here follows a classification system previously described, and is based on the type of information obtained, analytical or affective. This also is a reminder to the reader, albeit indirectly, that one should not mix and/or combine methods. Probably the most common method is asking for a preference judgment after a subject provides a discrimination judgment. This is an easy task to complete, but the results present the sensory professional with a dilemma in treating preference responses from the non-discriminating subject. We have more to say about these practices later in this section.

There are two categories of methods, analytical and affective, each providing different kinds of information:

1. *Analytical methods* provide analyses of products; e.g., are two product formulations perceived as different, what kinds of differences are perceived, and what are the magnitudes of the differences? Examples of these methods include discrimination and descriptive. The former identifies whether the difference was perceived and the latter determines the types of differences and their magnitudes.
2. *Affective methods* provide liking/preference information about products. Examples of these methods include paired preference and hedonic scoring. The former identifies which of two products is preferred and the latter measures the degree of liking/preference for each product.

Before discussing details about the analytical methods, the reader is reminded that no one method is superior or more sensitive than any other method. Claims of superiority for a method should be viewed with caution, as they are often based on the product tested not the method. By selecting an appropriate stimulus such as a pure chemical and pairings that are easily differentiated, one can demonstrate an effect; however, when applied to a complex food or beverage, the effect is often lost.

A well-known superiority claim is the triangle discrimination method. This claim is primarily based on the probability of 1/3 (there are three coded samples) for statistical significance vs the paired or duotrio method with a probability of 1/2 (there are two coded samples). While this is statistically correct, i.e., one needs fewer correct matches to achieve statistical significance, logically it follows that one would expect that a test involving four, five, or more products would be even more sensitive because of a concomitant reduction in probability!

Empirically we observe that as the number of samples increases, there is a decrease in sensitivity as evidenced by few tests in which significance is obtained. There are two possible explanations for this: the effect of sensory fatigue from sampling too many products and the associated effect of having too many choices. This should lead the sensory professional to be cautious about a claim of superiority for one method vs another.

Discrimination testing, as mentioned earlier, is at the core of sensory analysis. The task is to indicate whether two stimuli are perceived as different, nothing more. It is a special case of ranking. If an individual cannot order products based on which one is different or which one is stronger, then one should be skeptical of any judgment (from that individual) that involves a similar or related task such as judging the strength of a product. As noted in the discussion about qualifying, most people need guidance and training in how to use their senses and how to be a subject in a sensory test. Becoming a subject for analytical testing is an experience not unlike a school test involving choices. With some practice it is reasonable to expect that any volunteer will be able to demonstrate his or her sensory ability.

The number of subjects required for discrimination testing depends on the research objectives. In most instances using 25 qualified subjects is sufficient to determine whether a difference was perceived. The key is to use qualified subjects as described in the previous section. Some investigators propose that the number of subjects be based on statistical assumptions and using people recruited from the general population, not subjects who have met the screening criteria described previously. The use of naïve consumers makes no sense as products are marketed to specific population segments and these will be more sensitive than the general population. For more detail the reader is referred to Lawless and Heymann (1999) and Stone and Sidel (2004).

13.5.1 Analytical methods

13.5.1.1 Discrimination

Discrimination testing is one of two analytical methodologies available to the sensory professional. Discrimination methods can be further categorized as directional or non-directional. An example of the former would be the paired test in which two coded samples are presented and the subject is instructed to identify the sample that has "more of . . ." the designated attribute being tested (see Fig. 13.1). As the earliest form of discrimination testing (Cover, 1936), it enjoyed considerable popularity until processing technology and the use of more complex ingredients made it difficult to specify the nature of the difference and this probably encouraged development of other, non-directional methods. Researchers also came to realize that not all subjects understood the attribute

Paired comparison (version A)

There are two coded products; the task is to indicate which one is sweeter. Circle the code in the space provided

Which product is sweeter? ____ or circle   647   129

A/not-A   (version B)

There are two coded products; the task is to indicate whether they are the same or different. Indicate your choice by circling the word or placing a check on the line adjacent to the word that represents your decision.

Same   Different

Figure 13.1 Sensory analytical discrimination test – paired comparison and A/not-A.

These are two versions of the paired test. The choice of which to use is based on the specific objective. See text for more details as to the choice.

There are three coded products; the task is to indicate . . .

527   613   484

Which product is most different from the other two? ____

Figure 13.2 Sensory analytical discrimination test – triangle.

that was designated, further complicating both the decision and the interpretation of the results.

The *triangle* (Helm and Trolle, 1946) and *duo-trio* (Peryam and Swartz, 1950) methods represented the researchers' efforts to address these concerns. Both are non-directional methods, providing a simple resolution to the matter of specifying the difference. Both methods quickly eclipsed the paired method in terms of applications and popularity and remain more popular to this day. However, the methods are not equivalent; not only in terms of their respective probabilities but also in terms of the applications and the degree of complexity of the respective tasks.

The triangle method is a three-sample test (Fig. 13.2): all three samples are coded and the subject's task is to indicate which sample is most different from the other two. The duo-trio also is a three-sample test (Fig. 13.3); however, only two of the three samples are coded and the other is marked as reference or some similar designation. The task is to indicate which one of the coded samples is most similar to the reference. The triangle method can be viewed as three-paired comparisons (A vs B, B vs C, A vs C) while the duo-trio involves two-paired comparisons (A vs Ref and B vs Ref). For products with strong carry-over effects, i.e., adhering to the palate, it would make sense to minimize the number of samplings in order to reach a decision, and this should

There are two coded products and a third marked as a reference; the task is to indicate . . .

R   813   921

Which product is most similar to the reference?____

Figure 13.3 Sensory analytical discrimination test – duo-trio.

lead one to select the duo-trio as the method of choice.

As noted previously, the triangle method is usually cited as the most frequently used method and this choice is based on the statistical claim of greater sensitivity ($p = 1/3$ vs $p = 1/2$). The claim of statistical superiority is interesting in view of the evidence derived from an informal poll of approximately 15 companies in which it was reported that statistical significances were not often obtained. Since all products are different, by chance alone, one would expect significant differences to be obtained about one-third of the time. This may be due to use of experienced but not screened subjects, a lack of replication, or the increased potential for sensory fatigue when sampling the products to reach a decision. Sensitivity should not be an issue assuming that the subjects are qualified and replication is part of the design.

Replication should be an integral part of all analytical testing. By replication is meant that within a single session, a subject provides two judgments. A set of coded products is served and a decision is made, the products and scorecard removed, and following a timed interval of 2–3 minutes, another set of products and scorecard are presented and the process repeated. In a single session of about 10–15 minutes, two decisions are reached and the results provide the experimenter with substantially more information and confidence when making a recommendation to the requester. This increase in total number of judgments also increases the power of the test;, i.e., minimize ß risk by increasing the likelihood that if a difference can be detected, it will be. Replication enables one to examine response patterns, e.g., percentage correct for first trial vs second, and develop a better understanding of the results, and that leads to more confidence in the recommendation.

The discrimination model has remained surprisingly robust as a sensory methodology. This is most likely related to its simplicity and the ease with which responses can be obtained and results reported. It also has attracted the interest of researchers (see, for example, Sawyer *et al.*, 1962; Bradley, 1963), focused on issues such as subject selection, and increased the precision of the results and/or broadened the results beyond the core issue of whether the products were perceived as different. In more recent times, methods such as the R-index, the *n*-alternative forced-choice (AFC) method, and sameness testing have been described.

There are two coded products and two marked reference 1 and 2; the task is to indicate . . .

Which coded product is most similar to reference 1?____

Which coded product is most similar to reference 2?____

Figure 13.4 Sensory analytical discrimination test – dual standard.

Signal detection theory and the R-index require subjects to indicate their confidence or sureness as to their particular choices (O'Mahony, 1986; Ennis, 1990). The index is a probability value based on signal detection concepts that provides a measure of the degree of difference between two products. Like many psychophysical methods, these approaches require numerous evaluations and this directly impacts sensory fatigue and the concomitant reduction in sensitivity. The methods have had limited use, in part, because of this multi-sampling requirement. Finally and perhaps equally important is that no evidence has been presented to demonstrate it is more sensitive than typical methods.

The *dual-standard test* is another discrimination test used to compare two products (Fig. 13.4). Subjects are served two reference samples and two coded samples which are the same as the reference samples. The subject's task is to match each coded sample with its reference sample. The probability of choosing the correct answer is $p = 1/2$ or 50% and the duo-trio tables are referenced for significance.

A variety of approaches have been used with *multi-sample discrimination testing*, such as the *two-out-of-five test*. Again, like other discrimination methods, it is used to compare two products. In this test, subjects are presented with five coded products, two of one, and three of the other product. The subject's task is to sort the products into the two groupings, that is, to identify the two odd samples versus the remaining three that match. The probability of selecting the first of the two odd products correctly is 2 out of 5 (2/5). With the four remaining products, the probability of correctly picking out the second odd one is 1 out of 4 (1/4). The two selection events are dependent, so therefore the associated probabilities of each event are multiplied to calculate the overall probability of choosing the correct answer: $2/5 \times 1/4 = 1/10$.

The *A/not-A* discrimination test is used to compare two products, and prior to testing subjects are familiarized with each product ("A" and "not A") and informed that they may be served two "As" or two "Bs" in addition to the AB, BA orders. Subjects are then served a series of products and asked to identify them as "A" or "not A". The probability for this test is $p = 0.50$.

Another type of discrimination test is the *attribute difference test*. The question asked in this test is, "How does a given attribute differ between products?" These tests may be used to determine whether products differ in intensity in one sensory attribute – which one is more salty, has more vanilla flavor, etc. Depending on the null hypothesis, these tests can be either one tailed or two tailed; either you know that one product has more of an attribute (one tailed) or you don't (two tailed). Prior to testing, it is important to determine whether one-tailed or two-tailed probabilities will be used to analyze the data. The subject's task is to determine which product is stronger in the identified attribute. This method is called a 2-Alternative Forced Choice (2-AFC) test.

The *n-AFC methodology* has been actively promoted by Ennis and co-workers (Ennis, 1993; Bi *et al.*, 1997). In these methods, the attribute that varies must be identified prior to testing and this assumes that no other sensory attributes change. The method is an effort to extend the paired-comparison and still suffers from the same problem, that the attribute being evaluated is the only one changed and that the subjects can detect the attribute. The *3-Alternative Forced Choice test (3-AFC)* is a triangle test version.

Sameness or *similarity* testing is an approach that some have adopted without an appreciation of the consequences. Similarity testing is described (Meilgaard *et al.*, 1999) as a method to determine whether products in a difference test are the same. This is based on a difference test where there is no statistical difference between two products, leading to the conclusion that the products must be the same. To assume that products not significantly different from each other must be the same is incorrect. Concluding that products are similar would be incorrect vs concluding that the effect was so small as to be of minimal or no importance. Advising a brand manager that products are the same would be communicating misinformation because no two products are the same. A detailed discussion about this concept can be found in Cohen (1977).

Other modifications to the discrimination methodology include providing a preference judgment, providing a measure of the magnitude of the difference, and having the subject describe the basis for the difference or respond to a previously listed set of attributes. In each instance there are problems. Considering all of these modifications, the common feature is the treatment of responses from the subjects that did not make the correct discrimination choice. If a subject cannot make a correct match, then one would be reluctant to use any other information obtained from that individual. While statistical algorithms have been developed (Bradley and Harmon, 1964), the basic issue is not addressed, to wit, why collect information with the a priori knowledge that as much as 50% of it will not be usable (assuming that the differences are sufficiently small that as much as half of the product matches will be incorrect)? As mentioned for other proposed improvements in the discrimination model, no evidence has been presented that supports a claim of superiority in decision-making vs traditional discrimination procedures.

13.5.1.2 Descriptive analysis

Descriptive analysis methods represent the other analytical method available to sensory scientists. They are the most useful of sensory methods, enabling the researcher to identify which products are different and the specific basis for the difference(s) and their magnitude(s). For example, documenting the specific flavor attribute differences for an array of competitive products is essential to understanding preference differences and purchase intent differences. In non-food categories, e.g., personal care, home care, apparel, athletic equipment, electronics, automotives, descriptive analysis provides detailed descriptions of the product before, during, and after usage. This ability to measure the perceptions that take into account the entire use experience adds significant value to any brand strategy, hence the growing interest in the use of the methodology.

Much has been written about the methodology and the interested reader is directed to Lawless and Heymann (1999), Stone and Sidel (2003, 2004), and Sidel and Stone (2006) for descriptions of various methods, details about establishing panels, and recommended analyses of the results. Flavor Profile$^{®}$ was the first published method and was followed soon thereafter by Texture Profile®, a method based on Flavor Profile® focused solely on texture. This was followed by Quantitative Descriptive Analysis – QDA®. Since the introduction of the latter, other methods have been described in the literature; however, most appear to be based on these aforementioned methods. As one might expect, all descriptive methods share some common features, such as the limited number of subjects, usually less than 20, and the need for subjects to be trained to develop a language or learn to use an existing one. There are differences in the methods; for example, how subjects are screened, how the language is developed, and whether subjects can change a language, use of references and/or standards, whether the panel leader functions as a subject, how the strengths/intensities of the attributes are measured, and the types of analyses used.

Flavor Profile® and Texture Profile® are methods focused on specific modalities, with a typical panel of six subjects, each of whom takes turn serving as the panel leader. The number of attributes also is limited. For Texture Profile® the attributes are linked to standardized scales with various product anchors specified. Clearly, methods that focus on a single modality run risks that some perceptions will not be captured or will be embedded in other attributes. As has often been noted, dependency is the rule and independence is the exception; i.e., while the receptors are unique, the signals reaching higher centers in the brain interact with other signals and yield far more complex responses. The appearance of a product will influence the subject's taste expectation, and so forth. Interestingly enough, these methods also devote substantial amounts of time to subject training not unlike requiring subjects to recognize certain stimuli and to assign strengths based on those stimuli. We will have more to say about these procedures later in this discussion.

Descriptive analysis was first formally described in the literature by a method called *Flavor Profile®* (Cairncross and Sjöstrom, 1950; Sjöstrom and Cairncross, 1954; Caul, 1957). These investigators demonstrated that it was possible to select and train a group of individuals to describe their perceptions of a product in a consensus format and to report the results as a group effort. This method led to actionable results without dependence on the individual "expert". However, less sophisticated informal descriptive methods existed long before Flavor Profile® was published as a method. Early

chemists often used their senses to describe various chemicals, and perfumists and flavorists provide longstanding examples of people using descriptive words to characterize and communicate perceptions about the products with which they work. Experts in the wine, tea, coffee, spirits, chocolate, and a variety of other traditional industries have long used some form of descriptive language to characterize products, although it has not always been objective.

The Flavor Profile® method attracted considerable interest as well as controversy, but there is no question as to its historical importance to sensory science. Since then, other methods have been described and the application of descriptive analysis has increased substantially. The next descriptive method of importance was *Texture Profile®*, developed at the General Foods Research Center (Brandt *et al.*, 1963; Szczesniak, 1963; Szczesniak *et al.*, 1963). This method provided advancements in the development of terminology, included use of scales for recording intensities, development of intensity references with words, and specific anchors for each scale category. All other parts of the method were the same as for Flavor Profile®.

The objectives of these Profile methods were to eliminate reliance on the single expert by using a group of qualified people, through extensive training to eliminate subject variability, allow direct comparison of results with known reference materials, and to provide a direct link with instrumental measures. Having references for each scale may seem like an ideal means of focusing responses and reducing or eliminating variability entirely; however, these actions are counter productive in the light of knowledge of behavior. It is unrealistic to expect to eliminate variability without resorting to some form of behavior modification which would seem to be inconsistent with measuring behavior with minimal bias. People are variable from one moment to the next, from day to day, and, not too surprisingly, different from each other. In addition, use of references creates its own set of challenges because references themselves are a source of variability. If the references are commercial products, they are subject to change based on market conditions and related business issues. So, training subjects to be invariant is behaviorally unrealistic.

Profile methods also provide attributes to subjects based on their own knowledge and consultation with product technologists. Subjects are familiarized with these through training sessions over an extended period of time (usually about 3 months or more). A second concern with Profile methods was their separation of texture or flavor from the other sensory properties such as appearance, aroma, and aftertastes/aftereffects. As previously noted, the receptors are specific, but the transmission of the signals to other structures in the brain make clear the interactions that occur, leading to a more complex response. Failing to measure all perceptions makes it likely that useful information is lost. By measuring responses to all perceptions, the experimenter can derive a more complete picture of the product's sensory properties.

Profile methods and related "expert" systems treat subject differences as unwanted error and their training programs resort to behavior modification procedures in an attempt to eliminate this source of variance. The panel leader trains subjects to provide what the leader judges to be correct responses to a stimuli, and subjects are trained to agree and repeat these responses in the presence of the stimulus. The approach ignores the importance of individual differences that are reflective of consumers in general.

Interest in eliminating subject variability is, on the surface, a reasonable (although neither likely nor practical) idea, but it must not give way to procedures that sacrifice the validity of the measurement system. The human is a living and changing organism, influenced by physiological and psychological conditions that better support a concept of expected variability. Products also are variable. For these reasons, a panel of subjects is used, rather than a single subject (representing a move away from the expert), and replication has become a mainstay in most current descriptive analysis methods. Contemporary statistical practices enable the researcher to account for these sources of variation.

It is important for sensory scientists to understand the development of descriptive methods and to recognize these limitations. Sensory professionals must be able to adequately evaluate the merits of all methods relative to their business or research objective.

The next milestone in descriptive analysis was the development of the Quantitative Descriptive Analysis (QDA®) method by Stone et al. (1974). This method was developed in response to the aforementioned criticisms with the existing methods. At that time companies were experiencing growth in new products and increasing competition, and consumers were looking for new sensory experiences. Knowledge about behavior and its measurement and the use of computer-based systems for data capture and

analyses created an environment for the development of new methods. The QDA® method by the Tragon Corporation (Stone et al., 1974; Stone and Sidel, 1998; 2003) represented such a new method. It was a substantial departure from the aforementioned Profile methods, starting with how the subjects were selected, the source of the language, the use of replication, and the analysis of the responses. It addressed behavioral and measurement issues, weaknesses of the profile methods, and also responsiveness issues needed by the consumer products industry. Details about these differences are discussed later in this chapter.

As noted previously, all descriptive methods share some common features, but there also are substantial differences.

Recruiting

The QDA® method recommends recruiting subjects from outside of the technology center, with preference for individuals outside a company. This avoids use of technical experts or any individuals with prior technical knowledge of that product as their judgments are biased based on that knowledge. About 20–25 are recruited based on their product usage and related criteria. After screening, about 12–15 are identified as potential candidates for language based on their sensory acuity and availability to complete the project.

Screening

Screening is conducted with the recruited subjects using the discrimination model with products from the category to be tested. For QDA® it is important to screen subjects using actual products of interest, rather than simple stimuli in water solutions as used in the Profile methods. About 15–20 replicated product pairs are necessary, representing a range of increasing difficulty to discriminate from easy to very difficult. Product pairings are selected by the sensory team to represent all modalities and including any differences known to be important to consumer liking and purchase behavior. Subjects are selected based on their performance across all the trials, and those with more than 65% correct are contacted for the next stage, language development.

Language development

The QDA® language training focuses on using a common everyday language to describe the products, along with developing an evaluation protocol.

The panel leader functions as a facilitator, and keeps the conversation focused on these two tasks. For most product categories, about 40–50 attributes are sufficient to cover all modalities. In contrast, profile methods focus on a much smaller set of words, usually a technical-based language and a prescribed set of "universal references" obtained from the experimenter. "The choice of terminology and reference standards are factors too important to be left to the panelists, however well trained" (Meilgaard *et al.*, 1991). QDA® also uses references; however, they are selected directly from the product category and only introduced when it is beneficial to facilitate discussion and provide a common experience for the subjects. The QDA® attributes and definitions require less time to develop and the entire language development process requires between 5 and 10 hours, achieved in five 90-minute sessions, rather than over several months.

Data collection

Once the descriptive vocabulary is developed, subjects evaluate products individually in booths or, for an extended usage QDA®, in their own homes or by using the product under its normal circumstances (e.g., personal care, home care, functional apparel). Typically, subjects evaluate as few as 2 or 3 to more than 20 for an experimental design-type of test. Balanced block serving orders are used and each subject evaluates each product three or four times. This provides a sufficient data file for various analyses. Products are evaluated one at a time, with timed rest intervals and rinsing agents (e.g., for foods and beverages, water and unsalted crackers) between evaluations to minimize fatigue. When scoring, subjects use a graphic rating scale to rate the strength of each attribute for each product. The scale consists of a 6-inch line scale (~15 cm) line anchored at 1/2 inch from each end to identify the direction (e.g., slightly to very; weak to strong) for each attribute. The scale has no numbers and minimizes word biases. This helps preserve the interval character of the scale.

Analysis

From an historical perspective, QDA® was the first trained panel method to insist on, and provide for, statistical analysis of subject performance in addition to the primary focus on product differences. Panel responses on the scale are converted to numerical values (0–60) for analysis. QDA® also introduced the use of radar plots or "spider graphs" to readily com-

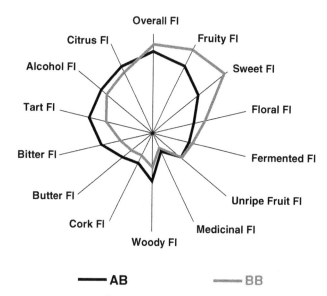

Figure 13.5 QDA® spider plot of sensory attributes. Each spoke in the wheel represents a sensory attribute. Entries are product mean values for each attribute. At the center of the wheel = low intensity; at the end of each spoke = higher intensity. Products that exhibit quantifiable sensory differences can be evaluated by target consumers. Then, by using a variety of multivariate analysis techniques, the key sensory characteristics that most influence target consumer liking can be designed into products, as well as providing a benchmark versus key competition. When consumer-based sensory evaluation techniques are used in conjunction with marketing and marketing research techniques, it provides an effective business strategy for brand management.

municate product similarities and differences, and this graphing system has become an industry norm (Figs 13.5 and 13.6). Once the data have been converted to numbers, a series of analyses examine all aspects of the subject's and attribute's "performance". The analyses include the one- and two-way Analysis of Variance models along with rank order analyses where required. In addition, there are specific algorithms for measuring subject variability, crossover and magnitude interactions, scale usage, and sensory mapping. A description of the various analyses can be found in Stone and Sidel (2004).

Other methods

Other methods have been introduced over the years; for example, the Spectrum Method by Gail Civille and collaborators in the late 1980s (Meilgaard *et al.*, 1991). This method closely follows Flavor Profile with intensive training (about 14 weeks or more), absolute ratings, and extensive use of references and calibration points. It is a technological/engineering

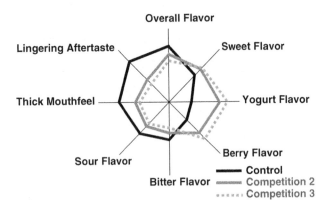

Figure 13.6 Example of a QDA® spider plot of fruit yogurt. Results illustrate product similarities and differences by attribute. Control was: *Higher/Stronger* in Overall Flavor, Bitter Taste, Sour Taste, Thick Mouthfeel, and Lingering Aftertaste; *Lower/Weaker* in Sweet Taste, Yogurt Flavor, and Raspberry Flavor.

approach to descriptive analysis. Training is lengthy, as previously noted, and separate training programs are required for texture, flavor, and other modalities. Subjects are taught about sensory processes and experimenter-assigned attribute vocabulary and rating scales. These rating scales are anchored at multiple points with experimenter-assigned reference standards. The method, however, does include category-specific discrimination tests, intensity scales, exposure to a broad array of products in the category, evaluation in individual test booths, and application of statistical analysis. However, the latter are not specified. To date, the method has not adequately demonstrated the claim of absolute scales nor has it recognized the concept of individual differences in perception.

Free-Choice Profiling (FCP) by Williams and Langron (1984) also has been described. The authors proposed a radically different procedure requiring no screening of the subjects and no language training. Unfortunately these researchers soon found that time was needed by the subjects to rationalize differences in language use, thus negating any time saved by not screening and not meeting as a group.

Descriptive analysis using a trained sensory panel continues to increase in popularity due to its usage in understanding product similarities and differences. This information is beneficial to businesses for a variety of applications. For the most part, many of the descriptive methods are more diagnostic in nature or they are variations on the two primary approaches to descriptive analysis – Profile methods and QDA®. A

few other methods have attempted to establish themselves as different.

Larson-Powers and Pangborn (1978) introduced the use of a reference product against which the other products were evaluated. With the reference product, the other products must be presented simultaneously, thereby reducing the number of products that could be evaluated in one session. The developers of Flavor Profile (Hanson *et al.*, 1983) introduced a new version of descriptive analysis called Profile Attribute Analysis (PAA) that included seven-point intensity scales to allow statistical treatment of the results. In addition, Stampanoni (1993) introduced "Quantitative Flavor Profiling" as a hybrid method based on Flavor Profile and QDA®. Several other authors have proposed methods that have elements of profiling and QDA®, often referred to as DA, or generic descriptive analysis (Einstein 1991; Gilbert and Heymann, 1995; Lawless and Heymann 1999).

Descriptive analysis methodology has evolved over time to become a strategic source of product information for companies. Developments in descriptive analysis can be traced from the use of product experts to the more formal and rigorous approach applied in the QDA® method. There will continue to be changes and developments in the growth of the methodology and its application, with particular emphasis on measurement theory, consumer behavior, psychology, quantitative data, and enhanced statistical procedures. Continued and successful application of descriptive analysis will result in greater awareness and the role of sensory evaluation within the business community as it continues to evolve into a strategic source of product information.

Further reading and references

Anonymous (1975) *Minutes of Division Business Meeting.* Institute of Food Technologists – Sensory Evaluation Division, Chicago.

Bi, J., Ennis, D.M. and O'Mahoney, M. (1997) How to estimate and use the variance of *d'* from difference tests. *Journal of Sensory Studies*, **12**, 87–104.

Bradley, R.A. (1963). Some relationships among sensory difference tests. *Biometrics*, **19**, 385–97

Bradley, R.A. and Harmon, T.J. (1964) The modified triangle test. *Biometrics*, **20**, 608–625.

Brandt, M.A., Skinner, E.Z. and Coleman, J.A. (1963) Texture profile method. *Journal of Food Science*, **28**(4), 404–409.

Cairncross, W.E. and Sjöström, L.B. (1950) Flavor Profile – a new approach to flavor problems. *Food Technology*, **4**, 308–311.

Caul, J.F. (1957) The profile method of flavor analysis. *Advances in Food Research*, **7**, 1–40.

Cohen, J. (1977) *Statistical Power Analysis for the Behavioral Sciences*, revised edn. Academic Press, New York.

Cover, S. (1936) A new subjective method of testing tenderness in meat – the paired- eating method. *Food Research*, **1**, 287–95.

Eggert, J. and Zook, K. (eds) (1986) *Physical Requirement Guidelines for Sensory Evaluation Laboratories*. ASTM Special Technical Publication 913. American Society for Testing and Materials, Philadelphia.

Einstein, M.A. (1991) Descriptive techniques and their hybridization. In: *Sensory Science: Theory and Applications in Foods* (eds H.T. Lawless and B.P. Klein). Marcel Dekker, New York, pp. 317–38.

Ennis, D.M. (1990) Relative power of difference testing methods in sensory evaluation. *Food Technology*, **44**(4), 114, 116–117.

Ennis, D.M. (1993) The power of sensory discrimination methods. *Journal of Sensory Studies*, **8**, 353–70.

Gambaro, A., Varela, P., Boido, E., Gimenez, A., Medina, K. and Carrau, F. (2003) Aroma characterization of commercial red wines of Uruguay. *Journal of Sensory Studies*, **18**, 353–66.

Gilbert, J.M. and Heymann, H. (1995) Comparison of four sensory methodologies as alternatives to descriptive analysis for the evaluation of apple essence aroma. *The Food Technologist (NZIFST)*, **24**(4), 28–32.

Hanson, J.E., Kendall, D.A., Smith, N.F. and Hess, A.P. (1983) The missing link: correlation of consumer and professional sensory descriptions. *Beverage World*, November, 108–15.

Helm, E. and Trolle, B. (1946) Selection of a taste panel. *Wallerstein Laboratories Communications*, **9**(28), 181–94.

Irigoyen, A., Castiella, M., Ordonez, A.I., Torre, P. and Ibanez, F.C. (2002) Sensory and instrument evaluations of texture in cheeses made from ovine milks with differing fat contents. *Journal of Sensory Studies*, **17**, 145–61.

Kuesten, K. and Kruse, L. (eds) (2008) *Physical Requirement Guidelines for Sensory Evaluation Laboratories: 2nd Edition*. ASTM Special Technical Publication MNL 60. American Society for Testing and Materials, Philadelphia.

Larráyoz, P., Mendia, C., Torre, P., Barcína, Y. and Ordóñez, A.I. (2002) Sensory profile of flavor and odor characteristics in Roncal cheese made from raw ewe's milk. *Journal of Sensory Studies*, **17**, 415–27.

Larson-Powers, N. and Pangborn, R.M. (1978) Descriptive analysis of the sensory properties of beverages and gelatins containing sucrose or synthetic sweeteners. *Journal of Food Science*, **43**, 42–51.

Lawless, H.T. and Heymann, H. (1999) *Sensory Evaluation of Food: Principles and Practices*. Aspen, Gaithersburg, Maryland.

Mackey, A.O. and Jones, P. (1954) Selection of members of a food tasting panel: discernment of primary tastes in water solution compared with judging ability for foods. *Food Technology*, **8**, 527–30.

Meilgaard, M., Civille, G.V. and Carr, B.T. (1991) *Sensory Evaluation Techniques*, 2nd edn. CRC Press, Boca Raton.

Meilgaard, M., Civille, G.V. and Carr, B.T. (1999) *Sensory Evaluation Techniques*, 3rd edn. CRC Press, Boca Raton.

Naes, T. and Risvik, E. (1996) *Multivariate Analysis of Data in Sensory Science*. Elsevier, Amsterdam.

O'Mahony, M. (1986) *Sensory Evaluation of Food: Statistical Methods and Procedures*. Marcel Dekker, New York.

O'Sullivan, M.G., Byrne, D.V., Martens, H. and Martens, M. (2002) Data analytical methodologies in the development of a vocabulary for evaluation of meat quality. *Journal of Sensory Studies*, **17**(6), 539–58.

Peryam, D.R. and Swartz, V.W. (1950) Measurement of sensory differences. *Food Technology*, **4**, 390–395.

Rainey, B.A. (1986) Importance of reference standards in training panelists. *Journal of Sensory Studies*, **1**, 149–54.

Sawyer, F.M., Stone, H., Abplanalp, H. and Stewart, G.F. (1962) Repeatability estimates in sensory-panel selection. *Journal of Food Science*, **27**, 386–93.

Schutz, H.G. and Cardello, A.V. (2001) A labeled affective magnitude (LAM) scale for assessing food liking/disliking. *Journal of Sensory Studies*, **16**, 117–59.

Sidel, J. and Stone, H. (2006) Sensory science: methodology. In: *Handbook of Food Science, Technology, and Engineering*, Vol. 2 (ed. Y.H. Hui). CRC Taylor & Francis, London, pp. 1–24.

Sjöström, L.B. and Cairncross, S.E. (1954) The descriptive analysis of flavor. In: *Food Acceptance Testing Methodology* (eds D.R. Peryam, F.J. Pilgrim and M.S. Peterson). National Academy of Sciences – National Research Council, Washington, DC, pp. 25–30.

Stampanoni, C.R. (1993) The quantitative flavor profiling technique. *Perfumer Flavorist*, **18**, 19–24.

Stone, H., Sidel, J.L., Oliver, S., Woolsey, A. and Singleton, R.C. (1974) Sensory evaluation by quantitative descriptive analysis. *Food Technology*, **28**(11), 24–34.

Stone, H. and Sidel, J.L. (1998) Quantitative descriptive analysis: developments, applications, and the future. *Food Technology*, **52**(8), 48–52.

Stone, H. and Sidel, J.L. (2003) Descriptive analysis. In: *Encyclopedia of Food Science*, 2nd edn. Academic Press, London, pp. 5152–61.

Stone, H. and Sidel, J. (2004) *Sensory Evaluation Practices*, 3rd edn. Academic Press, San Diego.

Stone, H. and Sidel, J.L. (2007) Sensory research and consumer-led food product development. In: *Consumer-led Food Product Development* (ed. H. MacFie). Woodhead Publishing, Cambridge.

Szczesniak, A.S. (1963) Classification of textural characteristics. *Journal of Food Science*, **28**, 385–9.

Szczesniak, A.S., Brandt, M.A. and Friedman, H.H. (1963) Development of standard rating scales for mechanical parameters of texture and correlation between the objective and the sensory methods of texture evaluation. *Journal of Food Science*, **28**, 397–403.

Williams, A.A. and Langron, S.P. (1984) The use of free-choice profiling for the evaluation of commercial ports. *Journal of the Science of Food and Agriculture*, 35, 558–68.

Supplementary material is available at www.wiley.com/go/campbellplatt

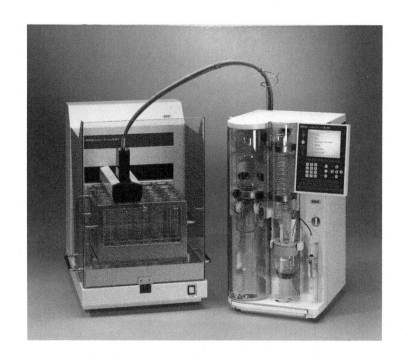

Plate 1 Modern Kjeldahl apparatus (printed with kind permission of Büchi Labortechnik, Essen, Germany).

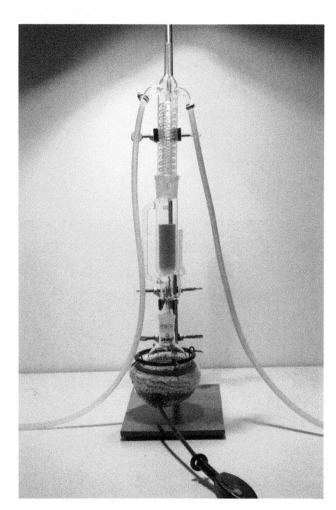

Plate 2 Soxhlet apparatus with extraction thimble, reflux condenser and electric heating jacket (photo: G. Merkh, University of Hohenheim).

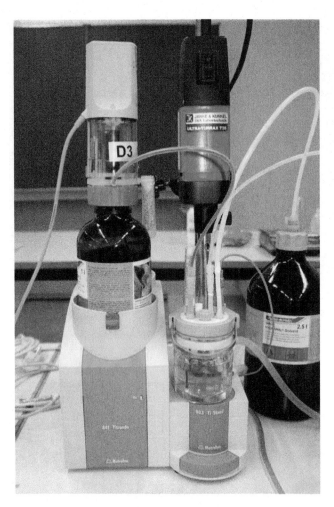

Plate 3 Karl Fischer titrator with a double-walled titration vessel for titrations at elevated temperatures and equipped with an internal homogeniser (photo: G. Merkh, University of Hohenheim).

Plate 4 View of a Karl Fischer titration cell with burette tip and platinum electrode pair (photo: G. Merkh, University of Hohenheim).

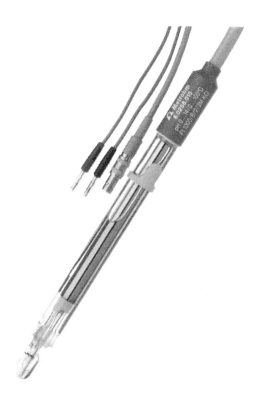

Plate 5 pH electrode with cables (printed with kind permission of Deutsche Metrohm, Filderstadt, Germany).

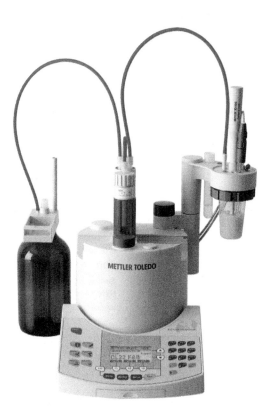

Plate 6 Automatic titrator (printed with kind permission of Mettler-Toledo AG, BU Analytical, Schwerzenbach, Switzerland).

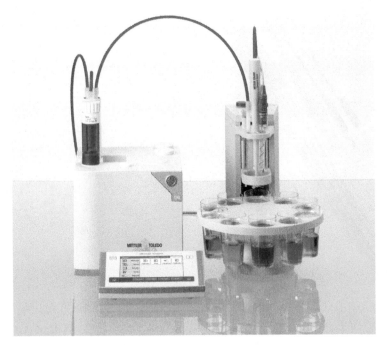

Plate 7 Titrator with sample changer (printed with kind permission of Mettler-Toledo AG, BU Analytical, Schwerzenbach, Switzerland).

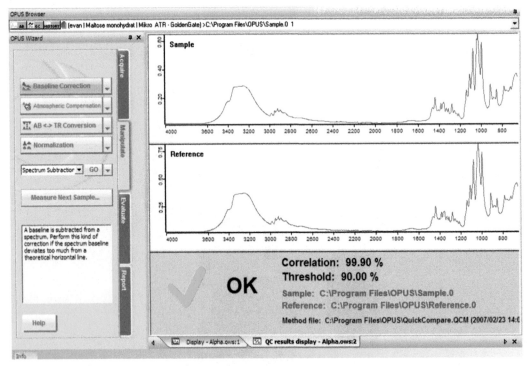

Plate 8 Comparison of IR spectra of a sample with the spectrum of a reference substance (here maltose monohydrate) (printed with kind permission of Bruker Optics, Ettlingen, Germany).

Plate 9 NIR measurement by reflection through the bottom of a Petri dish (printed with kind permission of Büchi Labortechnik, Essen, Germany).

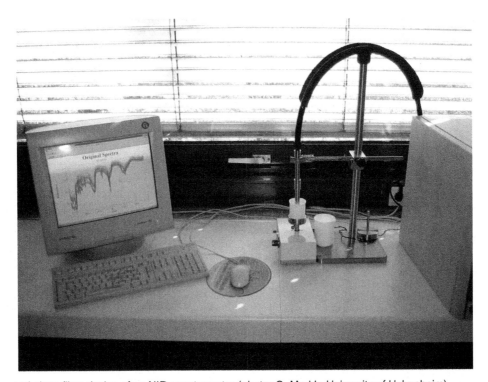

Plate 10 Probe and glass fibre device of an NIR spectrometer (photo: G. Merkh, University of Hohenheim).

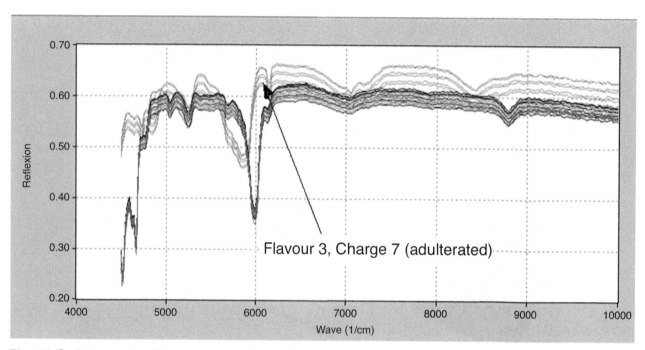

Plate 11 Qualitative control of cherry flavours by NIR spectroscopy: NIR-spectra of genuine cherry flavours and three spectra of adulterated products (from Köstler, M. and Isengard, H.-D. (2001) Quality control of raw materials using NIR spectroscopy in the food industry. *G.I.T. Laboratory Journal*, **5**, 162–4).

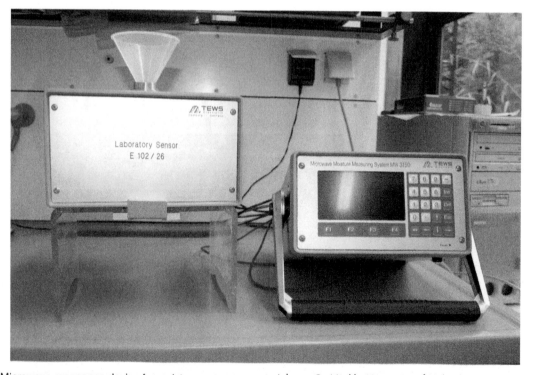

Plate 12 Microwave resonance device for moisture measurements (photo: G. Merkh, University of Hohenheim).

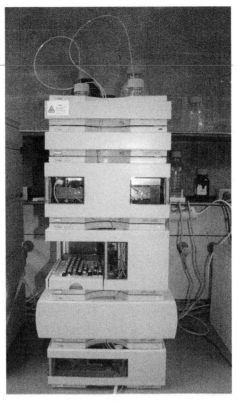

Plate 13 HPLC with (from top to bottom) eluent containers, degassing unit, high pressure pumps, sample vials and injection unit, column chamber, detector (photo: G. Merkh, University of Hohenheim).

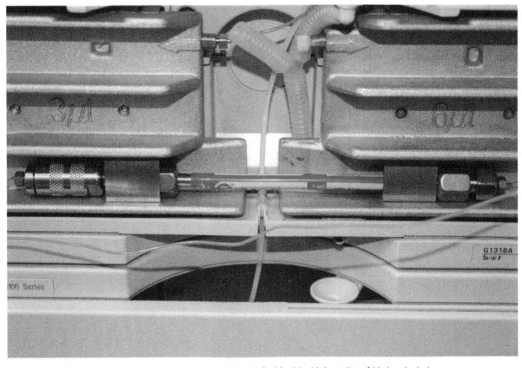

Plate 14 HPLC column in the opened column chamber (photo: G. Merkh, University of Hohenheim).

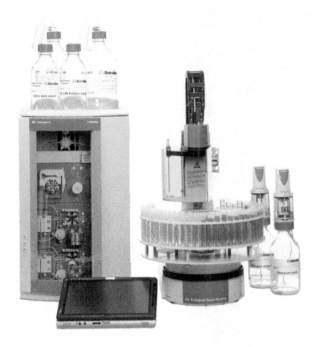

Plate 15 Ion-exchange chromatograph with automatic sample changer (printed with kind permission of Deutsche Metrohm, Filderstadt, Germany).

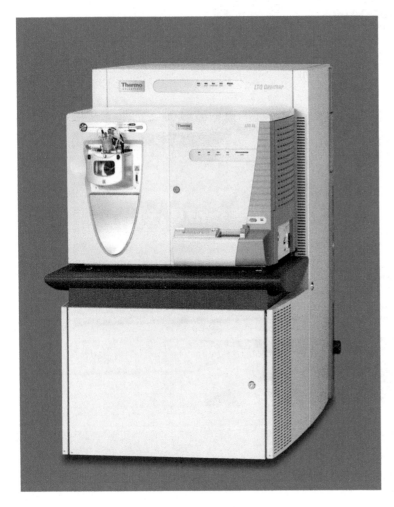

Plate 16 Example of a modern mass spectrometer: Thermo Scientific LTQ Orbitrap hybrid mass spectrometer (printed with kind permission of Thermo Electron, Dreieich, Germany).

Plate 17 Example of a modern atomic absorption spectrometer: PerkinElmer AAnalyst 400 (printed with kind permission of PerkinElmer LAS, Rodgau-Juegesheim, Germany).

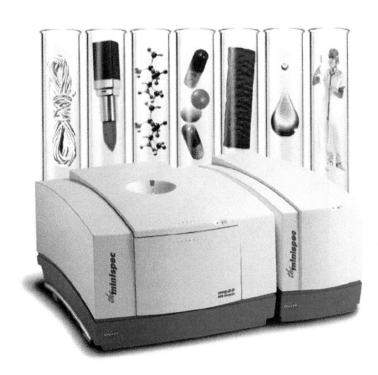

Plate 18 Time-domain NMR spectrometer (printed with kind permission of Bruker Optik, Rheinstetten, Germany).

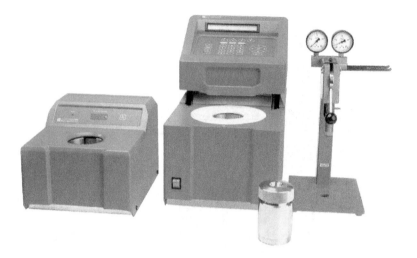

Plate 19 Isoperibolic calorimeter with calorimetric bomb, cooling station and pressure filling station (printed with kind permission of IKA-Werke, Staufen, Germany).

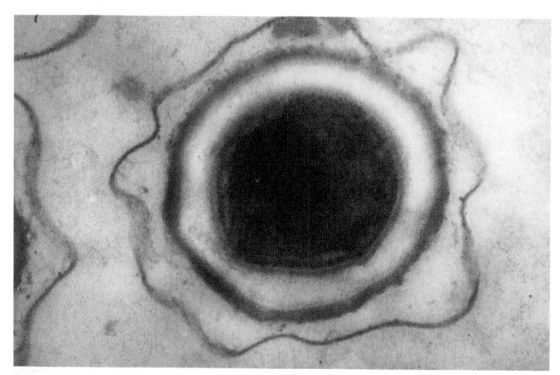

Plate 20 Electron micrograph of a cross section of a spore of *Clostridium bifermentans*. Layers (from outside) are exosporium, coat, cortex and proplast.

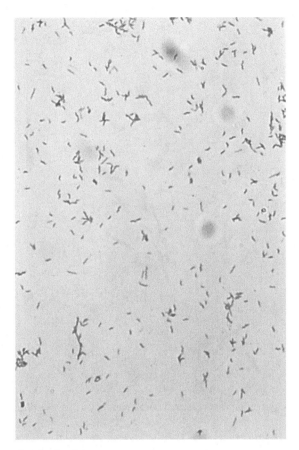

Plate 21 *Escherichia coli* Gram negative (red cells).

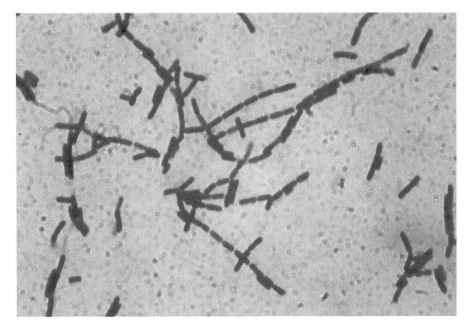

Plate 22 Exponential growing Gram positive cells of *Bacillus megaterium.* Note the long chains formed. These break up into much shorter chains when growth stops.

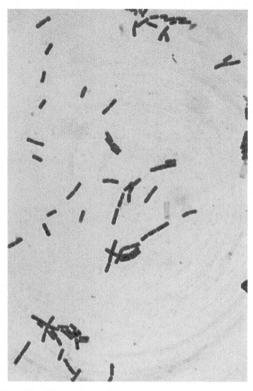

Plate 23 Sporulating Gram positive cells of *Bacillus cereus*. Spores appear as bright centres. Note short chains.

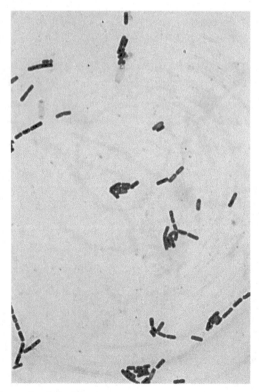

Plate 24 Late Gram positive sporulating cells of *Bacillus cereus*.

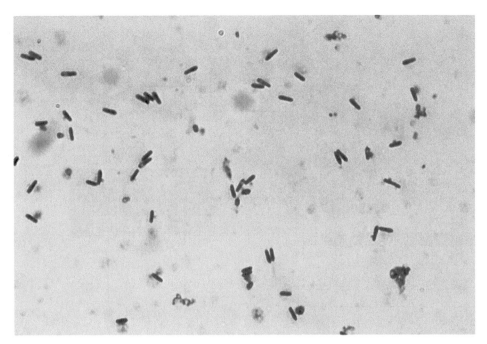

Plate 25 Gram positive sporulating *Clostridium perfringens* cells with "tennis racket" appearance produced by ends of cells swelling during spore formation.

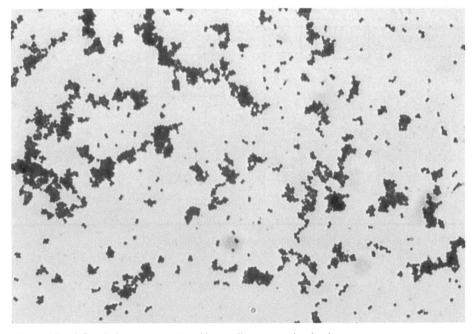

Plate 26 Gram positive cells of *Staphylococcus aureus*. Note cells are growing in clumps.

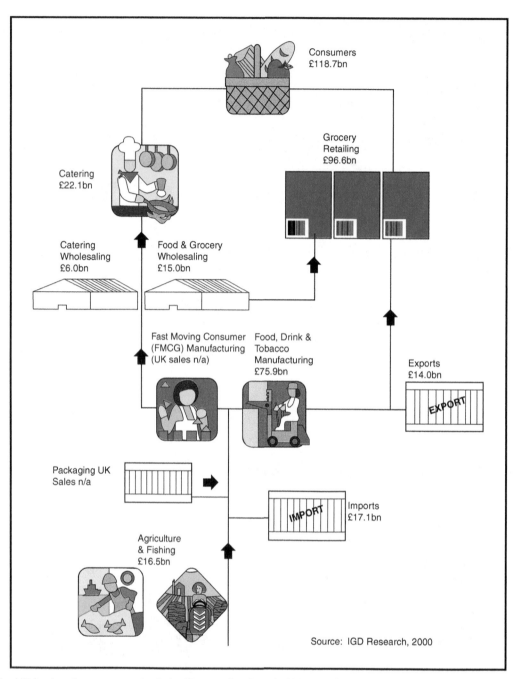

Plate 27 The UK food and grocery supply chain. (Source: Patel *et al.*, 2001, p. 5.)

Herbert Stone and Rebecca N. Bleibaum

Key points

■ This chapter provides an introduction to the design and analysis of data, and guides the food scientist in decision-making from an applied rather than a theoretical perspective.

■ Due to behavior aspects of the measuring instrument (people), the analysis of data is approached quite differently from the physical sciences (instrumental data, chemical data, etc.).

■ Food scientists must understand various types of scales and the types of analyses associated with each scale. They also must understand multivariate analysis methods and their wide variety of applications.

■ The analysis of variance (AOV) is the most common statistical test used in sensory research. It easily accommodates groups of subjects, multiple treatment conditions, and multiple products.

■ Correlations and regression equations are useful to examine sensory/instrumental relationships, and in product optimization research where sensory attributes are used to predict consumer liking, appropriateness, and other behavioral measures.

■ Cluster analysis is used extensively in sensory optimization research to identify unique consumer preference groups and is used in descriptive analysis to identify unique subject groups.

■ Visual presentation of sensory research is important to communicate data. Graphs and charts reveal insights not readily observed in other ways, and they simplify results for less technical audiences.

14.1 Introduction

In sensory evaluation, the process of measuring the strength of perceptions, product usage, attitudes, and preferences is approached differently from data from the physical sciences (instrumental data, chemical data, etc.) due to the behavioral aspects of the measuring instrument, i.e., the people. However, there is, and continues to be, much discussion about the types of statistical procedures used to analyze results from sensory tests, whether those results are part of a re-search project or are part of a product test before market introduction.

In most instances, sensory scientists rely on ordinal and/or interval scales to capture responses. Oftentimes, tests are designed and analyses completed without sufficient input from the sensory scientist, and, not surprisingly, results either lead to no definitive answer or in some cases, an incorrect conclusion. Sensory scientists must understand the type of scale being proposed vs alternatives and the types of analyses associated with that scale choice.

Once a scale has been selected, statistics becomes an essential part of the sensory evaluation process, providing a basis for summarizing information to reach conclusions about subject performance and product differences. There are numerous statistical procedures available and new ones being researched and reported regularly. It can be challenging to select the appropriate statistic to use in part because sensory scientists and statisticians often differ in their views about how responses should be treated, which analysis is more or less appropriate, the extent of any data transformation, and so forth. To add to this, there are a wide variety of relatively inexpensive statistical software packages that are menu driven; i.e., easy for the novice to use and, therefore, very attractive. However, this does not make them appropriate. Using appropriate statistics minimizes the risk of reaching a wrong decision, but it does reduce the likelihood of errors in decision-making. There is no excuse for the sensory scientist to lack sufficient knowledge about the design and analysis of a test.

The goals of this chapter are to provide an introduction to the design and analysis of sensory and related preference testing, and to guide the sensory scientist in decision-making from an applied rather than a theoretical perspective. There are many textbooks on statistics and several specifically focused on sensory evaluation; these are listed at the end of this chapter. Here we will provide principles and examples of practices that are best suited to behavioral information. However, it is not and cannot be an exhaustive discussion; that we leave to other texts that provide considerable detail (e.g., Lawless and Heymann, 1999; Stone and Sidel, 2004). In this chapter we focus on analyses that we consider especially useful for basic types of sensory tests.

14.2 Descriptive statistics

In most tests subjects typically use some form of scaling to register the strength of a product attribute such as sweetness. The type of scale is usually ordinal or interval and the responses are then summarized, and subjected to a series of analyses, often referred to as descriptive statistics. For example, one can compute a measure of central tendency such as the average (or mean), along a measure of the distribution or dispersion of the scores, e.g., standard deviation. These summaries are usually referred to as descriptive statistics. Results are often displayed in tables

and graphs, enabling the reader to easily compare results for a series of products, which is much easier to discuss as against using only the raw scores.

However, one must be careful to ensure that this summary does not mask or distort the information provided from individual or group responses.

14.2.1 Measures of central tendency

For measures of central tendency, the mean, mode, and median are useful in summarizing individual response data. The arithmetic mean is the most widely used measure in sensory evaluation. It is the sum of all ratings divided by the number of summed responses. Means can be distorted by extreme scores, in which case the harmonic or geometric mean may be a better choice. Another useful measure is the mode, which is not influenced by extreme scoring. The mode is the scale category that contains the highest number of responses. The median is another measure that is not influenced by extreme scores. It represents the category containing the 50th percentile, dividing an ordered distribution in half.

14.2.2 Measures of dispersion

Measures of dispersion, such as the range, the variance, and the standard deviation are the three most frequently used statistics when reporting results. Range refers to the absolute difference between the lowest and highest scores, and is easily influenced by extreme scores. By itself, range has limited value but provides much more information when combined with other measures such as central tendency, variance, and frequency distributions. Variance provides a measure of dispersion from the mean. Variance is calculated by subtracting each score from the mean, and then that number is squared to eliminate negative numbers. These squared differences are summed and then divided by the number of responses minus 1. This calculated variance is displayed as 'S2' to represent sample data rather than population data. Standard deviation is the square root of the variance and displayed as 'S' to represent sample data. The standard deviation will always be smaller than the variance, which reduces the influence of extreme scores.

14.2.3 Frequency distributions

A frequency distribution is the number of response for each category on the scale. Data from frequency distributions are typically displayed as tables or graphs, and provide valuable information on how well other statistics (e.g., mean and range) represent the response data. Sensory scientists should examine frequency distributions for each product and serving order, and for different groups of subjects, to discover trends in the data. Products with similar frequency distributions may be included in significance tests, whereas those with very different shape distributions may not be included.

14.3 Inferential statistics

Inferential statistics are used by sensory scientists to determine the risk associated with declaring that an observed difference in a test represents a real difference that can be generalized to other test populations, or to a larger consumer population. A variety of publications go into detail on inferential statistics. Therefore, our discussion will focus on the most common inference statistics used by sensory professionals, both non-parametric and parametric.

14.3.1 Non-parametric

Non-parametric statistical tests are appropriate for ordinal and nominal scales, when data from groups or products to be compared are in the form of counts (i.e., frequencies), percents, or ranks. Data from these scale types are not suited to parametric statistics such as means, t-tests, analysis of variance, and multiple means comparisons. There are a vast number of non-parametric statistical tests and further discussion can be found in the many useful references in the sensory statistics literature. In this chapter, we will discuss a few important and useful tests for the sensory professional.

Data from forced choice tests such as most discrimination tests (triangle, duo-trio, etc.) and paired preference tests can be analyzed for statistical significance using tables based on the binomial distribution. For small numbers of subjects, there are tables of the cumulative binomial distribution that provide the exact probability of the outcome. For larger numbers of subjects, one uses the normal distribution approximate calculations, expressed as Z scores.

Non-parametric statistics also include Chi-square (χ^2), which is useful for determining whether the observed number of responses is significantly different from an expected number of responses. In sensory research, Chi-square is primarily used in consumer testing to determine whether two response distributions (or populations) are significantly different. It also is used as an initial test to isolate and compare two response categories among several that were tested. Since there are several Chi-square formulations, the selected formula is based on how the data are collected and the number of categories being compared.

In addition to the above mentioned test, there are a variety of other non-parametric tests that may be useful for sensory evaluation. These include, but are not limited to, the Cochran Q test, Friedman test, Kruskal–Wallis test, Mann–Whitney U test, McNemar test, and the Wilcoxon test.

14.3.2 Parametric tests

Parametric tests are viewed as more powerful than non-parametric tests, and this is in part due to the types of scales appropriate for these measures. Parametric tests require data that satisfy the constraints of the normal curve, which includes measures on interval and ratio scales.

Most scales used in sensory evaluation have adequate interval characteristics to allow use of parametric statistics for analysis. Even if the measurement does not completely fit the statistical theory and assumptions, sensory results still may be reliable, valid, and useful. For sensory applied research, it may be more beneficial to modify the statistical theory and assumptions, rather than forgo the application of parametric analysis.

The first and most simple parametric statistic is the *t-test*, used to determine whether the means for two products are significantly different. The t-test formulae are slightly different, depending on experimental conditions, such as dependent or independent observations, small or large number, proportions, and equal or unequal numbers of observations. The t-test is most appropriate when two unique populations of consumers evaluate one product each, but this statistic is not the best choice when the same subjects evaluate both products. In this case, when one subject evaluates both products, the analysis-of-variance should be substituted. The t-test also is inappropriate where more than two products are evaluated and statistically compared. This constitutes multiple t-tests of the same data, which is a commonly observed error in sensory research.

The most common statistical test used in sensory research is the *analysis of variance (AOV)*. This is a useful statistical procedure that easily accommodates various groups of subjects, multiple treatment conditions, and multiple products. AOV is used to determine whether an effect (main or interaction) is statistically significant. When a difference is found, a multiple range test is then used to determine which means are different. A variety of AOV models are widely used in sensory research, including the one-way for independent means, the two-way for dependent means, treatment x levels for different treatment conditions (e.g., x levels of different variables), and split-plot analysis models for different subject groups evaluating the same and different treatments.

Challenges for selecting the best AOV model are to determine which model is most appropriate for the test conditions, to accurately account for all the sources of variance in the test, and to select the appropriate error term for testing the significance of an effect. Statistical significance in the AOV is based on a calculated F ratio consisting of an effect variance in the numerator and an error variance in the denominator. Probability tables for the significance of the calculated F are published in most statistics books; most statistical packages for sensory and behavioral data include the AOV analysis and exact probabilities for the calculated F.

Multiple range tests are used following an AOV to determine from among a set of means for a significant variable which ones are significantly different. Different range tests reflect what is to be compared and the respective authors' preference for controlling the various error rates possible when making multiple comparisons for a single data set. Too conservative a test will make it difficult to find statistical significance and result in more type II errors, whereas a less conservative test may result in false differences (i.e., type I error). The sensory scientist is cautioned here that the underlying theoretical assumptions for a multiple range test may not agree with practical experience using that test. Where this is the case, paying attention to the risk one can still proceed cautiously. Frequently used multiple range tests include: Fisher's LSD (least significance difference), Bonferroni, Duncan's, Dunnett's, some tests by Tukey, S-N-K (Student–Newman–Keuls), and Scheffe tests. Whichever test is selected, the results must be examined to determine whether they agree with the knowledge and expectation the researcher has about the products and the sensory test.

Sensory evaluation research typically produces large data sets, especially when consumer testing and analytical (chemical/physical) data are involved. The size of the data set depends on the number of subjects, attributes, products, treatments, etc. Other statistical treatments are needed with these large data sets, such as multivariate analysis methods.

14.3.3 Multivariate analysis

Multivariate data analysis methods are designed to illustrate the main structures and relationships in large data sets, producing relatively simple output graphs and tables that have a maximum of information and a minimum of repetition and noise. Relationships among the data are rarely univariate and multivariate data analysis achieves data reduction with minimal loss of information. These multivariate techniques provide graphical outputs, and allow a simple, yet in-depth understanding of relationships in the data. Sensory scientists must understand multivariate analysis methods and their wide variety of applications.

14.4 Correlation, regression, and multivariate statistics

Sensory scientists are often interested in the relationships between different sets of data and there are a variety of statistic procedures that analyze more than a single variable at a time. Responses to products may be influenced by a variety of factors or variables including different groups of subjects, products, attributes, questions, the context of the test, or combinations thereof. Correlation and multivariate methods are used to understand these relationships. Because correlation does not imply causation, summary statistics and graphs depicting association are useful for data reduction, substitution, prediction, and generally improving understanding about variables. Different statistics are useful for different types of measurement scales (e.g., nominal, rank order, or continuous) and the sensory scientist must understand the types of questions being asked, along with the types of analyses that are most appropriate.

Correlations are useful to examine and understand sensory and instrumental relationships, and in consumer-based product optimization research where sensory attributes are used as predictor variables for consumer liking, appropriateness, and other behavioral measures. The predictive measure is the regression equation (y = bX + a) which indicates the relationship between two variables. It also indicates the extent to which one measure can be predicted by knowing the other value (e.g., if caramel flavor strength score was 25, then overall liking would have a score of 7.2), or the extent to which two variables are associated. Regression fits the "best" line to the data by minimizing the sum of error terms:

$$y = bX + a$$

where

a   is the intercept,
b   is the slope or regression coefficient,
X   is the value of the independent variable.

14.4.1 Correlation

To easily illustrate a two-variable correlation one can graph results as a scatter plot with the data from one variable (e.g., caramel flavor) on the X-axis and data from the second variable (consumer liking) on the Y-axis. Results from a set of products can be graphed and a correlation calculated for the relationship between the two measures. The Pearson product moment correlation coefficient (r) measures the degree to which two variables are linearly related. It is used for continuous scaled data and is depicted by r, where r values can range from -1 to $+1$. The closer to 0, the weaker the association between the two variables. The sign depicts the direction of the relationship.

Pearson product moment correlation coefficient is useful only for linear data; curvilinear data can produce low values approaching 0, misleading the casual observer to conclude the variables are unrelated. Correlation can be determined for non-continuous scaled data as well; an example is Spearman's rank order correlation coefficient.

14.4.2 Multiple correlation (R)

Multiple correlation is used to determine the degree of association between a dependent variable and a set of predictor variables. This is useful in sensory science, especially in product optimization research where several sensory attributes (independent variables) may be included as important pre-dictors of consumer acceptance (dependent variable), and the multiple correlation (R) describes the degree of association between the dependent and independent variables. Several books have been published on the topic of multiple regression/correlation (MRC), and the interested reader is directed to Cohen and Cohen (1983) for an in-depth discussion of these techniques.

14.4.3 Regression

Regression is a general term applied to equations that fit a line to observed data points. Simple linear, non-linear, and multiple regression are possible, and the resulting regression line is often used to predict values of the dependent variable (y) from values of the independent variable (x). The sensory scientist will find regression equations useful to predict perceived intensity for an attribute based on ingredient concentration. In optimization research, multiple regression is used to predict consumer liking from a combination of sensory attributes and their intensities. Chemical/physical analytical measures may also be used as predictor variables for acceptance in optimization, alone or in conjunction with sensory measures.

14.4.4 Additional multivariate methods

In addition to the tests previously described, several other tests are used in sensory evaluation to simultaneously examine multiple variables. Multivariate analysis often provides understanding of important relationships not readily observable by other means. The book by Dillon and Goldstein (1984) provides further explanation of additional multivariate analysis methods, their usefulness, and application.

14.4.5 Multivariate analysis of variance (MANOVA)

MANOVA is an extension of analysis of variance (ANOVA) which applies to one variable, but it is for situations where there are several variables. It determines whether significant differences exist among treatments when compared against all dependent variables of interest. A good example would be quantitative descriptive analysis where several attributes are evaluated for a product set. Whereas ANOVA evaluates one dependent variable at a time, MANOVA analyzes all the dependent variables simultaneously. The correlation matrix of the sensory attributes, physical, and chemical measures provides useful insight into potential causal relationships. The matrix likely contains redundancies, i.e., there is overlap in what is being measured. Multivariate methods such as those described below enable us to identify these relationships.

MANOVA provides a single F-statistic, based on Wilks' lambda, which assesses the influence of all variables simultaneously. A significant F-statistic (due to a small Wilks' lambda) means that the samples are significantly different across the dependent variables and warrants individual ANOVAs for each variable. If the F-statistic is not significant, individual ANOVAs are not warranted. Performing a MANOVA first protects the statistician against an inflated overall type I error rate that can be brought about by a high number of individual ANOVAs. Finally, MANOVA also examines colinearity (through the covariance matrix) among the variables, which can point to a group of variables discriminating among the products, thus guarding against type II error.

14.4.6 Discriminate analysis

Discriminate analysis is a methodology for finding linear combinations of the independent variables that can act as scoring functions to estimate to which of several classification categories an observation belongs. Sensory scientists use this and related procedures (e.g., canonical correlation) in optimization research to identify lifestyle, attitude, and classification information that best identify membership to different consumer preference groups.

14.4.7 Principal components analysis (PCA)

PCA is a data reduction or simplification technique used to transform the original set of variables into as few linear combinations as possible to explain as much of its total variation. To better understand how this analysis functions, PCA first locates the center of the data. It then searches for a straight line through the data center that accounts for as much of the variation as possible. Next, it searches for a new line that accounts for as much of the remaining variation as possible, and then continues on with this approach. The so-called principal components are plotted as biplots with the objects and/or the variables. The linear combinations, identified as factors or components, are independent from all other factors. The method is often applied to sensory descriptive panel data to reduce the number of attributes, or to identify a smaller set of independent attributes to include as independent variables in developing multiple regression models for optimization research. It is also used in optimization research to identify products that form independent groups based on sensory attributes.

How do we figure out how many principal components we should examine and plot? Eigenvalues are associated with each principal component. A principal component with an eigenvalue above 1.0 typically has some relevance in the data set. Scree plots of decreasing eigenvalues are useful to better understand at what point the number of principal components are enough to describe the data and at what point they can be overlooked without losing much information.

Using biplots of the principal components allows a better understanding of the relationship among the variables (products and sensory attributes). These variables are plotted as vectors on the biplot and their placement reveals potential relationships in the data. A smaller angle between two vectors typically indicates a high positive correlation between them, a 90 degree angle suggests independence, whereas vectors that are plotted at an 180 degree angle typically have a high negative correlation between the two variables. It is important to understand that the position of one vector in the plot is determined by its relationship to ALL the other vectors, so the relationships may require careful study of the correlation or covariance matrix to verify relationships among variables.

To better understand the sensory attributes of the products from a PCA biplot, one must examine the placement of products relative to the location of the vectors. If a product has a high score on a principal component, it may be relatively high in the intensity of those attributes with high vector loadings on a principle component. Remember that PCA is a data reduction technique and it is important to verify findings using the original product and attribute mean values from the analysis of variance. However, if products are relatively far apart on the PCA plot, they are perceived as more different from products that are spatially more similar.

14.4.7.1 Factor analysis

Similar to PCA, factor analysis is a data reduction technique except it focuses on that part of the total variation that a variable shares with other variables in the set. It analyzes the interrelationships among a large number of variables and then groups them based on common underlying dimensions. Results are based on the correlation or covariance matrix. Factor loadings are the correlations of the variables with the factors. The initial solution can be rotated to facilitate interpretation. Oblique factors are correlated and orthogonal factors are uncorrelated. Additional statistics such as eigenvalues suggest the number of factors to extract. Factor solutions provide attribute groupings and variance explained by each factor. The communalities explain how well the factors account for an attribute. Some sensory researchers prefer to build optimization models using factor scores rather than attributes identified through PCA. However, such a result has limited value when applying the information to make product change, for example.

14.4.7.2 Factor analysis methods

Factor analysis methods attempt to describe a data matrix in terms of the sum of a few underlying latent factors (axes, dimensions, principal components, factors, or 'main tendencies of variation'). These factors constitute a simpler axis system, e.g., a space of lower dimensions (typically 2 or 3), than the dimensional space spanned by the original variables in the raw data.

By providing the location of each point on that first dimension, one does not lose too much information in the reduction process. Factor analysis methods reduce a matrix of sensory descriptive attributes to a few factors, without losing too much of the information in those original attributes.

Factor analysis methods typically consist of two independent steps. Rank reduction is carried out to discover the number of important factors in the data matrix; each factor is described by its factor scores for the objects (products) and then by the factor loadings (sensory attributes). Factor interpretation is the process of finding the linear transformation of the resulting factors that is most suitable for interpreting the sensory results. This may include expansion/contraction, rotation, or translation.

14.4.7.3 Cluster analysis

This data reduction technique is used to find a smaller number of groups whose members have elements more similar to one another than to members of another group. Hierarchical clustering produces a dendrogram from some measure of similarity or dissimilarity. Non-hierarchical clustering arbitrarily assigns n clusters to the data. Cluster analysis methods differ primarily in how the distance from a center point is calculated. It can be calculated to the nearest point in the cluster, to the farthest point in the cluster, or to some point near the cluster center.

What cluster analysis shows is that objects within the same cluster are more similar to each other than objects (e.g., products, subjects) found in different clusters. Cluster analysis is used extensively in sensory optimization research to identify unique consumer preference groups within and between markets or countries, and it is used in the same way in descriptive analysis to identify unique subject groups. In the latter situation, uniqueness may indicate the need for additional panel training.

14.4.7.4 Response surface methodology (RSM)

This is an experimental design approach adapted from the chemical industry. This methodology has specific requirements in that the important variables for the research must be known prior to the test and the products must be systematically varied based on the important variables. Output from these variables can be measured (limited to the products you manufacture). Optimization based solely on tests without any competitive products restricts the value of the results.

RSM is based on regression analysis and provides information on how the response (dependent variable) varies across the factors (independent variables) and their levels. RSM will allow for the determination of an optimum combination of the factor levels and also is useful to investigate the interaction effects between dependent variables

The mathematical model for RSM includes the linear effects of each of the factors, the second order or quadratic effects (squared values), and the interactions (the combination of the factors).

When using RSM in sensory and consumer-based optimization programs, there are several assumptions that must be met or thoroughly understood, as they impact the usefulness of the research:

- Factors critical to the product must be known.
- The region of interest where the factor levels influence the product must be known.
- The factors must vary continuously throughout the experimental range tested.
- There must exist a mathematical function that relates the factors to the measured response.
- The response that is defined by its function must be a smooth surface.

Depending on which components of the regression equation are significant, the response surface may be a plane (only the linear component of the equation is significant), a dome (with a maximum), a cradle (with a minimum), or a saddle (the quadratic and cross-product components are significant as well). F-ratios corresponding to the respective components of the equation are produced by the analysis and attest to their significance (or lack of significance).

A variety of other multivariate data analysis techniques are not covered here, including Partial Least Squares Regression (PLS), Generalized Procrustes Analysis (GPA), PCR, STATIS, and neural networks. There tend to be highly specialized applications for these analyses; however, the ones presented here are used most often with sensory data.

14.4.7.5 Data analysis software programs and visual presentation

Data analysis software programs and visual presentation of sensory research is an important part of understanding and communicating data. Graphs and charts reveal events and trends in data not readily observed any other way, and they simplify results for less technical and non-statistical audiences. The ready availability of computers in most companies and countries allows sensory scientists to manage large amounts of data. Techniques are available to stretch, shrink, rotate, organize, and reorganize large and small amounts of data with a simple keystroke or two, and in doing so examine countless relationships among subjects, attributes, and products, and to establish databases of information, develop normative values, and provide insight into products, subjects, and consumers. In addition to the large number of generally available analysis and graphing software programs, such as SAS® and JMP® (SAS

Institute Inc., Cary, NC), S-PLUS®(Insightful, New York, NY), SPSS (SPSS Inc., Chicago, IL), BMDP Statistical Solutions , Saugus, MA), custom programs have been developed by several companies for application in sensory evaluation. For the latter, the interested reader is directed to: The Unscrambler®, CAMO Software Inc., Woodbridge, NJ; Compusense, Guelph, Ontario, Canada; Sensory Computer Systems LLC, Morrisontown, NJ; and the Tragon Corporation, Redwood Shores, CA. A few examples of graphs and charts useful to the sensory scientist are described below.

14.4.7.6 Histograms

Frequency histograms and measures of central tendency and variance provide valuable information about sensory responses to products and for product comparisons. These measures also allow one to determine how well the data satisfy assumptions required for other statistical tests. Figure 14.1 illustrates a product by serving order distribution, where each subject evaluated two products. The means are lower for both products in the second serving order, demonstrating a possible order bias, a not uncommon finding in sensory studies, and the primary reason for using balanced serving orders. The bimodal distribution and variance differences for Product 1 when served in the second order would alert the sensory researcher that it may be risky, or even inappropriate, to conduct statistical tests for this result if homogeneity of variances is an issue.

14.4.7.7 Quantitative Descriptive Analysis (QDA®) spider charts

Figure 14.2 represents a popular way for graphing descriptive panel results. Examining this figure we readily see product differences for individual attributes, relationships among the attributes, and the overall tendency for a product to score low (Product A), high (Product B), or in the middle (Product X) compared to other products evaluated.

14.4.7.8 Sensory maps

Mapping is frequently used for displaying the results of product optimization research involving consumer acceptance and sensory descriptive panel data. The maps often group well-liked products together with sensory attributes and with marketing segments,

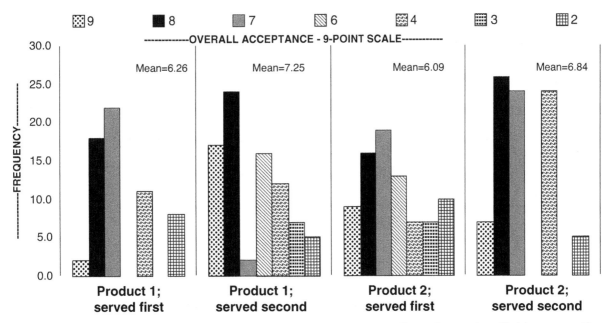

benefits, uses, consumer demographic, and psychographic information (e.g., lifestyle, income, gender, usage, and attitude) identified as highly related to those product preferences. In the case of descriptive panel data, it is often helpful to visually display products by attribute results.

Figure 14.3 is such a graph, and demonstrates how products may differ, or group, based on several attributes across different modalities. Here we see that

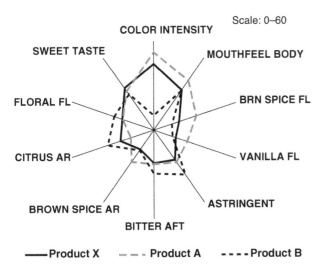

Figure 14.2 QDA spider plots. (Copyright 2008 Tragon Corporation. All rights reserved.)

Product A and X are scored high for several different attributes, whereas Product B falls between the two. We would conclude that Product X is a strong and high-impact product in several areas.

Figure 14.4 is an example of a density map obtained from a cluster analysis of consumer acceptance scores for a broad range of products of the type typically included in a category benchmarking or optimization program. The cluster map reveals three unique preference segments embedded within the total population. Typically, once stable segments are identified, the next step is to apply multiple regression statistical techniques to develop sensory attribute models that best fit each preference segment as well as one or more models for a 'bridge' style product that would best satisfy the total population.

Figure 14.5 is a sensory map that includes consumer benchmarking and sensory descriptive analysis results for an array of non-food products (hand lotions). The relative relationships presented here provide useful information to product developers and marketing researchers. These maps help identify the sensory attributes most closely associated with key products. When used effectively, these data displays can help reveal innovation opportunities and help companies provide products that are well liked by consumers. These types of sensory maps are typically derived from Factor Analysis (FA) or Principal Component Analysis (PCA) output.

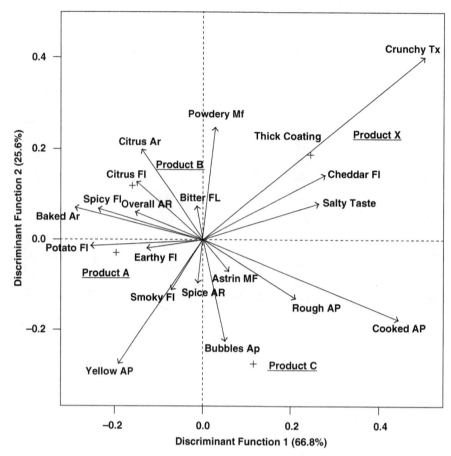

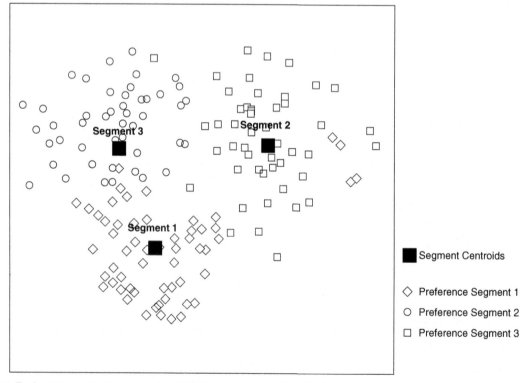

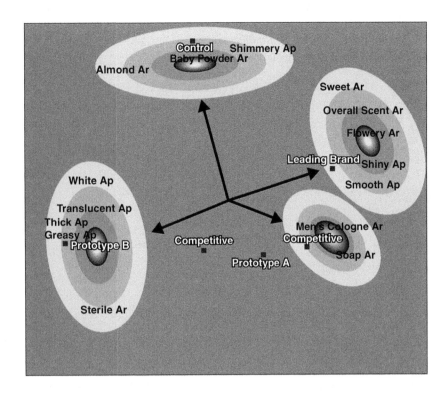

Further reading and references

Cohen, J. and Cohen, P. (1983) *Applied Multiple Regression/Correlation Analysis for the Behavioral Sciences*, 2nd edn. Lawrence Erlbaum Associates, Hillsdale, New Jersey.

Dillon, W.R. and Goldstein, M. (1984) *Multivariate Analysis: Methods and Applications (Wiley Series in Probability and Statistics)*. Wiley, New York.

Giovanni, M. (1983) Response surface methodology for product optimization. *Food Technology*, 37(11), 41–45, 83.

Gower, J.C. (1975) Generalized Procrustes analysis. *Psychometrika*, 40(1), 33–51.

Green, B.G., Shaffer, G.S. and Gilmore, M.M. (1993) Derivation and evaluation of a semantic scale of oral sensation magnitude with apparent ratio properties. *Chemical Senses*, 18, 683–702.

Green, B.G., Dalton, P., Cowart, B., Shaffer, G., Rankin, K. and Higgins, J. (1996) Evaluating the "labeled magnitude scale" for measuring sensations of taste and smell. *Chemical Senses*, 21, 323–34.

Guinard, J.-X. and Cliff, M.C. (1987) Descriptive analysis of Pinot noir wines from Carneros, Napa and Sonoma. *American Journal of Enology and Viticulture*, 38, 211–15.

Lawless, H.T. and Heymann, H. (1998) *Sensory Evaluation of Food. Principles and Practices*. Chapman and Hall, New York.

Lee, S.-Y., Luna-Guzman, I., Chang, S., Barrett, D.M. and Guinard, J.-X. (1999) Relating descriptive analysis and instrumental texture data of processed diced tomatoes. *Food Quality and Preference*, 10, 447–55.

Martens, H. and Russwurm H. Jr (1983) *Food Research and Data Analysis*. Applied Science Publishers, London.

Naes, T. and Risvik, E. (1996) *Multivariate Analysis of Data in Sensory Science*. Elsevier, Amsterdam.

Noble, A.C., Williams, A.A. and Langron, S.P. (1984) Descriptive analysis and quality ratings of 1976 wines from four Bordeaux communes. *Journal of the Science of Food and Agriculture*, 35, 88–98.

O'Mahony, M. (1986) *Sensory Evaluation of Food: Statistical Methods and Procedures*. Marcel Dekker, New York.

Pangborn, R.M., Guinard, J.-X. and Davis, R.G. (1988) Regional aroma preferences. *Food Quality and Preference*, 1, 11–19.

Parducci, A. (1965) Category judgment: a range-frequency model. *Psychological Review*, 72, 407–418.

Rummel, R.J. (1970) *Applied Factor Analysis*, Chapter 22. Northwestern University Press, Evanston, Illinois.

Schiffman, S.S., Reynolds, M.L. and Young, F.W. (1981) *Introduction to Multidimensional Scaling. Theory, Methods and Applications*. Academic Press, New York.

Schutz, H.G. (1983) Multiple regression approach to optimization. *Food Technology*, **37**(11), 46–8, 62.

Williams, A.A. and Langron, S.P. (1984) The use of free-choice profiling for the evaluation of commercial Ports. *Journal of the Science of Food and Agriculture*, **35**, 558–68.

Quality assurance and legislation

David Jukes

Key points

- Consumers expect a safe food supply and legislation is developed to help ensure this. Food scientists and technologists need to understand the system that operates in their own country.
- National food control systems are affected by regional and international agreements. One key element of this is the use of risk analysis and its role is considered.
- The Codex Alimentarius Commission develops international food standards which have a special status for the World Trade Organisation (WTO). The relationship between Codex standards and the WTO is described.
- Food scientists and technologists are key people in the implementation of quality management systems which enable food businesses to provide their customers with products matching their requirements.
- Quality management procedures are considered and key components described.
- The requirement to meet private standards has become an important element in the food supply chain and examples of these standards are considered.
- Quality assurance requires the application of valid statistical techniques. This chapter introduces key statistical concepts and introduces Shewhart control charts and acceptance sampling.

15.1 Introduction

Food scientists and technologists have a key role to play in the production of safe and wholesome food meeting customers' requirements. Within a factory, they are required to ensure that appropriate checks on supplies are conducted, to establish effective quality systems and to help maintain the safety and quality of the products. With a global food supply chain, their work will extend back through the chain to suppliers who may be thousands of miles away and will involve collaboration with customers who also may

be elsewhere in the world. With the complex nature of the food supply chain, controls have been established at national and international level to provide protection to consumers and to ensure fair trading practices. The maintenance and enhancement of these controls also requires a detailed understanding of food science and technology so professional food scientists and technologists often work for national food control organisations.

It is therefore necessary for food scientists and technologists to have an understanding of the key elements in food control – whether within a business or for a national authority. The controls may be legally established or they may be based on the more rigorous demands of the market. In this chapter three related topics are considered:

1   The fundamentals of food law where the basic elements of national controls are considered and the links to international developments are described.
2   Food quality management systems where the role of quality systems are described and their key elements are discussed.
3   Aspects of statistical process control which introduces some of the mathematical tools which can be used to ensure compliance with quality requirements.

A key point to make with regard to this chapter is that, unlike many of the scientific concepts described in other chapters, its content is based more on policies and practices. It is therefore subject to change based on newly identified hazards or on changed government priorities. This chapter has therefore tried to focus on key elements and has used current legislation and standards to illustrate them. Readers should be prepared to ensure that they identify any changes that have occurred before basing decisions on these examples.

15.2 Fundamentals of food law

15.2.1 Key objectives

When a society believes that it needs protection, it is usual for the national government to respond by passing legislation which provides that protection. Early laws on food control can be traced back many centuries, but the modern concepts of food law are mainly based on ideas developed in the mid-nineteenth century. At that time, advances in science, including microscopy and chemical analysis, made it clear that the food supply was frequently contaminated and often harmful.

With continuing advances in science and technology and the rapid urbanisation of much of the world's population, the food supply chain has become increasingly complex and globalised. Although the advances have enabled most people to enjoy a regular and varied food supply, many consumers, however, are still reluctant to accept the role of mod-

ern science and technology in the food supply. The resistance in some parts of the world to irradiated and genetically modified foods are clear examples where the advances have met resistance and the legislators attempt to adopt controls which assure safe food but also allow clear consumer choice.

Food legislation is only one part of the system used to try to provide consumer protection. Whilst the scientist or technologist may be focused on the content of the legislation – what level of additive is permitted, for example – the effective implementation of a national food control system depends upon the smooth and efficient operation of various different elements. A modern and well-drafted set of legal documents may appear to be sufficient, but in practice, without an effective administrative team within government, without a committed and professional enforcement team and without the support of a well-resourced laboratory service, the legislation is likely to be ineffective.

Food law can usually be identified as belonging to one of two key areas of protection – food safety and food quality. A third issue, the use of legislation to enhance the nutritional quality of the diet, is a more recent development.

■ *Food safety*
   To protect consumers from adverse health effects, legislation usually places responsibility on the food suppliers to ensure that the food is safe. To provide for ease of enforcement, detailed legislation has been developed to provide clear statements of what constitutes safe food. Examples of these include controls on the hygiene of production and distribution, limits on the levels of chemicals in the food (whether added on purpose or found as contaminants), approval systems for the control of new processes (food irradiation or genetically modified foods, for example) and specifications for packaging materials.
■ *Food quality*
   Food can be perfectly safe but may not be satisfactory. As an example to illustrate the difference, consider a container of milk. The milk may have been subject to detailed hygiene controls and may be entirely safe for consumption. However, if during the collection and/or distribution of the milk, some water had been added but the milk was still being sold as pure milk, the consumer buying the milk would in fact be paying for a mixture of milk and water. If the product was sold as milk, then the

consumer is being defrauded. Food law is therefore developed to provide consumers with protection in the compositional quality of the food they are buying. The manner of these controls varies and different approaches are possible.

The 'vertical approach': at one extreme, it is possible to provide detailed specifications for a wide range of food products. Manufacturers are then required to only produce products meeting these legal specifications. This method ensures that consumers can be confident that the foods they buy will meet nationally approved standards. However, this approach limits the manufacturer, stifles product development and restricts consumer choice.

The 'horizontal approach': at the other extreme, it is possible to allow manufacturers total freedom to manufacture any food product, but for the legislation to require fully informative labelling so that the consumer then can decide whether or not to buy the food from the information on the label.

In practice many countries have adopted a combination of these two approaches. It is common to provide a number of controls based on the 'vertical approach' for products which are considered of importance to the national diet – bread and milk, for example. When manufacturing other products, manufacturers have much greater freedom but must comply with labelling rules. The balance between these two approaches does vary. In developing countries with low literacy rates or a lack of consumer awareness, reliance on labelling may not provide sufficient consumer protection. In these countries there tends to be a greater reliance on product standards. In many economically developed countries or regions, the approach to recipe law which was prevalent during the mid-twentieth century has largely been replaced by the use of informative labelling controls.

Dietary control

Limited attempts to protect or improve health by food law have been adopted for some time. The main control has been the requirement for fortification of certain foods so as to enhance the consumption of key micronutrients. Examples include the addition of vitamins to margarine, of vitamins and minerals to flour and/or bread and the iodisation of salt. These targeted uses of dietary intervention have usually been successful in achieving their limited objective.

On a larger scale, though, many countries are now facing a significant public health problem due to changes in diet and lifestyle. Examples of these are the increasing levels of diabetes and cardiovascular disease. The issue though is how to tackle the causes. Encouraging people to adopt a healthier lifestyle may be possible but adopting policies that can lead to a healthier diet are also being considered and adopted. Possible options include a return to some vertical controls, but an alternative is the use of more detailed labelling and advertising controls including enhanced nutritional guidance and tighter controls on nutrition and health claims.

These three different objectives cover most issues that are normally regarded as food law. However, as food can have a great significance within a country's culture and economy, there are frequently issues that arise where laws are passed which have an impact on the food supply situation but which have their origins outside the objectives described above. Countries with a large agricultural economy may seek to protect their farmers or rural industries by adopting certain protective measures, whilst some countries use food law to ensure compliance with religious rules. Often general controls applied to protect the society from a wide range of risks will also impact businesses in the food chain – weights and measures, health and safety, customs and taxes or environmental controls are examples. These wider controls are not normally considered as 'food law' and will not be dealt with here. However, someone setting up a food business will need to be aware of all the laws affecting his or her business.

15.2.2 The role of risk analysis

Current approaches to food safety control are frequently based around the concept of 'risk analysis'. The different components of risk analysis are illustrated diagrammatically in Fig. 15.1.

The consideration of the scientific evidence is known as the 'risk assessment' process. However, risk assessment on its own is unsatisfactory. When the results of the 'risk assessment' are known, then others will need to consider what action is needed to provide protection – consideration may need to be given to various options. The process of assessing which option to apply is regarded as a separate component and is termed 'risk management'.

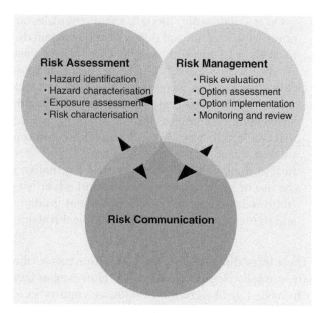

Structure of risk analysis.

for the control measures adopted by the national authority.

The application of modern scientific methods to the food supply chain has left many consumers concerned about the safety and integrity of their food. Health scares such as bovine spongiform encephalopathy (BSE), avian flu and dioxin contamination are major concerns and have occurred at a similar time to the development of new technologies such as genetic modification and food irradiation. It is perhaps not surprising that consumers often seek more natural foods (whether simply produced with no food additives or whether meeting more specific criteria such as 'organic'). These are perceived to represent a return to traditional food production methods. However, with the modern food supply chain, even 'natural foods' can pose risks and 'risk analysis' is still applicable to these situations.

Detailed working definitions linked to risk analysis have been adopted by the Codex Alimentarius Commission (discussed below) and are given in Table 15.1.

15.2.3 National food control systems

As described above, an effective national food control system requires the effective assembling and interconnection of a number of different elements. Five key components are often identified as being involved (FAO, 2006):

- food control management;
- food legislation;
- food inspection;
- food control laboratories;
- food safety and quality information, education and communications.

Finally, so as to ensure that the public are aware of the need for the controls and how they have been developed, both the risk assessors and the risk managers have to participate in a process of 'risk communication'. With the public often expecting food to be 100% safe, the realisation that this is never achievable can be a difficult concept to explain. The announcement of the results of scientific studies which may suggest a toxic hazard can make for good newspaper headlines, but the risk assessment and risk management stages may suggest a more complex situation in which some risk may be inevitable. Effective risk communication in which the situation is explained to the public is vital if their support is to be provided

Table 15.1 Some definitions of terms linked to risk analysis. (From Codex Alimentarius Commission, 2007.)

Hazard: a biological, chemical or physical agent in, or condition of, food with the potential to cause an adverse health effect.

Risk: a function of the probability of an adverse health effect and the severity of that effect, consequential to a hazard(s) in food.

Risk Analysis: a process consisting of three components: risk assessment, risk management and risk communication.

Risk Assessment: a scientifically based process consisting of the following steps: (i) hazard identification, (ii) hazard characterization, (iii) exposure assessment, and (iv) risk characterization.

Risk Management: the process, distinct from risk assessment, of weighing policy alternatives, in consultation with all interested parties, considering risk assessment and other factors relevant for the health protection of consumers and for the promotion of fair trade practices, and, if needed, selecting appropriate prevention and control options.

Risk Communication: the interactive exchange of information and opinions throughout the risk analysis process concerning risk, risk-related factors and risk perceptions, among risk assessors, risk managers, consumers, industry, the academic community and other interested parties, including the explanation of risk assessment findings and the basis of risk management decisions.

15.2.3.1 Food control management

Definition (FAO, 2006, p. 18)

Food control management is the continuous process of planning, organising, monitoring, coordinating and communicating, in an integrated way, a broad range of risk-based decisions and actions to ensure the safety and quality of domestically produced and imported food.

The provision of safe quality food within a country does not just happen – the process has to be managed and maintained. This task is normally assumed by one or more parts of government. As mentioned above, the political importance of the food supply system can vary and the allocation of responsibilities can vary as a result. For countries with an economy dominated by agriculture, food control management may be seen as an agricultural activity with a 'Ministry of Agriculture' taking a lead role. However, health issues may be more dominant in another country and the 'Ministry of Health' may take lead responsibility. In other countries, other priorities may lead to other structures and industry, trade or local government ministries may take the lead. The complex nature of the food supply system means that the right system for one country may not be right for another.

Failures in the control system due to either overlapping responsibilities, gaps between components or competing bodies not cooperating have led to a much greater focus on the overall system structure and the increasing use of a 'single agency' management structure to try to prevent and resolve issues.

15.2.3.2 Food legislation

Definitions (FAO, 2006, p. 39)

Food legislation (or food law) is the complete body of legal texts (laws, regulations and standards) that establish broad principles for food control in a country, and that governs all aspects of the production, handling and trade of food as a means to protect consumers against unsafe food and fraudulent practices.

Food regulations are subsidiary legal instruments (usually issued by a minister rather than by parliaments) which prescribe mandatory requirements that apply to various aspects of food production, handling, marketing and trade, and provide supplementary details that are left open in the main parliamentary-level legislation.

Food standards are nationally or internationally accepted procedures and guidelines (voluntary or mandatory) that apply to various aspects of food production, handling, marketing and trade to enhance and/or guarantee the safety and quality of food.

The creation of a legal basis for food control is vital if it is to be effective in protecting consumers. It is to be expected and hoped that the vast majority of food suppliers will wish to ensure that the food that they supply is safe and meets consumers' requirements. However, there will be others who are prepared, through either negligence or a desire for personal gain, to put these at risk. Legislation is necessary to provide a deterrent to minimise the number who are prepared to take this risk.

The written legal documents need to provide a comprehensive structure for food control. Key elements that are usually found in the main food legislation are:

- Introductory provisions: legislation works best if there are clear definitions and it is common to find key terms defined in the law. Common definitions are for 'food' and 'food business'.
- Enabling and administrative provisions: this will identify which public authorities have responsibility under the law and may establish an agency or board to act on behalf of government. The law will also specify the enforcement authority and the powers of the inspectors to enter premises and to take samples.
- Offences and penalties: key offences are usually created which provide general protection from unsafe food or food that is labelled in a misleading manner. The penalties have to be sufficient to deter potential offenders.
- Specific provisions on food: depending upon the priorities in a country, the law is likely to contain a number of specific controls. These could relate to import and export conditions or to requirements

relating to registration or licensing. More generally, the law is likely to provide the authority for the adoption of secondary food regulations.

Although the fundamental components of the main food law may not need to be changed very often, it is necessary to ensure that it remains an effective legal document. This requires a regular review of its provisions.

The secondary legislation in the form of food regulations will contain the main details required for effective food control. These can be categorised into three main types:

- Regulations affecting food products in general – for example, food hygiene and food labelling.
- Regulations affecting specific food products – for example, bread, chocolate, baby foods.
- Regulations for organisational or coordinating purposes – for example, the procedure for the issuing of licences or the taking of samples.

The more technical requirements contained in the food regulations are likely to be more regularly updated as new information becomes available on potential hazards or as technological advances introduce new substances or processes into the food supply chain.

15.2.3.3 Food inspection

Definition (FAO, 2006, p. 66)

Food inspection is the examination of food or systems for the control of food, raw materials, processing and distribution, including in-process and finished product testing, in order to verify that they conform to requirements. Food inspection can be operated by government agencies, as well as independent organizations that have been officially recognized by national authorities.

Inspection at all stages in the food supply by official enforcement officers is necessary to provide the public with an assurance that those who fail to meet the legal requirements are detected and prevented from continuing in such a manner. It is also helpful to the food businesses who are complying with the law to know that competitors who fail to meet the same standards will not be permitted to continue.

Modern inspection systems use a professional and systematic approach built upon a risk-based programme of inspections. Examples of key components of the food inspection system are:

- Documented policies and procedures for risk-based inspection.
- Database of food premises categorised by risk.
- Adequate professional food officers with appropriate training, qualifications and experience.
- Access to resources including facilities, equipment, transport and communications.
- Procedures for the collection and handling of food samples.
- Procedures for handling food emergencies, outbreaks of food-borne disease and consumer complaints.

15.2.3.4 Food control laboratories

Once the enforcement officers have taken a sample, it is essential that it is subject to the appropriate testing. This requires an official laboratory to be available which can provide the officer with a clear and accurate statement of the physical, chemical or microbiological status of the food sample. The result of the analysis should be capable of being used as evidence in any subsequent court proceedings. As such, the analysis needs to be conducted to a high level of integrity ensuring that the methodology is appropriate and meets national or international standards.

To help ensure that the laboratories are appropriately staffed and using appropriate techniques, it is common for them to participate in proficiency schemes in which their performance is independently judged in comparison to other official laboratories. The maintenance of high standards of analytical procedures linked to the use of standardised methods aids the accuracy of the results.

Although many routine analyses can be conducted with a fairly limited range of analytical equipment, advances in technology and the need to detect low levels of contaminants (e.g. pesticide residues or aflatoxins) have increased the range and sophistication of the equipment needed for official laboratories.

15.2.3.5 Food safety and quality information, education and communications

Although the government can provide the legislation and a well-resourced enforcement and laboratory service, the vast majority of food will not be seen by any official as it passes through the food chain. As the final step in the chain, consumers therefore play a key role in the maintenance and enhancement of food safety standards. For them to be an effective component of the food control system requires them to have access to information on food safety and quality. Informed consumers are more likely to identify unsafe food and take their custom elsewhere – they are also more likely to alert the authorities to its existence, allowing a swift response to food being sold illegally.

The education process goes beyond the consumer. In most countries the food supply system consists of a large number of small businesses which frequently do not have access to trained scientists or technologists. Education and communication also needs to identify suitable methods for providing relevant information to these small businesses so as to ensure, particularly, that they are producing and selling safe food in compliance with legal requirements.

15.2.4 International food standards

Although governments have responsibilities to their own populations for providing protection, the widescale adoption of different legislative requirements in different countries can be a major hindrance to trade. As the objective of the legislation was to provide consumer protection, it seemed likely that some agreement could be achieved at an international level to define the necessary controls and, with their acceptance by national governments, it should be possible to provide adequate protection but to allow unhindered trade. This was the origins of the international Food Standards Programme co-ordinated by the Food and Agriculture Organisation (FAO) and the World Health Organisation (WHO) involving the Codex Alimentarius Commission (CAC).

15.2.5 The Codex Alimentarius Commission

The CAC was set up in 1961 to act as the international forum for the adoption of food standards. The adopted standards could then be used by individual countries as the basis of their own national laws and regulations. By having a single basis, it was hoped that there would be greater uniformity and hence fewer barriers.

Although it did achieve much during the first 30 years of its life, the incorporation of the Codex standards into national legislation was limited. In particular, developed countries felt the Codex standards might dilute the controls they had developed nationally over many years. Developing countries that did adopt the Codex standards found that access to developed markets was still limited. More authority was, however, given to the CAC with the formation of the World Trade Organisation (WTO) in 1995. This provided a new recognition to the Codex which established Codex standards as the benchmark against which national legislation could be assessed for compliance with WTO rules (see below).

The Codex Alimentarius is financed and run by the FAO and WHO. Membership is open to any country that is a member of either the FAO or the WHO. Interested international organisations can also participate in the work of Codex and assist in the development of standards.

Currently the main meeting of the CAC is an annual meeting open to all member nations and participating organisations. At this meeting decisions are taken on the adoption of standards and on general issues affecting the work of the CAC. Most of the preparatory work takes place in meetings of the various committees which have responsibility for different areas of work (see Fig. 15.2). Some of these committees consider matters of a 'horizontal' nature (for example, food hygiene, food labelling or food contaminants) whilst others consider matters of a 'vertical' nature (for example, fats and oils, sugars, and fruit and vegetables. These committees are also open to all member nations and participating organisations. A defined 'step-wise' procedure governs the adoption of standards and allows all countries to participate even if they do not attend all the meetings.

The CAC also hosts meetings of regional committees which enables discussion of matters affecting different regions of the world. These regional committees have been particularly valuable for the developing parts of the world where the sharing of experiences encourages the adoption of approaches suitable to the regional situation.

Although the Codex process can seem quite lengthy and the time taken to adopt standards can be surprisingly long, achieving agreement on sensitive food issues can be difficult. The Codex tries to work on the basis of consensus in which standards are adopted by general agreement. Although voting has been used on occasions when there has been a

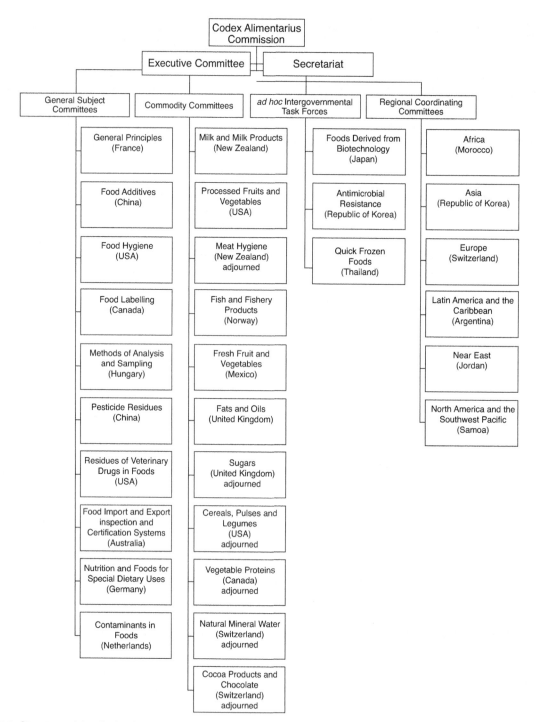

Figure 15.2 Structure of the Codex Alimentarius Commission.

clear difference of opinion, the resulting adoption of a controversial standard does not usually result in the matter being concluded. More detailed analysis of the differences and a more thorough understanding of the issues in dispute may eventually lead to a better agreed standard.

15.2.6 The World Trade Organisation (WTO)

The creation of the World Trade Organisation (WTO) in 1995 provided a new set of rules for the operation of world trade. Although the WTO represented a significant development, its origins can be traced back to attempts after World War II to put into place agreed international trade rules. This led to the operation of the General Agreement on Tariffs and Trade (GATT) and its progressive expansion with various rounds of negotiations.

With respect to food standards, the WTO provided more specific rules. These are contained in two agreements linked to the main WTO Treaty. The first, the 'Agreement on the Application of Sanitary and Phytosanitary Measures' (the SPS Agreement) considers, with respect to food, issues affecting human health. Potential barriers to trade caused by differences in other matters are generally covered by the 'Agreement on Technical Barriers to Trade' (the TBT Agreement). In addition, with the adoption of procedures for resolving disputes, there now exists a substantial framework for preventing and overcoming potential trade problems relating to food.

15.2.6.1 The Sanitary and Phytosanitary Agreement (SPS)

When a national food requirement is related to food safety, then it usually comes within the definition of a sanitary measure. Two key definitions are given in Table 15.2. The SPS Agreement covers issues relating to human food, animal feed and plant health. For food, the Agreement stipulates that the standards, guidelines and recommendations of the CAC shall be used as the basis for judging compliance with the requirements of the Agreement.

Article 3 of the Agreement sets out the basic concepts linked to harmonisation. The first three parts of Article 3 are given in full in Table 15.3. Article 3.1 provides that countries should adopt international standards, guidelines and recommendations unless alternatives are permitted under other Articles in the Agreement. By virtue of the definitions, for food this relates to the publications of the Codex Alimentarius Commission. Article 3.2 provides that any country which has adopted these standards is complying with the terms of the Agreement. Article 3.3 then provides that countries may introduce more demanding controls which provide a higher level of protection if their use is scientifically justified or where the chosen level of protection requires tighter controls. These however must be developed in accordance with other Articles in the Agreement, notably Article 5.

Table 15.2 Terminology used in the Sanitary and Phytosanitary Agreement.

Sanitary or phytosanitary measure

Any measure applied:

a to protect animal or plant life or health within the territory of the Member from risks arising from the entry, establishment or spread of pests, diseases, disease-carrying organisms or disease-causing organisms;

b to protect human or animal life or health within the territory of the Member from risks arising from additives, contaminants, toxins or disease-causing organisms in foods, beverages or feedstuffs;

c to protect human life or health within the territory of the Member from risks arising from diseases carried by animals, plants or products thereof, or from the entry, establishment or spread of pests; or

d to prevent or limit other damage within the territory of the Member from the entry, establishment or spread of pests.

Sanitary or phytosanitary measures include all relevant laws, decrees, regulations, requirements and procedures including, inter alia, end product criteria; processes and production methods; testing, inspection, certification and approval procedures; quarantine treatments including relevant requirements associated with the transport of animals or plants, or with the materials necessary for their survival during transport; provisions on relevant statistical methods, sampling procedures and methods of risk assessment; and packaging and labelling requirements directly related to food safety.

International standards, guidelines and recommendations

a For food safety, the standards, guidelines and recommendations established by the Codex Alimentarius Commission relating to food additives, veterinary drug and pesticide residues, contaminants, methods of analysis and sampling, and codes and guidelines of hygienic practice;

b for animal health and zoonoses, the standards, guidelines and recommendations developed under the auspices of the International Office of Epizootics;

c for plant health, the international standards, guidelines and recommendations developed under the auspices of the Secretariat of the International Plant Protection Convention in cooperation with regional organizations operating within the framework of the International Plant Protection Convention; and

d for matters not covered by the above organizations, appropriate standards, guidelines and recommendations promulgated by other relevant international organizations open for membership to all Members, as identified by the Committee.

Table 15.3 Key Articles from the Sanitary and Phytosanitary Agreement.

Article 3.1–3.3

1 To harmonize sanitary and phytosanitary measures on as wide a basis as possible, Members shall base their sanitary or phytosanitary measures on international standards, guidelines or recommendations, where they exist, except as otherwise provided for in this Agreement, and in particular in paragraph 3.

2 Sanitary or phytosanitary measures which conform to international standards, guidelines or recommendations shall be deemed to be necessary to protect human, animal or plant life or health, and presumed to be consistent with the relevant provisions of this Agreement and of GATT 1994.

3 Members may introduce or maintain sanitary or phytosanitary measures which result in a higher level of sanitary or phytosanitary protection than would be achieved by measures based on the relevant international standards, guidelines or recommendations, if there is a scientific justification, or as a consequence of the level of sanitary or phytosanitary protection a Member determines to be appropriate in accordance with the relevant provisions of paragraphs 1 through 8 of Article 5. Notwithstanding the above, all measures which result in a level of sanitary or phytosanitary protection different from that which would be achieved by measures based on international standards, guidelines or recommendations shall not be inconsistent with any other provision of this Agreement.

A key requirement for the adoption of alternative controls is that they are based on an assessment of risk. Countries are required to ensure the application of risk assessment methodology, taking into account guidance issued by the CAC. As discussed above, the process of risk assessment involves the evaluation of scientific data and the quantification of the risk. This then enables risk managers to identify the appropriate control measure.

Article 5.7 is particularly contentious. It states:

'In cases where relevant scientific evidence is insufficient, a Member may provisionally adopt sanitary or phytosanitary measures on the basis of available pertinent information, including that from the relevant international organizations as well as from sanitary or phytosanitary measures applied by other Members. In such circumstances, Members shall seek to obtain the additional information necessary for a more objective assessment of risk and review the sanitary or phytosanitary measure accordingly within a reasonable period of time.'

In some parts of the world, most notably the European Union, this Article has been used as the basic justification for the application of a concept known as the 'Precautionary Principle'. Following the crisis in confidence in food safety caused by the BSE outbreak in cattle, it was felt necessary that in future, action would need to be taken earlier in the event of other potential diseases where human health could be at risk. Where a significant risk to human health might exist, under the concept of the Precautionary Principle, action would be taken to implement precautions in advance of a full scientific understanding or a full risk assessment.

What is, however, in dispute is the basis of any action that is taken. The concern of those opposed to the application of the 'Principle', most notably the USA, is that as the scientific evidence may be limited, other factors will be used to reach a decision. These other factors could be influenced by political, economic or cultural concerns. This counter view therefore believes that even when the scientific evidence is limited, any decisions will still have to be based on that limited scientific evidence. With this in mind, the application of a special principle is not considered helpful – alternative terminology using terms such as 'a precautionary approach' is then regarded as more valid.

15.2.6.2 The Technical Barriers to Trade (TBT) Agreement

Technical requirements in national legislation which are not covered by the SPS Agreement are likely to be covered by the Technical Barriers to Trade (TBT) Agreement. Here there is less detail than in the SPS Agreement, but it is designed to try to achieve similar objectives. Countries are required to ensure that their country's technical regulations and standards (see Table 15.4 for definitions) do not create unjustifiable barriers to trade.

The Agreement recognises that countries may have different requirements based on, for example, their own specific level of development or environmental situation. These may lead to countries adopting

Table 15.4 Terminology used in the Technical Barriers to Trade Agreement.

Technical regulation
Document which lays down product characteristics or their related processes and production methods, including the applicable administrative provisions, with which compliance is mandatory. It may also include or deal exclusively with terminology, symbols, packaging, marking or labelling requirements as they apply to a product, process or production method.

Standard
Document approved by a recognized body, that provides, for common and repeated use, rules, guidelines or characteristics for products or related processes and production methods, with which compliance is not mandatory. It may also include or deal exclusively with terminology, symbols, packaging, marking or labelling requirements as they apply to a product, process or production method.

different regulations or standards; however, certain requirements are laid down. Any controls should not be more trade-restrictive than necessary to achieve the desired objective. The regulations and standards should also not discriminate between home produced and imported products. Governments are expected, where possible to achieve their objective, to use international standards as the basis of their own requirements.

15.2.6.3 The Dispute Settlement Understanding

The adoption of the detailed rules for the WTO was also used accompanied by the establishment of a specific procedure to try to resolve disputes. Detailed rules were agreed which are known as the Dispute Settlement Understanding (DSU). When a country considers that another country is failing to comply with the rules, it is entitled to initiate proceedings within the DSU to try to ensure compliance. Decisions are taken by the Dispute Settlement Body (DSB).

Countries are encouraged to try to resolve the dispute through negotiation. However, when this has clearly failed, then the complaining country can request the formation of a 'panel' to consider the evidence and to reach a conclusion. The panel usually consists of three specialists who are experts in the legal interpretation of the various WTO documents. A key component of the DSU is the tight schedule which is applied to the different stages in the process. This should mean that a dispute should normally be ruled upon within 15 months of the request for the formation of a panel.

Once a panel report has been adopted by the DSB, either side in the dispute has a right to appeal the decision. The appeal must, however, be based on matters of law or the legal interpretation developed by the panel. Again a tight schedule is applied such that

the decision of the appellate body is usually reached and adopted within 3 months. A country that has been found to be failing to comply with any of the WTO provisions is required to take action to remedy the situation promptly. Failure to act will allow other countries to seek compensation or to take agreed retaliatory action to try to encourage compliance.

Whilst many disputes are resolved by negotiation prior to the establishment of a panel, several major differences in national food laws have led to panels. In particular, the approaches adopted by the European Union (EU) towards restricting the use of hormones in meat and in only slowly approving genetically modified organisms (GMOs) led, in each case, to a detailed consideration by a panel. In both cases the EU policy was more restrictive and cautious that that of the USA which initiated the proceedings. Even when the panel reports had been adopted (1998 in the case of hormones and 2006 in the case of GMOs), the resolution of the problems has been difficult. In the case of hormones, the USA was authorised to take retaliatory action, but even this has led to a further challenge by the EU resulting in another panel being established. Currently, in 2009, the problem has not been resolved. Both these cases are linked to issues of safety and the scientific evaluation of risk. Despite the existence of the DSU and its procedures, public concern within the EU with regard to these two issues makes it very difficult politically for agreement to be reached.

15.2.7 Regionalisation: the working of regional blocks

As described above, governments of individual countries have responsibilities to their citizens to provide adequate protection. However, they also have responsibilities to promote economic development and the promotion of trade is seen as an effective way to achieve this. Adopting national legislation may be important for consumer protection, but, if every country adopts its own set of controls, the result can be to create additional barriers which make it complex and expensive for businesses to trade – whether for imports or for exports. One of the main objectives in working together as regional blocks is to try to adopt harmonised controls which enable barriers to be removed and for trade to take place smoothly and with only limited interruption when crossing frontiers.

Various different regional groupings now exist around the world and they have different roles and objectives – some go well beyond the desire to seek harmonisation, but they still incorporate the general objective of harmonisation and the abolition of barriers. One example, the EU, is described in more detail in Box 15.1.

15.3 Food quality management systems

It has been a long time since most people sourced their own food – whether by hunting or by farming. For the vast majority of people, their daily food supply comes through a huge interconnected web and, at each stage, there will be a supplier and a customer. Ensuring that the final consumers obtain the food that they desire – whether from the perspective of safety, of sensory characteristics, of shelf-life, of convenience or of value – requires detailed procedures.

Most industrialised processors have found it necessary to adopt a systematic approach to quality and have implemented management systems designed to deliver products that meet, or exceed, their customers' expectations. The complex nature of the food chain, the varied nature of potential difficulties, the variety of different raw materials and the high expectations of many consumers for absolute safety make it hard to identify and establish appropriate quality systems. The systems necessary to deliver the correct quality to consumers have been progressively developed and refined.

The driving force for the maintenance of quality standards comes from the ultimate food consumer. In developed economies, the multiple retailers have worked very hard to successfully identify consumer needs and to pass this information back up through the food chain. These requirements may be based on the minimum national legal requirements, but fre-

quently they may seek to go beyond this. More detailed requirements are often set out as 'private standards' which can be:

- specific to one retailer or business;
- agreed by a group of retailers or businesses at a national level;
- adopted as international standards, either by the International Organisation for Standardisation (ISO) or by another global body (Table 15.5).

In this section we explore the various building blocks of modern food quality management systems and see how some of the private standards work.

15.3.1 What is 'quality'?

The starting point for any discussion of quality management systems needs to be the definition of the term 'quality' itself. The different uses of the word create confusion as, for example, it is possible to use phrases such as 'a quality product' and 'poor quality' where the word has very different meanings. In addition, a food might be considered as 'good quality' by one purchaser but as 'bad quality' by another. The key point to note is that whether a product is correct, and of the right 'quality', is for the customer or consumer to determine.

When purchasing food, consumers will have very varied ideas as to what they require. Although many of these will relate to the issue of food safety, most food choices will be based on personal preferences relating to convenience, nutrition and sensory characteristics. Product development in many food companies attempts to create products that successfully match these preferences.

For the purposes of consistent terminology, the ISO has established a systematic approach to the terms used in quality management standards. Key definitions, including that for 'quality' itself, are given in Table 15.6.

15.3.2 Specifications

The starting point for any system of control must be a definition of what is actually required. In the case of the ultimate consumer, the requirement may never actually be expressed – either written or verbally. Consumers visiting their local supermarket may have only a vague concept of what they might want to eat or what they may wish to cook for a special meal. Although some people will know quite precisely what

Box 15.1 Food law in the European Union (EU)

The EU is one of the most influential trading blocks. Food law developed by the EU is often used by other countries seeking to develop modern controls. In seeking to protect its citizens, the EU also requires imports to be in compliance with internal requirements and this often drives legal developments elsewhere.

Key components of the EU

The EU currently consists of 27 Member States who have agreed to work together to achieve certain common objectives. The methods by which this is achieved are set out in a number of Treaties which have established the structures, procedures and rules which govern the operation of the Union. New members have been required to agree to the terms of the Treaties although it is possible to negotiate transitional arrangements. Changes or additions to the Treaties require all Member States to agree.

For the purpose of understanding food law within the EU, two key points need to be understood – the institutions and the legal documents. The Treaties have established various key bodies with powers and responsibilities. These are referred to as the 'Institutions'. For our purposes, four of them are of significance:

- The Council
  The Council of Ministers represents the views of the Member States. It is composed of the ministers from each government. Depending upon the topic under discussion, different ministers will attend meetings. In addition there are meetings of the Heads of Government which are known as 'European Council' meetings. This EU Institution should not be confused with the separate body known as the 'Council of Europe' which involves many more countries.
- The European Commission
  The Commission acts to develop the EU and to implement agreed policies. There is one Commissioner from each Member State, but, once appointed, he/she acts independently of his/her country. The Commission proposes new legislation and, in some cases, has authority to adopt legislation. The Commissioners are supported by civil servants who work for the Commission.
- The European Parliament
  The Parliament provides the democratic element within the EU. There are currently 785 MEPs who are elected for a period of 5 years. The Parliament has an influential role in the adoption of EU legislation.
- The European Court of Justice
  The Court of Justice provides a sound legal basis for all the activities of the EU – whether another institution, a Member State government, a business or an individual. The Court bases its judgements on the Treaties and ensures that agreed procedures have been followed. It ensures that EU law is interpreted and applied equally across the whole EU.

European legislation

European legislation can take several forms and can be adopted by a number of different procedures defined in the Treaties. The vast majority of European food law is adopted in the form of either a 'Regulation' or a 'Directive':

- An EU 'regulation' is directly applicable in all Member States and is therefore a legal document. Compliance with a regulation is necessary. Member States may, however, have to introduce national legislation to provide for the enforcement of the requirements and for appropriate sanctions in the event of non-compliance. This is now the preferred form of legal document as it provides for the effective application of any new legislation simultaneously in all Member States.
- An EU 'directive' is an agreement to apply its requirements, although the method of application may vary in each Member State. Most directives will result in an equivalent legal document being enacted nationally and, since the legislative process will vary between Member States, the overall impact of a directive can be delayed.

The main route for adopting food law, the co-decision procedure, is set out in Article 251 of the Treaty. This procedure ultimately requires both the Council of Ministers and the European Parliament to jointly agree the final text – hence the term 'co-decision'. However, the original concept for the legislation will be prepared by the Commission. This proposal will be drafted and discussed with a wide range of interested parties and with Member States before the Commission adopts its formal proposal. The co-decision procedure then allows the European Parliament to consider the proposal and to suggest amendments (known as the 'First Reading'). The Council considers the proposal, along with any amendments from the Parliament, and tries to reach agreement on a 'Common Position' – this can be by qualified majority voting. The Common Position is then considered by the Parliament (the 'Second Reading') and further amendments may be requested. The Parliament's views are then again considered by the Council – if any amendments are acceptable then the measure can be adopted. If the Council objects to any amendments, a period of negotiation is required involving a 'Conciliation Committee' which attempts to agree a joint text. Any agreed text will then be considered by both the Council and the European Parliament and, all being well, will then be adopted. During the whole process, the Commission keeps watch on the document and the amendments and can make its own proposals for amendments.

As will be seen from the above paragraph, the co-decision is complex and time consuming.

EU food law: risk analysis and scientific advice

Significant food safety problems developed within the EU during the late 1980s and into the 1990s. Combined with the need to show compliance with WTO rules, this led the EU to reassess the overall approach to food safety controls and, in particular, the allocation of the responsibilities for the components of risk analysis. Although it was clear that the European Commission had the lead role in risk management, the provision of scientific advice to the Commission ('risk assessment') had not been effectively established. The relationship between scientific advice given by national committees and that provided to the Commission by its own committees was not clear. To provide a more rigorous legal basis for food safety within the EU and restore public confidence in the safety of food, a new regulation was adopted in 2002 (Regulation (EC) No. 178/2002). To overcome the problem of scientific advice, the regulation also established a new body, the European Food Safety Authority (EFSA), which works according to procedures set out in the regulation.

it is they need for a meal, many will be influenced by what they find at the shop. The skill of the successful retailer is to identify emerging trends and provide a range of products matching different market needs. With good marketing, consumers will appreciate the selection that is available and will be able to prepare meals that are satisfying and enjoyable.

Retailers, however, have to ensure that they obtain consistent supplies of the right quality, in the right quantity and at the right time. Contracts have to be placed and this requires the precise nature of the product to be defined – the key document for this purpose is the specification. Any purchase is against a defined specification and any dispute about the quality of the product would be judged against the agreed specification.

There are some aspects in a specification that are relatively easy to agree – the weight or volume, the size and shape and the recipe, for example. However, other aspects may be harder – colour, flavour and texture.

Specifications can be agreed between two parties to a contract – a supplier and a customer – but they will also be needed within a production process. Process specifications provide the requirements for the product as it passes through the various stages of the production process. Compliance with the different requirements (whether temperature/time combinations for pasteurisation or frying temperatures for oil used in crisp manufacture, for example) will ensure that the product is passed on to the next process stage in the right condition.

Table 15.5 Selected ISO standards linked to quality management systems.

ISO 9000:2005 Quality management systems – Fundamentals and vocabulary
ISO 9001:2008 Quality management systems – Requirements
ISO 9004:2000 Quality management systems – Guidelines for performance improvements
ISO 10002:2004 Quality management – Customer satisfaction – Guidelines for complaints handling in organizations
ISO 10005:2005 Quality management systems – Guidelines for quality plans
ISO 10006:2003 Quality management systems – Guidelines for quality management in projects
ISO 10007:2003 Quality management systems – Guidelines for configuration management
ISO 10012:2003 Measurement management systems – Requirements for measurement processes and measuring equipment
ISO/TR 10013:2001 Guidelines for quality management system documentation
ISO 10014:2006 Quality management – Guidelines for realizing financial and economic benefits
ISO 10015:1999 Quality management – Guidelines for training
ISO/TR 10017:2003 Guidance on statistical techniques for ISO 9001:2000
ISO 10019:2005 Guidelines for the selection of quality management system consultants and use of their services
ISO 15161:2001 Guidelines on the application of ISO 9001:2000 for the food and drink industry
ISO 17000:2004 Conformity assessment – Vocabulary and general principles
ISO 17011:2004 Conformity assessment – General requirements for accreditation bodies accrediting conformity assessment bodies
ISO 17021:2006 Conformity assessment – Requirements for bodies providing audit and certification of management systems
ISO 17024:2003 Conformity assessment – General requirements for bodies operating certification of persons
ISO 17025:2005 General requirements for the competence of testing and calibration laboratories
ISO 19011:2002 Guidelines for quality and/or environmental management systems auditing
ISO 22000:2005 Food safety management systems – Requirements for any organization in the food chain
ISO/TS 22003:2007 Food safety management systems – Requirements for bodies providing audit and certification of food safety management systems
ISO/TS 22004:2005 Food safety management systems – Guidance on the application of ISO 22000:2005

15.3.3 Quality control (QC)

With knowledge of the requirements written in the specification, it is then possible to devise a strategy to try to ensure the delivery of the correct product. Within a manufacturing operation, there are three points to consider when establishing a quality control system:

- raw materials;
- processing conditions;
- finished product.

Although it might appear logical to consider these three in the order given above, it is more relevant

Table 15.6 Selection of key ISO definitions relating to quality assurance.

Word	Definition (words in bold are defined in the Standard)
Quality	Degree to which a set of inherent **characteristics** fulfils **requirements**
Requirement	Need or expectation that is stated, generally implied or obligatory
Characteristic	Distinguishing feature
Specification	**Document** stating **requirements**
Quality management	Coordinated activities to direct and control an **organization** with regard to **quality**
Quality planning	Part of **quality management** focused on setting **quality objectives** and specifying necessary operational **processes** and related resources to fulfil the quality objectives
Quality control	Part of **quality management** focused on fulfilling quality **requirements**
Quality assurance	Part of **quality management** focused on providing confidence that quality **requirements** will be fulfilled

to consider them in the reverse order as it is the nature of the finished product that determines the required processing and, from this, the nature of the raw materials that will be purchased.

Finished product

Manufacturers will want to ensure that the product is going to be acceptable to their customers and will therefore wish to monitor the final product. It is also often necessary to gather data on the nature of the finished products so as to be able to confirm compliance with legal requirements.

However, finished product testing is an expensive process and, if all the preceding steps in the process have been complied with, then the chances of finding unacceptable product at the end of the production line may be rather limited. The end of the production line is also too late to make changes if the product is incorrect.

Processing conditions

The value of control is much more evident when considering steps during the manufacturing process. Here it is likely that control will be needed to ensure that the production process goes according to the process specification. Quality control staff will monitor the different steps in the process and compare the results to the process specification. Where the results of the checks suggest that the process is deviating from the target values, action can be taken to adjust the process to keep it within the required performance parameters.

Raw materials

It is sensible to conduct checks on the incoming raw materials – no manufacturer would wish to start processing using unsatisfactory materials. However, as with finished product checks, the process can be expensive if sufficient testing is to be done to provide a statistically valid procedure. Depending upon the nature of the process and the potential for things to go wrong, testing will be carried out to try to ensure that only acceptable raw materials are used. Raw materials that do not meet the purchase specification can then be excluded.

Although the process of quality control is necessary, particularly within the production process to monitor production conditions and to ensure they remain on target, the testing of raw materials and finished product by the application of extensive quality control is harder to justify.

15.3.4 Quality assurance (QA)

The inspection systems necessary to ensure that out-of-specification material is not used or not produced are frequently incapable of providing the necessary protection. It is impossible to 'inspect quality into a product'; it is much better to try to build it in in the first place. Quality control on its own is insufficient – more extensive systems are necessary.

The use of techniques to create conditions that assure the production of food meeting the customer quality requirements is seen as a better way of managing a process. Looking at the three points described above, examples can be given:

- *Raw materials* – working with suppliers to ensure that they have met your quality requirements has long-term benefits which will reduce the need for raw material testing.
- *Processing* – analysing sources of variability in product quality from a production process may enable a better control system to be installed, allowing for a reduction in the frequency of checks.
- *Finished product* – gathering information from your customers about their reaction to your product will enable you to monitor their perception of your product quality and to identify improvements.

The term 'quality assurance' is therefore used to represent these wider activities used to help guarantee the delivery of the correct product to your customers. The effective application of advanced quality assurance techniques will minimise the need for extensive quality control.

The recognition that the quality of your products will frequently be dependent upon the quality of the products from suppliers led to the increasing use of checks on suppliers and, in the case of retailers, designation of approved suppliers. However, just as constant checking of raw materials is expensive, spending time checking suppliers requires considerable time and resources. The growing number of checks being conducted on food suppliers by retailers also led to complaints from the manufacturers. A more systematic way was needed and this led to the development of national and international standards against which quality systems could be judged.

The idea for an international standard giving requirements for quality management can be traced back many years, but the adoption of the British Standard BS5750 in 1979 marked a major move forward. It was this standard that in 1987 was taken as the basis for the three international standards – ISO 9001, ISO 9002 and ISO 9003. Of these, ISO 9001 was the most comprehensive, but ISO 9002, without the section on product development, could be applied to food production situations. ISO 9003 was a much reduced version with a more limited scope. In 1994 the standards were subject to some minor revisions, but a subsequent radical review led to a major relaunch of the standards in 2000. Further minor changes were made to the definitions in ISO 9000 in 2005 and a more significant review has recently led to the publication of a new version of ISO 9001. Although this new version of ISO 9001 contains no new requirements, it does provide some additional clarifications based on experience and some amendments to ensure consistency with other standards.

The three main standards that now exist are therefore:

- ISO 9000:2005 Quality management systems – Fundamentals and vocabulary
- ISO 9001:2008 Quality management systems – Requirements
- ISO 9004:2000 Quality management systems – Guidelines for performance improvements

ISO 9000 sets out the basic concepts and, importantly, establishes a detailed set of definitions. A selection of these is given in Table 15.6.

ISO 9001 provides the detailed requirements against which quality systems can be assessed. Organisations have to meet the general requirement – specifically they have to 'establish, document, implement and maintain a quality management system and continually improve its effectiveness in accordance with the requirements of this International Standard' (extract from Section 4.1 of the standard). The quality management system has itself got to be managed, with senior management of the organisation taking responsibility for this. Summary details of key documents are provided which should include:

- a quality policy statement;
- a quality manual;
- procedures;
- process control documents including work instructions.

Additional suggestions as to the content of these documents have been provided in another ISO publication – ISO/TR 10013:2001 Guidelines for quality management system documentation.

The standard then uses a process approach to identify the key aspects of a quality management system and provides four key components:

- management responsibilities;
- resource management;
- product realisation;
- measurement, analysis and improvement.

Each of these components is subdivided into several key aspects which must be incorporated into the organisation's quality system (see Table 15.7).

Although ISO 9001 has been widely accepted in many industrial sectors, its adoption by the food industry has not been easy. Major food companies that have fully functioning quality assurance departments have been able to ensure that their systems include all the required components. However, the supply of food to consumers in many countries is through a large number of small businesses with only the family or a small number of employees involved. It is not easy for these smaller companies to understand and implement the standard. Moreover, the food industry has already been adopting the Hazard Analysis and Critical Control Point (HACCP) approach to food control. Many countries now have food legislation that expects control to be based on HACCP. ISO 9001 does not discuss HACCP and the relationship between the two can be confusing.

In an attempt to encourage food businesses to adopt ISO 9001, another document was published by ISO which attempted to illustrate how companies who had effectively implemented HACCP could incorporate this into an overall quality management system meeting the requirements of ISO 9001. These guidelines were published in 2001 as ISO 15161 Guidelines on the application of ISO 9001:2000 for the food and drink industry.

Table 15.7 Section headings used in ISO 9001 and ISO 22000.

ISO 9001: 2008	ISO 22000: 2000
1 Scope	**1 Scope**
2 Normative references	**2 Normative references**
3 Terms and definitions	**3 Terms and definitions**
4 Quality management system	**4 Food safety management system**
4.1 General requirements	4.1 General requirements
4.2 Documentation requirements	4.2 Documentation requirements
5 Management responsibility	**5 Management responsibility**
5.1 Management commitment	5.1 Management commitment
5.2 Customer focus	
5.3 Quality policy	5.2 Food safety policy
5.4 Planning	5.3 Food safety management system planning
5.5 Responsibility, authority and communication	5.4 Responsibility and authority
	5.5 Food safety team leader
	5.6 Communication
	5.7 Emergency preparedness and response
5.6 Management review	5.8 Management review
6 Resource management	**6 Resource management**
6.1 Provision of resources	6.1 Provision of resources
6.2 Human resources	6.2 Human resources
6.3 Infrastructure	6.3 Infrastructure
6.4 Work environment	6.4 Work environment
7 Product realization	**7 Planning and realization of safe products**
7.1 Planning of product realization	7.1 General
7.2 Customer-related processes	7.2 Prerequisite programmes (PRPs)
7.3 Design and development	7.3 Preliminary steps to enable hazard analysis
7.4 Purchasing	7.4 Hazard analysis
7.5 Production and service provision	7.5 Establishing the operational prerequisite programmes (PRPs)
7.6 Control of monitoring and measuring equipment	7.6 Establishing the HACCP plan
	7.7 Updating of preliminary information and documents specifying the PRPs and the HACCP plan
	7.8 Verification planning
	7.9 Traceability system
	7.10 Control of nonconformity
8 Measurement, analysis and improvement	**8 Validation, verification and improvement of the food safety management system**
8.1 General	8.1 General
8.2 Monitoring and measurement	8.2 Validation of control measure combinations
8.3 Control of nonconforming product	8.3 Control of monitoring and measuring
8.4 Analysis of data	8.4 Food safety management system verification
8.5 Improvement	8.5 Improvement

15.3.5.2 ISO 22000

Although ISO 15161 gave some additional help for food businesses wishing to achieve accreditation, it was still a complex task to match food industry systems to the content of ISO 9001. Many food companies, including in particular the major retailers, were starting to establish private standards (discussed later) which their suppliers had to meet.

These standards are designed to provide the retailers with enhanced confidence in the ability of their suppliers to meet the needs of consumers for safe food.

ISO therefore recognised the need to provide a more focused quality management standard specifically addressing food industry needs. After much discussion, ISO 22000:2005 Food safety management systems – Requirements for any organization in the food chain was published in September 2005.

Although based upon the concepts and requirements set out in ISO 9001, the content of ISO 22000 is designed to directly address the special requirements of the food industry. The following points are worth noting:

- The standard relates to 'any organisation in the food chain' and can therefore be used by any suppliers to the food chain – including farmers, packaging suppliers, water companies and ingredient manufacturers – as well as the manufacturers of the finished product.
- The main topic addressed is 'food safety' and issues relating to more general food quality issues may not be directly affected by the standard.
- Although based substantially on the concepts of HACCP, some confusion can occur as the standard introduced an additional category of control. In most companies that have adopted HACCP, many of the potential food safety issues are minimised by routine procedures (usually known as 'pre-requisite programmes') and then key important controls are implemented at Critical Control Points (CCPs). In ISO 22000, an additional category, known as 'operational pre-requisite programmes' has been introduced which allows some of the important controls to be implemented by more routine procedures. This allows the management to maintain a high level of vigilance at selected control points designated as the CCPs. The definitions used by ISO 22000 for these three control elements are given in Table 15.8.

Table 15.8 Key definitions used within ISO 22000.

PRP/prerequisite programme: basic conditions and activities that are necessary to maintain a hygienic environment throughout the food chain suitable for the production, handling and provision of safe end products and safe food for human consumption

Operational PRP/operational prerequisite programme: PRP identified by the hazard analysis as essential in order to control the likelihood of introducing food safety hazards to and/or the contamination or proliferation of food safety hazards in the product(s) or in the processing environment

CCP/critical control point: step at which control can be applied and is essential to prevent or eliminate a food safety hazard or reduce it to an acceptable level

15.3.6 Private standards

The application of systematic and comprehensive quality management systems is seen by many companies as a valuable management tool. However, their use is not usually a legal requirement. As described earlier, the application of legislation is limited to areas where the potential for food safety failure and/or fraud requires national action to require food businesses to adopt minimum protective measures. Many retailers have been attempting to ensure that their suppliers meet additional requirements which have been set out in various different private standards. The successful agreement of a contract to supply that retailer can be dependent upon the supplier having been certified as meeting one or more of these additional private standards.

Although ISO 9001 and ISO 22000 have been discussed above, as their use is not legally required, they can also be classified as private standards.

The increasing number of other standards, many developed by national retailing organisations, was recognised as causing problems for suppliers around the world. In an attempt to provide a more consistent approach to these standards, the retailers established an organisation to co-ordinate their work. Under the auspices of the Comité International d'Entreprises à Succursales (CIES or International Committee of Food Retail Chains), a Global Food Safety Initiative (GFSI) has been established. The GFSI produced a benchmarking document which is used to assess other standards. If a food supplier meets the requirements of one of the benchmarked standards, it is to be hoped that that supplier's products will be acceptable to all the retailers who have established the GFSI.

Key standards that have been benchmarked by the GFSI include:

- *The British Retail Consortium (BRC) – Global Food Standard*. This was one of the first private standards adopted at a national level. Within the United Kingdom, changes in national legislation had put increased pressure on retailers to monitor their suppliers. In order to avoid excessive and repetitive checking by different retailers, the BRC published its first standard in 1998. This rapidly became accepted as a key standard for food companies to obtain – not just as suppliers to the major retailers. The standard has been progressively updated, with a fifth version published in January 2008.

- *IFS – International Food Standard*. In many ways the IFS is very similar to the BRC standard, but it has been developed by German and French retailers.
- *SQF 2000 – Safe Quality Food Scheme*. This scheme was developed to meet the needs of the Australian retail mark and actually consists of two standards – SQF 1000 is focused on the agricultural supply of food raw materials whilst SQF 2000 incorporates the more comprehensive requirement applicable to food manufacturing and distribution operations. The scheme was adopted by the Food Marketing Institute (FMI) in the USA where it has become the leading private standard.
- *Dutch HACCP Scheme*. Although developed originally to provide a clear means of demonstrating the successful application of the HACCP system within a company, the Dutch HACCP scheme has also been benchmarked by the GFSI.

Another private standard that has achieved international significance is that adopted by GLOBALGAP (formerly known as EUREPGAP) which has established standards linked to good agricultural practice (GAP). As certification to GLOBALGAP may be a requirement for trading with major retailers, particularly within Europe, the standard has been widely applied in many of the major food-producing regions. The standard promotes integrated crop management (particularly with regard to the application of pesticides) and consideration of worker welfare.

The increasing use of private standards by large retailers in the developed world has been raising some concerns. Exporters in developing countries have already been finding it difficult to meet the legal requirements established by developed countries, particularly Europe and North America. Access to these lucrative markets is seen as being one way for the developing world to improve its economic performance. The reality now is that frequently meeting the legal requirements is not sufficient. The need for a supplier to meet a private standard before a contract is placed adds a new set of requirements. These additional requirements can add to the financial costs to be met by suppliers before they are even able to compete for the contract. As there is no guarantee that they will be successful, the risks may be considered too great.

15.3.7 Meeting the requirements of the standards

The challenge of meeting compliance with the standards described above is often the responsibility of the food scientist or technologist employed by a food business. Many of the decisions about the nature and type of quality management that is needed will be based upon the science and technology of food and food manufacturing processes. The multidisciplinary nature of food science and technology enables food scientists and technologists to take the lead.

When faced with this task, the challenge can seem great. However, there are several key elements that can be used to make the development of quality systems a more structured process. A staged implementation process which allows sufficient time for each stage to be effectively completed is highly recommended.

A key element specified in most standards is the commitment of the senior management of the business to the overall quality policy. Attempts to implement a quality system as a simple marketing exercise without a belief in the inherent value of the system are likely to lead to failure. In addition, any attempt to devolve the development of the quality system to a lower level of management within a subdivision of the business will leave other parts of the business feeling that they have little role in the development and implementation of the quality system. Again, failure is likely.

Certain additional considerations are listed here, but there are specific publications that provide guidance on the creation of effective quality systems meeting the requirements of the standards.

15.3.7.1 Good manufacturing practice (GMP)

The concept of good manufacturing practice (GMP) can be related to the concept of 'prerequisite programmes' used within HACCP and ISO 22000. However, in some countries (and in some industries, notably pharmaceuticals) the maintenance and/or operation of GMP may be a legal requirement for the operation of the business. The phrase GMP will then be found in the relevant legal documents and there is likely to be detailed guidance given as to what constitutes GMP. Compliance with the specific requirements will then be essential.

Within the context of more voluntary quality systems, GMP can provide a foundation for the development of an effective quality system. One text that has received international recognition is that published by the Institute of Food Science and Technology in the UK entitled *Food and Drink: Good Manufacturing Practice – A Guide to its Responsible Management* (2006).

The Guide emphasises that good manufacturing practice requires the application of certain key activities, referred to in the Guide as 'effective manufacturing operations' supported by 'food control' (itself consisting of 'quality assurance' and 'quality control'). Whilst the description of the general requirements for effective GMP are explained in the Guide, the majority of the Guide consists of detailed practical advice on appropriate procedures and action for a wide range of quality systems and for different food sectors. The value of the Guide comes from the large amount of detail that is given relating to specific food control issues. Whilst much of the guidance goes beyond minimum legal requirements, by specifying procedures that can be considered 'good', the food scientists and technologists who contributed to the Guide have provided a very valuable tool for others to use to enhance their own operations and procedures.

15.3.7.2 HACCP

Fundamental to nearly all the food quality management systems is the application of the Hazard Analysis Critical Control Point (HACCP) system. The system itself is described elsewhere is this book with regard to its use in the control and prevention of microbiological hazards and readers should refer to those sections for full details.

It does need to be stressed, however, that the use of HACCP is not limited to microbiological issues and it is frequently stressed that there are three main types of hazard which need to be considered – microbiological, chemical and physical. The microbiological hazards are often seen as the most important and can often be controlled by appropriate processing and/or storage conditions. There may be a reliance on the control of microbial contamination prior to the arrival of raw materials at a factory and it will be necessary to ensure that the HACCP principles are applied throughout the food chain (as required, for example, by ISO 22000).

Chemical contamination may also occur at any point in the food chain. The contamination of animal feed by dioxin in Belgium in 1999 clearly demonstrated to the European regulators that the food chain extends backwards through many links. A hazard can enter the chain at any point and even though it might become diluted, some chemicals can still pose a hazard at very low levels. The application of HACCP principles should be used to identify these risks.

Another important hazard is the presence of physical contaminants – often termed 'foreign bodies'. These can arrive with some raw material (sticks and stones, for example) and many preparatory processing operations are designed to try to remove these contaminants.

Most food factories pay particular attention to the possible contamination of food by metal and glass. These can be the result of cutting blades snapping, allowing a sharp piece of metal to enter the food. The usual method of protecting against these hazards is to ensure that all packed product passes through a metal detector with an automatic reject device. In smaller manufacturing operations, a regular check on the state of knives and saw blades may be sufficient to demonstrate that contamination has not occurred.

Glass contamination could come from a number of different sources within a factory – windows, light fixtures or glass containers are examples. Many companies minimise the use of glass so that, in the event of a breakage, the risk of contamination is minimised. In factories where glass is used as a packaging method, more specific controls will be needed. The use of in-line X-ray equipment might be justified.

The implementation of an effective HACCP should be able to identify these hazards and will ensure that, either through appropriate prerequisite programmes or through the monitoring of the critical control points, the risks are minimised.

15.3.7.3 Documentation

A key requirement in most of the quality management systems described above is that the system is 'documented'. As the system is developed, it is only possible to propose, discuss, agree and amend parts of the system if they have been written down. The written documents also form a key part of the auditing process (see below) which allows external auditors to assess the components of the systems and consider how effective they are in meeting the requirements of a standard.

The excessive reliance on documentation has been one of the major criticisms against the standards. Preparing the complete set of documents considered

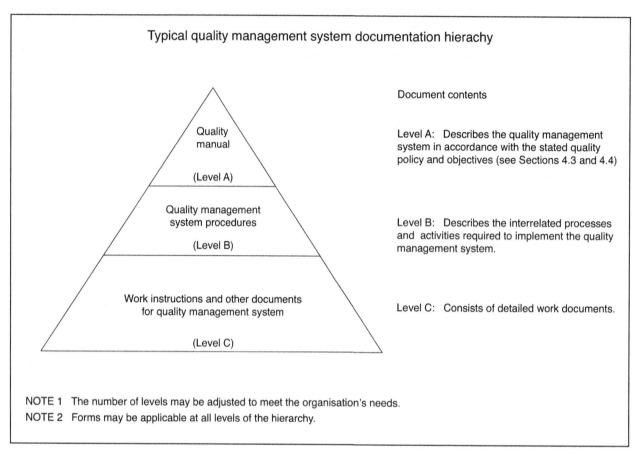

Typical quality management system documentation hierachy

Document contents

Level A: Describes the quality management system in accordance with the stated quality policy and objectives (see Sections 4.3 and 4.4)

Level B: Describes the interrelated processes and activities required to implement the quality management system.

Level C: Consists of detailed work documents.

Quality manual
(Level A)

Quality management system procedures
(Level B)

Work instructions and other documents for quality management system
(Level C)

NOTE 1   The number of levels may be adjusted to meet the organisation's needs.
NOTE 2   Forms may be applicable at all levels of the hierarchy.

Figure 15.3 Representation of documentation hierarchy described in ISO 10013.

necessary for an audit can lead employees to think that the documents are more important than the effective use of the information contained in them.

An indication of the likely structure of the documentation has been provided by ISO in the form of 'Guidelines for quality management system documentation' (ISO 10013:2001). Figure 15.3 gives an overview of the hierarchy of the documentation. The main components described by ISO are:

- *Quality manual*: the overall quality management system is likely to be set out in the main quality manual. It will contain a copy of the agreed quality policy of the business which commits the organisation to work in an effective manner to meet key quality objectives, as determined by customer requirements. The quality manual has sections that describe the overall organisation, responsibilities and authority for quality within the organisation. There will then be more detailed sections on different aspects of quality management. For most food businesses, the HACCP will form a major part

of the manual, but there will be many additional aspects of quality management which will be described (e.g. supplier quality assurance, complaint handling, prerequisite programmes). When developing a structure for the quality manual it is often helpful to base it upon the various components of the standard being used by the organisation. This will help ensure that the system does meet all the requirements of the standard and it will help demonstrate this at any subsequent audit.
- *Procedures*: the procedure is a more detailed document which describes how the main requirements of the quality management system are actually implemented.
- *Work instructions*: at the lowest level of the hierarchy are the work instructions which set out the specific tasks that are performed to achieve the set procedure. The amount of detail contained in the work instructions can be excessive and in writing them care must be taken to describe only those elements that are critical to achieve the requirements of the related procedure.

In addition to these three main elements there are likely to be a large number of other documents and records which are necessary to demonstrate the effective application of the quality management system.

15.3.7.4 Auditing, certification and accreditation

Fundamental to most quality management standards is the concept of internal auditing. For example, ISO 9001 specifies that (in Section 8.2.2):

'The organization shall conduct internal audits at planned intervals to determine whether the quality management system

a conforms to the planned arrangements, to the requirements of this International Standard and to the quality management system requirements established by the organization, and

b is effectively implemented and maintained.'

ISO 22000 contains an identical requirement (in Section 8.4.1) except that the word 'quality' is replaced by 'food safety'.

Internal auditing enables the organisation to assess for itself whether the planned management system is meeting its objectives. The audit can consider whether

- the established system is appropriate to meet the stated aims of the system (sometimes termed a 'systems audit'), and
- the system is being operated in practice (sometimes termed an 'adherence or compliance audit').

Auditing has, however, become a very important activity, demonstrating compliance with standards. A common terminology is based upon the following:

- *First party audits*: this is in practice an alternative terminology to internal audits. It covers the situation where the organisation is itself checking its own system to verify compliance.
- *Second party audits*: the second party in this context is usually the customer of the organisation. In order to check that the organisation is capable of meeting the requirements of the contract, the customer may decide to carry out an audit of the supplier – either before the contract is awarded or at intervals subsequently. Second party audits can be the key part

of a supplier quality assurance (SQA) programme established to ensure the safety of raw materials.

- *Third party audits*: if an independent audit is conducted, with the auditor having no direct connection with any contract, then it becomes a third party audit. Here the auditor has the sole task of checking whether the organisation meets the requirements of a particular standard (whether, for example, an ISO standard or one of those benchmarked by the GFSI). If the auditor is satisfied that this is the case, he/she can issue a certificate demonstrating compliance to the standard. The organisation is then able to use this to demonstrate to customers (current or future) that it is a reliable organisation meeting the requirements of the standard. It therefore removes the need for second party audits.

There is also a further issue. How can it be shown that third party auditors (or organisations employing them) are competent to undertake this work? Companies who offer third party audits are frequently referred to as 'certification bodies' as their aim is to provide certificates. It is necessary that these certification bodies should also meet certain standards. Within the ISO system of standards, this has been covered by the issuing of further standards. For certification bodies undertaking audits to ISO 9001, they will need to meet the requirement of ISO 17021:2006 Conformity Assessment – Requirements for bodies providing audit and certification of management systems. For certification bodies undertaking audits to ISO 22000, a more specific document has been published (currently as a Technical Standard): ISO/TS 22003:2007 Food safety management systems – Requirements for bodies providing audit and certification of food safety management systems.

To demonstrate compliance to these standards, the certification bodies will need to be subject to an audit by another organisation. These additional organisations are termed 'accreditation bodes'. Countries frequently have established a single accreditation body at the national level to oversee the effective operation of certification bodies. This may be the national standards organisation or another specially designated body. As a final check on the whole system, accreditation bodies have been issued with guidance. This is contained in ISO 17011:2004 Conformity assessment – General requirements for accreditation bodies accrediting conformity assessment bodies.

This whole succession of levels is illustrated in Fig. 15.4.

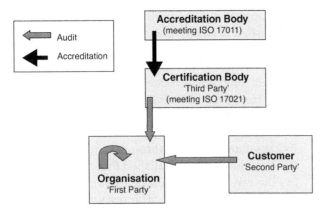

Figure 15.4 Relationship between auditing and accreditation.

It can be seen that the work of an auditor is fundamental to the successful operation of the whole process of certification and accreditation. ISO has recognised this and issued a further standard providing guidance on how an audit should be conducted and the training and competency for auditors. This is provided in ISO 19011:2002 Guidelines for quality and/or environmental management systems auditing.

As outlined by ISO, auditing has to be based on five strict principles, identified as the following:

1 *Ethical conduct*: the foundation of professionalism.
2 *Fair presentation*: the obligation to report truthfully and accurately.
3 *Due professional care*: the application of diligence and judgement in auditing.
4 *Independence*: the basis for the impartiality of the audit and objectivity of the audit conclusions.
5 *Evidence-based approach*: the rational method for reaching reliable and reproducible audit conclusions in a systematic audit process.

For a person to be an effective auditor it requires the development and application of personal attributes. A selection of these have been described by ISO as follows:

- ethical, i.e. fair, truthful, sincere, honest and discreet;
- open-minded, i.e. willing to consider alternative ideas or points of view;
- diplomatic, i.e. tactful in dealing with people;
- observant, i.e. actively aware of physical surroundings and activities;

- perceptive, i.e. instinctively aware of and able to understand situations;
- versatile, i.e. adjusts readily to different situations;
- tenacious, i.e. persistent, focused on achieving objectives;
- decisive, i.e. reaches timely conclusions based on logical reasoning and analysis;
- self-reliant, i.e. acts and functions independently while interacting effectively with others.

Beyond these, it is necessary for an auditor to have a range of generic skills linked to the audit process as well as suitable specific skills linked to food quality management issues. Trained food scientists and technologists possessing the right personal attributes and trained in generic auditing skills can make very competent auditors of food businesses.

15.3.7.5 Traceability, product recall and crisis management

When a food safety or quality problem arises, it is in the interests of everyone – consumers, retailers, distributors, manufacturers and raw material suppliers – to be able to identify the cause and to prevent the consumption or further distribution and sale of any affected stock. Failure to take prompt and effective action will result in increased damage to public confidence in the company and, more widely, in the whole food supply chain. Many companies now appreciate the importance of preparing for such problems and planning in advance the procedures that will be followed should they occur.

A key element in any such action will be the ability to identify (1) the source of the problem and (2) the extent of possible affected stock. Procedures to implement this are known as 'traceability' or 'product tracing'. At its most simple level, a food business will need to know where its raw materials have come from and where it has sold products to. This is referred to as 'one up, one down' and is the minimum now required under European law.

The value of this basic concept, however, can be significantly enhanced if the business is able to identify specific batches of raw materials which can be linked to specific codes of finished product – so-called internal traceability. When combined with data on recipes, the additional information can be invaluable in tracking problems through the food chain. Increasingly sophisticated systems are being employed which link the different components of the food chain so that the origins of many of the ingredients used in a manufactured food product can be traced back to their source. Following previous problems, notably arising from the cattle disease BSE, highly developed systems are frequently used in the tracing of meat so that its full history can be identified.

Despite the use of quality management systems, it can still be possible for things to go wrong and for a product to be sold that is not of the correct quality. The extent of the failure and the potential risks to consumers and to the reputation of the company all need to be evaluated and appropriate action taken to minimise the impact. Where consumer safety is at risk then rapid action will be needed to prevent further exposure. Businesses will need to consider whether to implement 'product withdrawal' or, where the products have reached consumers, a 'product recall'. Where appropriate, legal requirements will need to be considered and discussions with national food control authorities are almost certainly necessary.

Where either a product withdrawal or a product recall is necessary, a company needs to implement effective crisis management procedures. This occurs best if systems and procedures have been developed in advance as part of the quality management system. Planning for a crisis, combined with trials of the plans, will allow for their speedy implementation when a problem actually happens. Key components of a crisis management plan are likely to include:

- *Product recall policy*: if an incident occurs that has the potential to harm consumers or the company's image or brand, it is best if it can be assessed against criteria established in advance. Expressed in simplified form, the criteria can be drafted into a policy statement which, when an incident occurs, should enable quick agreement as to whether a product recall is needed.
- *Product recall plan*: once the need for a recall has been agreed, fast implementation is needed. A prepared plan enables everyone to know in advance what their role is during the recall.
- *Risk assessment*: a key component of the decision-making process is the evaluation of the scientific evidence related to the incident. In the early stages of the crisis, the level of risk may be unknown. For example, if consumers start reporting illness or off-flavours in a product, evidence will be needed to determine whether the problem is of microbiological or chemical origin. Access to laboratories with advanced analytical equipment and which offer a

rapid response service will enable any samples to be scientifically assessed. The analytical information will enable a risk assessment to be conducted to determine the best way to minimise the health risk to consumers.
- *Incident Management Team*: it is highly recommended that companies establish a group of key staff who together can form an Incident Management Team. When an incident does occur (which can be at any time of day or night), it is important to ensure that key staff can be brought together. As a product recall involves many different aspects of the business, it is helpful for all to be represented in the team. This could include staff from different departments including technical, quality control, sales, marketing, public relations, buying, production, distribution, logistics and legal. The team should not require particular individuals to be present as they may not be available when the incident occurs. A single person should be designated team leader and he/she will ultimately take responsibility for the handling of the recall.
- *Traceability and documentation*: as discussed above, traceability provides a key component in the analysis of the recall process and could enable it to be limited to a small number of batches or codes. With speed being a key factor, access to the appropriate documentation is essential. Much of this may be in the form of computerised data and the Incident Management Team will need to gain access to this (either directly or via other staff). It would obviously be helpful to the team if the development of the information system has been designed taking into account its potential use during a product recall.
- *Communication process*: speedy and accurate communication is vital for a successful recall. The Incident Management Team will need to be provided with access to telephones, fax machines and e-mail. Information provided verbally should be followed up by a documented confirmation (fax or e-mail). In addition to communication with people within the company, there will be extensive need to communicate outside. This will include, amongst others, suppliers, distributors, customers, national control bodies, local enforcement officers, media organisations, analytical laboratories, legal advisors and, in some cases, police forces. The provision of dedicated telephones will avoid any problems with the phone lines becoming blocked by incoming calls.

■ *Training*: an effective product recall, with minimal damage to the company, does not just happen by chance. It is necessary to prepare for the eventuality and to conduct trial product recall exercises so as to test the system in advance. By undertaking training, people become aware of their role in the recall and can quickly respond in the event that it occurs. Training exercises also allow for gaps or weaknesses in the procedures to be identified and eliminated. Regular testing of the plan will be needed to ensure that new staff can participate and to check the plan takes into account changes to other aspects of the company's systems.

15.3.7.6 Laboratory accreditation

Successful implementation of quality assurance procedures requires the sound application of scientific and technological principles. Control of processes relies upon the accurate measurement of process parameters (temperature, pressure, weight, etc.). More complex testing is usually carried out in laboratories where samples are analysed and the results used to reach decisions on the quality of raw materials, intermediate product or the finished product. The accuracy of the laboratory results is key in reaching the right decisions.

Another area where accuracy is vital is linked to the legal requirements. When enforcement officers take an official sample they will send it for analysis to a competent laboratory. The result from the analysis can then be used in any subsequent prosecution relating to the failure of the food to meet the legal requirement – whether owing to some aspect of chemical composition or the microbiological safety of the food. The accuracy of the analytical result is important for the safe operation of the legal process.

Procedures have been established to provide a means for establishing whether laboratories operate under recognised quality standards. Another ISO standard sets out the requirements – ISO 17025:2005 General requirements for the competence of testing and calibration laboratories. The first edition of ISO 17025 in 1999 followed earlier documents produced at an international level (ISO/IEC Guide 25) and at a European level (EN 45001), both of which it replaced.

ISO 17025 now contains all of the requirements that testing and calibration laboratories have to meet if they wish to demonstrate that they operate a management system, are technically competent and are able to generate technically valid results. The standard has

been written to be consistent with the requirements of ISO 9001.

15.4 Statistical process control

Effective quality assurance requires the analysis of data and the correct interpretation of such data. Statistical process control provides the tool for doing this.

The control of food safety hazards cannot be restricted to a system based on statistical process control. Although it is impossible to ensure 100% safety, the use of food safety management systems is designed to build safety into products rather than to rely on a statistical evaluation of the finished product. Modern systems, such as HACCP, focus attention on preventive measures.

There are, however, situations where sampling is appropriate and the use of valid statistical techniques provides guidance as to the correct action to be taken. The techniques used are based on an assessment of probabilities.

This section focuses on those controls that are based on sampling. Two situations frequently occur:

1 when monitoring a production process to ensure compliance with the requirements;
2 when deciding whether to accept or reject a delivery of food produced elsewhere.

In the first case there will be regular samples taken so as to check on the performance of the process. Using control charts, such as the Shewhart control charts described below, allows unexpected variation to be identified and action taken to correct it. In the second case, the assessment relates to the whole batch and different acceptance sampling techniques are used.

ISO has adopted a comprehensive set of standards designed to ensure consistent application of valid criteria. A selection of key standards is given in Table 15.9.

15.4.1 Background concepts

15.4.1.1 Variation

In any production process there will be variation. Modern machinery with sophisticated controls may be capable of very precise operations, but such

Table 15.9 Selected ISO standards linked to statistical process control techniques.

Terminology and symbols
ISO 3534 Statistics – Vocabulary and symbols:

- Part 1: General statistical terms and terms used in probability (2006)
- Part 2: Applied statistics (2006)
- Part 3: Design of experiments (1999)

Acceptance sampling
ISO 2859 Sampling procedures for inspection by attributes:
- Part 1: Sampling schemes indexed by acceptance quality limit (AQL) for lot-by-lot inspection (1999)
- Part 2: Specification for sampling plans indexed by limiting quality (LQ) for isolated lot inspection (1985)
- Part 3: Skip-lot sampling procedures (2005)
- Part 4: Procedures for assessment of declared quality levels (2002)
- Part 5: System of sequential sampling plans indexed by acceptance quality limit (AQL) for lot-by-lot inspection (2005)
- Part 10: Introduction to the ISO 2859 series of standards for sampling for inspection by attributes (2006)

ISO 3951 Sampling procedures for inspection by variables:

- Part 1: Specification for single sampling plans indexed by acceptance quality limit (AQL) for lot-by-lot inspection for a single quality characteristic and a single AQL (2005)
- Part 2: General specification for single sampling plans indexed by acceptance quality limit (AQL) for lot-by-lot inspection of independent quality characteristics (2006)
- Part 3: Double sampling schemes indexed by acceptance quality limit (AQL) for lot-by-lot inspection (2007)
- Part 5: Sequential sampling plans indexed by acceptance quality limit (AQL) for inspection by variables (known standard deviation) (2006)

ISO 8422 Sampling procedures for inspection by attributes: Specification for sequential sampling plans (1991)

ISO 8423 Sampling procedures for inspection by variables: Specification for sequential sampling plans for percent nonconforming. Known standard deviation (1991)

ISO/TR 8550 Guide for the selection of an acceptance sampling system, scheme or plan for inspection of discrete items in lots (1994)

Process control
ISO 7870 Control charts – general guide and introduction (1993)
ISO 7873 Control charts for arithmetic average with warning limits (1993)
ISO 7966 Acceptance control charts (1993)
ISO 8258 Shewhart control charts (1991)

machinery will still have minor variation. Older equipment with fewer controls will display greater variation. The variation will come from a range of minor fluctuations – temperatures, levels, pressures, position and density, for example. These fluctuations, when combined, are inherent to the production process and are referred to as 'chance causes'. Although it might be possible to identify their origin, this variation is usually accepted as part of the process. Reduction in the variation is likely to need investment in new controls or new machinery.

Process controls consider that the chance cause variation is part of the process but is used to identify variation from it. When something changes unexpectedly there will be additional variation on top of that produced by the chance causes. These additional variations derive from 'assignable causes'. Process control procedures are used to identify and eliminate these causes.

If the only variation present is due to chance causes, the process is considered to be in a 'state of statistical control'. If there is additional variation due to assignable causes it can be considered 'out of control'.

Some relevant definitions, taken from ISO 3534 Part 2, are given in Table 15.10.

15.4.1.2 Variable and attribute data

When analysing a situation, data can be generated in two different ways and the type of data will determine the statistical technique that is subsequently used.

Table 15.10 Selected ISO definitions linked to process control.

Process in control: a process in which each of the quality measures (e.g. the average and variability or fraction nonconforming or average number of nonconformities of the product or service) is in a state of statistical control.

State of statistical control: a state in which the variations among the observed sampling results can be attributed to a system of chance causes that does not appear to change with time.

Assignable cause: a factor (usually systematic) that can be detected and identified as contributing to a change in a quality characteristic or process level.
Notes:

1 Assignable causes are sometimes referred to as special causes of variation.
2 Many small causes of change are assignable, but it may be uneconomic to consider or control them. In that case they should be treated as chance causes.

Chance causes: factors, generally many in number, but each of relatively small importance, contributing to variations, which have not necessarily been identified.
Note: chance causes are sometimes referred to as common causes of variation.

Variable data represent observations obtained by measuring and recording the numerical magnitude of the characteristic. As the scale is continuous, the data can have any value and, depending upon the equipment, can be determined to any accuracy. Examples are length, pH and weight.

The alternative approach is to obtain observations based on the presence (or absence) of some characteristic. The data are counted and are therefore only in the form of integers. There are two types of attribute data:

- *Nonconforming units (defectives)*: counting of units which are either conforming (acceptable) or nonconforming (defective). Examples could be a damaged can, a broken biscuit, a bruised potato.
- *Nonconformities (defects)*: counting of the number of nonconformities (defects) in a unit. Examples could be the number of damaged cans in a case, the number of broken biscuits in a packet or the number of bruises on a potato.

15.4.1.3 The normal distribution

The generation of the overall process variation based on random chance causes most frequently generates data displaying the characteristics of the normal distribution. As a result, most of the techniques of statistical process control are based on an assumption that the data have a normal distribution.

As it is not necessary to have a detailed understanding of the mathematics of the normal distribution, it is not covered here. However, as a result of the mathematical basis, it is possible to identify certain key elements. Two components are used for this: the 'standard deviation' which provides a measure of the spread of the data and the 'mean' which provides the central point of it. (Note: the standard deviation can be most easily found by the use of a statistical function on a mathematical spreadsheet (e.g. using the MS Excel formula STDEV(A1:A10) would return the standard deviation of the 10 numbers in cells A1 to A10). Most scientific calculators also enable the easy calculation of the standard deviation of a set of entered numbers.)

The spread of values, as measured by the standard deviation, allows the following to be stated (see also Fig. 15.5):

- 68.3% of values lie within ±1 standard deviation of the mean
- 95.4 % of values lie within ±2 standard deviations of the mean
- 99.7 % of values lie within ±3 standard deviations of the mean

If the values of the mean and standard deviation are known, then it is possible to determine, using standardised values, the proportions of items that would lie beyond certain points in the distribution. See Box 15.2 and the tabulated values in Table 15.11.

15.4.1.4 Sampling risks

When taking decisions based on statistical probabilities, it is necessary to be aware of two potential errors.

In the first situation, the data may suggest that something has changed in a process or that a delivery

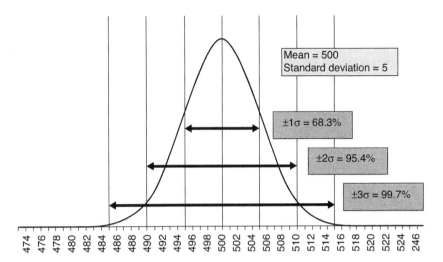

Figure 15.5 Concept of standard
deviation and the normal distribution.

may not meet the required specification. However, this may be due to random sampling variation when the actual population conforms to the requirements. In this case any decision taking to modify the process or reject the delivery will prove wrong and is likely to result in further adjustments. This is often referred to as an 'error of the first kind' or, more simply, a 'type I error'.

In the second situation, the data may suggest that the process is still in control or that a delivery does meet the specification. However, again due to random variation of the sampling process, the data may not have identified a change or a poorly produced batch. In this case no changes will be made to the process or a delivery may be accepted when it should have been rejected. This is referred to as an 'error of the second kind' or, more simply, a 'type II error'.

In both of the above cases, it is difficult to prevent the errors occurring. However, by basing the analysis on valid statistical techniques, any decision will be the best under the circumstances. The fact that it subsequently proves to have been wrong cannot be blamed on the person taking the decision. All parties, whether commercial, technical or production, need to accept that there will be occasions when a decision proves to have been wrong. On the other hand, failure to use valid statistical techniques can lead to significant arguments and potentially claims for recovery of losses. The use of recognised statistical standards is therefore strongly recommended.

15.4.1.5 Process capability

Consideration as to whether a process is in a state of statistical control is important when attempting to operate a production line. Customers, however, are not directly interested in the particular control issues but need to be confident that their requirements will be met – they are more interested in the potential for the process to meet the specification. This leads to the separate idea of 'process capability'.

Process capability is based on a measure of the spread of the process. It is therefore linked to the standard deviation which is the statistical measure of spread. Since 6 standard deviations (6σ) covers 99.7% of the population (see Fig. 15.5), it is common practice to calculate a 'process capability index' (PCI or C_p) based on 6σ. Where there are two specification limits, an Upper Specification Limit (USL) and a Lower Specification Limit (LSL), this can be calculated as the ratio of the difference between these limits divided by 6σ or:

$$C_p = (USL - LSL)/6\sigma$$

If only one specification exists or if the process is not centred, then it is more appropriate to consider the specification limits separately and to calculate the PCI, using the mean (μ), as the lesser of two alternatives:

$$C_{pk} = (USL - \mu)/3\sigma \quad \text{or} \quad (\mu - LSL)/3\sigma$$

Values of C_p or C_{pk} less than 1 indicate that the process as set up is not capable and will not meet the customer's requirements even when the process is in control. Values between 1 and about 1.33 are considered to have low to medium capability, whilst values greater than 1.33 are high and indicate that it should

Box 15.2 Example calculation using the normal distribution

During a production run, packed product is weighed and the mean weight was found to be 505 g (μ) and the standard deviation (σ) was 3.4 g. If the product specification demanded a weight of 500 $\pm$ 10 g, how much of the output would lie outside the permitted specification? If the process was centred at 500 g, with the same standard deviation, what would be the proportion outside the specification limits?

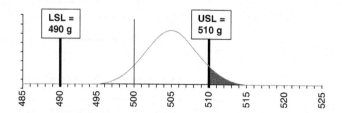

The figure shows that the major concern relates to the upper specification limit (USL) at 510 g. Proportions under the tail of a normal distribution can be found by reference to a standardised value (z) given by the equation:

$$z = (X - \mu)/\sigma$$

where

X is the value of interest,
μ is the mean of the normal distribution,
σ is the standard deviation of the normal distribution.

In this example, X = 510 g, μ = 505 g and σ = 3.4 g, giving:

$$z = (510 - 505)/3.4$$
$$z = 1.47$$

Consulting the tabulated values of z and P (see Table 15.11) gives a value for P of about 0.93 or 93%. The tabulated values of P are, however, the proportion below the value X (in this case 510, the value for the USL). In this example we are actually interested in the proportion greater than 510. Since the total proportions must add up to 1.00 (or 100%) the proportion above the USL is 1.00 − 0.93 = 0.07 or 7%.

It can also be noted that theoretically the normal distribution will also have a proportion below the lower specification limit. Although the diagram suggests that this value is likely to be much smaller than the 7% calculated for the USL, the actual value can be determined. Using the value of 490 g for the LSL gives X = 490 g, μ = 505 g and σ = 3.4 g, giving:

$$z = (490 - 505)/3.4$$
$$z = -4.29$$

From the table it can be seen that this gives a proportion of only about 0.00001 or 0.001%.

If the process had been centred with a mean of 500 g, the tail would be identical at both specification limits. A single calculation can therefore be done using one specification limit and the result doubled to obtain the overall proportion outside both limits, thus:

$$z = (490 - 500)/3.4$$
$$z = -2.94$$

From the table, this gives a proportion of about 0.0016 or 0.16%. The total outside both specification limits will then be double this – 0.0032 or 0.32%.

Table 15.11 Proportions under a normal distribution.

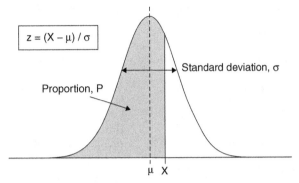

z	P	z	P	z	P	z	P	z	P
−5.00	0.000000	−3.00	0.001350	−1.00	0.158655				
−4.95	0.000000	−2.95	0.001589	−0.95	0.171056	1.05	0.853141	3.05	0.998856
−4.90	0.000000	−2.90	0.001866	−0.90	0.184060	1.10	0.864334	3.10	0.999032
−4.85	0.000001	−2.85	0.002186	−0.85	0.197663	1.15	0.874928	3.15	0.999184
−4.80	0.000001	−2.80	0.002555	−0.80	0.211855	1.20	0.884930	3.20	0.999313
−4.75	0.000001	−2.75	0.002980	−0.75	0.226627	1.25	0.894350	3.25	0.999423
−4.70	0.000001	−2.70	0.003467	−0.70	0.241964	1.30	0.903200	3.30	0.999517
−4.65	0.000002	−2.65	0.004025	−0.65	0.257846	1.35	0.911492	3.35	0.999596
−4.60	0.000002	−2.60	0.004661	−0.60	0.274253	1.40	0.919243	3.40	0.999663
−4.55	0.000003	−2.55	0.005386	−0.55	0.291160	1.45	0.926471	3.45	0.999720
−4.50	0.000003	−2.50	0.006210	−0.50	0.308538	1.50	0.933193	3.50	0.999767
−4.45	0.000004	−2.45	0.007143	−0.45	0.326355	1.55	0.939429	3.55	0.999807
−4.40	0.000005	−2.40	0.008198	−0.40	0.344578	1.60	0.945201	3.60	0.999841
−4.35	0.000007	−2.35	0.009387	−0.35	0.363169	1.65	0.950529	3.65	0.999869
−4.30	0.000009	−2.30	0.010724	−0.30	0.382089	1.70	0.955435	3.70	0.999892
−4.25	0.000011	−2.25	0.012224	−0.25	0.401294	1.75	0.959941	3.75	0.999912
−4.20	0.000013	−2.20	0.013903	−0.20	0.420740	1.80	0.964070	3.80	0.999928
−4.15	0.000017	−2.15	0.015778	−0.15	0.440382	1.85	0.967843	3.85	0.999941
−4.10	0.000021	−2.10	0.017864	−0.10	0.460172	1.90	0.971283	3.90	0.999952
−4.05	0.000026	−2.05	0.020182	−0.05	0.480061	1.95	0.974412	3.95	0.999961
−4.00	0.000032	−2.00	0.022750	**0.00**	**0.500000**	2.00	0.977250	4.00	0.999968
−3.95	0.000039	−1.95	0.025588	0.05	0.519939	2.05	0.979818	4.05	0.999974
−3.90	0.000048	−1.90	0.028717	0.10	0.539828	2.10	0.982136	4.10	0.999979
−3.85	0.000059	−1.85	0.032157	0.15	0.559618	2.15	0.984222	4.15	0.999983
−3.80	0.000072	−1.80	0.035930	0.20	0.579260	2.20	0.986097	4.20	0.999987
−3.75	0.000088	−1.75	0.040059	0.25	0.598706	2.25	0.987776	4.25	0.999989
−3.70	0.000108	−1.70	0.044565	0.30	0.617911	2.30	0.989276	4.30	0.999991
−3.65	0.000131	−1.65	0.049471	0.35	0.636831	2.35	0.990613	4.35	0.999993
−3.60	0.000159	−1.60	0.054799	0.40	0.655422	2.40	0.991802	4.40	0.999995
−3.55	0.000193	−1.55	0.060571	0.45	0.673645	2.45	0.992857	4.45	0.999996
−3.50	0.000233	−1.50	0.066807	0.50	0.691462	2.50	0.993790	4.50	0.999997
−3.45	0.000280	−1.45	0.073529	0.55	0.708840	2.55	0.994614	4.55	0.999997
−3.40	0.000337	−1.40	0.080757	0.60	0.725747	2.60	0.995339	4.60	0.999998
−3.35	0.000404	−1.35	0.088508	0.65	0.742154	2.65	0.995975	4.65	0.999998
−3.30	0.000483	−1.30	0.096800	0.70	0.758036	2.70	0.996533	4.70	0.999999
−3.25	0.000577	−1.25	0.105650	0.75	0.773373	2.75	0.997020	4.75	0.999999
−3.20	0.000687	−1.20	0.115070	0.80	0.788145	2.80	0.997445	4.80	0.999999
−3.15	0.000816	−1.15	0.125072	0.85	0.802337	2.85	0.997814	4.85	0.999999
−3.10	0.000968	−1.10	0.135666	0.90	0.815940	2.90	0.998134	4.90	1.000000
−3.05	0.001144	−1.05	0.146859	0.95	0.828944	2.95	0.998411	4.95	1.000000
				1.00	0.841345	3.00	0.998650	5.00	1.000000

not be difficult for the process to satisfy the requirements.

It is important to recognise that this can lead to four situations:

- *The process is capable and in control.* This is the ideal situation.
- *The process is incapable but is in control.* Here the production is being made according to expectation, but the chance cause variation is too big for the specification limits. Changes will usually be needed to the production line so as to reduce the overall variation.
- *The process is capable but out of control.* Here the production has some unexpected variation (probably due to an assignable cause), but, so long as it is not excessive, it is likely that the final product will still meet the customer's requirements. Tight control of the process is not needed and it may be possible to relax some of the control procedures.
- *The process is incapable and out of control.* Here the production has unexpected variation, but, even without this, there would be no chance of meeting the specification. At the same time as attempting to gain control of the process, it is necessary to consider what alterations are needed so as to make the process more capable.

Some examples of different distributions and their relationship to process capability are shown in Box 15.3.

15.4.2 Shewhart control charts (mean, range and standard deviation charts)

Control charts are the standard way of assessing the performance of a process. They are frequently referred to as Shewhart control charts after the person who first introduced the concept. The chart should be able to indicate whether the process is stable or 'in control' or whether something has changed and investigation is needed to identify and correct the cause. More generally, the charts provide a means of improving the performance of the process and provide information relevant to maintaining control and making process management decisions.

Detailed international standards have been written to cover several different possible types of control chart. Some may relate to a sequence of observations where there are no specified standard values. In a production situation it is, however, more common to have standard or target values and the charts are constructed with these as the basis. In addition there are types of charts linked to data that are in the form of variables and in the form of attributes.

The key steps in setting up control charts are as follows:

- Observations are made on the process to assess the extent of the chance cause variation (attempting to ensure that no assignable variation is present at the time of observation).
- From these data, the target distribution is set up. This will take into account the extent of the chance cause variation, any customer specifications and an appropriate process capability index.
- Based on the target distribution, the control chart is constructed with the expected values as the central line and additional control lines. The control lines may be set at a single value (an 'action limit') or may include additional lines ('warning limits') as well.
- Samples are then taken from the production and values calculated from the sample are plotted on the charts.
- If the plotted values lie beyond the action limit then action should be taken to identify the cause. Other rules can also be applied based on any warning lines used.

15.4.2.1 Variable data

Two problems can occur either together or separately: the mean may shift and/or the spread may change. Two charts are therefore used to identify these problems. To monitor the mean, a plot of sample means is used; to monitor the spread, a plot of sample ranges or standard deviations is used.

Means chart

For a stable process, most of the individual results (99.7%) will lie within 3 standard deviations of the target mean. When a sample is obtained, the mean of the sample is calculated. The value of this sample mean will vary around the target mean (μ) and the degree of variation will depend upon the standard deviation of the population (σ) and the size of the sample (n). The sample means will also form a normal distribution with the same mean as the population (μ) but with a spread (actually a standard deviation but commonly referred to as the 'standard error of the means' (SE)) which relates to the population

Box 15.3 Examples of the process capability index

Two specification limits with centred distributions

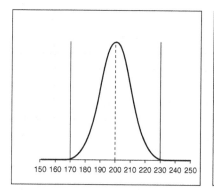 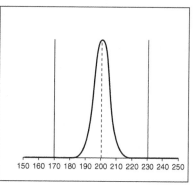 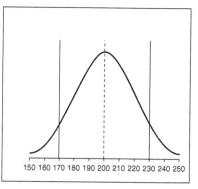

In this first set of figures, the USL is 230 and the LSL is 170. In all cases the mean is at 200:

- Case 1 (*left*): here the standard deviation is 10 so the calculation for the process capability index gives: $C_p = (USL - LSL)/6\sigma = (230 - 170)/60 = 1.0$. The process is just capable with the distribution fitting almost exactly between the specification limits. Any small additional variation will cause the production of out of specification product.
- Case 2 (*centre*): here the standard deviation is reduced to 5 which increases the value of Cp to 2.0. The process is very capable. Small additional variation will not be of any concern.
- Case 3 (*right*): here the standard deviation is now 20 which decreases the value of C_p to 0.5. The process is clearly incapable, with a significant proportion of out of specification product even though the process is in statistical control.

Use of C_{pk}

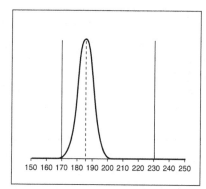 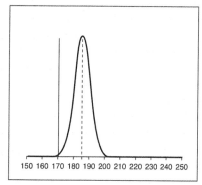 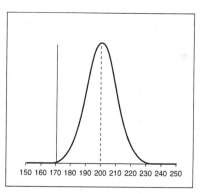

- Case 4 (*left*): here there are still two specification limits, but the process now has a mean of 185 and a standard deviation of 5. The calculation for the process capability index gives: $C_{pk} = (\mu - LSL)/3\sigma = (185 - 170)/15 = 1.0$. Although overall the process is potentially very capable (a calculation of C_p would give the same value as in Case 2 above), because of the positioning of the distribution, the process is only just capable. Any small additional variation of the mean downwards or an increase in the standard deviation will cause the production of out of specification product.
- Case 5 (*centre*): here the situation is the same as Case 4 except, as there is only the single specification limit, it is not possible to calculate a value for C_p.
- Case 6 (*right*): this situation is the same as Case 1 above but with no USL. The calculation of C_{pk} would give a value of 1.0.

standard deviation as follows:

$$SE = \sigma/\sqrt{n}$$

If the sample contains four items ($n = 4$), this would give a standard error of half the population standard deviation. Based on this idea, for a stable process, most of the sample means (99.7%) will lie within 3 standard errors of the target mean.

The action limits are set to represent this situation. Only about 3 samples in 1000 are likely to lie beyond action limit. Where warning limits are used, these are set at 2 standard errors for which about 1 sample in 20 will lie beyond these limits.

Range or standard deviation charts

Although it is mathematically more exact to use a standard deviation chart to monitor spread, the practical aspects often lead to a preference for range charts. A sample range, calculated as the largest result in the sample less the smallest result, is frequently easy to calculate and the resulting value is easy to interpret. The calculation of the standard deviation requires a calculator and any errors in the calculation may not be easy to identify. (As an example, consider the numbers 196, 200, 204 and 199. It can be quickly calculated and checked that the range is 8. The standard deviation has a value of 3.3 which cannot be easily verified or understood.)

In both cases the construction of the chart is not as easy as the means chart. The expected distribution of sample ranges (or standard deviations) does not follow a normal distribution and is asymmetrical about the mean range. It is therefore common practice to use predetermined factors with the value of the population standard deviation (or an estimate of it) to establish the position of the action limits – these factors will vary with the size of the sample. Example values for factors used to calculate the position of upper action limits are given in Table 15.12 and their use is illustrated in Box 15.4. When using larger sample sizes it is also possible to calculate lower action limits. However, since sample values below a lower action limit would represent a positive improvement in the process, they are not necessary to prevent potential problems in supplying customers. Using other factors, it is also possible to calculate the position of warning limits, although their use is less common and values are not provided in ISO 8258.

Tests for interpreting control charts

As indicated above, the action limits are set to represent situations that would be expected to occur with

Table 15.12 Factors for calculating control lines for range and standard deviation charts.

Sample size	Factor for range chart (D) (UCL = D × σ)	Factor for standard deviation chart (B) (UCL = B × σ)
2	3.686	2.606
3	4.358	2.276
4	4.698	2.088
5	4.918	1.964
6	5.078	1.874

a very low probability. If a sample result is beyond an action limit, it can be assumed that the process has shifted from the target distribution and that action should be taken to identify the cause and to eliminate it, returning the process to the target position.

When warning lines are used in addition to action limits, additional tests can be applied which represent situations where the probability of occurrence is also considered to be very low. The most common additional test is to check for two out of three points in a row beyond a warning limit.

An even more detailed use of the means chart is suggested in ISO 8258. This incorporates the use of lines that are positioned symmetrically about the central line at distances of 1, 2 and 3 standard errors (with the last two representing the warning and action limits respectively). The lines define six zones which are labelled as A, B, C, C, B, A, with zones C placed symmetrically about the central line. The following eight tests are then specified:

Test 1: One point beyond Zone A (i.e. beyond an action line)

Test 2: Nine points in a row in Zone C or beyond on one side of the central line

Test 3: Six points in a row steadily increasing or decreasing

Test 4: Fourteen points in a row alternating up and down

Test 5: Two out of three points in a row in Zone A or beyond (i.e. beyond a warning line)

Test 6: Four out of five points in a row in Zone B or beyond

Test 7: Fifteen points in a row in Zone C above and below the central line

Test 8: Eight points in a row on both sides of the central line with none in Zone C

Box 15.4 Setting up Shewhart control charts

A production line is set to give a product with a weight of 150 g. Data obtained from 20 samples, each containing four items, gave the following results:

Sample Number	Packet weights				Sample Mean	Sample Range	Standard Deviation
	1	2	3	4			
1	148	143	151	156	149.5	13.0	5.45
2	155	158	139	148	150.0	19.0	8.45
3	155	14	146	141	146.5	14.0	6.03
4	140	145	150	139	143.5	11.5	5.07
5	147	147	150	148	148.0	3.0	1.41
6	148	148	156	149	150.3	8.3	3.86
7	149	147	159	154	152.3	12.3	5.38
8	161	146	158	141	151.5	20.5	9.54
9	152	154	159	149	153.5	10.5	4.20
10	147	153	148	153	150.3	6.3	3.20
11	142	145	142	148	144.3	6.0	2.87
12	149	150	148	160	151.8	12.0	5.56
13	141	146	137	157	145.3	20.3	8.66
14	143	146	153	152	148.5	10.0	10.0
15	154	152	143	144	148.3	11.3	5.56
16	153	151	145	148	149.3	8.0	3.50
17	150	152	150	145	149.3	7.0	2.99
18	159	152	150	154	153.8	9.8	3.86
19	154	146	145	155	150.0	10.0	5.23
20	143	142	153	153	147.8	11.0	6.08

Grand Mean = 149.2
Grand SD = 5.49

Means		
149.2	11.0	5.08

The overall data standard deviation, based on all 80 items, is 5.49. The overall mean is 149.2, although, as the target mean is specified (150 g), the Shewhart control chart will be based on the target value.

Means chart

To calculate the position of the action and control lines, the standard error (SE) is first calculated. With samples of size 4, the value of SE is given by:

$$SE = \sigma/\sqrt{n} = 5.49/\sqrt{4} = 5.49/2 = 2.75$$

The control lines are then calculated at $\pm$ 3 SE and $\pm$ 2 SE as follows:

Upper Action Limit (UAL) $= \mu + 3 \times SE = 150 + 3 \times 2.75 = 150 + 8.25 = 158.25$ g
Upper Warning Limit (UWL) $= \mu + 2 \times SE = 150 + 2 \times 2.75 = 150 + 5.5 = 155.5$ g
Lower Warning Limit (LWL) $= \mu - 2 \times SE = 150 - 2 \times 2.75 = 150 - 5.5 = 144.5$ g
Lower Action Limit (LAL) $= \mu - 3 \times SE = 150 - 3 \times 2.75 = 150 - 8.25 = 141.75$ g

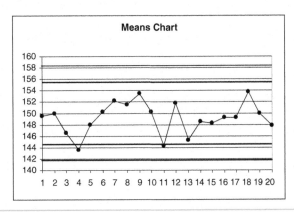

Range chart

Using the factors given in Table 15.12, the Upper Control Limit can be determined by multiplying the estimated population standard deviation by the appropriate factor – in this case based on the sample size of 4:

$$\text{Upper Control Limit} = D \times \sigma = 4.698 \times 5.49 = 25.79\,\text{g}$$

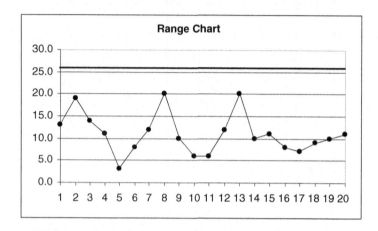

Standard deviation chart

Using the factors given in Table 15.12, the Upper Control Limit can be determined by multiplying the estimated population standard deviation by the appropriate factor – in this case based on the sample size of 4:

$$\text{Upper Control Limit} = B \times \sigma = 2.088 \times 5.49 = 11.46\,\text{g}$$

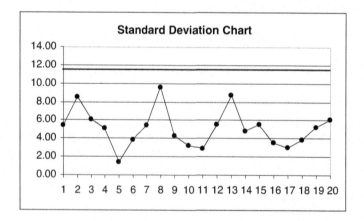

Note the similarities of the two charts for assessing the spread of the process – the range chart and the standard deviation chart. For all three charts, although the data are scattered, they are all within the action lines, suggesting that the data were all produced under similar conditions. If one or more points were beyond the action/control limits it would be appropriate to recalculate the position of the lines excluding the data from these 'out of control' points.

As indicated above, Tests 1 and 5 are rules used when both action and warning limits are used. However, an indication of any of the conditions stipulated in the tests is an indication of the presence of an unusual sequence of samples which should be investigated. In some cases, for example Test 7, the occurrence is likely to suggest an improvement in the process which, if it can be maintained, could lead to longer-term benefits.

15.4.2.2 Attribute data

Attribute data are derived for characteristics that can only be counted and not measured. A sample will therefore only generate a single value which is then plotted onto a single control chart. As described above, there are two kinds of attribute data and, based on these two types, two alternative charts can be used, but each has two types:

1 Charts showing non-conforming units (defectives):
  a number of non-conforming units (for samples of constant size) (the 'np chart')
  b fraction non-conforming (for samples of varying size) (the 'p chart')
2 Charts showing number of non-conformities (defects):
  a number of non-conformities (for samples of constant size) (the 'c chart')
  b non-conformities per unit (for samples of varying size) (the 'u chart')

Charts based on non-conforming units

Attribute data of defectives from a process in control will generate a binomial distribution. The distribution is mathematically defined on the basis of the fraction non-conforming units present in the population (p). When a sample is taken (of size n), the number of non-conforming units will vary on the basis of the binomial distribution. This distribution has a standard deviation which is calculated as:

$$\sigma = \sqrt{(n \cdot p \cdot (1 - p))}$$

This can then be used to establish action limits at $+3\sigma$ and, if required, warning limits at $+2\sigma$.

Charts based on non-conformities

Attribute data of defects from a process in control will generate a Poisson distribution. In this case the distribution is defined by the average proportion of non-conformities per unit present in the population (c). In this case the standard deviation of the distribution is calculated as:

$$\sigma = \sqrt{c}$$

As above, this can be used to establish action limits at $+3\sigma$ and, if required, warning limits at $+2\sigma$.

15.4.3 Acceptance sampling

When a batch has been manufactured or when a delivery is being made, it may be necessary to reach a decision on the suitability of the batch or delivery based on a sample taken from it. There are a number of possible approaches including:

- *One hundred percent inspection*: in this case everything is sampled. This can be very expensive and may be impossible where the sampling requires the opening of containers.
- *Sampling based on mathematical theories of probability*: in this case not everything is checked but a sample is taken. The size of the sample will have been based on probabilities and will allow the risks of incorrect decisions to be precisely calculated. The selected plan will be chosen to allow no more risk than can be tolerated.
- *Ad hoc sampling*: this case is not based on theory and only, for example, a fixed percentage or spot checking may be used. This may be considered appropriate where other quality assurance procedures are in place and the sampling is more for verification than as a formal acceptance sampling procedure.

15.4.3.1 Operating characteristic (OC) curves

An assessment of the suitability of any particular sampling plan will require the consideration of various different issues. From the mathematical perspective, the performance of a plan can be interpreted visually by the construction of an operating characteristic (OC) curve. For a particular sampling plan, each point on the curve shows the proportion of batches that may be expected to be accepted (on the y-axis) if batches of a particular percentage defective (on the x-axis) are offered for acceptance.

In an ideal sampling plan any batch that had a defect level less than a defined quality level would be accepted all the time and any batch that had a defect level greater that this level would be rejected all

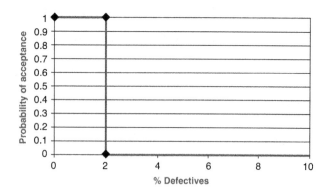

Figure 15.6 An ideal OC curve.

the time. The OC curve in this case is represented by three straight lines (see Fig. 15.6) including a vertical section at the defined quality level. This is only possible with 100% inspection. As the sample gets progressively smaller, the vertical part of the ideal curve twists to give a shallow slope, indicating an increasing risk of accepting batches with a higher number of defects and also an increasing risk of rejecting batches with a lower number of defects (which could have been accepted).

For the purposes of developing a consistent approach to acceptance sampling it is usual to define an 'acceptance quality limit' (AQL). This has been defined by ISO as the 'worst tolerable quality level'. In a note to this definition, ISO emphasises that: 'Although individual lots with quality as bad as the acceptance quality limit can be accepted with fairly high probability, the designation of an acceptance

quality limit does not suggest that this is a desirable quality level'.

Although a decision on the value for the AQL is important, it is linked to certain other important concepts which have also been defined by ISO. These are as follows and are illustrated in Fig. 15.7:

- Limiting Quality Level (LQL): 'quality level which, for the purposes of acceptance sampling inspection, is the limit of an unsatisfactory process average when a continuing series of lots is considered'.
- Consumer's Risk Point (CRP): 'point on the operating characteristic curve corresponding to a predetermined low probability of acceptance'.
- Consumer's Risk (CR): 'probability of acceptance when the quality level has a value stated by the acceptance sampling plan as unsatisfactory'.
- Producer's Risk (PR): 'probability of nonacceptance when the quality level has a value stated by the plan as acceptable'.
- Producer's Risk Point (PRP): 'point on the operating characteristic curve corresponding to a predetermined high probability of acceptance'.

For the purposes of developing suitable acceptance sampling plans, ISO has based its plans on a producer's risk of 0.05. This implies that a producer, submitting material of the quality specified by the AQL, will have a 5% chance that the delivery will be rejected (or a 95% chance that it will be accepted).

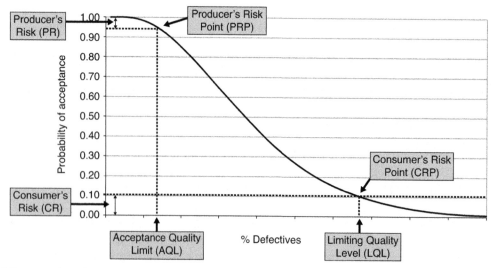

Figure 15.7 OC curve – key elements.

Although the consumer's risk is not used to establish the ISO sampling plans, in comparing the performance of different plans it can be considered that a consumer's risk of 0.1 would be appropriate. It is possible that when the OC curve for a plan is being evaluated it is found that material with the quality specified by the LQL gives a consumer's risk greater than 0.1. In this case consideration will need to be given to amending the plan, possibly by increasing the number of items in each sample.

15.4.4 ISO sampling plans

15.4.4.1 Acceptance sampling by attributes

In developing appropriate sampling systems, ISO provides a range of specific sampling plans which are linked together, by switching rules, into sampling schemes. A combination of sampling plans and sampling schemes can be collected together to create an overall sampling system.

The selected sampling scheme is usually made up of three different sampling plans which provide a higher (tightened inspection) or lower (reduced inspection) level of protection in comparison to a standard level (normal inspection). Specifically these are defined as follows:

- *Normal inspection*: 'inspection which is used when there is no reason to think that the quality level achieved by the process differs from a specified level'.
- *Reduced inspection*: 'inspection less severe than normal inspection, to which the latter is switched when inspection results of a predetermined number of lots indicate that the quality level achieved by the process is better than that specified'.
- *Tightened inspection*: 'inspection more severe than normal inspection, to which the latter is switched when inspection results of a predetermined number of lots indicate that the quality level achieved by the process is poorer than that specified'.

Switching rules which govern when the sampling plan can move from one inspection to another are defined in advance. In ISO 2859 Part 1, the rules are described and these are summarised in Fig. 15.8. Although most elements are self-explanatory, the decision to move from normal inspection to reduced inspection includes reference to a 'switching score'. Details of how this is calculated are given in ISO 2859 Part 1, but, in summary, each time a lot is accepted and meets certain conditions, the switching score is increased by either 2 or 3. If the conditions are not met, then the score is reset to zero. A minimum of 10 lots which have met the conditions will be required before any consideration is given to switching to reduced inspection. Therefore only very reliable suppliers producing to a consistently high quality are likely to be subject to reduced inspection.

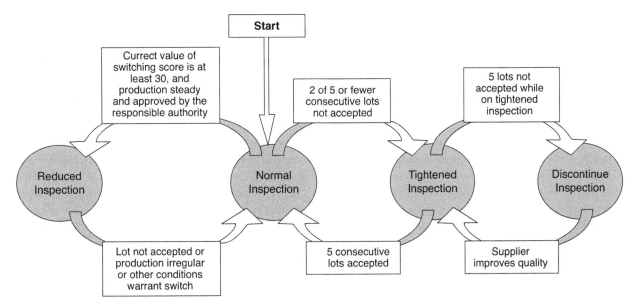

Figure 15.8 Switching rules applied to ISO sampling plans.

When using ISO sampling plans, the following sequence of steps is followed:

1 An AQL is selected which will preferably be one of the designated preferred AQLs used in ISO 2859. These values are:

0.01	0.015	0.025	0.040	0.065
0.10	0.15	0.25	0.40	0.65
1.0	1.5	2.5	4.0	6.5
10	15	25	40	65
100	150	250	400	650
1000				

The values can be either expressed as percent non-conforming items or non-conformities per 100 items.

2 ISO 2859 then provides an opportunity to select an 'inspection level'. There are three general levels (designated I, II and III) and some additional special levels (S-1, S-2, S-3 and S-4). These allow for some discrimination between different situations. The lower the inspection level, the smaller the number of samples taken, with the special levels being lower than the general levels. Level II would be regarded as the basic level unless a different level had been chosen.

3 For a given lot or batch that is to be inspected, the size of the lot/batch is then used to determine a sample size code letter.

4 The sample size code letter is used, in conjunction with the selected AQL, to determine the sample size and the acceptance and rejection numbers. Different tables are provided in ISO 2859 to allow the values of the sample size and the acceptance and rejection numbers to be obtained for normal, tightened and reduced inspection.

Single, double and multiple sampling

Another optional criterion that is built into the ISO procedures is the use of different numbers of samples to reach a decision. The simplest arrangement is to take a single sample from the lot and to base a decision on the result of that single sample in comparison to the specified acceptance and rejection numbers.

However, an alternative approach is to operate a double sampling plan in which, on the basis of a smaller initial sample, a rapid decision can be taken if the lot is identified as either very good or very bad. If, however, the lot is of an intermediate quality (close to the selected AQL) then a second sample is taken and a decision made on the basis of the combined total number of defectives in the first and second samples. If a double sampling scheme is used, the initial sample is typically only two thirds the size of the equivalent single sample. Although it may be necessary to take a second sample, making the total sampling about a third more, it will on average result in a saving in the total amount of sampling needed. The system is, however, more complex to operate and a decision will need to be taken based on the different circumstances.

The savings in inspection can be extended if an even more complex arrangement is introduced. ISO 2859 provides details for multiple sampling plans which allow for up to five separate samples to be taken. In this case each individual sample is only about a quarter of the equivalent single sample and, at its worst, the average sample size will only rise to about three quarters of the single sample. Even greater saving can therefore be made but at the expense of a more complex scheme.

An example of the establishment of an acceptance sampling plan based on the ISO tables is shown in the Box 15.5.

15.4.4.2 Acceptance sampling by variables

When sampling by variables it is possible to make use of two measures – the sample mean and the sample standard deviation. Statistically this enables a more accurate decision to be made or, alternatively, a smaller sample size can be used to give similar probabilities to those obtained when using the larger sample required when using an attribute measurement.

As with the ISO system for establishing sample plans for attributes, the ISO scheme for acceptance sampling by variables also contains many different alternatives. For example, there are schemes based around plans using sample standard deviations (the s method) or sample ranges (the R method). Alternatively, if the standard deviation is known and stable, the σ method is recommended. Another option to be considered is whether there is a single specification limit or whether there are both upper and lower specifications. In the latter case, it will be necessary to decide whether the proportion outside the specification can be applied to both limits or whether it is a combined total.

The basic concept used by ISO in establishing the criteria to accept or reject a lot is built around the concept of the normal distribution and the

(Note: access to a copy of ISO 2859 Part 1 will be needed to follow this example.)

The following sequence demonstrates the operation of ISO 2859 Part 1 to select a sampling plan based on a single sample.

1. Choice of an AQL

AQLs are selected on the basis of the characteristic under consideration. According to the 'General Guidelines on Sampling' adopted by the Codex Alimentarius Commission, the characteristics which may be linked to critical defects (for example to sanitary risks) shall be associated with a low AQL (i.e. 0.1% to 0.65%) whereas the compositional characteristics such as the fat or water content, etc. may be associated with a higher AQL (e.g. 2.5% or 6.5% are values often used for milk products)'. For this example a compositional characteristic will be assumed with an AQL of 2.5%.

2. Choice of inspection level

As, in the absence of any other reason, the standard inspection level suggested by ISO is II, Level II will be used in this example.

3. Determination of sample size code letter

The Sample Size Code Letter is based upon the size of a lot under consideration and the selected inspection level. In this case, it will be assumed that the lot size is 500. This generates a sample size code letter of H.

4. Determination of sample size and the acceptance and rejection numbers

Based on the above, the following sample scheme is then recommended by ISO:

Inspection	Sample size	Acceptance number	Rejection number
Normal	50	3	4
Tightened	50	2	3
Reduced	20	2	3

The following are the OC curves based on this scheme:

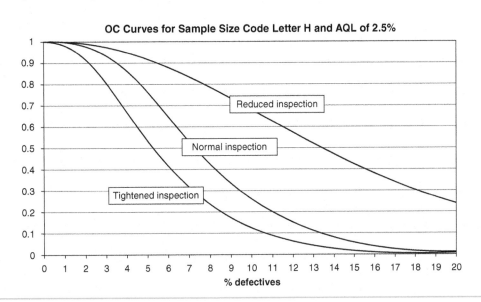

probabilities that a sample will represent the whole lot. It again uses the ideas described earlier of the proportion of a normal distribution lying beyond a certain value. The assessment of a sample therefore calculates a value known as the 'quality statistic' which has close similarities to the value of z used in earlier calculations (see Box 15.2). This quality statistic is then compared to an 'acceptability constant' selected from the relevant table in the ISO standard. The value of the acceptability constant incorporates the same allowances as that described earlier for attributes – a lot that is just at the AQL should only have a probability of 0.05 of being rejected.

In summary, the procedure specified by ISO for a variable acceptance sampling plan is as follows:

1 Before starting, check:

That the distribution can be considered to be normal and that production is considered to be continuous.

Whether the 's' (or 'R') method is to be used initially or whether the standard deviation is stable and known, in which case the 'σ' method should be used.

That the inspection level to be used has been designated. If none has been given, inspection level II shall be used.

That the AQL has been designated and that it is one of the preferred AQLs for use with the standard. Preferred AQLs in the ISO standard are:

0.1 0.15 0.25 0.4 0.65 1 1.5 2.5 4 6.5 10

If a double specification limit has to be met, whether the limits are separate or combined and, if the limits are separate, whether AQLs are determined for each limit.

2 Obtain the sample size code letter using Table I-A in ISO 3951.

3 From the relevant table (Tables II or III in ISO 3951) obtain the sample size (n) and acceptability constant (k).

4 Taking a random sample of this size, measure the characteristic x in each item and then calculate the mean (x) and the sample standard deviation (s). If the mean is outside the specification limit, the lot can be judged unacceptable without calculating s.

5 If single specification limits, either upper (U) and/or lower (L), are given, calculate the quality statistic:

$$Q_U = (U - x)/s \quad \text{and/or} \quad Q_L = (x - L)/s$$

6 Compare the quality statistic with the acceptability constant (k). If the appropriate quality statistic is greater than or equal to the acceptability constant, the lot is acceptable; if less, it is not acceptable.

i.e. If Q_U or $Q_L \geq k$ accept

If Q_U or $Q_L < k$ reject

Switching rules operate in this standard in a similar way to the attribute standard (see Fig. 15.8) although there are some slight differences. When switching to a tightened inspection, usually only the acceptability constant changes, not the sample size. Reduced inspection usually provides for a smaller sample and decreased values of k.

An example of the operation of a variable plan is given in Box 15.6.

15.4.5 Weight control

Perhaps the most common use of statistical techniques in food production is with weight control or, for liquids, volume control. There are sound business reasons for this emphasis. From an economic perspective, providing excessive weight puts a business at a competitive disadvantage. From a legal perspective, failure to put sufficient product in a package could mean non-compliance with weight control legislation. A manufacturer therefore will attempt to put the minimum necessary in a pack whilst still meeting legal obligations.

As discussed above, product from a manufacturing process tends to have weights that display a normal distribution. When a decision is taken to label a product with a certain weight, the resulting controls will need to take this into account. It is possible for legislation to require all packs to contain at least the weight stated on the pack (a minimum weight system). However it is also common practice for the legal requirements to incorporate statistical procedures and to recognise that some packs may be lower than the stated weight so long as, on average, the consumer obtains the stated weight (an average weight system).

The discussion that follows is based around the concepts used within the EU to establish a legal control system based on average weights (or volumes).

15.4.5.1 The European average weight system

Although the basic concept is that the weight of any pack shall, on average, be at least the stated weight,

Box 15.6 Operation of a variable sampling plan

(Note: access to a copy of ISO 3951 Part 1 will be required to follow this example.)
  The following are assumed:

- An AQL of 2.5%.
- The selected inspection level will be II.
- The lot of size is 500.
- A single sample will be used.
- The 's' method will be used.
- The Lower Specification Limit (L) is 12.5.

Determination of sample size code letter

The sample size code letter is based upon the size of a lot under consideration and the selected inspection level. With the above assumptions, the sample size code letter is I.

Determination of sample size

Using the 's' method, the sample size code letter (I) gives a sample size of 25 (for normal and tightened inspection) and 10 (for reduced inspection).

Determination of acceptability constant (k)

With an AQL of 2.5%, the acceptability constants are as follows:

  Normal inspection: 1.53 ($n = 25$)
  Tightened inspection: 1.72 ($n = 25$)
  Reduced inspection: 1.23 ($n = 10$)

Example

Assuming that 25 samples are taken and the sample mean (x) is found to be 13.3 and the sample standard deviation (s) is found to be 0.49. Calculation of the quality statistic gives:

$$Q_L = (x - L)/s = (13.3 - 12.5)/0.49 = 1.63$$

If the sample had been taken under normal inspection, since Q_L (1.63) $\geq$ k (1.53), the lot would be accepted. However, if the sample had been taken under tightened inspection, since Q_L (1.63) < k (1.72), the lot would be rejected.

it is also necessary to provide additional controls so that the actual spread of weights is not so great that consumers get excessively light (or excessively heavy) packs.

The system therefore includes provisions that recognise that there will be spread but that this must be controlled to within certain limits. Although the legislation does not specify absolute criteria, the legal requirement is that packed product shall be capable of passing a 'reference test' conducted by inspectors. The procedures for the reference test are laid down

in the legislation and are based on detailed statistical criteria.

As the reference test is carried out on a batch of packed product, it is not an appropriate technique to be used by manufacturers to assess and control their actual packing performance. What is required is a system that, when correctly implemented, will enable the manufacturer to be confident that a reference test will be passed if an inspector should conduct one on any batch of packed product.

By using the same statistical requirements that were used to establish the criteria for the reference test, it is possible to provide guidance to the manufacturer. This guidance is based on the 'Three Rules for Packers' which are as follows:

1 The actual quantity of the packages shall be not less, on average, than the nominal quantity.
2 Not more than 2.5% of the packages may be nonstandard.
3 No package may be inadequate.

Since the statistics of the reference test assume a normal distribution, Rule 3 is not mathematically possible. The statistics therefore use a value of 0.01% as the tolerance for inadequate packs. However, if such a package is found, its sale is not permitted by the legislation.

The interpretation of the Three Rules requires an understanding of certain additional terms (see example in Fig. 15.9):

■ *Nominal quantity (nominal weight or nominal volume* (Q_n): the quantity indicated on the package, i.e. the quantity of product that the package is deemed to contain.
■ *Target quantity (target weight or target volume)* (Q_t): the quantity selected by the packer to be the intended average quantity produced on the packing line.
■ *Actual contents*: the quantity (weight or volume) of product that it in fact contains. In all operations for checking quantities of products expressed in units of volume, the value employed for the actual contents shall be measured at or corrected to a temperature of 20°C, whatever the temperature at which packaging or checking is carried out. However, this rule shall not apply to deep frozen or frozen products, the quantity of which is expressed in units of volume.
■ *Negative error*: the quantity by which the actual contents of the package are less than the nominal quantity.
■ *Tolerable negative error* (TNE): an amount, specified in the legislation and based upon the nominal quantity, that defines the permitted spread of the production.
■ *Tolerance Limit 1* (T_1): the nominal quantity minus one tolerable negative error.
■ *Non-standard packages*: packages that have a negative error greater than the tolerable negative error (i.e. their quantity is below T_1).
■ *Tolerance Limit 2* (T_2): the nominal quantity minus one tolerable negative error.
■ *Inadequate packages*: packages that have a negative error greater than twice the tolerable negative error (i.e. their quantity is below T_2).

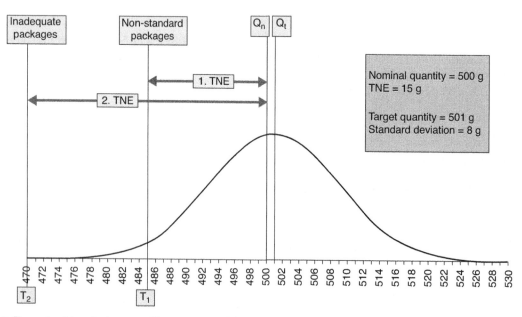

Figure 15.9 Example of terminology used for average weight control.

Detailed consideration will need to be given to a wide range of factors involved with a production line. The aim will be to identify a target quantity that needs to be established for a product line. Linked to this quantity there will be a control system based, for example, on Shewhart control charts. The combination of a target quantity and control system should give the producer confidence that a reference test, if conducted, will be satisfied.

The following sets out, for weight control, a sequence of steps which can be considered as the minimum necessary to identify an appropriate target weight (summarised from DTI, 1979):

1 *Collect data on the production process.*

Samples/sample size n will need to be collected over several days or shifts. For example, over a period of 3 days, from the end of a production line collect 40 sets (k) of 5 packages (n) (giving a total number of $k.n = 200$). The weight of each package is determined (x).

2 *Calculate the characteristics of the data.*

For each set of n results, calculate the standard deviation (s) and the average ($\bar{x}$). From the results:

Calculate the short-term variability, s_o:

$$s_o = \sqrt{\left(s_1^2 + s_2^2 + s_3^2 + s_4^2 + \cdots + s_n^2\right)/k}$$

Calculate the grand average of the means, $\bar{x}$:

$$\sum(\bar{x})/k$$

Calculate the standard deviation of all individual observations. The result is a measure of the medium-term variability, s_p.

3 *Interpret the results.*

Test A: Check for normality:

Calculate $\bar{x} - 2\,s_p$; if more than 4 out of 200 fall below this level, the indication is that the distribution is not normal.

Calculate $\bar{x} - 3.72\,s_p$; no quantity should be below this value.

If the process is not normal distributed, more detailed studies will be required.

Test B: Stability of average:

Divide s_p by s_o; for $n = 5$ and $k = 40$, if the result is more than 1.056, then the average is probably unstable and s_p should be used instead of s_o in setting the target quantity. For other values of n and k, see Table 15.13.

Test C: Comparison of variations with TNE:

Check to see if the value of s_p is equal to or less than TNE/2. If it is, then the packer should have little trouble in complying with Packers Rules 2 and 3 and Q_t can be set close to Q_n.

If the value of s_p is between TNE/2 and TNE/1.86, then there is a likelihood that the output will break Rule 2 (i.e. contain an unacceptable number of non-standard packages). The target quantity (Q_t) should be at least $2.s_p$ − TNE above the nominal quantity (Q_n).

If the value of s_p exceeds TNE/1.86, then the output will also break Rule 3 (i.e. contain an unacceptable number of inadequate packages). The target quantity (Q_t) should be at least $3.72.s_p$ − 2TNE above the nominal quantity (Q_n).

Table 15.13 Factors used in determining stability of average.

Number of samples (k)	Number of items per sample (n)									
	2	3	4	5	6	8	10	12	15	20
20	—	—	—	—	—	—	1.038	1.031	1.024	1.0181
25	—	—	—	—	—	1.044	1.035	1.028	1.022	1.0164
30	—	—	—	—	—	1.039	1.030	1.025	1.020	1.0145
35	—	—	—	—	1.048	1.035	1.028	1.023	1.0179	1.0133
40	—	—	—	1.056	1.045	1.033	1.026	1.021	1.0167	1.0124
50	—	—	1.065	1.049	1.040	1.029	1.023	1.0187	1.0147	1.0109
60	—	—	1.059	1.045	1.037	1.027	1.021	1.0174	1.0138	1.0102
70	—	1.077	1.053	1.041	1.033	1.024	1.0190	1.0156	1.0124	1.0092
80		1.071	1.050	1.038	1.031	1.023	1.0178	1.0147	1.0116	1.0086
100	1.114	1.064	1.044	1.034	1.028	1.020	1.0161	1.0133	1.0105	1.0078

4 *Deciding on the target quantity*

The target quantity is established by considering a number of factors. The most important is the nominal quantity to be used. However, additional factors will need to be considered:

Process variability including wandering average: this is the issue dealt with in Test C of Step 3.

Sampling allowance: if, in any production period, only a small number of items will be checked (i.e. less than 50) then an allowance will be needed. The actual amount will be a function of the process variability and the type of control system being used (e.g. Shewhart control charts with action and warning limits).

Storage allowance: where a product is known to lose weight during storage (e.g. bread might lose moisture), this needs to be taken into account.

Tare variability allowance: if the control system is based on the weighing of packed product, it is likely to assume a standard packaging weight. An allowance will be needed if the packaging weight varies.

The target quantity will then be calculated:

$$Q_t = Q_n + \text{process variability allowance} + \\ \text{sampling allowance} + \text{storage allowance} \\ + \text{tare variability allowance}$$

Further reading and references

Codex Alimentarius Commission (2007) *Procedural Manual*, 17th edn. WHO/FAO, Rome.

DTI (1979) *Code of Guidance for Packers and Importers*. Department of Trade and Industry, HMSO, London.

FAO (2005) *Perspectives and Guidelines on Food Legislation, with a New Model Food Law*. FAO Legislative Study 87. FAO, Rome

FAO (2006) *Strengthening National Food Control Systems: Guidelines to Assess Capacity Building Needs*. FAO, Rome.

FAO/WHO (2003) *Assuring Food Safety and Quality – Guidelines for Strengthening National Food Control Systems*. FAO Food and Nutrition Paper 76. FAO, Rome

FAO/WHO (2006) *Food Safety Risk Analysis: A Guide for National Food Safety Authorities*. FAO Food and Nutrition Paper 87. FAO, Rome.

Hubbard M.R. (2003) *Statistical Quality Control for the Food Industry*, 3rd edn. Springer, Berlin.

Institute of Food Science and Technology (2006) *Food and Drink: Good Manufacturing Practice – A Guide to its Responsible Management*, 5th edn. IFST, London.

Oakland, J.S. (2008) *Statistical Process Control*, 6th edn. Butterworth-Heinemann, Oxford.

Websites

British Retail Consortium (BRC):
http://www.brc.org.uk/

Codex Alimentarius Commission (CAC):
http://www.codexalimentarius.net/

Food and Agriculture Organisation (FAO):
http://www.fao.org/

GLOBALGAP: http://www.globalgap.org/

Institute of Food Science and Technology, UK (IFST):
http://www.ifst.org/

International Organization for Standardization (ISO):
http://www.iso.org/

World Health Organisation (WHO):
http://www.who.int/

World Trade Organisation (WTO):
http://www.wto.org/

Supplementary material is available at
www.wiley.com/go/campbellplatt

Regulatory toxicology

Gerald G. Moy

Key points

- The presence of potentially toxic chemicals in food is a global public health concern and most governments use regulatory toxicology to address these potential hazards.
- Regulatory toxicology characterizes hazards by evaluating animal and other studies and, when possible, establishes safe or tolerable levels based on appropriate levels of protection.
- Regulatory toxicology also estimates the likely dietary intake of chemicals by the population to assure that safe or tolerable levels are not exceeded.
- As chemicals arise in food from a number of different sources, regulatory toxicology approaches are different for various categories of chemicals, such as food additives, veterinary drug residues, contaminants, natural toxicants and adulterants.

16.1 Introduction

The presence in food of chemicals at potentially toxic levels is a public health concern worldwide. Contamination of foods may occur through environmental pollution of the air, water and soil, such as the case with toxic metals, polychlorinated biphenyls (PCBs) and dioxins. The intentional use of various chemicals, such as food additives, pesticides, veterinary drugs and other agro-chemicals can also pose hazards if such chemicals are not properly regulated or appropriately used. Other chemical hazards, such as naturally occurring toxicants, may arise at various points during food production, harvest, storage, processing, distribution and preparation. Furthermore, accidental or intentional adulteration of food by toxic sub-

stances has resulted in serious public health incidents in both developing and industrialized countries. For example, in Spain in 1981–82, adulterated cooking oil killed some 600 people and disabled another 20,000, many permanently (WHO, 1984, 1992). In this case, the agent responsible was never identified in spite of intensive investigations.

Over the past 50 years, the widespread introduction of chemicals in agriculture and in food processing has resulted in a more abundant and arguably safer food supply. To protect consumers, most governments have adopted a risk assessment paradigm, including regulatory toxicology, to scientifically estimate the potential risk to human health posed by chemicals in food. While risk assessment methods have been to a great extent harmonized, risk management approaches will necessarily vary depending on whether the chemical is intentionally added to the food supply or is present as the result of unavoidable or natural contamination. In addition, the choice of a risk management option may vary among countries depending on their desired level of health protection and technical, economic, sociocultural and other factors. In a number of cases, these differences have resulted in disruption of international food trade.

16.2 Regulatory toxicology

The regulatory control of potentially toxic chemicals in food is an essential responsibility of governments and their food control agencies. Up-to-date food legislation and enforcement, including monitoring programs, must support the work of those agencies. Much progress has been made in protecting the consumer from chemical hazards. However, with the incorporation of risk analysis principles into the development of international standards, it is becoming increasingly clear that risks must be approached in a more methodical and harmonized manner. Codex defines risk analysis as consisting of risk assessment, risk management and risk communication. Each of these has been the subject of an FAO/WHO expert consultation (FAO/WHO 1995, 1997, 1998). One of the outcomes of these consultations was the recognition that risk must be characterized more precisely and transparently by the scientific committees than has been done in the past. In addition to long-term risks, consumption of one meal or over one day of certain substances may pose acute risks. Examples include organophosphorus pesticides, pharmacologically active veterinary drugs and certain mycotoxins. Methods for evaluating these risks have been developed during the last few years, but more work needs to be done in this area.

To assist in the risk assessment process, a risk assessment paradigm has been developed consisting of four components, namely hazard identification, hazard characterization, exposure assessment and risk characterization. Hazard identification conducts a preliminary review of a chemical's inherent properties that may cause adverse health effects at likely levels of exposure. This preliminary assessment provides the basis for deciding if a full risk assessment is necessary. Hazard characterization assesses the dose-response relationship between expo-

sure and the onset of adverse effects and is used to establish a level of exposure that is considered acceptable or tolerable. Exposure assessment examines the actual or predicted exposure levels to the substance and risk characterization summarizes the risk assessment findings, including attendant uncertainties. These components are described in more detail below.

16.2.1 Hazard identification

The first step in a risk assessment process is hazard identification. Obviously, a chemical that produces toxic effects shortly after ingestion of small amounts can be easily identified as a hazard. However, at the beginning of the previous century, the potential long-term health risks posed by the many chemicals that were added to food were not well recognized. For example, the safety of benzoic acid and boric acid were hotly debated by physicians and scientists in the USA in the early 1900s. Today in most countries, most food safety legislation requires that any chemical intentionally added to food be shown to be safe through appropriate tests. Consequently, hazard identification has become automatic in the case of food additives, pesticides and veterinary drugs and such chemicals are required to undergo a risk assessment before they can be marketed.

For contaminants and other chemicals not intentionally added to food, hazard identification is not always a straightforward matter. For many contaminants, there is a paucity of reliable data on which to identify hazards. Often information on the toxic nature of contaminants will result from intoxication incidents caused by exposure to high levels of a chemical, as in the cases of methyl mercury and PCBs. However, the toxic effects of low-level, long-term exposure to many toxic chemicals often remain undetected and unknown. Each year around 1500 new chemicals are brought onto the market, adding to the approximately 70,000 existing ones. The United Nations Environment Program (UNEP) has estimated that production of chemicals is likely to increase by 85% over the next 15 years. In addition, there are an almost limitless number of naturally occurring toxicants that may be present in food. The presence of some toxicants in food has been revealed by improved analytical methods, which have steadily become more sensitive over the years.

Based on the results of hazard identification, risk managers may decide that the potential health risks warrant the completion of the risk assessment paradigm. The next step is hazard characterization, which may be considered the core of regulatory toxicology.

16.2.2 Hazard characterization

Hazard characterization may include the review carried out under the hazard identification component and it is perhaps worth pointing out here that the components of risk assessment are iterative and may even overlap. For example, hazard characterization and exposure assessment are often conducted simultaneously. The toxicological evaluation of a chemical for regulatory purposes is based upon a principle first identified in the sixteenth century by Philippus Paracelsus, a Swiss alchemist and physician. He recognized that "all things are toxic, it is only the dose which makes something a poison". The corollary to this is also true, namely that "all things are safe, it is only the dose which makes something innocuous". Besides understanding the dose-response characteristics of a chemical, the practical goal of regulatory toxicology is to find the dose at which no adverse health effects are expected to occur.

To accomplish this, hazard characterization relies on standard toxicity tests performed according to internationally accepted protocols such as those published by the Organization for Economic Cooperation and Development. Hazard characterization considers the dose levels at which no adverse effects occur in order to establish an intake level considered to be acceptable (intentionally used chemicals) or tolerable (contaminants and naturally occurring chemicals). The standard reference value used at the international level to indicate the safe level of intake of an intentionally used chemical is the "acceptable daily intake" (ADI) which is the estimate of the amount of a substance in food and/or in drinking water, expressed on a body weight basis, that can be ingested daily over a lifetime without appreciable health risk to the consumer.

For contaminants and naturally occurring chemicals, the corresponding reference intake value is the "provisional tolerable intake", which can be expressed on a daily, weekly or monthly basis. The tolerable intake is referred to as "provisional" since there are often insufficient data on the consequences of human exposure at low levels, and new data may result in a change to the tolerable level. For contaminants that may accumulate in the body over time,

such as lead, cadmium and mercury, the provisional tolerable weekly intake (PTWI) is used as a reference value in order to minimize the significance of daily variations in intake. For contaminants that do not accumulate in the body, such as arsenic, the provisional tolerable daily intake (PTDI) is used. For dioxins and dioxin-like PCBs, the reference is the provisional monthly tolerable intake (PTMI), which was used to emphasize the long half lives of these chemicals in the body. These tolerable intakes are primary health standards that apply to total ingested exposure.

At the international level, two joint FAO/WHO committees have, over a period of four decades, evaluated thousands of chemicals found in food. The Joint FAO/WHO Expert Committee on Food Additives (JECFA) evaluates food additives, contaminants and veterinary drug residues, and the Joint FAO/WHO Meeting on Pesticide Residues (JMPR) evaluates pesticide residues. The principles for the safety evaluations carried out by JECFA for food additives and contaminants are described in Environmental Health Criteria 70 published by WHO. Principles for the toxicological assessment of pesticide residues in food are described in Environmental Health Criteria 104. The use of animal studies for assessing toxicity and their consideration for public health purposes are also described by a number of national agencies. In most cases, a tiered approach is used based on the potential toxicity of the chemical and the likely levels of exposure of the population. For chemicals of lowest concern, only a genotoxicity study and short-term toxicity study in the rodent would be required. For chemicals of intermediate concern, a genotoxicity study, two subchronic toxicity studies in the rodent and non-rodent, a developmental/teratogenicity study and metabolism and toxicokinetics study would need to be performed. For chemicals of highest concern, chronic toxicity and carcinogenicity studies in two rodent species, usually the mouse and rat, would need to be performed in addition to those required for chemicals of intermediate concern. Other specific studies may also be required depending on the nature of the chemical and the specific requirements of the national authority, including neurotoxicity, immunotoxicity and acute toxicity studies.

The ability of certain chemicals to cause endocrine disruption in environmentally exposed animals is well documented and their potential health effects in humans are receiving great attention. Developmental neurotoxicity has not been evaluated for many chemicals and it is recognized that immunotoxicity may occur at levels previously thought to produce no adverse effects. The growing rates of breast cancer in women, testicular cancer in men and brain cancer in children all suggest that further research is needed to rule out the possible contribution of chemicals in food to these diseases.

16.2.3 Exposure assessment

Estimation of exposure to contaminants depends on knowledge of the level of the contaminant in the food coupled with knowledge of the amount of each food consumed, though there is a degree of uncertainty associated with both of these parameters. The level of contamination of food is influenced by a variety of factors such as geographical and climatic conditions, agricultural practices, local industrial activity and food preparation and storage practices. The level of contamination of food, as consumed, can be determined from food monitoring data when available. Different methods of dietary intake modeling combine data on contaminant levels in food with food consumption data in different ways to provide estimates of the daily, weekly or monthly dietary exposure. The models may be either deterministic or probabilistic. In the deterministic model, the mean is usually calculated, while with the probabilistic model, a more complete picture of the distribution of intakes is calculated to take into account all sections of the population for which food consumption data are available. In some models, distributions of levels of contaminants are also used.

Another approach is to use total diet surveys to estimate the exposure of selected contaminants in food. These surveys are a direct measure of the level of dietary exposure to contaminants in food as consumed. While cost-effective, these surveys cannot be used for chemicals that sporadically contaminate the food supply, such as aflatoxins. In this regard, the Global Environment Monitoring System/Food Contamination Monitoring and Assessment Program (GEMS/Food) of WHO collects, collates and disseminates information on the levels of contaminants in food and on time trends of contamination, enabling preventive and control measures. Data from GEMS/Food and from surveys undertaken in industrialized countries suggest that the food supply in developed countries is largely safe from the chemical viewpoint because of the extensive food-safety

infrastructure (i.e. legislation, enforcement mechanisms, surveillance and monitoring programs) and the cooperation of the food industry. However, data from developing countries are largely lacking. Accidental contamination or adulteration does occur in both industrialized and developing countries. Such contamination causes international concern because of extensive media coverage and disruption of global trade.

For intentionally added chemicals, methods have been developed to predict the likely exposure to populations if and when the chemical is permitted on the market. In the case of pesticides and animal drugs, exposure assessments are linked to recommendations on the maximum residue levels (MRLs) based on good agricultural and good veterinary practices, respectively. Using model diets and assuming maximum treatment doses and coverage, estimates of exposure are calculated and compared to the established ADI. If the exposure exceeds the ADI, further refinement of the exposure assessment is considered, for example processing factors, such as cleaning, peeling and cooking. In the case of veterinary drug residues, longer withholding periods from the time of administration of the drug to the time of marketing are considered.

For food additives, various screening methods have been devised based on the proposed maximum use levels (ML) and the foods or food categories in which the additive will be used. For some additives, such as flavors, the exposure is based on per capita production and assuming that 10% of the population consumes the chemical. If these screening methods indicate that the ADI may be exceeded, national exposure assessments are consulted because these offer more precise estimates of consumption and use levels in specific food.

16.2.4 Risk characterization

Risk characterization brings together the information on the level of exposure to the chemical for various population groups and compares this with the reference values for health effects. If the exposure does not exceed the reference value, then it is assumed that the chemical does not pose a health concern. In the case of contaminants and natural toxicants, this might be expressed in terms of a margin-of-safety between the tolerable level of intake and the known level of human exposure via the diet. This information allows a decision to be made regarding appropriate regulatory action, such as the issuance of a food standard, MRL or ML for a particular chemical.

16.2.5 Risk management

Based on the recommendations of JECFA and JMPR, the Joint FAO/WHO Codex Alimentarius Commission and its member governments may establish international food standards, guidelines and other recommendations. Since its inception in 1963, Codex has adopted more than 240 commodity standards, 3500 MRLs for various pesticide and veterinary drug/commodity combinations, 780 food additive standards and 45 codes of hygienic or technological practice. The World Trade Organization refers to Codex standards, guidelines and recommendations in the arbitration of trade disputes involving health and safety requirements. However, this is a necessary, but insufficient condition for assuring the safety of the food supply from potentially hazardous chemicals.

After appropriate legislation has been adopted, the production of food in which added chemicals are used must be within the limits prescribed by health-based legislation. For this purpose, the primary industry (producers of agricultural, animal and fishery products) and the processing industries have to comply with laws and regulations and must observe the principles of good agri/aquacultural, animal husbandry and manufacturing practices. Chemicals, such as preservatives, can help prevent spoilage and pathogenic microorganisms, but their use must strictly adhere to the law. Efforts to reduce the use of potentially toxic chemicals, such as integrated pest management, should be promoted.

In addition to industry compliance with legal requirements, the application of technologies that can prevent or reduce the use of chemicals in food, for example, by drying crops to prevent mold growth and thus the production of mycotoxins in food during storage, are encouraged. Food irradiation can replace the use of potentially harmful chemicals for insect disinfestations, sprout inhibition, microbial reduction and fumigation. Modern biotechnology also offers the possibility of reducing the need for chemicals, in particular insecticides, with potential health and environmental benefits.

Finally, monitoring programs for chemicals in food are necessary to verify industry compliance, assess the impact of interventions and identify unsafe foods. In addition to the monitoring of contaminants, infor-

mation is also needed on the total dietary exposure of the population and vulnerable subpopulation groups to ensure that public health is protected and to reassure the public that the food supply is safe.

16.3 Chemical hazards in food

For the purpose of this chapter, the reader is referred to the definitions of the Codex Alimentarius Commission, which are used to describe the various chemicals found in food. A key definition of Codex is its definition of "contaminant", which includes the important phrase *"not intentionally added to food"* (Codex, 2006):

"Any substance not intentionally added to food which is present in such food as a result of production (including operations carried out in crop husbandry, animal husbandry and veterinary medicine), manufacture, processing, preparation, treatment, packing, packaging, transport or hold of such food or as a result of environmental contamination. The term does not include insect fragments, rodent hairs and other extraneous matters."

Consequently, this definition excludes food additives and pesticide and veterinary drug residue chemicals. Certain pesticides, such as DDT, which are no longer intentionally applied to crops but are still found in foods, would be regarded as "contaminants". However, compounds that occur naturally in foods derived from plants are not considered to be contaminants as they are inherent components of the plants. Such substances are referred to by Codex as "natural toxicants". The various definitions are important in the context of national food safety legislation as they often determine regulatory requirements. In the context of this chapter, however, the Codex definitions are used as convenient reference points to facilitate communication.

16.3.1 Food additives

Food additives comprise a large and varied group of chemicals, which have a long history of use or are thoroughly tested to assure their safety prior to marketing. They are added to food to improve keeping quality, safety, nutritional quality, sensory qualities (taste, appearance, texture, etc.) and certain other properties required for processing and/or storage. Food additives evaluated by JECFA and used in accordance with Codex recommendations are considered to present no appreciable risk to health. Some traditional measures, such as salt curing and smoking are, however, considered to be risk factors for certain diseases, for example, hypertension and some cancers. If possible, other methods of preservation should be used (WHO, 1990a). The illegal use of banned or unapproved food additives, such as boric acid and textile dyes, continues to be a problem in many developing countries.

16.3.2 Veterinary drug residues

Veterinary pharmaceuticals have been a key element in increasing the production of animal-derived foods. Vaccines and therapeutic drugs are essential to protect the health of confined animals, which are under more stress and are more at risk of communicable diseases. Antibacterial drugs are also given to animals in less than therapeutic doses to promote weight gain and to improve feed efficiency. The use of antibiotics in this way has, however, contributed to problems with antibiotic-resistant microorganisms (Shah et al., 1993; WHO, 1995). As a means of intensifying meat production, use is also made of hormonal anabolic agents. These are widely used in some parts of the world for promoting growth, especially in ruminants. The application of such anabolics can yield a net increase in muscle meat of 5–10% or more. In the past decade, new hormones have been developed for other purposes such as an increase in milk production. Made by modern biotechnology, species-specific purified protein pharmaceuticals, most notably bovine somatotropin (BST), may be particularly important in developing countries. JECFA has evaluated the safety of many veterinary drug residues, including several anabolics and BST, and concluded that, under good agricultural and veterinary practices, such substances do not present an appreciable risk to the consumer (WHO, 1998a, 2000a). Consequently, the Codex Alimentarius Commission has adopted numerous MRLs for veterinary drug residues, including several anabolics. However, BST has not yet been approved by Codex because of risk management concerns due to factors not related to food safety.

The estimation of the risks associated with residues from illegal and uncontrolled treatment of food animals with veterinary drugs is, however, a different

matter. Continued monitoring is therefore necessary to ensure that only approved veterinary drugs are used at permitted doses, and withdrawal periods, if defined, are observed.

16.3.3 Pesticide residues

With regard to pesticides, laboratory animal studies and accidental contamination of food, as well as occupational and intentional exposure to pesticides, have provided evidence that these chemicals may cause serious health problems following excessive exposure. The reported effects range from acute fatal poisoning to neurotoxicity, immunotoxicity, teratogenicity and carcinogenicity. Poor nutrition and dehydration can further aggravate these effects and, therefore, lower the toxic threshold of certain pesticides (WHO, 1990b).

For these reasons, the use of good agricultural practices is extremely important when these substances are employed. In a number of situations, foods have been found to contain high levels of pesticide residues, for example, when the crops had been harvested too soon after applications of pesticides or when excessive amounts of pesticides had been applied.

In industrialized countries, there is little evidence that approved pesticides, used according to good agricultural practices, have caused harm to human health. In most recorded cases where food had been implicated in pesticide poisoning, the contamination was due to inappropriate or illegal use. Sometimes, food had been contaminated because of accidental contact with pesticides during storage or transport. In other cases, seeds for planting that had been treated with fungicides were inadvertently consumed. However, potential acute effects of certain pesticides, particularly in children, warrant further study.

In developing countries, poor food safety infrastructures have hindered an accurate assessment of the problem of pesticides in food, although acute outbreaks are reported periodically in the media. Indirect information suggests that consumers may be frequently exposed to high levels of pesticides in their diets. For example, the very large number of acute poisonings among agricultural workers implies a poor knowledge of the handling and application of pesticides. Indeed, monitoring data on food imported from developing countries by industrialized countries indicate that these foods are sometimes highly contaminated at their points of origin. Information on organochlorine pesticide residues in the breast milk of women in developing countries is further evidence of significant cumulative exposure to these chemicals (WHO, 1998b). Further assessments are needed in view of the acute and chronic health hazards involved.

16.3.4 Contaminants

A number of chemical substances may occur in the food supply as a result of environmental contamination. Their effects on health may be extremely serious and have caused great concern in past years. Serious consequences have been reported when foods contaminated with toxic metals such as lead, cadmium or mercury have been ingested.

16.3.4.1 Lead

Lead affects the hematopoietic, nervous and renal systems. It has been shown to reduce mental development in children in a dose-related manner with no apparent threshold. When lead pipes or lead-lined water storage tanks are used, the lead exposure from such water may be appreciable. Similarly, processed food and beverages may be contaminated by lead in pipes or other equipment. Food packed in lead-soldered cans may also contain significant amounts of lead. During recent years, many countries, especially industrialized ones, have initiated efforts to reduce lead in drinking water systems and food equipment and containers and these efforts have led to a significant decrease in lead exposure (WHO, 1988a, 2000b). The elimination of lead-containing additives in gasoline resulted also in the elimination of lead contamination of foods grown along motorways as well as the reduction of airborne exposure.

16.3.4.2 Mercury

Methylmercury, the most toxic form of mercury, has been shown to have serious effects on the nervous system, which in severe cases may be irreversible. The fetus, infants and children are particularly sensitive (WHO, 2000b). The tragic incident of methylmercury intoxication in Minamata Bay, Japan, in the late 1950s was caused by industrial discharge of mercury-containing compounds that were subsequently taken up by fish and shellfish. Fish are usually the major dietary source of methylmercury, although certain marine mammals can also contain high levels.

Therefore, several countries have recommended that pregnant women restrict their intake of certain predatory fish and marine mammals in order to protect the developing fetus (Rylander and Hagmar, 1995).

While mercury occurs naturally in the environment, the level of mercury in fish may be influenced by industrial pollution. For instance, in the mid-1960s it was found in Sweden that the use of mercury compounds in the paper and pulp industry, as well as other industrial discharges of mercury compounds into the environment, significantly increased the level of methylmercury in freshwater and coastal fish. Following a series of interventions, including prohibition of the use of phenyl mercury acetate in the wood industry and alkyl mercury in agriculture, the level of mercury contamination gradually decreased (Oskarsson *et al.*, 1990).

16.3.4.3 Cadmium

Cadmium is a naturally occurring contaminant that often is present in certain volcanic soils. Cadmium also occurs as an industrial pollutant and at high levels in bird guano fertilizers. The first case of mass cadmium poisoning was documented in Japan in 1950 and was known as Itai-itai disease (literally ouch-ouch disease) because of the severe pain in the joints and the spine that it produces. However, long-term low-level exposure to cadmium is associated with renal failure. While molluscs and especially oysters contain the highest levels of cadmium, consumption of grains results in the highest exposure. Presently, the estimated total dietary intake of cadmium is about 40–60% of the provisional tolerable weekly intake, which JECFA has established at 7 μg/kg body weight/week (WHO, 2006).

16.3.4.4 Polychlorinated biphenyls

Other environmental chemicals of interest are polychlorinated biphenyls (PCBs) used in varied industrial applications. Information on the acute effects of PCBs in humans has been obtained from two large-scale incidents that occurred in Japan (1968) and Taiwan (1979) after the consumption of contaminated edible oil. In the first case, rice oil was contaminated with PCBs due to a leaking pipe in a factory plant (Howarth, 1983). Experience from these outbreaks showed that, as well as their acute effects, PCBs may have carcinogenic and other long-term effects. Levels of PCBs in adipose tissue of women in the United States have been correlated with developmental and behavioral deficits in their infants (Jacobson *et al.*, 1990). Drastic restrictions in the production and use of PCBs have been introduced in many countries since the 1970s (WHO, 1988a). In 1998, animal feed contaminated with PCBs, including high amounts of dioxins (see below), resulted in a wide-scale contamination of poultry, eggs and, to some extent, beef in Belgium. However, no acute effects in humans were reported.

16.3.4.5 DDT

DDT was widely used between 1940 and 1960 as an insecticide for agricultural purposes and for the control of vectorborne diseases. DDT and its degradation products are still present as environmental contaminants in many countries. While banned in all countries for agricultural use, in many tropical countries DDT is still an important chemical used for the control of malaria. Besides its adverse effects on wildlife, DDT has been associated with several adverse health effects in humans, including cancer (Ahlborg *et al.*, 1995).

16.3.4.6 Dioxins

Dioxins are among a group of toxic chemicals known as persistent organic pollutants (POPs). In the environment their stability and fat solubility allow POPs to biomagnify in the food chain. The name dioxin applies to a family of structurally and chemically related dioxins and dibenzofurans, which are mainly by-products of industrial processes and waste incineration. Dioxins are found at low levels throughout the world in practically all foods, but especially dairy products, meat, fish and shellfish. Incidents involving elevated levels of dioxins in animal-derived foods have recently occurred in Belgium and the US. In the latter case, the source of the dioxins was a type of natural clay used as a binding agent in the formulation of animal feed. Acute effects of exposure to high levels of dioxins include skin lesions, such as chloracne, altered liver function and a shift in the sex ratio of progeny to favor girls. Long-term exposure is linked to impairment of the immune system, the developing nervous system, the endocrine system and reproductive functions, and cancer.

In 2002, JECFA considered the tolerable intake of dioxins to which a human can be exposed without harm. Based on human epidemiological data and animal studies, JECFA established a PTMI of 70 pg/kg body weight expressed as WHO Toxic Equivalence Factors (WHO, 2002a). This is one of the lowest tolerable intakes established for a substance. Levels of exposure to dioxins in several countries are estimated to be in the range of the WHO recommended PTMI. In a number of industrialized countries, the trend in exposure is downward as source-directed measures have reduced environmental emissions.

16.3.5 Natural toxicants

16.3.5.1 Mycotoxins

Mycotoxins, the toxic metabolites of certain microscopic fungi (molds), may cause a range of serious adverse effects in humans and in animals and have been of growing national and international concern since the 1970s (Moy, 1998). Animal studies have shown that besides serious acute effects, mycotoxins are capable of causing carcinogenic, mutagenic and teratogenic effects (European Commission, 1994).

Currently several hundred mycotoxins have been identified. Aflatoxin is the most well known and important mycotoxin from an economic point of view. As fungi producing aflatoxin prefer high humidity and temperatures, crops such as maize and groundnuts grown in tropical and subtropical regions are more subject to contamination. Epidemiological studies show a strong correlation between the high incidence of liver cancer in some African and Southeast Asian countries (12–13 per 100,000 annually) and exposure of the population to aflatoxin. Certain studies suggest that aflatoxins and hepatitis B virus are co-carcinogens and the probability of liver cancer is higher in areas where both aflatoxin contamination and hepatitis B are prevalent (Pitt and Hocking, 1989). Aflatoxins are found in peanuts, maize, tree nuts and some fruits such as figs. Besides adverse weather conditions (both too wet and too dry), post-harvest handling plays an important role in the growth of molds (WHO, 1979; FAO, 1987). In this respect, compliance with good agricultural/manufacturing practice is of the utmost importance. Aflatoxin-contaminated animal feed is also of human health concern as it appears in tissues that are used as human food. This is of particular importance in relation to dairy cows as aflatoxin B in feed is metabolized by the animals and excreted in milk as aflatoxin M.

Other mycotoxins of concern include ergot alkaloids, ochratoxin A, patulin, fumonisin B and the trichothecenes. JECFA has established very low provisional tolerable intakes for ochratoxin A, patulin, fumonisin B and some of the trichothecenes (WHO, 2002b). In view of their presence in many foods and their stability during processing, mycotoxins must be considered a major public health concern.

16.3.5.2 Marine biotoxins

Intoxication by marine biotoxins is another area of concern. In many parts of the world this type of poisoning is a major public health problem, affecting many thousands of people. The most common type is ciguatera, also known as "fish poisoning". In severe cases, symptoms may last for several weeks, months or years, and the case/fatality rate may range from 0.1 to 4.5%. Ciguatera has been associated with the consumption of a variety of tropical and subtropical fish, mainly coral fish, feeding on toxin-producing dinoflagellates, or predatory fish consuming such coral fish.

Another group of marine biotoxins produces acute intoxication after consumption of contaminated shellfish. Toxins causing shellfish poisoning are produced by various species of dinoflagellates. Under certain conditions of light, temperature, salinity and nutrient supply, these organisms may multiply and form dense blooms that discolor the water, as in the case of so-called red tides. Shellfish feeding on these algae accumulate the toxins, without being affected. The shellfish most often implicated are clams, mussels and occasionally scallops and oysters. Depending on the symptoms, different types of intoxications have been described as a result of the consumption of contaminated shellfish. These include paralytic shellfish poisoning (PSP), diarrheal shellfish poisoning (DSP), neurotoxic shellfish poisoning (NSP), amnesic shellfish poisoning (ASP) and azaspiracid poisoning (AZP). Recent warming of the world's oceans has altered the distribution and range of the dinoflagellates, introducing the problem to previously unaffected areas (Kao, 1993).

16.3.5.3 Plant toxicants

Toxicants in edible plants and poisonous plants that resemble edible plants are important causes of ill health in many areas of the world (WHO, 1990a). In some places, the poorer sections of the population eat grains known to be potentially toxic (e.g., *Lathyrus sativus*) out of hunger. Seeds of plants producing pyrrolizidine alkaloids have accidentally contaminated wheat and millet, leading to acute and chronic liver disease (WHO, 1988b). In Europe, misidentification of toxic mushrooms is by far the leading cause of illness and death in this category. In various Asian countries, such as India, economic adulteration of mustard seeds with similar looking but poisonous seeds is reported repeatedly.

16.3.5.4 Biogenic amines

Biogenic amines, including histamine, tyramine, cadaverine and putrecine, are decarboxylation products of amino acids, which are formed during fermentation, such as during cheese ripening, wine fermentation and decomposition of protein, most notably certain species of fish. Symptoms appear a few minutes to an hour after ingestion and include a strange taste, headache, dizziness, nausea, facial swellings and flushing, abdominal pain, rapid and weak pulse and diarrhea. Histamine is not destroyed by cooking.

16.3.5.5 Chemicals produced during processing

While most processing of food is intended to make the food safer over longer periods of time, occasionally a process will produce chemical residues of health concern. For example, smoking of food using traditional methods can result in high levels of benzopyrenes and other polycyclic aromatic hydrocarbons, which are known human carcinogens. Several chemicals with toxic properties are produced during the cooking process. Frying of fish can produce heterocyclic amines and frying of nitrite-cured meat can produce nitrosamines. Even cooking oils themselves can decompose into toxic degradation products. Recently acrylamide has been discovered in a range of high carbohydrate foods when cooked at high temperatures (WHO, 2006).

16.3.6 Adulterants

The adulteration of food is of concern for both health and economic reasons. The use of unapproved chemicals to make the food appear to be of higher quality, to mask an inferior or decomposed product or to simply increase the weight or volume of a product have been practised by unscrupulous producers and traders since ancient times. Societies have passed laws to protect consumers against such practices and often imposed severe punishments for violations. At the international level, Codex has adopted more than 240 food standards to help ensure the identity, quality and safety of common food commodities. However, incidents of adulteration continue to occur. These are mostly motivated by greed and often abetted by a complete ignorance of, or disregard for, the risk to human health.

As mentioned earlier in regard to the cooking oil incident in Spain (WHO, 1984, 1992), the health consequences of adulteration can be extremely serious, including death and permanent disability. Other adulterants can pose a variety of health problems, including cancer, birth defects and organ failure. Common adulterants include talc, formaldehyde, boric acid, unapproved colouring agents and even stones. Water is a common adulterant of liquid products, which can be harmful if it contains pathogens or toxic chemicals. The addition of melamine to food and feed has caused an international recall of contaminated products. In this case, the contamination of milk powder with melamine resulted in widespread contamination of secondary products that use milk powder as an ingredient.

In most countries, the addition of any chemical to food that has not been approved for use is considered adulteration. While such adulteration is illegal, the detection and prosecution of violations is extremely difficult. On the other hand, the intentional contamination of food with the intent to alarm consumers is often communicated by the perpetrators. Because of its criminal dimension, this will involve police and security personnel. Motivations may be political (terrorism), economic (extortion) or personal (revenge). Besides the usual contaminants known to occur in food, a range of possible threat agents could be used. The possibility of such incidents needs to be assessed based on perceived threat levels and resources allocated appropriately (WHO, 2008).

When an adulteration is threatened or discovered, the response decisions of risk managers, e.g. public warnings and recalls, will depend, in part, on the nature and extent of the risk posed by the adulterated product. Unfortunately, information initially available may be sparse and unreliable. Furthermore, relevant toxicology and adequate chemical data for the adulterant are often insufficient to support a reliable risk assessment. In such situations, appropriate com-

munication with all stakeholders, and especially the food industry, is essential to ensure that the public concern is similar in scale to the public risk.

Given the attendant uncertainties, risk managers should err on the side of caution, but should not neglect the potential implications for the food supply. Consequently, preparedness for adulteration incidents, including terrorist threats, is essential for assuring a timely, appropriate and coordinated response. Preparedness, including developing capabilities for rapid assessment and tracing of food, is essential to respond to emergency situations regardless of their origins. Most importantly, establishing clear lines of authority and communication among all relevant government agencies as well as with key contact points in the food industry is of greatest importance. Finally, simulations and exercises are necessary to test systems for robustness and effectiveness before a serious incident occurs (WHO, 2008).

16.4 Conclusions

While the food supply in developed countries is generally regarded as safe, certain chemicals can still cause long-term public health problems. For example, human exposure to acrylamide and various mycotoxins is well documented, but their potential health effects in humans need urgent study. In addition, developmental neurotoxicity has not been evaluated for many chemicals and even fewer chemicals have been tested for immunotoxicity. The growing rates of certain cancers suggest that further research is needed to rule out the possible contribution of chemicals in food to these diseases.

Periodic emergency incidents involving chemical hazards also point to the need for more effective approaches for ensuring that such events do not occur and that when they occur, appropriate action is promptly taken, including rapid, transparent and accurate communication with the public and the international community.

In developing countries, the situation regarding chemicals in food is largely unknown. Most of these countries do not have detailed legislation to control chemicals in food or lack food control capacities to enforce such legislation. In addition, most of these countries have no monitoring capabilities and little information about the dietary exposure of their populations to chemicals in food. Developing countries must develop risk assessment and management capabilities to effectively deal with chemical hazards in food. Key to this is the development of national capacities to conduct health-oriented, population-based monitoring programs to assess exposure of populations to chemicals in food, including conducting total diet studies.

Research into the potential adverse health effects of chemicals should include refinements of our knowledge about both hazard characterization and exposure assessment in order to provide the best scientific assessments of the risks posed by these hazards.

Further reading and references

Ahlborg, U.G., Lipworth, L., Titus-Ernstoff, L., et al. (1995) Organochlorine compounds in relation to breast cancer, endometrial cancer, and endometriosis: an assessment of the biological and epidemiological evidence. *Critical Review of Toxicology*, **25**, 463.

Codex (2006) *Codex Alimentarius Commission Procedural Manual*, 16th edn. Joint FAO/WHO Food Standards Programme, Codex Secretariat, FAO, Rome.

European Commission (1994) *Mycotoxins in Human Nutrition and Health*. Agro-Industrial Research Division Directorate General, XII Science, Research and Development, European Commission, Brussels.

Food and Agricultural Organization (1988) *Nairobi + 10, Mycotoxins 1987*. Report of the 2nd Joint FAO/WHO/UNEP International Conference on Mycotoxins, Bangkok, 28 September to 2 October 1987. FAO, Bangkok.

Food and Agricultural Organization/World Health Organization (1995) *The Application of Risk Analysis to Food Standards Issues*. Report of a joint FAO/WHO consultation. WHO, Geneva.

Food and Agricultural Organization/World Health Organization (1997) *Risk Management and Food Safety*. Report of a joint FAO/WHO expert consultation. FAO Food and Nutrition Paper 65. FAO, Rome.

Food and Agricultural Organization/World Health Organization (1998) *The Application of Risk Communication to Food Standards and Safety Matters*. Report of a joint FAO/WHO expert consultation. FAO Food and Nutrition Paper 70. FAO, Rome.

Howarth, J. (1983) *Global Review of Information on the Extent of Ill Health Associated with Chemically Contaminated Foods*. World Health Organization, Geneva (document WHO/EFP/FOS/EC/WP/83.4).

International Atomic Energy Agency (1991) *International Chernobyl Project Assessment of Radiological Consequences and Evaluation of Protective Measures – Conclusions and Recommendations of a Report by an International Advisory Committee*. IAEA, New York, p.3.

Jacobson, J., Jacobson, S. and Humphrey, H. (1990) Effect of the exposure to PCBs and related compounds on growth and activity in children. *Neurotoxicology and Teratology*, **12**(4), 319–26.

Kao, C.Y. (1993) Paralytic shellfish poisoning. In: *Algal Toxins in Seafood and Drinking Water* (ed. I.R. Falcone). Academic Press, London.

Moy, G.G. (1998) The role of national governments and international agencies in the risk analysis of mycotoxins. In: *Mycotoxins in Agriculture and Food Safety* (eds K.K. Sinha and D. Bhatnager). Marcel Dekker, New York, pp. 483–96.

Oskarsson, A., Ohlin, B., Ohlander, E.M. and Albanus, L. (1990) Mercury levels in hair from people eating large quantities of Swedish freshwater fish. *Food Additives and Contamination*, **7**, 555.

Pitt, J.I. and Hocking, A.D. (1989) *Mycotoxigenic Fungi. Foodborne Microorganisms of Public Health Significance*. Australian Institute of Food Science and Technology, North Sydney, New South Wales.

Rylander, L. and Hagmar, L. (1995) Mortality and cancer incidence among women with high consumption of fatty fish contaminated with persistent organochlorine compounds. *Scandinavian Journal of Work Environment and Health*, **21**, 419.

Shah, P.M., Schafer, V. and Knothe, H. (1993) Medical and veterinary use of antimicrobial agents: implications for public health: a clinician's view on antimicrobial resistance. *Veterinary Microbiology*, **35**, 269.

World Health Organization (1979) *Mycotoxins. Environmental Health Criteria 11*. WHO, Geneva.

World Health Organization (1984) *Toxic Oil Syndrome: Mass Food Poisoning in Spain*. WHO Regional Office for Europe, Copenhagen.

World Health Organization (1988a) *Assessment of Chemical Contaminants in Food. Report on the Results of the UNEP/FAO/WHO Programme on Health-Related Environmental Monitoring*. UNEP/ FAO/WHO, Geneva.

World Health Organization (1988b) *Pyrrolizidine Alkaloids. Environmental Health Criteria No. 80*. WHO, Geneva.

World Health Organization (1990a) *Technical Report Series No. 797. Diet, Nutrition, and the Prevention of Chronic Disease*. Report of WHO Study Group, WHO, Geneva.

World Health Organization (1990b) *Public Health Impact of Pesticides Used in Agriculture*. WHO, Geneva.

World Health Organization (1992) *Toxic Oil Syndrome: Current Knowledge and Future Perspective.* WHO Regional Publication, European Series No. 42. WHO, Copenhagen.

World Health Organization (1995) *Report of the WHO Scientific Working Group on Monitoring and Management of Bacterial Resistance to Antimicrobial Agents.* WHO/CDS/BVI/95.7. WHO, Geneva.

World Health Organization (1998a) *Evaluation of Certain Veterinary Drug Residues in Food.* 50th Report of the Joint FAO/WHO Expert Committee on Food Additives. WHO, Geneva.

World Health Organization (1998b) *Infant Exposure to Certain Organochlorine Contaminants from Breast Milk: A Risk Assessment.* GEMS/Food International Dietary Survey. WHO, Geneva.

World Health Organization (1999) *Evaluation of Certain Food Additives and Contaminants.* 53rd Report of the Joint FAO/WHO Expert Committee on Food Additives. WHO, Geneva.

World Health Organization (2000) *Evaluation of Certain Veterinary Drug Residues in Food.* 52nd Report of the Joint FAO/WHO Expert Committee on Food Additives. WHO, Geneva.

World Health Organization (2002a) *Evaluation of Certain Food Additives and Contaminants.* 57th Report of the Joint FAO/WHO Expert Committee on Food Additives. WHO, Geneva.

World Health Organization (2002b) *Evaluation of Certain Mycotoxins in Food.* 56th Report of the Joint FAO/WHO Expert Committee on Food Additives. WHO, Geneva.

World Health Organization (2006) *Evaluation of Certain Food Contaminants.* 64th Report of the Joint FAO/WHO Expert Committee on Food Additives. WHO, Geneva.

World Health Organization (2008) *Terrorist Threats to Food.* Revised 2008. WHO, Geneva, http://www.who.int/foodsafety/publications/fs_management/terrorism/en/index.html.

Food business management: principles and practice

Michael Bourlakis, David B. Grant and Paul Weightman

Key points

- To introduce some of the important aspects of the role and importance of business and management in selected parts of the food industry.
- To illustrate the food business environment and to focus on the food chain system and its key members.
- To pay particular attention to the key management functions of food firms including operations management and human resource management.
- To analyse the role of both finance and accounting in the context of their contribution to the management of food businesses.

17.1 Introduction

Food is fundamental to health, happiness and political stability (Bourlakis and Weightman, 2004). The need to obtain food, water and shelter are the basic physiological needs according to Abraham Maslow's hierarchy of needs (Jobber, 2004). It is therefore always a challenge to ensure that the processes of food production, processing and distribution are well managed.

This chapter introduces some of the important aspects of the role and importance of business and management in selected parts of the food industry. The first section illustrates the food business environment and is followed by a section on the food chain system. Food operations management is analysed in another section whilst key academic fields such as human resource management, finance and accounting are brought together in a separate section and are

examined in the context of their contribution to the management of food businesses.

17.2 The food business environment

Fine *et al.* (1996) argued that food studies are in disarray and have become fragmented due to several external factors. These factors can be categorised into two constituents. First, technological changes in food freezing and preservation, and even the development of the microwave oven, have helped alter the nature of food processing as well as consumption patterns and consumer preferences. Second, differing lifestyles and culture have developed into a divide between unlimited choices of many foods, especially "fast food" which has seen the erosion of traditional meal patterns in western industrialised societies, and limited choices in many third world countries whereby people are unable to feed themselves. Both these constituents affect the nature of food production, processing and packaging and final consumption, and thus the shape of the food chain.

Additionally, Strak and Morgan (1995) identified five environmental dimensions affecting the food industry:

1 globalisation;
2 market structure and power;
3 consumer tastes and lifestyles;
4 technological change; and
5 regulatory effects.

They argued that "food industry activities may perhaps be better understood as a 'web' not a chain" (Strak and Morgan, 1995, p. 337), as shown in Fig. 17.1, with these five dimensions surrounding such activities.

For example, changing consumer tastes and busy lifestyles have led to a demand for convenience foods, i.e. prepared in advance, but also healthy and different and discerning foods. Thus, this dimension is linked to the dimensions of technological change, i.e. the microwave oven allows for faster food preparation of packaged meals, and globalisation, i.e. the search for and acquisition of world foods.

Whilst these five dimensions have been introduced elsewhere in terms of change factors for logistics and the food industry, Strak and Morgan's approach has located customers at the centre of the web à la marketing thoughts of Jobber (2004).

Tansey and Worsley reported that the development of "canning, freezing and chilling technologies" in the nineteenth century dramatically "transformed preservation and distribution" of foodstuffs and developed a further time and distance separation between farmers and consumers (1995, p. 43). An extreme example of such a transformation is the growth of Australia's exports of canned meat to the UK from

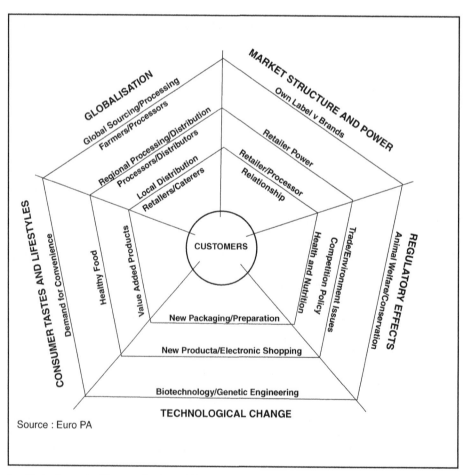

Figure 17.1 The food industry web. (Source: Strak and Morgan, 1995, p. 337.)

"16,000 lb in 1866 to 22,000,000 lb in 1871 at half the price of fresh meat in England" (ibid).

These technologies further stimulated firms in the food processing industry to grow from being pre-servers of food to manufacturers in their own right. Tansey and Worsley cited OECD 1992 statistics that global production of processed food "amounts to some U.S.$1.5 trillion, making it one of the world's largest industries" (1995, p.111). Further, the industry has become very concentrated with "the 100 largest OECD-based companies" accounting for "about 20 per cent of global production" (ibid).

In addition, Tansey and Worsley argued that the larger, global food manufacturers have put new pres-sures on suppliers to provide standardised products of superior quality at fixed prices. They presented the following grim perspective on the nature of the food industry and the motivations of actors within in it:

"It uses an industrial approach to agriculture and food production, is highly productive in response to high inputs and overcomes seasonality for all foods. It draws on produce from around the world and, by using a mixture of trading and preserva-tion techniques, enables a wide range of foodstuffs always to be available. In the development of this food system, foods became more and more like commodities, rather than matters of life and death, or of religious and cultural meaning.

Commodities are produced, traded and trans-formed, bought and sold, in a market whose reach has extended from a largely local level to an in-creasingly global stage. It is a market in which ac-tors seek to control their costs, their production or marketing practices, as closely as they can. They want to minimize their uncertainties and costs and maximize their returns. It is a market in which each actor is thrown into competition with others, both within their areas of operation and outside them" (1995, pp. 47–48).

The nature of competition in the food industry does not appear conducive to good relationship develop-ment and the maintenance considered necessary in supply chains.

Fine *et al.* (1996) undertook a UK Economic and Social Research Council (ESRC) study on food con-sumption and proposed that food chains are diverse systems of provision (SOP) that are determined by consumption of a certain food type. This contradicted usual consumption studies examining horizontal factors across consumer groups or broad product ranges. They argued that "consumption must be in-vestigated within a vertical framework in which each commodity or groups of commodities is differenti-ated from others" (1996, p. 6). The latter factor in-cludes the growth of bulk purchases of meat, ready meals and eating out or catering.

The British spend about £23 billion a year eating out, with about £6 billion spent on fast food (Hogg, 2001). This is an increase of 32% since 1996. In the US over $110 billion was spent on fast food in 2000 com-pared to $6 billion in 1970 (ibid). Such increases are related to general economic growth and total UK con-sumer expenditure that has grown 104% from £171 million in 1963 to £348 million in 1993 (Strak and Mor-gan, 1995, p. 3). And yet, whilst these values are sig-nificant, the purchase of food only accounted for 12% of average UK consumer expenditure in 1993 com-pared to 20% in 1963 (ibid). This percentage decrease may be evidence of food "commoditization" as ar-gued by Tansey and Worsley.

17.3 The UK food chain system

The UK food chain (UKFC) comprises the agriculture, horticulture, food and drink manufacturing, food and drink wholesaling, food and drink retailing, fish-eries and aquaculture, and catering industries (Food Chain Group, 1999). Comparable and detailed sec-tor statistics are difficult to amass and "just present-ing the data in a consistent format ... is a significant task" (Strak and Morgan, 1995, p. 1). This difficulty is demonstrated by presentation of aggregate food chain data available from several sources.

The Food Chain Group, a UK Government work-ing group, reported the UKFC accounts for gross added value of £56 billion to the UK economy, or 8% of GDP (1999, p. 12). The UKFC, excluding the fish-eries and aquaculture and catering sectors, also em-ploys 3.3 million people or 12% of the UK's workforce (ibid). Patel *et al.* (2001), writing for the Institute of Grocery Distribution (IGD), have presented the UK food and grocery supply chain in Colour Plate 27. They report sector values of £16.5 billion for agricul-ture and fishing and £75.9 billion for food and drink manufacturing; however, the latter includes tobacco in the sector value.

The UKFC has evolved significantly since the end of the World War II. Four factors dominated the sup-ply and distribution of food in the post-war period:

- commonplace rationing;
- local or regional product sourcing and provision;
- lack of a national distribution system; and
- consumers' low expectations.

Supply chains as such were non-existent, and manufacturers and wholesalers controlled food distribution (Patel *et al.*, 2001). As farming yields and consumer prosperity increased and road transportation developed through the building of motorways and transport deregulation, retailers grew in importance during the 1970s and 1980s.

Power in the food supply and grocery chain started to shift to large multiple retailers such as Tesco, Morrison's, Sainsbury's and Asda (owned by Wal-Mart). This shift in power enabled these large multiple retailers to realise operating profit margins of 7–8%, which is much more than margins of 2% in other EU countries or 1% in Australia. Moreover, whilst the total food industry "profit pie" has grown from £1 billion in 1981 to £5 billion in 1992, the retailers' share of the "pie" has increased from 20% to 40% at the expense of the manufacturers and processors (Tansey and Worsley, 1995, p. 124).

The concentration of power amongst the large multiple retailers has led them to integrate supply chains and develop and own regional distribution centres (RDCs). They have also outsourced logistics and supply chain activities (Bourlakis and Bourlakis, 2001) and introduced technological tools such as efficient consumer response (ECR). Dawson and Shaw (1990) examined changes in the supplier–retailer dyad and proposed a continuum of relationships that runs from transactional to fully integrated. Relationships towards the latter end of the continuum may develop as a result of a changing business environment, emerging techniques for SCM and the development of distribution technology. It is clear that retailers are the progenitors of such integration, particularly in the food processing industry.

The IGD and others (Fernie *et al.*, 2000; Alvarado and Kotzab, 2001) have promoted the benefits of closer supplier–retailer integration, technological advancements and relationships in the UK food chain resulting from increased retailer concentration. However, other authors have criticised this concentration on the grounds of coercive power and retailer motives (Tansey and Worsley, 1995; Food Chain Group, 1999; Grant, 2005; Vlachos and Bourlakis, 2006).

Tansey and Worsley argued that:

"Small farmers and workers must compete with large and powerful users of their products and services. Large manufacturers, especially in the UK, have found themselves supplying increasingly powerful retailers who are able to set terms and drop their products if they fail to meet retailers' sales standard. Retailers themselves might find their role changing, however, with the use of interactive technology now becoming available in the store and home. This may raise the question of who is the middleman. Whatever happens there is a fascinating battle going on for who processes – in the factory, home or small business – the food that goes into people's stomachs world-wide" (1995, p. 141).

However, the UK Competition Commission in 2000 (Blythman, 2005) found the consumer marketplace to be competitive and retailer profits not excessive. However, it also recommended a new Code of Practice on Supermarkets' Dealings with Suppliers. The new Code was meant to be "important in rebalancing the relationship between supermarket and supplier" to reduce "extreme practices, such as retailers imposing retrospective price cuts to contracts, asking suppliers to meet the costs of shop refurbishment or staff hospitality" (Blythman, 2005, pp. 152–154). The UK Office of Fair Trading undertook a review of the Code in early 2004, principally to assess the Code's impact on supermarket and supplier relationships. The review "revealed a widespread belief among suppliers that the code is not working effectively and that most believed that the code had not brought about any change in the behaviour of the supermarkets" (Fearne *et al.*, 2005, p. 572).

17.4 Characteristics of UK food retailers

The structural activities of UK food retailers today are characterised by high levels of concentration and own branding, and the development of horizontal alliances (Strak and Morgan, 1995). Browne and Allen argued that these characteristics stem from UK food retailers being in the fourth "advanced retailing" stage of European retail development where "retail markets have a high degree of market concentration, segmentation, capitalisation, supply chain integration and use of information technology" (1997, p. 36). Strategic approaches utilised within an advanced retailing stage include the formation of vertical or horizontal alliances notwithstanding retailer concentration levels, an increased use of information management available from the new technology, and development of own brands to capitalise on strengths in segmentation and consumer awareness. The two former approaches are considered to represent the future of logistics and distribution systems in the UK food chain, where a "network information-driven value chain" replaces the "traditional linear supply chain" (Mathews, 1997).

Browne and Allen (1997) have also argued that retailer control of the food chain is linked to the degree of market concentration and difference between retailers' own label and manufacturers' brands. The more retailers are able to acquire control in the food chain, the more they are able to organise logistics activities such that products flow to their own RDCs via third-party logistics (3PL) service providers. With greater control, retailers are also able to impose more stringent demands on their suppliers such as "greater reliability of supply, consistently high quality levels, favourable price levels and sufficiently flexible production control to ensure variety of product" (ibid, p. 34).

Moreover, levels of trust between retailers and suppliers in the UKFC may not be as collaborative or as cordial as some authors would like due to retailer control and perceived power. Some of the issues affecting logistics relationships generally have been discussed above. Specific examples from the UKFC follow. For example, P-E International (1991) surveyed 54 grocery suppliers and 9 grocery retailers, amongst other industry sectors, regarding partnership development during the 1990s. Grocery retailers were "very keen on mutual objectives and two-way communication, but less enthusiastic about full involvement in each others' businesses" leading to the proposition that "mutual objectives will be set by retailers" (1991, p. 14), "perhaps because they are expected to be the principal beneficiaries" (1991, p. 18). Grocery suppliers on the other hand were "less enthusiastic about two-way communication, mutual objectives and many of the technological developments" but were keener for "full involvement in each others' businesses" (1991, p. 15). Thus, P-E International concluded there was "widespread doubt and suspicion ... about retailer moves to develop relationships into partnerships" which supported the "notion of one-sided partnerships, and the need for reciprocity" (ibid).

Robson and Rawnsley (2001) supported the original P-E International contentions whilst interviewing food industry managers in a qualitative study some 10 years later. They found that although supermarkets should be leading the way in developing vertical relations in the UKFC, in practice partnerships and relationships have not fully developed unless they are on retailers' terms. Their study focused on ethical considerations in UKFC relationships in terms of retailers conforming to the notions of consumer sovereignty and *caveat venditor* and extending such behaviour and attitudes towards suppliers. Robson and Rawnsley found little evidence of partnering based on resource sharing in a trusting environment. Indeed, they noted that the IGD's model of ethical behaviour "excludes supply chain relationships ... in favour of ... product safety and manufacturing efficiency" (2001, p. 47).

Lastly, Fearne (1998) examined partnerships in the UK beef supply chain. He noted they "have been difficult to establish and slow to develop" but argued they "are the only sustainable form of trading relationship in the long term" (1998, p. 214). He saw four key drivers behind the evolution of partnerships in this sector:

- changing attitudes and purchasing behaviour of meat consumers;
- competitive strategies of supermarket chains;
- the 1990 Food Safety Act; and
- the effects of the BSE crisis.

These drivers will lead actors in the UK beef supply chain to develop the partnerships Fearne calls for. Fearne recognised that the "development of partnerships requires hard work, commitment and a fair degree of trust in the long-term intentions of partners", but he also admitted "partnerships, in certain circumstance, may offer no improvement in returns to producers over the open market" (1998, p. 230).

This scepticism of suppliers towards retailers was recently reinforced by a retailer-driven initiative known as factory gate pricing (FGP). Finegan (2002) reported on FGP and backhauling initiatives in an IGD report designed to be an objective and impartial account of the processes surrounding these initiatives. FGP is where retailers ask suppliers to provide product costs at the suppliers' factory gate, i.e. excluding transport costs to retailers' RDCs or other receipt points. Backhauling is the "movement of goods from supplier to the retailer via vehicles which have previously made a delivery in the local area" (2002, p. 25), i.e. retailers will collect for themselves, possibly using 3PL service providers, as opposed to have

Table 17.1 UK consumer expenditure on food by sector at current prices, 1999–2003.

	1999	2000	2001	2002	2003
Meat and meat products	11,883	12,265	12,397	12,535	13,917
Fish and fish products	2,063	2,152	2,284	2,375	2,395
Fruit and vegetables	12,189	12,400	12,533	12,734	13,342
Dairy products, eggs, oils and fats	8,486	8,631	8,719	8,567	8,924
Bread, cakes, biscuits and cereals	8,065	8,346	8,743	9,046	9,149
Miscellaneous foods	3,543	3,634	3,759	3,955	4,125
Total (£ millions)	**£46,229**	**£47,428**	**£48,435**	**£49,212**	**£51,852**
Percentage change year-on-year	—	2.6	2.1	1.6	5.4

Source: Key Note, 2005.

suppliers deliver. Suppliers see these initiatives as an unbundling of suppliers' prices into constituent components of transport and production costs. Finegan noted that food manufacturers, processors and suppliers fear retailers will "use this information ... to query manufacturing cost structures and ultimately put pressure on manufacturers to reduce the price of goods" (2002, p. vi). Further, manufacturers may incur increased transport costs due to sub-optimisation of delivery loads to other customers and increased operations costs responding to a retailer's timetable and priorities (Gannaway, 2001).

17.5 Characteristics of UK food processors

The Food Chain Group (1999) noted the food and drink manufacturing (FDM) sector, net of alcoholic drink, accounts for gross added value of about £16.2 billion or 2.2% of GDP. The sector employs 455,000 people, mostly full-time. Thus, the FDM sector is a major factor in the total UK food chain and represents about 25% of value added and employment. The Food Chain Group recorded about 8000 firms classified as food and drink manufacturers, but the sector is highly concentrated with the ten largest manufacturers accounting for 21% of sector turnover or revenue and three firms accounting for over 75% turnover for many products (1999, p. 44).

Despite the concentration, the number of firms in the FDM sector has grown by 43% since 1977. At that time there were about 5600 firms classified as food and drink manufacturers and the ten largest manufacturers then accounted for 60% of employment and

value added in the food sector (Tansey and Worsley, 1995). Thus, whilst there is substantial concentration in the FDM sector, there are still many relatively small firms. Browne and Allen reported that "around 85% of food, drink and tobacco companies had less than 50 employees and 60% had fewer than 10 employees in 1995" (1997, p. 35).

There are many sub-sectors within the FDM sector as shown in Table 17.1.

17.5.1 UK grocery logistics

Fernie *et al.* examined retail grocery logistics and noted that "the old inefficient manufacturer and supplier-led practices have been swept away by the modern, technologically rich, retailer-led and customer-focused ways of ensuring product availability" (2000, p. 83). They also argued that further progress in retail logistics and further improvements in supply chain efficiency can "only be achieved through collaboration between supply chain partners" (ibid).

With the cooperation of IGD, Fernie *et al.* surveyed grocery retailers, manufacturers and logistics service providers on four main questions:

"What factors are likely to have most impact on cost, service or structure within the grocery supply chain over the next three years?
What technologies would facilitate benefits throughout the supply chain?
Where would inventory be held in the supply chain in the next three years?
How would warehouse and transport practices change in this period?" (2000, p. 86)

They found that the most important factors, based on means of scores that will impact costs, are traffic congestion, transport taxation, 24-hour trading and home shopping/home delivery. Respondents reinforced these factors as regards applicable technologies, with in-vehicle communications and automated sorting systems topping the list. Respondents also noted inventory reductions taking place at the retail and RDC levels, thus moving inventory upstream to manufacturers' facilities. Cross-docking, shared user services and communication and load consolidation and backhauling featured highly as regards the last question.

Fernie et al. concluded that radical changes are unlikely in the near future and that relationship development "will continue as supply chain members seek to take further costs out of the system" (2000, pp. 88–89). Ultimately, they considered that grocery supply chains must become more efficient to ensure existing profit margins. This will be achieved through "greater collaboration between retailers and suppliers through sharing of information, the use of global retail exchanges and the implementations of Collaborative Planning Forecasting and Replenishment" (2000, p. 89).

17.6 Marketing in food business management

Stank et al. (1998) discussed logistical service capabilities of personal product and prepared food supply chains. They cited a former president of American Express as stating "in a commodity-like business, service is the only way to create product differentiation" (1998, p. 78), which they included to mean logistics or distribution service. They interviewed restaurant managers in their role as food service customers and retailers of personal products to develop a continuum of logistical capabilities that is presented in Fig. 17.2. The continuum suggests that capabilities such as cost minimisation or TQM lead to operational effective-

ness, but do not lead to "customer closeness" or relationships.

Stank et al.'s contention that "as clichéd as it sounds, business really does begin and end with the customers" (1998, p. 79) also suggests that existing conceptual research in food chains is unsuitable for understanding customer service and ultimately customer satisfaction. They noted that "identifying core operational service elements is a minimum requirement for competing, but it will certainly not be enough to distinguish a service provider from the pack, or guarantee that customers will be loyal" (ibid).

Flanagan (1992) conducted a study of customer service requirements of UK food processing buyers. He found that decision factors influencing buyers to purchase from a supplier were, in order of importance, product quality, price, reliability of supply, response to problems, and delivery lead times. He also found that the essential elements of customer service from the buyer's perspective were, again in order of importance, continuity of supply, advice on non-availability, delivery on the day required, condition of goods on arrival, and emergency deliveries.

17.7 Food operations management

Most organisations have an operations function that transforms various inputs into products and services to meet customer requests (Slack et al., 2004). Inputs include materials and information, while the transformation process makes use of the organisation's facilities and staff. The food supply chain can be characterised as a complicated system consisting of several stages from agricultural production and industrial processing through to marketing or retailing and consumption (Yakovleva and Flynn, 2004). Sophisticated consumers now demand and major food retailers are currently delivering high-quality

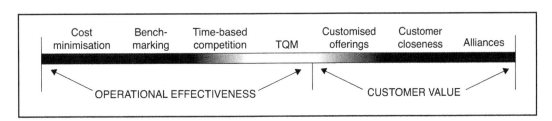

Figure 17.2 Continuum of logistical capabilities. (Source: Stank et al., 1998, p. 79.)

food products in various forms at competitive prices throughout the entire year (Apaiah *et al.*, 2005).

The operations function and its management in food supply chain organisations is no different to those in other sectors; however, the food processing sector also faces issues of seasonality, perishability, quality and traceability that are unique to the sector and of concern to consumers and governments. An efficient and effective food supply chain design that includes efficient operations within organisations can satisfactorily address these issues (Van der Vorst, 2000). However, food supply chains are also becoming more global and complex and thus contain many more business relationships. Knowledge from many disciplines, such as food processing technology, operations research, environmental science, marketing and business economics, needs to be combined to enable the efficient and effective design of such supply chains (Apaiah *et al.*, 2005).

From a general operations management perspective, the design of operational processes to produce products and services must consider four areas:

1 the *supply chain design*;
2 the *layout and flow* within the production facility;
3 what *process technology* to incorporate; and
4 the various *jobs and work design* to facilitate the process (Slack *et al.*, 2004).

17.7.1 Supply chain design

Supply chain design decision issues include:

■ whether to *make or buy* products;
■ *supplier sourcing*;
■ the *location* of facilities and operations relative to others in the supply chain; and
■ the *processes* involved in *production*.

17.7.1.1 Make or buy

The make or buy decision is also termed the outsourcing decision, i.e. should organisations outsource their production to third parties? Reasons to make instead of buy include small quantities that would be uneconomic for third parties, a need for quality, technological secrecy, capacity utilisation and smooth continuous production flow, and to avoid sole source dependency (Leenders *et al.*, 2002). Conversely, reasons to buy include lack of expertise within the organisation, the process or product is non-core to the organisation, uncertainty about long-term costs

and viability, more flexibility in source selection, and less administrative and overhead expenditures (ibid). Food organisations tend to outsource production less than in other industries; however, a lot of logistical activities such as warehousing, transportation and information processing are outsourced, especially by large food retailers such as Tesco in the UK or Carrefour in France (Van Hoek, 1999).

17.7.1.2 Supplier sourcing

Supplier sourcing issues include single or multiple sourcing, the size of the supply base, supplier locations, propensity for relationships, and transactional considerations such as price, quality and delivery (Gadde and Håkansson, 2001). The latter transactional elements are very important to UK food processing organisations (Pecore and Kellen, 2002; Grant, 2004). Also, international and global sourcing have become important as consumer demand for different and exotic foods has increased. Lastly, while food processors are desirous of relationships with suppliers and retailers the propensity for doing so is affected by the imbalance of power that has accrued to food retailers (Grant, 2004; Fearne *et al.*, 2005; Hingley, 2005).

All these issues include elements of sourcing or purchasing risk. One classic approach to deal with such risk is the portfolio matrix developed by Kraljic that maps the value of goods against risk. Low value and risk purchases are termed non-critical, high value and low risk purchases are termed leverage, high value and high risk purchases are termed strategic, and low value and high risk purchases are termed bottleneck. Organisations can use this matrix to categorise their purchases into one of the four quadrants and make appropriate purchasing decisions (Leenders *et al.*, 2002, p. 245).

17.7.1.3 Location

The location of facilities can be approached from macro and micro perspectives (Grant *et al.*, 2006b). The macro perspective examines the issue of where to locate facilities geographically within a general area so as to improve the sourcing of materials and the company's market offering (improve service and/or reduce cost). The micro perspective examines factors that pinpoint specific locations within the large geographic areas.

One macro approach includes the combined theories of a number of well-known economic geographers. Many of these theories are based on distance and cost considerations and use a strategy based on cost minimisation. One well-known strategy was devised by Alfred Weber, a German economist. According to Weber, the optimal site was one that minimised total transport costs, assuming that these costs varied in direct proportion to the weight of goods transported multiplied by the distance moved (traditionally expressed as ton mileage). Weber classified raw materials into two categories according to their effect on transportation costs: location and processing characteristics. Location referred to the geographical availability of the raw materials.

Few constraints would exist on facility locations for items that had wide availability. Processing characteristics were concerned with whether the raw material increased, remained the same or decreased in weight as it was processed. If the processed raw material decreased in weight, facilities should be located near the raw material source because transportation costs of finished goods would be less with lower weights. Conversely, if processing resulted in heavier finished goods, facilities should be located closer to the final customers. If processing resulted in no change in weight, a location close to raw material sources or to markets for finished goods would be equivalent (ibid).

Another approach, the centre-of-gravity approach, is more simplistic in scope, and locates facilities based solely on transportation costs. This approach locates a facility centre at a point that minimises transportation costs for products moving between suppliers and the markets. The centre-of-gravity approach provides general answers to the facility location problem, but it must be modified to take into account factors such as geography, time and customer service levels (ibid).

From a micro perspective, more specific site-selection factors must be examined. A company must consider:

- the quality and variety of transportation carriers serving the site;
- the quality and quantity of available labour;
- labour rates;
- the cost and quality of industrial land;
- potential for expansion;
- tax structure;
- building codes;
- the nature of the community environment;
- costs of construction;
- cost and availability of utilities;

- cost of money locally;
- local government tax allowances and inducements to build.

The site-selection process is interactive, progressing from the general to the specific. It may be formalised or informal, centralised at the corporate level, decentralised at the divisional or functional level, or some combination of each. It is important that management follow some type of logical process that recognises many trade-offs when making a location decision (ibid).

17.7.1.4 Production processes

There are several types of production processes but they all have one thing in common. Processes need to be planned to show what an organisation is going to do over the next period, and a useful place to start looking at this effect is through capacity (Waters, 2003). The capacity of any process is set by a restricting bottleneck and must be planned so that available capacity matches forecast demand. There are standard approaches to planning, which iteratively develop and compare solutions. The goal is for production to remain constant even through periods of varying demand. There are many ways of designing aggregate plans, ranging from negotiations through to mathematical modelling.

A master schedule "disaggregates" an aggregate plan and shows the number of individual products to be made, typically, each week. Material requirements planning (MRP) starts with a master schedule and bill of materials and uses these to develop a timetable for gross material requirements. MRP "explodes" the master schedule to find the gross requirements for materials, then adds related information about stocks to give net requirements; finally, it adds information about suppliers and operations to give schedules and details of ordering policies for orders and related internal operations. This approach has a number of benefits, particularly relating stocks to known demand. However, there are weaknesses, such as the complexity of systems and reduced flexibility. As a result, it is only really suitable for certain types of process. At the moment it is, perhaps, most successful for batch manufacturing.

MRP can be extended from its original function of scheduling materials to scheduling other resources. There are many ways of extending the basic MRP approach. The simplest extensions add more information about, say, suppliers or operations, typically using a batching rule to combine smaller orders into larger, more economical ones. More widespread extensions include the use of the MRP methods to plan more resources.

Manufacturing resource planning or MRP II is such a major extension for manufacturing organisations, extending MRP to other functions within the organisation. MRP II provides an integrated system for synchronising all functions within an organisation and connects schedules of all activities back to the master schedule (Waters, 2003). More recently, enterprise resources planning or ERP extends MRP to other organisations in a supply chain, giving highly efficient, integrated operations. MRP is a dependent demand method that finds demand for materials directly from the master schedule. Another process that is a dependent demand method is just-in-time (JIT) operations. JIT eliminates waste by organising operations to occur at exactly the time they are needed. In this sense, stock becomes a waste of resources that should be eliminated.

Organising operations to occur just as they are needed seems an obvious idea, but achieving it has proven difficult (Waters, 2003). While some organisations have had great success with JIT, many others have hit serious problems and have simply given up trying to implement it. The message is that JIT – like MRP – can work well in some circumstances, but it should not be considered a universal tool for all organisations.

JIT generally works best for specific types of organisation such as large-scale manufacturers using continuous processes. JIT works by pulling materials through a supply chain, rather than the traditional methods that push them through. Customer demand triggers operations, with a message passed backwards through the supply chain by kanbans, which are systems recording the levels of stock in the container used to introduce materials just-in-time. There are many different forms and ways of using kanbans, most usually electronic tags. JIT minimises rather than eliminates stock. The overall stock of any item depends on the number of kanbans and size of containers used.

JIT has its roots in the lean principle developed by Toyota, which has spawned lean production. Lean production and lean thinking principles evolved from Toyota's efforts to eliminate waste and non-value-adding activities to derive a competitive advantage. Most automakers have adopted the lean concept whilst other supply chains (e.g. food) move to that direction too. Food supply chains have also developed as lean processes at the producer and processor end (Simons and Zokaei, 2005). However, food retailers operate on an independent demand basis driven by consumer needs and are thus more flexible in their operations to be responsive to that demand. Such a process system is agile in its ability to be flexible and respond (Christopher, 2005). Decisions concerning production processes may involve choosing between, or incorporating both, lean and agile production. These two types of production are based on different scheduling principles: push (lean) and pull (agile). In a push supply chain, goods are produced to stock before the actual orders are placed, based on a forecast generated in advance.

There are many methods of forecasting, none of which is always the best. These methods can be classified as judgmental, causal and projective forecasts. Judgmental forecasts are based on opinions, knowledge and skills rather than a more formal analysis. The most widely used methods of judgmental forecasting are personal insight, panel consensus, market surveys, historical analogy and the Delphi method. Although they all have their advantages, their main problem is unreliability. They are the only means of forecasting when there is not historical data (Waters, 2003; Grant et al., 2006a).

Historical data often appear as a time series, i.e. a series of observations at regular periods of time. These observations usually follow a regular pattern but contain random "noise" observations that make forecasting difficult due to an ever-present error. The most important measures of this error are mean error or bias and mean absolute deviation (MAD). Causal forecasting is a method that uses a cause and effect relationship to produce forecasts. A general approach uses linear regression, which finds the straight line of best fit through a set of data, and uses coefficients of correlation to determine how good the fit is. Projective forecasting extends historical patterns into the future by methods based on moving averages and exponential smoothing. These forecasts can be "fine-tuned" by choosing an appropriate smoothing constant or using a tracking signal (Waters, 2003; Grant et al., 2006a).

In a pull or agile supply chain products are not made until a customer gives a signal, by placing an actual order, in the upstream information flow (Harrison and van Hoek, 2005). The materials are produced just-in-time as needed by the customer, which helps eliminate stock and reduce costs. Inventory reduction together with higher quality, flexibility of manufacturing and volume, and rapid response to customer demands are the goals of an agile or pull strategy. However, if the response is not in time then out-of-stocks (OOS) can occur and production is affected. At the food retail level this problem translates into less on-shelf availability (OSA), which is important to consumers. Consumers react to OOS and OSA by purchasing substitutes, delaying their purchase, shopping at other stores or not purchasing at all, any of which affects their satisfaction (Corsten and Gruen, 2003). Thus, too lean a system that is not sufficiently flexible can cause problems for organisations and reduce revenues and profitability.

17.7.2 Layout and process flow

Layout and process flow is concerned with the physical location of resources such as facilities, machines, equipment and staff in the operation; i.e. the type of "production line" used (Slack *et al.*, 2004). It has been said that Henry Ford based his ground-breaking automobile production line process on the process used in Chicago slaughterhouses (Simons and Zokaei, 2005).

Basic facility layout types are:

- fixed-position layout, where the product is stationary, such as an aircraft assembly;
- process layout, where like products are grouped together, such as frozen or chilled foods in a supermarket;
- cell layout, where complementary products are grouped together, such as lunch products in a supermarket; and
- product layout, where products are processed for the convenience of the resources being used, such as an assembly line or self-service cafeteria

(Slack *et al.*, 2004). The food supply chain will use three types of layouts and food processors will primarily use a product layout.

For example, bread and bakery product manufacturers often buy entire automated production lines from a single food product machinery manufacturer (Liberopoulos and Tsarouhas, 2005). A typical line consists of several workstations in series integrated into one system by a common transfer mechanism and a common control system. Material moves between stations automatically by mechanical means,

and no storage exists between stations other than that for material handling equipment such as conveyors.

Food product machinery manufacturers usually design all the workstations in a production line based on the bottleneck workstation, i.e. the station with the lowest nominal production rate. The bottleneck workstation is important because it determines the nominal production rate of the entire line. In bread and bakery products manufacturing, the bottleneck workstation is almost always the baking oven (ibid). One method of overcoming bottleneck stations is to use the theory of constraints (TOC) to focus on the bottleneck constraint using an optimised production technology (OPT) software package (Slack *et al.*, 2004).

The use of standardisation in a product layout leads to efficient flow and production. In a lean food processing situation the two practices of takt-time and standardised work exist at an operational level (Simons and Zokaei, 2005). The German word "takt" refers to beat of music. Takt-time is used to communicate and synchronise the rate of the production process with customer demand to prevent overproduction waste and is the elapsed time between units of production output. Work standards can be defined as the best way of doing a job and work standardisation refers to operational procedures on the shop floor that ensure customer satisfaction.

Essentially, producing to takt-time is a type of standardisation. Takt-time is about standardising the production cycle times and work standardisation is standardising the procedure of tasks on the shop floor, both of which lead to efficient production systems (ibid). However, environmental issues in food processing such as production regime change to accommodate seasonal products and demand shifts, and high setup and changeover costs due to equipment cleaning can inhibit the efficiency of a lean production system (Houghton and Portougal. 2001).

17.7.3 Process technology

There are two areas in the food supply chain where technology is important. One is in the food production area and the other is at the retail level as regards stock replenishment. In both areas information and information flow is critical to the efficient use of technology; however, the use of technology has lagged in the processing end of the food supply chain (Mann *et al.*, 1999; Grant, 2004). However, information technology has been used at the food retail level to

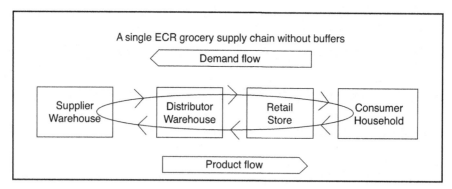

Figure 17.3 The ECR model. (Source: Kotzab, 1999, p. 367.)

effect efficient replenishment to overcome OOS and increase OSA (Bourlakis and Bourlakis, 2005, 2006). The efficient consumer response (ECR) initiative began in the early 1990s by US consultants Kurt Salmon Associates (Kotzab, 1999). ECR is defined as a grocery industry strategy in which distributors and suppliers are working closely together, i.e. in partnership to bring better value to the grocery consumer through a seamless delivery of products at a total low cost (Kotzab, 1999; Whipple *et al.*, 1999). The Salmon ECR model is shown in Fig. 17.3.

This seamless delivery is consumer-driven through a paperless information flow initiated by a retailer's electronic point-of-sale (EPOS) that also sets and manages production levels for suppliers (Kotzab, 1999). Expected benefits from ECR include lower to-

tal system inventories and costs, enhanced consumer value in terms of choice and quality of products, and more successful development of new consumer-driven products (Kotzab, 1999; Whipple *et al.*, 1999). The ECR global scorecard is shown in Fig. 17.4 and highlights the process enablers and integrators that affect the supply and demand management strategies of suppliers and retailers.

Leading European retailers and manufacturers founded ECR-Europe in the mid-1990s to consider ECR for the European business situation (Kotzab, 1999; Fearne *et al.*, 2005). Much was expected in terms of short-term results despite Salmon and other ECR theorists claiming it was a long-term strategy (Kotzab, 2000). In the UK, the top five grocery retailers – Tesco, Sainsbury, Morrison's, Somerfield and

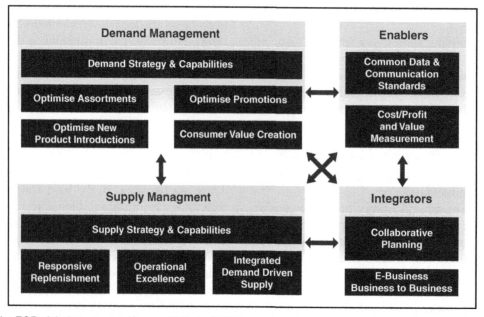

Figure 17.4 The ECR global scorecard. (Source: Kotzab, 1999.)

Asda – account for more than 70% of the UK retail food market. This has led to the UK grocery supply chain being declared "amongst the most efficient in the world"; thus the potential impact of ECR in the UK "may not be as significant as in the U.S. or Europe" (Patel *et al.*, 2001, p. 140).

However, implementation of ECR in the US and Europe, whilst easy in theory, has proved difficult in practice, and early results have been disappointing (Mathews, 1997; Whipple *et al.*, 1999; Kotzab, 2000). Implementation of an ECR system means firms must decide how to vertically coordinate various supply chain actors. Thus, issues regarding adversarial power and channel control have also been barriers to successful implementation (Whipple *et al.*, 1999; Grant, 2005).

An early ECR pilot programme at Somerfield saw inventory levels reduced by up to 25% but service levels improved by only about 2.5%. Despite integration difficulties some "soft" benefits occurred, such as improved management of seasonal events (Younger, 1997). Other ECR pilot programmes have benefited primarily dyadic relationships between firms as opposed to the entire supply chain (Kotzab, 2000).

Stock-outs continue to be a problem in some settings, product category management that is a feature of some ECR applications has been criticised as being too time and data intensive, and ECR is still perceived as a technique only suitable for large manufacturers and retailers (Mitchell *et al.*, 2001; Corsten and Gruen, 2003). Other implementation issues to be solved as ECR continues to unfold include:

- "Who identifies and allocates costs and benefits in the supply chain?
- Who resolves an actor benefiting at the cost of another actor?
- What supply chain performance standards are appropriate? and,
- What sanctions should apply to actors who do not perform to these standards?"

(Patel *et al.*, 2001, p. 142).

Notwithstanding these issues and its lack of early success, the concept of ECR as a prescriptive management technique will never go away and will continue to evolve (Mathews, 1997). This evolution is intended to incorporate all actors in the food and other fast moving consumer goods (FMCG) supply chains. One such evolution is the concept of collaborative planning, forecasting and replenishment (CPFR).

CPFR was developed by the Voluntary Interindustry Commerce Standards (VICS) group in the US to "minimise out-of-stocks by synchronising forecasting and planning between retailers and manufacturers" (Corsten and Hofstetter, 2001, p. 62). This enhancement is therefore a "step beyond ECR" or other automatic replenishment programmes (ARP) that rely on "inventory restocking triggered by actual needs rather than relying on long-range forecasts and layers of safety stock just in case" (Stank *et al.*, 1999, p. 75). CPFR, as presently configured between only manufacturers and retailers, is currently unsuitable for every firm as firms require sufficient revenue and product volumes to be economically feasible and real-time information sharing on a common platform such as the internet (Stank *et al.*, 1999; Marzian and Garriga, 2001). This will require collaboration and technological sophistication throughout the entire supply chain.

17.7.4 Jobs and work design

Professor Hans-Christian Pfohl of the European Logistics Association has commented that business is people and that the success or failure of a business depends on management's ability to harness the willing participation and creativity of its people. However, a study in the early 1990s of over 200 European and US organisations found that the top six barriers to instituting a high-quality business programme were related to employees or organisational issues. In order of importance they were:

- changing the corporate culture;
- establishing a common vision throughout the organisation;
- establishing employee ownership of the quality process;
- gaining senior executive commitment;
- changing management processes;
- training and educating employees

(Grant *et al.*, 2006b).

Further, in the mid-1990s the UK government supported a Food and Drinks Industry Benchmarking and Self-Assessment Initiative to enable organisations to assess their management systems and business performance programmes against the European Business Excellence Model (Mann *et al.*, 1999). A report on the feedback from the first 50 organisations to take part found that only a minority of food and drinks organisations were developing their management systems along the lines of business excellence.

A majority of organisations were applying traditional methods of management, were not learning from the experiences of best-in-practice organisations, and did not apply a systematic approach to achieving business improvement. The prime criteria of weakness were people (i.e. employees and staff) management and satisfaction, customer satisfaction, impact on society, and policy and strategy. These findings were supported by data from a cross-industry comparison that showed that the food industry was performing the least well (Mann *et al.*, 1999). The result was that financial results and overall competitiveness of the industry were reduced, thus highlighting the importance of instituting a proper job and work design process.

Job and work design is about structuring an "individual's job, the workplace or environment they work and their interface with the technology they use" (Slack *et al.*, 2004, p. 284). The structuring of jobs is termed scientific management or Taylorism, after Frederick Taylor's book on worker time and motion studies published in the early 1900s (ibid). Scientific management is a well-established discipline but is sometimes criticised as it does not include behavioural considerations.

A typical behavioural job design model will combine techniques of the job design and core job characteristics together with mental states of the worker and performance and personal outcomes. Other behavioural considerations include job rotation, enlargement and enrichment, empowerment, flexible working and team working. The latter considerations, the work environment and technological interface are also known as ergonomic or human factor considerations. It is important that individuals in the workplace are treated with respect and dignity regarding their job and the environment in which they fulfil it. For example, proper working temperatures, illumination and noise levels and anthropometric and neurological aspects are important physical human factor issues (Slack *et al.*, 2004).

17.8 Human resource management

Human resource management is concerned with people at work and relationships with the firm aiming to bring together and develop people in an organisation. The key objective is to develop and apply policies in key areas such as (Needham, 2001):

- workforce planning, recruitment, selection
- induction, education, training
- terms of employment, remuneration
- working conditions, health and safety
- negotiation of agreements (wages, conditions, holidays, sickness leave)
- establishing procedures for avoidance and settlement of disputes.

These areas may be broadly divided into three parts:

1. Utilisation – recruitment, selection, training, etc.
2. Motivation – job design, remuneration, participation, etc.
3. Protection – working conditions, safety, etc.

The manager commands a critical role in every firm function and the Human Resource Management functions are, according to Cole (1997):

1. Set objectives for each work area including the decision making.
2. Plan how to reach the objectives.
3. Organise work, analyse and allocate to groups and individuals.
4. Motivate, communicate and provide incentives and information.
5. Measure results, check against plans and control function.
6. Develop people (or manage them to enable self development).
7. Delegate, but do not abdicate, and provide clear instructions/rules and support for those responsible for delegated work.

The successful manager should also be effective, efficient and should ensure that the efforts of subordinates are directed towards the organisation.

Management also evolves and adapts to changing economic, social and technical circumstances. The situation in which managers operate is dynamic and it is always necessary to adapt and improve. There are, however, basic principles and/or precepts that are well established. The Scientific Management School typically asks: What's the job to be done? How best to organise it? Their response is to measure the task, design the system and then repeat the action. The workforce was in those early years treated as mere components of the system, more or less as robots not expected to think. As social conditions changed management developed and another school of management emerged with foundations of approach shaped by education and training in sociology and psychology. These are known as 'Human Behaviourists' (HB). They ask: What do people want from their job? and What leads to a contented work force? Central to the HB school are:

- motivational theory which rests upon the assumption that all rational behaviour is attributable to a cause;
- assumptions about people: McGregor pointed to the two basic assumptions about human behaviour and labelled them Theory X and Theory Y. These are two possible differing perspectives of managers. The first is that people at work are basically lazy, requiring coercion and control, whereas the other perspective is that people naturally like their work, are eager to learn, committed to the organisation and do not need coercion and rigorous control. Under good management people will seek responsibility and be able to exercise imagination and ingenuity for mutual benefit (Cole, 1997).

17.9 Finance and accounting for food firms

All businesses have a financial management component. The responsibility may be delegated to specialists who have the essential tasks of ensuring there are adequate controls, conformance to standards and conventions and integrity within the financial system. Managers must ensure earnings are duly received by the firm, and wages, salaries, expenses and taxes are paid at the due time in correct amounts. Managers must have a financial awareness of the accounting systems, terminology and the strengths and weaknesses of the measurement of income and wealth of the business.

According to Weightman (2006), the purpose and functions of financial management and accounting are as follows:

- Reporting
    - Internal to the firm
    - External to relevant interests
- Planning and controlling
    - Financial budgeting and follow up (control)
    - Capital investment appraisal (resource allocation)
- Decision making where the costs and returns are important
- Fiscal compliance and reporting
    - Value Added Tax
    - Income/Corporation Tax
- Company law compliance
    - Returns to the Registrar of private and public companies at Company House
    - The existence of records and supporting systems
    - Auditing – internal and external

The accounting discipline, on the other hand, deals with different groups/stakeholders including shareholders, company's management, creditors, suppliers, customers and employees. It involves managing assets (what the entity owns), liabilities (what the entity owes) and capital (the owners' interest in the business) and should conform to the following formula: Capital + Liabilities = Assets.

In the accounting field, there are two main branches. First, financial accounting is of basic importance. It is concerned with accounting for external providers of finance (investors) and creditors such as banks and finance sources. It also covers all aspects of published financial reports usually on an annual basis. Second, management accounting aims to provide information for internal direction and control. It will, for example, provide costs of production, margins on sales of specific products or services, budgets to portray the expected levels of income and earnings for each month or other accounting period.

Basic accounting incorporates the following accounting terms (Weightman, 2006):

- *Entity*: try to restrict the amount of data available only to the business. The accounts should exclude the private affairs of individuals; this is the case for small food companies where owners may charge other or personal expenditures to the company.
- *Periodicity*: accounts should be prepared at the end of a defined period of time and this period should be adopted as the regular period of account.
- *Going concern*: this assumes that the business entity will continue in existence for the foreseeable future unless there is strong evidence to suggest that this is not the case.

- *Quantitative*: only data that are capable of being easily quantified should be included in an accounting system.
- *Historical cost*: this requires transactions to be recorded at their original/historic purchasing or selling cost, with subsequent changes such as arising through price inflation being ignored.
- *Matching*: this endeavours to match expenditure with income generated by those expenditures such as wages, raw materials and purchased services in the accounting period.
- *Prudence/conservatism*: the rule is, if caution is in doubt, overstate losses and understate profits; apply to budgets and valuations of stocks of products for sale, work in progress and the value of fixed assets, which are things that the business aims to keep.
- *Consistency*: once specific accounting policies have been adopted, they should be followed in all subsequent accounting periods in order to compare different time periods. Consistency may not be upheld due to changes in accounting rules and conventions made by the accountancy profession and/or by government regulation. When these occur they should be disclosed in the notes to the accounts.
- *Objectivity*: personal prejudice must be avoided in the interpretation of the basic accounting rules.
- *Double entries*: for every transaction, there is always a twofold effect giving rise to double-entry book keeping.

Management accounting should determine and predict the flows, balances and requirements of short-term financial resources, should provide financial information and predict the economic condition of the accounting entity, and provide financial information that is useful for monitoring performance under legal and contractual requirements. The key financial reports should provide information useful for planning the budgeting and for predicting the impact of the allocation of resources on the achievement of operational objectives and information that is useful for evaluating managerial performance. These reports should be also governed by specific principles as set out above, such as (Weightman, 2006):

- Objectivity – do the controllers of the company report in a fair and objective manner to the owners?
- Consistency – how consistent are financial reports over time? Do the same rules apply?

- Comparability – reports of an organisation may be compared with another, although public sector accounts are not comparable with private sector accounts and an inter-sector analysis may not be appropriate (e.g. supermarkets versus manufacturers).
- Timeliness – the time it takes for an organisation to produce the financial reports.

17.9.1 The essential financial reports for external reporting

The three key financial reports are the balance sheet, profit and loss account and cash flow statement.

17.9.1.1 The balance sheet

The balance sheet may be defined as a "statement at a point in time of the assets and liabilities of the business". Assets are proprietary rights and all are owned by someone or some entity/organisation. They may be tangible such as buildings, machinery, stocks of product and raw materials or intangible such as patent rights, customer lists and brand names which are much more difficult to value. Liabilities are claims against those assets and are such things as bank loans, creditors to whom the business owes money and shareholders' funds. Assets must equal liabilities, or "balance" as if measured on scales; hence the balance sheet is so named. It is not a true measure of the market value of the business, but rather a statement of where the money in a business has come from and where it is currently located. The balance sheet is regarded with reservations because it contains estimates of value and omissions. The most evident omission is assets that are not acquired by a transaction; for example, the skills of the workforce and management, the value of loyal customers and much more.

One of the most useful parts of the balance sheet is the entry "changes in reserves" between two years. Reserves are identified in the balance sheet and arise from several sources in a business, but usually from profits or surpluses that have not been distributed to owners or shareholders. These are the profits that have been ploughed back into the firm to hopefully allow it to grow.

Here is a generic example expressed in columnar format and in "traditional" side by side format (Weightman, 2006). Assets must balance with liabilities, or claims on the value of assets of the business.

CXYZ Co. Balance Sheet at 31.12.20xy

ASSETS	£
Fixed assets	1000
Current assets	1000
Investments	1000
	3000

LIABILITIES/CLAIMS or financed by	
Share capital	1000
Loan capital	1000
Reserves	1000
	3000

OR ALTERNATIVELY PRESENTED AS
(*traditional format*)

Liabilities or claims
on the firm   Represented by

	£		£
Share capital	1000	Fixed assets	1000
Loan capital	1000	Working capital	1000
Reserves	1000	Investments	1000
	3000		3000

The following accounting principles are used in the preparation of a balance sheet:

- Measured in money terms
- The business in separate entity
- Claims = Assets
- Value as a going concern
- Plant and equipment measured at cost less depreciation

In a balance sheet the assets and claims should conform to the following formula:

Assets = Current Liabilities + Long-Term Liabilities + Proprietorship Interest

A balance sheet gives only a snapshot at a single moment and is prepared on the same accounting principles every year.

Reserves arise primarily from success in business. Reserves are not usually held in cash, although there may be a small element of cash held by the firm. They arise secondly from "holding gains", for example from assets held by the firm that appreciate (rather than depreciate). They arise thirdly when shares are sold by the company above their nominal value (named value such £1 or $1).

17.9.1.2 The profit and loss account

In contrast to the balance sheet, a statement at a point in time is the profit and loss account that determines the profit and loss a business incurs over a trading period. It complements the use of the balance sheet. However, it is unwise to rely entirely upon the profit and loss account and the balance sheet as they cannot provide information about where the cash has come from and where it went.

A simplified example is shown in below (Weightman, 2006):

Example plc: Profit and loss account for the year ended 31 December 200x

	£'000	
Turnover	5590	
Cost of sales	4100	
Gross profit	1490	
Other expenses	840	
Profit before interest and taxation	650	Operating profit
Interest payable	50	Net of interest received
Profit before taxation	600	
Taxation	135	
Net profit for the year	465	
Dividends	230	
Retained profit for the year	235	Transferred to balance sheet reserves

- *Turnover* is another word for sales, but excludes VAT and is adjusted for stocks held at the start and end of the year. Problems can arise in the basis of valuation of these. Sales are not necessarily cash. Many businesses sell on credit terms, e.g. 30, 60, 90 days, yet these are counted as a sale before the cash is received.
- *Cost of sales* is the cost of goods or services used to generate turnover, including wages, materials and depreciation on assets used. The account includes a factual record of expenditure, but it also makes estimates. These include depreciation which is a notional charge made for the use and/or obsolescence of assets used over more than one accounting period.
- *Depreciation* is a charge for the use of assets (such as machinery, buildings). There are several ways of

calculating this; the most frequently used method is the "Straight Line Method" where:

$$D = \frac{\text{Purchase Price} - \text{Scrap Value}}{\text{Number of years of use of asset}}$$

It may alternatively be calculated by the reducing balance method where a fixed percentage is used to depreciate the capital items. For example, 20% for machinery and vehicles is applied to the purchase price or acquisition value in year 1, then to the "written down value" thereafter (the depreciated value). Depreciation is not a fund to generate replacement of the asset, although it indirectly does so.

Retained profit allows the business to grow. Profit can be alternatively defined as an increase in net assets.

17.9.1.3 The cash flow statement

Managers will look to the cash flow statement that focuses on the true movement of funds through the business in the accounting period. It begins with operating profit from the profit and loss account to which adjustments are made to eliminate non-cash items referred to above such as depreciation and changes in value of stocks. When these and other adjustments are made, the standard format of the cash flow statement will show where the cash has been spent or allocated. This includes to providers of finance as interest, shareholders as dividends, new capital investment and acquisition and disposal of assets.

17.9.1.4 Analysis of financial information

In terms of conducting a financial analysis, there is a range of relevant ratios dealing with the raising of capital, the use of that capital to earn profits and the cash management in order to remain solvent. In general, ratios provide vital signs of a company's health and have little meaning on their own and have to be compared with previous years' figures to identify trends, the figures with similar firms in the same industry and the sector/industry average. The key performance measurement ratios include the profitability ratios, the short-term liquidity ratios, the long-term solvency ratios and the efficiency ratios (Weightman, 2006).

In relation to profitability ratios, a key ratio is profit earned by capital employed which measures the success of operations management in using the total assets available.

$$\text{Profit earned by capital employed} = \frac{\text{Profit before taxation} + \text{Interest expense}}{\text{Current assets} + \text{Unquoted investments} + \text{Fixed assets}}$$

For the liquidity ratios it is worth stressing that a business can be profitable and insolvent at the same time. Liquidity is the ability of a company to pay its bills and to stay in business and requires careful cash flow management. The current ratio should not be read in isolation with the rest of the balance sheet and it emphasises that different industries work on different working capital.

$$\text{Current ratio} = \frac{\text{Current assets}}{\text{Current liabilities}}$$

The liquid ratio (acid test ratio) demonstrates the organisation's ability to pay in the short term and those assets that may be readily turned into cash. A safe ratio is 1:1 but this depends on the sector. A major drawback is that it can be easily manipulated (an aspect of creative accounting).

$$\text{Liquid ratio} = \frac{\text{Current assets} - \text{Stocks}}{\text{Current liabilities}}$$

The debtor turnover period ratio measures the average length of time that it takes customers to pay what they owe. In general, many food businesses depend on credit (e.g. food retailers) only, others depend only on cash, and others on both. If the company depends only on cash, the ratio is not applicable.

$$\text{Debtor turnover ratio} = \frac{\text{Debtors}}{\text{Sales turnover}}$$

The stock turnover period ratio examines stock stated at cost price and sales at selling price and is influenced by changes in profit margin and the rate at which the physical stocks are sold.

$$\text{Stock turnover period ratio} = \frac{\text{Average stock}}{\text{Materials element of cost of goods sold}}$$

or

$$\frac{\text{Average stock}}{\text{Sales (turnover)}}$$

The long-term solvency ratios deal with borrowing where too much borrowing causes big changes in the return to ordinary shareholders. In principle, the fixed assets should be bought out of long-term capital and companies that have a high proportion of fixed interest loan capital to equity are highly geared. The capital gearing ratio indicates that the organisation should consider some borrowing to improve its liquidity.

$$\text{Capital Gearing Ratio} = \frac{\text{Equity capital}}{\text{Fixed investment borrowing}}$$

The debtor's ratio is an efficiency measure and assesses the average time to collect a debt as more companies fail daily to pay their way.

$$\text{Debtor's ratio} = \frac{\text{Debtors}}{\text{Average daily credit sales}}$$

The creditor's ratio measures the average time taken to pay a debt and depends on the industry average, the company's market power and the relationship with its suppliers.

$$\text{Creditor's ratio} = \frac{\text{Creditors}}{\text{Average daily credit purchases}}$$

The stock turnover ratio ensures that the company does not tie up large volumes of working capital in stocks, and the faster the stock is turned over the better. This ratio is very important for food retailers as a high turnover is essential for their profitability.

$$\text{Stock turnover} = \frac{\text{Cost of goods sold} \times 360 \text{ days}}{\text{Average stock of finished goods}}$$

Finally, another relevant ratio is gearing or leverage, which is the proportion of profits needed to meet interest on loan capital. The greater the level of gearing, the greater the degree of risk for both lenders and ordinary stockholders.

$$\text{Gearing} = \frac{\text{Interest}}{\text{Profit before taxation} + \text{Interest}}$$

Highly geared businesses are when borrowed money exceeds the value of equity, or owners' funds. This is satisfactory when profits are high and rising. Interest can easily be paid out of earnings. When profits fall or if interest rates rise these highly geared businesses are exposed to the danger of creditors foreclosing to recover the monies owing.

Last but not least, considering that the business environment is under continuous change, budgetary planning becomes very crucial for companies' survival. There are two types of budgetary planning: physical (showing the expected scale of production) and financial (estimating the costs of the physical plan). In principle, a budget is the representation of the policy of a business that is to be pursued. It starts with forecasts of sales, production and costs of capital, labour and materials. The budget needs to be co-ordinated by the top management in order to avoid variances between various units within the company. Budgets should be agreed ahead of time since they become the target to be achieved, or the limit for expenditures. Variance always occurs and it is a need for the analysis of results between the initial budget and the actual out-turn. Results can be either favourable (positive) or unfavourable (negative). The key causes of variance can be related to different prices for bought-out materials and components, changes in product design altering the cost of inputs, policy decisions of various kinds, inflation, productivity issues, external environment issues such as competitors' actions, strikes, power failures and new technologies. There are different types of variance (Weightman, 2006). Specifically, there are four direct material variances:

1 price variance that is related to price fluctuations;
2 usage variance due to materials' deterioration from poor storage conditions, purchasing with a substandard quality or using wrong materials;
3 mix variance since different materials have different costs; and
4 yield variance due to unexpected outcomes such as the processing of a standard mix of standard materials at the wrong temperature.

The direct labour variances are related to the different grades of direct labour and the lack of availability at a particular point in time of a particular type of labour. The manufacturing overhead variances are due to the standard costs of overhead items and actual costs, the standard usage of overhead items and actual usage and the normal capacity and the actual level of capacity utilisation experienced during a given period. Normally, unfavourable variance arises when the actual cost is higher than the budgeted cost, the actual revenue is less than the budgeted revenue and the actual profit is less than the budgeted profit.

17.10 Conclusions

This chapter covered the critical role of business and management in key sectors of the food industry. We analysed key business and management functions including food operations management, human resource management, finance and accounting. We envisage that our analysis will prove beneficial to, inter alia, academics, students and practitioners and will expand their understanding of that specialised field of study.

Further reading and references

Alvarado, U.Y. and Kotzab, H. (2001) Supply chain management: the integration of logistics in marketing. *Industrial Marketing Management*, **30**, 183–98.

Apaiah, R.K., Hendrix, E.M.T., Meerdink, G. and Linnemann, A.R. (2005) Qualitative methodology for efficient food chain design. *Trends in Food Science and Technology*, **16**(5), 204–214.

Blythman, J. (2005) *Shopped: The Shocking Power of British Supermarkets*. Harper Perennial, London.

Bourlakis, M. and Bourlakis, C. (2006) Integrating logistics and information technology strategies for sustainable competitive advantage. *Journal of Enterprise Information Management*, **19**(2), 389–402.

Bourlakis, C. and Bourlakis, M. (2005) Information technology safeguards, logistics asset specificity and fourth-party logistics network creation in the food retail chain. *Journal of Business and Industrial Marketing*, **20**(2), 88–98.

Bourlakis, M. and Bourlakis, C. (2001) Deliberate and emergent logistics strategies in food retailing: a case study of the Greek multiple food retail sector. *Supply Chain Management: An International Journal*, **6**(3/4), 189–200

Bourlakis, M. and Weightman, P. (2004) *Food Supply Chain Management*. Blackwell, Oxford.

Browne, M. and Allen, J. (1997) The four stages of retail. *Logistics Europe*, **5**(6), 34–40.

Christopher, M. (2005) *Logistics and Supply Chain Management: Creating Value-Adding Networks*, 3rd edn. FT Prentice Hall, Harlow.

Cole, G.A. (1997) *Management Theory and Practice*. DB Publications, London.

Corsten, D. and Gruen, T. (2003) Desperately seeking shelf availability: an examination of the extent, the causes, and the efforts to address retail out-of-stocks. *International Journal of Retail and Distribution Management*, **31**(12), 605–617.

Corsten, D. and Hofstetter, J.S. (2001) An interview with VICS Chairman, Ron Griffen. *ECR Journal – International Commerce Review*, **1**(1), 60–67.

Dawson, J.A. and Shaw, S.A. (1990) The changing character of retailer–supplier relationships. In: *Retail Distribution Management* (ed. J. Fernie), pp. 19–39. Kogan Page, London.

Fearne, A. (1998) The evolution of partnerships in the meat supply chain: insights from the British beef industry. *Supply Chain Management*, **3**(4), 214–31.

Fearne, A., Duffy, R. and Hornibrook, S. (2005) Justice in UK supermarket buyer–supplier relationships: an empirical analysis. *International Journal of Retail and Distribution Management*, **33**(8), 570–82.

Fernie, J., Pfab, F. and Marchant, C. (2000) Retail grocery logistics in the UK. *International Journal of Logistics Management*, **11**(2), 83–90.

Fine, B., Heasman, M. and Wright, J. (1996) *Consumption in the Age of Affluence: The World of Food*. Routledge, London.

Finegan, N. (2002) *Backhauling and Factory Gate Pricing*. Institute of Grocery Distribution, Watford.

Flanagan, P. (1992) Customer service requirements in the UK food processing industry. *Focus: The Journal of the Institute of Logistics and Distribution Management*, **11**(10), 22–4.

Food Chain Group (1999) *Working Together for the Food Chain: Views from the Food Chain Group*. Ministry of Agriculture, Fisheries and Food, London.

Gadde, L.-E. and Håkansson, H. (2001) *Supply Network Strategies*. Wiley, Chichester.

Gannaway, B. (2001) Issues of control. *The Grocer*, **224**, 32–4.

Grant, D.B. (2004) UK and US management styles in logistics: different strokes for different folks? *International Journal of Logistics: Research and Applications*, **7**(3), 181–97.

Grant, D.B. (2005) The transaction-relationship dichotomy in logistics and supply chain management. *Supply Chain Forum: An International Journal*, **6**(2), 38–48.

Grant, D.B., Karagianni, C. and Li, M. (2006a) Forecasting and stock obsolescence in whisky production. *International Journal of Logistics: Research and Applications*, **9**(3), 319–34.

Grant, D.B., Lambert, D.M., Stock, J.R. and Ellram, L.M. (2006b) *Fundamentals of Logistics Management: First European Edition*. McGraw-Hill, Maidenhead.

Harrison, A. and van Hoek, R. (2005) *Logistics Management and Strategy*, 2nd edn. FT Prentice Hall, Harlow.

Hingley, M.K. (2005) Power imbalanced relationships: cases from UK fresh food supply. *International Journal of Retail and Distribution Management*, **33**(8), 551–69.

Hogg, C.D. (2001) Fast food: some facts and figures to make you lose your appetite. *The Independent*, 5 September, Wednesday Review, 8.

Houghton, E. and Portougal, V. (2001) Optimum production planning: an analytic framework. *International Journal of Operations and Production Management*, **21**(9), 1205–1221.

Jobber, D. (2004) *Principles and Practice of Marketing*, 4th edn. McGraw-Hill, Maidenhead.

Key Note (2005) *Food Market (UK)*. www.keynote.co.uk, viewed September 2006.

Kotzab, H. (1999) Improving supply chain performance by efficient consumer response? A critical comparison of existing ECR approaches. *Journal of Business and Industrial Marketing*, **14**(5/6), 364–77.

Kotzab, H. (2000) Managing the fast moving goods supply chain – does efficient consumer response matter? *Proceedings of the Logistics Research Network 5th Annual Conference*, September, Cardiff Business School, Cardiff, pp. 336–43.

Leenders, M.R., Fearon, H.E., Flynn, A.E. and Johnson, P.F. (2002) *Purchasing and Supply Management*, 12th edn. McGraw-Hill, New York.

Liberopoulos, G. and Tsarouhas, P. (2005) Reliability analysis of an automated pizza production line. *Journal of Food Engineering*, **69**, 79–96.

Mann, R., Adebanjo, O. and Kehoe, D. (1999) An assessment of management systems and business performance in the UK food and drinks industry, *British Food Journal*, **101**(1), 5–21.

Marzian, R. and Garriga, E. (2001) *A Guide to CPFR Implementation*. ECR Europe, Brussels.

Mathews, R. (1997) A model for the future. *Progressive Grocer*, September, 37–42.

Mitchell, A., Corsten, D., Jones, D.J. and Hofstetter, J.S. (2001) A platform for dialogue. *ECR Journal – International Commerce Review*, **1**(1), 8–17.

Needham, D. (2001) *Business for Higher Awards*. Heinemann, Oxford.

Patel, T., Sheldon, D., Woolven, J. and Davey, P. (2001) *Supply Chain Management*. Institute of Grocery Distribution, Watford.

P-E International (1991) *Long-Term Partnership – Or Just Living Together?* P-E International, Leinfelden-Echterdingen.

Pecore, S. and Kellen, L. (2002) A consumer-focused QC/sensory program in the food industry. *Food Quality and Preference*, **13**(6), 369–74.

Robson, I. and Rawnsley, V. (2001) Co-operation or coercion? Supplier networks and relationships in the UK food industry. *Supply Chain Management: An International Journal*, **6**(1), 39–47.

Simons, D. and Zokaei, K. (2005) Application of lean paradigm in red meat processing. *British Food Journal*, **107**(4), 192–211.

Slack, N., Chambers, S. and Johnston, R. (2004) *Operations Management*, 4th edn. FT Prentice Hall, Harlow.

Stank, T.P., Daugherty, P.J. and Ellinger, A.E. (1998) Pulling customers closer through logistics service. *Business Horizons*, **41**, 74–81.

Stank, T.P., Daugherty, P.J. and Autry, C.W. (1999) Collaborative planning: supporting automatic replenishment programs. *Supply Chain Management*, **4**(2), 75–85.

Strak, J. and Morgan, W. (1995) *The UK Food and Drink Industry*. Euro PA and Associates, Northborough.

Tansey, G. and Worsley, T. (1995) *The Food System: A Guide*. Earthscan Publications, London.

Van der Vorst, J.G.A.J. (2000) *Effective food supply chains: generating, modelling and evaluating supply chain scenarios*. PhD Thesis, Wageningen University, Wageningen.

Van Hoek, R. (1999) Postponement and the reconfiguration challenge for food supply chains. *Supply Chain Management*, **4**(1), 18–34.

Vlachos, I. and Bourlakis, M. (2006) Supply chain collaboration between retailers and manufacturers: do they trust each other? *Supply Chain Forum: An International Journal*, **7**(1), 70–81.

Waters, D. (2003) *Inventory Control and Management*, 2nd edn. Wiley, Chichester.

Weightman, P. (2006) *Lecture Notes*. Newcastle University, Newcastle.

Whipple, J.S., Frankel, R. and Anselmi, K. (1999) The effect of governance structure on performance: a case study of efficient consumer response, *Journal of Business Logistics*, **20**(2), 43–62.

Yakovleva, N. and Flynn, A. (2004) Innovation and sustainability in the food system: a case of chicken production and consumption in the UK. *Journal of Environmental Policy and Planning*, **6**(3–4), 227–50.

Younger, R. (1997) *Logistics Trends in European Consumer Goods: Challenges for Suppliers, Retailers and Logistics Companies*. Financial Times Management Report, London.

Takahide Yamaguchi

- Marketing activities are implemented by companies and organizations in the food industry 'to create value for customers' and 'to build strong customer relationships'.
- Marketers have to understand their markets and customers to formulate a marketing strategy for gaining competitive advantages.

18.1 Introduction

Marketing activities are generally carried out in all types of industries. To a large extent, the kinds of marketing activities carried out are common across all industries. However, the nature of the industry does lead to some differences in the activities. This implies that the methodology of marketing can be applied to many industries, albeit with some modifications. Marketing can also be applied to the 'food' industry. There exist marketing books that offer information about the general methodologies and theories of marketing based on studies of food companies. Basic textbooks will contain case studies on Nestlé, Unilever, Coca-Cola, etc. McDonald's is a popular case in recent years. Food companies contribute to the development of marketing theory as objects of case study.

However, discussions on a common marketing methodology for the entire food industry are confronted with one difficulty. 'Food' is a general term for edible products that human beings ingest in daily life. The companies mentioned above produce different kinds of food. Nestlé is famous for dairy products, chocolate, coffee, etc. Unilever makes margarine, Coca-Cola is a soft drink company and McDonald's is a hamburger chain. All these companies are included under the broad classification of the 'food industry' despite the fact that they produce different kinds of food. In addition, companies that produce perishable food, canned food, frozen food, processed food, drinking water, alcoholic beverages, etc. are also classified under the same umbrella. The food industry comprises plural sub-industries. Therefore, there is a need to consider many types of marketing applications that correspond to the different products in the food industry.

In this chapter, the marketing activities implemented by companies and organizations in the food industry are described as 'food marketing'. As mentioned above, the food industry covers a wide range of products, and each product needs the application of different marketing activities. It is impossible to explain all the activities in this chapter. However, while explaining basic food marketing, we will describe the marketing activities of some food companies. Food marketing will be explained in the following manner. We will begin by describing the marketing concept and the marketing process in order to explain the definition of marketing. In the following sections, we will explain marketing research, the formulation of a marketing strategy and the devising of a marketing plan.

18.2 Marketing principles

In this section we:

- discuss the nature and intent of the marketing concept;
- determine the marketing process;
- apply social, legal, ethical and environmental principles to marketing situations.

18.2.1 The marketing concept

Our first question is 'What is marketing?' You are aware that TV commercials and big signboards on the roofs of buildings are a part of marketing activities. When you read newspapers and magazines, you come across a huge amount of advertising. Advertising is an important part of marketing. Sometimes, you are requested to sample new products, such as a new flavour of cheese or a new taste of beer, in a supermarket. Sales people explain to you how the new sample differs from existing products. This selling activity is also a part of marketing. Thus, we see some examples of marketing activity in our daily lives. However, advertising and selling constitute only a part of marketing.

Today, marketing is considered to have a broader and deeper meaning than in the past. Kotler and Armstrong (2006) define marketing thus:

'Marketing is the process by which companies create value for customers and build strong customer relationships in order to capture value from customers in return.'

This definition comprises two parts. The first part is 'to create value for customers'. Marketing begins with understanding the customer's needs and wants.

Finding potential needs and wants that the customer is still unaware of also plays a key role. In order to create value for customers, it is necessary to realize customers' needs and wants. Knorr's soup stock was accepted by customers since it realized one such customer need: it not only reduced the soup to powder but also met the demand for quick and easy soup preparation.

The second part of the definition is 'to build strong customer relationships'. This part suggests that the goal of marketing is to capture value from the customer. In order to deliver the value created to the customer, many kinds of activities are carried out, such as advertising and sales promotion. Further, once a customer buys a product, some measures must be taken to ensure that he/she makes repeat purchases. It is important to establish the brand name for the customer to recognize the product.

18.2.2 The marketing process

As mentioned above, the aim of marketing is to capture value from the customer. To arrive at this value, marketers go through five steps, which are collectively referred to as the 'marketing process' (Kotler and Armstrong, 2006):

1 Understand the marketplace and customer needs and wants.
2 Design a customer-driven marketing strategy.
3 Construct a marketing programme that delivers superior value.
4 Build profitable relationships and create customer delight.
5 Capture value from customers to create profits and customer equity.

The first step of the marketing process requires marketers to obtain an understanding of their marketplace and customers. Marketing research is one method to understand the market. Marketers need to understand the main influences on the marketing activities within the macro and micro environments and on customer behaviour. Further, they need to collect, analyse and evaluate information and data by using different methods.

The second step is to design a customer-driven marketing strategy. Marketing strategy formulation depends on the targeted customers; therefore, marketers are forced to decide who their customers are. In other words, it is important for marketers to carry out market segmentation and to deliver their goods to their target customers.

The third step is to prepare a practical marketing programme. The preparation of the marketing programme begins with planning the details of the marketing mix. The marketing mix is a combination of the firm's marketing tools – product, price, promotion and place. These marketing tools are known as the 'four Ps'. By controlling these four Ps, firms need to draft a marketing programme that will meet the wants of the target market.

The fourth step consists of two parts. The first part is customer relationship management, and the second part is partner relationship management. The former pertains to the manner in which firms build relationships with their customers, while the latter pertains to how firms build relationships with their trade partners.

The final step of the marketing process is aimed at capturing value from the customer. Firms attempt to build customer loyalty by enhancing customer satisfaction. They focus not only on increasing their market share but also on increasing customers' mind share. By building customer loyalty, firms are able to increase their profits substantially.

In the following sections, the marketing process for food marketing is discussed in detail.

18.2.3 Application of social, legal, ethical and environmental principles to marketing situations

Before explaining the marketing process, it is important to discuss corporate social responsibility (CSR). The concept of CSR is not new. McGuire (1963) defined it as follows:

> 'The idea of social responsibility supposes that the corporation has not only economic and legal obligations, but also certain responsibilities to society which extend beyond these obligations.'

The economic responsibilities demand that the corporation should produce goods and services that society wants and sell them at fair prices (Carroll, 1996). Fair prices do not naturally imply low prices. Rather, a fair price is one that includes dividends to the investors and sufficient profits for the continuance of the business. In recent years, legal responsibilities have come to the fore due to the keen market competition. Compliance with laws is an important aspect of a business's responsibilities. However, a company's ethical responsibilities toward society go beyond its economic and legal obligations. Ethical responsibilities embrace those activities and practices that are expected or prohibited by societal members even though they are not codified into law (Carroll, 1996).

The application of CSR to marketing situations is linked to producing a positive image that the company contributes to society and that its existence is valuable to society. The positive image is not a tangible asset of the company but is stored in the minds of customers. Thus, neglecting CSR can lead to major problems. For example, when Snow Brand Milk failed to prioritize CSR, it not only destroyed its favourable reputation in the Japanese market but also incurred huge losses in its milk business. In 2000, low-fat milk produced by Snow Brand Milk led to rampant food poisoning. The ensuing investigation revealed that the production process, although compliant with the existing laws, was unacceptable to consumers. However, Snow Brand Milk failed to take appropriate measures to regain consumer confidence in its products. As a result, its brand value plummeted and it lost market share.

A discussion on CSR includes various topics, including the important topic of environment protection. The significance of this topic came to light with the environmental pollution of the 1970s. In the 1980s, there emerged the so-called global environmental problems, such as global warming, depletion of the ozone layer, reckless deforestation and sea pollution. However, environment protection has rarely been discussed in the context of marketing. In 1992, Peattie suggested the concept of 'green marketing'. Green marketing proposes the development of a new marketing method to cope with both the pursuit of benefits and a reduction in the environmental load. For companies to build their reputation in the market, it is important that they carry out green marketing. In addition, customers who are very sensitive to global environmental problems constitute an important market that firms cannot afford to ignore.

18.3 Marketing research

In this section we look at how to:

- understand the main influences on the marketing process within the macro and micro environments;

- analyse and respond to the issues associated with buying behaviour in both consumer and organizational markets;
- evaluate the need for information in marketing and understand the different methods of collecting and analysing data;
- apply the marketing research process in selected food markets.

As the first step of the marketing process, it is important for marketers to understand their marketplace and customer needs and wants. Therefore, marketers research markets and customers, analyse consumer behaviour and establish methods of collecting and analysing many kinds of data. This is explained in the following section. Later, we present an example of the entire marketing research process in a food market.

18.3.1 The macro and micro environments of a company

For effective marketing, marketers need to understand all relationships that surround the company. These relationships are called the 'marketing environment'. The marketing environment consists of the actors and forces outside marketing that affect marketing management's ability to build and maintain successful relationships with target customers (Kotler and Armstrong, 2006, p. 60). It is imperative that companies adapt to the changing environment.

The marketing environment comprises a macro environment and a micro environment. A company's macro environment is formed of some societal forces that influence its marketing activities. Generally, the following forces are considered to constitute the macro environment (Kotler and Armstrong, 2006):

- Demographic environment: demography is the scientific study of the human population. It includes the study of the size, structure and distribution of populations and how populations change over time. For example, the baby boomers, who were born between 1946 and 1964, have become the most powerful market. Developing products for this market has gained prominence.
- Economic environment: this comprises the level and distribution of the income effect and consumers' purchasing power and purchasing patterns. In the case of a company that approaches a foreign market, the price of the imported food is

higher than the price of the local food. Therefore, it is important to know the size of the market for the high-priced product.

- Natural environment: concern for the global environment has been increasing steadily. Companies need to develop products that do not lead to global warming and air and water pollution. Moreover, the production processes should be modified to suit this need.
- Technological environment: new technologies create new markets and opportunities. In addition, new technology replaces older technology. Marketers should not ignore technological changes. New technology also creates new regulations. For example, it is mandatory to check the safety of genetically modified organism (GMO) foods before they are delivered to the market.
- Political environment: laws, government agencies and lobby groups form the political environment. These factors influence businesses in various ways. The number of laws and regulations is increasing year by year.
- Cultural environment: people in a society have common basic beliefs and values, namely, culture. These core beliefs and values affect their perceptions, preferences and behaviours. Eating habits are especially influenced by the food culture.

On the other hand, the micro environment consists of the actors close to the company that affect its ability to serve its customers: it comprises the company, suppliers, marketing intermediaries, customer markets, competitors and the public (Kotler and Armstrong, 2006):

- The company: a marketing plan that requires the coordination of many intra-firm functions – top management, finance, research and development (R&D), purchasing, operations and accounting – is formulated. These interrelated functions constitute the internal environment.
- Suppliers: companies need many kinds of resources to produce goods and services. These resources are provided by suppliers. If there are shortages and late deliveries of these resources, marketing activities are affected. Moreover, the prices of resources influence the prices of products and services.
- Marketing intermediaries: in making products accessible to the final customer, marketing intermediaries, for example, resellers, physical distribution

firms, marketing services agencies and financial intermediaries, play an important role. Companies need to build good relationships with their marketing intermediaries.

- Customer markets: the three well-known types of customer markets are personal consumption markets, business markets and government markets. Personal consumption markets are subdivided into individuals and households. Marketers today need to recognize both the international market and the domestic market.
- Competitors: companies produce goods in order to satisfy the needs of their target consumers. These goods must realize greater customer satisfaction than those of their competitors.
- Public: public relations influence marketing activities. A company's relationships with investors, the media, government, consumer and citizen groups and local communities play an extremely important role. Moreover, in a broad sense, workers, managers and the board of directors constitute the internal public. It is important to make employees feel good about their company.

18.3.2 Buying behaviour

In this section we describe the marketing environments that must be taken into account when carrying out marketing activities. Next, we discuss the influence of marketing activities on buying behaviour. As markets are classified into consumer markets and business markets, buying behaviour is also classified into consumer buyer behaviour in the consumer market and business buyer behaviour in the business market. These two types are explained in this section.

In order to understand consumer buyer behaviour, marketers have to consider two things: factors influencing consumers and the buyer decision process. As for factors influencing consumers, the following four are well known:

- cultural
- social
- personal
- psychological

The details of and the relationships among these factors are shown in Fig. 18.1. These four factors are presented in a kind of progression, from factors that have a broad influence to those that have a personal influence.

Marketers need to understand the three components of the buyer decision process. Specifically, they need to understand the buying decision maker, types of buyer decisions and the steps in the buyer decision process. First, the buying decision maker has a role in the buying decision process. Consider the example of a man who purchases daily food items from a supermarket by referring to a list. He is the purchaser. His partner, who made the list, is the decision maker of these purchases. We can picture the partner asking the man what he would like to have for dinner while drawing up the list. In this case, the purchaser plays the role of the influencer. Many kinds of roles exist in the buying process. Marketers have to identify the buying decision maker and approach him/her.

Second, the decision-making process of a purchaser depends on what he/she wants to buy. The types of buyer decisions are classified along two axes: the commitment level to the buying process and the differences between brands (Assael, 1987). There is a

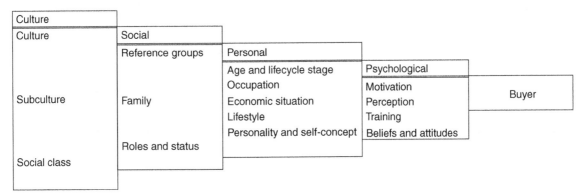

Figure 18.1 Factors influencing consumer buyer behaviour. (Source: Kotler and Armstrong (2006), p. 130, Figure 5.2.)

difference between the wine purchased for a Christmas dinner and that purchased for daily consumption. In the case of the latter, the consumer customarily buys the usual wine. The buying process seldom takes a long time. Moreover, in the process, the buyer emphasizes custom over the brand. However, in the process of buying wine for Christmas, the purchaser takes a long time and selects a good brand.

Third, the following model depicts the five steps in the buyer decision process (for example, Engel, *et al.*, 1982). The buyer:

1 recognizes a need;
2 seeks information;
3 evaluates alternatives;
4 decides the purchase; and
5 makes a post-purchase assessment.

The buying process begins long before the actual purchase and continues long after it is completed. Therefore, marketers need to focus on the entire buying process rather than on merely the purchase decision (Kotler and Armstrong, 2006, p. 147).

We now discuss business buyer behaviour. The business market consists of organizations that purchase other companies' products and services in order to produce their goods. Some characteristics of a business market differ from those of a consumer market (Kotler, 2000):

1 The business market has fewer buyers than a consumer market.

2 The buyers in a business market are bigger than those in a consumer market.

3 There exists a close relationship between the supplier and the customer.

Similar to consumer buyer behaviour, business buyer behaviour is influenced by various factors. The most important of these are environmental, organizational, interpersonal and individual factors (see Fig. 18.2). These factors include many factors that differ from those that influence the general public and those that are related to individual transactions.

In a business market, marketers also need to understand the three components of the buyer decision process, that is, the buying decision maker, types of buyer decisions and the steps in the buyer decision process.

First, in business purchasing, the decision is made in the buying centre (Webster and Wind, 1972). The buying centre includes organization members who play five roles in the purchase decision process: 'users' use the product or service; 'influencers' help to prepare the specifications; 'buyers' select the supplier for routine buying; 'deciders' have the final authority to select suppliers; and 'gatekeepers' control the information flow into the buying centre.

Second, there are three orientations in the case of the buyer decision (Anderson and Narus, 1998). The 'purchasing orientation' is a market transaction; the buyer aims to buy goods that are cheaper. The 'sourcing orientation' builds more cooperative relationships with suppliers; by building close relationships, the buyer aims to simultaneously achieve

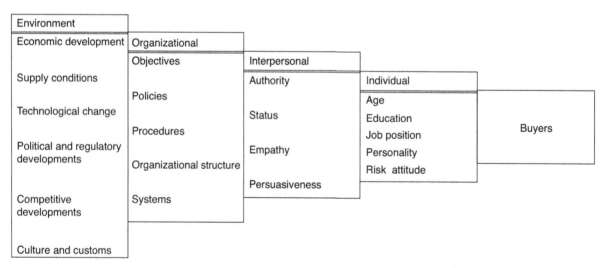

Figure 18.2 Major influences on business buyer behaviour. (Source: Kotler and Armstrong (2006), p. 169, Figure 6.2.)

quality improvement and lower costs. Further, in the case of 'supply management orientation', the buying centre plays a broader role than in other orientations; the company aims to enhance value through the entire value chain that it formulates.

Third, the following eight steps are suggested to constitute the buyer decision process. Business buyers:

1 recognize problems;
2 describe the general needs;
3 specify the products;
4 seek suppliers;
5 solicit proposals;
6 select the supplier;
7 order the desired specification;
8 review the performance.

This is a general description of the process.

18.3.3 Information in marketing

It is essential for marketing decision makers to have access to timely and accurate information. This information is based on a system that gathers, analyses and distributes information. This system is called the 'marketing information system (MIS)' (Kotler and Armstrong, 2006). In MIS, information users, usually marketing managers, recognize the need for marketing information. Next, they specify the information that they require from internal company databases, marketing intelligence activities (systematic collection and analysis of information about competitors and the marketplace) and marketing research. The system provides the kind of information that marketing managers seek for their information analysis

and subsequent actions. Appropriate information is distributed through the system and helps marketing managers in their decision-making. MIS need not necessarily be a computer system. Marketing managers should be able to utilize MIS in any form.

18.3.4 The market research process in the food market

As a background for marketing managers' utilization of MIS, we can point out that the managers need not only general information for decision-making but also information relevant to specific marketing situations that the organization is facing. This specific marketing information is obtained through marketing research.

The marketing research process has four steps (Kotler and Armstrong, 2006):

1 defining the problem and research objectives;
2 developing the research plan;
3 implementing the research plan;
4 interpreting and reporting the findings.

The first step, defining the problem and research objectives, is the most difficult step in the process. Marketing managers are usually aware of the occurrence of some event in their market, but they are not always able to identify the specific cause. Marketing managers must think deeply about the causes through reflection and discussions with organization members. Defining the problem is the most important aspect. For example, Japanese sake companies are seeing a decrease in demand (see Fig. 18.3). The production volume of Japanese sake has been

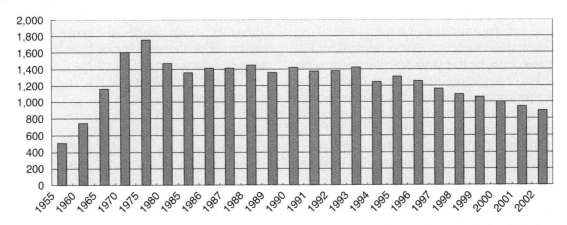

Figure 18.3 The shift in the annual production of Japanese sake (taxable volume). (Source: Japanese National Tax Agency (2003) *Sake no Shiori* (Guide for Japanese sake).)

dwindling since 1996. The marketing managers of these companies pursued many possible causes, but they were unable to identify the specific determinants. However, they discovered that it was important to increase sake consumption among young people. Thus, the marketers defined the problem that they needed to solve.

Next, in order to define the problem, marketing managers must set the research objectives. There are three types of marketing research. The type selected depends on the research objectives (Kotler and Armstrong, 2006). Exploratory research is selected when the objectives are to gather preliminary information that will help define the problem and suggest hypotheses. Descriptive research is opted for when the objectives are to better describe the marketing problems, situations or markets; this is equivalent to researching a potential market for new products, demographics, etc. In the third type, causal research, hypotheses on the cause-and-effect relationship are tested. Let us return to the case of Japanese sake. The problem that the sake companies face is the estrangement of young people from Japanese sake. In order to solve this problem, the marketing managers of Japanese companies can think of various ideas, such as developing new tastes and containers, finding new distribution channels and creating advertisements that portray a new image. Marketing managers set the research objectives after deciding which points they want to focus on.

The first step in the marketing research process is to define the problems and objectives. Therefore, the definition influences the final results of the marketing research. The second step is to develop the research plan for collecting information. The research plan outlines the sources of existing data and spells out the specific research approaches, contact methods, sampling plans and instruments that researchers will use to gather new data (Kotler and Armstrong, 2006). The research objectives are translated into the information required. For example, the marketing managers of the Japanese sake companies require information regarding the drinking behaviour, trends, fashions, etc. of young people. The research plan is presented in a written proposal. The proposal must cover the objectives of the marketing research and provide information that will help the managers in decision-making.

Information is classified into secondary and primary data. Secondary data are information that has been collected for another purpose. Such data are available free or for a fee. Government statistics are also secondary data. Secondary data must be collected first because such data offer a clue to the research and can be collected within a short period. However, since secondary data are collected for another purpose, managers do not always find the information that they seek from secondary data. On the other hand, primary data are collected for a specific purpose. They are needed data. However, collecting such data involves the expenditure of an immense amount of time and money. Observational research, survey research and experimental research are the well-known methods of primary data collection (Kotler and Armstrong, 2006). In the case of observational research, primary data are gathered by observing relevant people, actions and situations. This method is used to gather information from the daily behaviour of test subjects. For example, consider a Japanese sake company's marketing manager attempting to create opportunities to drink for young people. The manager can extract information about the drinking behaviour of Japanese youths from such observations. The second method of obtaining primary data, survey research, is the best-suited approach for gathering descriptive information. The company can learn about people's attitudes, preferences, buying behaviour, etc. by asking questions directly to them. The marketing manager of the Japanese sake company can explore the possibility of producing bottles of sake cocktail. This different kind of a bottled cocktail could become popular in the Japanese market. The young generation will find sake cocktail more appealing than ordinary Japanese sake. Thus, the manager can conduct a questionnaire survey to gather information on the preferences of bottled cocktail buyers. The third method of gathering primary data, experimental research, involves a comparison between two groups. Each group is treated differently, which clarifies the cause-and-effect relationship. The above-mentioned marketing manager can test the sake cocktail that his company has developed among different groups. This will clarify the preferences of each age group.

The third step of marketing research is to implement the research plan. This includes the collection, processing and analysis of information. Information collection has already been explained in the research methods described earlier. In the processing and analysis of information, it is essential to isolate the important information and findings easily.

The fourth step is to interpret and report the findings. It is possible for a marketing manager to interpret some findings incorrectly. Past experience may prevent managers from accepting a new perspective. Marketing managers must discuss their interpretations with internal staff and external experts. After these discussions, they should be prepared to face the consequences of their decisions.

18.4 Strategic marketing and the marketing plan

This section explains the approach to strategic marketing. The first two subsections discuss the marketing mix and the components in devising a marketing plan. The third subsection complements the previous two. The tools used to formulate a competitive strategy are explained. In the final subsection, the strategic actions based on the marketing strategy are considered in the case of the Japanese sport drink market. This section looks at how to:

- assess the importance of market segmentation to the overall process;
- apply the key concepts to issues associated with the marketing mix variables of product, price, promotion and distribution;
- research and produce a basic marketing plan for a firm in the food market;
- analyse food marketing programmes and opportunities and offer appropriate alternative solutions/programmes;
- compile information on food companies' products and marketing strategies and make rational judgments on marketing strategy;
- apply marketing concepts and principles to real cases in the food market.

18.4.1 Market segmentation, target marketing and market positioning

Many kinds of people participate in the buying and selling process in a market. The preferences of the consumers that compose a market are reflected in the goods that they want to buy. These preferences are based on geographic, demographic, psychographic and behavioural factors. Consumers are grouped by this difference of preference. The process of dividing one market into distinct groups of buyers is called market segmentation (Kotler and Armstrong, 2006).

Every market has segments. Many kinds of criteria are used to segment a market. In the chocolate market, a man may eat KitKat (Nestlé) when he takes a break from his work. However, he may buy chocolates at Marks and Spencer when he invites friends for dinner. It is difficult to make one particular kind of chocolate the first choice of every customer in every situation. A product should focus on meeting the needs of a market segment.

After a company carries out market segmentation based on its chosen criteria, it must decide which segment it wants to cater to. Thus, the marketer implies a focus on the specified market segments as the target market. It involves evaluating the attractiveness of each market segment and selecting one or more segments to enter (Kotler and Armstrong, 2006). The criterion in segment selection is whether the segment can generate the greatest customer value for a long period of time. If a company has limited resources, it can select a niche market. By serving market segments that are ignored by major competitors, a company can choose to be special by catering to people with special needs. Thirty years ago, it was virtually impossible for Japanese working mothers to prepare dinner every day. Traditionally, Japanese women had to go shopping every day in order to prepare meals. In those days, working mothers were a minority. A company began to offer daily delivery of foodstuffs for dinner. This niche market has grown in recent years, and, with it, the company too has grown. Thus, focusing on niche markets has possibilities for the future.

After deciding the market segment, the company must decide the position that it wants to occupy in the minds of the target consumers. Market positioning is arranging for a product to occupy a clear, distinctive and desirable place relative to competing products in the minds of target consumers (Kotler and Armstrong, 2006). Marketers plan for their products to occupy a unique position in the minds of target consumers. In positioning a product, the product's competitive advantages must set an available position. In the chocolate market segment, in order to compete with KitKat, Marks and Spencer will need to capitalize on its reputation and sell high-end chocolates. If it succeeds in doing so, it will avoid a direct confrontation with Nestlé and secure its benefit.

The marketing strategy is formulated through the above-mentioned process. Strategy formulation enables marketing managers to present their business domain to company members.

Once a company has decided its target segment and positioning, it needs to consider how to implement its marketing strategy. The marketing mix is the framework through which companies implement their marketing strategies. The marketing mix is the set of controllable, tactical marketing tools that the firm blends to produce the response it wants in the target market (Kotler and Armstrong, 2006). It comprises four groups of variables known as the four Ps: product, price, place and promotion (see Fig. 18.4):

- Product: goods, services or goods-and-service combinations that the company offers to the target market.
- Price: the sum of money that the customer has to pay to obtain the product.
- Place: activities through which companies deliver their products to the target customers, for example, the distribution channel, shop location and transport.
- Promotion: company activities that communicate the advantages of the product to the target customers.

For the successful implementation of the marketing strategy, the four Ps of the marketing mix must be well coordinated and designed as a marketing action plan.

Consider the case of a health beverage developed by a food company. The company was looking for a new market for the special vegetables produced in the district. The health food market was increasing year by year. The company decided to process the juice extracted from the vegetables. Since it had strong connections with local farmers, it was able to source pesticide-free vegetables. The company decided that its target segment would be middle-aged women. It expected this segment to be interested in health and beauty and therefore to accept the special vegetable juice. In order to increase the appeal of the product, its price was set higher than the prices of other juices. Further, as the company focused on middle-aged women, it used the shop channel to distribute and promote its product. The company was successful in its attempts to gain market share.

This case substantiates the need to coordinate the components of the marketing plan in a consistent manner.

Formulating a marketing plan is a complicated and difficult process. One reason for this is the presence of competitors. Marketing managers analyse their competitors' behaviours in the market and use these findings to reflect on their own marketing strategies and plans. This analytical framework is called 'SWOT analysis'. A SWOT analysis enables marketers to understand both the opportunities and threats presented by the external environment, which surrounds the company, and its strengths and

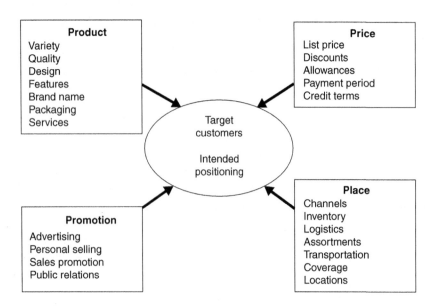

Figure 18.4 The four Ps of the marketing mix. (Source: Kotler and Armstrong (2006), p. 48, Figure 2.5.)

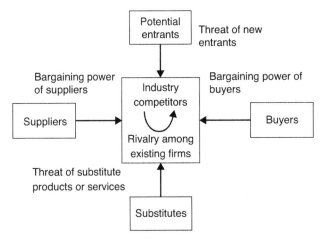

Figure 18.5 The five competitive forces that determine industry profitability. (Source: Porter (1985), p. 5, Figure 1.1.)

ing of the external environment by analysing these five industry forces:

- industry competitors
- supplier
- buyers
- potential entrants
- substitutes

The strengths and weaknesses of companies are identified from their managerial resources. However, some guidelines are needed to analyse these resources. In the vegetable juice case, the company's good connections with local farmers are considered as a strength. Although no weakness has been pointed out in this case, a high production cost can be inferred from the high price of the product. In the absence of guidelines, it is difficult to discuss managerial resources.

The VRIO (value, rarity, imitability and organization) framework is suggested as a guide to analyse resources. The answers to the four questions in Table 18.1 determine whether a particular firm resource is a strength or a weakness (Barney, 2002). If a resource is valuable, rare and costly to imitate, its use will be a continuous source of strength. However, if a resource does not meet these conditions, its advantage will be only temporary.

When marketing managers consider market competition, they must understand the competitive environment that the competitors create. The SWOT analysis framework helps to understand a company's environment. This framework clearly introduces comparison with competitors into the process of marketing strategy formulation. In other words, it plays the role of reinforcing marketing strategy formulation.

weaknesses, which depend on the internal environment based on the company's resources.

An analysis of the external environment yields many kinds of facts, even if the facts are limited to the view of opportunities and threats. Let us revert to the case of the vegetable juice company. The marketing managers of this company can identify certain factors as opportunities, for example, enhancing health consciousness and increasing demand for health food. On the other hand, as threats, they can identify factors such as the entry of a big food company. The more the number of factors identified, the better is the understanding of the external environment. In order to obtain a more comprehensive understanding of the external environment, the five competitive forces model has been suggested (see Fig. 18.5). This model was developed to identify the forces that determine industry profitability (Porter, 1980, 1985). Marketing managers can acquire a comprehensive understand-

Table 18.1 Questions that need to be answered when conducting a resource-based analysis of a firm's internal strengths and weaknesses.

	Questions
The question of value	Do a firm's resources and capabilities enable the firm to respond to environmental threats or opportunities?
The question of rarity	Is a resource currently controlled by only a small number of competing firms?
The question of imitability	Do firms without a resource face a cost disadvantage in obtaining or developing it?
The question of organization	Are a firm's other policies and procedures organized to support the exploitation of its valuable, rare and costly-to-imitate resources?

Source: Barney (2002), p. 160, Table 5.1.

18.4.4 Application to a case

The following case is based on Nonaka and Katsumi (2004). Consider the marketing strategy formulation process in the case of Suntory's soft drink. (Suntory is Japan's leading producer and distributor of alcoholic and non-alcoholic beverages.)

The soft drink market in Japan is valued at approximately four trillion five hundred billion yen (twenty-two billion five hundred million pounds) based on the retail price. One thousand new products are introduced into the market every year. However, of these, only three items continue to be sold in the next year. Thus, the market survival rate is 0.3%. Furthermore, since the sales volume of each of these products exceeds 15 million cases (one case = 24 × 350 ml) per year, they can be considered as major brands and standard items. Suntory 'Dakara' was launched in March 2000. The total sales volume of Dakara exceeded 15 million cases in 2000, reached 24.7 million cases in 2001 and finally reached 34 million cases in 2002.

Dakara is classified as a sport drink. Sport drinks were initially designed to help athletes rehydrate and replenish electrolytes, sugar and other nutrients. Gatorade was introduced in 1966 and is a well-known sport drink in the world. The sport drink segment of the soft drink market was a niche market. However, at present, this segment has expanded to include non-athletes. It is one of the important segments of the soft drink market. Until 2000, the sport drink market of Japan had two major brands: 'Pocari Sweat' by Otsuka Pharmaceutical and 'Aquarius' by Coca-Cola Japan. These two brands had occupied more than 90% of the Japanese sport drink market for the past 20 years. The annual sales volume of Pocari Sweat was about 60 million cases and that of Aquarius was about 50 million cases. It was difficult for other companies to penetrate this market.

The marketing manager of Suntory conducted a survey research in which he asked the sampled consumers questions regarding the occasions when they drank Pocari Sweat and Aquarius. Seventy-six percent of the respondents answered 'during sport activities' or 'after sport activities'. Thus, the survey research did not yield conclusive results. Therefore, the members of the marketing department decided to observe the daily lives of the selected subjects. They recorded the details in a diary. The members termed this observational research 'diary research'. From the diary research it was found that most of the sub-jects drank sport drinks when they had a hangover or when they desired a break, rather than when they engaged in sporting activities. In this manner, the manager realized that a sport drink was not always consumed as one. Thus, the size of the sport drink segment expanded.

As the next step, the managers analysed the strengths of Pocari Sweat. The customers had a medicinal image of this product. The manager believed that this image was the element that consumers desired in a sport drink at the time. This image was transformed into an image of a nurse and then communicated to the marketing department. In addition, the members of the marketing department worked in a convenience store and collected data from real consumers. The data revealed that there was an imbalance in people's daily diet. Hence, the contents of Dakara were developed such that they emphasized the product's role in the excretion of surplus from the body rather than as a supplement for inadequate nutritional intake. Thus, the positioning of Dakara was different from that of the existing sport drinks, which were projected as nutritional supplements.

In the Japanese market, the prices of soft drinks are fixed. Suntory used its existing distribution channel for Dakara. The marketing department sought a new challenge in the promotion of Dakara. Soft drink television commercials typically show famous celebrities drinking the product. However, the marketing manager of Suntory wanted to emphasize the cleansing aspect of Dakara. Therefore, Manikin Piss from Belgium was used as the main character in the commercial. Through projecting Manikin Piss in a humorous manner, the company was able to elegantly convey its message to the consumers. It was essential for the marketing manager to ensure consistency between the product's concept and promotion strategy.

Further reading and references

Anderson, J.C. and Narus, J. A. (1998) *Business Market Management*. Prentice Hall, Englewood Cliffs, New Jersey.

Assael, H. (1987) *Consumer Behavior and Marketing Action*, 3rd edn. Kent Publishing, Brisbane.

Barney, J.B. (2002) *Gaining and Sustaining Competitive Advantage*, 2nd edn. Prentice Hall, Englewood Cliffs, New Jersey.

Carroll, A.B. (1996) *Ethics and Stakeholder Management*, 3rd edn. South-Western College Publishing, Florence, Kentucky.

Engel, J.F., Blackwell, R.D. and Miniard, P.W. (1982) *Consumer Behavior*, 3rd edn. Dryden Press, New York.

Japanese National Tax Agency (2003) *Sake no Shiori* (Guide for Japanese sake) (in Japanese). Japanese National Tax Agency, Tokyo.

Kotler, P. (2000) *Marketing Management*, 10th edn. Prentice Hall, Englewood Cliffs, New Jersey.

Kotler, P. and Armstrong, G. (2006) *Principles of Marketing*, 11th edn. Prentice Hall, London.

McGuire, J.W. (1963) *Business and Society*. McGraw-Hill, New York.

Nonaka, I. and Katsumi, A. (2004) *The Essence of Innovation* (in Japanese). Nikkei BP, Tokyo.

Peattie, K. (1992) *Green Marketing*. Pitman, London.

Porter, M.E. (1980) *Competitive Strategy*. The Free Press, Cambridge.

Porter, M.E. (1985) *Competitive Advantage*. The Free Press, Cambridge.

Webster, F.E. and Wind, Y. (1972) *Organizational Buying Behavior*. Prentice Hall, Englewood Cliffs, New Jersey.

Supplementary material is available at www.wiley.com/go/campbellplatt

Ray Winger

19.1 Introduction

The aim of this chapter is to facilitate the application of food science and technology to food product development.

Food product development (FPD) is the ultimate capstone course in food science and technology. It requires the integration of the comprehensive knowledge obtained in the degree programme. FPD is the cornerstone of all industrial manufacturing processes, as this discipline links all parts of a factory: from marketing (and the consumer), through production (recipe sheets, quality assurance), commer-

cial and financial activities (profitability, sustainability) and purchasing.

Yet systematic product development remains a relatively poorly utilised function within most food companies. Even technologists tend to overlook the powerful opportunities associated with the FPD techniques.

The key features of FPD are integrated, systematic qualitative and quantitative techniques used in developing new or modifying existing food products from conceptual ideas through to successful and sustainable products in the marketplace. The customers may be the ultimate consumers, or they may be intermediate food manufacturers, institutional food providers (restaurants, fast food chains, etc.) or – in the case of waste raw materials – they may be other animals and pets.

However viewed, FPD requires knowledge of food science (chemistry, biochemistry, biology, microbiology), food processing, packaging, sensory and consumer sciences, marketing, quality assurance, legislative matters, research methods and experimental design, commercial and business aspects (particularly finance), environmental issues, purchasing (ingredients and raw materials) and management.

On successful completion the student will be able to carry out the following tasks in relation to:

1 New product development strategy:
  prepare a defined product specification;
  understand the role of market research in product design;
  understand the roles of different professional disciplines in product development;
  identify the steps in the development process.
2 Product development at experimental kitchen/ pilot plant scale:
  make, and appraise, proposals for the small-scale manufacture of a product to an agreed specification;
  design and carry out appropriate experimental work at this level;
  evaluate product characteristics using appropriate analytical procedures.
3 Scale up to commercial production:
  evaluate the progression of a new food product from the development stage to commercial production.

19.2 Background

19.2.1 The product development process

The most widely referenced normative product development models are those of Booz, Allen and Hamilton Inc. (Anonymous, 1982) and that of Cooper and Kleinschmidt (1986). While there is a variation in the number of stages performed in these models there are essentially four basic stages in these models for every product development process. These are product strategy development, product design and process development, product commercialisation, and product launch and post-launch. Each stage has

activities that produce outcomes (information), and based on these outcomes, management decisions are made (Fig. 19.1).

In practice, some of the activities performed in the product development process can be truncated,

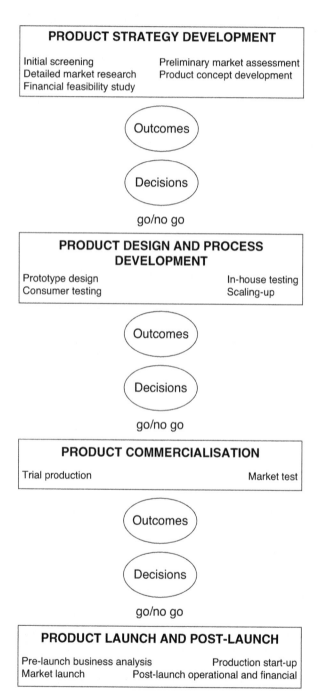

Figure 19.1 Schematic of the overall product development process. (Source: Siriwongwilaichat (2001); adapted from Earle and Earle (2000).)

or some stages can be omitted or avoided based on a company's accumulated knowledge and experience. This is determined by the degree of product newness and the product life cycle. Products, which are incrementally new, such as product improvement and line extension, require less time to develop. For short-lifecycle products or fads, speed of new product introduction is important, as changes in consumer needs are rapid. In this instance, relevant product requirements rather than new scientific results or advanced technology are sought. For long-lifecycle products, process innovation and cost control are important foci.

As companies grow and develop, the product development process also matures. In early generations of the product development process, the research and development (R&D) project was not integrated into the business strategy of a company. The next stage was the recognition that new product development should begin with the business strategy, working through the product strategy, to the new product area definition. In the final generation, new products were strategically planned at company level. Technology was considered a weapon for competitive advantage.

The modern innovation model is viewed as an interactive model in which stages are interacting with and interdependent on each other. The innovation process is a complex net of communication paths, both intra-organisational and extra-organisational, linking together the various in-house functions and the firm to the broader scientific and technological community, and to the marketplace.

Organisational knowledge creation is a key driver of innovation and organisational success. The new product development process deals with the transition from *organisational knowledge* to *embodied knowledge* in both new products and human resources. The importance of this cumulated knowledge source becomes evident when the "timing" of product launches is concerned. Highly desirable and consumer-friendly products will fail in the market if they are introduced at the wrong time. The stock of cumulated knowledge allows timely, rapid and effective development of a product to suit a market opportunity. In many companies, especially in the food industry, cyclic innovation success often follows the turnover of experienced staff – suggesting the company lacks an effective permanent database of knowledge which can be shared from one generation of staff to the next.

19.2.2 Innovation in the food industry

The product development process in the food industry has some specific characteristics. Product innovation is determined by the interaction between consumer expectations and demand, and the role of technological opportunities. Innovation in the food industry may be different from that in other industries due to the following characteristics:

- In the food industry true "never seen before" innovations are rare. Food products innovation tend to be incrementally innovative.
- Consumers of food products tend to show "risk aversion". They want new products but the new product has to be familiar, or similar to those they are used to. Food consumers have a long experience with certain tastes and eating habits, which are usually difficult and time consuming to change.
- Because foodstuffs are easy to imitate, the time for an innovator to obtain monopolistic gains is short. Thus the food market is more likely to be oligopolistic.
- The food industry is often defined as a low-technology sector, and R&D is usually dominated by a few large companies. In the food industry, compared to other industries, most companies are small and internal product development activities usually focus on product design. Only a few large multinational companies are dealing with modern and advanced technology.
- In the food industry, major advances in technology occasionally occur, but diffuse rapidly across the industry. Important technical advances in food innovation often come from embodied forms of technology from outside upstream industry, such as equipment, ingredient and packaging suppliers.
- It is also important to note that there are differences between sub-sectors within the food industry. The large-scale intermediate foods manufacturers (e.g. flour milling or sugar refining) have been characterised as scale-intensive firms in which a high proportion of own process technology was made. On the other hand, the processed convenience food manufacturers generally conduct little in-house product and process development and tend to be characterised as supplier-dominated firms.

R&D in the food industry is a market-driven, application-oriented effort, directed toward specific products and based more on incremental technological innovations than on major changes from basic research. R&D expenditure in the food industry is quite low compared to some other higher science and technology based industries such as the chemical industry and pharmaceutical industry. For multinational companies, Nestle spent 1.2% of its sales on R&D in 1992 and Unilever about 2% in 1993. According to Food Processing's 1998 top 100 R&D, no company reported its R&D budget higher than 1.5% of its food sales (Meyer, 1998). R&D budgets spent in chemical companies ranged from 4% to 10% of sales (Moore, 2000). In the pharmaceutical industry, the R&D budget was as high as 15–17% of sales (Cookson, 1996; Scott, 1999; Mirasol, 2000).

Whilst these R&D figures indicate a degree of commitment to new product development, they are also potentially misleading. The above-mentioned companies invest heavily in new growth markets and/or products, while they spend relatively little on well-established products with poor growth prospects. Technical support for the latter products focuses on cost reduction strategies, in the main.

R&D strategies employed in food industries can be summarised as follows:

- R&D activities in food industries are dictated by opportunities in markets. Consumer preference, market size, profit margin and capacity for effective response to competition are important criteria used to direct the R&D function.
- R&D activities are differentiated according to goals and situation, and allocated to strategic business units (SBUs) and central facilities. R&D employing process and product technology to develop new products or line extensions is usually assigned to SBUs. This allows immediate response to specific markets. In some cases, R&D is also done at the corporate level. This would aim at new markets, long-term product development and the development of technology with a major impact on the entire corporation or which could support multiple SBUs.
- Firms distinguish the purpose and impact of technologies and aim for internal or external development. Development of base or key technology is frequently done in-house. Pacing technology is frequently employed by means of joint ventures or acquisitions, rather than developed in-house.
- Successful product lines can be sustained in the market by R&D efforts focused on cost reduction, line extensions and product renewal. The key is continued improvement based on consumer preference and new technology.

19.2.3 Product characteristics

A new food product can be categorised by its key product attributes. These are:

- form (size, shape, density, packaging, stability);
- content (nutrition, additives, contaminants);
- palatability (taste, odour, colour, texture, social and identification culture);
- cost (raw materials, conversion, overhead costs).

Each attribute should be considered as follows.

- Determine the consumer perception of the attribute, e.g. what the consumer perceives as the embodiment of the attribute in the product.
- Determine the actual or potential value of the attribute, e.g. the nutritional value to each person may differ by individual diet and nutritional needs.
- Estimate market value in order to estimate market attractiveness. This requires consideration of questions such as: How large is the segment? How stable is consumer demand? What alternatives are there for a certain attribute? How does the attribute affect the other products sold?

An accurate technical characterisation of the attribute must be achieved, e.g. specification, process conditions, stability, packaging and method of measuring the attribute for quality control purposes.

19.2.4 Company acquisitions

Nowadays, food companies often simply acquire or merge with companies that have product potential. Such mergers and acquisitions lead to a decreasing R&D investment. For example, Groupe Danone acquired Nabisco; Unilever acquired Bestfoods in 2000; General Mills acquired Pillsbury.

Rapid economic liberalisation and relaxation of barriers to foreign investment have attracted international companies, which have gained strong positions in Asian markets through mergers and acquisitions.

In the USA and Japan, external technology acquisitions rather than in-house development is more likely when many other rivals are expecting to develop a similar product.

19.2.5 Linkages between marketing and technical operations

While the consumer need is a marketing input, technology is the ability to fulfil that need by producing the desired product. The key combination of these two dimensions of knowledge provides product differentiation in the competitive market and changing consumer demand. Technical knowledge appears to be important as an input and as a potential marketing strength. Technology strategy is therefore important to the firm's long-term profitability and growth.

19.2.6 Knowledge sources

The food industry also requires technical people to deal with the complexity of food legislation, food ingredients and their behaviour during processing. In-house technical people provide:

- strong motivation to do the hard work needed;
- sufficient in-house influence to get the necessary allocations of time, money and facilities; and
- mastery of the technological knowledge needed.

Internal knowledge plays a vital role for four main reasons:

- the need of firms to utilise appropriate technology related to specific artefacts to gain short-term profits from radical innovation;
- the cumulative nature of technological development requires a company to build on existing capability to facilitate the new knowledge generation;
- to catch up an unfamiliar technology a firm needs in-house knowledge to identify, digest and utilise external knowledge;
- internal knowledge enables a firm to use specific, as opposed to general, knowledge for its product differentiation in the market.

19.2.7 Corporate strategies

The single most critical aspect of a successful product innovation strategy is corporate commitment. The introduction of new products is expensive, very risky and requires a special corporate culture. A corporation must be prepared to commit to a business plan and remain with that plan with a resolve that at times, under normal business methods of review, may seem to be irrational.

Any executive manager has a natural human desire for self-preservation. Executives want to know that activities they are supporting will keep them in their job. New product development (NPD) often requires an executive to support a concept despite insufficient information on which to base sound business judgements. The formal NPD process provides a systematic method to provide the executive team with information on progress against milestones, cost control and external factors impacting upon the company. This clearly requires a special sort of senior executive. It also requires an insistence upon regular reporting of progress on any NPD project.

Without exception, successful NPD companies from any manufacturing sector have one common feature. Their NPD process is multidisciplinary and integrates *all* parts of the corporation into the business planning and NPD process itself.

NPD should be treated at the most senior executive level as a calculated gamble. It has to be recognised at the onset that the NPD process is more likely to fail than to succeed. The typical "success rate" for unstructured, random product development is of the order of 1 success per 200 products introduced. For the most sophisticated and well-run product development processes currently available, the success rate is still only in the order of 1 product in 3 to 1 product in 5 launched. Success in this context means not only survival in the marketplace for the product lifecycle, but also a positive return on investment.

Every NPD project requires a careful strategic review, identifying the relevance and "fit" to the corporation and its culture, assessing the market opportunity and potential, considering the technical feasibility and operational capability and the potential financial and legal implications of the work. Business planning *must* incorporate all aspects of the company. Failure to have an integrated approach from the inception of the project is identified as the biggest cause of NPD failure.

The corporation *must* know what business it is in. This defines the type of products that would be considered.

One of the best food examples of this vision was the move by Mars Corporation into frozen novelty bars. The two major players in frozen confectionery, Unilever and Nestle, while watching each other closely, assumed Mars was in pet food and confectionery bars. Mars, on the other hand, defined themselves as being in the novelty confectionery bars business (and in that scenario, whether something was frozen or not was irrelevant). The technology they developed in frozen novelty bars was unique and it took Unilever and Nestle over 2 years to be able to successfully produce competitive products. Unilever and Nestle did not consider Mars a competitor because they wrongly identified their market position as being in ice cream.

19.2.8 Organization for new product development

A clearly defined product development strategy and an awareness of the contribution of innovation are not the only requirements for successful product development. New product development is, by its very nature, a cross-disciplinary process and cannot therefore be a segregated functional activity. To be successful, it has to occur with the participation of varied personnel drawn from across the company, working as a team.

The "sequential" approach to product development with each function taking it in turn to take responsibility for a project and make its contribution before passing it on to the next, has often been used throughout the food industry and indeed most industries. Marketing generally identifies a new product opportunity, R&D investigates and develops the technology and generates concepts which are then designed and handed onto manufacturing to be produced, who in turn hand the product onto sales and marketing for distribution and ultimate sale. Though this is clearly logical and, on the surface, easy to understand and therefore manage, this sequential approach has several faults.

With this sequential system individual roles and responsibilities are relatively clear and compartmentalised within functions. Risk should be controlled when projects are transferred between stages. However, there is little requirement for communication between functions whilst the project is within handover periods. Consequently omissions and mistakes are frequently made.

As a means to overcome these problems, an improved approach to the organisation of product development which had been advocated for a number of years was adopted and introduced into a range of manufacturing companies in the late 1980s. Variously termed simultaneous engineering, concurrent engineering or parallel working, this has focused attention on the project as a whole rather than on the individual stages. By enabling different operations to be undertaken in parallel not only is the overall development period reduced, but also the needs of the project as a whole are better satisfied. Primarily by involving all functions throughout the project from concept to production, all aspects and implications of the design can be addressed as it develops.

The concurrent approach has been found to work best when used in partnership with interdisciplinary project teams.

The key benefits that are available from the use of multifunctional teams include:

- elimination of iterations between functions because they are all working together on the same project, and hence development periods can be reduced;
- overcoming hierarchical structures within organisations, with the consequence that project decision making can become decentralised and more appropriate;
- better team member project focus so that relevant information is more effectively filtered from the mass of market statistics, technical data and manufacturing figures which are obtained from within departments and outside;
- ownership of the project objective and project goals, which overcomes the traditional problems associated with division of labour;
- more frequent exchange of ideas, opinions, knowledge, etc. which leads to faster learning by inexperienced team members.

In successful companies, team members represent the various corporate functions, which in turn depends upon the size and organisational matrix of the company. Effectively the following functions need to be involved at all stages of the NPD cycle:

- marketing
- sales, distribution and logistics
- R&D
- quality assurance
- engineering
- manufacturing/operations
- finance and legal
- procurement/purchasing
- packaging and design

It is worth noting that many of these functions would consider themselves irrelevant in the NPD team. If they are not included in the team at all stages of the development phases, then the effectiveness of the NPD process will be reduced. All of these functions *have* a role to play at some stage during the NPD. In some companies, one person may cover several of these roles.

19.2.9 The impact of retailers and supermarkets

The literature has begun focusing upon the importance of the intermediate groups in the product development process, and for good reason. Food manufacturers normally have no direct linkages to the final consumers of foods. The key direct link is the supermarket or hospitality trade chef.

These vertical alliances (note this is not acquisition) and close associations with external organisations (such as research institutions and universities) is the feature of so-called fourth generation R&D (Miller and Morris, 1999). This is, in essence, the integration of production, marketing and internal R&D groups with external organisations.

19.2.10 Key features for product failure

A variety of studies have tried to isolate the issues for product failure; however, the lists are often contradictory and confusing. One of the reports that seems to cover most of the issues and to rank them in some order was published in *Food Technology* (Hollingsworth, 1994). The key reasons and their ranking, based on the percentage frequency they were included in the top three reasons for failure by the respondents, were as follows:

strategic direction	44
product didn't deliver promise	35
positioning	33
competitive point of difference	32
price/value relationship	30
management commitment	29
packaging	20
misleading research	19
development process	19
creative execution	18
marketing/trade support	18
branding	15
consumer input	14
marketing group experience	9
advertising message	8
promotions	8
pay-back period	8

19.3 Class protocols

19.3.1 New product development strategy

19.3.1.1 Planning and timing of NPD project

Systematic NPD is a carefully defined, timed and structured discipline. The introduction of new products will be successful only if the project is carefully planned and timelines properly scheduled. This is the first element of any NPD process and is an important discipline for all students.

There are essentially six components to an NPD project:

- concept design and idea generation;
- preliminary product screening;
- formal economic analysis;
- experimental food product development;
- product testing;
- commercialisation.

Normally, a launch date would have been identified (this may be determined by the length of the teaching period in this exercise) and this date determines the total length of the project. It *must* be finished by that date.

A job progress path and critical path analysis is a constructive exercise for all students to better understand the overall NPD process, the amount of work involved and how to structure their time management procedures to ensure they meet all the required deadlines. Many of the product development texts provide examples of these procedures and there are project management databases and software programmes available to prepare the critical paths.

Students should be encouraged to identify where activities can be conducted concurrently (e.g. a literature review can be conducted while the initial screening and economic analysis are being performed). Each element of the NPD process needs to have a realistic time estimate (e.g. product formulation requires ordering ingredients which can sometimes take a week or more to obtain).

The timeline must fit to the critical end date. If the timing is too long, parts of the project must be truncated so the total project time is reduced. During the project, these timelines will be regularly checked for compliance. *Failure to meet a key milestone is sufficient reason to terminate the project*. If a product becomes too difficult to develop in the laboratory, it is better to stop further work on that product (no matter how great its potential) and focus on another product which will be easier and more timely to prepare.

Because there are several "*go/no go*" review points throughout the NPD process, it is important to ensure that more than one product is being designed at any stage. As a general rule of thumb, the initial concept design and idea generation stage should identify several hundred possible products. By the time experimental product development starts in the laboratory, there should be about five products being considered. By the time products are being launched, there will only be one or two products left. This progressive culling of products is essential to maximise the success of NPD activities.

19.3.1.2 Concept design and idea generation

Fundamental to successful product development is the need for a careful, strategically defined, relevant set of product ideas which fit a company's business plan. It is mandatory that the exercise involves a company. There needs to be a very clear vision of the business strategy, the company's expertise and capabilities and a clear set of constraints surrounding the NPD exercise. Ideally, NPD exercises should be run in conjunction with a real food company. Alternatively the teacher must ensure that a virtual company is defined in considerable detail.

Relevant company details might include:

- the business and food industry sector the company operates within;
- size, type of staff available (e.g. technical, marketing);
- financial and related issues that may constrain NPD (e.g. willingness to invest in new capital);
- processing capability and capacity;
- distribution and retail (or industrial ingredients) chain;
- domestic or export market focus (plus pertinent regulations and food law).

There are many ways to generate ideas and identify possible concepts. The specific drivers will vary amongst different NPD projects. For example:

- Wish to introduce a totally new product to the company.
- Wish to copy a competitor's product.

- Wish to improve an existing product (e.g. add functional or nutritional ingredient).
- Wish to provide a line extension (e.g. new flavour in an existing range).
- Desire to better utilise a waste material.
- Opportunity to utilise a new ingredient.
- Desire to reduce the cost of an existing product.
- Desire to use surplus capacity on a particular machine.

Some ideas will be driven by novel process, packaging or ingredient opportunities (i.e. process driven) while many will be driven by new or emerging market opportunities. In *all* cases, the fundamental objective of this stage of the NPD process is to generate as many possible ideas as possible (no matter how silly they may seem). It is worth remembering that from this stage, about 1 product idea in 200 will be successful. Most product ideas will be eliminated throughout the NPD process, so always start with a large number.

An essential part of this stage is marketing. This has been covered in other chapters in this book.

A further critical need at this stage is a clear definition of the target consumer. For some NPD projects (e.g. utilisation of a currently waste material) the target market may be difficult to define. However, for the majority of products, the target market is known and marketing staff can define their consumers accurately.

At the end of this stage, you will have:

- a clear picture of the company you are working for;
- a defined understanding of any company constraints (equipment, financial investment, budget);
- a broad picture of the market and targeted consumer;
- a general understanding of the supply chain (distribution, etc.) required;
- a general recognition of any specific product constraints (e.g. shelf life, special storage conditions, food legislation);
- a very large number of product ideas.

19.3.1.3 Preliminary product screening

The next stage in the NPD process is to systematically reduce the number of product ideas to a manageable group. This requires a multidisciplinary team review of all products identified. This is effectively a preliminary economic analysis. Each product needs to be scored (to the best or the individual's existing knowledge) on a range of criteria. These criteria need to be defined specifically for a given NPD project. Criteria would cover issues such as:

- potential for market success;
- relevance to the company's existing products and distribution channels;
- ability of the company to make the product;
- technical complexity of the product (hence the relative cost of technical development);
- potential cost of the product (hence profitability).

Each product is scored by each team member (it is possible the specific skills would be used, as appropriate: e.g. marketing would score the market-based criteria) and the scores totalled. Products would be ranked and the top scoring products chosen for the next stage. Clearly, success at this stage requires people to have a reasonable understanding of the various criteria being screened. Some preliminary market research might be required, but this should be kept relatively broad at this stage. *Note*: this is not a process of "guesswork". If information is not clearly understood, some preliminary literature research may be required. An experienced team will be able to screen these products very quickly, whereas a team with limited knowledge will require more background investigation to make sound judgements.

As a guideline, the number of products remaining should be about 15–20% of the original list.

At this stage, each product should include a brief description that describes the unique features to be developed. For example, the product is a ready-to-drink, high-protein, fruit-based beverage, shelf-stable, in a 100-ml UHT processed carton plus straw. It is important that all team members have a clear, consistent vision and understanding of the product.

19.3.1.4 Formal economic analysis

Market research

Critical to the success of the NPD process from this point is a detailed understanding of the market and the consumer. The end-use market must be clearly defined (note there may be different markets for different products being reviewed). Information required includes:

- end-market user needs, wants and demands;
- market trends and key issues being targeted (e.g. nutritional needs);

- pricing;
- distribution and promotional strategies required to succeed;
- market size and market share estimates (current and future);
- competitive pressure (hence influence on market potential);
- economic life of the product.

Consideration of the end-users' actions should be evaluated in terms of factors, such as the following:

- Product demand will be driven by . . . ?
  necessity
  indulgence
  health, etc.
- How well will the product satisfy these needs?
- What is our product's unique ability to satisfy a need?
- How easy is it for competitors to copy this unique feature?
- Is competition going to grow, or decline?
- How important is price-cutting going to be in this market segment?
- Will the number of customers be large or small?
- Will the geographic distribution of customers be advantageous or an obstacle?
- Do customers have strong brand loyalty or not?
- Does the product fit in with current company brands?

These criteria can then be scaled more objectively than the preliminary screening, as much more detailed data have been collected and analysis undertaken at this stage for each of the products. These criteria should be scaled (e.g. on a 10-point scale) and marked accordingly – product by product. It is often desirable to have a similar scale (with different anchors) for each attribute being reviewed.

Note: at this stage, these various topics indicated with bullet points above would normally be grouped into, say, four or five scoring criteria. These might be market size, level of competition, product uniqueness and potential market share, for example.

Company potential

In conjunction with the market potential, the impact on the company must also be evaluated carefully. This requires a sound technical knowledge of the probable product to be developed.

Information required includes the following:

- potential safety of the product (risk to company);
- legal issues relevant to manufacture, export, market food law;
- are there already patents on this product or product type?
- potential to patent the new technology/product;
- how easy is it for competitors to develop similar products?
- ability of existing distribution channels to cope with product;
- generation of by-products or waste materials;
- technical difficulty in formulating/developing the product for the company;
- ability to manufacture the product on existing equipment (capability and capacity);
- unique advantages for the company (e.g. specialised equipment, unique raw materials);
- ability of technical staff to support the development;
- potential conflicts within the company;
- potential expenses:
    - development costs;
    - inventory or related costs associated with unique ingredients;
    - probable capital investment;
    - marketing start-up costs/promotion, etc.;
    - production start-up costs;
    - depreciation and related financial matters.
- potential for adding value (related to potential product cost).

As was done for the marketing potential, these criteria should be scored on a suitable scale (e.g. 10-point) with appropriate anchors. These company criteria may be grouped into five or six criteria. These might be shelf life, processing cost, extra capital requirements, technical feasibility and running costs, for example.

Final product screening

Once the products have been scored for marketing and company-based attributes, the data can be combined using a weighted average. To do this, each marketing and company criterion would be assigned a "weighting grade". An example might be:

- Market size — 10 marks
- Level of competition — 5 marks
- Product uniqueness — 10 marks
- Potential market share — 15 marks
- Shelf life — 5 marks
- Processing cost — 15 marks
- Extra capital requirements — 15 marks
- Technical feasibility — 10 marks
- Running costs — 15 marks

To summarise the process: the scoring criteria are identified (examples in the list above) and each is given a weighting. Each team member then scores each product against these criteria, normally using, say, a 10-point scale. The score for each product for each criteria is then multiplied by the weighting factor for that criteria. Finally, the result for each criterion is summed across all criteria to end up with a single number. In the sample above, the maximum value for each product is 1000. The example is shown in greater detail in Table 19.1.

The products would then be ranked and the top 3–5 products selected for further NPD activity. *Note*: Do *not* end up with one product! At this stage of the NPD process you have about 1 chance in 20 that you have identified a successful product.

Product specification

The final step for this initial stage of NPD is to define the accurate product specifications for each of the products that have successfully emerged from this screening step. This specification should be as detailed as possible, remembering there is still the development step to come. However, there must be no ambiguity in the specifications at this point. These must include:

- target market and end-user;
- product attributes:
    - colour
    - packaging
    - product size
    - unique features
    - special storage requirements
    - product quality and brand image
- product composition issues:
    - raw materials to be used
    - labelling needs (e.g. genetically modified ingredients usage, organic)
    - flavours and tastes
    - cultural or other issues (e.g. halal, kosher)
    - legal requirements
    - linkages to existing products (line extensions, etc.)

Table 19.1 Example of an individual team member's scoring sheet for final product screening. Scores are out of 10.

Criterion		ms		loc		pu		pms		sl		pc		ecr		tf		rc	Total score
	Score	Weight	Score	Weight	Score	Weight	Score	Weight	Score	Weight	Score	Weight	Score	Weight	Score	Weight	Score	Weight	
Product 1	5	10	7	5	3	6	15	6	5	7	15	9	15	8	10	7	15		660
Product 2	9	10	7	5	3	5	15	6	5	4	15	3	15	3	10	9	15		530
Product 3	1	10	2	5	9	7	15	6	5	6	15	7	15	2	10	3	15		505
Etc.		10	5		5		15		5		15		15		10		15		

ms, Market size; loc level of competition; pu product uniqueness; pms product market share; sl shelf life; pc processing cost; ecr extra capital requirements; tf technical feasibility; rc running costs.

- processing constraints:
    - equipment needs
    - equipment capacities
    - specialised handling (e.g. aseptic, process line design and equipment layout)
- shelf life issues:
    - shelf stable, refrigerated, frozen, dry, etc.
    - special packaging needs
    - ability to formulate for shelf life (e.g. can you use antioxidants, preservatives?)
    - special safety issues (microbiological, biological, chemical, etc.)

19.3.2 Product development at experimental kitchen/pilot plant scale

This stage requires a multidisciplinary approach (which is typical of all stages in the NPD process). Consumer surveys would be desirable to better define the end-user expectations for these products. Literature reviews and discussions with ingredient or equipment suppliers will also provide recipes and processing procedures for the five or so products that have been identified.

The first stage is to obtain all the ingredients to be used, including ensuring a wide range of appropriate food additives are acquired. A modern NPD laboratory would have these ingredients at hand, but students will usually need to contact suppliers for various specialist materials. *Note*: ingredients for industrial manufacture of food cannot normally be sourced from supermarkets! Technical industry magazines are replete with advertisements for ingredients and identify suppliers. Web-based searches for ingredients (even using function names, such as gums, thickeners, flavours, colours, etc.) will quickly identify suppliers. There are very few books on food ingredients (one of the more comprehensive is Branen *et al.*, 2002), but food chemistry texts provide some valuable key words for searching.

The most valuable sources of recipes and food ingredients are the ingredient suppliers themselves. Industrial trade shows also include ingredient manufacturers. The vast majority of food ingredients and additives can *only* be sourced through these suppliers.

It is important that specification sheets be obtained with the ingredients. These include detailed compositional analyses (chemical, purity, microbiological, stability, etc.), methods for storing and using these ingredients (you won't find this information in any

books!) and special tricks for use, ideal food conditions (pH, ionic strength, etc.) to be used, or conditions to be avoided, and other information that would be essential for NPD (e.g. halal, kosher, GE-free, organic, treated with ionising radiation).

Identify special processing conditions and how you plan to achieve this (e.g. UHT processing can only be done with appropriate high pressure equipment – can you access this somewhere?). Are there some effective steps you can take in the laboratory that you might not use in the commercial operation (e.g. size reduction with a hammer, where you might use a pin-mill in industry).

Plan and prepare your shelf life testing. This must start as soon as you develop your first prototype product. Shelf life testing takes *time*. You invariably don't have time . . . so start shelf life with the very first product you make and continue these trials on every batch of product you make. Don't wait until you have your final product . . . it's too late to find you have a problem then.

Invariably the experimental process revolves around formulating products with several unique variables to be assessed. For example, a product might have to be a specific colour, contain certain flavours, or have a certain taste and texture. The ingredients that you would use for any one of these attributes will most likely interact with many of the other ingredients – the result is unpredictable. Thus experimental work in NPD invariably involves several (often 6–10) variables, which in turn requires the use of sophisticated and specialised experimental designs (such as Plackett and Burman). These designs allow meaningful results to be obtained from a limited number of experiments, the outcome being an ability to identify which of the variables are the most important. Once this has been identified, more routine, balanced experimental designs can be used with the smaller number of variables.

Once a suitable set of products has been designed, the products need to be tested. First and foremost, chemical and microbiological testing should be completed to ensure the products are safe to consume. These tests would be designed suitably for the specific products in question.

If the products are safe, sensory tests would be performed to ascertain product organoleptic characteristics. Sometimes, unexpected flavours or tastes develop in foods. Alternatively, if you are imitating an existing competitor's food product, you may wish to test for similarity (or difference). It is often important to describe the sensory characteristics of the products, so they can be related to your product as it is scaled up (through the pilot plant) or even help explain consumer acceptability. These sensory tests are normally run with trained or semi-trained panellists in the laboratory. They are designed to ensure the product specifications are matched and that there are no surprises.

In making the product in the laboratory, a clear set of process steps would be finalised. Some of these may be special to the pilot plant/commercial operation (for example, pasteurisation using a plate-and-frame rig would be unlikely in the laboratory, but a routine operation in the factory). Use of pasteurised milk in the laboratory might be replaced by use of fresh milk followed by pasteurisation in the factory. A formal process flow – from receipt of raw materials to the final packaging of the finished product – would be defined at this stage. This would be compared to the capability and capacity of the company's manufacturing operation. The outcome of this stage is a schematic flow diagram, showing every step – operation, transfer, inspection – in the conversion of the raw material to the finished product.

Product costing is essential. Ingredient costs can be obtained from suppliers and the recipe cost can be readily defined. However, there are matters of wastage to consider, people's time (how many man-hours of labour is required in the factory?), possible packaging costs, distribution costs and the company's overhead and profit contributions need to be included. This is best predicted using computer spreadsheets, to allow "what-if" scenarios and therefore an indication of the risk associated with producing these products.

19.3.2.1 'Go/no go' decision time

The secret of successful NPD is the regular review of every project being run. Given the extreme difficulty in identifying successful products that will work both in the market place *and* can be manufactured in an appropriate commercial manner by a particular company, it is vital to always ensure that continued work on an ineffective product is terminated as soon as it is no longer viable (and as early as possible in the NPD process).

At this stage, the senior management team needs to review each product for continuing, or not. They will have:

- a sample of the product (and variations) to taste;
- the product specification in terms of the original brief (detail to ensure the product is exactly what was originally envisaged);
- the probable product cost and market price (and therefore profit/degree of added-value);
- samples of competitive products (if they exist);
- the ability of the company's production unit to manufacture the product, as well as any special conditions (e.g. new capital investment required);
- an updated forecast of expected market share (i.e. price times volume sales analysis);
- identification of any special issues of concern (e.g. key milestones, such as desired launch date will be missed).

The outcome of this review is a defined list of products for further development work. This may be a selection of specific product variants (flavours, etc.). It may have resulted in the rejection of products for further development. Without exception, only those products that still show real positive benefit to the company will be identified for further work. From this point, development work becomes very expensive.

19.3.2.2 Food pilot plant scale-up

The progression from laboratory scale to pilot plant is beset by numerous problems. Scale-up is rarely straightforward and the equipment being used is normally very different from anything in the laboratory. The full process flow needs to be properly designed and set up in the pilot plant. If external contracting is required for some specialist activities, the methods necessary to prepare raw materials and ingredients for off-site manufacture need to be designed.

Products need to be prepared and checked against the original laboratory products. Ensure safety testing (microbiological, chemical, etc.) is completed before sensory testing. If products differ from the laboratory products, the NPD team must approve any change before this stage is completed. It is critical that marketing, management and operational staff are intimately involved with the assessment of the product throughout this stage. Shelf life studies should be routinely started at the completion of every batch of product.

Accurate product costings can now be completed. Exact commercial processes can be defined and the specific parameters around every process step can be elucidated. Critical control points for safety and key product characteristics need to be identified and quality control sheets prepared. Product specifications need to be accurately delineated and target variances predicted.

At this stage, consumer testing can be completed on the products. This is a valuable exercise using pilot plant product. However, this should not be confused with market testing which occurs after commercial scale-up. Consumer testing should focus on product acceptability, comparisons with competitors (if they exist) and identification of characteristics that need to be adjusted or optimised. Ideally, a variety of product examples should be prepared: these may vary in flavours, or colour, or texture, or some other key attribute. With a good knowledge from chemical, physical and in-house sensory testing, the reaction of consumers to these products will provide valuable insight to the desirable and optimal qualities for these products. This consumer testing should be part of an iterative development process to optimise product qualities to maximise consumer acceptability (within limits of time and money).

19.3.2.3 "Go/no go" decision time

With the newly acquired knowledge, a further review is required. The final development stage (industrial scale-up) is expensive and should be completed only on those products that show definite promise. New data for review include:

- consumer acceptability;
- updated product costing;
- potential to meet company specifications (shelf life, market share, profit);
- knowledge of operational (manufacturing) capability (can it be done?);
- prediction of volumes (can these be met by manufacturing – i.e. capacity?);
- ability to meet launch deadlines.

Successful products (hopefully now only 1–3 should remain) will be scheduled for commercial scale-up.

19.3.2.4 Commercialisation

The final stages in the technical part of NPD development is the scale-up (make sure it works in the factory) and the market testing.

In this stage, a trial commercial run is planned. This would normally manufacture several hundred kilograms of product using the actual processing equipment that would ultimately make the product. Factory trials are difficult to plan – most commercial equipment is scheduled for continuous use more than a month in advance. Equipment availability, process flow needs, raw material ordering and packaging decisions (temporary packaging can be used, as final packages would not yet be designed) all need attention.

Issues related to scale-up would be identified (invariably the scale-up offers a high degree of unpredictability), special parameters and quality control practices would be checked (for feasibility, practicality) and the process flow would be confirmed or adjusted.

The product resulting from this trial commercial manufacture would be used for market testing. Once again, assure safety is achieved (chemical, microbiology testing) as is conformity to product specification (sensory testing). Pre-market testing involves not only consumer sensory testing (acceptability), but also information about customer willingness to try the new product, plus repeat purchasing behaviours. Products have no branding information (usually), but there are legal requirements to label ingredients used, ensure clear product descriptions and so forth. It is possible to show advertising materials, various options in messages that might be used – and numerous marketing/promotional issues that will provide the marketing staff with invaluable perspectives on consumer behaviour. Assessment might include focus groups, large consumer surveys or in-home testing regimes (with accompanying questionnaires). Typically for industry the survey would involve 300–600 consumers and will take 6–14 weeks to properly complete and analyse. It may cost $100,000–500,000 – so this is a very expensive exercise.

The appropriate use of these techniques of pre-market consumer testing needs careful consideration. They are not widely used as they are expensive. Their accuracy is questionable and the data are mathematically complex to evaluate. For seasonal products, clearly the time factor is relevant: it is not realistic to test one year and sell the product the following year! The techniques are of use only if products are sold to mass-market end-users.

Test marketing does allow the marketing staff to adjust their volume forecasts, assess promotional and related tactics and to better predict consumer buying habits and patterns. However, it is time-consuming and may alert competitors to your plans.

Many companies use product "roll-outs" to launch their new products. They would tend to skip market testing and systematically roll-out products progressively. For example, they might sell first to influencer users. They might sell to a specific geographic location first and "see how it works there". They could use specific market segments, or sell through a special market channel (e.g. service stations).

The final stage of this activity is another review. This review is very detailed and comprehensive. All the data are now fully available: costing, pricing, profitability, feasibility to manufacture, promotional requirements, distribution and planning for launch, consumer acceptability and the risks associated with launching each of the products developed. This review needs to be very systematic and products must be carefully scrutinised at this point.

Final product specifications can be defined. These need to be clearly written and laid out in a logical sequence. They need to be realistic and relevant, proven in practice (you've had all the practice you need now!), approved by all relevant people and authorised in writing. They also need to be user friendly (often meaning there are several versions – for operations, marketing, R&D and QC) and they need to have clear delegations and authorities: in other words, who is authorised to make changes. Raw materials should be specified (so purchasing understands any specific technical constraints, e.g. GE-free and QC specifications on quality).

Specification headings might include:

- ingredients and their individual specifications;
- formulation (bills of materials, recipe);
- manufacturing process (including process flow chart);
- product, process and raw material testing;
- packaging specifications (materials and labelling);
- product costing;
- product weights/volumes;
- shelf life and special storage conditions;
- authorities (who can alter the specifications).

19.3.2.5 Product launch

Product launch delimits the end of new food product development and the beginning of the product lifecycle. Activities here include merchandising, advertising, sales recording, buyers' diaries, competition studies and marketing costing.

Following product launch, NPD switches to different activities closely linked to the product lifecycle. There is a need to look at production efficiencies, improving (reducing) product quality variations, reducing the cost of production (alternative processes, ingredient substitution, packaging optimisation), and checking and adjusting shelf life. Total Quality Management activities need to be reviewed and adjusted, and raw material purchasing practices should be evaluated (can you gain cost reductions through economies of scale purchasing?). More often, there is a regular need to provide "line extensions" – similar products with, say, different flavours. Eventually, most products will reach a mature marketing stage and then sales will decline until the product is no longer viable to produce.

There is also a real need for post-launch monitoring and evaluation. What is the true production and marketing efficiency? Has the market result been as expected? Does the product actually fit the company? Are there issues around food safety, nutrition or other key attributes that were identified for this product? Are there environmental, social, legal or physical issues that need to be addressed? The economic analysis can now review the true costs and evaluate payback time, cash-flow, return on investment (ROI) and related matters.

Further reading and references

Anonymous (1982) *New Product Management in the 1980s*. Booz-Allen and Hamilton, Inc., New York.

Branen, A.L., Davidson, P.M., Salminen, S. and Thorngate, J.H. (2002) *Food Additives*, 2nd edn. Marcel Dekker, New York.

Cookson, C. (1996) International R&D. *Chemical Week*, S15.

Cooper, R.G. and Kleinschmidt, E.J. (1986) An investigation into the new product process: steps, deficiencies and impact. *Journal of Product Innovation Management*, **3**(2), 71–85.

Earle, M.D. and Earle, R.L. (2000) *Building the Future on New Products*. Leatherhead Publishing, Leatherhead.

Hollingsworth, P. (1995) Food research cooperation is the key. *Food Technology*, **49**(2), 65, 67–74.

Meyer, A. (1998) The 1998 top 100 R&D survey. *Food Processing*, August, 32–40.

Miller, W.L. and Morris, L. (1999) *4th Generation R&D: Managing Knowledge, Technology, and Innovation*. John Wiley and Sons, New York.

Mirasol, F. (2000) Pfizer bags Warner-Lambert to form #2 global pharma giant. *Chemical Market Reporter*, **257**(7), 1, 20.

Moore, S.K. (2000) R&D management: finding the right formula. *Chemical Week*, **162**(16), 29–33.

Scott, A. (1999) Aventis, tech giant. *Chemical Week*, **161**(1), 46.

Siriwongwilaichat, P. (2001) *Technical information capture for food product innovation in Thailand*. PhD Thesis. Massey University, Albany.

Supplementary material is available at www.wiley.com/go/campbellplatt

Sue H.A. Hill and Jeremy D. Selman

Key points

- How information technology can be best exploited.
- What computer hardware and software are available.
- How to manage and store information.
- Mechanisms for electronic communication.
- How best to search and interrogate the Internet and the World Wide Web.

In this chapter we will consider how information technology (IT) can be exploited to the advantage of students of food science and technology.

20.1 PC software packages

Scientists and technologists need to be able to work as efficiently and effectively as possible. Therefore, in order to take maximum advantage of the opportunities afforded by IT, they must know how to go about making themselves aware of different software packages that are available and commonly used in science and by the food community. They may need to be capable of establishing and comparatively evaluating the features of different packages for a given set of user requirements, and select the most suitable computer systems and software packages based on their evaluation. Other factors may be important that can influence selection, such as the viability of the supplier, the

quality of support, software compatibility, the maturity of the product in terms of the latest version, and the quality of any accompanying documentation.

20.1.1 Computer hardware and software

Personal computers (PCs) and laptops are manufactured by a number of international companies such as Dell, Macintosh, Toshiba, Viglen, Compaq and Hewlett-Packard (see http://www.pcworld.com). Each of these manufacturers will utilise an operating system (see sections 20.2.1 and 20.2.2), and the three main operating systems available are Microsoft Windows, Macintosh and Linux. Essentially these operating systems are designed to run specific software, and so there tends to be no cross compatibility for software, although there may be for files. So, for example, the current operating system for general PCs is Microsoft Windows Vista; for Macintosh computers it is Apple Mac OS X; and for Linux computers it is Linspire from Lindows Inc. There are many Linux 'flavours', common ones being 'Red Hat' and 'Suse'. Linspire claims it can run Windows software.

Depending on the choice of specification, a typical high-performance PC in 2008 will have a central processing unit (CPU), or processor, with a speed of about 3 GHz (gigahertz). The CPU interprets computer program instructions and processes data. The computer memory is a form of solid state storage known as random access memory (RAM). This fast but temporary storage would typically comprise about 2 GB (gigabytes) of memory. The hard disk is also a device for storing digitally encoded data, but is slower than RAM and of a more permanent nature. The hard disk will typically have about 160 GB of memory. A suitable screen will be required, and there is a choice of screen sizes available, and options for curved CRT (Cathode Ray Tube) screens and the generally more expensive LCD (Liquid Crystal Display) or plasma flat screens. At the time of writing, Eee PCs or Netbooks (mini-PCs) are becoming increasingly popular too.

Before using a computer, the user needs to consider some important security issues. Computers can be damaged or become infected with a virus, and laptops can be stolen, so it is necessary to consider what back-up of the data is to be done. Regular back-up of data (daily, weekly) is essential for business. Although the latest computers include a firewall to prevent illegal external access via the Internet (see section 20.3.2) to the computer files, it is essential to subscribe to a reliable anti-virus software program, and regularly update this. Only when these are in place should access to the Internet be attempted.

In order to access the World Wide Web (Internet) (see sections 20.3.2 and 20.3.3) the computer will need a modem with a bandwidth (amount of data that can flow in a given time) of 56 Kbps (kilobits per second) to link to a cable telephone line. For a faster connection, a link to Broadband, via cable or satellite, with a bandwidth typically from 512 Kbps to 10 Mbps (megabits per second) or more is available, and this is preferred by many businesses. To enable such a connection, an Assymetric Digital Subscriber Line (ADSL) is needed. Searching the Internet requires a browser, and again there are different browser software packages for different computer operating systems such as Internet Explorer or Firefox for the Windows system, and Netscape for the Windows,

Macintosh and Linux operating systems. Safari is the common option for Macintosh.

It will be necessary to print out documents and worksheets from time to time and so a suitable printer will be required. For home use, inkjet printers are cheaper and adequate; however, for frequent and professional use, the more expensive laser printers are best. The cost of printers has reduced significantly in recent times. Removable data storage systems are to be recommended for ease of transporting and sharing data, such as CD-ROMs (Compact Disc Read Only Memory) and USB (Universal Serial Bus) port memory sticks. New computers no longer tend to come with floppy drives as standard. (A floppy disk is a data storage device that is composed of a disk of thin, flexible – hence 'floppy' - magnetic storage medium encased in a square or rectangular plastic shell. It is read and written by a floppy disk drive.)

A typical computer will cost at today's prices about £500–800 sterling (and due to much cheaper prices in the USA, probably a similar amount or less in US dollars). However, computer hardware and software are being developed and improved all the time. Therefore, if the individual wishes to be able to continue to communicate with colleagues and business contacts using compatible software and hardware systems, it is likely that it will be necessary to upgrade the computer every 3–4 years.

The most commonly used software package worldwide is Microsoft Office, and considerable information about this package and its use is available in many languages (Brown and Resources Online, 2001; Bott *et al.*, 2007; Pierce, 2007). For the food scientist the package offers an invaluable and efficient tool for many aspects of work. However, it is important to be aware that it is not the only office software package available which will run on the Windows operating system. Wordperfect Office from the Corel Corporation, Ability Office from Ability Plus Software, Star Office from Sun and the free OpenOffice software are just some of several other options that may be suitable for the requirement. Most of these office software packages offer at least four general features: word processing, spreadsheets, databases and presentations, which are discussed below.

20.1.2 Word processing

The efficiency of creating a document in a word processing environment is that it can be edited as you go along, and then printed or e-mailed when required.

Text files, pictures, graphs and diagrams can be inserted with relative ease.

20.1.3 Spreadsheets

Spreadsheets provide the tools to work with numbers simply and efficiently. They consist of cells which are organised by columns and rows, and these form the building blocks. A cell can contain words, numbers or a formula. Almost all of this type of work will involve entering or manipulating information in cells. By formatting the cells with borders, colours or special fonts, tables can be created that help the audience to understand the data. Spreadsheets, or Worksheets, are like individual sheets of paper within a notebook, but they can hold much more information than a real piece of paper. Worksheets are, therefore, the main organisers of the data. Workbooks contain a set of related information located on one or more Worksheets. Data can then be presented in the form of graphs and bar charts.

20.1.4 Databases

Essentially a database is an organised collection of information. A database enables the information to be looked at from various angles, and a variety of reports can be generated to show relationships between disparate pieces of information in the database. A database contains tuples (tables). Each table contains a number of rows of information, and each row can have a set number of columns defined at the table level. Querying this information is done using SQL (Structured Query Language). There are many 'out of the box' tools in databases that mask SQL and make it easy for users to query data. Examples include the report forms in the Microsoft database software Access, and the report form tool Crystal Reports which can be used with a variety of database types (see http://www.businessobjects.com/products/reporting/crystalreports). A report can also be designed to show the relationships between the records in different tables.

20.1.5 Presentations

A presentation graphics program is software that enables the creation of a slide show presentation. Such a show presentation can be made up of a series of slides containing charts, graphs, bulleted lists, eye-catching text, multimedia video and sound clips. Design templates include colours and graphics for use, whilst content templates contain both design and content. There will be many opportunities where such a program is invaluable, from presenting the results of a project, to presenting the structure and work of a department.

20.1.6 Tables, graphs and diagrams

The use of tables and graphs (or charts) to present information is often required. A graph often makes numerical data more visual and easier to grasp. Numbers and appropriate labels can be entered into a datasheet or spreadsheet. Then the data can be presented in, for example, a bar, line, pie, surface or bubble chart. Graphics applications can be used for producing diagrams, or for editing pictures.

20.1.7 Creation of compound documents

A compound document is, for example, a Word document that includes, say, tables, pictures and graphs as well.

20.1.8 Transferable skills in computing

If an individual wishes to be able to demonstrate some measure of this transferable skill, then he/she can study and sit for the International Computer Driving Licence (ICDL) or the European Computer Driving Licence (ECDL) (http://www.ecdl.co.uk, and http://www.ecdl.com). This syllabus comprises the following modules:

Basic concepts of IT

To have an understanding of some of the main concepts of IT at a general level. It is a requirement to understand the make-up of a PC in terms of hardware and software and to understand some of the concepts of IT such as data storage and memory. It is necessary to understand how information networks are used within computing and be aware of the uses of computer-based software applications in everyday life. An appreciation is required of health and safety issues as well as some environmental factors involved in using computers. There needs to be an awareness of some of the important security and legal issues associated with using computers.

2 Using the computer and managing files

This requires the demonstration of knowledge and competence in using the common functions of a PC and its operating system. There is a need to be able to adjust main settings, use the built-in help features and deal with a non-responding application. It is essential to be able to operate effectively within the desktop environment and work with desktop icons and windows. The ability is required to manage and organise files and directories/folders, and know how to duplicate, move and delete files and directories/folders, and compress and extract files. You must understand what a computer virus is and be able to use virus scanning software. This part includes being able to demonstrate the ability to use simple editing tools and print management facilities available within the operating system.

3 Word processing

It is necessary to be able to use a word processing application on a computer. This includes being able to accomplish everyday tasks associated with creating, formatting and finishing small-sized word processing documents ready for distribution. You must be able to duplicate and move text within and between documents. You must also demonstrate competence in using some of the features associated with word processing applications such as creating standard tables, using pictures and images within a document, and using mail merge tools.

4 Spreadsheets

It is important for many data handling applications to be able to understand the concept of spreadsheets, and to demonstrate the ability to use a spreadsheet application on a computer. Then you need to be able to accomplish tasks associated with developing, formatting, modifying and using a spreadsheet of limited scope ready for distribution. You also have to be able to generate and apply standard mathematical and logical formulas using standard formulas and functions, and demonstrate competence in creating and formatting graphs/charts.

5 Database

You must understand some of the main concepts of databases and demonstrate the ability to use a database on a computer. This includes being able to create and modify tables, queries, forms and reports, and prepare outputs ready for distribution. You must be able to relate tables and to retrieve and manipulate information from a database by using query and sort tools available in the package.

6 Presentation

You must demonstrate competence in using presentation tools on a computer, and be able to accomplish tasks such as creating, formatting, modifying and preparing presentations using different slide layouts for display and printed distribution. You must be able to duplicate and move text, pictures, images and charts within the presentation and between presentations. You must demonstrate the ability to accomplish common operations with images, charts and drawn objects and to use various slide show effects.

7 Information and communication

The first section, Information, requires you to understand some of the concepts and terms associated with using the Internet, and to appreciate some of the security considerations. You will also be able to accomplish common web search tasks using a web browsing application and available search engine tools. You should be able to bookmark websites, and to print web pages and search outputs. You should be able to navigate within and complete web-based forms. In the second section, Communication, you are required to understand some of the concepts of electronic mail (e-mail), together with having an appreciation of some of the security considerations associated with using e-mail. You shall also demonstrate the ability to use e-mail software and to send and receive messages, and to attach files to mail messages. You should also be able to organise and manage message folders/directories within e-mail software.

From the http://www.ecdl.com website it can be seen that there are certification contacts in 135 countries in the regions of Africa, Asia Pacific, Europe, the Middle East and the Americas. There are other certification programmes, including for people with disabilities, such as ECDL Advanced, ECDL CAD for computer aided design and ECDL Certified Training Professional, and more courses are being planned.

20.1.9 Statistical packages

The application of statistics is an extremely important part of studying and understanding foodstuffs and food manufacturing operations. Foods are biological materials and, therefore, there is natural variability between samples of both raw materials and final products. Key aspects include the design of experiments and the subsequent analysis of results. For manufacturing operations, sampling, statistical process control, sensory analysis and consumer testing are important (Hubbard, 2003; De Veux and Velleman, 2004).

For the general applications of statistics there are a variety of choices. In higher education the *Minitab* package, and cut down versions (http://www .minitab.com), have a long-standing reputation for a variety of statistical tests, and also design and control charts. More powerful than Minitab is *Genstat* (UK) which again has been used in agricultural and food research for many years.

Design Expert (http://www.statease.com) offers a wide functionality. The use of traditional factorial designs and of Taguchi designs (for example, using Qualitek-4 (http://www.rkroy.com) from Nutek Inc.) enable the good practice of weeding out the known weak relationships, or the faulty products, as soon as possible.

SAS/STAT software (originally the Statistical Analysis System – SAS) (http://www.sas.com) is very powerful and has a large market share in the USA, being well known especially in the pharmaceutical sector. However, if used to its full power it is necessary to type in appropriate syntax, which may be slightly offputting to someone who is not a statistician.

SPSS (http://www.spss.com) has been around a long time, especially in the social science and healthcare communities. It is especially good for survey type work and for tabulation. Analysis of data using up to version 11.5 is based on the assumption that the sample data are simple and random. If, in reality, the sample data are not simple and random, but for example stratified, then the analysis is limited. However, in the more recent version 12, there is a module for complex samples. *SPSS* also has a good and user friendly windows interface. The family of Matlab and Simulink software from http:// www.mathworks.com also provides opportunities for powerful manipulation of data.

For government survey type work *STATA* may be used. This software from the USA (http://www .stata.com) is popular in economics circles, and version 8 has a windows interface. Originally developed in conjunction with the Norwegian Food Research Institute, a particularly good package relevant to the processing of biological materials that has more recently been applied to solving food-related problems is *Unscrambler* (see also section

20.1.10), which has been commercialised by CAMO (http://www.camo.com).

For many simpler applications of statistics there are options for *Microsoft Excel* add-ons. For example, *Analyse-it* (http://www.analyse-it.com), *StatTools* (http://www.palisade-europe.com) and the website http://www.statistics.com give a review of over 100 packages including, for example, Statserv (http://www.statserv.com) and XLStat (http://www. xlstat.com). Another useful list of statistics software is given at http://www .freestatistics.info.

20.1.10 Software tools for food science and technology applications

The availability of computers now permits an ever-increasing array of predictive tools and models, and also training opportunities for food applications. Some examples of these tools are given in this section.

- *Sensory analysis and sampling*: these are the two areas that have traditionally required the use of statistics. For sensory analysis and consumer tests, two leading packages are *FIZZ* (http://www .biosystemes.com) and the suite of software from Compusense Inc, Canada (http://www. compusense.com). *Unscrambler* (see also section 20.1.9) from CAMO (http://www.camo.com) with its multivariate analytical functions has also been applied successfully in this field. Some of the Campden and Chorleywood Food Research Association (CCFRA) reviews on and guides to the application of statistics to sampling in food manufacture include a statistics tool, or spreadsheets, for use with *Microsoft Excel* (CCFRA, 2001, 2002, 2004).
- *Microbial growth and shelf-life*: there is increasing availability of predictive models, for example, for predicting shelf-life of food (see http://www .foodrisk.org). The following are examples of these:
  ComBase: microbiological safety and likely spoilage of a range of food formulations can be predicted using this Internet-based free database. A collection of predictive models using the ComBase data, such as Microfit and Growth Predictor, is available from the Internet (http:// www.combase.cc).
  Forecast: the Campden and Chorleywood Food Research Association (CCFRA) has developed a collection of bacterial spoilage models which account for fluctuating temperatures, dynamic

processing environments, modified atmospheres and new product types (http://www.campden .co.uk/scripts/fcp.pl?   words=forecast&d=/ research/features_06_3.htm).

Food Spoilage Predictor: developed by researchers in Australia, this program can be used to predict the rate of microbial spoilage in a wide range of chilled, high protein foods such as meat, fish, poultry and dairy products. The system uses a small data logger that is integrated with the software containing the models. It can predict remaining shelf-life at any time in the cold chain (http://www.arserrc.gov/cemmi/FSPsoftware .pdf).

Seafood Spoilage Predictor: developed by the Danish Institute for Fisheries Research, this free Internet software can be used to predict the shelf-life of seafood stored under either fluctuating or constant temperature conditions (http://www.dfu.min.dk/micro/sssp/).

FARE Microbial™: (http://www.foodrisk.org/ exclusives/FARE_Microbial/) developed by Exponent Inc. together with the US Food and Drug Administration. It consists of a contamination and growth module, and an exposure module to enable probabilistic microbial risk assessment. Sym'Previus (http://www.symprevius.net) is a French language predictive microbiology tool and database.

- *Water activity and mould-free shelf-life (Water Analyzer Series)*: this series of programs can be used to predict the water activity (A_w) of component products under a range of different conditions, including the efficacy of packaging films (http://www .users.bigpond.com/webbtech/wateran.html).

- *ERH-CALC*: users can input basic recipe formulations and the software calculates the theoretical equilibrium relative humidity (ERH). From these data, the model then predicts the mould-free shelf-life (MFSL) of the ambient stored product. (http://www.campden.co.uk/publ/pubfiles/ erhcalc.htm).

- *Hazard Analysis Critical Control Point (HACCP)*: HACCP documentation software (http://www .campden.co.uk) has been widely used in both Europe and the USA.

- *Chill Chain (Coolvan)*: developed by the Food Refrigeration and Process Engineering Research Centre, Coolvan can predict the temperature of food during a single/multi drop journey in a refrigerated van. Knowing the temperatures in a food can help in predicting shelf-life as well as enabling a producer to ensure that a chilled food will be at the correct temperature when it reaches the retailer (http://www.frperc.bris.ac.uk/pub/pub13.htm).

- *Heat processing*: CTemp for calculating heat processes (http://www.campden.co.uk); and the spreadsheet for process lethality determination from the American Meat Institute (http://www .amif.org).

- *Packaging*: Mahajan *et al.* (2007) discuss the development of the user-friendly software PACKinMAP for the design of modified atmosphere packaging for fresh and fresh-cut produce. A download is available from the Swiss Federal Office of Public Health (http://www.bag.admin.ch) which can be used to estimate the amount of a substance that migrates from a plastic material into a food product during a given time.

- *Nutrition*: Leake (2007) discusses currently available databases and software providing information about the nutritional value of foods and beverages, ingredients and E-numbers, for example (see http://www.nutricalc.co.uk). The usefulness of such information for the preparation of nutrition fact labels by food producers is considered. There are of course a variety of nutrition packages for diet control applications (for example, http://www.nutribase.com).

- *Enzyme analysis*: Enzlab (http://www.ascanis. com/Enzlab/enzlab.htm) ensures the correct method and calculations are used for enzymatic food tests.

- *Bread and cakes*: Bread Advisor and the Cake Expert System Software have been developed by Campden BRI (formerly Campden and Chorleywood Food Research Association – CCFRA) together with the food industry (http://www .campden.co.uk).

20.1.11 Distance learning

Software developments aimed at the Internet training and education sector are increasingly being used for distance learning programs which include the food science area. For example, for food legislation at Michigan State University (http://www .iflr.msu.edu), for food science at Kansas State University   (http://www.foodsci.k-state.edu/Desktop Default.aspx?tabid=709), and for courses for the food industry at the University of Guelph (http: //www.open.uoguelph.ca/offerings/). The USDA/

FDA food and nutrition information center lists a number of links to distance learning, on-line courses and curriculums relevant to nutrition and dietetics at http://riley.nal.usda.gov/nal_display/index.php?info_center=4&tax_level=2&tax_subject=270&topic_id=1326&&placement _default=0 A survey of distance courses offered both in the USA and internationally was recently done by the International Union of Food Science and Technology (IUFoST) and this is at http://www.iufost.org/education_training/distance_education/ and http://www.iufost.org/education_training/.

More recently, there have been approaches to presenting virtual experiments in food processing on the Internet (http://rpaulsingh.com/virtuallabs/virtualexpts.htm). A free educational website (http:// www.foodinfoquest.com) has been produced by the International Food Information Service (IFIS Publishing). This website is targeted primarily at students and helps to develop the skills required to find and use food science information effectively. It explains the types of information resources available, suggests where they may be found, demonstrates basic searching techniques and discusses methods for writing up research.

20.2 Managing information

20.2.1 Introduction

When you have amassed (created or retrieved or received) data on your PC (see Chapter 21 and section 20.3 of this chapter) it becomes increasingly necessary to manage or organise this information in some way in order for you to be able to refer to it quickly and easily on an 'as needs' basis. This becomes especially important when the volume of the data held inevitably increases with time, and is true for both individuals and organisations, although the scale of the task obviously varies and can require correspondingly varying levels of sophistication.

In this chapter we shall only focus on electronic data and, in doing so, it is important to remember that the formats of electronic records and the platforms on which they are stored not only vary but also can become unreadable in a very short space of time. This happens because computing systems frequently have short life spans and information can be readily corrupted. The extent of the problem becomes more apparent when one considers other factors influencing the storage and management (archiving) of electronic data. For example:

- it is possible for electronic data to be subject to undetectable changes if proper precautions are not taken;
- electronic records may fail to be captured because most record keeping processes are based on printed records;
- the context of an electronic record and its relation to other records can be lost;
- capture of all the relevant contextual information can be expensive; and
- systems for managing electronic records are not routinely designed to incorporate archiving as part of their functionality.

At the level of the PC these problems are addressed by the operating system which is installed on your particular PC (see below); organisations wanting to archive (preserve) large amounts of data for prolonged periods of time do not have such a simple remedy for their problems. We shall first address the situation with respect to PCs and then go on to briefly consider the problems and solutions employed by organisations in their attempts to electronically archive and access large amounts of diverse scientific information over prolonged time periods.

20.2.2 Managing and storing (archiving) information on a personal computer

The development of operating systems has negated the need for computer programmers to write routines for commonly used functions and has provided a uniform method for all application software to access the same resources. The majority of operating systems are related to the type of machine on which they are installed. PCs make use of an operating system that undertakes the following tasks:

- initialising the system;
- providing routines for handling input and output requests;
- memory allocation; and
- providing a system for handling files (which are the basis for the storage model on PCs).

Examples of operating systems for PCs include UNIX, Microsoft Windows, Mac OS X and Linux. Online courses are offered on understanding PCs and operating systems (for example, http://www .helpwithpcs.com) and the different producers of the operating systems available offer their own 'help screens' in association with their products to explain how to use them and enable users to take maximum advantage of their functionality. It is, therefore, more appropriate to consult the 'help screens' pertaining to your particular system than to try to describe the functionality of all the available systems here. Instead we will focus on the generalities common to all the systems currently available. Operating systems are responsible for the provision of essential services within a computer system to include initial loading of programs and transfer of programs between secondary storage and main memory, supervision of input and output devices, file management and protection facilities. It is the file management service that is of interest to us in the context of managing and storing information. Operating systems provide a file management service that allows the user to locate and manipulate the various programs and data files which are stored on the hard disk of a PC.

A file is essentially an electronic repository (a digital storage point) for information. A file can contain anything from picture images to word processor documents. Files can be renamed, copied and deleted; they can also be manipulated in a variety of ways, for example, they can be hidden from view or cannot be altered (read only). A collection of files is called a directory. Directories (or folders) are used to divide the files into logical (possibly subject) groups that are easy to work with; files can easily be moved from one directory to another.

Directories can have subdirectories which, in turn, can also have subdirectories which, in turn, can also have subdirectories, and so on; in principle there is no limit to the number of subdirectories that you can create. However, it is important to remember that directories and subdirectories take up space on the hard drive of your PC such that you are limited by the memory capacity of the hard drive. In practice, it is not a good idea to create too many directories and subdirectories because they can complicate the situation and spread subject areas too thinly, making it difficult to retrieve all related topics in their entirety (you could miss out on key information stored in a remote and thinly populated subdirectory). The key is to keep the number to the minimum that will allow good organisational control of your files, but in essence you should do what works best for you. The collection of directories and subdirectories on a hard

disk drive (a device inside a PC where the bulk of the information is stored) is sometimes referred to as the directory tree. The root or base directory is the starting point from which all subdirectories are generated.

It is important to name your files, directories and subdirectories in a way that will help you to easily identify what they contain, for example:

Food Science (*directory*)
Dairy Science (*subdirectory*)
Dairy Products (*subdirectory*)
Butter (*file*)
Buttermilk (*file*)
Cheese (*subdirectory*)
Cheese varieties (*subdirectory*)
Brie (*file*)
Cheddar (*file*)
Edam (*file*)
Mozzarella (*file*)
Etc. (*files*)
Cream (*file*)
Dairy beverages (*subdirectory*)
Lactic beverages (*file*)
Milkshakes (*file*)
Whey beverages (*file*)
Yoghurt (*file*)
Lactose (*file*)

By adopting this type of model it is possible to organise the information stored on your PC in a logical and easily identifiable and retrievable fashion, thereby creating your own electronic archive.

20.2.3 Problems and solutions associated with large-scale electronic archiving and providing permanent access to scientific information

During their undergraduate careers, most food science and technology students are unlikely to encounter the problems faced by organisations attempting to electronically archive, and provide permanent access to, often quite large amounts of scientific information. It is, however, appropriate to address this subject here, albeit briefly, in order to give some idea of the enormity of the problem and to provide the student with some understanding of the range and depth of information that is stored. On this grander scale, it is now considered more appropriate to replace the term 'electronic' with the word 'digital'; similarly, there is also a trend to replace 'archiving' with 'preservation' (see *Digital Preservation and Permanent Access to Scientific Information: The State of the Practice* by Gail Hodge and Evelyn Frangakis (2004) at http://www.icsti.org/digitalarchiving/getstudy .php).

At this level the operating systems which cope with managing and storing information on a PC are not able to resolve the problems encountered (see section 20.2.1). So-called off the shelf systems do not, as yet, adequately satisfy all the requirements that an organisation may have with respect to 'archiving'. Therefore, custom (purpose)-built systems are usually developed to ensure that all requirements are met. In developing a customised solution it is important that the system caters for the maintenance of the necessary records and the accessibility of record content over prolonged periods of time. Therefore, the system should have modular architecture, with different modules defining different requirements of the system; the interfaces between the modules should be well defined.

Communication (transmitting messages) between the various modules is essential and necessitates using a format that is unlikely to become obsolete. Developing a system that is composed of a number of separate modules is of benefit because it allows work to be done on an individual module (or for it to be replaced) without any effect on the other modules of the system. The system must be protected from damage (hardware damage, physical damage, attack by computer viruses and hackers) which means that it must be backed up in such a way that it can be rebuilt in its entirety; it must also be protected from unauthorised access.

The foregoing gives only a brief and introductory understanding of what is required in developing customised systems and is sufficient for the purposes of this book. However, one of the first public organisations to set up an operational digital preservation system was the Australian State of Victoria and if you are interested in pursuing the subject further, a detailed explanation of how they created the Victorian Electronic Records Strategy can be found on their website at http://www.prov.vic.gov.au.

20.2.4 Copyright issues

When storing information it is important to keep in mind the constraints imposed by copyright legislation. Copyright is a very complex issue and has been made even more complicated by the advent of the electronic information era and the increasing need to apply international standards. Essentially, copyright (the ownership of a piece of original work albeit for a limited but defined period of time) is given to the creator of an 'intellectual property' and whether the 'creator' is an original author of, for example, a scientific article in a journal, or the publisher of the journal in question (authors often assign copyright of their articles to the publisher of the journal in which their article appears), legislation is in place both to protect their work and to allow the right of the public to have access to knowledge. For example, in creating a paper collection of literature on food science, it is important to remember that the UK Copyright, Designs and Patents Act 1998 does not permit multiple photocopying of articles without payment, or copying more than one article from a single journal issue.

In the electronic environment the position is compounded by the concept of downloading (the transfer and storage of references from a database to a computer). In practice, downloading is often covered by the licensing agreements imposed by database producers (see Chapter 21) and agreed by libraries when they subscribe to a particular database. The majority of database producers will allow downloading to set up personal files (for use by an individual) but will forbid the subsequent transfer of the downloaded material by any means to third parties (for example, to students in other colleges or universities or to corporate concerns). Therefore, in practice, there should be no copyright problems if you are creating a single copy literature file for your sole personal use. In fact the situation has been impacted by, and in some respects become more relaxed, with the arrival of open access initiatives which are based on the premise that all scholarly publications should be freely available on the Internet and accessible at no charge. Open access (see section 21.2.2) permits any user to read, download, copy, distribute, print, search or link to the full text of any scientific articles. Although open access is increasing in momentum, it is important to remember that it has not been universally adopted and a lot of material available via the Internet still has costs associated with its use.

The issue of copyright remains complex and continues to become increasingly so – it is something of a minefield. If you are in any doubt at all about what you can legally download onto your PC, it would be very worthwhile to consult your librarian or information specialist/centre (if your university or college has one). References are included for further information on copyright, but it would be wise to seek direct advice for specific issues (Wall, 1998; Armstrong, 1999; Wall *et al.*, 2000).

20.3 Electronic communication

20.3.1 Introduction

Electronic communication (the sending and receiving of information electronically via connected computer systems or the Internet) is still relatively new but tremendously popular; it has rapidly become almost indispensable in both the personal and working lives of a huge number of people (see also Chapter 21). It is important to remember that electronic communication differs from more traditional communication mechanisms in a number of key respects. These relate to:

■ speed (information can be delivered electronically in seconds);

■ permanence (electronic information is sometimes transient and can be interfered with by third parties);

■ cost (electronic communication permits widespread distribution of information with only the equivalent effort and cost of sending a single message by more traditional means);

■ security and privacy (electronically received information may be intercepted by third parties, i.e. not the sender or recipient of the information);

■ authenticity of the sender (the sender of the information is verified).

In this chapter we shall consider electronic communication with respect to the Internet, the World Wide Web (WWW), online and local (intranet) networks, the use of search engines and the use of e-mail.

20.3.2 The Internet

The Internet is the name given to an enormous, worldwide network of computer systems which links computers to other computers using TCP/IP (Transmission Control Protocol/Internet Protocol) network protocols. TCP/IP software was originally designed for the UNIX operating system but is now available for every major operating system and continues to evolve. A valuable source of information about all matters pertaining to the Internet is the Internet Society which can be accessed at http://www.isoc.org.

The term 'internet' may also be used to describe networks of linked computers which are not part of the Internet proper (hence the use of lower case rather than a capital letter to start the word) – membership of the Internet proper demands the use of TCP/IP protocols. The term 'intranet' is used to describe a private network of computers owned by a company or organisation. Intranets make use of the same kinds of software used on the public Internet but are only available for internal use within the company or organisation.

To access the Internet you will need a computer, a modem or other telecommunications device and software to connect you to an Internet Service Provider (ISP). There are a great many ISPs to choose from; each is a company that sells Internet connections via a modem. It is probably best to shop around for the best deals if buying for yourself, having first sought the advice of your university/college librarian or information centre.

Each computer that is part of the Internet network carries software which enables information to be provided and/or accessed and viewed via the World Wide Web (see section 20.3.3); these computers are known by a variety of names such as servers, web servers and host computers. Every computer that is on the Internet has a unique Internet Protocol (IP) number or address which consists of four parts separated by dots, for example, 123.123.123.1; most also have one or more domain names which refers to the initial part of a Uniform Resource Locator (URL) which, in turn, represents the unique address of a web document. Domains are a human-readable way of mapping IP addresses to Internet locations. Essentially, domain names indicate who produced the page you are looking at.

URLs have a logical layout as shown in the following fictitious example for a university library:

> http://www.lib.xxx.edu/Guide/FoodScience/
> Catalogue.html

where

http:// refers to the type of file (hyper text transfer protocol, or http, is the protocol used to transmit web pages), www.lib.xxx.edu/ refers to the domain name (edu signifying an educational site),

Guide/FoodScience/ usually refers to the path or directory the file is stored in on the computer (see *Note*),

Catalogue.html usually refers to the name of the file and its extension (see *Note*).

(*Note*: the URL enables the user to quickly find a file. As indicated above, parts of the URL *usually* refer to the actual file, and most of the time this is the case. However, sometimes the URL may not refer to the actual file itself. This is because the file may have been built in a different dynamic format, for example as an active server page, or because the structure of the files may have been designed differently, for example for the needs of the directory subject matter.)

The Internet itself does not contain any information. Instead, it provides access to the information contained on one of the many computers which make up the Internet network. Internet-linked computers offer access to one or more of the following services:

- the World Wide Web (WWW or the Web);
- electronic mail (e-mail);
- Telnet or remote login;
- FTP (File Transfer Protocol); and
- gopher.

We will consider two of these services (the WWW and e-mail) in more detail. For the other three, it is sufficient for our purposes to have a brief but working definition of their functionality as follows:

- Telnet allows your computer to log on to a second computer and use it as if you were located at the site of the second computer.
- FTP permits your computer to retrieve complete complex files from a remote computer and to view and/or save them on your computer.
- Gopher provides a text-only method for access to Internet documents but has been virtually superseded by the advent of the WWW; however some gopher documents may still be encountered.

20.3.3 The World Wide Web (WWW, the Web)

The WWW is essentially the information store for services available via the Internet. It consists of web-sites and web pages which, for example, enable you to:

- retrieve documents;
- view images, animation and video;
- listen to sound files;
- send and receive oral messages; and
- use programs which run on virtually any type of software that is available.

A web page is found by browsing to a Uniform Resource Locator (URL – see section 20.3.2.); a website is a collection of related pages which you can link to from that site. For example, the website for the International Food Information Service (IFIS) is called Food Science Central and can be accessed at http://www.foodsciencecentral.com. All of the pages associated with Food Science Central branch out from there. The range of services available to you is controlled by the types of hardware and software that you have access to.

The WWW was originally developed by Tim Berners-Lee in 1989; a year later he went on to create the first WWW server, the first web browser, the URL addressing system and HTML (hypertext markup language). Browsers are software programs that allow you to view WWW documents (pages); they convert HTML-encoded files into the text and images that you see (sometimes referred to as page rendering). A number of different browsers are available, for example, Microsoft Internet Explorer, Netscape, Firefox, Opera and Safari.

HTML provides the basis for WWW functionality. HTML is a standardised language of computer code which is used to define the content of web pages and provides formatting instructions for display of same on your computer screen; it works in conjunction with your browser which is programmed to interpret HTML for display. In addition to providing the content of web pages, HTML has a feature called hypertext built into it. Hypertext enables web pages to be linked to other web pages which need not necessarily be housed on the same computer. Linking is limited by the ability of the computer linking in to display the content it does not house – it needs software that will enable it to do this which is often built into browsers or can be added as so-called plug-ins.

In 1994, Berners-Lee went on to found the World Wide Web consortium (W3C) which is the main international standards organisation for the WWW (see http://www.w3.org). W3C is working with another mark-up language – XML (extensible mark-up language). XML is a simple, flexible text format derived from another mark-up language – SGML (standard generalised mark-up language). XML was originally designed to meet the challenges posed by large-scale electronic publishing but is now also being used for data exchange in an increasingly wide variety of applications both on the WWW and in other areas.

Searching for information on the WWW cannot be undertaken directly due to its sheer size and because it is not indexed using a standard vocabulary. Therefore, instead, you should use one or more of the intermediate search tools currently available. We shall now go on to consider the types of search tools that are available.

20.3.4 Search engines

Search engines are enormous databases of web pages that have been put together automatically by machine and enable you to search subsets of the WWW. There are two types of search engines – individual search engines which compile their own searchable databases, and metasearch engines which do not compile their own databases but simultaneously search the databases of a number of individual search engines.

Examples of individual search engines include AlltheWeb, AltaVista and Google. Individual search engines compile their databases by using computer robot programs known as spiders which roam the WWW identifying and indexing the pages they visit; the findings can then be included in the databases used by the various search engines currently available. When you use a search engine you are not searching the WWW either in its entirety or as it exists at the particular moment in time that you conduct the search – you are searching a subset of the WWW which has been captured using a fixed index system at some time in the past. It is difficult to say how far in the past the databases being searched were created. Spiders regularly return to the web pages that they index to keep the databases up to date, but, unfortunately, the update process can take quite a long time depending on how regularly the spiders revisit the WWW and then how quickly any changes can be made to the databases. Although search engines are not completely up to date in terms of the status of the WWW, some have formed partnerships with news databases which are state of the art in terms of currency; for example, AlltheWeb (Fast) News, Al-taVista Breaking News and Google Breaking News; such search engines carry a 'news tab' which can be used to access the most recent information.

Search engines work by using selected software programs to search their indexes for matching keywords and phrases (see section 21.2.4); the results are presented in a defined order known as relevance ranking (with the most relevant references appearing towards the top of the list of results and the least relevant towards the bottom). The software programs employed by the various search engines are often similar, but no two engines are exactly the same either in terms of size, speed of retrieval or content. They also offer different search options and different ranking schemes. Therefore, no two search engines will give you exactly the same results – the differences may be small but they could be significant. However, search engines are undoubtedly the best way currently available for searching the WWW, although, it is important to remember that they often produce enormous numbers of references to simple search requests, many of which may be irrelevant to your needs, and so-called information overload can then kick in.

Metasearch engines (examples include Ixquick, Metor, Profusion and Vivisimo) do not roam the WWW compiling their own searchable databases – instead they simultaneously search the databases of a number of individual search engines (see above). In this way, metasearch engines provide a quick and easy mechanism of discovering which search engines are capable of doing the best job in satisfying your particular search requirements. Metasearch engines are very fast and present their results in one of two ways: as a single, deduplicated merged list or as multiple lists exactly as they are obtained from each of the individual search engines used (duplicate references may, therefore, occur). Metasearch engines are handy tools for conducting simple searches and when you are in a hurry – they will give you a quick overview of your area of interest and are also helpful if searching using individual search engines is not picking up any useful references.

20.3.5 Subject directories, portals and vortals

Unlike search engines, subject directories are created manually and maintained by people rather than computer robot programs. Sites are reviewed and selected for inclusion in directories according to a defined policy or set of criteria. There is a tendency for directories to be smaller than search engine databases and usually only the key pages of a website are indexed. A search engine for searching a given directory is sometimes incorporated. Subject directories come in a variety of forms, for example, general directories, academic directories, commercial directories, portals and vortals. Portals (for example, Excite, MSN and Netscape) are created or acquired by commercial concerns and reconfigured as gateways (see section 20.3.6) to the WWW; they link to popular subject areas and offer additional services such as e-mail and access to current news. Vortals (vertical portals) differ from portals in being subject specific; they are routinely created by subject experts/specialists/professionals. Examples of vortals include Educator's Reference Desk (educational information), SearchEdu (college and university sites) and WebMD (health information).

It should be noted that the dividing line between search engines and subject directories is starting to get hazy. While the majority of subject directories offer search engines to interrogate their content, search engines are now either acquiring existing subject directories or creating their own. On balance, however, because of the way in which they are structured, subject directories can frequently deliver a higher quality of content and fewer irrelevant references in terms of search results than can search engines. However, because most subject directories do not compile their own databases and rely on pointing to web pages rather than storing them, there is a danger, because of the transient nature of information on the WWW, that changes could go unrecognised and users might be directed to pages that no longer exist. In practice, subject directories are probably most valuable for browsing information and undertaking simple, general searches; they are useful sources of information on popular subject areas, organisations, commercial/corporate sites and products. Examples of subject directories include CompletePlanet, LookSmart, Lycos and Yahoo!

20.3.6 Gateways

There are two types of gateways – library gateways and portals (for portals see section 20.3.5 above). Library gateways are collections of databases and information sites which are arranged by subject content; they are created, regularly reviewed and recommended by specialists. Gateway collections, therefore, represent high-quality information sites on the WWW and, consequently, are important pointers to high-quality information sources. Examples include Academic Information, Digital Librarian, Internet Public Library, Librarians' Index to the Internet and WWW Virtual Library.

20.3.7 The Deep Web

There is a large area, estimated to be around 70% of the total WWW, which cannot be accessed and indexed by search engine spiders. This area is known as the Deep Web or the Invisible Web; it contains information that is protected by passwords or firewalls and transient information which is created only on demand. Although search engines continue to improve and are getting better at searching the Deep Web, you really need to point your browser directly at Deep Web sites in order to access them. Because this is essentially what a number of library gateways and subject-specific databases (vortals) do, they are particularly useful tools for searching the Deep Web.

20.3.8 Approaches to searching

Although there are no right or wrong ways to search for information there are approaches to searching that can improve the efficiency of the exercise. In section 21.2.4 we describe how to interrogate bibliographic databases using Boolean logic and the Boolean operators (AND, OR and NOT). The efficiency of searching using Boolean logic can sometimes be improved by so-called nesting which involves the use of parentheses (and sometimes inverted commas) to combine several search requests into a single request. For example, keying in the search request (*Salmonella* OR *Listeria*) AND (chocolate OR cheese) would retrieve reference sources about *Salmonella* and chocolate, *Salmonella* and cheese, *Listeria* and chocolate, and, *Listeria* and cheese as separate entities.

In addition to Boolean operators, proximity or positional operators can be used to create search requests. Not all search engines recognise proximity operators, but a few can accept some if advanced search functionality is employed. Examples of proximity operators are:

- NEAR (probably the most commonly recognised), allowing you to search for terms or keywords which are located within a specified distance of each other in any order – the closer the terms are the higher the reference appears on the list of sources retrieved;

- ADJ (adjacent to – rarely recognised by search engines), allowing you to search for phrases, for example, low ADJ calorie, which will retrieve 'low calorie' and 'calorie low' references;
- SAME, for retrieving terms located in the same field – see below;
- FBY (followed by).

Currently, search engines do not recognise either the SAME or FBY proximity operators.

Electronic records have a field structure. A typical web page comprises a number of major fields, for example, the title, domain, host (site), URL and link are located in separate fields. Examples of other searchable fields include object, text, language, sound and pictures. Not all websites have the same fields available for search purposes – you need to be aware of what is available at your chosen WWW location and this can usually be found somewhere on the web page or via a linking button. Some search engines will permit you to interrogate individual fields in combination with your chosen term(s) or keyword(s). Field searching is a precise and powerful mechanism for searching the WWW, but its use is limited by its availability in terms of both the fields available for searching and the ability of your search engine to retrieve the same.

20.3.9 E-mail – electronic mail

E-mail is a system of sending and receiving messages using computers; it has rapidly become one of the most popular means of communication for personal, professional and business purposes. In order to send and receive e-mail messages you will need an e-mail address which will be unique to you. To this end you will need to open an account with an e-mail provider. Your e-mail address will take the following form:

student@e-mailprovider.com

where 'student' is your chosen name.

When you have set up your account you will be able to send and receive electronic messages via your computer. There is no cost for sending or receiving e-mail messages other than your telephone charges (usually the price of a local call). Some e-mail providers offer a Post Office Protocol (POP) server with their e-mail package. A POP server enables you

to receive your messages via an e-mail client program, for example Outlook or Netscape Messenger, which permits you to connect to your e-mail provider to retrieve your messages and then read them offline; thus your online time is reduced and so, correspondingly, are your telephone charges.

Similarly, some e-mail providers offer a Simple Mail Transfer Protocol (SMTP) server with their e-mail package. An SMTP server enables you to send e-mails which have been written offline within your e-mail client program; you then only have to go online briefly to send them, which can, again, reduce your telephone charges. These savings are important factors to consider when choosing an e-mail provider. You have to be connected to the Internet to receive an e-mail, but you do not have to remain connected in order to read it.

ISPs (see section 20.3.2) normally offer a package which includes an e-mail address together with a POP and SMTP server. This makes life very easy but can be a problem if you want to change your ISP, because moving means that you will have to change your e-mail address. Therefore, it can be a good idea to set up an account with a free permanent e-mail provider, such as Yahoo® or Hotmail®, as a backup.

In the following Chapter (21) we shall go on to consider how IT can be employed to foster and improve communication and transferable skills.

Further reading and references

(Note: all URLs cited were correct in November 2008.)

Armstrong, C.J. (ed.) (1999) *Staying Legal – A Guide to Issues and Practice for Users and Publishers of Electronic Resources.* Library Association Publishing, London.

Armstrong, C.J. and Bebbington, L.W. (2003) *Staying Legal – A Guide to Issues and Practice Affecting the Library, Information and Publishing Sectors*, 2nd edn. Facet Publishing, London.

Bott, E., Siechert, C. and Stinson, C. (2007) *Windows Vista Inside Out.* Microsoft Press, Redmond, Washington.

Brown, C. and Resources Online (2001) *Microsoft Office^XP Plain and Simple.* Microsoft Press, Redmond, Washington.

CCFRA (2001) *Designing and Improving Acceptance Sampling Plans – A Tool.* Review No. 27. Campden and Chorleywood Food Research Association, Chipping Campden.

CCFRA (2002) *Statistical Quality Assurance: How to Use Your Microbiological Data More Than Once.* Review

No. 36. Campden and Chorleywood Food Research Association, Chipping Campden.

CCFRA (2004) *Microbiological Measurement Uncertainty: A Practical Guide.* Guideline No. 47. Campden and Chorleywood Food Research Association, Chipping Campden.

De Veaux, R.D. and Velleman, P.F. (2004) *Intro Stats.* Pearson Education, Upper Saddle River, New Jersey.

Hubbard, M.R. (2003) *Statistical Quality Control for the Food Industry*, 3rd edn. Springer, New York.

Leake, L.L. (2007) Software automates nutrition labelling and more. *Food Technology*, **61**(1), 54–7.

Mahajan, P.V., Oliveira, F.A.R., Montanez, J.C. and Frias, J. (2007) Development of user-friendly software for design of modified atmosphere packaging for fresh and fresh-cut produce. *Innovative Food Science and Emerging Technologies*,8(1), 84—92.

Pierce, J. (ed.) (2007) *2007 Microsoft Office System Inside Out.* Microsoft Press, Redmond, Washington.

Wall, R.A., Norman, S., Pedley, P. and Harris, F. (2000) *Copyright Made Easier*, 3rd edn. Europa Publications, London.

Supplementary material is available at www.wiley.com/go/campbellplatt

Communication and transferable skills

Jeremy D. Selman and Sue H.A. Hill

Key points

- Study skills (effective learning techniques and resources, and time management).
- Information retrieval (library resources, databases, the Internet and World Wide Web and competitive intelligence).
- Communication and presentational skills.
- Team and problem solving skills.

There is a growing awareness in higher education worldwide of the need for students to have good information literacy skills, and to be able to communicate clearly. This is essential to help with accessing, manipulating and efficiently using the many information resources now available to everyone. Standards for information literacy competency have been drafted in order to try to improve the relevant skills (Society of College, National and University Libraries, 1999; American Library Association, 2000; Bundy, 2004). The Society of College, National and University Libraries (SCONUL) paper summarises seven key skills:

1 the ability to recognise the need for information;
2 the ability to distinguish ways in which the information 'gap' may be addressed;
3 the ability to construct strategies for locating information;
4 the ability to locate and access information;
5 the ability to compare and evaluate information obtained from different sources;
6 the ability to organise, apply and communicate information to others in ways appropriate to the situation;
7 the ability to synthesise and build upon existing information, contributing to the creation of new knowledge.

These seven key skills build upon the knowledge that students should have about basic library skills, as well as Information Technology (IT) skills. Effective communication is based on the ability to access and organise knowledge, and to work efficiently and professionally with colleagues.

21.1 Study skills

21.1.1 Use of effective learning techniques

As a student you need to pose your own queries to probe, discover, explore, clarify and understand new ideas and concepts. This involves independent study and deciding how to spend your time which includes ensuring that learning is part of your daily rhythm, i.e. just as sleeping, eating and exercise should be. Learning needs to be done in surroundings that make you feel positive and comfortable. (Marshall and Rowland, 1998). Decide when you find it easiest to concentrate well. Some people work intensively for short periods, followed by relaxation; others can concentrate for longer times.

It is possible to learn almost anything if there are goals that require it. Anyone can learn faster by structuring the information, so taking some time to organise information can increase effectiveness. It also helps to be in the right state of mind, and to have a personal reason why you wish to learn some material. Build anticipation about learning the material by imagining being able to answer an exam question well, or being able to answer correctly a technical question at interview. Have positive expectations that the material will be interesting, useful and easy to understand. To get the most out of the different kinds of information, it is better to understand the structural features of each type. So a book comprises contents, chapters, references, index and so on. It will be worthwhile taking notes in your preferred style to aid the absorption and understanding of what is being read. This may involve jotting down keywords, maps of related ideas, highlighting or underlining words, or making written or pictorial notes, or perhaps using a collage of post-it notes. Sometimes it can help to paraphrase (in your own words) an idea, or for example précis (a shorter summarised version) an argument. Ultimately, the goal is to try to use what you have read as a framework to support your own ideas and arguments (Northedge et al., 1997).

Sometimes the amount of material that needs to be consulted, for example in a reading list, can appear daunting. So break it down into smaller elements. A key issue is the chunk size that you naturally choose when exploring something new. If something seems hard, then you may need to move up to a higher level. Conversely, if something seems easy, then you may need to move down to a more detailed level. Find a size that seems manageable and with which you can

make progress. It may be a good idea to put things into a framework and to connect them to things you already know. This can reduce the feeling of being lost, and can reduce the amount of explicit content you have to remember.

Learning and remembering the key things needs to be done at your pace, in your own way, and often it helps to build on existing skills and knowledge, starting with principles and concepts with which one is already familiar. It is important to be systematic in the approach to remembering. Some people will use images, sound links, emotions, mnemonics or mind maps to help them remember facts and arguments. Mind maps comprise a joined up landscape of issues, facts and arguments, using colour, capital letters and different sizes of writing to make up the landscape (Turner, 2002). Finally, remember that you can also share reading with colleagues, and share views.

Regarding lectures, it helps to have given some advance thought to what the next lecture will be about. Then as you listen, question and evaluate what is being said, and make some notes. Minimise distractions to maintain concentration. Leave gaps to fill in notes later if there is a point that you did not understand. List any questions and ask the lecturer or colleagues at the end. Later, reflect on and evaluate the lecture to see if you really understand and have learned the critical points of what it was about.

Turner (2002) discusses learning techniques and summarises the approach as 'ascertain':

A – actively seeking out information
S – select information from different sources
C – classify the information in various ways
E – evaluate the information and exercise judgement about it
R – reflect on your learning and your study process
T – target: set goals and meet them
A – again: re-read, re-draft and hone your knowledge and arguments
I – integrate the information
N – negotiate: by interrogating the meaning of a text, and weighing it against other sources of information, approaches and arguments.

21.1.2 Effective use of learning resources

Learning resources come in many forms ranging from textbooks, reference books, journals, e-journals, the World Wide Web, databases, and the media such as newspapers, radio and television. Libraries are of course the hub of many learning resources and, therefore, an understanding of how these can be used to advantage is critical. How to handle the various types of information is also important in terms of being able to evaluate usefulness, how to navigate, using manipulation techniques and the presentation of issues. Laboratory-based work, including language and scientific laboratory work, may require a variety of learning resources comprising understanding written instructions, planning, understanding how to use specific equipment, using your practical skills, and then understanding how to analyse and report the outcomes of the learning experience (Northedge *et al.*, 1997).

Common to all of these is reading. We need to read to get ideas, to expand our knowledge, to improve our own writing style, to be able to back up ideas and arguments one wishes to make, and to be able to demonstrate evidence of reading to examiners. In whatever way you approach reading, it is important to read at a pace that allows you to engage and absorb ideas. Sometimes one may read for keywords by skimming, scanning or sampling the text. However, it is always important to be clear why we need to look at a certain text, and what we are expecting to gain from it. It is essential to develop the ability to discriminate between the presentation of knowledge from ideas, from observations, and from results and arguments. Equally, it is important to be wary of being over-reliant on the apparent authority and reliability of academic texts (Fairbairn and Fairbairn, 2001).

So when investigating a topic, prepare by selecting materials, and evaluate them in terms of their importance. Decide whether the materials should be bought, copied or borrowed. As you work with these materials question and evaluate the facts and ideas, just as you would with a book, a lecture or when watching a film or exploring the Internet. Select the key points, order and record them systematically, and organise and integrate the ideas with factual information. Towards the end of your researching, give time to reflect and review your conclusions against your initial brief.

21.1.3 Time management

Whatever you wish to do, there is always something you have to do first.

Better time management can help in achieving goals, whether it is the development of a new prod-

uct, or achieving higher exam grades. Time is not variable, and everyone has the same amount. However, time can be lost due to the telephone, meetings, bad planning, too much attention to detail and many other reasons. So there is a need to prevent and reduce such time wasting. Initially, the setting of goals is required to establish a context for managing time. Goals can be immediate, short term, medium term and so on. It is then useful to subdivide these into manageable steps or tasks. This will help in tackling them one task at a time. In the case of studying for a degree, each sub-goal might be the successful completion of one year of the programme. This approach may seem a challenge, but it does emphasise that there is a connected path linking what is done today with the achievement of a longer-term goal (Northedge *et al.*, 1997).

Sometimes you just do not feel motivated to work. Try working for a short time and see if you can 'get into it'. It might help to realise that when you are not motivated to work, you are not necessarily out of motivation, you are just motivated to do something else. However, remember the saying: procrastination (delaying; postponing; putting off) is the thief of time (Edward Young). Life should not be all routine or habit, but developing good time management habits will make it easier to be more effective.

It is better to work to priorities and, therefore, there has to be a list of tasks in order to see how much there is to be done. Each task has to be marked important or urgent or both, and there is a good case for doing the more disliked jobs first. The difference between an organised person and a disorganised person is that an organised person knows what still has to be done. When making a task list, it is important to indicate the level of priority of each task. One way to do this is to consider those tasks with a time deadline (urgent) and those that are directly relevant to your objectives (important), and then categorise each task:

	Urgent	*Non urgent*
Important	1	2
Less important	3	4

It is important, therefore, to be efficient with time, and that means being in control of your activities such as managing paperwork, delegating effectively and making meetings as efficient as possible. Being effective relates to the effort and efficiency you put into achieving your objectives, whatever they are. Two key problem areas that must be managed are paperwork and e-mails. These must be actively managed not just postponed. So papers/e-mails should be deleted, actioned, held for future action, filed or delegated/redirected. Avoid marking anything as pending.

In order to improve time management it helps to know where time is being spent. One approach is to track your time, which is like making a schedule but in reverse. At the end of every hour briefly write down how the time was spent. If how the time was spent did not match an already planned activity, enter a comment about what was done during that time. There are of course several different ways that this time awareness and time tracking can be achieved. The results will enable some modification of work patterns so that time is more efficiently spent on the goal-related activities.

To assist with time planning, use tools such as a monthly planner, a weekly objectives list, a weekly planner and a time log. The monthly planner can be used as a time-based memory aid, tracking major deadlines and appointments and so on. However, more can be made by recording interim deadlines and forecasting busy periods prior to final deadlines. This will help show whether there is room for new tasks, and whether you are on target to achieve your goals. The weekly objective list is a to-do list with additional features to further break down tasks into smaller units, and to record time estimates for the task. Having entered an activity it is important to assign a time estimate (Cottrell, 1999).

One way of making more time for yourself is to ask the question: why should I not delegate this task? However, delegation must be done sensibly and you must bear in mind that it will usually involve some decision making by the person to whom you have delegated, even though you remain accountable for the final outcome. The benefit is that it saves you time, and can positively provide to the delegated person motivation, recognition and greater involvement.

Remember to plan tasks flexibly, including making fixed appointments with yourself, and leaving realistic margins, such as half an hour, between jobs or appointments. Insert time-filler jobs – jobs of less than 10 minutes that can be slotted in between larger tasks. For the very busy days, a daily action plan can be very helpful. Most people spend much of their time working with or through others, and so it is essential to sensitively handle the conflict of needing to manage your own time efficiently and the obligation to consider others' needs, priorities and time pressures.

Meetings can take up large amounts of time ineffectively, not just for you, but for everyone attending. Where possible meetings should be decision orientated rather than discussion orientated, and should be planned in advance. The agenda should be a list of things to be done, not just discussed, listing the important issues first, and should not really include 'Any Other Business' which can open up the meeting to unplanned topics. Effective chairmanship is needed to keep the meeting time efficient but effective too. Similarly the management of a very large project will dictate the need to break it down into a series of component elements, and a list of tasks to enable control and simple critical path analysis may be helpful.

21.2 Information retrieval

21.2.1 Introduction

In reality there are no right or wrong ways to search for information; the key is to carefully select your information sources in order to optimise the quality and quantity (and hence cost) of the information you retrieve. It, therefore, becomes increasingly important to understand how information is communicated before one sets about trying to retrieve it. Information relevant to food science and technology is generated by a wide variety of organisations including publishers, industrial and commercial companies, academia and government institutions; it reaches the people who need this information (the users) by a number of different mechanisms.

Probably the most obvious way of spreading and receiving information is by word of mouth. However, it is now undoubtedly the Internet that is rapidly becoming the world's most popular and widely used information source. Yet, despite the ever-increasing popularity of the Internet, libraries continue to be a vital and invaluable source of information and can provide access to a myriad of information sources including journals, electronic archives, theses, conference proceedings, reports, patents, trade literature, standards, books, reviews, factual and bibliographic databases, and the Internet itself.

21.2.2 Using library resources

In order to make best use of the services and facilities that any given library, whether real or virtual, has to offer, especially when visiting it for the first time, it is a good idea to consult the library catalogue which will list the library's stock of information sources (journals, books, etc.) in a defined order, for example, alphabetically for individual authors or editors of books or alphabetically by journal title. Not all library catalogues are organised in the same way, but the majority are now offered in electronic format as an online database and their appearance and searchability is governed by the library management software used to create them. Library staff are always willing to give help and advice, and user documentation should be routinely available. Staff will also explain the layout of the library, the inter-library loan system, any available networks, special subject guides, and how to make recall requests.

Every book in a library is given a classification number or code which represents the subject area covered. A number of different classification systems exist, but the most widely used is the Dewey Decimal Classification system. It is estimated that this system, which was conceived by Melvil Dewey in 1873 and first published in 1876, is used by more than 200,000 libraries in 135 countries. The system has ten main subject divisions and further subdivisions (see http://www.oclc.org/dewey/). The main subject divisions are given a three-figure classification number, for example, food technology has the classification number 664; this three-figure number may be followed by a decimal point and further numbers can be assigned depending on the specific subject area – the more numbers, the more specific the subject area. Books are arranged on the library shelves by classification number or by code order to facilitate their retrieval. Library catalogues should detail this information, and, ideally, also record publisher details, publication dates, edition/issue numbers (when appropriate) and in the case of books and journals, respectively, ISBN (International Standard Book Number) and ISSN (International Standard Serial Number) numbers.

As well as books and journals, libraries may contain graphic materials including:

- illustrations and models;
- audio-visual sources such as DVDs, video tapes and slides;
- microfilm and/or microfiche of, for example, back issues of newspapers;
- collections of ephemeral material such as newsletters;

- repositories and reference collections including dictionaries, encyclopaedias, yearbooks, handbooks, directories, bibliographies, indexes, abstracts and e-materials.

Because libraries are used by many people, books may be out on loan at the time of searching. It is therefore important to know how the library loan system works, in order to ensure that when a book is returned, it can be made available to you immediately. In addition there may be a need to find references to other books not in the library. Reference works, such as encyclopaedias, can be a useful starting point because they provide good summaries and useful references to other books and articles. Periodicals (also known as journals, serials or magazines) publish the latest research on a topic. If a particular one is not available in a particular university, then use may be made of the inter-library loans service.

Some journals are freely available via the Internet (see section 20.2.4 in Chapter 20 and section 21.2.4 below); to readily access these visit the websites for the Directory of Open Access Journals (DOAJ) at http://www.doaj.org, or Highwire Press at http://highwire.stanford.edu (see also section 21.2.4). Other sources of information may include theses, conference proceedings and patents. Legislation, standards and statistics may be relevant to a particular project, and specialised sources of information may be searched for these.

21.2.3 Factual databases

Factual databases routinely contain basic facts and figures which are of potential value to users. For example, they may contain melting, boiling or freezing points of various substances, conversion tables or, of specific interest to food scientists and technologists, food composition data. Good examples of the latter type of factual database may be found on the Internet at http://www.nal.usda.gov/fnic/foodcomp/search/ (produced by the United States Department of Agriculture – Agricultural Research Service) and http://www.fao.org/infoods/directory_en.stm (produced by the Food and Agriculture Organisation of the United Nations).

21.2.4 Bibliographic databases

Bibliographic databases play a very important role in information retrieval by collating information from a variety of different sources. Providers (producers) of bibliographic databases essentially do the first phase of searching for their users and provide bibliographies (records) of relevant source material; this saves their users time and money and can ensure that important source material is not overlooked (something which can easily be done if you are conducting your own search). Bibliographic databases may cover quite broad subject areas (for example, the Web of Knowledge produced by Thomson Reuters which covers all subject areas in science and technology – see http://www.isiwebofknowledge.com/), or they may be very specialised; those specialising in food science and technology information will be described after we have considered the properties of bibliographic databases in general.

The records in bibliographic databases may just be references or citations, but more commonly they also contain summaries (abstracts) of the source material. Sources can include journal articles, books, theses, patents, standards, legislation, reports, conference proceedings, lectures and reviews. Records in bibliographic databases are indexed to enable searching and subsequent retrieval of relevant records which may then be consulted in their own right or used to track down the source material from which any given record has been prepared. The full text of the source material may be held by your library; alternatively, it might be necessary to make use of the services of a document delivery service. There are a large number of document delivery services offered both across the Internet and via traditional postal means. Accessing full text can be expensive, so it is sensible to do some independent research and/or seek the advice of your librarian about the best service to use for your particular needs. As previously described, some documents are freely available via the Internet (see also sections 20.2.4 and 21.2.2).

A sophisticated and relatively recent development designed to facilitate and manage information exchange in the electronic environment has been seen with the advent of the Digital Object Identifier (DOI) System. Amongst other things, the DOI System provides a mechanism for linking customers with content (information) suppliers. DOIs are persistent (unchanging) names given to intellectual property such as electronic journal articles, e-books, images – in fact any kind of content; they consist of characters and/or digits and can be used to direct you to the relevant Internet locations. The DOI System has not yet been universally adopted but is an exciting prospect for the

future; its impact is rapidly increasing. For more information about the System please visit the website http://www.doi.org.

Different producers vary in the way in which they index their databases although, essentially, the underlying principles remain the same. The subject index is usually created by extracting keywords (terms) from the abstract (some producers create the index from the source material) and listing them separately, either in a new field in the case of electronic formats of the database or as an alphabetised list at the back of printed versions. The keywords are not selected randomly – they are chosen to accurately reflect the important topics described in an abstract and are usually taken from a defined word list which is, in turn, derived from a thesaurus. A thesaurus is a collection of terms chosen to describe a particular subject area and reflects the relationship between the terms. For example, the term yoghurt has a relationship to the less specific (broader) terms fermented milk, fermented dairy products, dairy products, animal foods, foods, fermented foods, processed foods and milk; it also has a relationship to the more specific (narrower) terms buffalo yoghurt, drinking yoghurt, flavoured yoghurt, frozen yoghurt, fruit yoghurt and yoghurt beverages (International Food Information Service, 2007). By using the relationships between keywords described in a thesaurus it is, therefore, possible to refine your search of the electronic format of a bibliographic database to create a general (broad) search for records on a particular subject or a more refined or specific (narrower) search. A search can also be further refined by combining key words to describe precisely your subject area of interest. This is achieved by applying the principles of Boolean logic (see also section 20.3.8).

Boolean logic essentially breaks a subject area down into concepts. For example, if you were interested in conducting a combined search on *Salmonella* and chocolate, both terms should be keyed into the search box joined by the word 'and'. The information retrieval software incorporated by bibliographic databases to conduct searches will find all the records about *Salmonella* and all the records about chocolate and then combine the two searches such that only records containing both terms are identified – hence 'and' logic narrows down a search. There are three Boolean logic operators which are used to link search terms – these are 'and', 'or' and 'not'. In our example, use of 'or' in the search strategy would broaden the search and identify all records containing the terms

Salmonella and all the records containing the term chocolate. If 'not' logic is employed the retrieval software will identify all the records containing the term *Salmonella* and then exclude all those records that also contain the term chocolate. Search strategies can be developed still further by specifying the particular fields that a record contains in the search conditions; alternatively, the entire record can be searched (free text). By developing search strategies in this way the number of false drops (irrelevant records) retrieved is minimised and only those records you are really interested in are identified.

Some bibliographic databases are available free of charge and others charge for access on either a subscription or a 'pay as you use' basis; their update frequency is variable. Bibliographic databases are offered in a variety of formats including printed journals, CD-ROM and online (via intranets and the Internet). Internet access may be obtained direct from the database producer or via a host or vendor, for example, Ovid (see http://www.ovid.com), Dialog (now part of ProQuest – see http://www.dialog.com) and Thomson (see http://www.thomsonreuters.com/business_units/scientific/ or http://www.thomsonreuters.com/). There are many vendors and a great many bibliographic databases – far too many to describe in this chapter; instead please refer to the references Lee (2000), Hutchinson and Greider (2002), and Lambert and Lambert (2003) and/or consult your librarian. An overview of current publishing trends and technologies has been published by the International Food Information Service (2005).

We will now focus on two major bibliographic database producers, both of which specialise in information on food science and technology. The first of these is Leatherhead Food International (see http://www.leatherheadfood.com/lfi/). They produce a collection of databases bringing together technical, market and legal information. These include Foodline News, Foodline Product, Foodline Market, Foodline Science, and Foodline Legal (see http://services.leatherheadfood.com/foodline/index.aspx).

The second database producer specialising in food science and technology is the International Food Information Service (see http://www.foodsciencecentral.com). Their website offers a variety of news, reports, articles and links to other websites of interest to food scientists and technologists; it also provides an access point to the world's largest food science, food technology and food-related nutrition bibliographic database, i.e. *FSTA – Food Science and Technology Abstracts*®. The offering of *FSTA* on the Food Science Central website is called *FSTA Direct*™; it comes directly from the producers – the International Food Information Service (IFIS). *FSTA Direct*™ can also be accessed independently of the Food Science Central website (see http://www.fstadirect.com.). *FSTA Direct*™ is offered with access to a companion thesaurus and an online dictionary of food science and technology – both produced by IFIS. The thesaurus and dictionary are also available in printed format (International Food Information Service 2007 and 2009, respectively). *FSTA*™ is also available via a variety of vendors, on CD-ROM and as a printed journal. The database is sourced from all the world's literature (journal articles, conference proceedings, books, reports, theses, patents, legislation, reviews and standards) of relevance and significance to food science and technology and has been produced since 1969. A sample record from *FSTA Direct*™ is shown in Fig. 21.1.

Some bibliographic databases are offered to universities and colleges via service providers known as data centres. For example, in the UK the first data centre to be set up and probably the best known is *BIDS* (Bath Information Data Service) which is based at the University of Bath. The service is free at the point of use, but university libraries pay for subscriptions to the databases which allows staff and students open access. There are a number of different services available throughout the world and, therefore, it is a good idea to check with your librarian for data centre availability at your university or college.

21.2.5 Use of the World Wide Web as an information source

As mentioned previously (section 20.3.3) the Internet is rapidly becoming the world's most popular and widely used information source. An obvious starting point when searching the Internet is to look for relevant websites, but it is likely that there will be a great many of these that appear to be of interest – the World Wide Web is a daunting place for the unwary or novice searcher. It should be borne in mind that while some of these websites may be excellent (as evidenced by the web addresses quoted as references in this chapter), others may be of poor quality and

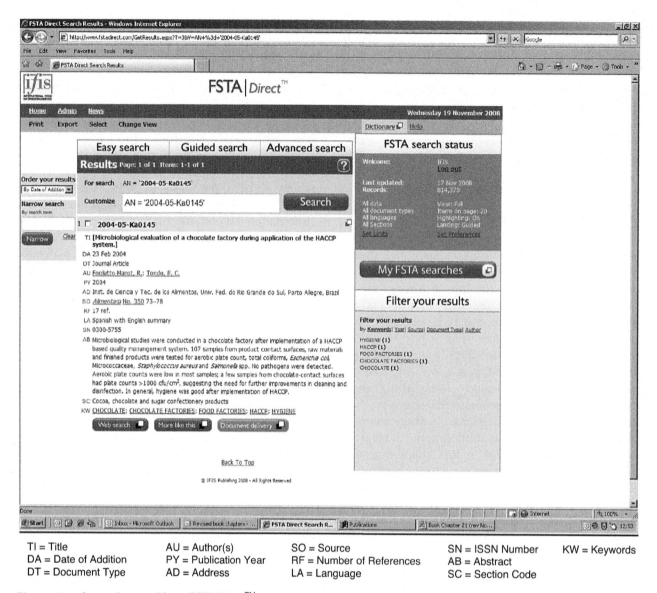

TI = Title   AU = Author(s)   SO = Source   SN = ISSN Number   KW = Keywords
DA = Date of Addition   PY = Publication Year   RF = Number of References   AB = Abstract
DT = Document Type   AD = Address   LA = Language   SC = Section Code

Figure 21.1 A sample record from *FSTA Direct*™.

information overload can quickly become a big problem. An equally large problem also looms with respect to key information which may be missed when surfing the web. In essence the message is to use the World Wide Web with caution and not to rely on it entirely. We also discuss this topic in Chapter 20.

21.2.6 Competitive intelligence

A different, but related, area of information to that of science and learning is that of competitive intelligence. The commercial world, including the food industry, makes use of competitive intelligence. This can be defined as the systematic process of obtaining and analysing publicly available competitor information to facilitate organisational learning, improvement, differentiation and competitor targeting in industries, markets and with customers (Hasanali *et al.*, 2004). The Society of Competitive Intelligence Professionals (http://www.scip.org) describes it as remaining cognisant of competitors' intentions and unanticipated marketplace developments by: scanning public records; monitoring the Internet and mass media; and speaking with customers, suppliers, partners, employees, industry experts and other knowledgeable parties.

21.3 Communication and presentational skills

If language is not correct, then what is said is not what is meant; if what is said is not what is meant, then what ought to be done remains undone (Confucius).

21.3.1 Writing essays, reports and abstracts

Writing an essay or a dissertation on a technical subject is a useful exercise in being able to order and present ideas and arguments to a reader, to clearly get across the messages about the understanding of the issues involved. Practice at doing this, of course, provides experience in the art of writing to get across ideas, facts, pros and cons, and conclusions. Having chosen or been given the essay title and the objective for writing about this subject, a draft plan of the key sections needs to be drawn up, followed by some structure within each of the various sections. A very similar approach can be used to that for writing a report. The essay should be the author's own work and it is essential to try to avoid plagiarism. This means that other people's work and ideas must be clearly acknowledged. Never download essays from the Internet and try to pass them off as your own work.

During the preparation and planning of the structure of the report, it is essential to bear in mind who will read it, who will need it, what they will want from it and what are their terms of reference. In response, one should constantly be asking the question: how can I best meet their needs? For the benefit of the reader, information should be presented simply, clearly, logically and systematically. As a guide, the report layout should include the following sections and have a structure similar to the following:

Title page
Abstract
Acknowledgements
Contents list
Introduction
Literature review
Materials and methods
Results
Discussion
Conclusions and recommendations
Bibliography and references
Glossary
Appendices

The Introduction section should introduce the general subject area and its significance, and show how the topic in the title fits in. The terms of reference should be referred to, and the overall objective of the report should be clearly spelt out. The Literature review should clarify why the current work being reported was needed. Any important and related studies should be compared and contrasted *critically*. This means that it is not enough merely to refer to the fact that there have been other workers who have previously studied a similar topic. Critical detail must be referred to where, for example, there have been differences in the analytical methods, differences in the raw materials used, agreements in some findings, differences in the parameters studied and so on. The discussion of the work of others and the conclusions drawn about the unanswered questions, or remaining gaps in the knowledge and understanding, should then lead to a clear statement of the objective of the work and the reasons for it.

The Materials and methods section should list and specify all the materials used. Use chemical nomenclature correctly. It is essential to report exactly what was done (as opposed to what one might have liked to have done) and to be accurate. In the writing of scientific papers, there should be enough detail for someone elsewhere in the world to read and then repeat exactly what was done (coupled with referenced material). This section should always be written in the third person and the past tense. A key element to the approach of studying foods is the use of statistics. The application of statistics to the study of foods is important because foods are biological materials which are inherently variable, but the use of statistics can save time and money (see also section 20.1.9). So report here on the experimental design used, for example to reduce the number of experiments needed. Analyse the results using statistics in order to be certain what differences are real rather than just due to biological variation, or due to errors in sampling and experimentation. It is essential to draw out the certainty of the meaning of your results, and employ appropriate controls to ensure validity of your data and subsequent conclusions.

The Results section may include a variety of diagrams, tables, graphs and photographs, and it is important to ensure that each one has the correct title, correct and clear labels for the parts or graph axes, and the correct units. Always check that the numbering of the tables and graphs has been done precisely, and that these numbers match those referred to in the text of the Discussion section. Do not present both a table of data and a graph of the same data. Generally, it is better to do one or the other. Considerable effort may go into producing tables and graphs, and because as the author you know what they all mean, it is sometimes easy to forget that the reader may not understand what is meant. So remember to discuss all your results – they cannot speak for themselves. During this discussion, be quantitative. State the relevance of each result to, for example, related research, or to some aspect of food manufacturing. Discuss the effectiveness of your approach to studying the problem. To what extent did you achieve the objective? All the time try to simplify, justify and quantify the results in the Discussion.

A report is a kind of story; therefore punctuation, grammar and spelling are critical. The language used is important in ensuring that the story is clear, but the style should be much more concise than that of a novel. So use simple sentences, use one word instead of two where possible, and use the past tense. Do not refer to yourself, use the third person. Avoid using what are sometimes referred to as incomprehensible sentences, or sentences that are ambiguous. Examples of incomprehensible sentences might be one including the phrase *having completed the observations, the microscope.* . . . The point here is that microscopes do not make observations. Or, *using a meter, the current* . . . . The point here is that it is not the current that uses the meter. Finally, *after standing in boiling water for an hour, examine the flask* . . . . The concern here might be for the well-being of the experimenter's feet! An ambiguous sentence may, for example, include the phrase *the starch yielded more glucose than maltose* . . . . Make clear what is greater than what. Write either: . . . *than did maltose* . . . or . . . *starch produced a greater yield of glucose than of maltose* . . . (Booth, 1977).

The Conclusions section is the most important part. Here it is important to make the final deductions and summaries of the findings, and to draw the inferences from the results. Draw all the related aspects together. Finally, make suggestions for further work and any relevant recommendations that are appropriate.

References and bibliographies are dealt with in the next section, and these are a formal part of the report. However, the Appendix is usually reserved for additional and supporting information, so the report itself should be complete without any Appendices.

The Abstract should be written last as it should reflect what has been written in the report. Whether the Abstract is in the form of an Executive Summary or a detailed scientific abstract, it is critical to your reader. This is because it will usually be read first and will, therefore, determine whether the reader decides to read the main body of the report. So it is essential to summarise what has been done in terms of objectives, results and conclusions. Reflect the requirements of the original brief. Be quantitative, and record key data from the results and ranges of the key parameters studied. Give the main conclusion and its relevance. Time is often limited for reading so it is desirable to grab the reader's attention and make the right impact. So be precise, concise, relevant and complete.

Larger reports may take considerable time to draft, and comprise a number of significantly different sections. It is, therefore, important to set aside the overall draft for some days, and to re-read it with a fresh pair of eyes. Then make further modifications and improvements, leave again and then repeat the editing process to finalise the report.

21.3.2 References and bibliographies

The general rule is that a citation or a reference should only be used when necessary to acknowledge the ideas of others, or to criticise in order to build your own argument. Sometimes references are included at the bottom of the page on which they occur. However, in scientific writing it is normal to present references at the end of the article or book.

There are two principal formats for quoting references: the Harvard system and the numeric system. In both cases the objective is to link ideas in the text with their source. Using the Harvard system, the surname of a single author, or two authors, and year of publication would be quoted in the text. For more than two authors, the citation is written as the first author followed by *et al*. Examples of how to write the full reference are:

- From books: author(s) (surname followed by initials); year of publication (in parentheses); title of book (italics or underlined), edition of the book (if not the first); the publisher; place of publication; volume number (if applicable); page number(s), and section (if applicable).
- From journals: author(s) (surname followed by initials); year of publication (in parentheses); title of article; name of journal (italics or underlined); volume number (in bold or underlined); part number (if applicable); page number(s).

In the reference section, references should be listed in strict alphabetical and chronological order. Where more than one reference has been published by the same author(s) in the same year, then an 'a', 'b', etc. is placed after the date. This approach is popular in the research world, because it is so much easier to identify familiar researchers.

With the numeric system, where the first reference needs to be made in the text, the number (1) in parentheses is given. Where the next reference appears, (2) would be written, and so on for all the references. In the reference section itself, the references are listed in numerical order.

The World Wide Web presents both opportunities and problems. Access is relatively fast and easy, but the information can be of transient nature, and there is a lack of peer refereeing of much information published there. Where citation of a web-based source is necessary, the citation in the text should be by author and year (although problems can occur if there is no obvious author). Then in the bibliography also quote the article title, followed by the URL: http://internet address/remote path, written as one line, followed by [date visited] in square brackets as shown. It is advisable to retain a printed copy of that site on the day visited, because the site could be changed without notice.

21.3.3 Oral presentations and the use of visual aids

Initial experience of orally presenting ideas usually comes through involvement in informal discussion or learning groups associated with parts of a course being studied. Participation in these can help build confidence in both the learning process and in speaking.

Unlike a report, an oral presentation allows the speaker to present him-/herself, and something of his/her character, to the audience. In the first instance an oral presentation must be audible to your audience. If there is no microphone then it is essential to speak up and speak clearly. If there is a microphone, then let the amplifier do the amplification for you and speak normally. Because you are standing up in front of an audience, people in that audience are bound to spend some time looking at you. So your appearance in terms of dress and smartness will be important to consider. The purpose of an oral presentation is to get across one or more messages, and it is important to remember not to become a distraction to that purpose. So it is better to be still, rather than move about a lot. Use gentle rather than rapid gestures, and whether you feel like it or not, smile and be upbeat rather than the opposite. Do catch the eye of some members of the audience from time to time, and briefly hold their attention, and generally try to ensure you are 'connected' to them.

Structure your talk and take some note of what is required for a report. What is the brief, and how long have you got to speak for? There is some truth in the old adage – if you can't get your key message across within 10 minutes, you are not likely to do so in 1 hour. It helps to forewarn the audience what you are going to talk about, so indicate at the beginning what structure you intend to follow. Then after presenting each point, summarise it before moving on to the next point. Finally, draw all the key points together in a concluding summary. The repetition of key points can be useful to the audience. If you are not clear, then sometimes what you think you have said can be received as something else.

It is important to have the structure and the key messages in your mind as you speak. Usually the use of visual aids can help to illustrate and reinforce what is being said. It is equally important not to be seen to be obviously reading a script, with head down, and speaking down. Sometimes it is necessary to present a detailed table or complex graph as a visual aid. Do not keep speaking with the slide on view, and then simply move on to the next slide. In these cases it is important to stop and explain clearly what the graph consists of, and what the results are showing. It is the same principle as writing a report, in that graphs and tables cannot speak for themselves. Allow time for this, and unless you are very familiar with your presentation, this means you will need some form of rehearsal.

The visual aids themselves are very important because if they cannot be read by the audience, there is no point in using them. This means that the use of ordinary typescript will never be suitable. Usually seven or eight lines give the minimum font size (typically 32 or 28 for text, and 44 for titles in say Arial typescript) that can be read comfortably by the audience, especially if being projected in a large meeting room. Use appropriate colour combinations for the backgrounds of slides, to ensure that the text contrasts well and stands out clearly. With transparent view-foils for overhead projection, colour is either not available or is too expensive. However, using a computer, packages such as Microsoft Powerpoint provide a range of options to help make a presentation interesting and stimulating through the use of animation and the addition of graphics and images.

21.3.4 Group presentations and the use of posters

For the presentation of work done by a group, it may be decided that two or more members of the group will present different parts of the work. In such a case it must be agreed what format each will present in, so that the various contributions appear obviously related.

At some conferences, or to provide highlights of the work of an organisation, there may be the opportunity to present on the walls or suitable boards posters summarising a project or an area of work. The content of a poster should broadly follow the structure of any written report (see section 21.3.1). Although there can be much variety, these posters may be typically from 65×90 cm to 90×120 cm in size, or based on A1 size paper. The posters may be laminated or printed on plastic for long life. People viewing the posters will stand at a distance of about 1–2 m and, therefore, need to be able to read the text comfortably at that distance. Therefore, the title of the poster needs to be in letters 2–3 cm high. Other text needs to use a font size of about 1 cm high, and this means that the total number of words that can be used will be 300–500. In addition, there will usually be two or three photos or graphs, and typically these will need to be 15–20 cm square, and be very clearly labelled using an appropriately large font. Having prepared the basic text and charts, the material is usually given to a graphic designer who will be responsible for the design and publishing of the poster.

21.4 Team and problem solving skills

21.4.1 Effective team working

At some stage, in all walks of life, you may find yourself working as part of not only a group but also a team (Procter and Mueller, 2000). A work group is a team if all the members of the group share at least one goal that can be accomplished only through the joint efforts of all. Having a shared goal or goals makes the difference between a group and a team. Teams are often put together in order to meet specific needs at a certain time, and so they may exhibit a kind of life cycle. In other words, teams may go through a form-

ing phase, a developing phase, a mature performing phase and then an ending phase. This team could be a group of students put together to carry out laboratory or project work or it could, for example, be a cricket team.

In comparing a cricket team with a work team, it is possible to see some of the issues that may arise, and why teams need special attention (Hardingham, 1998):

Cricket Team	Work Team
Clearly defined role for everyone in the team	Many people not sure where they or others fit
Concrete measurable goal	Often the goals of the team have never been spelt out – different people have different ideas of what they are
Visible competition for the team to unite against	As much competition inside the team as outside it
Has a coach	Is left to its own devices

An effective team is one that achieves its objectives. As teams are made up of people, then clear communication with everyone in the team is essential to ensure that everybody understands the role of the team and his or her individual role within the team (Widdicombe, 2000). In trying to understand a team, people need to understand the competitive pros and cons, the need or otherwise for clearly defined job descriptions in the team, and issues of team loyalty, team size and time span. Team spirit is an important element to the success of a team, and individuals need to enjoy their role and feel confident about where the team is going. A mental check-list for a quick appraisal for team functioning is summarised by the acronym PERFORM (Hardingham, 1998):

P Productivity: is the team getting enough done?
E Empathy: do the team members feel comfortable with one another?
R Roles and goals: do they know what they are supposed to be doing?
F Flexibility: are they open to outside influence and contribution?
O Openness: do they say what they think?
R Recognition: do they praise one another and publicise achievement?
M Morale: do people want to be in this team?

By answering these simple questions, it will soon become clear if there are any indicators of problems.

Working in teams can release creativity and energy, especially when the communication is genuinely interactive, with people building on one another's suggestions. People can enjoy work more, because we all like, and need, to 'belong'. Teamworking can result in reduced costs and improved productivity. Sometimes teamworking is the only way to do a job. Neither a concert nor a play can be performed without teamwork. Neither can many essential and mundane organisational tasks.

Sometimes things can go wrong with teams. A major example of teamworking is meetings. People can turn up late; meetings can overrun; they can be boring; people may not contribute much to the discussion; the first item on the agenda takes forever, so there is insufficient time for the other items; people leave feeling frustrated or exhausted; people who should be there don't turn up; one or two people dominate the whole meeting; meetings are used as a forum for settling private scores; decisions are either not made or are arbitrarily imposed by the team leader after inconclusive discussion. As meetings are where the whole team get together, then such symptoms can be very destructive, and represent a lost opportunity. As good teamworking depends on human relationships, then these can also be badly damaged (Widdicombe, 2000). Planning and discipline are, therefore, needed and this includes circulating the agenda and related papers well in advance, and ensuring that meetings start and finish on time.

The importance of professionalism rather than individualism is discussed by Widdicombe (2000), together with the role of the Chairman with respect to the various ways in which he or she will need to listen, or intervene, and steer the debate to a conclusion. Sometimes teams can suffer because, although the meetings are well run and enjoyable, nothing happens between the meetings and no one delivers. The better approach to improve this is to ensure that every action decided by the team has an individual's name attached to it. Check that the named individual has understood and agreed to the action. Then review the progress of all actions at the next meeting, and explore helpfully why they may not have been completed, checking what other team members can do in support. Remember to publicise and celebrate achievements.

'Groupthink' can be adverse to a team's and an organisation's well-being because it undermines the effectiveness of the team's or organisation's decision making. The symptoms are: when there is little or no debate about issues; little or no challenging of a decision; little or no self criticism, and defensiveness towards criticism; an absolute conviction that the team is right; and declining interest in facts or opinions from outside the team. The best solution is an external force with impact and credibility to challenge the group. Some teams use external consultants; others keep their leader at some distance from much of the team's activity so that he/she is less likely to get sucked in to 'Groupthink'.

Team players are usually thought to be the sort of people who are willing to share opportunity and credit, to be open, direct, and ready communicators, and to be likeable. How people approach problems, weigh and use information and make decisions are components of 'thinking style'. These styles are:

- Action-orientated thinker, or a reflection-orientated thinker
- Facts-based thinker, or an ideas-based thinker
- Logic-focused thinker, or a values-focused thinker
- Ordered thinker, or a spontaneous thinker

Team leaders have a key role in teamworking, and whilst personality is difficult to change, behaviour is not. Team leaders may have certain responsibilities, and hence the team may look for the person with the best skills to meet this responsibility. Examples of such responsibilities may be: for organising the team to meet its goals; for the quality of the team's output; for developing the team; or for the interface between the team and the organisation. The ideal leadership style is whichever works best for the team, and the team may be starting up, or competent, or unsure, or in a high-risk situation. Typically there can be directive, delegating, supporting and inspirational styles of leadership.

21.4.2 Problem solving strategies and techniques

Many problems can only be solved by effective group working. However, it is important to recognise that there are just as many that are far better tackled by an individual. Group problem solving is the best way to tackle issues where the problem concerns more than one person, or where there is no straight-forward single answer, so an amalgam of different views is needed. Finally, there are situations where it is important that those involved are committed to the solution.

The problem solving process may require combining a number of steps (Robson, 2002). Brainstorming is a fast-paced method of getting a group of people to generate many ideas in a short space of time. For success there are several key rules:

1 There should be no criticism of any ideas thrown up during the meeting.
2 Everyone's minds should be allowed to roam free in order to allow any ideas to come out, however impractical they may seem at the time.
3 Brainstorming is very much about quantity rather than quality, with the goal being to generate as many ideas as possible in a short space of time.
4 All ideas must be written down, even if the difference between two ideas is just the way in which they are expressed. So the whole group can see these ideas, they are usually put on to flip charts and stuck on walls around the room.
5 Some time is required (maybe a few hours, or up to a week) to allow the ideas to incubate, especially to ensure that nothing is rejected out of hand.
6 Then the ideas can be evaluated at the next group meeting, starting by grouping the ideas into themes.

The problem must be defined clearly. A key challenge is that members of a problem solving group often have quite different perceptions about what they want to get out of the group in terms of solutions. Sometimes the problem itself is couched in terms of a possible solution, and this can limit the solution-finding process (e.g. we need a new machine, instead of, the existing machine is inefficient). Another challenge is that the problem statement can be so broad and general that the group struggles to get its mind around the issue. In order to avoid getting the right answer to the wrong question, it is essential to keep focused on the core issue. It needs to be clear that the group can influence the issue in a reasonable time, and that data can be collected about it to ensure that problems are solved in terms of facts and not opinions.

In trying to analyse any problem, the more alternatives the group considers, the better will be the solution that it eventually decides upon. There are several possible approaches that can be taken, but it is essential that the problem is looked at from all angles, and from a broad view. One such approach is

the cause and effect diagram (or fishbone diagram) and, in summary, this comprises writing down the precise (rather than general) effect on the right-hand side of a large piece of paper. Second, draw in the main ribs of the fish and write down the headings of the main problem areas. These will typically be about people, the environment, methods, plant, equipment and materials. Brainstorming is used to generate the actual list of causes which relate to the effect that has been written down. Time needs to be allowed to reflect on the suggestions made, and even to allow further input from other people. Finally the whole diagram needs to be analysed. The Pareto principle (80:20 rule) points to the typical situation where only a few causes will be responsible for most of the effect.

All of us have so many opinions about so many things, and our opinions make so much sense to us, that it is often difficult to appreciate how anyone else could not agree. Thus, it is necessary to collect data in order to have factual information. Typically, data can be collected on a simple check sheet. These sheets should be the same for everyone involved, have a clear title and be dated when used. Ensure that the data collected are really those that are needed to help solve the problem. Think also about what other factors, inside or outside the company, are influencing the data being collected, and make sure that sufficient data are collected to show the full picture and not just part of it. Then tools such as Pareto diagrams and histograms may be applicable for helping to interpret the data collected. In many instances a simple line graph or pie chart or some other simple way of visualising the figures will suffice.

In addition to using techniques for analysing problems and opportunities, creative and analytical approaches are required for finding the possible solutions. For example:

- the solutions fishbone technique involves writing down the problem statement, agreeing the main 'ribs', drawing the diagram, brainstorming solution possibilities, reflecting on the ideas, and then evaluating the diagram for the best possible solution ideas;
- the force-field analysis technique comprises defining the worst and best possible situations, identifying and ranking the restraining forces, identifying and ranking the driving forces, assessing the possibility of influencing the forces, highlighting the

priority areas, and finally devising an action plan to solve the problem;

- the Delphi approach involves reviewing the problem analysis and data, generating solution ideas individually, combining the ideas into one list, individual ranking and recording of solution ideas, and finally discussing the rankings and agreeing the solutions by consensus.

Different methods will be appropriate to different circumstances (Robson, 2002).

The preferred solution must then be subjected to a rigorous cost-benefit analysis, since solutions are not likely to be welcomed if the costs exceed the benefits. Sometimes a solution will yield benefits that carry on year after year, but only cost money at the start, so the group needs to be able to calculate the payback period for its solution. Some solutions cannot be quantified in such a way, e.g. improvements in working conditions, so it may be important to examine if there are any other cost-benefits that may be realised at the same time. The key point is that the cost-benefit relation does not have to be expressed in terms of money: it could be time saved, relationships between people getting better, or communications between departments improving.

Finally, the solutions will have to be presented to the management and other decision makers, and so a very clear and logical presentation must be prepared and delivered. Monitoring and evaluating the implementation of the action plan should normally be the responsibility of the group itself.

Further reading and references

(Note: all URLs cited were correct in November 2008.)

American Library Association (2000) *Information Literacy Competency Standards for Higher Education*. The Association of College and Research Libraries, Chicago, Illinois.

Booth, V. (1977) *Writing a Scientific Paper*, 4th edn. The Biochemical Society, London.

Bundy, A. (2004) *Australian and New Zealand Information Literacy Framework: Principles, Standards and Practice*, 2nd edn. Australian and New Zealand Institute for Information Literacy, Adelaide.

Cottrell, S. (1999) *The Study Skills Handbook*. Palgrave, Basingstoke.

Fairbairn, G.J. and Fairbairn, S.A. (2001) *Reading at University*. Open University Press, Buckingham.

Hardingham, A. (1998) *Working in Teams*. Institute of Personnel Development, Management Shapers Series, London.

Hasanali, F., Leavitt, P., Lemons, D. and Prescott, J.E. (2004) *Competitive Intelligence: A Guide for Your Journey to Best-Practice Processes*. American Productivity and Quality Center, Houston, Texas.

Hutchinson, B.S. and Greider, A.P. (2002) *Using the Agricultural, Environmental and Food Literature*. Marcel Dekker, New York.

International Food Information Service (2005) *Food Science Information Discovery and Dissemination – Current Publishing Trends and Technologies Enabling Access to Essential Knowledge*. IFIS Publishing, Shinfield.

International Food Information Service (2007) *FSTA Thesaurus: Eighth Edition*. IFIS Publishing, Shinfield.

International Food Information Service (2009) *Dictionary of Food Science and Technology*, 2nd edn. Wiley-Blackwell, Oxford.

Lambert, J. and Lambert, P.A. (2003) *Finding Information in Science, Technology and Medicine*. Europa Publications, London.

Lee, R. (2000) *How to Find Information: Genetically Modified Foods*. British Library, London.

Lumsdaine, E. and Lumsdaine, M. (1995) *Creative Problem Solving – Thinking Skills for a Changing World*. McGraw-Hill, New York.

Marshall, L. and Rowland, F. (1998) *A Guide to Learning Independently*, 3rd edn. Open University Press, Buckingham.

Northedge, A., Thomas, J., Lane, A. and Peasgood, A. (1997) *The Sciences Good Study Guide*. The Open University, Milton Keynes.

Procter, S. and Mueller, F. (eds) (2000) *Teamworking*. From the series Management, Work and Organisations. MacMillan Press, Basingstoke.

Robson, M. (2002) *Problem Solving in Groups*, 3rd edn. Gower Publishing, Aldershot.

Society of College, National and University Libraries (1999) Briefing Paper: *Information Skills in Higher Education*. SCONUL Advisory Committee on Information Literacy, SCONUL (http://www.sconul.ac.uk/groups/information_literacy/papers/Seven_pillars2.pdf).

Turner, J. (2002) *How to Study – A Short Introduction*. Sage Publications, London.

Widdicombe, C. (2000) *Meetings that Work – A Practical Guide to Teamworking in Groups*. Lutterworth Press, Cambridge.

World Wide Web, 473–6
worms, 138–9

xanthan gum, 10
xanthenes, 24
XML, 474
xylans, 10

yakult, 93
yeasts, 118–19
    beer, 89–90, 163–4

bread, 97
    "killer" strains, 164
    wild, 164
    wine, 89, 164
Yersinia enterocolitica, 126–7
Yersinia pestis, 127
yogurt, 93, 94, 166–7
    bacterial contaminants, 141–2
Young's modulus, 202

zinc, 320

9 780632 064212